Lecture Notes in Networks and Systems 1001

The series "Lecture Notes in Networks and Systems" publishes the latest developments in Networks and Systems—quickly, informally and with high quality. Original research reported in proceedings and post-proceedings represents the core of LNNS.

Volumes published in LNNS embrace all aspects and subfields of, as well as new challenges in, Networks and Systems.

The series contains proceedings and edited volumes in systems and networks, spanning the areas of Cyber-Physical Systems, Autonomous Systems, Sensor Networks, Control Systems, Energy Systems, Automotive Systems, Biological Systems, Vehicular Networking and Connected Vehicles, Aerospace Systems, Automation, Manufacturing, Smart Grids, Nonlinear Systems, Power Systems, Robotics, Social Systems, Economic Systems and other. Of particular value to both the contributors and the readership are the short publication timeframe and the world-wide distribution and exposure which enable both a wide and rapid dissemination of research output.

The series covers the theory, applications, and perspectives on the state of the art and future developments relevant to systems and networks, decision making, control, complex processes and related areas, as embedded in the fields of interdisciplinary and applied sciences, engineering, computer science, physics, economics, social, and life sciences, as well as the paradigms and methodologies behind them.

Indexed by SCOPUS, EI Compendex, INSPEC, WTI Frankfurt eG, zbMATH, SCImago.

All books published in the series are submitted for consideration in Web of Science.

For proposals from Asia please contact Aninda Bose (aninda.bose@springer.com).

David Pastor-Escuredo · Imene Brigui ·
Nishtha Kesswani · Sushanta Bordoloi ·
Ashok Kumar Ray
Editors

The Future of Artificial Intelligence and Robotics

Proceedings of 5th International Conference on Deep Learning, Artificial Intelligence and Robotics, (ICDLAIR) 2023 - Progress in AI-Driven Business Decisions & Robotic Process Automation

Springer

Editors
David Pastor-Escuredo
UNHCR
Geneva, Switzerland

Imene Brigui
Emlyon Business School
Écully, France

Nishtha Kesswani
Department of Computer Science
Central University of Rajasthan
Tehsil Kishangarh, Rajasthan, India

Sushanta Bordoloi
National Institute of Technology Mizoram
Mizoram, India

Ashok Kumar Ray
NERIST, North Eastern Regional Institute
of Science and Technology
Nirjuli, India

ISSN 2367-3370 ISSN 2367-3389 (electronic)
Lecture Notes in Networks and Systems
ISBN 978-3-031-60934-3 ISBN 978-3-031-60935-0 (eBook)
https://doi.org/10.1007/978-3-031-60935-0

This Springer imprint is published by the registered company Springer Nature Switzerland AG
The registered company address is: Gewerbestrasse 11, 6330 Cham, Switzerland

Contents

A Comparative Study of Machine Learning Algorithms for Predicting Cardiovascular Disease

Anu Lohachab(✉) and Kuldeep Kumar

Department of Computer Engineering, National Institute of Technology, Kurukshetra, Kurukshetra, India
anulohachab@gmail.com, kuldeepkumar@nitkkr.ac.in

Abstract. Cardiovascular diseases (CVDs) are one of the primary causes of global morbidity and mortality, presenting significant healthcare challenges. The critical need for accurate prediction of CVD risk is paramount for timely and proactive intervention, yet it remains a challenge. To enhance CVD risk prediction, this study delves into a range of machine learning algorithms, encompassing supervised, and ensemble algorithms. Furthermore, recognizing the literature's limitations, our focus has been on enhancing model performance through hyperparameter tuning, implementing robust feature selection methods, and conducting thorough model evaluations. Besides, for feature selection, we utilize chi-squared tests and correlation analysis to ensure the relevance and significance of the features. Moreover, our comprehensive evaluation, spanning three diverse datasets, assesses both supervised and ensemble learning algorithms for their accuracy and generalizability. The results indicate that the K-Nearest Neighbors-based model excels, achieving 97.82% accuracy. By enhancing predictive accuracy and model robustness, our study not only contributes to improved patient-specific interventions but also aids in shaping more effective and efficient public health strategies in cardiovascular care.

Keywords: Machine Learning in Healthcare · Predictive Modeling for CVD · Supervised Learning Algorithms · Ensemble Learning Models · Healthcare Data Analysis · Patient Care Strategies

1 Introduction

Cardiovascular disease (CVD), one of the leading causes of morbidity and mortality worldwide, poses significant challenges in healthcare [2]. The complexity of CVD stems from its multifactorial nature, where factors such as age, gender, smoking habits, and genetic predispositions interplay to influence an individual's risk. Indeed, recent advances in data mining and neural network applications have paved the way for more sophisticated assessments of CVD risks [1,3]. However, accurate risk quantification is complicated by the heterogeneity of these

D. Pastor-Escuredo et al. (Eds.): ICDLAIR 2023, LNNS 1001, pp. 1–11, 2024.
https://doi.org/10.1007/978-3-031-60935-0_1

contributing factors, underscoring the need for more nuanced and robust predictive models that can integrate and analyze a broad spectrum of risk indicators.

Nevertheless, building upon the intricate dynamics of cardiovascular disease, the emphasis on early detection and the creation of predictive models is pivotal in reducing CVD-related fatalities. To this, employing artificial intelligence, particularly in cardiac disease detection systems, significantly bolsters the efficacy of established models for CVD detection and prediction [15]. For instance, longitudinal studies using multivariate regression analysis are instrumental in constructing these risk prediction models [13]. Furthermore, in the literature (e.g., [6,10,12,14]), numerous studies have explored various classification techniques, spanning supervised, unsupervised, hybrid, and deep learning methodologies for CVD prediction. Despite these, a critical analysis reveals significant areas for improvement, particularly concerning hyperparameter optimization, the implementation of more effective feature selection strategies, and the thoroughness of model evaluation processes. These underlying issues are key contributors to the reduced accuracy observed in existing predictive models, underscoring a need for methodological enhancements in this domain. Specifically, such improvements are essential for achieving higher precision in CVD risk prediction and advancing the overall effectiveness of these models.

Accordingly, in this study, a variety of machine learning algorithms are utilized to predict CVD risk from historical data. To ensure the robustness and generalizability of our findings, our analysis includes three distinct datasets: the Statlog Heart Disease Database [11], the UCI Cleveland Database [5], and the Kaggle Database. The selection of each dataset is based on its unique attributes and relevance to CVD risk factors, facilitating a thorough evaluation of the predictive capabilities of our models in diverse clinical contexts. The primary contributions of this paper are delineated as follows:

- This paper presents a comparative study that evaluates a diverse range of machine learning algorithms across three distinct datasets. Such a holistic approach aims to provide insights into the strengths and weaknesses of each algorithm in different clinical contexts and suggest an approach for relatively more robust CVD risk prediction models.
- This study addresses the complexities of CVD risk factors, resulting in an enhancement of the accuracy of the underlying algorithms; specifically, with the K-Nearest Neighbors (KNN)-based model, we achieved an accuracy of 97.82%, a relatively better performance.

The remainder of this paper is organized as follows: Sect. 2 reviews literature in the field of cardiovascular disease prediction. Section 3 details the methodology and study design for CVD prediction. In Sect. 4, we discuss experimental outcomes, highlighting the significant improvements achieved in model accuracy and, subsequently, predictive capabilities. Finally, Sect. 5 concludes the paper and outlines potential future directions.

2 Related Works

This section briefly discusses some of the existing works dealing with CVD prediction, encompassing supervised, unsupervised, and hybrid approaches. A brief comparison is provided in Table 1.

2.1 Supervised Machine Learning-Based Approaches

Supervised machine learning algorithms have been widely employed for heart disease prediction, demonstrating promising results. Shah et al. [12] investigated the effectiveness of various supervised learning algorithms, including Random Forest, KNN, Decision Trees, and Naive Bayes, using the Cleveland database from the UCI Machine Learning Repository and selected a subset of 14 attributes. Among the employed algorithms, KNN exhibited the highest accuracy rate of 90.78%. Katarya et al. [6] summarized supervised machine learning algorithms, including Naive Bayes, Random Forest, Support Vector Machines (SVMs), and Decision Trees, and emphasized feature selection before prediction. In another study, Tasnim et al. [14] employed two datasets, the UCI Cleveland dataset, and the Statlog Hungarian dataset, to evaluate the performance of supervised machine learning algorithms. They implemented Principal Component Analysis (PCA) and Chi-Square feature selection techniques to reduce the dimensionality of the data. The combination of the Random Forest Classifier and PCA yielded the most favorable accuracy, reaching 92.85%. Mohan et al. [8] investigated the performance of supervised machine learning algorithms. They applied a range of algorithms, including K-Nearest Neighbors, Random Forest, Naive Bayes, and Logistic Regression(LR). LR emerged as the relatively better algorithm, achieving an accuracy rate of 90.2%.

2.2 Unsupervised Machine Learning-Based Approaches

Though notable unsupervised machine learning-based approaches for CVD prediction exist, limited accessibility restricts our review to a select few. In their work, Islam et al. [4] addressed the challenge of dataset dimensionality reduction using PCA and further enhanced their approach by incorporating Genetic Algorithms with unsupervised k-means clustering. Their experimentation, conducted on the UCI Machine Learning heart disease dataset, yielded an accuracy of 94.06% . In a separate study, Rajalakshmi et al. [10] proposed a novel approach to cardiovascular disease prediction. Their methodology combined K-Means Clustering, Weighted Associative Classifier, and Decision Tree C5.0 algorithms applied to a carefully curated dataset containing 11 features. The outcome of their research demonstrated a high accuracy of 94.54% in predicting cardiovascular disease.

2.3 Hybrid Machine Learning-Based Approaches

Several studies have explored hybrid machine learning approaches for cardiovascular disease prediction. Kavitha et al. [7] implemented machine learning

Table 1. A Brief Comparative Analysis of ML Approaches for CVD Prediction

Reference	Dataset	Algorithms	Feature Selection/Engineering	No. of Features	Accuracy
Shah et al. [12]	UCI Cleveland	Random Forest, KNN, Decision Trees, Naive Bayes	Not specified	14	90.78%,
Tasnim et al. [14]	UCI Cleveland, Statlog	NB, KNN, SVM, LR, DT, Xgboost	PCA, Chi-Square	14 & 12 respectively	92.85%
Islam et al. [4]	UCI heart disease	Hybrid Genetic algorithm with K-Means	PCA	2(after transformation)	94.06%
Rajalakshmi et al. [10]	UCI Cleveland	K-Means with Decision Tree C5.0 and Weighted Associative Classifier	Not specified	11	94.54%
Kavitha et al. [7]	UCI Cleveland	Random Forest, Decision Tree and a Hybrid Model	Not specified	14	88.7%
S.Mohan et al. [9]	UCI Cleveland	HRFLM (Random Forest and Linear Model)	Not specified	13	88.7%
Our Study	UCI Cleveland, Statlog & Kaggle	Supervised Models(Random Forest, KNN, Logistic Regression, Decision Tree) & Ensemble Models(AdaBoost, CatBoost)	Correlation Analysis& Chi Square test	13, 11 & 11 clinical feautures respectively	97.82%

algorithms using the UCI Cleveland dataset, employing decision trees, random forests, and a hybrid model combining random forests and decision trees. The hybrid model exhibited an accuracy rate of 88.7% in predicting cardiovascular disease. Mohan et al. [9] also employed machine learning-based models on the UCI Cleveland dataset, utilizing a hybrid HRFLM approach that combines random forest (RF) and linear model (LM). This hybrid approach demonstrated an accuracy of 88.7% in predicting heart disease.

Nevertheless, a brief review of the selected works suggests that there is potential for further exploration in various aspects, including hyperparameter optimization, more effective feature selection techniques, and more comprehensive model evaluation.

3 Methodology and Study Design

In this section, we elaborate on the essential phases of our study, as depicted in Fig. 1. These phases align with the typical stages found in machine learning-based research. However, our study features distinct adaptations, notably in hyperparameter tuning, dataset diversity, and a wide range of algorithm considerations, making it distinctive from the highlighted existing works (Table 1).

1. **Data Acquisition:** In this phase, we collect datasets from various sources, including the UCI Cleveland dataset [5], Statlog heart disease datasets [11], and the Kaggle database.
2. **Data Preprocessing:** In this phase, the datasets undergo preprocessing. The Interquartile Range (IQR) method is utilized for outlier detection and removal. This method segments the data into quartiles. The first quartile (Q1) indicates the value below which 25% of the data lies, while the third quartile (Q3) marks the value below 75% of the data. The IQR is calculated as:

$$IQR = Q3 - Q1 \tag{1}$$

Subsequently, the lower and upper bounds for identifying outliers are computed using Eqs. 2 and 3, respectively.

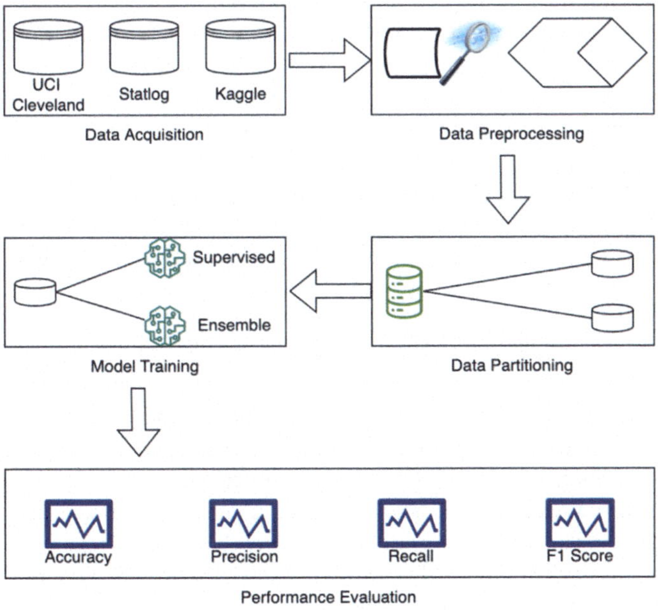

Fig. 1. Overview of Study Methodology Phases

$$\text{Lower\_limit} = Q1 - 1.5 \times IQR \tag{2}$$

$$\text{Upper\_limit} = Q3 + 1.5 \times IQR \tag{3}$$

For feature selection, correlation analysis, and the Chi-Square test are employed. These methods assess the dependency between attributes and the output class label. A higher Chi-Square value suggests a stronger association. The most relevant features are selected based on these methods.

3. **Data Partitioning:** The typical 80:20 split ratio is chosen to provide a substantial amount of data for training while reserving a significant portion for model validation. Ensuring this representative split helps mitigate the risks of overfitting and underfitting, thus enhancing the model's ability to generalize to unseen data.
4. **Model Training:** The training phase involves instructing machine learning algorithms-based models, which include both supervised and ensemble learning methods, to learn from the training data. The supervised methods employed include:
   - Random Forest (RF): An ensemble of decision trees, employing bagging to enhance model robustness against overfitting.
   - K Nearest Neighbours (KNN): A non-parametric method that classifies data points based on the majority vote of their neighbors, with k representing the number of neighbors considered.
   - Logistic Regression (LR): A regression model where the dependent variable is categorical, ideal for binary classification tasks.
   - Decision Trees (DT): A model that segments the data into subsets based on feature values, forming a tree-like structure of decisions.

   Similarly, ensemble learning algorithms such as AdaBoost and CatBoost are included. These methods aggregate multiple base models to enhance predictive performance, often resulting in improved accuracy and stability. A typical formulation of an ensemble-based model can be written as follows:

$$\text{Ensemble Model} = \sum_{i=1}^{N} \alpha_i \times \text{Base Model}_i \tag{4}$$

   where α_i are the weights assigned to each base model, determined through the ensemble learning algorithm. These weights reflect the importance or confidence of each base model within the ensemble.
5. **Performance Evaluation:** The evaluation phase is critical in assessing the effectiveness of the trained models. This involves analyzing various metrics, each providing different insights into the model's predictive capabilities. The metrics employed in this study are:

Table 2. Attributes Availability Across Datasets

Attribute	Cleveland Dataset	Statlog Dataset	Kaggle Dataset
ID (Patient's ID)	×	×	✓
Age (Patient's Age)	✓	✓	✓
cp (Chest Pain Type)	✓	✓	×
Sex (Patient's Gender)	✓	✓	✓
Hypertension (presence of Hypertension)	×	×	✓
oldpeak (Exercise induced ST Depression)	✓	✓	×
Heart Disease (History of Heart Disease)	×	×	✓
trestbps (Blood Pressure while Resting)	✓	✓	×
Ever_married (marital status)	×	×	✓
thalach (Maximum Heart Rate attained)	✓	✓	×
work_type (Employement status)	×	×	✓
chol (Serum Cholesterol)	✓	✓	×
exang (Exercise-Induced Angina)	✓	✓	×
Residence_type (Rural or Urban)	×	×	✓
restecg (Resting Electrocardiographic Results)	✓	✓	×
avg_glucose_level (average glucose level in blood)	×	×	✓
fbs (Blood Sugar Level in the time of Fasting)	✓	✓	×
thal (Blood Disorder i.e. Thalassemia)	✓	×	×
bmi (body mass index)	×	×	✓
Slope (ST segment's slope during peak exercise)	✓	✓	×
smoking_status	×	×	✓
ca (Major Vessel's count)	✓	×	×
Target/Class/Stroke (CVD Prediction)	✓	✓	✓

- Precision: It is calculated as the proportion of correctly predicted positive observations (TP) to the total predicted positives (TP + FP).
- Recall (Sensitivity): Recall quantifies the fraction of TP that the model successfully identifies from all actual positive cases (TP + FN).
- Accuracy: It assesses the overall accuracy of the model by comparing the sum of correct predictions (TP and TN) with the total number of predictions made.
- F1-Score: It is the harmonic mean of precision and recall and becomes particularly insightful for datasets with imbalanced class distributions or when the impact of false positives and negatives is distinct.

Table 3. Performance Metrics of Supervised and Ensemble ML Models on Various Datasets

Dataset	Model Type	Model Used	Accuracy	Recall	Precision	F1-Score
UCI Cleveland	Supervised	KNN	0.978	0.94	0.97	0.98
UCI Cleveland	Supervised	DT	0.82	0.94	0.96	0.83
UCI Cleveland	Supervised	RF	0.87	0.87	0.93	0.87
UCI Cleveland	Supervised	LR	0.86	0.87	0.93	0.87
UCI Cleveland	Ensemble	AdaBoost	0.89	0.94	0.96	0.89
UCI Cleveland	Ensemble	CatBoost	0.80	0.94	0.93	0.91
Statlog Heart	Supervised	KNN	0.95	0.95	0.94	0.95
Statlog Heart	Supervised	DT	0.94	0.94	0.94	0.94
Statlog Heart	Supervised	RF	0.96	0.97	0.96	0.97
Statlog Heart	Supervised	LR	0.92	0.92	0.89	0.92
Statlog Heart	Ensemble	AdaBoost	0.932	0.95	0.92	0.93
Statlog Heart	Ensemble	CatBoost	0.938	0.94	0.94	0.94
Kaggle	Supervised	KNN	0.95	0.95	0.96	0.95
Kaggle	Supervised	DT	0.93	0.94	0.93	0.94
Kaggle	Supervised	RF	0.96	0.96	0.98	0.97
Kaggle	Supervised	LR	0.92	0.92	0.96	0.92
Kaggle	Ensemble	AdaBoost	0.92	0.93	0.95	0.93
Kaggle	Ensemble	CatBoost	0.93	0.94	0.94	0.94

Overall, our approach in this study begins with the meticulous collection of data, followed by its systematic partitioning. We then train a diverse range of machine learning-based models, emphasizing both accuracy and robustness. Finally, these models are rigorously evaluated using established metrics, ensuring their reliability and effectiveness.

4 Experiments and Discussions

4.1 Experiments: Setup and Datasets

Due to resource limitations, we utilized Google Colab notebooks for our experimental setup. Besides, as stated before, our study focuses on analyzing three distinct datasets, each aimed at detecting cardiac disease. For a comparison of attribute distribution and availability, refer to Table 2, which provides an overview of each dataset's attributes. Basically, we employed the UCI Cleveland Dataset, which comprises 13 clinical features and a binary target variable, to signify the presence (1) or absence (0) of cardiac disease. Similarly, the Statlog Heart Disease Dataset utilizes 11 clinical attributes, along with a binary target variable that follows the same coding scheme as the UCI dataset. Finally, the dataset sourced from Kaggle also employs 11 clinical features and a binary target variable, consistent with the previous datasets.

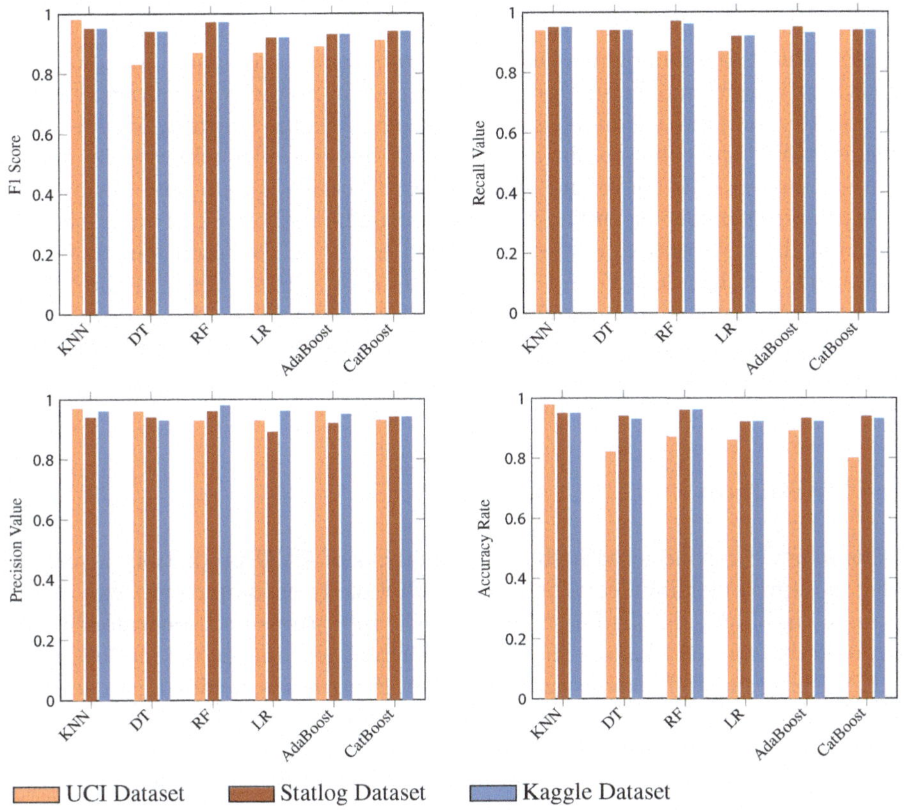

Fig. 2. Comparison of Machine Learning Models: F1 Score, Recall, Precision, and Accuracy Across Multiple Datasets

4.2 Discussions: Performance Evaluation

The metrics considered for evaluating the performance of our study are detailed in Sect. 3. Based on these metrics, Table 3 presents the results of our analysis. A key finding is that the KNN-based model, specifically configured with a leaf size of 45, Minkowski distance metric (with $p = 1$), uniform weights, and considering 5 neighbors, achieved the highest accuracy in the UCI Cleveland Dataset. This suggests that for this dataset, the KNN-based model's specific configuration plays a crucial role in its predictive performance. In contrast, the Random Forest Model initialized with a random seed value of 760, demonstrated superior accuracy on the Statlog and Kaggle Heart Disease datasets. This highlights the model's robustness and effectiveness in handling different types of heart disease data. Further, in the realm of Ensemble Learning Models, the AdaBoost-based model, with a base Random Forest Classifier consisting of 100 trees, achieved an 89% accuracy rate on the UCI Cleveland dataset. However, it was the CatBoost-based model, trained with a learning rate of 0.9, that outperformed AdaBoost, reaching an impressive accuracy rate of 93.8% on the Statlog and 93% on the Kaggle dataset. This indicates that CatBoost's advanced algorithms and learning rate optimization offer significant advantages in predictive accuracy. Figure 2 provides the comparative analysis of the various Machine Learning Models under consideration. According to our results, KNN-based model achieves the highest accuracy for the UCI Cleveland Dataset. Additionally, this model also demonstrates the best precision for the same dataset. The highest F1 Score in the UCI Cleveland dataset context is also attributed to the KNN-based model. On the other hand, the Random Forest-based model shows its strength in the Statlog Heart Disease dataset by achieving the highest recall value. To sum up, it can be said that these findings provide important insights into the performance of different models across various datasets, highlighting the need for careful model selection based on specific performance metrics.

5 Conclusion

In this paper, we conducted a thorough analysis of CVD prediction, leveraging widely accepted machine learning models. Our study encompassed three diverse datasets, enabling a detailed assessment of the robustness and generalizability of our predictive models. Among the various machine learning-based models tested, the KNN-based model demonstrated notable effectiveness, achieving an accuracy rate of 97.82%. From the Ensemble Learning Models standpoint, the CatBoost-based model also exhibited significant efficacy, with an accuracy rate of 93.8%. These results underscore the potential of our models as highly effective tools in supporting clinical decision-making, contributing valuable insights to the field of CVD prediction. However, it is also important to acknowledge that our study was conducted under certain assumptions and with relatively clean data sets. While the findings indicate a promising direction, they may not fully capture the complexities encountered in more realistic clinical scenarios. Therefore, further investigation might be needed. Future research should aim to explore more diverse and less curated datasets and consider the impact of real-world variables and conditions on the performance of these models. Additionally, future efforts may focus on feature engineering and the integration of additional clinical data sources to enhance model performance and utility. This approach will not only validate the effectiveness of the algorithms in practical settings but also refine their applicability in predicting CVD with greater accuracy and reliability. Our study, therefore, serves as an exploratory step towards more nuanced and comprehensive cardiovascular disease prediction research, acknowledging the ongoing journey toward achieving truly robust and generalizable models.

References

1. Durairaj, M., Revathi, V., et al.: Prediction of heart disease using back propagation MLP algorithm. Int. J. Sci. Technol. Res. **4**(8), 235–239 (2015)
2. Franco, M., Cooper, R.S., Bilal, U., Fuster, V.: Challenges and opportunities for cardiovascular disease prevention. Am. J. Med. **124**(2), 95–102 (2011)
3. Gandhi, M., Singh, S.N.: Predictions in heart disease using techniques of data mining. In: 2015 International Conference on Futuristic Trends on Computational Analysis and Knowledge Management (ABLAZE), pp. 520–525. IEEE (2015)
4. Islam, M.T., Rafa, S.R., Kibria, M.G.: Early prediction of heart disease using PCA and hybrid genetic algorithm with k-means. In: 2020 23rd International Conference on Computer and Information Technology (ICCIT), pp. 1–6. IEEE (2020)
5. Janosi, A., Steinbrunn, W., Pfisterer, M., Detrano, R.: Heart disease. https://archive.ics.uci.edu/dataset/45/heart+disease, Accessed July 2023. https://doi.org/10.24432/C52P4X
6. Katarya, R., Srinivas, P.: Predicting heart disease at early stages using machine learning: a survey. In: 2020 International Conference on Electronics and Sustainable Communication Systems (ICESC), pp. 302–305. IEEE (2020)
7. Kavitha, M., Gnaneswar, G., Dinesh, R., Rohith Sai, Y., Sai Suraj, R.: Heart disease prediction using hybrid machine learning model. In: 2021 6th International Conference on Inventive Computation Technologies (ICICT), pp. 1329–1333. IEEE (2021)
8. Mohan, N., Jain, V., Agrawal, G.: Heart disease prediction using supervised machine learning algorithms. In: 2021 5th International Conference on Information Systems and Computer Networks (ISCON), pp. 1–3. IEEE (2021)
9. Mohan, S., Thirumalai, C., Srivastava, G.: Effective heart disease prediction using hybrid machine learning techniques. IEEE Access **7**, 81542–81554 (2019)
10. Rajalakshmi, K., Nirmala, K.: Heart disease prediction with mapreduce by using weighted association classifier and k-means. Indian J. Sci. Technol. **9**(19), 1–7 (2016)
11. UCI Machine Learning Repository. Statlog (Heart). https://archive.ics.uci.edu/dataset/145/statlog+heart, Accessed July 2023. https://doi.org/10.24432/C57303
12. Shah, D., Patel, S., Bharti, S.K.: Heart disease prediction using machine learning techniques. SN Comput. Sci. **1**, 1–6 (2020)
13. Stevens, D., et al.: Modelling of longitudinal data to predict cardiovascular disease risk: a methodological review. BMC Med. Res. Methodol. **21**(1), 1–24 (2021)
14. Tasnim, F., Habiba, S.U.: A comparative study on heart disease prediction using data mining techniques and feature selection. In: 2021 2nd International Conference on Robotics, Electrical and Signal Processing Techniques (ICREST), pp. 338–341. IEEE (2021)
15. Yan, Y., Zhang, J.-W., Zang, G.-Y., Jun, P.: The primary use of artificial intelligence in cardiovascular diseases: what kind of potential role does artificial intelligence play in future medicine? J. Geriatr. Cardiol. JGC **16**(8), 585 (2019)

Deep W-Net: DNN for Spatial Saliency Prediction in Video Frames

Gaurav Sharma(✉) and Maheep Singh

Computer Science and Engineering, NIT Uttarakhand, Srinagar, India
{gauravsharma.phd2020,maheepsingh}@nituk.ac.in

Abstract. Deep neural networks (DNN) have recently been utilized to recognize saliency in images and movies. Various models have been put forth to anticipate spatial, temporal, and spatiotemporal saliency. This paper proposes a new approach for estimating spatial saliency in video frames. Our suggested approach, which achieves excellent results on various measuring parameters, is inspired by U-net architecture. We tested our model using video frame extracts and achieved satisfactory outcomes. Results on data sets with still imagery are also examined. The proposed model does not utilize any transfer learning approaches during any part of training or testing.

Keywords: U-net · Spatial saliency · Video data sets · Transfer learning · Deep Neural Network

1 Introduction

Video saliency is identifying the area that draws human attention for observation. While some pixels are in the background, most conspicuous components are concentrated in foreground regions. In image/video compression, analysis, summarization, Etc., traditional image/video saliency prediction/ recognition is utilized. While visual saliency varies on the area of the subject and may sometimes be different for everyone, detecting it in films and images has proven difficult.

Our work presents a DNN trained on frame level using a range of datasets. The model receives an associated frame for each frame. The classic U-Net architecture utilized for medical image segmentation served as an inspiration for this work's architecture. The notion of U-net architecture was intended to be applied to border domain [1,2]. To improve the precision and efficiency of video saliency detection in dynamic and complex visual environments, our work is motivated by the need to understand and model human visual attention using deep learning, as well as to advance various multimedia applications, improve user experiences, and contribute to advancements in fields like autonomous systems, security, and healthcare.

The paper is structured as follows. In Sect. 2, related works are reviewed. The proposed architecture is explained in Sect. 3. The experimental findings and data sets used for the training and testing portion are given in Sect. 4. The conclusion is presented in Sect. 5.

D. Pastor-Escuredo et al. (Eds.): ICDLAIR 2023, LNNS 1001, pp. 12–21, 2024.
https://doi.org/10.1007/978-3-031-60935-0_2

2 Related Work

For still pictures, a method for predicting saliency by comparing a multi-scale center-surrounding feature comparison with color, orientation, and intensity was developed by Itti et al. [3]. Harel et al. [4] applied graph theory to forecast picture saliency, following Itti et al. Background and foreground clues are crucial for predicting saliency. Yang et al. [5] employed graph-based manifold ranking to determine how comparable different image elements were using background and foreground signals. Their connection to the seed determined the saliency. It is advantageous to use image details and region-based attributes to determine picture saliency. Zhang et al. employed a model based on a multiple graph where the graph was constructed using picture attributes, and visual rarity was calculated for saliency assessment [6]. Zhang et al. [7] a model in which a local tree-structured low-rank representation technique obtained background. To improve salient object detection, some studies attempt to create universal frameworks from frameworks that already exist [8]. Wang et al. [9] mention where attention stimuli were utilized to examine saliency in stereoscopic images using fixation maps to identify salient objects.

Video saliency prediction considers motion information compared to saliency prediction in still images. Liu et al. created a super pixel-based model, also known as a saliency tree for video sequences [10]. According to Leboran et al. [11] computational approach, sensitive traits can be described by high-order statistical structure. Fang et al. proposed a video saliency prediction model that combined spatial and temporal saliency utilizing statistical uncertainty indicators [12]. To determine spatiotemporal saliency, Kin et al. used a steady-state distribution of walkers to represent random walks [13]. Li et al. [14] calculated the video saliency in videos by calculating the multivariate Gaussian data's lossy encoding length. Seo et al. employed a space-time local steering kernel to extract features from a video sequence [15]. Motion and spatial edges were considered by Wang et al. [16] for saliency assessment and treated as characteristics. Using object segmentation, Tu et al. model [17] included the object signature they discovered.

There are two aspects to the video saliency challenge: semantic segmentation and salient object identification. Semantic segmentation finds numerous things in videos and treats each equally, but salient object detection only finds one object in a video. Shen et al. [18] introduced a trajectory clustering approach employing sub-modular optimization for video saliency prediction; later, Shen et al. [19] employed Taylor expansion to create a lower-order function for segmenting images.

Many studies have been conducted due to the rapid progress of deep learning techniques in computer vision. Zhao et al. [20] created a multi-context DNN that considered local and global context information and used them to model a unified network. A fully convolutional neural network-based multitask deep learning network was proposed by Li et al. [21]. By extending their technique for films with handcrafted low-level features and deep contrast characteristics, Li et al. [22]. A DNN model that learned from both local and global features was proposed by Wang et al. [23]. Zang et al. [24] developed an approach by expressly penalizing the inaccuracy of the localization. For segmenting videos, Li et al. presented a neighborhood reversible flow and a complementary convolutional neural network [25]. The author of [26] introduced a methodology for employing visual saliency to address the issue of photo cropping. Wang et al. [27] employed hierarchical saliency information to address the saliency prediction issue.

2.1 Recent Work

A DNN was introduced by Wong et al. [28] to account for spatial and temporal saliency. Using an auto-encoder, Li et al. [29] suggested a DNN framework. RNN with ConvolutionalLSTM was utilized by Song et al. [30] to learn spatiotemporal information. The authors of [31] suggested an RNN that is flow-guided for video saliency detection. Saliency identification in images utilizing object proposal and omitting long-term temporal information was introduced by Guo et al. [32]. Shi et al. [33] mention a bi-directional method for extracting context information. Yang et al. [34] presented a bi-directional FCN to gather skeleton context information. W-Net architecture proposed by [35] resembles the English alphabet “W” and it uses a multi layer encoder and decoder for spatial saliency detection. The proposed model's idea and these investigations significantly differed based on diverse tasks. Our study suggests a Deep W-Net architecture influenced by conventional U-net architecture to determine spatial saliency in video frames. To learn spatial saliency, we created multi-layer encoders and decoders. In the first layer, all of the frame's pixels were used as features, and transfer learning was not used. In Sect. 4, experimental findings demonstrated that our model performs satisfactorily over various measuring parameters and data sets.

3 Proposed Model

This Section outlines the proposed Deep W-Net for predicting spatial saliency in videos. The proposed architecture is based on U-net [1] and W-Net [35]; nevertheless, we made several structural adjustments and evaluated the experimental findings to develop the proposed model. Google Colab is used for implementation, while Python was used on a local machine for frame extraction, resizing, and renaming of videos and image processing. Without applying transfer learning, every video frame pixel was considered input in our model. On any testing frame, no training has been conducted. Before modeling, testing videos were manually segregated.

3.1 Proposed Architecture

We employ an encoder and decoder to down-sample and up-sample the frames under U-net [1]. In contrast to U-net's concatenation, which has been performed between up-sampling and second down-sampling, multiple steps of down-sampling and up-sampling are performed similarly. Two stages of down-sampling

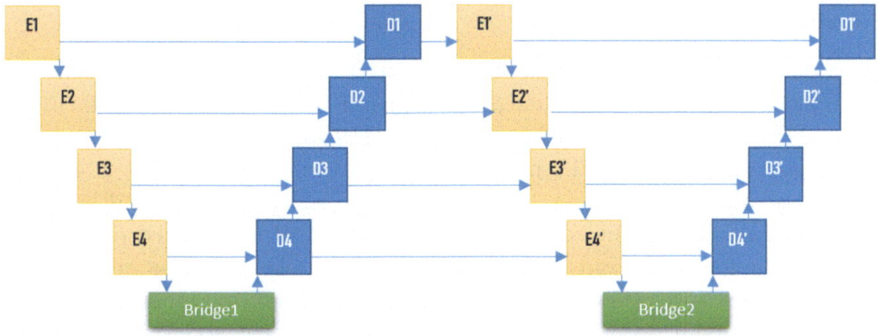

Fig. 1. Proposed Architecture (Deep W-Net)

and two of up-sampling make up the suggested model. E1, E2, and so on stand for encoder blocks, and the D series stands for decoder blocks, as seen in Fig. 1. Bridge1 is the convolution of the output of the E4 encoder performed by convBlock, as seen in Fig. 2. The convolution performed by E4's encoder is Bridge2. Convolutional transpositions of Bridges 1 and 2 are taken by D4 and D4', respectively. The convolution results of convBlock's (Fig. 2) output are concatenated to the convolution transpose output of D blocks; convolution from D blocks are concatenated to convolution from E1'–E4'.

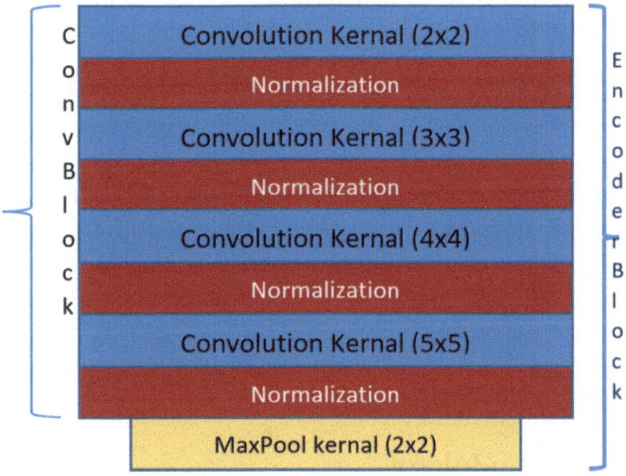

Fig. 2. Encoder Block

3.2 Encoder Block

Six encoder blocks are present in the suggested model. Four convolution layers and a normalizing layer follow each convolution in the encoder block. Figure 2 shows a convolution block. Each convBlock is followed by a Maxpooling layer to downsample. The input for the first convolution layer, C1, is 224 × 224 pictures (frames), and the kernel size is 2 × 2. C2 receives input from C1 and has a 3 × 3 kernel. Using a 4 × 4 kernel size in C3, the output of C2 is used as the input. C4 accepts the features from C3 as input. In C4, the kernel measures 5 × 5. The ability to extract low-level to high-level features is aided by kernel size variation. For all convolutional layers, padding is the same, but the stride is 1. Bridgel is accomplished from E4; however, there is no Maxpooling. Similarly, Bridge2 is produced by E4's output. In various phases of our model, 2 × 2 kernel size max pooling is performed.

3.3 Decoder Block

The up-sampling decoder component is shown in Fig. 3. Convolutional transposition layer, concatenation, and convBlock are included in this block. ConvBlock is the encoder block's convBlock. Concatenation is used to map feature information from the subsequent convolutions of the encoder block when convolutional transpose is used for up-sampling. Convolutional transposition uses a 2 X 2 pixel kernel with a 2 pixel stride.

3.4 Training Process

The proposed model was trained using the DAVSOD, SegTrack, and ViSal data sets. 80 % data is used for training, while 20 % is used for testing. Learning parameters involves minimizing the loss function. The loss function is the binary cross entropy. The proposed model, Deep W-Net, uses the Adam optimizer to identify the optimized network parameter while minimizing the loss function. End-to-end training is conducted using a batch size of 3 and a learning rate 0.001.

4 Experimental Results

4.1 Evaluation Methodology

We conducted experiments on DAVSOD [36], SegTrack [37], and ViSal [38] datasets to assess the performance of our model Deep W-Net. Two hundred twenty-six videos with 23938 frames can be found in the DAVSOD data set. 1000 frames make up SegTrack, while 17 videos from various sources are found in ViSal. F-measure, S-measure, and Mean Absolute Error (MAE) are utilized for performance comparison. F-measure is calculated as in Eq. 1.

$$F - measure = 2 \times P \times R \div (P + R) \tag{1}$$

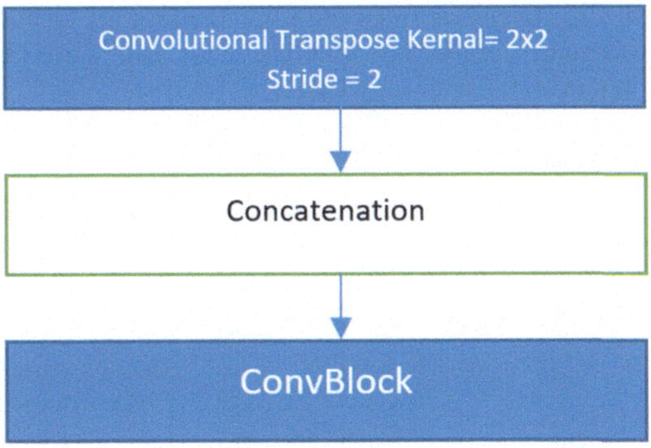

Fig. 3. Decoder Block

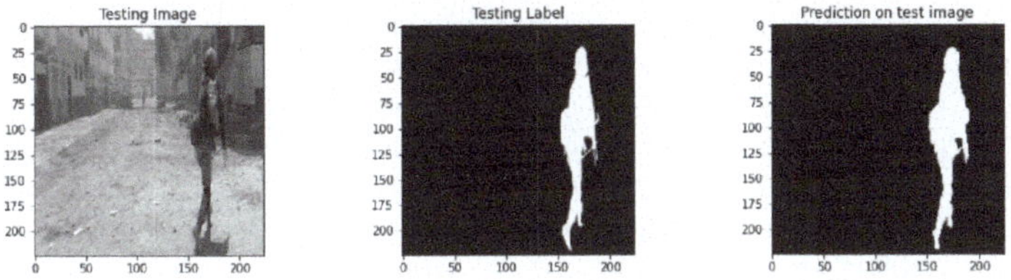

Fig. 4. Visual Comparison

where Precision is calculated as in Eq. 2.

$$Precision(P) = TP/(TP + FP) \tag{2}$$

where FP stands for false positive and TP for true positive. Recall is calculated as in Eq. 3.

$$Recall(R) = TP/(TP + FN) \tag{3}$$

where FN stands for false negative and TP for true positive. The region-aware structural similarity (S-measure) metric is calculated as:

$$Sr = \sum_{k=1}^{K} w_k * ssim(k) \tag{4}$$

wk is the weight assigned to each block, K is the total number of blocks in the sub-image, and ssim is the structural similarity metric. MAE is calculated as,

$$MAE = 1/n \sum_{i=1}^{n} |x_i - x| \tag{5}$$

where n is number of samples(frames) and absolute error is
$|x_i - x|$ where x and x_i are ground truth and model's result respectively.

Table 1. Quantitative Comparisons Between Our Model and Other Sota Vsod Models Using The F-Max, S-Measure, and Mae Metrics. The Top Three Results are Respectively Highlighted in RED, GREEN, and BLUE.

Dataset	SegTrackV2			ViSal			DAVSOD		
Metric / Model	F-Max	S-Measure	MAE	F-Max	S-Measure	MAE	F-Max	S-Measure	MAE
Ours	0.871	0.901	0.015	0.892	0.924	0.021	0.688	0.762	0.066
ESTI 2021	0.860	0.891	0.017	0.952	0.952	0.013	0.651	0.746	0.086
MGA 2019	0.821	0.865	0.030	0.933	0.936	0.017	0.640	0.738	0.084
CPD 2019	0.778	0.841	0.023	0.941	0.94	0.016	0.608	0.724	0.092
DLVS 2018	-	-	-	0.851	0.881	0.048	0.521	0.657	0.121
SCOM 2018	0.764	0.815	0.030	0.831	0.762	0.122	0.464	0.599	0.220
SFLR 2017	0.745	0.804	0.037	0.779	0.814	0.062	0.478	0.624	0.132
SGSP 2017	0.673	0.681	0.124	0.677	0.706	0.165	0.426	0.577	0.207
STBP 2017	0.640	0.735	0.061	0.622	0.629	0.163	0.410	0.568	0.160
MSTM 2016	0.526	0.643	0.114	0.673	0.749	0.095	0.344	0.532	0.211
GFVM 2015	0.592	0.699	0.091	0.683	0.757	0.107	0.334	0.553	0.167
SAGM 2015	0.634	0.719	0.081	0.688	0.749	0.105	0.370	0.565	0.184
MB+M 2015	0.554	0.618	0.146	0.692	0.726	0.129	0.342	0.538	0.228
RWRV 2015	0.438	0.583	0.162	0.440	0.595	0.188	0.283	0.504	0.245
SPVM 2014	0.618	0.668	0.108	0.700	0.724	0.133	0.358	0.538	0.202
TIMP 2014	0.573	0.644	0.116	0.479	0.612	0.170	0.395	0.563	0.195
SIVM 2010	0.581	0.605	0.251	0.522	0.606	0.197	0.298	0.486	0.288

4.2 Comparative Research Employing Techniques that Are Already Proposed

The proposed model is contrasted with other cutting-edge models in this section. Comparison with ESTI [50], MGA [49], CPD [39], DLVS [28], SCOM [40], SFLR [41], SGSP [42], STBP [43], MST [44], GFVM [38], SAGM [16], MB+M [45], RWRV [13], SPVM [46], TIMP [47], SIVM [48] is shown in Table.1. In this work, F-measure, S-measure, and MAE are used to compare experimental results. Figure 4 presents our results for comparison between prediction and ground truth. The proposed model Deep W-Net performs best on DAVSOD and SegTrack Dataset with F-Max as 0.688 and 0.871, S-Measure as 0.762 and 0.901, and MAE as 0.066 and 0.015, respectively. Overall, the effectiveness of the proposed model, Deep W-net, is convincingly evidenced by the F-Max, S-measure, and MAE values shown in Table 1.

5 Conclusion

This paper presented a novel approach (Deep W-Net) for detecting salient video objects. The proposed model uses an encoder and decoder block to map information from the encoder to the decoder and from the decoder to the encoder. We used each pixel from the normalized frames as a feature to train the proposed model. Extensive experiments have been done on three benchmark datasets, and the outcomes show that the proposed Deep W-Net has achieved a significant performance compared to other related models. In future work, we would explore cross-modal fusion, weakly supervised learning, and integrating temporal-spatial information for precise and adaptable models.

References

1. Ronneberger, O., Fischer, P., Brox, T.: U-net: convolutional networks for biomedical image segmentation. In: Navab, N., Hornegger, J., Wells, W., Frangi, A. (eds.) Medical Image Computing and Computer-Assisted Intervention – MICCAI 2015, MICCAI 2015, LNCS, vol. 9351, pp. 234–241. Springer, Cham (2015). https://doi.org/10.1007/978-3-319-24574-4_28
2. Sharma, G., Singh, M., Kumain, S.C.: W-net plus: DNN for spatial saliency prediction in videos. In: 2023 1st International Conference on Innovations in High Speed Communication and Signal Processing (IHCSP), pp. 263–267. IEEE, March 2023
3. Itti, L., Koch, C., Niebur, E.: A model of saliency-based visual attention for rapid scene analysis. IEEE Trans. Pattern Anal. Mach. Intell. **20**(11), 1254–1259 (1998)
4. Harel, J., Koch, C., Perona, P.: Graph-based visual saliency. In: Advances in Neural Information Processing Systems, vol. 19 (2006)
5. Yang, C., et al.: Saliency detection via graph-based manifold ranking. In: Proceedings of the IEEE Conference on Computer Vision and Pattern Recognition (2013)
6. Zhang, J., et al.: A novel graph-based optimization framework for salient object detection. Pattern Recogn. **64**, 39–50 (2017)
7. Zhang, Q., et al.: Salient object detection employing a local tree-structured low-rank representation and foreground consistency. Pattern Recogn. **92**, 119–134 (2019)
8. Lei, J., et al.: A universal framework for salient object detection. IEEE Trans. Multimedia **18**(9), 1783–1795 (2016)
9. Wang, W., et al.: Stereoscopic thumbnail creation via efficient stereo saliency detection. IEEE Trans. Vis. Comput. Graph. **23**(8), 2014–2027 (2016)
10. Liu, Z., Zou, W., Le Meur, O.: Saliency tree: a novel saliency detection framework. IEEE Trans. Image Process. **23**(5), 1937–1952 (2014)
11. Leboran, V., et al.: Dynamic whitening saliency. IEEE Trans. Pattern Anal. Mach. Intell. **39**(5), 893–907 (2016)
12. Fang, Y., et al.: Visual attention modeling for stereoscopic video: a benchmark and computational model. IEEE Trans. Image Process. **26**(10), 4684–4696 (2017)
13. Kim, H., et al.: Spatiotemporal saliency detection for video sequences based on random walk with restart. IEEE Trans. Image Process. **24**(8), 2552–2564 (2015)
14. Li, Y., Zhou, Y., Yan, J., Niu, Z., Yang, J.: Visual Saliency Based on Conditional Entropy. In: Zha, H., Taniguchi, Ri., Maybank, S. (eds.) Computer Vision - ACCV 2009, ACCV 2009, LNCS, vol. 5994, pp 246–257. Springer, Berlin, Heidelberg (2010). https://doi.org/10.1007/978-3-642-12307-8_23
15. Seo, H.J., Peyman, M.: Static and space-time visual saliency detection by self-resemblance. J. Vis. **9**(12), 15–15 (2009)
16. Wang, W., Shen, J., Porikli, F.: Saliency-aware geodesic video object segmentation. In: Proceedings of the IEEE Conference on Computer Vision and Pattern Recognition (2015)

17. Tu, Z., et al.: Fusing disparate object signatures for salient object detection in video. Pattern Recogn. **72**, 285–299 (2017)
18. Shen, J., Peng, J., Shao, L.: Submodular trajectories for better motion segmentation in videos. IEEE Trans. Image Process. **27**(6), 2688–2700 (2018)
19. Shen, J., et al.: Higher order energies for image segmentation. IEEE Trans. Image Process. **26**(10), 4911–4922 (2017)
20. Zhao, R., et al.: Saliency detection by multi-context deep learning. In: Proceedings of the IEEE Conference on Computer Vision and Pattern Recognition (2015)
21. Li, X., et al.: Deepsaliency: multi-task deep neural network model for salient object detection. IEEE Trans. Image Process. **25**(8), 3919–3930 (2016)
22. Li, G., Yizhou, Yu.: Visual saliency detection based on multiscale deep CNN features. IEEE Trans. Image Process. **25**(11), 5012–5024 (2016)
23. Wang, L., et al.: Deep networks for saliency detection via local estimation and global search. In: Proceedings of the IEEE Conference on Computer Vision and Pattern Recognition (2015)
24. Zhang, Y., et al.: Improving object detection with deep convolutional networks via bayesian optimization and structured prediction. In: Proceedings of the IEEE Conference on Computer Vision and Pattern Recognition (2015)
25. Li, J., et al.: Primary video object segmentation via complementary CNNs and neighborhood reversible flow. In: Proceedings of the IEEE International Conference on Computer Vision (2017)
26. Wang, W., Shen, J., Ling, H.: A deep network solution for attention and aesthetics aware photo cropping. IEEE Trans. Pattern Anal. Mach. Intell. **41**(7), 1531–1544 (2018)
27. Wang, W., Shen, J.: Deep visual attention prediction. IEEE Trans. Image Process. **27**(5), 2368–2378 (2017)
28. Wang, W., Shen, J., Shao, L.: Video salient object detection via fully convolutional networks. IEEE Trans. Image Process. **27**(1), 38–49 (2017)
29. Li, J., Xia, C., Chen, X.: A benchmark dataset and saliency-guided stacked autoencoders for video-based salient object detection. IEEE Trans. Image Process. **27**(1), 349–364 (2017)
30. Song, H., et al.: Pyramid dilated deeper convlstm for video salient object detection. In: Proceedings of the European Conference on Computer Vision (ECCV) (2018)
31. Li, G., et al.: Flow guided recurrent neural encoder for video salient object detection. In: Proceedings of the IEEE Conference on Computer Vision and Pattern Recognition (2018)
32. Guo, F., et al.: Video saliency detection using object proposals. IEEE Trans. Cybern. **48**(11), 3159–3170 (2017)
33. Shi, H., et al.: Bidirectional long short-term memory variational autoencoder. In: BMVC (2018)
34. Yang, F., et al.: Multi-scale bidirectional FCN for object skeleton extraction. In: Proceedings of the AAAI Conference on Artificial Intelligence, vol. 32. no. 1 (2018)
35. Sharma, G., Singh, M., Kumain, S.C.: W-net plus: DNN for spatial saliency prediction in videos. In: 2023 1st International Conference on Innovations in High Speed Communication and Signal Processing (IHCSP), pp. 263–267. IEEE (2023)
36. Fan, D.-P., et al.: Shifting more attention to video salient object detection. In: Proceedings of the IEEE/CVF Conference on Computer Vision and Pattern Recognition (2019)
37. Li, F., et al.: Video segmentation by tracking many figure-ground segments. In: Proceedings of the IEEE International Conference on Computer Vision (2013)

38. Wang, W., Shen, J., Shao, L.: Consistent video saliency using local gradient flow optimization and global refinement. IEEE Trans. Image Process. **24**(11), 4185–4196 (2015)
39. Wu, Z., Li, S., Qingming, H.: Cascaded partial decoder for fast and accurate salient object detection. In: Proceedings of the IEEE/CVF Conference on Computer Vision and Pattern Recognition (2019)
40. Chen, Y., et al.: SCOM: spatiotemporal constrained optimization for salient object detection. IEEE Trans. Image Process. **27**(7), 3345–3357 (2018)
41. Chen, C., et al.: Video saliency detection via spatial-temporal fusion and low-rank coherency diffusion. IEEE Trans. Image Process. **26**(7), 3156–3170 (2017)
42. Liu, Z., et al.: Saliency detection for unconstrained videos using superpixel-level graph and spatiotemporal propagation. IEEE Trans. Circ. Syst. Video Technol. **27**(12), 2527–2542 (2016)
43. Xi, T., et al.: Salient object detection with spatiotemporal background priors for video. IEEE Trans. Image Process. **26**(7), 3425–3436 (2016)
44. Tu, W.-C., et al.: Real-time salient object detection with a minimum spanning tree. In: Proceedings of the IEEE Conference on Computer Vision and Pattern Recognition (2016)
45. Zhang, J., et al.: Minimum barrier salient object detection at 80 fps. In: Proceedings of the IEEE International Conference on Computer Vision (2015)
46. Liu, Z., Zhang, X., Luo, S., Le Meur, O.: Superpixel-based spatiotemporal saliency detection. IEEE Trans. Circuits Syst. Video Technol. **24**(9), 1522–1540 (2014)
47. Zhou, F., Bing Kang, S., Cohen, M.F.: Time-mapping using space-time saliency. In: Proceedings of the IEEE Conference on Computer Vision and Pattern Recognition (2014)
48. Rahtu, E., Kannala, J., Salo, M., Heikkilä, J.: Segmenting salient objects from images and videos. In: Daniilidis, K., Maragos, P., Paragios, N. (eds.) Computer Vision - ECCV 2010, ECCV 2010, LNCS, vol. 6315, pp. 366–379. Springer, Berlin (2010). https://doi.org/10.1007/978-3-642-15555-0_27
49. Li, H., et al.: Motion guided attention for video salient object detection. In: Proceedings of the IEEE/CVF International Conference on Computer Vision (2019)
50. Chen, C., et al.: Exploring rich and efficient spatial temporal interactions for real-time video salient object detection. IEEE Trans. Image Process. **30**, 3995–4007 (2021)

Optimal Pose Estimation with Particle Filters Using Unpowered Wheels

Aprajit Singh Jamwal, Arushi Singhal, Atul Saini, Mishthi Jain, and Ravinder Singh(✉)

Netaji Subhas University of Technology, New Delhi, India
engg.ravindersingh@yahoo.com

Abstract. Accurate localization of the mobile robot is an essential element of autonomous vehicles that guarantees reliable navigation and secure passage along predetermined paths. Wheel odometry data from the powdered wheels and GPS-based location are the two main information sources for the pose estimation However, depending entirely on GPS might result in subpar performance owing to inherent errors and signal disturbances, particularly in metropolitan settings with tall buildings or tunnels. In addition, wheel odometry is used to increase precision and real-time responsiveness however, the reliability of the wheel odometry is reduced as the surface becomes slippery or the torque generation is very high. This resulted in the wheel slipping which is target as the problem statement in this research paper. In this proposed approach, the wheel odometry data is taken from the wheel encoder with unpowered wheel is taken that only change if the actual position is changed. As the unpowered wheel can only rotate when the mobile robot is moved however, the powered wheel may rotate even the robot is on stand still in the slippery surface. This proposed technique is implemented in various trails and the resulted are compared with various technique to check the reliability of the proposed technique. It has been experimentally obtained that the accuracy in the localization of the mobile robot is increased by 33.16% as compared to the conventional technique. This proposed technique may be implemented in path tracking, path planning, autonomous navigation, trajectory optimization, etc., to achieve accurate localization.

Keywords: particle filter · GPS · wheel odometry · path tracking · localization

1 Introduction

In recent years, there have been notable breakthroughs in the field of autonomous vehicle development without human guidance. Path tracking is one of the essential elements that has contributed to the success of autonomous cars. It involves accurately following a particular route while taking road geometry into consideration. Autonomous vehicles utilize various techniques such as Multi-Data Sensor Fusion, Kalman Filters/ Extended Kalman filter, Bayesian probabilistic approach, etc., to fuse the data from numerous sensors and minimize the possibility of accidents. These methods have made it possible

D. Pastor-Escuredo et al. (Eds.): ICDLAIR 2023, LNNS 1001, pp. 22–32, 2024.
https://doi.org/10.1007/978-3-031-60935-0_3

to improve trajectory tracking applications by fusing data from IMU (Inertial Measurement Units), GPS (Global Position System) -based positions, and wheel odometry [1–4]. Wheel odometry data improves accuracy in the near term, but it can accumulate inaccuracies over time, in the form of wheel slippage. For precise tracking, these independent data sets are generally employed together using data fusion techniques. These techniques, however, are costly. This research work is primarily focused on cost-effective localization. The paper begins with Sect. 2 and explores the various techniques employed for path tracking throughout the years. Section 3 provides the essential background on GPS and Wheel Odometrical Based Data that forms the core foundation of the paper. Section 4 focusses on acquiring data based on GPS and Wheel Odometry to provide a comparative analysis. Section 5, then proposes an efficient and cost-effective methodology to provide measurements with minimized error. The data is corroborated by the test simulations performed. Conclusions are provided then.

2 Literature Review

Earlier attempts in path tracking by autonomous cars employed a variety of techniques to ensure that tracking was accurate even in low-signal environments. One of the methods uses inexpensive vision sensors and probabilistic methods for multiple sensor fusion. Using a low-depth sensor, an Extended Kalman Filter is merged with another EKF algorithm [3]. Another method improved pose estimation by applying an Unscented Kalman filter (UKF) fusing data from odometry, IMU, and GPS using the new uncertainty measure [5–14]. The development and implementation of a high-integrity navigation system based on the joint use of the Global Positioning System (GPS) and an inertial measurement unit (IMU) for autonomous land vehicle applications is one of the most well-known algorithms for autonomous navigation [1]. These methods are employed singly or collectively to follow the course of an autonomous vehicle in poor signaling regions, guaranteeing precise and dependable navigation in such cases. One of the drawbacks that was still seen: the robot can successfully track the desired trajectory with 0.3 m accuracy only [3]. Another well renowned method for the accurate localization of mobile robot is to get information from the wheel odometry. This local movement information gives the vehicle more precise and rapid feedback on its motion, enabling it to identify deviations from the desired course and make the appropriate corrections. However, this approach favors low complexity environment with small acceleration. If the acceleration is very high, effects of wheel slipping dominates which results in inaccuracies [10]. This problem statement is targeted in this study to reduce the effect of wheel slipping and attain accurate localization.

The research significance of the proposed technique is as follows:

- Error analysis of the particle filter with GPS data for the location of the mobile robot.
- Error analysis of the particle filter with wheel odometry w.r.t the wheel slippage
- Reduction of error w.r.t wheel slipping by acquiring data from unpowered wheels.

3 Problem Statement

The accurate localization of mobile robot is an essential requirement for efficient trajectory tracing. As per literature, various techniques such as GPS, IMU, wheel odometry, hybrid version of GPS with wheel odometry, etc., has been used for the reliable navigation of autonomous mobile robot. GPS is very reliable in terms of accuracy and precision for tracking and localization-related problems. The drawbacks usually occur in the strength of the signal on the receiver's end. This has a significant impact during the utilization of GPS measurement datasets. As soon as a low signal strength area is encountered, the robot pose gets distorted due to the presence of large errors. The aim of the paper is to provide a reliable methodology to be used during such cases. At preliminary stage, proposed solution includes the usage of wheel odometrical data to compensate for the loss of poses. But the proposed solution is itself very troublesome due to a complex phenomenon, known as wheel slipping that produces erroneous data [10]. Taking the setup to the next step, the paper, then, offers a unique technique to overcome the shortcoming proposed by the preliminary solution. The measurements from odometrical data are sourced from unpowered wheels. The above problem statement has been briefly explored in three stages: Effect on tracking accuracy when GPS data is only used, when GPS and wheel odometrical data accommodated with wheel slipping and the technique proposed to minimize error due to wheel slipping by considering the wheel odometry from unpowered wheels.

4 Error Analysis in Conventional Approaches

The capacity to travel precisely and securely along planned paths is crucial for autonomous cars. Utilizing data from the GPS becomes an essential factor for this complex operation of route tracking. Wheel Odometry, on the other hand, is a technique that relies on internal parameters like wheel rotations to provide the position information of the autonomous vehicles [10]. These conventional approaches are explored in the subsequent sections.

4.1 GPS Based Path Tracking Analysis

A network of satellites called the Global Positioning System continually transmits signals with exact timing and position information from its orbit around the Earth. These signals may be picked up by autonomous cars equipped with GPS receivers, allowing them to instantly identify their position, speed, and direction. This global positioning data provides the coordinates of the vehicle inside a global reference frame, which forms the basis for planning and executing precise courses [4]. 5000 particles are employed by the particle filter, which means that it has 5000 likely positions at any instance of time. Particle Filter chooses from, among those 5000 estimated poses, the best feasible predictions based on the input. These estimations are computed using the linear and angular velocities given as control input. The hybrid solution of GPS and particle filter provide quite an accurate pose estimation however, as the GPS information gets distorted because of roofed area (or low signal strength area), the error comes into play as shown in Fig. 1. The type of receiver used is generally different with varying signal strength requirements. Thus, the strength of the signal becomes a key factor for high accuracy. Generally, the threshold for signal strength is taken to be 25 dB [11]. Below this, the error accumulation is very high in the measurements. The program is set up such that we receive accurate readings above this threshold and readings are discarded below it.

When signal strength is less than 25 dB threshold, GPS measurements won't be provided to the particle filter due to large accumulation of error as seen in Fig. 1. In this case, "Dead Reckoning" aids in estimation. Here, the previous pose and the general motion model of the wheeled robot predicts the future poses in an iterative process [12], this is mathematically represented in Eq. 1 by m

$$\mathbf{m} = \mathbf{p}(\mathbf{x_{t+1}} | \mathbf{x_t}, \mathbf{u_{t+1}}) \tag{1}$$

where

$\mathbf{x_{t+1}}$ = Pose at time t + 1

$\mathbf{x_t}$ = Robot Pose measurement at current time t

$\mathbf{u_{t+1}}$ = Control Input of the robot at time t + 1

When GPS signal strength is restored, particle filter continues depending on GPS. Signal Strength is often reduced by thick structures, or tunnel causing errors, this is known as Signal Interference. GPS signal often falls short in applications requiring centimeter-level precision [4]. Thus, GPS alone is not enough, as demonstrated in Fig. 1 and Table 1, hence, Wheel Odometrical datasets are used in weak reception areas.

4.2 Wheel Odometry Data-Based Path Tracking Analysis

Wheel odometry measures wheel revolutions using encoders and is an effective method for figuring the movement of vehicle [10] as shown in Eqs. 2, 3, 4 and 5. The net displacement mentioned in Eq. 6 is accounted in the past position to determine the present pose using linear and angular velocities of the wheeled robot.

$$\mathbf{v^{lin}_{leftwheel}} = (\mathbf{v_{vehicle}} + \boldsymbol{\omega}_{\mathbf{vehicle}}\ \mathbf{L})/2 \tag{2}$$

where

$\mathbf{v_{vehicle}}$ = represents the linear velocity of the vehicle.

$\boldsymbol{\omega}_{\mathbf{vehicle}}$ = represents the angular velocity of the vehicle.

$\mathbf{L}$ = signifies the wheelbase.

$\mathbf{v^{lin}_{leftwheel}}$ = represents the linear velocity of left wheel.

Similarly,

$$\mathbf{v^{lin}_{rightwheel}} = (\mathbf{v_{vehicle}} - \boldsymbol{\omega}_{\mathbf{vehicle}}\mathbf{L})/2 \tag{3}$$

*This negative sign ensures the slowing down of one wheel when there is change in direction.

$$\mathbf{d_{leftwheel}} = \mathbf{v^{lin}_{leftwheel}}/\mathbf{time} \tag{4}$$

where, $\mathbf{d_{leftwheel}}$ = represents the displacement of the left wheel.

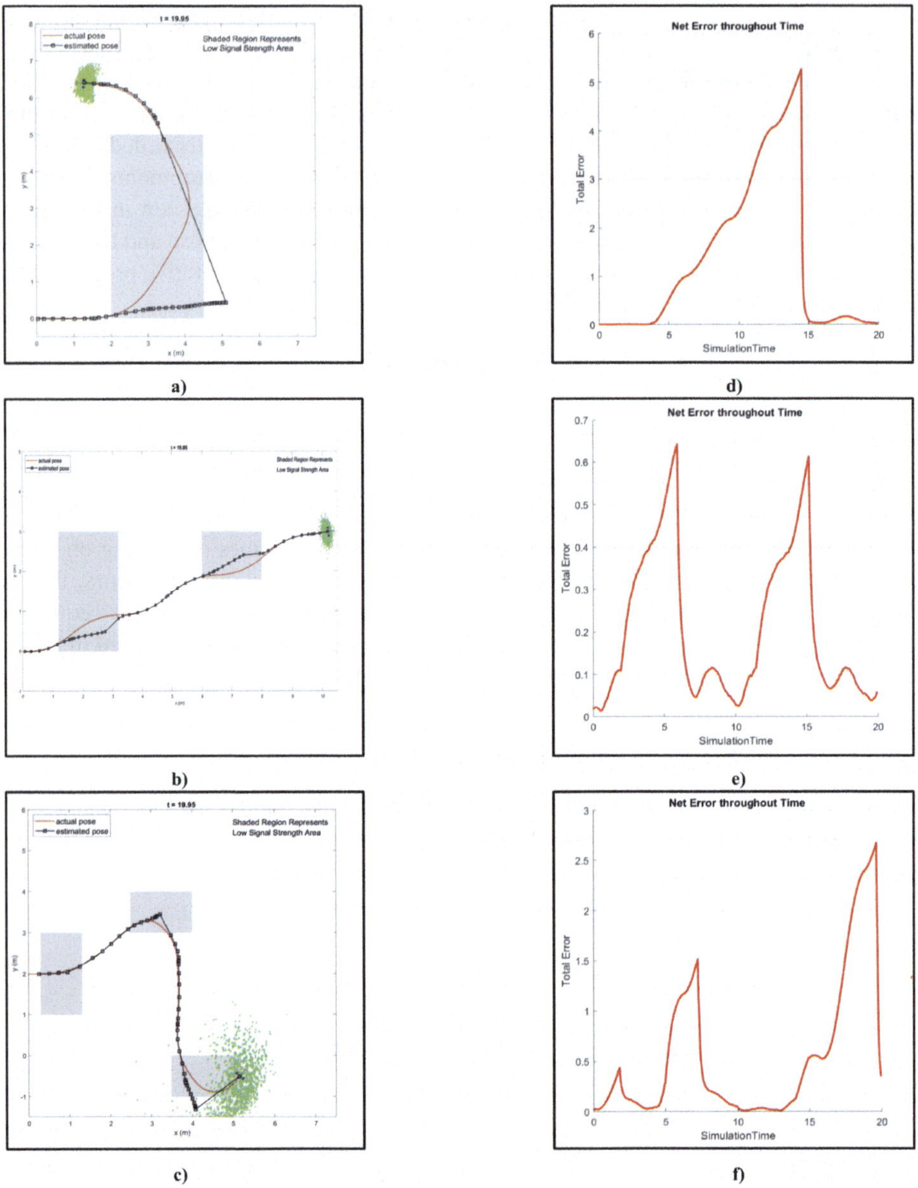

Fig. 1. Simulation of a wheeled robot in Trajectories (a), (b) and (c). Figures (d), (e) and (f) depicts the net error due to positional and angular parameters. The GPS tracking efficiency, decreases with increased steepness or large weak-reception area. Otherwise, the error plot stays near 0.1 as shown in Fig. 1e.

Similarly,

$$\mathbf{d_{rightwheel}} = \mathbf{v^{lin}_{rightwheel}/time} \tag{5}$$

Table 1. Stage 1: Accuracy in Localization vs Time based on GPS data. The simulation consists of 5 trials per trajectory with different input parameters. Constant parameters include simulation time (t = 19.95 s), threshold factor for net error ($\gamma = 0.1$) and the net number of particle for measurement data. ($\delta_t = 400$).

	Trial 1	Trial 2	Trial 3
$\boldsymbol{\delta_\gamma}$	144	166	172
$\boldsymbol{\alpha^{(x,y,\theta)}{}_{(t,1)}}$	36%	41.5%	43%

Where $\boldsymbol{\delta_\gamma}$, signifies the net number of particles that have net error below the threshold factor and $\boldsymbol{\alpha^{(x,y,\theta)}{}_{(t,1)}}$ represents the accuracy maintained by the particles that contain information x, y, θ during the computational time for stage 1. It is computed as the ratio of $\delta_{\gamma,1}$ and δ_t.

$$\mathbf{NetDisplacement} = (\mathbf{d_{leftwheel}} + \mathbf{d_{rightwheel}})/2 \tag{6}$$

Wheel odometry dataset is acquired using Eqs. 3, 4, 5, and 6. A vehicle's turning angle or change in orientation may also be predicted by comparing the speeds of the wheels on its two sides [13].

Wheel Odometry poses certain challenges in the form of wheel slippage. Its nature is unpredictable and is majorly dependent upon acceleration of the vehicle and the steepness of the turn [13]. The car covers an amount of distance which is considerably lower than the real distance due to wheel slipping. This erroneous data is then provided to particle filter. Due to these reasons, wheel odometrical data is not reliable for a large time interval. Table 2, depicts the results including wheel slippage (Fig. 2).

The amount of slippage significantly impacts the accuracy of the calculated position as shown in Table 2. It depends on many variables, such as topography, the acceleration of the wheeled robot (see Eq. 4), the size of the wheels, the wheelbase of the wheeled robot, etc. [13]. This paper has a streamlined accommodation for wheel slippage.

$$\boldsymbol{\omega^{s}_{wheel}} = \boldsymbol{\beta}(\boldsymbol{\omega^{o}_{wheel}}) \tag{7}$$

where

$\boldsymbol{\omega^{s}_{wheel}}$ = represents the angular velocity of wheel with slip.
$\boldsymbol{\omega^{o}_{wheel}}$ = represents the angular velocity of wheel without slip.
$\boldsymbol{\beta}$ = signifies the factor by which angular velocity is amplified.

The individual angular velocities of the wheels are predicted to rise by a certain factor during bends as shown in Eq. 7. Wheel slippage is introduced for a split second [13]. This approximation is crucial to show the efficacy of the proposed technique.

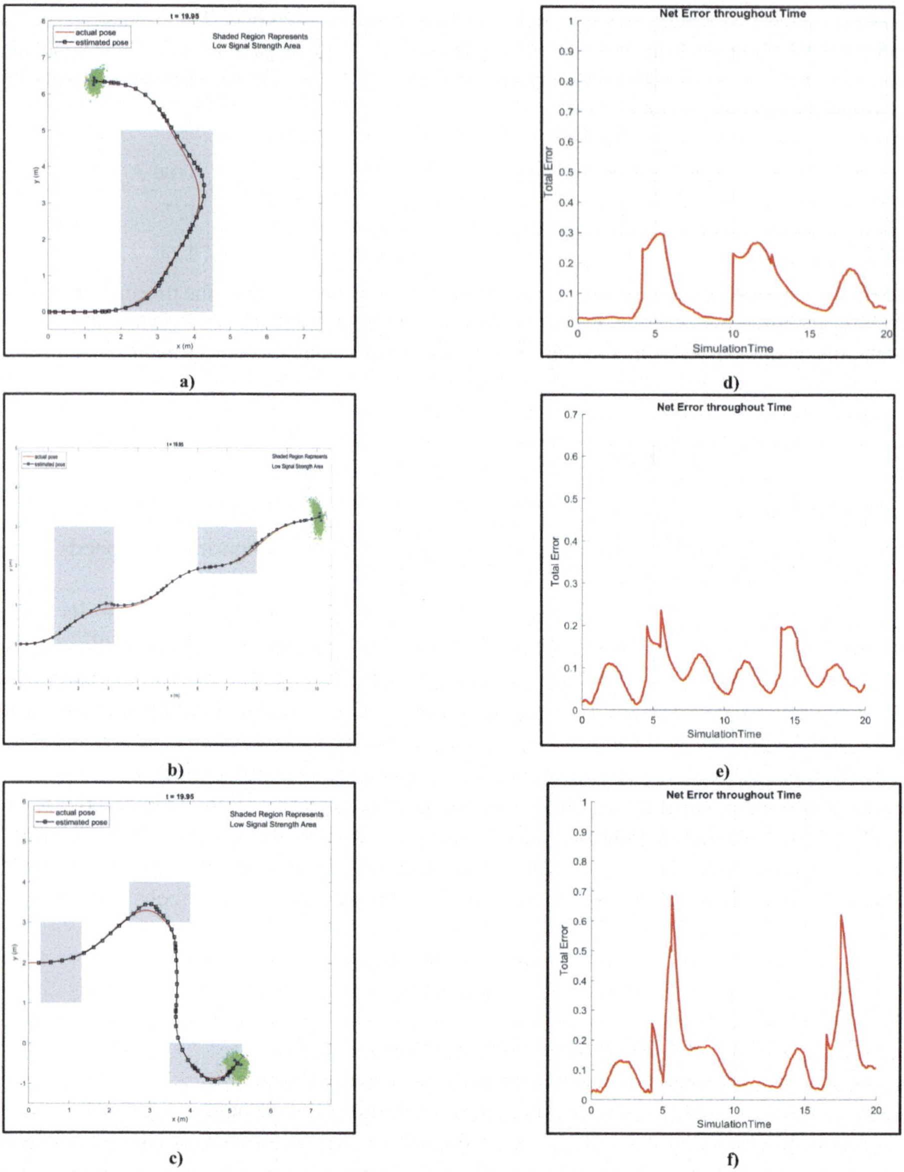

Fig. 2. Figures 2a, 2b and 2c represents the trajectories and 2d, 2e and 2f are the error plots.

5 Improved Path Tracking Technique

Improved Path Tracking Technique offers the solution to minimize errors accumulated due to wheel slipping. Wheel Odometrical data is often provided by the powered wheels (wheels movement is dependent upon torque produced by the engines). In terms of relative wheel slippage, unpowered wheels experience some wheel slippage, but it is

Table 2. Stage 2: Accuracy in Localization vs Time based on GPS and false odometrical data

	Trial 1	Trial 2	Trial 3
δ_γ	241	256	247
$\alpha^{(x,y,\theta)}{}_{(t,2)}$	60.25%	64%	61.75%

Where, $\alpha^{(x,y,\theta)}{}_{(t,2)}$, represents the accuracy maintained by the particles during the simulation in stage 2.

little compared to the wheel slipping experienced by driven wheels [14]. The unpowered wheels provide the necessary data for odometrical calculations; thus, the actual path is tracked and slippage minimized. Figure 3 accurately depicts this flow.

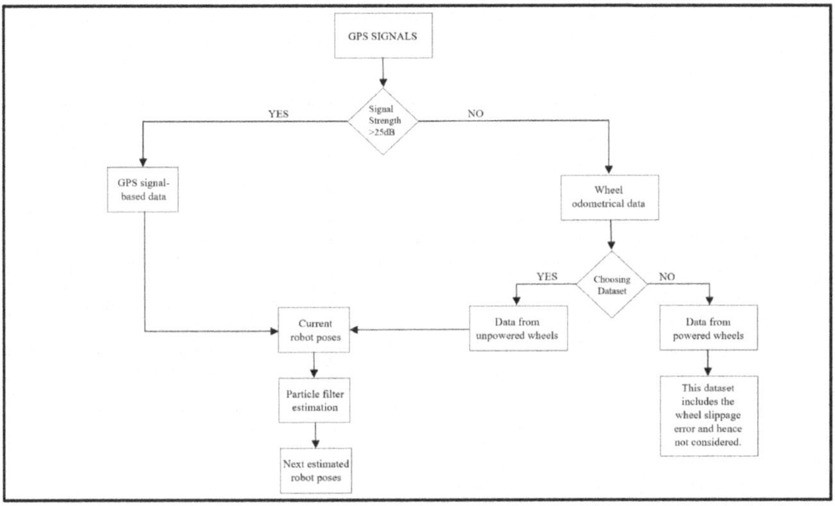

Fig. 3. Flowchart for Improved Path Tracking Technique using powered wheels

Unpowered wheels mean that there is no share of control of the wheels over change in direction of the wheeled robot. To address this issue, steering is taken into consideration. It aids in providing direction to robot's wheels, as necessary during turns [14] (Fig. 4).

Figure 4 demonstrates that wheel slippage error is significantly reduced, thereby improving accuracy by an average of 11.34% from Stage 2. Results are shown in Table 3.

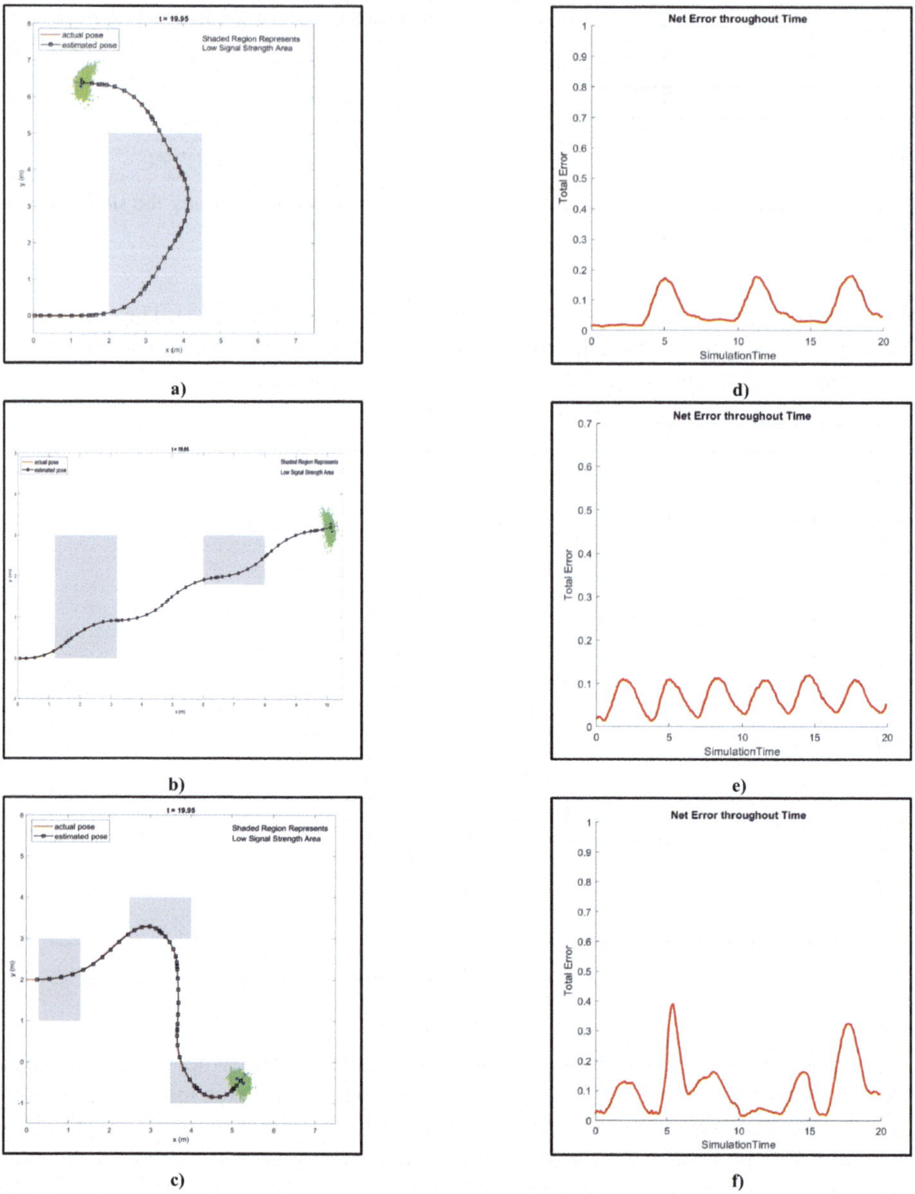

Fig. 4. Figures 4a, 4b and 4c represents the trajectories and 4d, 4e and 4f are the error plots. More error due to greater number of turns is seen in Fig. 4c and 4f.

Table 3. Stage 3: Accuracy in Localization vs Time based on GPS and Improved Technique

	Trial 1	Trial 2	Trial 3
δ_{γ}	287	304	289
$\alpha^{(x,y,\theta)}{}_{(t,3)}$	71.75%	76%	72.25%

Where, $\alpha^{(x,y,\theta)}{}_{(t,3)}$, represents the accuracy maintained by the particles during the simulation in stage 3.

6 Conclusion

This paper tries to eliminate the limitations of the conventional approaches and draws together the key considerations. The goal was to offer a highly accurate, low-cost method for trajectory tracking. For this, it was essential to provide the particle filter with inputs that are as noise-free as possible. The paper outline is

- GPS based path tracking algorithms yield low accuracy result when GPS reception is less than 25 dB. To overcome this, wheel odometry was proposed. The improvement in accuracy was seen about 21.83%. Wheel Slippage is still present.
- To counter wheel slippage, an Improved Path Planning Technique is proposed utilizing the concept that unpowered wheels have considerably less slippage than driven wheels. The improvement obtained from this model was around 33.16%.

In future the proposed technique will be implemented for the dynamic environment along with the multidata sensor approach w.r.t inertial measurement unit and LiDAR odometry for the reliable navigation of mobile robot.

References

1. Yousuf, S., Kadri, M.B.: Information fusion of GPS, INS and odometer sensors for improving localization accuracy of mobile robots in indoor and outdoor applications. Robotica **39**(2), 250–276 (2021)
2. Yang, J.-A., Kuo, C.-H.: Integrating vehicle positioning and path tracking practices for an autonomous vehicle prototype in campus environment. Electronics **10**(21), 2703 (2021)
3. Al Khatib, E.I., Jaradat, M.A.K., Abdel-Hafez, M.F.: Low-cost reduced navigation system for mobile robot in indoor/outdoor environments. IEEE Access **8**, 25014–25026 (2020)
4. Maier, D., Kleiner, A.: Improved GPS sensor model for mobile robots in urban terrain. In: 2010 IEEE International Conference on Robotics and Automation, pp. 4385–4390. IEEE (2010)
5. Hayajneh, M.R.: Interacting multiple model adaptive unscented Kalman filter for accurate localization of a high-dynamic motion mobile robot using wireless sensor network (2022)
6. Eldesoky, A., Hamed, A., Kamel, A.M.: Improved position estimation of real time integrated low-cost navigation system using unscented Kalman filter. J. Phys.: Conf. Ser. **1447**(1), 012017 (2020)
7. Lai, X., Zhu, G., Chambers, J.: A fuzzy adaptive extended Kalman filter exploiting the student's t distribution for mobile robot tracking. Meas. Sci. Technol. **32**(10), 105017 (2021)

8. Anbu, N.A., Jayaprasanth, D.: Integration of inertial navigation system with global positioning system using extended Kalman filter. In: 2019 International Conference on Smart Systems and Inventive Technology (ICSSIT), pp. 789–794. IEEE (2019)
9. Lv, M., Wei, H., Xinyu, F., Wang, W., Zhou, D.: A loosely coupled extended Kalman filter algorithm for agricultural scene-based multi-sensor fusion. Front. Plant Sci. **13**, 849260 (2022)
10. Ojeda, L., Borenstein, J.: Methods for the reduction of odometry errors in over-constrained mobile robots. Auton. Robot. **16**, 273–286 (2004)
11. Bamford, W., Naasz, B., Moreau, M.C.: Navigation performance in high earth orbits using navigator GPS receiver. In: 29th Annual AAS Guidance and Control Conference, no. AAS-06-045 (2006)
12. Farag, W.: Bayesian localization in real-time using probabilistic maps and unscented-Kalman-filters. J. Eng. Res. **10**(3A) (2022)
13. Klomp, M., Gao, Y., Bruzelius, F.: Longitudinal velocity and road slope estimation in hybrid electric vehicles employing early detection of excessive wheel slip. Veh. Syst. Dyn. **52**(sup1), 172–188 (2014)
14. Ani, O.A., Xu, H., Xue, K., Liu, S.-G., Zhang, Z.-Y.: Analytical modeling and multi-objective optimization (MOO) of slippage for wheeled mobile robot (WMR) in rough terrain. J. Cent. South Univ. **19**(9), 2458–2467 (2012)

A Short Survey on Comparative Study of Modern Cryptography Approach

Anurag Choudhary(✉)

Gorakhpur, Uttar Pradesh, India
anuragcho25@gmail.com

Abstract. Organizations all over the globe produce a lot of data every day because of the development of e-commerce apps, social networks, the World Wide Web, and other technologies. An intrusion occurs when hackers breach the data and use it to their benefit. Securing information is difficult in the internet era since terabytes of data are being generated daily and an online platform is available every second for user demands. Therefore, the current world's information security depends heavily on cryptography algorithms, which make the online environment safer. In terms of using the cryptography algorithm, you can ensure confidentiality, authentication, integrity, availability, and identification for genuine users, as well as protect sensitive data and information from unauthorized persons. The cryptography methods (encryption and decryption) use mathematics to secure data, and there are various cryptography algorithms available, but appropriate algorithm selection is vital for secure communication that delivers heightened security, reliability, and performance to meet the requirements of a user. This survey paper aims to provide an overview of popularly used cryptography algorithms and show their significance when deploying security in the required environment.

Keywords: Cryptography · Cryptography Algorithms · Symmetric · Asymmetric · Hashing

1 Introduction

Cryptography, stemming from the Greek word for "secret writing," is a fundamental strategy for upholding information security, encompassing data confidentiality, integrity, authentication, and non-repudiation while enabling secure communication in the presence of adversaries [2,6]. In the realm of modern technology, cryptography plays a pivotal role in safeguarding sensitive information in various applications like eCommerce, e-banking, e-mail, and medical databases. Preserving information security becomes paramount as unauthorized access to private data must be prevented at all costs [1,11]. Cryptography achieves this by transforming data into an unrecognizable and incomprehensible form, ensuring that unauthorized individuals cannot access it. The process of converting plaintext into ciphertext or encrypted text, rendering it unreadable, is known as

D. Pastor-Escuredo et al. (Eds.): ICDLAIR 2023, LNNS 1001, pp. 33–48, 2024.
https://doi.org/10.1007/978-3-031-60935-0_4

encryption. The encrypted form is then transmitted across an insecure medium, posing a challenge for unauthorized persons to decipher. Decryption, on the other hand, reverses this process, converting the unreadable ciphertext back into plaintext for the intended recipient to comprehend [8,15]. In Figure 1, the field of cryptology encompasses two disciplines: cryptography and cryptanalysis. Cryptography focuses on the art of securing messages through encryption, while cryptanalysis involves the skill of breaking ciphers. The study of cryptosystems, known as cryptology, acknowledges the importance of both divisions in ensuring information security and privacy in today's technology-driven world [4,13].

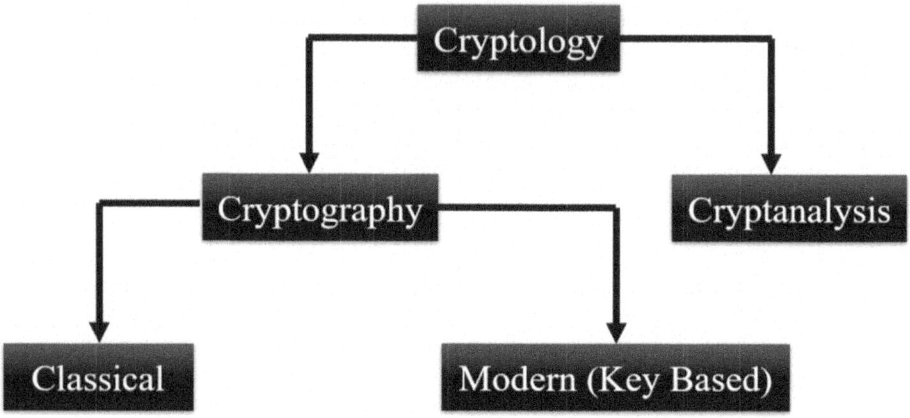

Fig. 1. Context of Cryptography

1.1 Cryptography Goals

Cryptography has some goals. These goals can be either achieve all at the same time in one application, or only one of them in one application. These goals are:

(a) **Authentication:** Concerned with assuring that a communication is authentic.
(b) **Confidentiality:** Keeping unauthorized users away from important information.
(c) **Data Integrity:** Protecting the data from being changed by unauthorized users.
(d) **Non-repudiation:** The impossibility of a transmitted message being denied by either the sender or the receiver.

1.2 Cryptography Algorithm

It is generally accepted that modern cryptography falls into three categories: Symmetric Key, Asymmetric Key, and Hashing. Table 1 contains some useful modern cryptography algorithms that are the major part of this survey paper and are discussed further in Sect. 3 along with the appropriate table.

Table 1. Modern Cryptography

Category	Algorithm
Symmetric Key Encryption	Block Cipher: DES, Double DES, Triple-DES, AES, IDEA, Blowfish, Twofish, RC2, RC5, RC6
	Stream Cipher: RC4, Seal, FISH, ISAAC
Asymmetric Key Encryption	RSA, DSA, ECC, El-Gamal, Diffie-Hellman
Hashing	MD2, MD4, MD5, MD6 (224-256-384-512), SHA-0, SHA-1, SHA-2 (224-256-384-512), SHA-3 (224–256-384–512)

1.3 Applications of Cryptography

These are some of the most common cryptography applications:

(a) **Digital Signatures:** Digital signatures, relying on cryptographic methods, verify documents' authenticity by being data dependent on both the document and the signer's private key.
(b) **Identification/password protection:** Identification is the ability to recognize active system users, safeguarding computer-accessible information from unauthorized users through password protection.
(c) **Secret Sharing:** Secret sharing systems are helpful because they allow for more secure storage of very sensitive data, such as encryption keys, missile launch codes, and digital bank accounts.
(d) **Zero-Knowledge Techniques:** A zero-knowledge proof (ZKP), used in cryptography, is a mathematical method to confirm the accuracy of information without disclosing it.

1.4 Aims and Objectives of the Study

(a) To weigh the strength of modern cryptography over classical cryptography.
(b) To provide an overview of the modern cryptography approaches with the appropriate table, and indicating the potential work for future work.

1.5 Structure of the Paper

Several sections make up the framework of this paper. The background is covered in Sect. 2, the comparison of various cryptography algorithms is covered in Sect. 3, and the conclusion is covered in Sect. 4.

2 Background

2.1 History of Cryptography

Egypt likely started employing cryptography around 2000 B.C. The method and examples from three different eras in the history of cryptography are shown in

Table 2. The history of cryptography has seen remarkable developments, attracting top mathematicians and computer scientists. Encryption's role in securing sensitive information has been crucial in business and military success. While some countries restrict cryptography to protect national interests, the internet has made innovative cryptosystems more accessible worldwide.

Table 2. Era of Cryptography

Era	Period	Method	Examples
Ancient	Before 500 AD	In this era, cryptography was conducted using pen and paper.	Scytale, Atbash, Scytale, Caesar cipher
Medieval	500–1500 AD	The invention of cipher machines marks the beginning of this era.	Playfair cipher, Vigenere cipher, Japanese Red and Purple Machines
Modern	After 1500 AD	The modern era of cryptography is characterized by the use of computers.	ElGamal, Rijndael, RSA, DES

2.2 Short Overview of Classic to Modern Cryptography Path

The path from classic to modern cryptography represents a long and fascinating evolution of techniques and algorithms for securing communication and data. When humanity needed to exchange secrets thousands of years ago, cryptography was invented (for example, in a war). Before moving on to the modern technique, classical cryptography offers an excellent illustration. Character-oriented ciphers are the classical ones [1,3]. The objective is to demonstrate how modern ciphers developed from them, even if they are now outdated. With the help of Fig. 3, transposition ciphers and substitution ciphers are the two main divisions of classical symmetric-key ciphers.

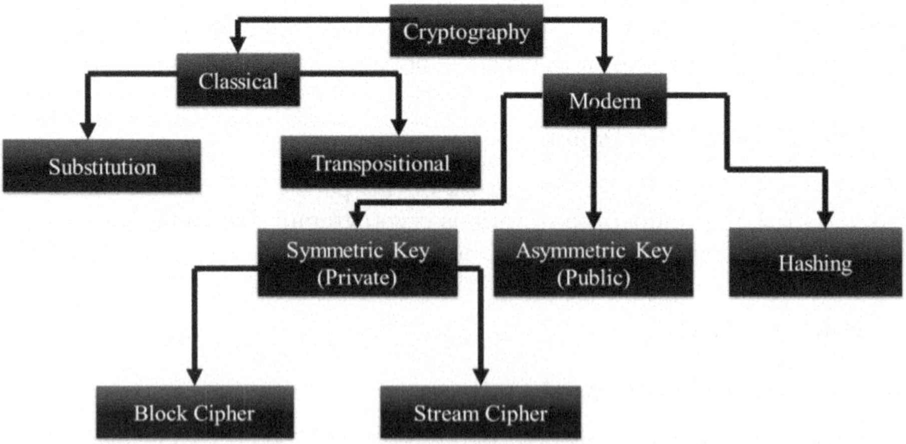

Fig. 2. Modern Cryptography

(a) **Substitution ciphers:** With the prescribed condition, changed one letter for another. For example with a shift of 3, A would be replaced by D, and so on.
(b) **Transposition Cipher:** Columnar transposition cipher converts plaintext into ciphertext using a predetermined permutation. For example, "GOOD-MORNING" becomes "GMIOONORGDN" in the vertical format Fig. 2.

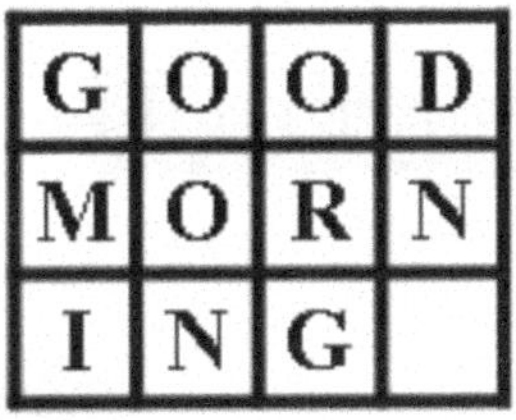

Fig. 3. Transposition Cipher

The journey from classical to modern cryptography reflects innovative progress in meeting evolving digital security needs. Classical cryptography's ease of encryption and decryption with pencil and paper also make it vulnerable to attacks, especially with computer technology [1,9]. Modern cryptography, using mathematical concepts like number theory and probability theory, is the cornerstone of computer and communications security. Figure 3 illustrates how modern cryptography relies on cryptographic keys and algorithms, categorized as symmetric key, asymmetric key, and hashing, offering much higher security compared to the classical approach.

2.3 Basic Terminology of Cryptography

Given the above scenario, the basic terminology in cryptography are in Table 3:

Table 3. Basic Terminology

Terms	Approach
Plaintext	This is the original message
Ciphertext	A message that is encoded
Cipher	Transformation of plaintext into ciphertext using an algorithm
Encipher	A method of changing plaintext into ciphertext
Decipher	The recovery of plaintext from the ciphertext
Encrypt	A method of encoding or enciphering
Decrypt	A method of decoding or deciphering
Key	A cipher contains information that is only known by the sender or receiver
Cryptography	Art or science of keeping messages secure
Crptoanalysis	Art or science of breaking ciphers
Cryptology	An area that combines cryptography and cryptanalysis
Cryptographers	Those who specialize in cryptography
Cryptanalysts	Those who specialize in cryptanalysis

2.4 Cryptography Attacks

As we move into the 21st century, information plays a growing role in almost every aspect of our lives. Accordingly, the attacks are typically classified based on when the attacker performs the action [13]. Figure 4 shows that cryptography attacks can be passive and active.

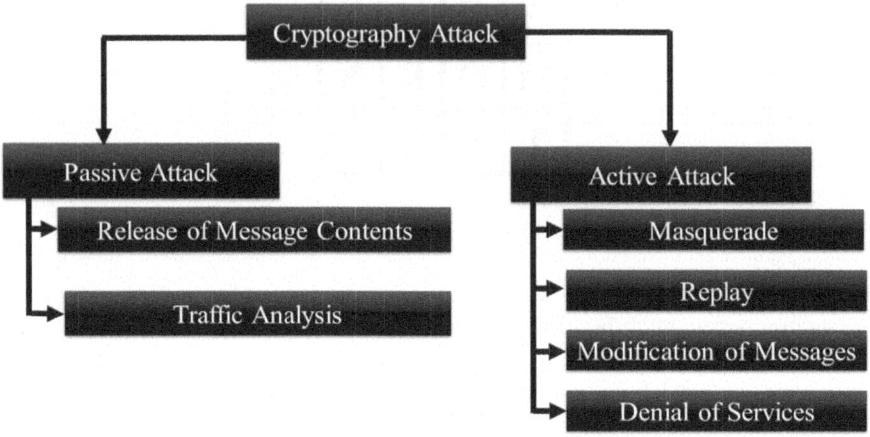

Fig. 4. Cryptography Attacks

Passive Attacks: In a passive attack, communications or data can only be observed. Passive attacks are of two types:

(a) **Release of message contents:** It is simple to access a phone call, an email message, or a file transfer without affecting the message.
(b) **Traffic analysis:** to look for message patterns coming from both the sender and the recipient.

Active attacks: Communications or data can be actively changed by an active assault. These attacks can be classified in to four categories:

(a) **Masquerade:** A single entity impersonates another.
(b) **Replay:** The replay process entails passively capturing a data unit and reusing it, which produces an authorized effect following the retransmission of the data unit.
(c) **Modification of messages:** Occurs when someone who is not authorized obtains access to an asset and tampers with it. Integrity is under threat in this situation.
(d) **Denial of service:** Hinders or prohibits the regular usage of communications facilities or their administration.

2.5 Related Work

In 2009, Alejandro Jiménez-Horas et al. [1] round-based block ciphers are proposed as a new method of protecting hardware cryptosystems. Online error detection is used in this system, which results in time performance loss and tiny area overheads due to partial time redundancy.
In 2014, S. Gomathi et al. [2] advanced the Substitution Technique (AST), a relatively new form of substitution that is more secure, to be used to encrypt the message, and suggested using a brand-new architecture to securely encrypt the text.
In 2016, Nikola Luburic et al. [3] introduced Crypto-Tutor, an educational tool for modern cryptography algorithms, offering students interactive learning with background, function, structure, and benchmarking on various datasets.
In 2017, Faiqa Maqsood et al. [4] the effectiveness of symmetric and asymmetric cryptography techniques has been examined in this paper. When compared to asymmetric methods, the performance findings demonstrate that symmetric techniques are computationally less expensive.
In 2017, Muhammad Faheem Mushtaq et al. [5] studied the encryption and decryption times of various algorithms with random data packet sizes. They present essential cryptography concepts and survey significant algorithms like RSA, ECC, and El-Gamal.
In 2019, Mohammed M. Alani et al. [6] proposed a current survey of how machine learning techniques are used in cryptography and cryptanalysis. They also talked about ways that systems using machine learning could be attacked from a security standpoint.
In 2019, Suzan Sallam et al. [7] analyze lightweight cryptography methods, including block ciphers and stream ciphers, and examine their hardware performance and security features.
In 2020, Shcherbakov A. Yu et al. [8] proposed that today's cryptography is mainly concerned with protecting the external perimeter of business processes in corporations, independent of the user, and providing cryptography services to all corporations.
In 2021, Meiran Galis et al. [9] present two modern concepts for symmetric cryptography systems' key setup, focusing on the decentralized establishment of keys without a trusted third party's assistance.
In 2022, S. K. Pal et al. [10] proposed an effective, portable, and dynamic artificial neural network (ANN) based cryptography algorithm to secure data during network transmission, and considered the CIA triad principles of confidentiality, integrity, and availability.

3 Comparative Study of Modern Cryptography

Figure 5 categorizes modern Cryptography into three types: Symmetric Key, Asymmetric Key, and Hashing. Table 4 compares them based on Method, Objective, Number of Keys, Key Length, Complexity, Security, Info Length, Process, Common Use, and Examples. Table 5 list Key Benefits and Drawbacks.

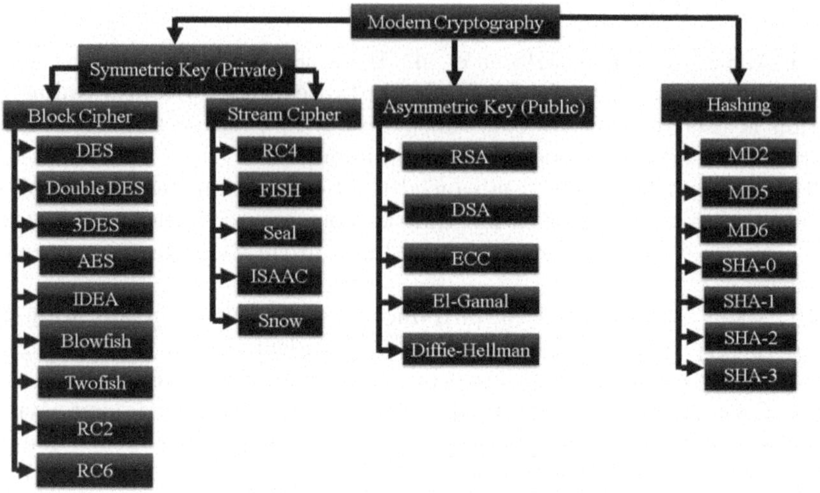

Fig. 5. Modern Cryptograph

Table 4. Comparison of Modern Cryptography

	Symmetric Encryption	Asymmetric Encryption	Hashing
Method	Two ways	Two ways	One Way
Objective	Confidentiality	Confidentiality, Authenticity, and Non-repudiation.	Integrity
Number of Key	1	2	0
Key Length	56-256 bit	1024-4096 bit	128-512 bit
Complexity	Medium	High	Medium
Security	Less Secure	More Secure	Highly Secure
Length of Information	Not Fixed	Not Fixed	Small and Fixed
Process	Reversible	Reversible	Irreversible
Commonly Used	Protecting top secret documents	Digital Signatures	One way password management
Examples	AES, 3 DES, etc.	RSA, DSA, etc.	MD5, SHA-2, etc.

Table 5. Key Benefits and Drawbacks

Types	Key Benefits	Drawbacks
Symmetric Encryption	There is just one key (symmetric key) used, and it is the same key used to encrypt and decrypt the message	The receiver must get the secret key from the sender
Asymmetric Encryption	Asymmetric encryption is that it does not force the user to share (secret) keys	Asymmetric encryption is that it is time-intensive, and it requires considerably more effort
Hashing	A hash function for any file can be generated to create a digital fingerprint	There are several problems with one-way hash functions, including the weakening of their security after repeated use

3.1 Symmetric Encryption

According to Fig. 6 symmetric cryptosystems are also known as private key cryptosystems since there is one key used for both encryption and decryption. Encryption and decryption are performed with the help of a private key, which is only accessible to those with the proper authorization [5]. A few examples of symmetric encryption include Blowfish, AES, RC4, DES, RC5, and RC6. AES-128, AES-192, and AES-256 are the three most used symmetric algorithms.

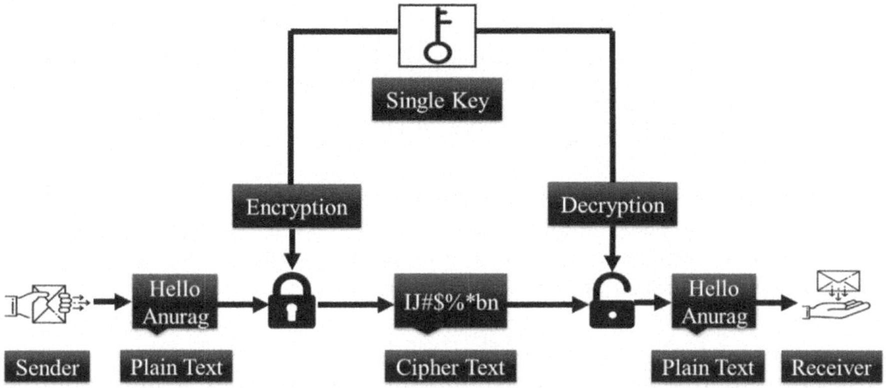

Fig. 6. Symmetric Key Encryption

Properties of Symmetric Encryption: These are some of the most common symmetric encryption properties:

(a) **Key:** It is sufficient to have only one key to encrypt as well as decrypt data.
(b) **Security:** It just offers confidentiality.
(c) **Process:** It takes only a little time to encrypt data.
(d) **Text Size:** Sizes are usually the same or smaller than those in the original clear text.
(e) **Usage:** Cannot be used for digital signatures; mostly used for encryption and decryption.

Applications of Symmetric Encryption: These are some of the most common symmetric encryption applications:

(a) **Banking:** In order to facilitate transactions, personal information (PII) such as credit card numbers must be encrypted.
(b) **Data storage:** Encrypting data that is being stored on a device when it is not being sent.

Types of Symmetric Encryption Algorithm: Symmetric encryption algorithms come in two forms: Block Ciphers, Stream Cipher:

- **Block Ciphers:** Bits are encrypted in discrete chunks of corresponding sizes [7]. Most popular and prominent block ciphers are listed below with appropriate Table 6.
  (a) **Data Encryption Standard (DES):** The National Institute of Standards and Technology (NIST) developed the Data Encryption Standard (DES), a block cipher with 48-bit keys that encrypts 64-bit plaintext using a 16-round Feistel structure. It's a symmetric key algorithm, using the same key for encryption and decryption [5].
  (b) **Double Data Encryption Standard:** Double DES uses two instances of DES with different keys for encryption. The first DES encrypts plaintext to middle text, and the second DES encrypts middle text to ciphertext [12].
  (c) **Triple Data Encryption Standard:** Triple DES employs three DES instances on a single plaintext with various key-choosing techniques. It provides stronger encryption than DES but is less secure than AES [12].
  (d) **Advanced Encryption Standard (AES):** Advanced Encryption Standard (AES) is a popular symmetric encryption algorithm based on an SP network, much faster than triple DES. It offers various key lengths (128-bit, 192-bit, or 256-bit), making it significantly more secure than DES [2,12].
  (e) **International Data Encryption Algorithm (IDEA):** IDEA is a 64-bit symmetric block cipher using 128-bit keys. It employs six subkeys for encryption and four for output, with eight identical rounds, ensuring diffusion and confusion [19].
  (f) **Blowfish Algorithm:** Blowfish, a symmetric block cipher, is an alternative to DES and IDEA. Created by Bruce Schneier in 1993, it allows key lengths from 32 to 448 bits, gaining acceptance as a strong, freely accessible encryption method without patents or licenses [5,12].
  (g) **Twofish Algorithm:** Twofish, a quick symmetric encryption method replacing Blowfish, requires 128, 192, or 256-bit keys. It uses a block cipher, suitable for software and hardware environments [18].
  (h) **Rivest Cipher 2 (RC2):** In the late 1990 s, RC2, a block symmetric cipher, gained popularity. Designed by Ron Rivest of RSA Security in 1987, it was kept secret until its exposure in 1996 [16].
  (i) **Rivst Cipher 5 (RC5):** Ron Rivest created RC5, a fast symmetric key-based block encryption. Its advantages include variable key length, variable bit size, and low memory consumption. Ideal for desktop and smart card applications.
  (j) **Rivst Cipher 6 (RC6):** RC6, a symmetric key block cipher derived from RC5, supports a wide variety of word lengths, key sizes, and round numbers, with a block size of 128 bits and a key size of 128, 192, or 256 bits. It combines two simultaneous RC5 processes, resulting in faster performance.

Table 6. Some Common Block Ciphers

Algorithm	Created By	Year	Key Size	Plain Text/ Cipher Text	Rounds	Attack	Encryption Speed	Structure	Level of Security
DES	IBM and NSA	1975	56-bit	64-bit	16	Brute Force Attack, Differential cryptanalysis	Very Slow	Feistel structure	Vulnerable
Double DES	IBM	1978	112-bit	64-bit	32	Meet-in-the-Middle attack	Very Slow	Feistel structure	Vulnerable
Triple-DES	IBM	1998	168-bit	64-bit	48	Meet-in-the-Middle attack	Very Slow	Feistel structure	Moderate
AES	Joan Daemen and Vincent Rijmen	2001	128-bit,192-bit,256-bit	128-bit	10,12, or 14	Side-channel Attack	Faster	Substitution-Permutation Structure	Highly Secure
IDEA	Xuejia Lai and James Massey	1991	128-bit	64-bit	8	Differential cryptanalysis, Luby-Rackoff attacks	Slow	Substitution-Permutation	Moderate
Blowfish	Bruce Schneier	1993	32–448 bit	64 -bit	16	Dictionary Attack	Fast	Feistel structure	Moderate
Twofish	Bruce Schneier	1998	128-bit,192-bit,256-bit	128-bit	16	Related-key attacks	Fast	Feistel structure	Secure
RC2	Ron Rivest	1987	40–128 bit	64-bit	18	Brute force	Slow	Feistel structure	Vulnerable
RC5	Ron Rivest	1994	0–2040 bit	32–128 bit	12,16, or 20	Slide attack	Fast	Feistel structure	Moderate
RC6	Ron Rivest	1998	128-bit,192-bit,256-bit	128-bit	20	Meet-in-the-Middle attack	Fast	Feistel structure	Secure

- **Stream Ciphers:** Since it is not stored in memory, data is encrypted as it comes in. [7,17]. Most popular and prominent stream ciphers are listed below with appropriate Table 7.
  (a) **Rivest Cipher 4 (RC4):** RC4 is one of the most extensively used software stream ciphers and is used in protocols like Secure Socket Layer (SSL), Transport Layer Security (TLS), and IEEE 802.11 wireless LAN standard. It was created in 1987 by Ron Rivest. In terms of software, it is said to be quick and straightforward [16,17].
  (b) **Software-Optimized Encryption Algorithm (Seal):** SEAL (Software-Optimized Encryption Algorithm) is a stream cipher that continuously encrypts data, generating arbitrary keystream segments quickly. This feature makes it suitable for encrypting hard drives.
  (c) **FIbonacci SHrinking (FISH):** FISH stream cipher is a fast software-based one, combining shrinking generator cipher principles with Lagged Fibonacci generators. It offers high speed and a long key length.
  (d) **Indirection, Shift, Accumulate, Add, and Count (ISAAC):** ISAAC is a cryptographically secure pseudo-random number generator (CSPRNG) with stream ciphers and bitwise operations. Created by Robert Jenkins in 1996, it's fast, adaptable, and highly effective when optimized.

Table 7. Some Common Stream Cipher

Algorithm	Created By	Year	Key Size	Clock/Byte of Output	Attack	Round	Speed
RC4	Ron Rivest	1987	40-2048 bit	Variable	Fluhrer-Mantin-Shamir (FMS), PTW, and related key recovery attacks	Not defined	Very fast
Seal	Philip Hawkes, James Massey	1990	128-256 bit	Variable	Integral cryptanalysis and differential cryptanalysis	8	Fast
FISH	Ross Anderson	1993	128-256 bit	Variable	Related-key attack	64	Fast
ISAAC	Robert Jenkins	1996	0-256 bit	Variable	None (not widely used)	8	Fast

3.2 Asymmetric Encryption

Accoding to Fig. 7 an asymmetric cryptosystem, often known as a public key cryptosystem, encrypts and decrypts data using two different keys. The encryption key is known as the public key since it is maintained in the open [4].

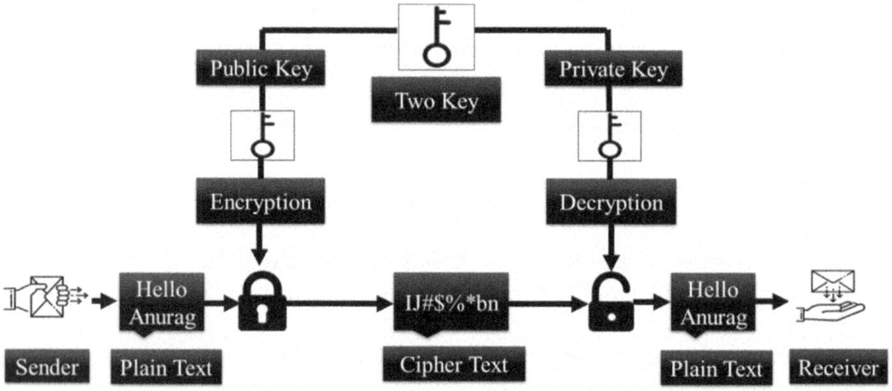

Fig. 7. Asymmetric Key Encryption

Properties of a Asymmetric Encryption: These are some of the most common asymmetric encryption properties:

(a) **Key:** One key is needed to encrypt data, while the other key is needed to decrypt data.
(b) **Security:** It offers non-repudiation, secrecy, and authenticity.
(c) **Process:** The encryption procedure takes a long time.
(d) **Text Size:** Greater than the original clear text size.

Applications of Asymmetric Encryption: These are some of the most common asymmetric encryption applications:

(a) **Digital signatures:** Verifying a person's identity before they sign a document.
(b) **Blockchain:** Authentication of identification is required to approve cryptocurrency transactions.

Types of Asymmetric Encryption Algorithm: Table 8 shows asymmetric encryption algorithms come in different forms: Rivest-Shamir Adleman (RSA), Diffie-Hellman, Digital Signature Algorithm (DSA), Elliptical Curve Cryptography (ECC):

(a) **Rivest-Shamir Adleman (RSA):** It is highly challenging for an attacker to do the reverse operation and generate the private key from the public key [12]. Consequently, this approach is exceedingly secure, but key creation is difficult and time-consuming [6].
(b) **Digital Signature Algorithm (DSA):** Compared to other asymmetric algorithms, this one is quicker and more advantageous, but the sharing procedure is difficult and the digital signatures have a limited lifespan.
(c) **Elliptical Curve Cryptography (ECC):** Elliptic Curve Cryptography (ECC) is a popular, efficient asymmetric encryption method, smaller and quicker than alternatives. Bitcoin uses it as its asymmetric cryptosystem due to its lightweight nature.
(d) **ElGamal:** ElGamal cryptosystem is related to Diffie-Hellman key exchange. While Diffie-Hellman shares a secret key, ElGamal enables secure communication over an insecure channel using public key encryption.
(e) **Diffie-Hellman:** The procedure moves along more quickly since the private key is so short. The method is more susceptible to assaults since the private key is short, and man-in-the-middle attacks may also be used against it.

Table 8. Some Asymmetric Encryption Algorithm

Algorithm	Created By	Year	Key Size	Security Attack	Performance
RSA	Ron Rivest, Adi Shamir, and Leonard Adleman	1977	1024–4096 bit	Factoring, Timing, Fault Injection, and Side-Channel Attacks	Slow for key generation, moderate for encryption and decryption
DSA	National Institute of Standards and Technology (NIST)	1991	1024–3072 bit	Quantum Attacks, Side-Channel Attacks, and Hash Function Attacks	Fast for key generation, moderate for signing and verification
ECC	Neal Koblitz and VictorMiller	1985	128–521 bit	Quantum Attacks, Side-Channel Attacks, and Pairing Attacks	Fast for key generation, encryption, and decryption
El-Gamal	Taher Elgamal	1985	1024–2048 bit	Discrete Logarithm Problem(DLP) and Side-Channel Attacks	Slow for key generation, moderate for encryption and decryption
Diffie-Hellman	Whitfield Diffie and Martin Hellman	1976	2048–3072 bit	Man-in-the-middle Attacks and Computational Attacks	Fast for key generation, moderate for encryption and decryption

3.3 Hashing

Acording to Fig. 8 any particular piece of material may be uniquely identified using a hashing. It also transforms plaintext into a particular length of distinctive ciphertext based on the amount of plaintext [13,14].

Fig. 8. Hash functions

Properties of a Hash Function: These are some of the most common hasing properties:

(a) **Deterministic:** The hash function produces an output that is significantly smaller than its input.
(b) **Not reversible:** Unpredictable unique hash values are created by a hash algorithm at random.
(c) **Collision resistant:** The outcome from two inputs is not the same.

Applications of Hashing: These are some of the most common hashing applications:

(a) **Verifying the password:** When connecting, your password's hash is used for verification, confirming your legitimacy as a user. If the submitted password matches, authorization is granted.
(b) **Storage of passwords:** Passwords are stored as hash values using hashing, ensuring security. The hash table keeps hash values; hackers can't access plaintext passwords.

Types of Hashing Algorithm: Hashing algorithms come in two forms: Message Digest (MD), Secure Hash Algorithm (SHA) are shows in Table 9:

- **Message Digest (MD):** MD2, MD4, MD5, and MD6 are among the hash functions in this family. MD5 was formerly thought to be the go-to hashing technique, however, it's now deemed broken due to collisions in the wild [11]. The Message Digest algorithm is listed below:
  (a) **MD2:** The MD2 algorithm was developed by Ronald Rivest. It is probably Rivest's most secure message digest function, but its computation takes the longest. MD2 is rarely used due to this reason. Using MD2, you can generate a 128-bit digest.
  (b) **MD4:** MD4, invented by Ronald Rivest as a rapid replacement for MD2, was later found to have a potential flaw allowing the identification of a second file with the same MD4 without brute-force search. It generates a 128-bit digest.
  (c) **MD5:** In order to generate a 128-bit string value from a string of any length, one can utilise the cryptography hash method MD5. Despite the fact that MD5 contains vulnerabilities, it is nevertheless frequently used. MD5 is most typically used to validate file integrity.
  (d) **MD6:** Cryptography hash functions encompass the MD6 Message-Digest Algorithm (MD6-224, MD6-256, MD6-384, and MD6-512). Its unique characteristics, including the use of a Merkle tree structure, enable efficient computation of hashes for long inputs in parallel.
- **Secure Hash Algorithm (SHA):** SHA includes SHA-0, SHA-1, SHA-2 (with SHA2-224, SHA2-256, SHA2-384, and SHA2-512), and SHA-3 (with SHA3-224, SHA3-256, SHA3-384, and SHA3-512). SHA-256 is widely used [11]. The Secure Hash Algorithm is listed below:

(a) **SHA-0:** SHA-0, a 160-bit or 20-byte hash function released in 1993, was quickly removed due to a severe defect, leading to SHA-1's introduction [11]
(b) **SHA-1:** SHA-1 is a cryptography hash algorithm that accepts an input and outputs a hash value of 160-bits. They also convert to 40-digit hexadecimal values [11,14]. It also succeeds SHA-0.
(c) **SHA-2:** SHA-224, SHA-256, SHA-384, and SHA-512 are part of the suite with varying output lengths. SHA-256 is the most common and is the successor to SHA-1, offering improved security [11,15].
(d) **SHA-3:** SHA-224, SHA-256, SHA384, and SHA-512, and it allows the same hash lengths as SHA-2, but its underlying structure differs dramatically from the rest of the SHA family [11,14].

Table 9. Some Hashing Algorithm

Algorithm	Created By	Year	Word Size	Digest sizes	Block sizes	Round	Used
MD2	Ronald Rivest	1989	32-bit	128-bit	64-bit	48	Used in digital signatures, but is now considered insecure
MD4	Ronald Rivest	1990	32-bit	128-bit	512-bit	3	Used in some protocols and applications, but is now considered insecure
MD5	Ronald Rivest	1991	32-bit	128-bit	512-bit	64	Password hashing, digital signatures, checksums
MD-6 (224)	Ronald Rivest	2008	64-bit	224-bit	256-bit	42	Hash-based message authentication codes (HMACs)
MD-6 (256)	Ronald Rivest	2008	64-bit	256-bit	256-bit	44	Hash-based message authentication codes (HMACs)
MD-6 (384)	Ronald Rivest	2008	64-bit	384-bit	512-bit	64	Hash-based message authentication codes (HMACs)
MD-6 (512)	Ronald Rivest	2008	64-bit	512-bit	512-bit	72	Hash-based message authentication codes (HMACs)
SHA-0	National Security Agency	1993	32-bit	160-bit	512-bit	80	Obsolete and no longer used
SHA-1	National Security Agency	1995	32-bit	160-bit	512-bit	80	Message integrity checks, digital signatures, password hashing
SHA-2 (224)	National Security Agency	2001	32-bit	224-bit	512-bit	64	Various security protocols and applications, including SSL/TLS, SSH, IPsec, SSH, HMACs, and others
SHA-2 (256)	National Security Agency	2001	32-bit	256-bit	512-bit	64	Various security protocols and applications, including SSL/TLS, SSH, IPsec, SSH, HMACs, and others
SHA-2 (384)	National Security Agency	2001	64-bit	384-bit	1024-bit	80	Various security protocols and applications, including SSL/TLS, SSH, IPsec, SSH, HMACs, and others
SHA-2 (512)	National Security Agency	2001	64-bit	512-bit	1024-bit	80	Various security protocols and applications, including SSL/TLS, SSH, IPsec, SSH, HMACs, and others
SHA-3 (224)	Keccak Team	2015	64-bit	224-bit	1152-bit	24	Securing embedded subsystems
SHA-3 (256)	Keccak Team	2015	64-bit	256-bit	1088-bit	24	Digital signatures, password hashing, integrity checking
SHA-3 (384)	Keccak Team	2015	64-bit	384-bit	832-bit	24	Digital signatures, password hashing, integrity checking
SHA-3 (512)	Keccak Team	2015	64-bit	512-bit	576-bit	24	Digital signatures, password hashing, integrity checking

4 Conclusion

The primary reason for surveying the most widely used modern cryptography algorithms is to comprehend their prevalence and significance, given that cryptography has become an essential part of the digital era. Modern techniques include symmetric encryption for efficiency, asymmetric encryption for stronger security, and hashing for authentication. Block and stream ciphers are standard symmetric algorithms, while RSA and ECC dominate asymmetric encryption. Additionally, SHA-256 is one of the hashing algorithms belonging to the SHA-2 family, which is the most widely used and considered the best hashing algorithm. However, there is a need for continuous development to meet emerging challenges, ensure the security of digital communications, and address new threats.

References

1. Jiménez-Horas, A., et al.: Pseudo-random number generation applied to robust modern cryptography: a new technique for block ciphers. In: 2009 15th IEEE International On-Line Testing Symposium, IOLTS 2009, pp. 203–205 (2009)
2. Gomathi, S.: A cryptography using advanced substitution technique and symmetric key generating algorithm. In: 2014 IEEE 8th International Conference on Intelligent Systems and Control: Green Challenges and Smart Solutions, ISCO 2014 - Proceedings, pp. 224–228, May 2014

3. Luburić, N., Stojkov, M., Savić, G., Sladić, G., Milosavljević, B.: Crypto-tutor: an educational tool for learning modern cryptography. In: 2016 IEEE 14th International Symposium On Intelligent Systems and Informatics (SISY), pp. 205-210, August 2016
4. Maqsood, F., Ahmed, M., Ali, M.M., Shah, M.A. Cryptography: a comparative analysis for modern techniques. In: IJACSA International Journal of Advanced Computer Science and Applications, vol. 8 (2017)
5. Mushtaq, M.F., Jamel, S., Disina, A.H., Pindar, Z.A., Shakir, N.S.A., Deris, M.M.: A survey on the cryptographic encryption algorithms. Int. J. Adv. Comput. Sci. Appl. **8** (2017)
6. Alani, M.: Applications of machine learning in cryptography: a survey. In: Proceedings of The 3rd International Conference On Cryptography, Security And Privacy, pp. 23–27 (2019). https://doi.org/10.1145/3309074.3309092
7. Sallam, S., Beheshti, B.D.: A survey on lightweight cryptographic algorithms. In: IEEE Region 10 Annual International Conference, Proceedings/TENCON, vol. 2018-October, pp. 1784–1789, February 2019
8. Shcherbakov, A.Y.: Prospects for modern cryptography. In: Proceedings of the 3rd International Conference "Futurity designing. Digital reality problems", vol. 3, pp. 227–233 (2020)
9. Galis, M., Unkasevic, T., Milosavljevic, M., Banjac, Z., Milosav, P.: Modern techniques for decentralized key establishment in symmetric cryptographic systems. In: 2021 29th Telecommunications Forum, TELFOR 2021 - Proceedings (2021)
10. Pal, S., Datta, B., Karmakar, A.: An artificial neural network technique of modern cryptography. J. Sci. Res. **14**, 471–481 (2022)
11. Rachmawati, D., Tarigan, J.T., Ginting, A.B.: A comparative study of message digest 5(md5) and sha256 algorithm. J. Phys. Conf. Ser. **978**, 012116 (2018)
12. Patil, P., Narayankar, P., Narayan, D.G., Meena, S.M.: A comprehensive evaluation of cryptographic algorithms: DES, 3DES, AES. RSA Blowfish. Procedia Comput. Sci. **78**, 617–624 (2016)
13. Sindhu, S., Sindhu, D.: Cryptographic algorithms: applications in network security. Int. J. New Innov. Eng. Technol. **7**, 11 (2017)
14. Iyer, N.C., Mandal, S.: Implementation of secure hash algorithm-1 using FPGA. Int. J. Inf. Comput. Technol. **3**, 757–764 (2013)
15. Sumagita, M., Riadi, I.: Analysis of secure hash algorithm (SHA) 512 for encryption process on web based application. Int. J. Cyber-Secur. Digit. Forensics **7**, 373–382 (2018)
16. Charbathia, S., Sharma, S.: A comparative study of rivest cipher algorithms. Int. J. Inf. Comput. Technol. **4**, 1831–1838 (2014)
17. Hammood, M.M., Yoshigoe, K., Sagheer, A.M.: RC4 stream cipher with a random initial state. In: Park, J.J., Barolli, L., Xhafa, F., Jeong, H.Y. (eds.) Information Technology Convergence, LNEE, vol. 253, pp. 407–415. Springer, Dordrecht (2013). https://doi.org/10.1007/978-94-007-6996-0_42
18. Haq, T.U., et al.: Improved twofish algorithm: a digital image enciphering application. IEEE Access **9**, 76 518–76 530 (2021)
19. Sahu, H.K., Jadhav, V., Sonavane, S., Sharma, R.K.: Cryptanalytic attacks on international data encryption algorithm block cipher. Def. Sci. J. **66**, 582–589 (2016)

Advances in Computer-Aided Detection and Diagnosis of Retinal Diseases: A Comprehensive Survey of Fundal Image Analysis

Tushar Taggar(✉), Subhag Sharma, and Sanjay Sharma

School of Computer Science and Engineering, Shri Mata Vaishno Devi University (SMVDU), Katra, India

tushartaggar2804@gmail.com, sanjay.sharma@smvdu.ac.in

Abstract. Diseases pertaining to the Retina are the leading causes of blindness comprising globally 22–25% of people suffering from it. The go to first line of diagnosis for ophthalmology processes to detect retinal diseases comprise of fundoscopy in which the fundus of image is dilated and subsequent imaging is performed in order to identify the probable diseases and predict possible outputs. The paper is a survey of latest accomplishment made in the defied in CADe-CADx domain with respect to fundal images and their associated diseases. It encompasses various image processing and classification techniques proposed by researchers in recent times. The current research scenario covers a variety of topics ranging from traditional CNNs to modern techniques like Visual Transformers. It explores multiple models used for classifying different numbers of diseases. Various performance metrics as used by various researchers are also discussed.

Keywords: ARMD · CNN · DR · Fundus · Glaucoma · Retina · Transformers

1 Introduction

Out of 2.2 billion people reported to be suffering from near or distance vision impairment as reported by WHO's report of 2022 approximate 1 billion of vision ailments could have been averted. The number is still insufficiently addressed and under-counted for it is yet to accommodate data from low income, unevolved and developing nations. Vision impairment presents a substantial economic burden on a global scale, as evidenced by the estimated yearly cost of productivity up to US$ 411 billion.

The primary factors contributing to visual impairment and blindness include diabetic retinopathy (DR),refractive errors, cataracts, age-related macular degeneration (ARMD), glaucoma (GLC) and The treatment for common impairments like refractive errors can be corrected with help of glasses prescribed by a certified ophthalmologist, cataract is also a wide spread alignment which requires a particular surgery involving interchanging of cloud lens with an artificial intraocular lens, highly complex and specific diseases

D. Pastor-Escuredo et al. (Eds.): ICDLAIR 2023, LNNS 1001, pp. 49–60, 2024.
https://doi.org/10.1007/978-3-031-60935-0_5

Diabetic retinopathy largely varies from person to person depending upon speed of progression and severity of disease.

Early diagnosis in all of the diseases mentioned in the document following can be well controlled if detected in time for instance, early diabetic retinopathy which include mild or moderate nonproliferative diabetic retinopathy one might have to go under treatment right away whereas advance diabetic retinopathy which is indicated by proliferative Diabetic retinopathy or macular edema might require prompt treatment included photocoagulation, vitrectomy and injecting medication into the eye even though Diabetic retinopathy can reversed using medication and advance treatment option diseases like glaucoma, ARMD can't be reversed however with help of advancement in the field of computer diagnosis these diseases if caught on early can be slowed down in progression under guidance of eye doctor example in case of intermediated ARMD in one or both eyes special dietary supplements have shown significant results in stopping from turning into a more severe form a late ARMD also if one suffer from Late ARMD in one eye treatments are available to stop progression in other.

This paper examines the utilisation of deep neural networks in the prognostication of various eye-related diseases by analysing fundus images obtained through pupil dilation, enabling a comprehensive assessment of the entire retina. It uses an ophthalmoscope for capturing the images. The algorithm uses multiple layers of interconnected convolution layers and detects unique patterns specific to each disease which further helps in identifying and localizing the diseased area of retina however this paper only deals with detection part.

The paper is split into four sections, of which Sect. 2 describes various literature reviews. Section 3 gives insight about critical evaluation techniques used. In conclusion, this study presents the findings and implications of the research, and identifies potential areas for future investigation in Sect. 4.

2 Literature Review

This section of the paper gives a detailed overview on various techniques and methodologies applied in order to achieve acceptable performance metrics.

2.1 Data Acquisition

Different researchers have relied upon a number of datasets both available in public domains and private acquisitions by researchers. The datasets are as elaborate as Retina 1M, Retina 100K consisting of 1 million and 100K fundus images respectively with 50 diseases [1], Cen L.P et al. in [2] uses a dataset under the title Picture Archiving and Communication Systems (PACS) consists of 88,702 color fundus images, fundus images of 35000 images available publicly on kaggle as used by Lam Carson et al. in [3, 4] is a public dataset comprising of 5000 images along with patient detail like age, doctors' diagnostic keywords etc. released under the title Ocular Disease Recognition (ODIR).Other datasets include Retinal Fundus Multi Disease Image dataset (RFMiD) consisting of 3200 images in color and their associated ground truth labels abnormal categories (comprising of 45 different type of diseases/ pathology stored in a csv format)

[5], Pan Yuhang et al. in [6] use a customized dataset of 1032 fundus images, The Methods to Evaluate Segmentation and Indexing Techniques in the field of Retinal Ophthalmology (MESSIDOR) dataset is a research resource that has received funding from the French Ministry of Research and Defence [7]. Additionally, there are several other smaller datasets available for evaluation purposes in the field of retinal ophthalmology, such as DRISHTI-GS, ORIGA, DRIVE dataset, STARE dataset, and CHASE-DB dataset.

Since most of the datasets cover diabetic retinopathy only other fields are not explored to the extent. Out of all the datasets the most common ones are ODIR and RFMiD. Some researchers also used a combination of multiple datasets like in [8] Chea N. et al. uses a combination of 5 datasets under the umbrella name NOISE-STRESS, Rodríguez A. et al. in [9] uses a collection of 8 datasets under the name of MuReD Dataset.

2.2 Data Pre-processing and Feature Selection

Researchers commonly employ standard pre-processing techniques, such as histogram equalisation, in order to enhance the performance metrics. N. Chea et al. in [8] extracts Region of Interests (ROI) and applies ISO-CLAHE (Contrast Limited Adaptive Histogram Equalisation). Also it was common observation that most of the peers normalized the fundus image pixels to fall in range of 0 and 1.Augmentation techniques included methodologies like rotation of the images disc, zooming in and out. R. Sarki et al. in [10] describes the ways to manage the class imbalance problem which is prominent in majority datasets given randomness of sampling.

As normalization is believed to reduce the number of features it is worth noting that Kadan et al. in [11] convert RGB images into HSI (High Spectral Imaging) format to preserve features. It further uses the I-band to process the image for the classification.

2.3 Techniques Used

Of all the techniques used, Convolution neural networks (CNN) were dominantly used though in multiple forms like RCNN, pre trained models like RESNET-50, VGG, etc. Other techniques used included ViT (Visual Transformers) which are comparatively newer domains. Some papers mentioned the ensemble methodology to enhance the classification scores.

Y. Pan et al. in [6] used a pretrained model with fine tuning of hyperparameters in order to achieve high accuracy scores greater than 90% the fine tuning process is known as adding subsequent layers in the end of the pretrained model. In the study conducted by the authors [6], Resnet-50 and Inception V3 models were employed and fine-tuned to effectively classify three distinct categories, specifically Normal, Macular Degeneration, and Tessellated Fundus. Paper reports an accuracy score of 93.81% while using Resnet-50 after fine tuning, it also reports an accuracy of 91.76% while using fine tuned version Inception V3 model.

Z. Li et al. in [12] gives an approach called BFPC-Net (Binocular Fundus Photographs Classifying Network). The BFPC-Net comprises 3 parts namely Residual Attention Module (RAM), Feature Fusion Module (FFM) and Image Augmentation Model (IAM). The Fig. 1 refers to model structure as represented in [12] and gives a brief architecture of BFPC-Net. The paper classified the ODIR dataset into 8 classes including normal class and DR being the major of all ailments. The paper reports a baseline accuracy of 80.82% which is boosted to 94.23% when the basic model is used along with FFM. The author special mentions that the accuracy is bound to fall significantly if the number of pathologies/ailments is increased.

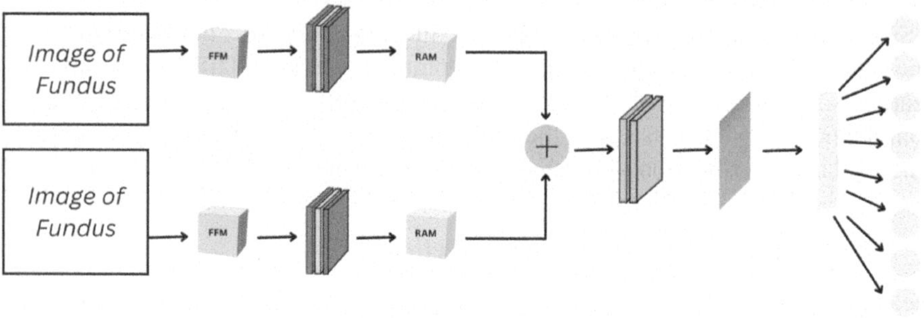

Fig. 1. Use of Feature Fusion Module use to extract features from images

In [13] N. Gour et al. have given a concept of 2 models the first model comprises of 2 input CNN and the second model comprises of concatenated CNN's, both the model make use of transfer learning the pretrained models being Resent, Inception V3, Mobile -Net and VGG 16.The model 1 demonstrates superior performance for VGG 16 architecture with stochastic gradient descent (SGD optimiser), model 2 although shows high training accuracy with these architectures it has shown overfitting phenomena with SGD Optimiser. The paper reports a validation accuracy 87.16% for model and 89.06% for model 2 while classifying ODIR data into 8 subsequent classes. The paper however reports a very high training accuracy model.

N. Chea et al. in [8] have used pretrained models and compared the performance of the model while using the standard dataset and augmented dataset. The authors in [8] compare RestNet-50, RestNet-101 ResNet-152, VGG 16 and VGG 19. It reports an accuracy in the range 63% to 77% Table 1 gives a better description of better scores. The model classifies data into Normal, GLC, ARMD and DR classes.

E. Ho et al. in [14] uses 5 different architectures to create an ensemble comprising SE-ResNeXt, DenseNet-12, Inception V3, EfficientNet-B4 and EfficientNet-B5.The study employed a 5-fold cross-validation technique to validate the models. The scores generated by each model were averaged, and this process was repeated for various architectures. The final ensemble was obtained by averaging the results. The paper reports a final AUROC score of 96.13% for healthy versus 28 different pathologies.

D. Müller et al. in [15] proposes a disease risk detector comprising of DenseNet-201 and EfficientNet-B4, along with disease label classifier comprising of ResNet-152, InceptionV3, DenseNet-201 and EfficientNet-B4.The utilisation of a 5-fold cross-validation technique and a stacked logistic regression model is acknowledged by the author along with ensemble techniques in which bagging and stacking are performed in respective order. The author reports an AUROC score of 97% for the classifier model and 99% for the detector model for 28 different pathologies.

M. T. Islam et al. in [16] have proposed a model to classify the data of ODIR dataset into 8 classes. The average F-score reported was 0.85, Kappa score reported 0.31and AUC value of 0.805.

In a study conducted by Lam et al. [3], pretrained GoogLeNet and AlexNet models were utilised for binary model classification. The authors employed the Contrast Limited Adaptive Histogram Equalisation (CLAHE) technique as a preprocessing step. The results indicated that the GoogLeNet model exhibited the highest sensitivity of 95% and specificity of 96%. The paper classifies large dataset of 35000 images into 5 class labels on the basis of severity the model makes the use of multiple optimiser including Adam, RMS Prop,SGD, etc. and reports a The maximum achieved accuracy in testing of 57.2%, 68.8%, and 74.5% on 4-ary, 3-ary, and 2-ary classification models, respectively.

In [17] the author uses an ensemble network of 4 networks under the name of DENet. The author uses RESNET-50 as backbone model and modified unit for guided segmentation at global fundus image level, it further uses RESNET model to classify the extracted Optic Disc Region. The paper reports an accuracy of 74.95% on the SINDI dataset and 84.285 on the SCES dataset.

R. Thanki in [18] suggests a SqueezeNet for feature extraction using a deep learning network as shown in Fig. 2. The features extracted are passed through a ML based classifier and at last the screening is performed. The SqueezeNet model consists of a total of 14 layers, including 2 convolution layers, 8 fire layers, 3 max pooling layers, 1 global average pooling layer, and a softmax activation function. The paper differentiates between normal and glaucomatous images. It makes use of DRISHTI-GS as its training data and ORIGA as testing data. The results are well compiled under Table 1.

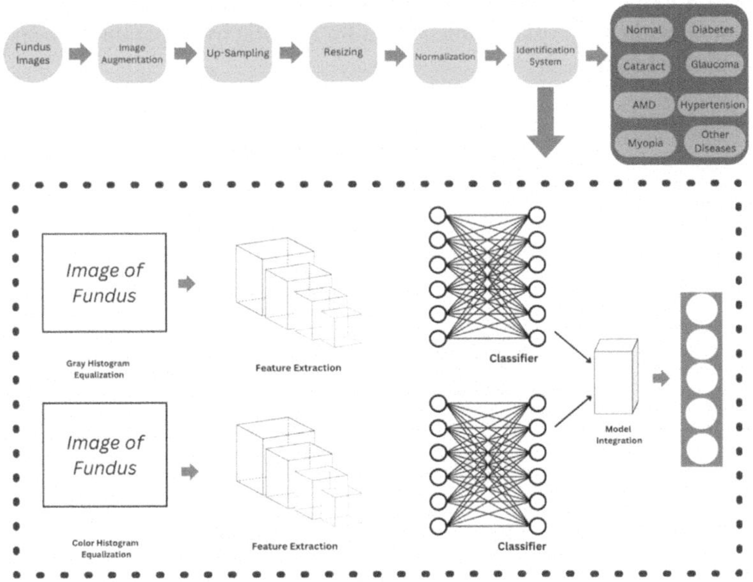

Fig. 2. Standard WorkFlow of Model using ML classifier

In [10] R. Sarki et al. uses DCNN VGG 16 to classify data according to 2 scenario. Scenario one being mild severity of DR, Diabetic macular edema (DME) GLC and Normal. And Scenario 2 being Normal, DR, GLC, Cataract. The author reports 88.3% accuracy for multi class Eye Disease (DED) and 85.95% for Mild Multi Class DED using VGG 16 Model.

S. Kumar et al. in [19] makes use of DenseNet-201 while EfficientNet-B4 was used for disease risk detection ResNet-105 is added for multi-disease classification. It makes 14 classification while being trained on RFMiD Dataset and being tested on ODIR dataset. It reports an accuracy of 97.42% using a stacked model approach.

In [9], M. A. Rodrguez et al. make use of C-Tran, which incorporates adam as an optimizer while being backed by multiple pretrained models, out of which DenseNet-161 gave the most promising results. C-Tran is supported by multiple models that have been pretrained. The C-tran is used for extracting features in order to generate a set of feature embeddings. Following this, the embeddings of the labels are combined with the embeddings of the states to facilitate the training of the model using incomplete information. In the end, the transformer's output is employed to provide the multi-layer perceptron (MLP) head with the requisite information for generating the probabilities linked to the unidentified classes. The representation of their conceptual framework is depicted in Fig. 3.The researcher conducted a trilogy of experiments. The initial series of experiments involved the utilisation of both conventional and cutting-edge convolutional neural network (CNN) models as backbones, specifically for the purpose of extracting features within the C-Tran architecture. The aim of this study was to identify the optimal and most effective model for the specific task at hand. The second set of experiments aimed to mitigate the problem of class imbalance by employing resampling techniques. Finally, the third set of experiments aimed to determine the optimal configuration of the model by identifying the most suitable values for specific hyperparameters, specifically image size and batch size.

$$AP = \sum_{i=0}^{|T|-1} [recall_i - recall_{i+1}] \times precision_i \tag{1}$$

$$ML\_mAP = \frac{1}{|T|} \sum_{i=1}^{|T|} AP_i \tag{2}$$

$$ML\_F1 = \frac{1}{|T|} \sum_{i=1}^{|T|} F1_i \tag{3}$$

$$ML\_AUC = \frac{1}{|T|} \sum_{i=1}^{|T|} AUC_i \tag{4}$$

$$ML\_Score = \frac{ML\_mAP + ML\_AUC}{2} \tag{5}$$

$$Model\_Score = \frac{ML\_Score + Bin\_AUC}{2} \tag{6}$$

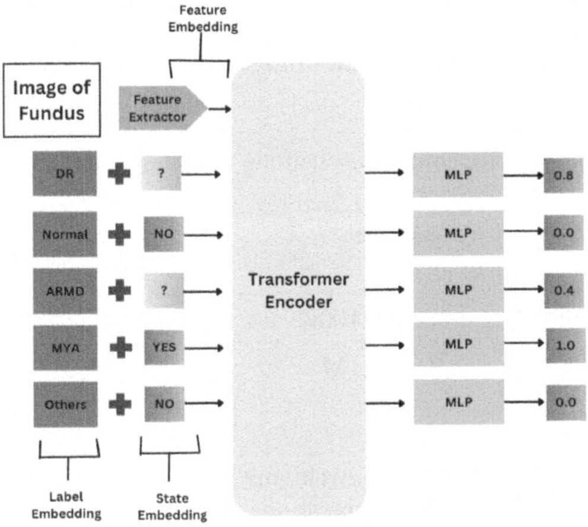

Fig. 3. C-Tran using MLP head to classify data

where T is set of labels and |T| is the total number of labels of set

L.-P. Cen et al. in [2] describes the role of a cluster of CNNs and a Mask RCNN and tries to establish relevant performance using the help of clustering which aims to cluster the similar type of diseases together based on characteristics and features such as maculopathy and disc swelling and elevation. The researcher presents the mean, a sensitivity of 97.8%, a specificity of 99.6%, and an Area under the receiver operating characteristic curve (AUC) of 0.9984, F1 score of 0.923.

In [20] the author adopts a DCNet consisting of a backbone Convolutional Neural Network (CNN), a Spatial Correlation Module (SCM), and a classifier as shown in Fig. 4. The Sparse Correlation Matrix (SCM) algorithm effectively captures strong correlations among the features extracted from the paired Colour Fundus Photographs (CFPs) at a pixel level. Furthermore, it integrates pertinent feature representations to enhance the overall fusion process. The author reports an increase of 7.7%, 1.3%,1.3% and 3.4% for the observed Kappa, F1, AUC and Final scores.

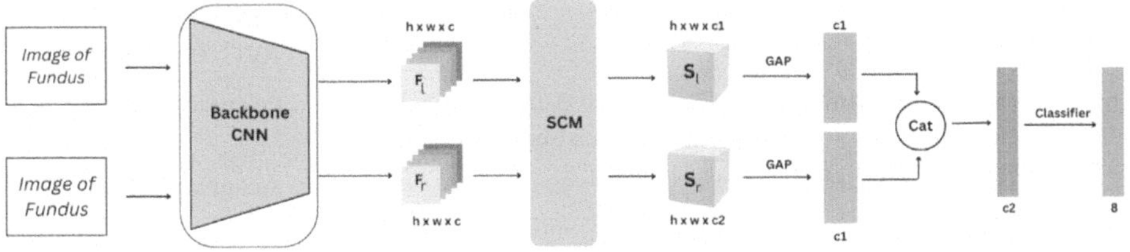

Fig. 4. DCNet using SCM backbone

J. Wang et.al. in [21] compares multiple pretrained models out of which Efficient-B3 Comes out be the best performing network. The author uses pre-trained model as backbone for the purpose feature extraction and adds customized top layer for multi class classification. The reported accuracy score is 92% while the running model standard ODIR dataset. The top layer classifies the images into 8 classes.

Table 1. Summary of Techniques used along with associated performance Score

Ref ID	Technique	DataSet	No of Classes	Results
[1]	ResNet-50 with pre-trained weights from ImageNet and ADAM Optimiser	Retina-100K	5	64.14 mAP
		Retina-1M		48.23 mAP
[2]	3 groups consisting of CNNs and Mask-RCNN	Picture Archiving and Communication Systems (PACS) consists of 88,702 color fundus images	39	F1 score:0.923
				Sensitivity: 97.8%
				Specificity: 99.6%
				AUC: 0.9984
[3]	Pretrained GoogleNet/AlexNet with contrast limited adaptive histogram equalization as pre-processing	35000 images available publicly on kaggle and Messidor -1	2	Peak test accuracies: 74.5% (2- ary with adam optimiser) with Data Augmented and CLAHE Images
				Peak test accuracies: 68.8% (3-ary with AdaGrad optimiser)
				Peak test accuracies:57.2% without data augment and CLAHE images (4-ary with Adam optimiser)
[4]	ResNet-101	ODIR	8	Kappa: 0.637 ± 0.007 F1: 0.913 ± 0.002 AUC: 0.930 ± 0.004 AVG: 0.827 ± 0.003
[6]	ResNet-50 (fine tuned)	Customized dataset of 1032 fundus images	3	Accuracy: 93.81%
	Inception V3 (fine tuned)			Accuracy: 91.76%

(*continued*)

Table 1. *(continued)*

Ref ID	Technique	DataSet	No of Classes	Results
[8]	ResNet-50 (Augmentation)	NOISE-STRESS	4	Accuracy: 85.79%
[9]	DenseNet161	MuReD dataset	20	AUC: 0.962
[10]	VGG16 model	Messidor, Messidor-2, DRISHTI-GS, and retina datasets publicly available	6	Accuracy: 85.95%
[12]	BFPC-Net	ODIR	8	Accuracy: 94.23%
[13]	VGG16	ODIR	8	Accuracy: 89.06%
[14]	VGG16	RFMiD	28	AUROC: 0.9613
[15]	Ensemble of ResNet152, InceptionV3, DenseNet201 and EfficientNetB4	RFMiD	28	AUROC: 0.97
[16]	Proposed Model	ODIR-5K	8	F-score: 0.85
[17]	DENet	ORIGA for training, SCES and SINDI for testing	2	SCES: 84.28% (Accuracy) SINDI: 74.95% (Accuracy)
[18]	SqueezeNet	DRISHTI-GS,ORIGA	2	Accuracy: 76.2%
[19]	ResNet105	RFMiD Training and ODIR Testing	14	Accuracy:97.42%
[20]	ResNet-101 backbone with SCM	ODIR	8	AUC: 0.827
[21]	EfficientNetB3	ODIR	8	Accuracy: 92%

3 Critical Evaluation

As discussed regarding various techniques used in subsection A of same section it is noteworthy that the field of transformers, RCNNs are yet and unexplored terrain. The authors widely make use of standard pretrained model as a backbone further adding customized layer on the top which created the scope for multiple scenarios. The Best accuracies were achieved in [19] which mounts 97.42%. Some authors have also made use of other performance metrics such as Kappa score (K), F1 score, AUC, Specificity, Sensitivity.

$$K = K\frac{P_0 - P_e}{1 - P_e} \tag{7}$$

$$Precision = \frac{True\ Positive}{True\ Positive + False\ Positive} \tag{8}$$

$$Sensitivity = \frac{False\ Positive}{True\ Negative + False\ Positive} \tag{9}$$

$$F1\ Score = \frac{2 \times Precision \times Sensitivity}{Precision + Sensitivity} \tag{10}$$

$$Specificity = \frac{True\ Negative}{True\ Negative + False\ Positive} \tag{11}$$

where P_0 - Probability of Agreement Observed and P_e - Probability of Agreement by Chance

A prevalent issue encountered in paper analysis involves the identification of an inverse relationship between the quantity of classes and the performance score. Specifically, as the number of output classes increases, a decrease in the model's performance is observed. The identical phenomenon was documented by Z. Li et al. in their publication [12].

Many authors also reported the common problem of class imbalance amongst various datasets publicly available, however only a few of the authors/researchers have given the means to eradicate the same. The class imbalance problem impacts significantly by training as the actual performance of the model might just be an overshoot. This is why in classification problems especially images related to medical data the class distribution must be in equity.

4 Conclusion

As described earlier the use of ViT and other images segmentation and classification model like YOLO and U-Net are worth trying for the above said task, given the complexity of the task it is advised for a trained ophthalmologist to segment the dataset and annotated it according so it can be fed to advance model and faster model like YOLO so that the computation time can be reduced substantially.

Also the inclusion of other features like age, region, gender, family history might also be promising step towards a better prognosis and early detection of any eye related or fundal disease pretrained model used by the author in section make use of large image datasets like that of Image Net but the problem with use of such dataset arise as they have no class separate for the medical images or specifically fundal image disease.

Hence it can conclude that research work for the sake of retinal image classification has reached satisfactory levels but slight improvements can be made on level of computation time and space. The introduction of edge computing and cloud servers integrated ophthalmoscope can also be a boon as an auto detection system can be formed.

References

1. Zhou, Q., Zou, H., Wang, Z.: Long-tailed multi-label retinal diseases recognition via relational learning and knowledge distillation. In: Wang, L., Dou, Q., Fletcher, P.T., Speidel, S., Li, S.

(eds.) MICCAI 2022. LNCS, vol. 13432, pp. 709–718. Springer, Cham (2022). https://doi.org/10.1007/978-3-031-16434-7_68
2. Cen, L.-P., et al.: Automatic detection of 39 fundus diseases and conditions in retinal photographs using deep neural networks. Nat. Commun. **12**(1) (2021). https://doi.org/10.1038/s41467-021-25138-w
3. Lam, C., Yi, D., Guo, M., Lindsey, T.: Automated detection of diabetic retinopathy using deep learning. AMIA Jt. Summits Transl. Sci. Proc. **2018**, 147–155 (2017). PMID: 29888061; PMCID: PMC5961805
4. He, J., Li, C., Ye, J., Qiao, Y., Gu, L.: Biomed. Sig. Process. Control **63**, 102167 (2021). https://doi.org/10.1016/j.bspc.2020.102167
5. Pachade, S., et al.: Retinal fundus multi-disease image dataset (RFMiD): a dataset for multi-disease detection research. Data **6**(2), 14 (2021). https://doi.org/10.3390/data6020014
6. Pan, Y., et al.: Fundus image classification using Inception V3 and ResNet-50 for the early diagnostics of fundus diseases. Front. Physiol. **14** (2023). https://doi.org/10.3389/fphys.2023.1126780
7. Messidor database. Image Anal. Stereol. **33**(3), 231–234 (2014). ISSN 1854-5165. http://www.ias-iss.org/ojs/IAS/article/view/1155 or https://doi.org/10.5566/ias.1155
8. Chea, N., Nam, Y.: Classification of fundus images based on deep learning for detecting eye diseases. Comput. Mater. Continua **67**(1), 411–426 (2021). https://doi.org/10.32604/cmc.2021.013390
9. Rodríguez, M.A., AlMarzouqi, H., Liatsis, P.: Multi-label retinal disease classification using transformers. IEEE J. Biomed. Health Inform. **27**(6), 2739–2750 (2023). https://doi.org/10.1109/jbhi.2022.3214086
10. Sarki, R., Ahmed, K., Wang, H., Zhang, Y.: Automated detection of mild and multi-class diabetic eye diseases using deep learning. Health Inf. Sci. Syst. **8**(1) (2020). https://doi.org/10.1007/s13755-020-00125-5
11. Kadan, A., Subbian, P.: Classification of Fundus Images Using Neural Network Approach Classification of Fundus Images Using Neural Network Approach (2019). https://doi.org/10.4018/978-1-7998-1192-3.ch006
12. Li, Z., Xu, M., Yang, X., Han, Y.: Multi-label fundus image classification using attention mechanisms and feature fusion. Micromachines **13**(6), 947 (2022). https://doi.org/10.3390/mi13060947
13. Gour, N., Khanna, P.: Multi-class multi-label ophthalmological disease detection using transfer learning based convolutional neural network. Biomed. Sig. Process. Control **66**, 102329 (2021). https://doi.org/10.1016/j.bspc.2020.102329
14. Ho, E., et al.: Deep ensemble learning for retinal image classification. Transl. Vis. Sci. Technol. **11**(10), 39 (2022). https://doi.org/10.1167/tvst.11.10.39
15. Müller, D., Soto-Rey, I., Kramer, F.: Multi-Disease Detection in Retinal Imaging Based on Ensembling Heterogeneous Deep Learning Models (2021)
16. Islam, M.T., Imran, S.A., Arefeen, A., Hasan, M., Shahnaz, C.: Source and camera independent ophthalmic disease recognition from fundus image using neural network. In: 2019 IEEE International Conference on Signal Processing, Information, Communication & Systems (SPICSCON), Dhaka, Bangladesh, pp. 59–63 (2019). https://doi.org/10.1109/SPICSCON48833.2019.9065162
17. Fu, H., et al.: Disc-aware ensemble network for glaucoma screening from fundus image. IEEE Trans. Med. Imaging **37**(11), 2493–2501 (2018). https://doi.org/10.1109/tmi.2018.2837012
18. Thanki, R.: A deep neural network and machine learning approach for retinal fundus image classification. Healthc. Anal. **3**, 100140 (2023). https://doi.org/10.1016/j.health.2023.100140
19. Kumar, S., Bindu, S.: MDCF: multi-disease classification framework on fundus image using ensemble CNN models. J. Jilin Univ.

20. Li, C., Ye, J., He, J., Wang, S., Qiao, Y., Gu, L.: Dense correlation network for automated multi-label ocular disease detection with paired color fundus photographs. In: 2020 IEEE 17th International Symposium on Biomedical Imaging (ISBI), Iowa City, IA, USA, pp. 1–4 (2020). https://doi.org/10.1109/ISBI45749.2020.9098340
21. Wang, J., Yang, L., Huo, Z., He, W., Luo, J.: Multi-label classification of fundus images with EfficientNet. IEEE Access **8**, 212499–212508 (2020). https://doi.org/10.1109/access.2020.3040275
22. Meena, G., Mohbey, K.K., Indian, A., Khan, M.Z., Kumar, S.: Identifying emotions from facial expressions using a deep convolutional neural network-based approach. Multimed. Tools Appl. (2023). https://doi.org/10.1007/s11042-023-16174-3

Driver Safety and Drowsiness Detection in Internet of Vehicles with Federated Learning

Satadal Banerjee, Mulagala Sandhya(✉), and Y. Sreenivasa Rao

National Institute of Technology Warangal, Hanamkonda, India
SBCS21205@student.nitw.ac.in, {msandhya,ysr}@nitw.ac.in

Abstract. Rising road accidents have been a serious issue to any nation, whatsoever developed it has been. The primary reasons all attribute to one sole reason- Driver negligence and unsafe driving habits. There are many factors that can impact driver safety, including the condition of the vehicle, the driver's physical and mental state, and the road and weather conditions. Considering there is nothing much to do about road and weather conditions, our primary goal here is to assess the driver of their state and take necessary precautions. We assume all the vehicles to be a part of an Internet of Vehicles (IoV), apply a federated learning based approach to get the driver behavioral patterns and alert them or nearby drivers/authorities. While the exact precision and value for this approach is still to be found out, we are hopeful about the goodwill and the safety it will bestow upon the road and transport.

Keywords: Internet of Vehicles · Deep Learning · Federated Learning

1 Introduction

The community has always tried to reduce the number of road accidents happening. The efforts have not been totally futile though, over the years, technological advances have proven to be lifesaver. Even machine learning algorithms and integrating them with Internet of Vehicles has been on the latest edge. Talking about technological advances, there are Advanced driver assistance system like Vehicle-to-vehicle (V2V) and vehicle-to-infrastructure (V2I) communication [1], Driver monitoring systems [2], Automatic emergency braking (AEB), Intelligent transport systems [3]. Though our technique involves a hybrid of Vehicle-to-Vehicle Communication and Driver Monitoring system in a simulated Internet of Vehicles (IoV), we additionally make use of Federated Learning. We focus on privacy preserving mechanisms to collaboratively train a machine learning model while keeping their data on the devices themselves, rather than sending it to a centralized server. In federated learning, each device taking part in the training process receives a copy of the machine learning model from the central server first. The model is then locally trained on each device's own data without transferring the data to the server. Each device transmits the model, updates back to the server after training. The server then compiles these updates to enhance the model

D. Pastor-Escuredo et al. (Eds.): ICDLAIR 2023, LNNS 1001, pp. 61–70, 2024.
https://doi.org/10.1007/978-3-031-60935-0_6

as a whole. Federated learning has an advantage that it allows the training of machine learning models on a much larger scale, since data from many devices can be used to train the model. Another advantage is that it can preserve the privacy of users, since their data never leaves their devices. Additionally, federated learning can be useful in scenarios where data is distributed across a large number of devices that have limited connectivity or computational resources.

2 Related Work

Yuan et al. [4] proposed a federated bidirectional connection broad learning scheme (FeBBLS) to solve the data sharing issues. A similar approach with federated learning was discussed by Zhang et al. [5] by training the model with transfer learning with drowsiness database. A CKKS-based privacy preserving communication protocol (CPA secure) was followed, which is also a fully homomorphic encryption. In [6], Chen et al. devised a driver behavior monitoring and warning (DBMW) framework encapsulated within Internet of Vehicles for road safety. In particular the DBMW incubates a wearable head motion sensor. The image sensor is used to detect the current located lane and determine the deviation degree of a vehicle. The wearable device is used to detect the head motion and measure the anomaly level of the driver by signal sampling, normalization and Feature Vector Transforms. In [7], One of the emblematic issues with federated learning, the non-IID problem of data distribution, was addressed by Shin et al. If the data distribution is not IID, the suggested approach can effectively remove those that lead to client covariate shifts from training. It has been demonstrated that, even when compared to the well-known FedProx algorithm, there is no performance loss until Fully-IID is used. Instead, it was demonstrated that it performs better in a wholly non-IID data distribution. In [8], Witt et al. overcame the challenge of heterogeneity of Federated Learning Framework in terms of use cases, applied focus, design choice, and thoroughness by offering a comprehensive and holistic comparison framework. The FL design was made client truly decentralized. This opened roads to other decentralized networks such as smart contracts in Blockchain and Distributed Ledgers. In [9] Guerra et al. designed Centralized Federated Learning (CFL), that used a simple FedAvg meachanism on a batch of clients, to update the central server periodically with a randomized batch of clients. Also, a Blockchain-Enabled Federated Learning (BFL) was designed that handled the data as a Distributed Ledger Technology and added update and edge weights as transactions in the block that could only happen through advance cryptographic functions. Fan et al. [10] took a step ahead and introduced Federated Learning Architecture to the Industrial Internet of Things (IIOT). A Federated Learning-based Privacy-preserving Data Aggregation scheme (FLPDA) for IIoT was suggested.

3 Proposed Method

The proposed methodology mainly comprises of four steps, server initialization and system startup, selection of random nodes/ clients for training at a particular epoch, local nodes training, and finally aggregation by the server.

System Initialization, Server startup and client setup
The model and the related networking framework need to be chosen first. Preferably a centralized server will be set up.

1. The server state must be stateless and have support for good load balancing.
2. It should also have an authentication and suitable encryption mechanism. The client or edge nodes should be authenticated regularly.
3. It should be powerful, power-efficient and be able to scale well on decently large networks.

The server might additionally consume some network resources, but those can be optimized as well. In [11], Dai et al. designed model optimization techniques for centralized federated architectures. The client setup, on the other hand, is simple. Preferably they should be lightweight, so as to they can allow the server to congregate a maximum number of them, within given deadlines.

Client Selection
The main motto here would be to select a randomized and maximum number of clients as possible, so that the overall aggregated model can achieve the desired performance faster. The choice whether the server will select randomized clients or select clients specifically that can help converge the model faster is totally up to us. Here we choose the clients that best suit the server. Let us assume there are a total of S clients with indices $S = \{1, .., S\}$ and a subset $S^i \subseteq S$ be a set of S^i best suited clients chosen at this step. We could also additionally offer client optimization algorithms [12]. The clients, in the given order of S, begin to upload their models respectively to the server. Assuming all the clients have the same deadline to upload their model, let $T_{S_i} \in R^+$ where T denotes the individual deadline for each client S_i, and R^+ denote the set of non-negative real numbers. Also, let T_i^{upload} denote the time taken by the ith client to upload the model to the central server and T_i^{update} denote the time taken by the ith client to update the edge weights in accordance with the servers updated weights, then the elapsed time from the beginning when the server sends initial weights, to the end, from receiving updated edge weights, for the ith client, can be formulated as-

$$\delta_i := \begin{cases} 0 & \text{if}\, i = 0, \\ T_i^{\text{Update}} + T_i^{\text{Upload}} & \text{otherwise}, \end{cases} \tag{1}$$

$$T_i^{\text{Update}} = \sum_{j=1}^{i} \max\left\{0, t_{i_j}^{\text{Update}} - \delta_{j-1}\right\}, \tag{2}$$

$$T_i^{\text{Upload}} = \sum_{j=1}^{i} t_{i_j}^{\text{Upload}}. \tag{3}$$

where T_{total} represents the total elapsed time, which is $T_{\text{upload}} + T_{\text{update}}$ and T_{avg} is the time taken for the federated averaging step.

Training in Local Nodes

A Convolutional Neural Network is able to detect and classify the drivers drowsiness. Patterns like facial expression, eye movement, mouth position, etc. can be extremely necessary to these. Considering the drivers have dashboard cameras fitted onto them that constantly keep on recording the faces, or take periodic snapshots of the drivers facial images, we have two choices, whether to use a pre-trained model or train now itself. Ali et al. [13] suggested using an Eye Aspect Ratio and Mouth Aspect Ratio as parameters to judge for drowsiness as in Fig. 1. The eye aspect ratio can be calculated from the formulae-

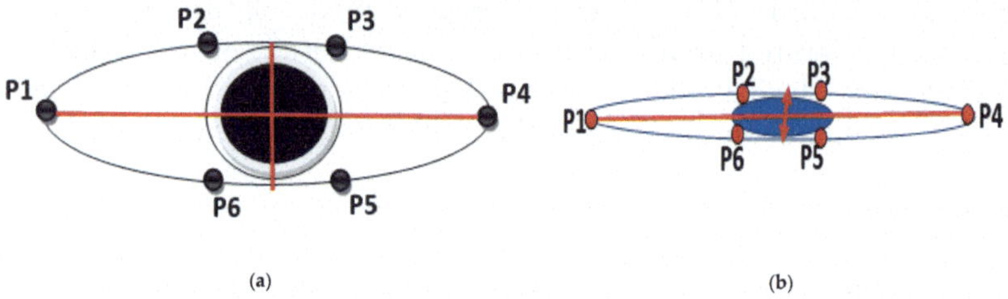

Fig. 1. Calculation of Eye Aspect Ratio

$$Aspectratio = \frac{(p6 - p2) + (p5 - p3)}{2X(P4 - P1)} \tag{1}$$

As little are the eyes, smaller is the aspect ratio. Wide open eyes denote a larger value of aspect ratio. So preferably, if the aspect ratio remains dangerously low for a certain threshold period of time, it will be safe to declare that the driver is drowsy. The exact opposite is for the mouth aspect ratio; a larger value is dangerous, it denotes the driver is yawning or talking loudly, and is definitely distracted. Personally, the choice of model used can differ, however, the core remains the same. Calculate the aspect ratios from the sequence of images and decide the best, that is, choose an option, drowsy or not drowsy. A VGG19 architecture is a modest one for the case, is handy and gives quite optimal results [14]. A simplified VGG19 architecture consists of a series of convolution and pooling layers, followed by a Fully Connected Neural Network and softmax activation function that provides the output, which in this case, is a classification.

Server Aggregation

The final step is the aggregation from the server, on the received edge weights from the clients. We perform a Federated Averaging process, also known as FedAvg, on the received edge weights.

Algorithm 1: FedAvg Algorithm proposed by K = no. of clients, B is local minibatch size, E = no. of local epochs, n_k is the number of data points for *kth* client, n is total number of data points and η = learning rate.

Server executes:
  Initialize w_0
  for each global round t=1,2,... **do**
    m ← max(c.K,1)
    Choose a random set of users S_t
    for each **client** $k \in S_t$ **do**
      $w^k_{t+1} \leftarrow$ ClientUpdate(w_t)
      $w_{t+1} \leftarrow \frac{1}{n} \sum_{u \in K} n_k w^k_{t+1}$
ClientUpdate(k,w):
  B ← split P_k into batches of size B
  for each local epoch i from 1 to E **do**
    for batch $b \in B$ **do**
      $w \leftarrow w - \eta \nabla l(w, b)$
  return w to **server**

Fondling with the parameters can result in some different but interesting results. If we change the local mini batch size B as ∞ then the full local dataset will be treated as a minibatch. The heart of the FedAvg algorithm lies in the aggregation step. A fraction of clients are chosen and with each of the received edge weight updates, it calculates an aggregate value by simply performing $\frac{n_k}{n}$ $\times$ the edge weight updates from the previous round, and then calculates the updated edge weights and send those values to all the clients in the new round.

4 Results and Discussion

We broadly classify the system into two parts. In first part, we mainly discuss about the Internet of Vehicles and the Vehicular Ad-Hoc Network (VANET) architecture and its implementations using Eclipse SUMO (Simulation of Urban MObility) and *OpenStreetMap.org*. In the second part, the main approach regarding how the drivers sequential images are being fed and trained using the federated model.

4.1 Tools Used and Dataset

Tools used are as follows:-

1. SUMO or Simulation Of Urban MObility which has been used to simulate a vehicular network.
2. *Openstreetmap.org* which generates .osm files of map roads of real places.
3. OMNET++ is an API or framework designed exclusively with the libraries of C++. Mostly used for simulation of network architectures for inter connected devices, which in this case, is simulation of an Internet of Vehicles [15].
4. Google Colab for federated learning implementation.

5. Keras and Tensorflow Federated Library.
6. Datasets used are - DrivFace dataset [16] and challenges-in-representation-learning-facial-expression-recognition-challenge dataset [17]

The DrivFace dataset is a multivariate image sequence of subjects while driving a car in real scenario. There are a total of 4 driver with several facial features and a total of 606 images of 640×480 pixels. There are about 6400 attributes associated to it. All the facial key points (eyes, nose and mouth) are visible (Fig. 2).

Fig. 2. Sample images from DrivFace dataset

4.2 Vehicle Network Simulation

Vehicle simulation is critical to understanding the topology to where the exact location and architecture using federated learning holds. Now, we really have two choices,

- use existing topology from map, say current city or any other city.
- Create custom map and nodes and recreate topology and routes.

SUMO road networks are encoded as XML files. There are various other options which are customizable like number of roundabouts, terrain, density of vehicles, etc. Adjusting such parameters open up more possibilities to more complex road architectures and vehicular network models. Some road traffic simulations were created using the Simulation of Urban Mobility tool where we want to integrate federated learning concept. Some notable parameters that were set are as follows-

- Traffic scale-100
- Delay-100 ms
- Location and map-Kharagpur, West Bengal, India

SUMO's feature in allowing users to import road network simplified the process for developing mobility model. Through Java Open Street Map (JOSM) Editor, real world road network can be easily downloaded and edited. In the editor, modification such as inserting traffic lights, configure road connections, define number of lanes on a road, etc. can be made to suit the users' need. The map is then further processed through SUMO's command netconvert to generate simulation network containing heuristically computed values. After that a trip file is generated to simulate vehicle flows on the network. Figure 3 shows the flux vs density plot for the given simulation of IOV we did.

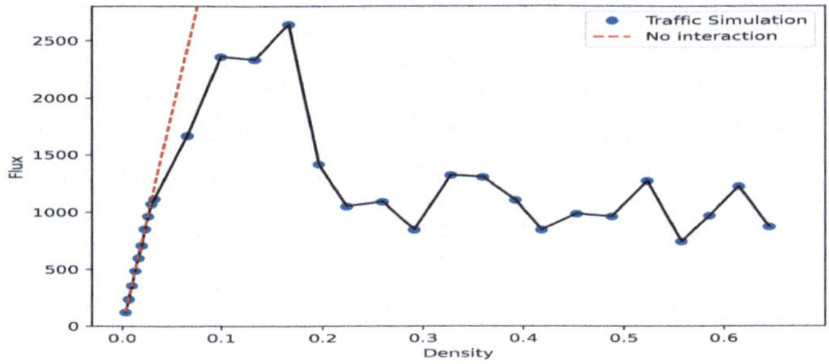

Fig. 3. Flux vs Density plot for the given Simulation of map

The aspect ratios are calculated and those values are back propagated throughout the model. The results are boundary boxes that specify the regions of eyes that in turn, calculate a score that decides whether the driver is drowsy or not. The classifier first checks whether the drivers eyes are open or not. It then calculates a drowsiness score. A simple ResNet50 model with 13 convolutional layers with 3×3 filters isused. We have used a CNN model with optimizer as Stochastic Gradient Descent, learninig rate as 1e-3, loss function as *Categorical CrossEntropy* and a simple *Softmax* activation function in the output layer.

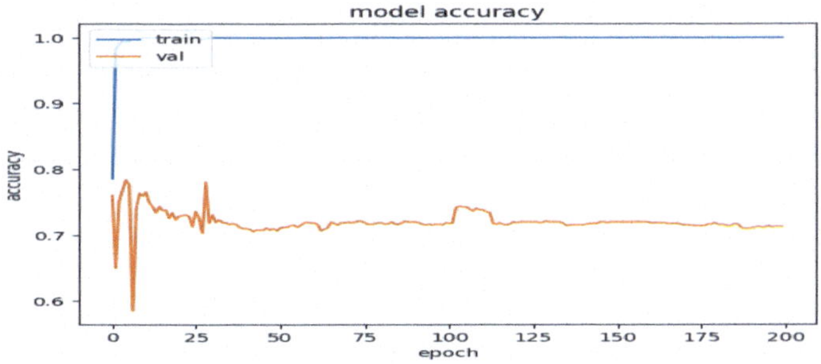

Fig. 4. epoch vs accuracy for 200 epochs

From Fig. 4, it can be seen that even though the training accuracy is close to 99%, the validation accuracy ranges on an average of 85% for the standard SGD model.

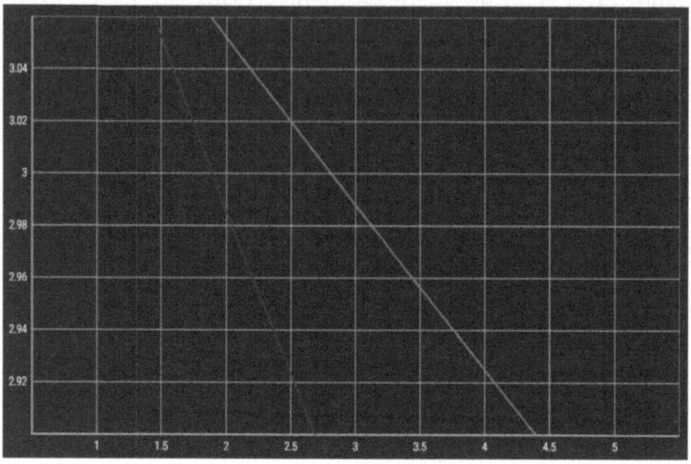

Fig. 5. time vs sparse categorical cross entropy for 200 epochs, down scaled by 20 where x-axis is time, y-axis is categorical loss

The images are spliced into Tensors of dimensions (1×784). Since the Time series graphs are long, we have down scaled the values by a factor of 20. Figure 5 defines the categorical loss for the batch we defined. It can be seen that the loss shoots down to approximately 59.272% in initial 4.5 s. That is the time when more than 10 batches of images were fed. The smoothing factor has been set to 0.63. Additionally we performed the Federated training for 5 clients, each with 3 images. We tabulated the testing accuracy (Sparse Categorical Accuracy) and Categorical Cross Entropy as Loss for 10 clients, with shuffled data separated folder wise. The results are shown in Table 1.

Table 1 gives us some insights on the ability of our Federated Model to classify the emotions correctly. The highest Sparse Categorical Entropy accuracy attained is 90.005% for some given client (number 8) and the corresponding minimum categorical loss attained is 48.931% for client number 10. The overall Categorical accuracy ponders between 89–90% with the mentioned model that uses the 7 layer VGG19 model with extra layers in the output layer, with the mean accuracy of 10 clients being 89.482%. We concluded that this model is quite flexible with respect to the FedAvg algorithm. We have shown that all the parameters are interchangeable and provide the same results, with our parameters producing the most optimal results. The Flux vs Density plot has no correlation with the models overall performance for the time being, but may affect the servers load as more and more traffic and vehicles get encapsulated in a single IOV network.

Table 1. Table of testing accuracies and losses for 10 clients

Client number	Sparse categorical Accuracy (In Percentage)	Loss(In Percentile)
1	89.341	49.126
2	89.277	49.329
3	89.279	49.112
4	89.388	48.951
5	89.306	49.443
6	89.582	48.973
7	89.681	49.384
8	90.005	49.218
9	89.201	49.087
10	89.766	48.931

5 Conclusion and Future Work

It seems a very much possibility to leverage local resources and minimize power consumption and dissipation at local edge devices. Moreover, with the repetition of the iterations, the communication cost is fairly high, that has to be minimized too. A cheaper and more economic way to achieve critical communication between the IoV network modules or the vehicles can be devised. The security concerns are thoroughly covered, but some issues still need to be looked after. The privacy is achieved through secure multi-party communications but at the cost of some model efficiency. There is some trade-off between these two factors, which need to be balanced. Some model compression schemes like sparsification and quantization are also on demand and sought after [18]. Overall this architecture might seem simple and straightforward, but has a lot of scope for improvement. The cost of the vehicle hardware could be reduced. In this work we have presented a model that can fairly work on sequence of facial images and predict whether they are distracted or not. However, this opens a larger scope of work, like we can feed a continuous stream of digital video of the driver to the model. We would also look forward to integrate the SUMO and the Internet of Vehicles with the Federated model, for a more robust performance on a large scale.

References

1. Malik, R.Q., Ramli, K.N., Kareem, Z.H., Habelalmatee, M.I., Abbas, H.: A review on vehicle-to-infrastructure communication system: Requirement and applications. In: 2020 3rd International Conference on Engineering Technology and its Applications (IICETA), pp. 159–163 (2020)
2. Liu, Y., Zhang, H., Liu, J.: Driver fatigue monitoring method based on eyes state classification. In: 2008 Chinese Control and Decision Conference, pp. 2257–2260 (2008)

3. Macheret, P.D., Savchuk, R.R., Shkuratov, G.I.: Intelligent transport systems: analysis of the current state and prospects of development. In: 2021 International Conference on Quality Management, Transport and Information Security, Information Technologies (IT&QM&IS), pp. 234–237 (2021)
4. Yuan, X., Chen, J., Zhang, N., Fang, X., Liu, D.: A federated bidirectional connection broad learning scheme for secure data sharing in internet of vehicles. China Commun. **18**(7), 117–133 (2021)
5. Zhang, L., Saito, H., Yang, L., Jiajie, W.: Privacy-preserving federated transfer learning for driver drowsiness detection. IEEE Access **10**, 80565–80574 (2022)
6. Chen, L.-W., Chen, H.-M.: Driver behavior monitoring and warning with dangerous driving detection based on the internet of vehicles. IEEE Trans. Intell. Transp. Syst. **22**(11), 7232–7241 (2021)
7. Shin, W., Shin, J.: Fedvar : federated learning algorithm with weight variation in clients. In: 2022 37th International Technical Conference on Circuits/Systems, Computers and Communications (ITC-CSCC), pp. 1–4 (2022)
8. Witt, L., Heyer, M., Toyoda, K., Samek, W., Li, D.: Decentral and incentivized federated learning frameworks: a systematic literature review. IEEE Internet Things J. **10**(4), 3642–3663 (2023)
9. Guerra, E., Wilhelmi, F., Miozzo, M., Dini, P.: The cost of training machine learning models over distributed data sources. IEEE Open J. Commun. Soc. **4**, 1111–1126 (2023)
10. Fan, H., Huang, C., Liu, Y.: Federated learning-based privacy-preserving data aggregation scheme for IIoT. IEEE Access **11**, 6700–6707 (2023)
11. Dai, H., Hong, Y.: Research on model optimization technology of federated learning. In: 2023 IEEE 8th International Conference on Big Data Analytics (ICBDA), pp. 107–112 (2023)
12. Zirui, G., Wei, W., Bingyong, T.: Research on resource optimization allocation of e-business system based on client-class. In: 2007 International Conference on Service Systems and Service Management, pp. 1–6 (2007)
13. Ali, N., Hasan, I., Özyer, T., Alhajj, R.: Driver drowsiness detection by employing CNN and DLIB. In: 2021 22nd International Arab Conference on Information Technology (ACIT), pp. 1–5 (2021)
14. Kavitha, S., Dhanapriya, B., Vignesh, G.N., Baskaran, K.R.: Neural style transfer using vgg19 and alexnet. In: 2021 International Conference on Advancements in Electrical, Electronics, Communication, Computing and Automation (ICAECA), pp. 1–6 (2021)
15. Abid, M.A., Belghith, A.: Modeling efficiency of mobile ad hoc networks in OMNeT++. In: 2011 4th Joint IFIP Wireless and Mobile Networking Conference (WMNC 2011), pp. 1–5 (2011)
16. Lpez, A.H.-S., Aura, Katerine, D.-C.: DrivFace. UCI Machine Learning Repository (2016). https://doi.org/10.24432/C5XC7Q
17. Will, C.Y.B.D., Ian, G.: Challenges in representation learning: facial expression recognition challenge (2013)
18. Zhang, W., Xi, X.: The innovation and development of internet of vehicles. China Commun. **13**(5), 122–127 (2016)

Privacy Preserving Fingerprint Classification Using Federated Learning

Ashok Soni, Mulagala Sandhya(✉), and Y. Sreenivasa Rao

National Institute of Technology Warangal, Warangal, India
ASCS21218@student.nitw.ac.in, {msandhya,ysr}@nitw.ac.in

Abstract. Fingerprint recognition is a widely used biometric authentication method that faces challenges due to the increasing sophistication of fake fingerprint creation techniques. This work focuses on the task of real and fake fingerprint classification using federated learning, leveraging a custom dataset derived from the CASIA fingerprint dataset. The goal is to develop a model capable of accurately distinguishing between real and fake fingerprints for different clients. The methodology involves training individual client models on their respective data using Convolutional neural network (CNN) architectures. A global model is updated using a federated learning technique that ensures collaborative model training while protecting data privacy. Using a separate test dataset, the performance of the trained global model is assessed, and several performance metrics, including accuracy, precision, recall, and F1-score, are produced. The results demonstrate the effectiveness of federated learning in addressing the real and fake fingerprint classification task. The custom dataset derived from the CASIA fingerprint dataset provides a focused analysis of the binary classification problem, enabling insights into the model's ability to accurately classify real fingerprints while minimizing false positives and false negatives. The findings contribute to the development of robust fingerprint recognition systems and highlight the potential of federated learning in addressing biometric authentication challenges.

Keywords: Fingerprint Recognition · Real and Fake Classification · Federated Learning

1 Introduction

Due to their ease and robustness when compared to traditional procedures like a password, biometric recognition systems are now used in a range of identifying sectors. Systems for biometric identification rely on a person's physiological and behavioural characteristics [1]. Biometrics finds applications in various domains such as access control, border security, forensic investigations, healthcare, and financial services. Biometrics are vulnerable to two different forms of attacks: direct and indirect [2]. The most frequent type of attack is a direct one because

D. Pastor-Escuredo et al. (Eds.): ICDLAIR 2023, LNNS 1001, pp. 71–80, 2024.
https://doi.org/10.1007/978-3-031-60935-0_7

it requires no prior knowledge to execute. Simple and practical tools like silicon, wood glue, etc. can be used in the sensor device to execute the fingerprint recognition system. In contrast, an indirect assault forces the system's module to provide deep knowledge. Researchers have been drawn to create a system that can evaluate and offer a solution for detection of liveliness of fingerprint systems as a result of the proliferation of attack tools. Accurate classification of real and fake fingerprints is crucial to ensure the security and integrity of biometric systems [3]. However, the availability of large-scale fingerprint datasets, coupled with privacy concerns, poses significant challenges for traditional machine learning approaches. In this paper, we explore the application of federated learning, a decentralized machine learning paradigm, for real and fake fingerprint classification using a custom dataset derived from the CASIA fingerprint dataset. Federated learning has gained attention as a privacy-preserving approach that allows model training without centralizing sensitive data [4]. In the context of fingerprint classification, we leverage the CASIA dataset, which contains fingerprints of various clients. We construct a dataset from this dataset so that each client contains two sub-folders containing fingerprint images belonging to that client and not belonging to that client. The main advantage of federated learning lies in its ability to preserve data privacy [5]. With sensitive biometric information at stake, it is crucial to ensure that the raw fingerprint data remains secure. In our approach, each client retains their data locally, and model training is performed on individual devices. By decentralizing the data and performing computations locally, we minimize the risk of privacy breaches and unauthorized access to personal information. Moreover, federated learning offers scalability and computational efficiency benefits. This decentralized approach enables the utilization of large and diverse datasets without the need for data transfer and central storage, making it suitable for scenarios with numerous clients or limited network bandwidth. Additionally, federated learning addresses the challenges of concept drift and data heterogeneity. In the context of fingerprint classification, clients may exhibit variations in fingerprint characteristics due to differences in acquisition devices or individual variations. Through our work, we aim to demonstrate the benefits of federated learning over traditional machine learning methods for real and fake fingerprint classification. In the context of fingerprint classification, concept drift can occur due to changes in fingerprint acquisition techniques, variations in the population being classified, or the emergence of new fingerprint spoofing techniques. Traditional machine learning models trained on a static dataset may become obsolete or less effective in the presence of concept drift, requiring frequent retraining to adapt to the changing data distribution. Scalability is another consideration when dealing with large-scale fingerprint datasets. Centralizing the entire dataset for training purposes can result in resource-intensive computations and storage requirements [4,5]. Federated learning enables training the model without the need for centralizing sensitive data. Instead, the learning process takes place on local client devices, and only model updates, in the form of model weights, are shared and aggregated to improve the global model. Initially global model will send the weights to each client and clients will update their weights. This decentralized approach minimizes privacy risks and ensures that sensitive data remains on the client's device. Federated learning offers several advantages over traditional machine learning approaches in the context of fingerprint classification Taheri R. et al. [6]. Firstly, it preserves data privacy by keeping the raw fingerprint data local to the client's device. This eliminates the need for centralizing sensitive biometric information, reducing the risk of privacy breaches and unauthorized access. Secondly, federated learning enables the utilization of large and diverse datasets without the need for data transfer and central storage. Each client retains its data locally, and model training is performed on individual devices. This decentralized approach allows for parallel computation, reducing the overall training time and resource requirements. Thirdly, federated learning addresses the challenges of data heterogeneity and concept drift. By training client-specific models that capture local patterns, and aggregating them into a global model, federated learning leverages the collective knowledge from diverse sources.

2 Related Work

Through collaborative model learning, federated learning (FL) extracts knowledge from decentralised data, improving data security, privacy, and secrecy [4] [5]. This paradigm is helpful for the transition from centralised fault diagnosis to intelligent diagnosis that protects privacy. For the purpose of diagnosing bearing faults, Zhang et al. [7] offer a cooperative training framework based on the CNN and Federated Learning. Zhang et al. [8] done validation of the FL with dynamic validation methods through two mechanical diagnosis scenarios shows that the isolated data island problem can be solved. To solve the issue of non-independent and identically distributed decentralised fault data, Zhang offers a federated transfer learning approach. FL still has a number of privacy problems, nevertheless. For instance, Fredrikson et al. [9] use a model-inversion attack to reassemble the instance by reverse-engineering the changed parameters. The distribution of privacy data can still be deduced by eavesdroppers from the updated gradient, despite the fact that only biometric authentication organisations can access each client's data. Zhang et al. [10] develop a FL method for defect diagnostics in the Internet of Ships that is cryptography-based. For industrial big data analysis, anonymous uploading and privacy differential techniques are incorporated into Federated Learning to avoid local modifications from being tracked [11].

In recent years, there has been significant research in the field of real and fake fingerprint detection. Several studies have proposed various techniques and algorithms to distinguish between genuine and spoofed fingerprints [12]. Accurate and dependable automatic person identification is becoming increasingly essential in various applications such as national ID cards, border crossings, access control, and payments. Biometric recognition technology emerges as a highly capable solution to meet this demand. In fact, biometric recognition systems have reached a level of maturity where they can surpass accuracy of human recognition under specific conditions. However, there are still several unresolved challenges that need to be addressed before biometric recognition systems can be fully trusted. Jain et al. [13] highlights five key areas of research that require further attention to establish the trustworthiness of biometrics. Performance Robustness and Scalability, Security, Privacy, Explainability and Interpretability as well as Biasness and Fairness. For each of these areas, the paper provides a clear problem definition, emphasizes the significance of the problem, references existing research, and concludes with suggestions for further exploration and study. Kumar et al. [14] developed a fingerprint liveliness detection system utilizing two distinct approaches: a 3-layered Convolutional NN model and a combination of this CNN architecture with various machine learning classifiers. They assessed the performance of these models and compared their results. Interestingly, when working with limited data, they observed that the 3-layered Convolutional NN architecture and the model employing features extracted from CNN and classified using Gaussian Naïve Bayes yielded the most favorable outcomes. Both architectures exhibited similar high F1-scores, indicating their effectiveness in liveliness detection of fingerprints. Convolution neural networks, which are suitable for image recognition, are one of the key tools for deep learning. The class labels or numerical responses can be predicted using a trained network that was built using a ConvNet architecture. Additionally, we can take characteristics from a pretrained network and utilise them to train a classifier [15].

3 Proposed Solution

Dataset Description

The dataset used in this work is derived from the CASIA Fingerprint dataset, which contains fingerprints of multiple clients. The dataset is organized into subfolders, with each subfolder representing a client. Each client's subfolder further contains two subfolders: one for real fingerprints and another for fake fingerprints.

- **Real Fingerprints**: The subfolder for real fingerprints contains genuine fingerprint images belonging to the respective client. These fingerprints are collected using standard fingerprint acquisition techniques and represent the legitimate biometric samples.
- **Fake Fingerprints**: The subfolder for fake fingerprints contains fingerprint images that do not belong to the respective client.

Few sample images of client 000 from the dataset is shown in the Fig. 1. The dataset aims to provide a balanced representation of both real and fake fingerprints for each client, allowing the development and evaluation of algorithms for real and fake fingerprint classification. The labels for the images indicate whether they are real (1) or fake (0) fingerprints. Using this dataset, the proposed method uses federated learning, where multiple client models are trained on their respective subsets of the data, and the global model is updated by aggregating the weights of the client models. To address the task of real and fake fingerprint

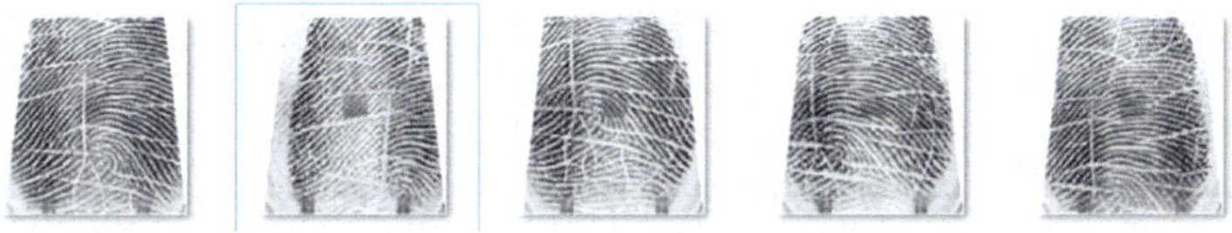

Fig. 1. Sample Fingerprints from Client 000

classification, we propose a federated learning-based approach that leverages the collaborative efforts of multiple clients to train a robust global model. The proposed approach aims to achieve accurate classification while maintaining data privacy and security.

Data Collection and Preprocessing

- Obtain a dataset containing fingerprints from multiple clients.
- Preprocess the fingerprint images by resizing them to a uniform size and converting them to grayscale.

Model Architecture

- Design a Convolutional Neural Network (CNN) model as the global model and client models.
- The CNN model combination of convolutional layers, max-pooling layers and fully linked layers.
- The number of layers and their configurations can be adjusted based on the complexity of the dataset.

Federated Learning Setup:

- Initialize the global model with random weights.
- Define the number of clients participating in the federated learning process.

Iterative Training Process:

- Perform a fixed number of rounds for the federated learning process.
- Randomly select a subset of clients for each round.
- Send the global model weights to the selected clients.
- On each client, Update the local model by training it on the client's training data. Perform local model updates using gradient descent optimization.

Global Model Aggregation:

- Aggregate the updated weights of the client models on the server using Algorithm 1 given below.
- Average the weights based on the number of samples used for training on each client.
- Update the global model weights with the aggregated weights.

Algorithm 1: FedAvg Algorithm K = no. of clients, B is localMinibatchSize, E = no. of local epochs, and η = learning rate.

Server execution: Initialize w_0
 for each global round t=1,2,... **do**
  m $\leftarrow$ max(c.K,1)
  Choose a random set of users S_t
  for each client $k \in S_t$
   $w_{t+1}^k \leftarrow \text{ClientUpdate}(w_t)$
   $w_{t+1} \leftarrow \frac{1}{n} \sum_{u \in K} n_k w_{t+1}^k$

ClientUpdate(k,w):
 B $\leftarrow$ split P_k into batches of size B
 for each local epoch i from 1 to E **do**
  for batch $b \in B$ **do**
   $w \leftarrow w - \eta \nabla l(w, b)$
 return w to server

---

where,
n_k is total number of data points of client k,
w_t is model weights on communication round #t,
c is fraction of clients performing computation in each round.

By leveraging federated learning, the proposed approach enables the collaborative training of a global model while respecting data privacy. The iterative training process and global model aggregation facilitate knowledge sharing among clients and the creation of a powerful model capable of accurately distinguishing between real and fake fingerprints.

4 Results and Discussion

4.1 Network Architecture

The network architecture used in our work for real and fake fingerprint classification consists of a convolutional neural network (CNN) model. The architecture is defined as follows:

Global Model:

- Input Layer: Conv2D layer with 32 filters with kernel size of 3X3, and ReLU activation function. Input shape is (128, 128, 1).
- Max Pooling layer with pool size of 2X2.
- Flatten layer to convert the 2D feature maps to a 1D feature vector.

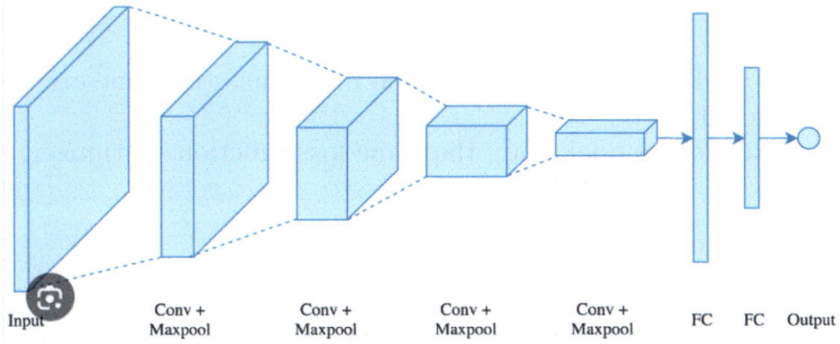

Fig. 2. A CNN Architecture

- Dense layer with 64 units and ReLU activation function.
- Dense layer with 1 unit and sigmoid activation function (output layer).

Client Model(Local Model):

- Same as Global Model.

The CNN architecture in Fig. 2 follows a typical design for image classification tasks. It starts with convolutional layers to extract local features from the input fingerprint images. Following the convolutional layers are the max pooling layers, which shrink the feature maps' spatial dimensions while capturing the most significant features. The flattened feature maps are then passed through fully connected (dense) layers, which learn higher-level representations and make the final classification decision. This network architecture allows the model to learn and differentiate between real and fake fingerprints by capturing the relevant patterns and features present in the input images. Through the federated learning process, the client models train on their respective subsets of the data, and the weights of global model is updated by applying FedAvg algorithm (aggregating the weights of the client models), leading to improved performance in real and fake fingerprint classification tasks.

4.2 Training Setup

The training setup follows the federated learning approach, where multiple client models collaborate to train a global model. The training setup can be described as follows:

Define the Global Model:

- Create a sequential model using the Keras API.
- Specify the layers and architecture of the model, including convolutional layers, pooling layers, and dense layers.
- Compile the model by specifying the loss function, optimizer, and evaluation metrics.

Define the Client Model:

- Create a list of client models, where each model has the same architecture as the global model.
- Compile each client model using the same loss function, optimizer, and evaluation metrics as the global model.

Create a Subset of Data for Each Client:

- Iterate over the number of clients.
- Randomly select a subset of the data from the dataset for each client.
- Divide the data into images and corresponding labels (real or fake).

Perform Federated Learning:

- Follows the federated learning paradigm, where the global model is collaboratively trained using local data on multiple client models. The process of aggregating client model weights ensures that the global model learns from the collective knowledge of the clients while preserving data privacy and security.

Evaluation of Global Model:

- Split the data of each client into training and test sets.
- Load the trained global model.
- Predict classes for the test set using the global model.
- Calculate performance by calculating metrics like accuracy, precision, recall, and F1-score.

4.3 Authentication Results

To test the performance of the proposed real and fake fingerprint detection approach, we conducted authentication experiments on a dataset containing fingerprints from multiple clients. The dataset was split into training and test sets with a ratio of 4:6, and the global model was trained using federated learning for 10 global rounds. Each clients trains their models for 10 times with a batch size = 10 and epoch = 5. The trained model was then tested on the test sets of each client to assess its authentication capability. The authentication results are summarized below in *Table* 1.

The F1 score ranges between 0 and 1, with 1 being the best possible score indicating perfect precision and recall. The model will perform better overall in terms of precision and recall if the F1 score is higher. Table 1 represents four performance metrics of 11 clients (client 000 to client 010). We have calculated the mean values of 11 clients and getting Accuracy is equal to 96%, Precision is equal to 95.13%, Recall is equal to 97.81% and F1 score is equal to 96.37% we are getting similar scores of performance metrics for each client. For example, the value Accuracy, Precision, Recall and F1 score of client 000 are 96.25%, 94.87%, 97.36% and 96.10% respectively. These values are very close to the mean value of scores. By this we can justify that this method can be adopted for any number of clients.

Table 1. Performance Results working with 11 clients on the CASIA database

Client Id	Accuracy	Precision	Recall	F1 score
000	0.9625	0.9487	0.9736	0.9610
001	0.9375	0.9130	0.9767	0.9438
002	0.975	1.0	0.9534	0.9761
003	0.975	1.0	0.9534	0.9761
004	0.9375	0.8958	1.0	0.9450
005	1.0	1.0	1.0	1.0
006	0.975	0.95	1.0	0.9743
007	0.975	1.0	0.9523	0.9756
008	0.9625	0.9487	0.9736	0.9610
009	0.9375	0.9130	0.9767	0.9438
010	0.9375	0.8958	1.0	0.9450

5 Conclusion and Future Direction

In this work, we presented a real and fake fingerprint classification approach using federated learning. By training the model collaboratively, we aim to enhance the authentication capabilities and overcome the limitations of traditional machine learning approaches. Through the experiments conducted on a dataset containing fingerprints from multiple clients, we demonstrated the effectiveness of the proposed approach in accurately classifying real and fake fingerprints. The authentication results showed promising performance, with high accuracy, precision, recall, and F1-score values across different clients. These results indicate that the proposed approach can effectively differentiate between genuine and spoofed fingerprints, thus enhancing the overall security and reliability of fingerprint-based authentication systems. Although the proposed approach achieved satisfactory results, there is still scope for further research and improvement. One of these is robustness to adversarial attacks. Investigating the robustness of the model against adversarial attacks is crucial for real-world deployment. Future work can explore techniques to enhance the model's resilience to attacks and ensure reliable authentication even in the presence of sophisticated spoofing attempts. The current approach relies solely on fingerprint images for classification. Future work can explore the integration of additional features, such as texture analysis, minutiae extraction, or multi-modal biometrics, to further improve the accuracy and robustness of the authentication system. As the number of clients and data size increase, scalability and efficiency become critical factors. Future work can explore methods to optimize the federated learning process, such as model compression, communication-efficient algorithms, or distributed training strategies, to handle large-scale fingerprint datasets effectively.

References

1. Wahhab, H.I., Alanssari, A.N.: Survey of primary methods of fingerprint feature extraction. Comput. Technol. Autom. Control Radioelectronics **18**, 140–147 (2018)
2. Jain, R., Kant, C.: Attacks on biometric systems: an overview. Int. J. Adv. Sci. Res. **1**, 283 (2015)
3. Jain, A.K., Ross, A., Prabhakar, S.: An introduction to biometric recognition. IEEE Trans. Circuits Syst. Video Technol. **14**(1), 4–20 (2004)
4. Kaissis, G., et al.: End-to-end privacy preserving deep learning on multi-institutional medical imaging. In Nature Mach. Intell. **3**(6), 473–484 (2021)
5. Ramage, D., Hampson, S., McMahan, B., Moore, E., Arcas, B.A.Y.: Communication-efficient learning of deep networks from decentralized data. In Proc. Artif. Intell. Statist. **54**, 1273–1282 (2017)
6. Taheri, R., Shojafar, M., Alazab, M., Tafazolli, R.: Fed-iiot: a robust federated malware detection architecture in industrial IoT. IEEE Trans. Industr. Inf. **17**(12), 8442–8452 (2020)
7. Zhang, Z., Xu, X., Gong, W., Chen, Y., Gao, H.: Efficient federated convolutional neural network with information fusion for rolling bearing fault diagnosis. Control Eng. Pract. **116**, 104913 (2021)
8. Zhang, W., Li, X., Ma, H., Luo, Z., Li, X.: Federated learning for machinery fault diagnosis with dynamic validation and self-supervision. Knowl.-Based Syst. **213**, 106679 (2021)
9. Fredrikson, M., Jha, S., Ristenpart, T.: Model inversion attacks that exploit confidence information and basic countermeasures. In: Proceedings of the 22nd ACM SIGSAC Conference on Computer and Communications Security, pp. 1322–1333 (2015)
10. Zhang, Z., Guan, C., Chen, H., Yang, X., Gong, W., Yang, A.: Adaptive privacy-preserving federated learning for fault diagnosis in internet of ships. IEEE Internet Things J. **9**(9), 6844–6854 (2022)
11. Zhao, B., Fan, K., Yang, K., Wang, Z., Li, H., Yang, Y.: Anonymous and privacy-preserving federated learning with industrial big data. IEEE Trans. Industr. Inf. **17**(9), 6314–6323 (2021)
12. Sandhya, M., Prasad, M.V.N.K.: Cancelable fingerprint cryptosystem based on convolution coding. In: Thampi, S., Bandyopadhyay, S., Krishnan, S., Li, KC., Mosin, S., Ma, M. (eds.) Advances in Signal Processing and Intelligent Recognition Systems. Advances in Intelligent Systems and Computing, vol. 425, pp. 145–157. Springer, Cham (2016). https://doi.org/10.1007/978-3-319-28658-7_13
13. Jain, A.K., Deb, D., Engelsma, J.J.: Biometrics: trust, but verify. IEEE Trans. Biometrics Behavior Identity Sci. **4**(3), 303–323 (2022)
14. TK, A.K., Vinayakumar, R., VV, S.V., Sowmya, V., Soman, K.P.: Convolutional neural networks for fingerprint liveness detection system. In: 2019 International Conference on Intelligent Computing and Control Systems (ICCS), pp. 243–246 (2019)
15. Lazimul, L.T., Binoy, D.L.: Fingerprint liveness detection using convolutional neural network and fingerprint image enhancement. In: 2017 International Conference on Energy, Communication, Data Analytics and Soft Computing (ICECDS), pp. 731–735 (2017)

Comparative Study of Ensemble Learning Models for Smart Meter Load

Jatinder Kumar[1(✉)], Kapil Gupta[1], and Ashutosh Kumar Singh[1,2]

[1] National Institute of Technology Kurukshetra, Thanesar, India
jatinderkumar2851@gmail.com, {kapil,ashutosh}@nitkkr.ac.in
[2] Indian Institute of Information Technology Bhopal, Bhopal, India

Abstract. Load Forecasting plays a crucial role in the operation and management, such as demand response, energy management and resource optimization in the smart grid. Ensemble learning-based load forecasting techniques can significantly improve the accuracy and robustness of load prediction models, which combine the predictions from multiple individual models to create a more accurate and reliable prediction. In this paper, four ensemble learning-based load forecasting models are discussed. Experiments are performed on the Open Energy Data Initiative dataset, which shows that the Extended Gradient Boosting Regressor model has superiority in terms of mean square error.

Keywords: Smart meter · Ensemble learning · Data preprocessing · Demand response management

1 Introduction

The modern world is experiencing an unprecedented increase in energy consumption driven by rapid urbanization, industrialization, and the proliferation of electronic devices. This surge in energy demand has not only put immense pressure on the existing electrical grid infrastructure but has also necessitated a paradigm shift in the management and monitoring of energy consumption. One of the key advancements in this domain is the deployment of smart meters, which provide real-time data on energy consumption patterns at a granular level [1]. Smart meters play a pivotal role in the transformation of traditional power distribution systems into intelligent and responsive grids, commonly known as smart grids. These devices enable utilities and consumers alike to monitor, control, and optimize electricity usage efficiently. In recent years, the application of machine learning techniques to smart meter data has garnered significant attention [2]. These techniques offer the promise of accurate load forecasting, anomaly detection, and energy consumption prediction, among other vital tasks. Among the various machine learning approaches, ensemble learning has emerged as a powerful tool for harnessing the full potential of smart meter data [3]. Ensemble learning methods combine multiple base models to achieve superior predictive performance, robustness, and generalization capabilities. They have demonstrated their effectiveness in a wide range of applications, including classification, regression, and anomaly detection [4]. However, the application of ensemble

D. Pastor-Escuredo et al. (Eds.): ICDLAIR 2023, LNNS 1001, pp. 81–90, 2024.
https://doi.org/10.1007/978-3-031-60935-0_8

learning to smart meter load analysis and management remains an area ripe for exploration [5].

The system model of the smart grid is depicted in Fig. 1. The Region has various localities, such as homes, buildings, and industrial areas. These smart meter collects the power consumption of electrical appliances and transfers it to the cloud environment for preprocessing. The preprocessing phase involves the data interpolation method, which removes the bad, missing, and "not a number" values from the data. Thenafter, aggregation is performed of live and historical data stored by the cloud environment. Normalization steps include scaling the input data in a specific range, such as [0, 1]. After that, normalized data is fed into the prediction model to train. The predicted load is computed as the desired accuracy or minimum error rate is achieved by the prediction model; otherwise, the prediction model is retrained iteratively. The electricity company provide electricity based on the predicted load of the smart meter load of regions and takes the decision of power generation and supply.

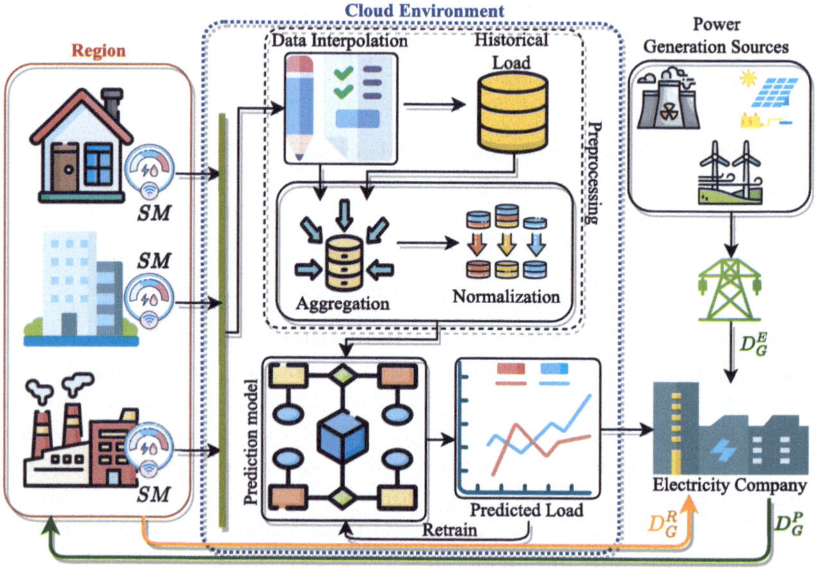

Fig. 1. System model

This paper embarks on a comprehensive comparative study of ensemble learning models for smart meter load analysis. We aim to investigate and evaluate the performance of various ensemble learning techniques, such as Random Forest (RF), Gradient Boosting Regressor (GBR), Bagging Regressor (BR), and Extended Gradient Boosting Regressor (XGBR), in the context of load forecasting, anomaly detection, and energy consumption prediction. By harnessing the power of ensemble learning, we aspire to contribute to the advancement of intelligent and efficient energy management systems, paving the way for a sustainable and resilient energy future.

2 Literature Review

Oprea and Bara [6] proposed a big data solution based short-term load forecast. Seven regression algorithms were executed simultaneously and the most suitable algorithm was opted automatically using the best performant algorithm based on the root mean square error. Jagait et al. [7] presented adaptive online ensemble learning-based load forecasting under the concept of drift. Ensemble learning is embedded with the recurrent neural network and autoregressive integrated moving average. A Long Short-Term Memory-based load forecasting model for the smart meter was implemented [8]. Missing and bad values were removed and replaced by the data interpolation method, while normalization was performed to transform the load into a similar scale of load. Khan et al. [9] proposed an enhanced support vector machine and an enhanced convolutional neural network for short and mid-term electric load forecasting. Feature selection was performed by extreme gradient boosting and random forest. Krannichfeldt et al. [10] proposed an online ensemble learning-based short-term load forecasting method. A modified passive-aggressive regression model was developed to implement the online ensemble forecasting. Ensemble learning-based power consumption forecast model developed by Kumar et al. [11]. Five base predictors, such as bagging regressor, gradient boosting regressor, linear regressor, random forest, and artificial neural network, were taken and a voting engine was utilized to select the most suitable algorithm. Kell et al. [12] used random forests, neural networks, long short-term memory neural networks and support vector regression algorithms to predict the 30-minute load. The k-means clustering was utilized to categorize similar types of smart meter customers. Short-term load forecasting for the industrial customers was proposed by Wang et al. [13]. An adaptive SVMD to decompose the raw electricity load and a Bayesian optimization algorithm were utilized to tune the hyperparameters of the XGBoost regressor. A weather identification and short-term load forecasting model was proposed by Deng et al. [14]. The Bagging-XGBoost algorithm was implemented for high-accuracy short-term load forecasting. A 48-hour-ahead load forecasting algorithm was implemented for campus university using the ensemble methods [15]. Four different algorithms, such as bagging, random forest, conditional forest, and boosting regression algorithms, were taken for short-term load forecasting.

3 Ensemble Learning Based Forecasting Model

In this section we will discuss about the four most popular ensemble learning based regression algorithms, i.e., Bagging Regressor, Gradient Boosting Regressor, Random Forest, and Extend Gradient Boosting Regressor, which are discussed below:

1. **Bagging Regressor:** A machine learning ensemble approach called bagging regression improves the predictability and stability of regression models. It works by averaging the predictions of many basic regression models that have been trained on bootstrapped subsets of the training data. Through this method, model variance is decreased and overall prediction accuracy is increased. In order to describe the Bagging Regressor mathematically, predictions from many base regression models are combined and designated as $h_i(x)$, where i stands for a different basis model. The average of these distinct guesses yields the final prediction, computed using Eq. 1,

$$F(x) = N_1 \sum_{i=1}^{N} h_i(x) \tag{1}$$

   where, the ultimate forecast for input x is represented by $F(x)$, N stands for the ensemble size or the total number of base models, the prediction of the ith base regression model for input x is represented by $h(x)$.
   This ensemble method makes use of the variety across base models to reduce overfitting and increase the resilience of the model, making it an effective tool for regression problems.
2. **Bagging Regressor:** A strong and precise prediction model is produced by successively combining a number of weak regression models using the effective ensemble machine learning approach known as boosting regression. By assigning more weight to data items that are challenging to forecast, it focuses on optimising model performance by successfully reducing mistakes and increasing predicting accuracy. The weighted mixture of predictions from many weak regressors represented mathematically as $h(x)$, where i stands for each weak learner, makes up the boosting regressor. Iteratively modifying the weights of each of these distinct forecasts yields the final forecasting, $F(x)$ is computed with Eq. 2, where α_i denotes weight given to the ith weak learner. prediction is represented by the integer i.

$$F(x) = \sum_{i=1}^{N} \alpha_i h_i(x) \tag{2}$$

   Boosting adjusts the weights of data items that are incorrectly categorised or forecasted in each iteration, working iteratively. The model's performance is gradually enhanced by training the weak learners to concentrate on the previously misclassified data points. A boosted ensemble model that performs well in regression tasks is produced by this iterative procedure, which is carried out until a predetermined stopping threshold is satisfied.
3. **Random Forest:** A reliable ensemble machine learning technique known as Random Forest is employed for both classification and regression problems. In order to produce a highly accurate and reliable prediction model, it combines the strength of decision trees with ensemble learning. By combining the predictions of many decision trees, each trained on a distinct sample of the data, Random Forest reduces overfitting and improves predictive accuracy.

By choosing a random subset of characteristics for each tree throughout the tree-building process, Random Forest adds unpredictability in addition to aggregation. The danger of overfitting is decreased by this randomization, which increases model variety. A strong and reliable predictive model is produced as a result of the consensus of the individual tree predictions in the final forecast.
Random Forest is a useful tool for machine learning research and practical applications since it has shown effective in a variety of applications and is especially well-suited for handling complicated and high-dimensional datasets.
4. **Extend Gradient Boosting Regressor:** A cutting-edge gradient boosting technique created for both classification and regression applications is called XGBoost. Its extraordinary prediction accuracy, effectiveness, and scalability have helped it become incredibly popular in machine learning. By building a collection of decision trees, XGBoost is able to systematically improve model performance by minimising a certain loss function. By using second-order gradient information and gradient descent, XGBoost effectively optimises the loss function. Additionally, regularisation techniques are used to reduce overfitting and improve model resilience.
For its outstanding performance in several machine learning contests and practical applications, XGBoost is well-known. It is an essential tool for cutting-edge research and real-world data science activities thanks to its variety, speed, and predictive capability.

4 Results and Discussion

4.1 Experimental Setup and Dataset

The experiments are conducted on 64 64-bit Ubuntu machine having 32GB RAM and Intel(R) Core(TM) i5-2520M @2.50GHz. Algorithms are implemented with Python of version 3.8.4. The experiments are conducted on the Open Energy Data Initiative (OEDI) dataset [16]. The train data is taken 80% of the dataset, while the remaining is taken for the testing of the regression model.

4.2 Performance Analysis

Figures 2, 3, 4 and 5 show the actual and predicted load over the OEDI smart meter dataset by the Bagging Regressor, Gradient Boosting Regressor, Random Forest, and Extend Gradient Boosting Regressor, respectively. The predicted load is compared with 200 time intervals and each time interval is comprised of 30 min. From these figures, it is observed the predicted load is too close to the actual load of the smart meter.

From Fig. 6(a)–(c), the performance of BR, GBR, RF and XGBR regression model are computed in terms of mean square error (MSE), median absolute error (MedAE), mean absolute error (MAE), is shown over the dataset. The smallest error of MSE is 0.0016 with 0.1 learning rate and at 300 estimators, while the

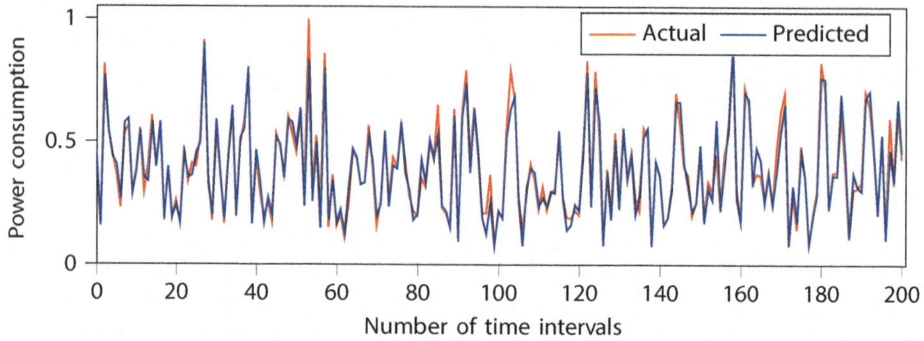

Fig. 2. Bagging Regressor

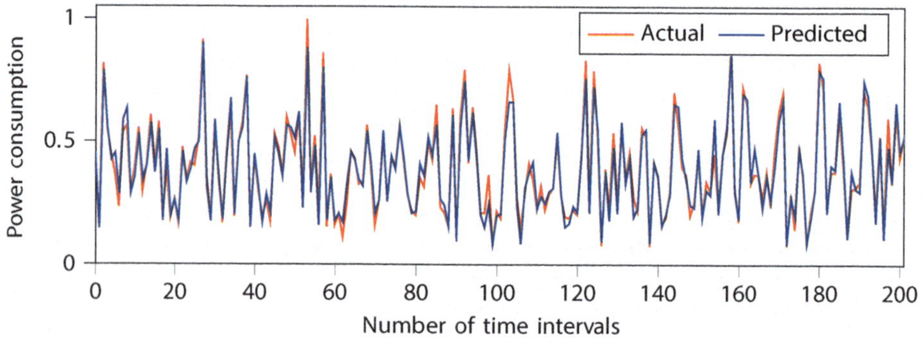

Fig. 3. Gradient Boosting Regressor

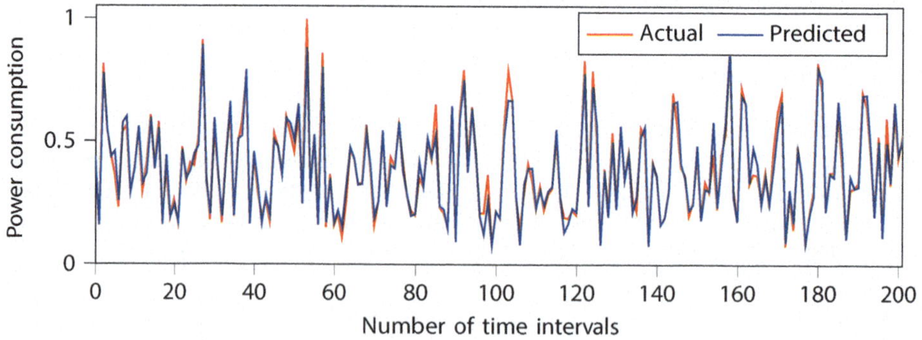

Fig. 4. Random Forest

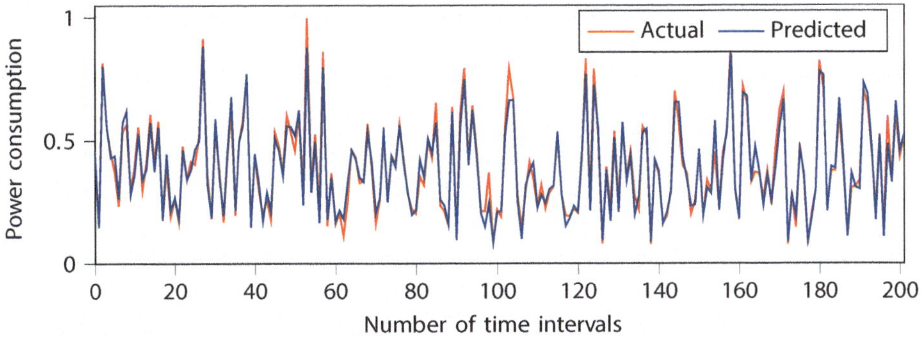

Fig. 5. Extend Gradient Boosting Regressor

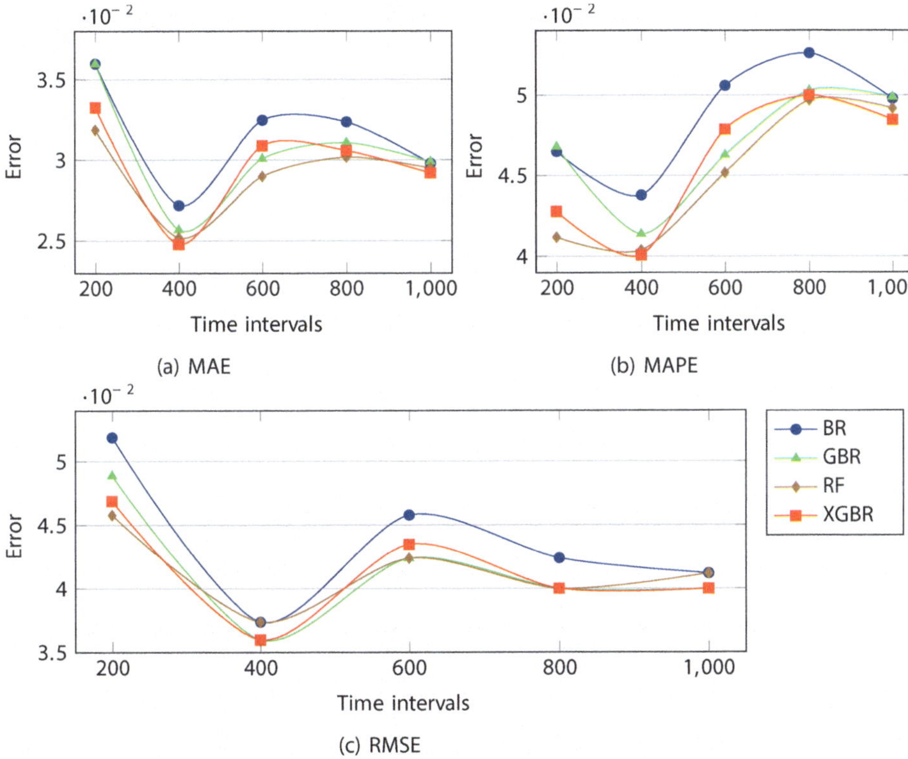

Fig. 6. Evaluation parameters with respect to number of time intervals

highest error is 0.0086 with 0.01 learning rate and at 100 estimators. The smallest error of MAE is 0.0285 with 0.1 learning rate and at 300 estimators, while the highest error is 0.0741 with 0.01 learning rate and at 100 estimators. The smallest error of MSE is 0.0887 with a 0.075 learning rate and at 100 estimator, while the highest error is 0.3356 with 0.01 learning rate and at 100 estimator.

Table 1. Comparison of ensemble learning regressors

Ensemble Model	Errors	Time Intervals				
		200	400	600	800	1000
Bagging Regressor	MedAE	0.022	0.0193	0.0195	0.0243	0.0195
	R-Square	0.6097	0.8415	0.8665	0.9287	0.9319
	NRMSE	0.1617	0.0749	0.0776	0.068	0.0618
Gradient Boosting Regressor	MedAE	0.0267	0.0168	0.0195	0.0236	0.0215
	R-Square	0.6601	0.8575	0.886	0.9367	0.9355
	NRMSE	0.1509	0.0711	0.0717	0.064	0.0601
Random Forest	MedAE	0.0196	0.0149	0.0173	0.0193	0.0185
	R-Square	0.7003	0.8505	0.8889	0.9346	0.9334
	NRMSE	0.1417	0.0728	0708	0.0651	0.0611
XGB Regressor	MedAE	0.0211	0.0156	0.0173	0.0228	0.0194
	R-Square	0.682	0.8578	0.8792	0.9354	0.936
	NRMSE	0.146	0.071	0.0738	0.0647	0.0599

Table 2. Performance of XGB Regressor model

n_estimators	Learning rate	Error metrics		
		RMSE	MAE	MAPE
100	0.01	0.0086	0.0741	0.3356
	0.05	0.0027	0.0346	0.1108
	0.075	0.0024	0.0317	0.0887
	0.1	0.0023	0.0331	0.1051
200	0.01	0.0039	0.0471	0.1896
	0.05	0.0025	0.0322	0.0961
	0.075	0.0034	0.0362	0.0992
	0.1	0.0031	0.0365	0.1096
300	0.01	0.0029	0.0381	0.1289
	0.05	0.0024	0.0324	0.0961
	0.075	0.0025	0.0352	0.1075
	0.1	0.0016	0.0285	0.0993
400	0.01	0.0033	0.0387	0.1146
	0.05	0.0029	0.0363	0.1126
	0.075	0.0027	0.0345	0.1027
	0.1	0.0029	0.0353	0.1047
500	0.01	0.0027	0.0349	0.1059
	0.05	0.0022	0.0321	0.1019
	0.075	0.0021	0.0318	0.1025
	0.1	0.0037	0.0368	0.1126

The MedAE, R-Square and NRMSE are computed over the time intervals 200, 400, 600, 800 and 1000 for the Bagging Regressor, Gradient Boosting Regressor, Random Forest and XGB regression algorithms as shown in Table 1. The Lowest MedAE for the Bagging Regressor is 0.0193 at interval 400, while the minimum MedAE achieved for the Gradient Boosting Regressor is 0.0168 at interval 400. Similarly, the MedAE are computed for RF and XGBR at time intervals of 400. The minimum R-Square error and normalized root mean square error (NRMSE) are achieved by bagging regressor and XGB regression algorithm, respectively. Table 2 shows the performance of the XGB Regressor model with the different learning rate and n_estimators in terms of root mean squared error (RMSE), MAE, and mean absolute percentage error (MAPE). The learning rate range is taken at 0.01, 0.05, 0.075 and 0.1 with the n_estimators of 100–500. The XGB Regressor model performs better at the learning rate of 0.1.

5 Conclusion

A discussion and comparative study are demonstrated about the ensemble learning-based load forecasting models. Four types of regression algorithms are taken, including bagging regressor, boosting regressor, random forest and extended boosting regressor. Simulation results show that the XBGR model has better accuracy as compared to other ensemble learning-based load forecasting models. Future work includes the mid-term and long term load forecasting with the weather input variable.

References

1. Singh, A.K., Kumar, J.: A privacy-preserving multidimensional data aggregation scheme with secure query processing for smart grid. J. Supercomput. **79**(4), 3750–3770 (2023)
2. Ibrahim, M.S., Dong, W., Yang, Q.: Machine learning driven smart electric power systems: current trends and new perspectives. Appl. Energy **272**, 115237 (2020)
3. Kumar, J., Saxena, D., Singh, A.K., Vasilakos, A.V.: A quantum controlled-NOT Neural network-based load forecast and management model for smart grid. IEEE Syst. J. **17**, 5714–5725 (2023)
4. Wang, L., Mao, S., Wilamowski, B.M., Nelms, R.M.: Ensemble learning for load forecasting. IEEE Trans. Green Commun. Netw. **4**(2), 616–628 (2020)
5. Singh, Ashutosh Kumar, and Jatinder Kumar. "A secure and privacy-preserving data aggregation and classification model for smart grid." Multimedia Tools and Applications (2023): 1-19
6. Oprea, S.-V., Bâra, A.: Machine learning algorithms for short-term load forecast in residential buildings using smart meters, sensors and big data solutions. IEEE Access **7**, 177874–177889 (2019)
7. Jagait, R.K., Fekri, M.N., Grolinger, K., Mir, S.: Load forecasting under concept drift: online ensemble learning with recurrent neural network and ARIMA. IEEE Access **9**, 98992–99008 (2021)

8. Kumar, J., Singh, A.K.: A demand and response management model using load forecasting technique for smart grid. In: 2023 5th International Conference on Energy, Power and Environment: Towards Flexible Green Energy Technologies (ICEPE), pp. 1–6. IEEE (2023)
9. Khan, S.U.R., et al.: Enhanced machine-learning techniques for medium-term and short-term electric-load forecasting in smart grids. Energies **16**(1), 276 (2022)
10. Von Krannichfeldt, L., Wang, Y., Hug, G.: Online ensemble learning for load forecasting. IEEE Trans. Power Syst. **36**(1), 545–548 (2020)
11. Kumar, J., Gupta, R., Saxena, D., Singh, A.K.: Power consumption forecast model using ensemble learning for smart grid. J. Supercomput. **79**, 11007–11028 (2023). https://doi.org/10.1007/s11227-023-05096-4
12. Kell, A., McGough, A.S., Forshaw, M.: Segmenting residential smart meter data for short-term load forecasting. In: Proceedings of the Ninth International Conference on Future Energy Systems, pp. 91–96 (2018)
13. Wang, Y., et al.: Short-term load forecasting of industrial customers based on SVMD and XGBoost. Int. J. Electr. Power Energy Syst. **129**, 106830 (2021)
14. Deng, X., et al.: Bagging-XGBoost algorithm based extreme weather identification and short-term load forecasting model. Energy Rep. **8**, 8661–8674 (2022)
15. Ruiz-Abellón, M.D., Carmen, A.G., Guillamón, A.: Load forecasting for a campus university using ensemble methods based on regression trees. Energies **11**(8), 2038 (2018)
16. Brodt-Giles, D.L., Rossol, M.N.: Open energy data initiative: advancing analytics and research innovation through improved data access. No. NREL/FS-6A20-73632; DOE/GO-102019-5171. National Renewable Energy Lab.(NREL), Golden, CO (United States) (2019)

Social-Media Video Summarization Using Convolutional Neural Network and Kohnen's Self Organizing Map

Shamal Kashid[1(✉)], Lalit K. Awasthi[1], Krishan Kumar[2], and Parul Saini[1]

[1] Computer Science and Engineering, NIT Uttarakhand, Srinagar, India
{kashid.shamalphd2021,parulsaini.phd2020}@nituk.ac.in

[2] Department of Computer Engineering, NIT Kurukshetra, Kurukshetra, India
k2b@ieee.org

Abstract. Video Summarisation (VS) techniques provide concise representations of original videos. This research presents a novel approach to extracting keyframes based on Convolutional Neural Network and Self-Organizing Map (SOM). The technique analyses videos frame by frame. Convolutional Neural Network is used to extract the deep-level feature vectors. Using Self-Organizing Map clustering, the feature descriptors corresponding to the frames are divided into keyframes and non-key frames. This SOM-based video summarization method effectively selects the most representative frames from the extracted feature vector. The proposed approach produces better results than other state-of-the-art video summarising techniques. In the comprehensive experimental analysis of two benchmark datasets, our methodology produced average F-scores of 0.81 and 0.82, respectively. The outcomes demonstrate that the method consistently generates high-quality summaries for videos across various categories.

Keywords: Convolutional Neural Network · Clustering · Key-frames · Self Organizing Map · Video Summarization

1 Introduction

The rapid advancements in communication technology, such as the 5G network and optical fiber, have led to a record-high growth in multimedia content, including text, audio, image, and video. This data is generated in various domains, including sports, surveillance, medical, and social media, primarily through optical instrument devices. The majority of the data on the internet is video data [1]. In enormous volumes, video data holds immense potential for various applications, including sports administration, medical analysis, and security monitoring, but its impact on hardware processing devices necessitates VS technology [2].

The widespread availability of affordable electronic devices with advanced capabilities primarily drives the rapid growth of multimedia data on the World Wide Web. This growth in video content produces a potential risk to network

D. Pastor-Escuredo et al. (Eds.): ICDLAIR 2023, LNNS 1001, pp. 91–100, 2024.
https://doi.org/10.1007/978-3-031-60935-0_9

infrastructure due to the continuous uploading and downloading of large video files. Researchers are actively working on solutions to reduce video data handling to overcome this issue. One such solution is Video Summarization, a procedure within video analysis [6,7]. Video summarization aims to enable efficient and effective retrieval of video data. This approach involves representing lengthy videos with a more concise and informative version while maintaining the core meaning of the original content. Video summarization creates brief abstracts that allow users to grasp the essential message of the video in a significantly shorter time than watching the full-length video [8].

The generation of concise and intelligent VS, also known as EVS (Equal Frames Partition based VS) [4], is proposed using an equal partition-based clustering technique with two approaches. The entire video is clustered into keyframe groups in the first variation. However, the second variant divides the video into equal-sized frames and then clusters them into keyframes. In both forms, the Davies-Bouldin Index, a cluster validation approach, is used to determine the ideal collection of clusters. ESVS (Eratosthenes Sieve-based VS) [5] has been suggested for VS to provide concise and intelligent video abstraction, commonly known as event summary.

VS has been extensively studied since the 1990s, and it can be broadly categorized into two classes: static and dynamic video summarization [3]. Static VS typically contains a storyboard with keyframes representing the video content. It primarily considers visual information while skipping audio messages. On the other hand, dynamic video summarization involves creating a video clip that integrates image, text, and audio information. Compared to dynamic video summarization, static video summarization is easier to browse and helps reduce the computational complexity involved in video retrieval and analysis [6,7]. The VS problem can be converted into a clustering problem, in which some keyframes are chosen to represent the video.

This paper investigates a novel method for creating summaries of videos from various categories. The study discusses the drawbacks of the current methods and focuses on deep features to identify important frames in a video. The following are the significant contributions of the proposed work as follows.

1. To create a VS system for producing high-quality summaries that consistently work on videos from many categories.
2. Application of deep feature vectors extracted using CNN's transfer learning technique.
3. A technique for summarizing social media videos is introduced by leveraging the CNN model and the Kohonen Self-Organizing Map (SOM).

The remaining paper is organized as follows: Sect. 2 outlines the existing video summarization techniques. In Sect. 3, the proposed VS model with details is given. Section 4 discusses the experimental results and performance analysis of the proposed framework. In Sect. 5, the proposed framework is concluded with future work.

2 Related Work

VS based on deep learning DeepReS [1] has been proposed for resource-constrained devices in industrial-based surveillance scenarios. It is best suited for contexts with restricted resources since it refined the data on a resource-constrained appliance to decrease the content volume for transmission via wireless network areas in an IIoT, the Industrial Internet of Things. In [2], a two-stream method that captures motion and visual features from the video is proposed for extracting keyframes using an unsupervised static VS technique. From different CNN layers, feature maps are obtained and fused. Redundant frames are removed by building a similarity network. This method performs better than other related unsupervised methods.

Video summarization techniques aim to extract essential keyframes from videos. Various approaches have been employed, including feature-based methods like color, motion, objects, gesture, speech, audio-visual cues, etc. Clustering-based techniques have also been utilized, such as k-means clustering, partitioned Clustering, and spectral Clustering. Other approaches include shot selection-based VS, event-based summarization, and trajectory-based VS [3]. The Generic VS (GVSUM) [11] is proposed in which keyframes are extracted whenever there is a change in the cluster number of the frame. Keyframes based on feature vectors extracted from Multi CNN [12] are proposed to be fed into a sparse autoencoder, which can also generate summaries of high quality from videos of all categories or genres.

In a VS framework, a novel strategy combines memorability and entropy scores for keyframe selection. This method evaluates the memorability of frames alongside their information entropy to ascertain their significance in the summarization process [12]. The Cost-Effective VS approach in [13] integrates aesthetic features and deep Convolutional Neural Networks (CNN) to generate video summaries. Additionally, Muhammad et al. [17] introduce an architecture for surveillance video summarization, explicitly addressing the challenges posed by devices with limited resources and lower computational complexity. This approach successfully navigates resource-constrained environments while maintaining effective video summarization capabilities.

Gharbi et al. [14] provide an effective keyframe extraction method. Their methodology is based on defining local interest spots and clustering repeatability graphs using approaching modularity values. The technique will only be used on a set of candidate frames chosen using a windowing rule to reduce the amount of data to be handled. Extracting interest points from a group of potential frames is, in fact, the initial step. The repeatability values between each pair of frames from the candidate set are then calculated. A repeatability straight graph is used to depict these data. Crucial frame selection is carried out using the modularity principle to graph clustering. Parihar et al. [15] proposed a unique method for multi-view summarization. Their approach focuses on eliminating static and redundant frames by employing the BIRCH clustering algorithm on the initial collection of frames. Similarly, Raut and Vrushal et al. [16] present a method for video summarization that leverages unsupervised learning.

3 Proposed Model

An effective VS should preserve the elements of the original video that captivate the users. This paper introduces a novel technique for creating a static VS of a given video, consistently producing high-quality summaries across various video categories. The method begins by extracting frames from the input video and eliminating redundant frames. Next, it processes the frames by employing CNNs to extract feature vectors. These feature vectors, corresponding to individual frames, are then classified into key-frames and non-key-frames using SOM clustering. Figure 1 provides an overview of this proposed approach. Subsequent sections will delve into the distinct steps comprising this method in detail. The following Sections describe the various phases applied in the proposed approach.

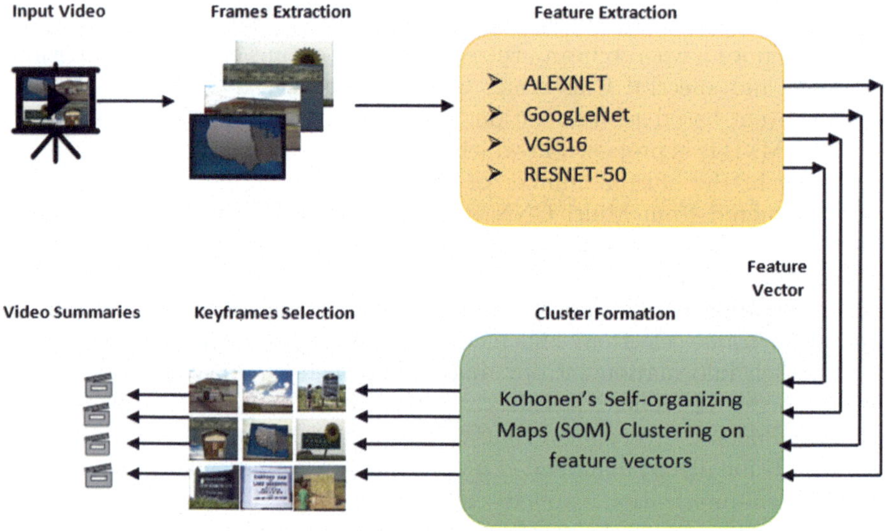

Fig. 1. Proposed Model

3.1 Pre-processing

This step involves video-to-frame conversion, frame feature extraction, and frame feature mapping to pre-process videos. We converted these videos into frames. This approach maintains the video's originality with the help of not converting the colored frames into grayscale ones, and no pre-sampling is done.

3.2 Feature Extraction

After the input video preprocessing, features are extracted from each frame using CNN based on transfer learning. Deep convolutional neural networks (CNN)

[11,12] are networks built using the deep learning technique and have interconnected layers. CNN has established itself as one of the best feature extractors for processing video [17]. Due to the lack of comprehensive datasets available for training purposes, our approach utilizes pre-trained CNN models. The dataset is preprocessed to feed it to the model, including the Feature extraction using the AlexNet [19], GoogLeNet [20], RESNET-50 [18], and Vgg-16 [21] models. These networks are trained using the ImageNet dataset, which contains millions of various types of images. The feature extraction process focuses on the fully connected layer known as 'predictions.' ResNet-50 is a hybrid model that combines residual connections with an inception network [18]. Residual connections have proven to be highly efficient when training deep architectures. Inception networks, characterized by their considerable depth, replacing the filter concatenation stage with residual connections led to a significant and steep improvement in network performance [13].

3.3 SOM Clustering

Using four pre-trained neural networks, extract the deep-level features from the frames as described in Sect. 3.2. Provide the extracted feature vectors as input to the Kohonen SOM model. Train the model using the batch SOM technique to discover the topology and distribution of the input data. The training continues for a fixed number of epochs, refining the SOM's representation. After training, identify and retrieve clusters of frames that indicate similar patterns on the SOM. These clusters show collections of frames with comparable traits, suggesting visual similarity. This approach can be highly effective in identifying and grouping visually similar frames, which is valuable for VS.

3.4 Keyframe Selection

This step aims to identify frames within each cluster that best represent the cluster's content. The purpose of this is to identify frames that are significantly different from the cluster's centroid frame. Firstly, calculate the centroid frame for each frame in a cluster. After that, calculate the dissimilarity between each frame and the centroid frame. Select the frame with maximum dissimilarity as the representative frame. Combine all the keyframes to represent VS.

4 Experimental Results and Discussion

In this Section, we have evaluated our proposed method, and for validation, we have compared our results with several existing approaches. The experimental environment is introduced in Sect. 4.1. Two datasets are used to check the efficacy of the proposed VS models. The overall description of the datasets is explained in Sect. 4.2. We evaluated the proposed VS through performance metrics, as discussed in Sect. 4.3. Finally, Sect. 4.4 highlights the result and analysis of the proposed VS model.

4.1 Experimental Settings

Our experiment used an Intel i7 processor and 32 GB of RAM to run the Windows 11 operating system. We conducted our work within the Python environment, utilizing the Jupyter Notebook framework, and leveraged various Python libraries, such as pandas, numpy, TensorFlow, and sklearn, to streamline our processes.

4.2 Dataset

Various procedures for VS assess their experimentation results using different datasets. We have used two publically available datasets for performing experiments for the proposed VS approach selected from a benchmark VSUMM (Video Summarization) dataset [10] due to their videos' diversity and different categories. The details of two video datasets with 100 videos of 1231 MB are detailed below.

1. *Open Video (OV)* Dataset includes 50 videos taken from OVP-Open Video Project (30 fps frame rate). The size of the video frame is 352 × 240 pixels. Every video in MPEG-1 forms in color and has sound. The total size of the OV dataset is 763 MB. These videos span various genres (documentary, lecture instructive, ephemeral, historical), span from 1–4 minutes, and total around 75 min.
2. *Youtube (YT)* Dataset comprises 50 videos taken from Youtube (30 fps frame rate). Every video in MPEG-1, in this Dataset, is also in color with sound. The total size of the YT dataset is 468 MB. These video elements are shared into several categories, including cartoons, sports, news, marketing, a series collection of TV, and homemade videos.

4.3 Evaluation Metrics

Significant research efforts have been dedicated to developing techniques for extracting keyframes to create effective video summaries. However, there is a need for a standardized or optimal method to evaluate the performance of these keyframe extraction techniques. To achieve quantitative measurements various methods are compared to user-generated summaries, considered the desired summaries (ground truth summaries). Different algorithms evaluation of video summaries typically relies on three primary evaluation metrics: Precision, Recall, and F-measure, as defined by Eqs. 1, 2, and 3, respectively [13,22].

Precision and Recall are widely used evaluation metrics for assessing the performance of keyframe extraction algorithms. These metrics are calculated based on the ratio between the algorithm's selected keyframes and the frames specified in the ground truth summaries, as shown in Eqs. 1 and 2. F-measure is represented by the harmonic mean of Precision (P) and Recall (R), calculated as in Eq. 3.

$$P = \frac{A \cap B}{B} \tag{1}$$

$$R = \frac{A \cap B}{A} \tag{2}$$

$$F - Measure = \frac{2.P.R}{(P + R)} \tag{3}$$

A is the number of frames in the ground truth or summary, and B is the number of frames in the proposed summary. The maximum value of the F-Measure indicates a more accurate approach. While estimating the P and R requires comparing the keyframes received from the proposed summary and the reference summary with which we compare our results.

Table 1. Results of ALEXNET, GOOGLENET, VGG-16 and RESNET-50 on YT Dataset

Model	Precision		Recall		F-score	
	[9]	Ours	[9]	Ours	[9]	Ours
AlexNet	0.66	**0.69**	0.86	**0.91**	0.74	**0.78**
GoogLeNet	0.61	**0.72**	0.81	**0.85**	0.69	**0.77**
Vgg-16	0.67	**0.69**	0.73	**0.74**	0.70	**0.71**
RESNET-50	0.71	**0.78**	0.78	**0.85**	0.75	**0.81**

Table 2. Results of ALEXNET, GOOGLENET, VGG-16 and RESNET-50 on OV Dataset

Model	Precision		Recall		F-score	
	[9]	Ours	[9]	Ours	[9]	Ours
AlexNet	0.66	**0.67**	0.86	**0.89**	0.74	**0.76**
GoogLeNet	0.61	**0.70**	0.81	**0.86**	0.69	**0.77**
Vgg-16	0.67	**0.70**	0.73	**0.75**	0.70	**0.72**
RESNET-50	0.71	**0.79**	0.78	**0.87**	0.75	**0.82**

4.4 Results Analysis

As discussed in Sect. 3, SOM clustering is done on the extracted features from the pre-trained CNN models to produce the video summary. The experiments are performed using four pre-trained models of CNN, namely AlexNet, GoogLeNet, Vgg-16, and RESNET-50 on the OV and YT datasets. Tables 1 and 2 show the ALEXNET, GOOGLENET, VGG-16, and RESNET-50 results on the YT and OV datasets. The following observations have been made from Table 1, 2 and 3:

Table 3. Results of Various Categories of Videos in VSUMM Dataset

Category	Model	Precision		Recall		F-score	
		[9]	Ours	[9]	Ours	[9]	Ours
Cartoons	AlexNet	0.77	**0.79**	0.87	**0.89**	0.82	**0.83**
	GoogLeNet	0.60	**0.75**	**0.82**	0.77	0.69	**0.76**
	Vgg-16	**0.65**	0.64	0.71	**0.72**	**0.68**	**0.68**
	RESNET-50	0.70	**0.78**	0.60	**0.82**	0.64	**0.80**
Sports	AlexNet	0.58	**0.63**	**0.90**	0.89	0.71	**0.73**
	GoogLeNet	0.57	**0.81**	**0.80**	0.77	0.67	**0.79**
	Vgg-16	0.69	**0.71**	0.73	**0.76**	0.71	**0.73**
	RESNET-50	0.81	**0.83**	0.70	**0.85**	0.75	**0.84**
News	AlexNet	0.62	**0.66**	0.82	**0.86**	0.71	**0.74**
	GoogLeNet	0.65	**0.77**	**0.80**	0.79	0.72	**0.78**
	Vgg-16	**0.67**	0.66	0.75	**0.79**	**0.71**	**0.71**
	RESNET-50	0.77	**0.79**	0.73	**0.80**	0.74	**0.79**

1. Table 1 shows that Resnet-50 gives better results in terms of Precision, Recall, and F-score as compared to [9] on the YT dataset. Also, it gives good results as compared to ALEXNET, GOOGLENET, and VGG-16 networks.
2. Table 2 shows good results compared to ALEXNET, GOOGLENET, and VGG-16 networks. Additionally, Resnet-50 gives better results in Precision, Recall, and F-score compared to [9] on the OV dataset.
3. Table 3 shows the results of categories (Cartoons, Sports, and News) of videos in VSUMM Dataset. From Table 3, we can conclude that Resnet-50 outperforms in all categories of the VSUMM dataset compared to [9], ALEXNET, GOOGLENET, and VGG-16 networks.

5 Conclusion

The video summarization technique uses lengthy videos to summarise the informative content from that video. In this work, we developed video summarization as a Clustering problem and proposed a SOM-based modified framework with CNN feature extractions. The dataset is pre-processed to feed to the model, including the Feature extraction using the AlexNet, GoogLeNet, RESNET-50, and Vgg-16 models. We have used a dataset that provides target summary frames for each video. Then, we formulated an unsupervised clustering algorithm to extract summary frames. According to experimental findings, the suggested technique performs competitively on the VSUMM dataset using all four models compared to other approaches. According to comparison data, The top two models that have given significantly better results on our proposed model are GoogLeNet and RESNET-50. Our future research will emphasize identifying the optimal number and fusion of pre-trained CNN models to enhance the overall quality of the generated video summaries.

Acknowledgment. The authors would like to thank to the DST GoI for sponsoring the work under DST/ICPS/General/2018.

References

1. Muhammad, K., et al.: DeepReS: a deep learning-based video summarization strategy for resource-constrained industrial surveillance scenarios. IEEE Trans. Ind. Inform. **16**(9), 5938–5947 (2019)
2. Khurana, K., Deshpande, U.: Two stream multi-layer convolutional network for keyframe-based video summarization. Multimedia Tools Appl. **82**(25), 38467–38508 (2023)
3. Saini, P., Kumar, K., Kashid, S., et al.: Video summarization using deep learning techniques: a detailed analysis and investigation. Artif. Intell. Rev. **56**(11), 12347–12385 (2023)
4. Kumar, K., Shrimankar, D.D., Singh, N.: Eratosthenes sieve based key-frame extraction technique for event summarization in videos. Multimedia Tools Appl. **77**(6), 7383–7404 (2018)
5. Kumar, K., Shrimankar, D.D., Singh, N.: Equal partition based clustering approach for event summarization in videos. In: 2016 12th International Conference on Signal-Image Technology & Internet-Based Systems (SITIS), pp. 119–126. IEEE (2016)
6. Dhiman, A., Deshmukh, M.: Optimized approach for video summarization using transfer learning and LSTM. In: 2023 International Conference on Computational Intelligence and Sustainable Engineering Solutions (CISES), Greater Noida, India, pp. 26–31 (2023). https://doi.org/10.1109/CISES58720.2023.10183585
7. Negi, A., Kumar, K., Chauhan, P., Saini, P., Kashid, S.: Resource utilization tracking for fine-tuning based event detection and summarization over cloud. In: Troiano, L., Vaccaro, A., Kesswani, N., Díaz Rodriguez, I., Brigui, I., Pastor-Escuredo, D. (eds.) ICDLAIR 2022. LNNS, vol. 670, pp. 73–83. Springer, Cham (2022). https://doi.org/10.1007/978-3-031-30396-8_7
8. Negi, A., Kumar, K., Saini, P., Kashid, S.: Object detection based approach for an efficient video summarization with system statistics over cloud. In: 2022 IEEE 9th Uttar Pradesh Section International Conference on Electrical, Electronics and Computer Engineering (UPCON), pp. 1–6. IEEE (2022)
9. Nair, M.S., Mohan, J.: Video summarization using convolutional neural network and random forest classifier. In: TENCON 2019-2019 IEEE Region 10 Conference (TENCON), pp. 476–480. IEEE (2019)
10. De Avila, S.E.F., Lopes, A.P.B., da Luz Jr, A., de Albuquerque Araújo, A.: VSUMM: a mechanism designed to produce static video summaries and a novel evaluation method. Pattern Recogn. Lett. **32**(1), 56–68 (2011)
11. Basavarajaiah, M., Sharma, P.: GVSUM: generic video summarization using deep visual features. Multimedia Tools Appl. **80**, 14459–14476 (2021). https://doi.org/10.1007/s11042-020-10460-0
12. Nair, M.S., Mohan, J.: Static video summarization using multi-CNN with sparse autoencoder and random forest classifier. SIViP **15**, 735–742 (2021). https://doi.org/10.1007/s11760-020-01791-4

13. Otani, M., Nakashima, Y., Rahtu, E., Heikkila, J.: Rethinking the evaluation of video summaries. In: Proceedings of the IEEE/CVF Conference on Computer Vision and Pattern Recognition, pp. 7596–7604 (2019)
14. Gharbi, H., Bahroun, S., Zagrouba, E.: Key frame extraction for video summarization using local description and repeatability graph clustering. SIViP **13**, 507–515 (2019)
15. Parihar, A.S., Pal, J., Sharma, I.: Multiview video summarization using video partitioning and clustering. J. Vis. Commun. Image Representation **74**, 102991 (2021)
16. Raut, V., Gunjan, R.: Transfer learning based video summarization in wireless capsule endoscopy. Int. J. Inf. Technol. **14**(4), 2183–2190 (2022)
17. Fan, Y., Lu, X., Li, D., Liu, Y.: Video-based emotion recognition using CNN-RNN and C3D hybrid networks. In: Proceedings of the 18th ACM International Conference on Multimodal Interaction, pp. 445–450. ACM (2016)
18. Szegedy, C., Ioffe, S., Vanhoucke, V., Alemi, A.A.: Inception-v4, inception-ResNet and the impact of residual connections on learning. In: Thirty-First AAAI Conference on Artificial Intelligence (2017)
19. Srivastava, N., Hinton, G., Krizhevsky, A., Sutskever, I., Salakhutdinov, R.: Dropout: a simple way to prevent neural networks from overfitting. J. Mach. Learn. Res. **15**(1), 1929–1958 (2014)
20. Szegedy, C., et al.: Going deeper with convolutions. In: CVPR (2015)
21. Simonyan, K., Zisserman, A.: Very deep convolutional networks for large-scale image recognition. arXiv:1409.1556 (2014)
22. Kashid, S., Awasthi, L.K., Kumar, K., Saini, P.: NS4: a Novel Security approach for extracted video keyframes using Secret Sharing Scheme. In: 2023 International Conference on Computer, Electronics and Electrical Engineering and their Applications (IC2E3), Srinagar Garhwal, India, pp. 1–6 (2023). https://doi.org/10.1109/IC2E357697.2023.10262778

Machine Learning and Deep Leaning in Predicting Coronary Heart Disease

Seda Demir(✉) and Harun Selvitopi

Erzurum Technical University, 25050 Erzurum, Turkey
{seda.demir,harun.selvitopi}@erzurum.edu.tr
https://erzurum.edu.tr/personel/4666/seda-demir,
https://erzurum.edu.tr/personel/493/harun-selvitopi

Abstract. Coronary heart disease is the leading cause of morbidity and mortality worldwide. Current diagnostic tools for this disease are not well suited to monitoring response to treatment. In this article, machine learning model k-nearest neighbor (k-NN) and deep learning model artificial neural network (ANN) were used and compared to evaluate the risk of coronary heart disease. The Framingham dataset was used to apply the techniques mentioned above. In addition, the Hotdecking missing data filing method was also taken into consideration to prepare the dataset before applying machine learning and deep learning techniques. At the end of the data analysis, 85.50% accuracy was obtained for k-NN and 85.85% accuracy for ANN.

Keywords: Coronary heart disease · k-nearest neighbour · machine learning · artificial neural network

1 Introduction

Heart diseases are among the leading fatal diseases. Approximately 30% of all deaths worldwide occur from cardiovascular diseases. More than 18 million people die every year from cardiovascular (CVD) disease due to risks such as obesity, hyperglycemia, cholesterol, etc. The most typical type of cardiovascular disease is coronary heart disease. CVD is characterized by many clinical symptoms that manifest as myocardial infarction, angina, and/or sudden cardiac death. Coronary heart disease accounts for 38% of deaths in women and 46% in men [1]. The high percentage of deaths is mainly due to the high prevalence of coronary heart disease. It is imperative to mitigate these increases in the prevalence of CVD and strengthen the prevention and treatment of CVD by conducting an appropriate scientific risk assessment to reduce the risk of risk groups.

For this purpose, the cloud-random forest (C-RF) model combining cloud model and random forest model has been applied to observe the risk of coronary heart disease in [2]. In [3], early detection of the coronary heart disease was studied using the random forest(RF), Logistic Regression (LR), and Support Vector Machine (SVM) methods. The polynomial-SVM with the use of the grid

D. Pastor-Escuredo et al. (Eds.): ICDLAIR 2023, LNNS 1001, pp. 101–108, 2024.
https://doi.org/10.1007/978-3-031-60935-0_10

search optimization method has given the better prediction than logistic regression (LR), decision tree (DT), linear discriminant analysis (LDA), k-nearest neighbors (k-NN), artificial neural network (ANN) in [4]. In [5], the machine learning techniques has been applied the personal data to detect the coronary heart diseases. Four different machine learning techniques has been applied the data set obtained in eleven years from 7350 patients between the ages 21–90 years using computerized tomography (CT) coronary angiography in [6]. In [7], machine learning method has been developed to identify a marker for detect the coronary artery disease. The six different machine learning techniques has been applied date set for comparison of the effectiveness to predict the heart disease in [8]. In [9], the authors find that the random forest methods has the 100% accuracy to predict the heart disease with comparison of different machine learning methods.

The data mining tools are also important to obtain the high percentage accuracy, sensitivity and specificity when using machine learning methods. In [10], the authors have considered six different data mining tools as: Orange, Weka, RapidMiner, Knime, Matlab, and Scikit-Learn and six different machine learning methods as: Logistic Regression, Support Vector Machine, k Nearest Neighbors, Artificial Neural Network, Naïve Bayes, and Random Forest to compare the aforementioned importance and find the best tool is Matlab and the best method is Matlab's Artificial Neural Network model. The data preparation is one of the main stage of the data analysis techniques. Therefore, a new normalization technique has been proposed comparing with ten traditional machine learning techniques in [11]. There are lots of techniques to prepare the missing data set before applying machine learning methods as mean flling, mode flling, and Lagrangian interpolation. The most useful method to fill the missing values of data set is Hotdecking method [12], because it takes values similar to the complete data.

2 Research Methodology

In our study, we first prepare our Framingham dataset, complete the missing data with the Hotdecking method, then divide this data into two parts as training set and test set. Then, the data from both sets are preprocessed and prepared to be entered into the model. Then ANN, k-NN models were trained with the data in our training set to classify coronary heart disease. We used ANN, k-NN models to classify our data and evaluate its performance. The accuracy of the model was measured by its ability to classify the data in the test set.

2.1 Collection and Properties of Dataset

There are a total of 4238 data in the Framingham CHD dataset. This dataset was taken from Kaggle (https://www.kaggle.com/datasets/navink25/framingham/code). Both models contain 80% of the training set data and the testing set contains 20% of the data to test the model.

2.2 Data Pre-processing

Data pre-processing is a common method for processing data in order to apply the model. In preprocessing, we take raw data and prepare the input data for the next stage by applying different techniques. The main purpose of data pre-processing is to improve the quality of the information in the data so that the machine can easily understand and analyze that data. Framingham CHD data set, consisting of a total of 4238 samples, covers 15 risk factors that cause disease, including sex, age, education, currentSmoker, cigsPerDay, BPMeds, prevalentStroke, prevalentHyp, diabetes, totChol, sysBP, diaBP, BMI, heartRate and glucose. Training indices such as cigsPerDay, BMI, heartrate, totChol, and glucose contain missing values. Therefore, before making a risk assessment, missing values in the data set must be transformed. Filling missing values is selected. Common data filling methods include Lagrange interpolation, mode filling, and mean filling, but these methods cannot be applied to the data set for various reasons. To fill in the missing values, the Hotdecking method was applied due to the characteristics of the CHD data set. Imputation techniques that use observed values from the sample to impute (fill in) missing values are known as hotdeck imputation. This method is to find an object that is most similar to it in all the data through the correlation relationship between the data and to use the value of this similar object as a substitute. After filling in the missing values, the target variable must be confirmed. In the CHD dataset, the target variable is TenYearCHD. This field indicates whether the patient will suffer from CHD. (0 = no disease; 1 = disease present).

3 Experiment and Output

In this section, we will show which model is best for classification and which model performs best on the dataset.

k-NN and ANN models are used in this paper. The k-nearest neighbor (k-NN) algorithm is the most commonly used machine learning algorithm. This supervised learning method was presented by Cover and Hart in 1967 [13]. The aforementioned algorithm is a supervised machine learning algorithm generally used for classification purposes. For the prediction of diseases, it is commonly employed. Using the features and labels of the training data, the supervised algorithm k-NN predicts the classification of unlabeled data. This approach classifies an example by looking at its previously classified neighbor examples. Commonly, the k-NN algorithm classifies datasets using a training model similar to the test query by considering the k closest training data points (neighbors) that are closest to the query it is testing. The algorithm applies the majority voting rule to control which classification is finalized.

Since the 1980s, artificial neural networks (ANNs) have become a popular topic in artificial intelligence. The human brain constructs a basic model, abstracts the neural network from its perspective of information processing, and generates distinct networks based on various connections. Our goal is to replicate the information processing of the brain's neural network, which processes memory data. A neural network is a computational model in which a large number of nodes (or neurons) are interconnected. According to the interconnectivity, the output of one node is used as the input by another node for additional processing. Nodes are typically arranged according to the transformations they carry out into a matrix called layers. An ANN framework can have input and output layers as well as one or more hidden layers. Nodes and edges have weights that allow to adjust the signal strengths of the communication, which can be strengthened or weakened through repeated training. ANNs are capable of predicting the test data based on the training and subsequent adaption of the matrix, node, and edge weights. Performance measures are used to analyze the results of binary classification models in the medical diagnosis of coronary heart disease. The use of confusion matrix and ROC curve was examined as performance measurement.

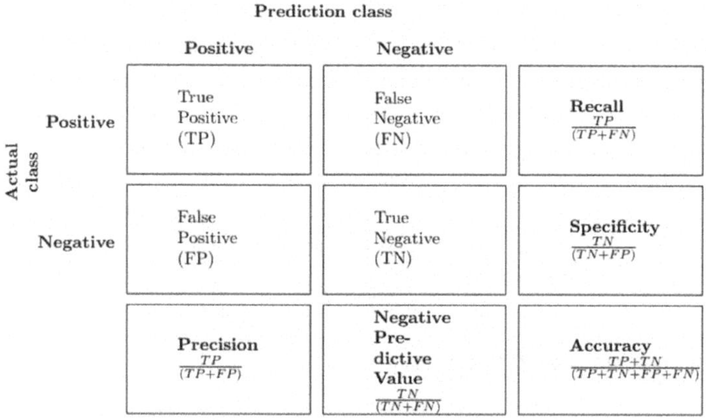

Fig. 1. Confusion Matrix

Confusion matrix is another name for error or contingency matrix in the field of machine learning research. Figure 1 shows the basic structure of the confusion matrix. True positives (TP) are positive cases in this framework that are successfully identified by the classifier. In the same way, true negatives (TN) are instances of negative data that the classifier correctly detects. Negative examples that are incorrectly recognized by the classifier are known as false positives (FP). Positive cases that the classifier incorrectly labels as negative are known as false negatives (FN) or positive and false negatives. The confusion matrix was used for several performance measures such as accuracy, precision, specificity, recall, F1-score. Below are the equivalents of a few of them.

$$\text{Accuracy} = \frac{TP + TN}{TP + TN + FP + FN}$$

$$F_1 \text{ score} = \frac{2 \times TP}{2 \times TP + FN + FP}$$

$$\text{Precisioin} = \frac{TP}{TP + FP}$$

$$\text{Recall} = \frac{TP}{TP + FN}$$

$$\text{Specificity} = \frac{TN}{TN + FP}$$

The accuracy measure is calculated by taking all correct predictions and dividing them by all predicted values. True positive values are divided by both true and false positive values to determine the precision measure. True positive values are divided by true positive and false negative values to get the recall measure.

		Predicted Class		
		Positive	Negative	Total
Actual Class	Positive	724	1	725
	Negative	122	1	123
	Total	846	2	848

		Predicted Class		
		Positive	Negative	Total
Actual Class	Positive	719	3	722
	Negative	117	9	126
	Total	736	12	848

Fig. 2. Confusion Matrix for k-NN and ANN

Figure 2 shows the confusion matrix for the machine learning model k-NN and the deep learning model ANN.

Evaluation indices for k-NN and ANN models are given in Table 1. As can be seen here, the accuracy value of ANN, which is a deep learning model, is higher than k-NN, which is a machine learning model.

Table 1. k-NN and ANN models evaluation indexes

Evaluation indexes	k-NN	ANN
Accuracy	0.8550	0.8585
Recall	0.8558	0.8600
Precision	0.9986	0.9958
Specificity	0.5000	0.7500
F1-score	0.9217	0.9230
Negative Predictive Value	0.0714	0.1111
False Positive Rate	0.5000	0.2500
False Discovery Rate	0.0014	0.0042
False Negative Rate	0.1442	0.1400

Plotting the real positive rate versus the false positive rate at different threshold levels gives the Receiver Operating Characteristic (ROC), one of the most important tools for evaluating diagnostic tests. The predictability of a classifier is also frequently evaluated using the area under the ROC curve (AUC). A classifier's superiority is shown by a larger AUC score, although the opposite is also true.

The sensitivities and specificities for various values of a continuous test measure are first tabulated in order to create a ROC curve. In essence, this yields a list of different test values along with the test's sensitivity and specificity at each value. Plotting sensitivity (true positive rate) on the y-axis versus 1-specificity (false positive rate) on the x-axis for the different tabular values subsequently creates the graphical ROC curve.

We find the AUC value obtained present study is very near the AUC values available in the literature [6]. That shows the k-NN and ANN has been applied correctly in this study.

Figure 3 shows the best k parameters in the k-NN model applied to the CHD data set. It is seen that k = 33 should be chosen here. Figure 4 shows the ROC curve for the k-NN model and the AUC value for this model appears to be 0.7019.

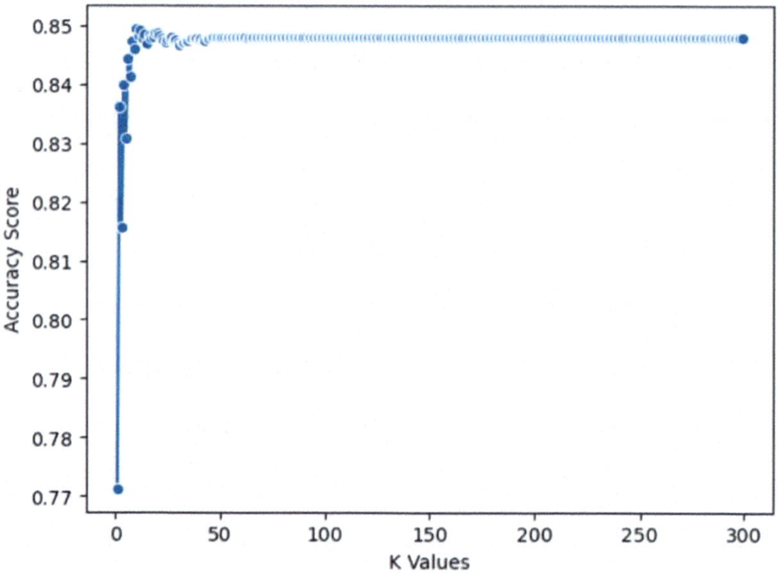

Fig. 3. This is the title of the best fitting k value figure for the k-NN model

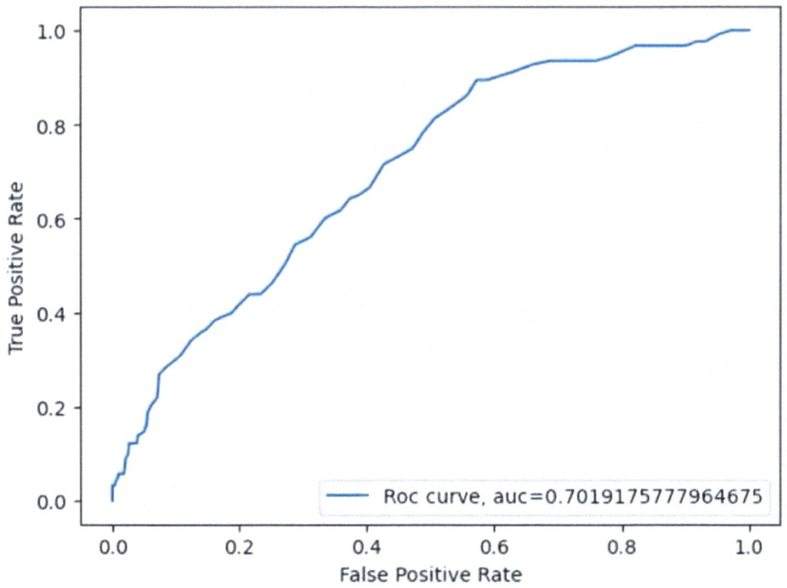

Fig. 4. This is the title of the Roc Curve figure for the k-NN model

4 Conclusion

The machine learning and deep learning methods are very keen on to predict the coronary heart disease using some risk factor from the date sets. This study deal with to examine the accuracy of the k-nearest neighbour and artificial neural network methods considering the Framingham data set. One of the main aim of this study is to compare the predictability of the machine and deep learning methods. To apply the aforesaid methods to considered data set we have use the Hotdeck imputation to fill the missing data. We find that the predictability accuracy of the k-nearest neighbour and artificial neural network 85.50% and 85.85% respectively. All of them gave high accuracy for our dataset. All of them gave high accuracy for our dataset.

Notes and Comments. In [2], a cloud random forest (C-RF) model combining cloud model and random forest was proposed for the Framingham dataset to evaluate the risk of CHD. The classification accuracy of this model was found to be 85%. This model was compared with CART, SVM, CNN and RF models and their accuracy values were examined. The accuracy values of other models are 0.77, 0.76, 0.81 and 0.82, respectively. As seen in this research, C-RF has the highest accuracy value. In our study, k-NN and ANN models were applied to the Framingham dataset and showed higher accuracy than previous models.

5 Future Works

In addition to the data preprocessing method and modeling used in this article, we will use new data preprocessing algorithms such as SMOTE, k- fold cross validation, min-max normalization, feature selection, and we will combine k-NN with a new model and compare it with ANN.

References

1. Rafiq, M., et al.: Competing endogenous RNA regulatory networks of hsa_circ_0126672 in pathophysiology of coronary heart disease. Genes **14**(3), 550 (2023). https://doi.org/10.3390/genes14030550
2. Wang, J., Rao, C., Goh, M., Xiao, X.: Risk assessment of coronary heart disease based on cloud-random forest. Artif. Intell. Rev. **56**(1), 203–232 (2023). https://doi.org/10.1007/s10462-022-10170-z
3. Yilmaz, R., Yağın, F.H.: Early detection of coronary heart disease based on machine learning methods. Med. Rec. **4**(1), 1–6 (2022). https://doi.org/10.37990/medr.1011924
4. Chen, X., Fu, Y., Lin, J., Ji, Y., Fang, Y., Wu, J.: Coronary artery disease detection by machine learning with coronary bifurcation features. Appl. Sci. **10**(21), 7656 (2020). https://doi.org/10.3390/app10217656
5. Bertsimas, D., Orfanoudaki, A., Weiner, R.B.: Personalized treatment for coronary artery disease patients: a machine learning approach. Health Care Manag. Sci. **23**, 482–506 (2020). https://doi.org/10.1007/s10729-020-09522-4
6. Johnson, K.M., Johnson, H.E., Zhao, Y., Dowe, D.A., Staib, L.H.: Scoring of coronary artery disease characteristics on coronary CT angiograms by using machine learning. Radiology **292**(2), 354–362 (2019). https://doi.org/10.1148/radiol.2019182061
7. Forrest, I.S., et al.: Machine learning-based marker for coronary artery disease: derivation and validation in two longitudinal cohorts. The Lancet **401**(10372), 215–225 (2023). https://doi.org/10.1016/S0140-6736(22)02079-7
8. Dwivedi, A.K.: Performance evaluation of different machine learning techniques for prediction of heart disease. Neural Comput. Appl. **29**, 685–693 (2018). https://doi.org/10.1155/2022/7529472
9. Ali, M.M., Paul, B.K., Ahmed, K., Bui, F.M., Quinn, J.M., Moni, M.A.: Heart disease prediction using supervised machine learning algorithms: performance analysis and comparison. Comput. Biol. Med. **136**, 104672 (2021). https://doi.org/10.1016/j.compbiomed.2021.104672
10. Tougui, I., Jilbab, A., El Mhamdi, J.: Heart disease classification using data mining tools and machine learning techniques. Health Technol. **10**, 1137–1144 (2020). https://doi.org/10.1007/s12553-020-00438-1
11. Abdar, M., Ksiazek, W., Acharya, U.R., Tan, R.S., Makarenkov, V., Pławiak, P.: A new machine learning technique for an accurate diagnosis of coronary artery disease. Comput. Meth. Prog. Bio. **179**, 104992 (2019). https://doi.org/10.1016/j.cmpb.2019.104992
12. Mander, A., Clayton, D.: Hotdeck imputation. Stata Tech. Bull. **9**(51), 156–166 (2000)
13. Cover, T., Hart, P.: Nearest neighbor pattern classification. IEEE Trans. Inform. Theory **13**(1), 21–27 (1967). https://doi.org/10.1109/TIT.1967.1053964

Augmented Super Resolution GAN (ASRGAN) for Image Enhancement Through Reinforced Discriminator

C. Bagavathi(✉), Uday Kiran Gadde, Ruthin Gupta Immadisetty, and Ananth Krishna Vasireddy

Department of Computer Science and Engineering, Amrita School of Computing, Amrita Vishwa Vidyapeetham, Coimbatore, India
c_bagavathi@cb.amrita.edu

Abstract. Image super-resolution improves an image resolution and provides better detail. Super resolution achieved through Generative Adversarial Networks has been found to produce good resolution but has few challenges such as discrimination ability, stability and convergence of SRGAN. To address these challenges, we propose an Augmented Super resolution Generative Adversarial Network (ASRGAN) that incorporates reinforcement learning (RL) techniques. In ASRGAN, we introduce an RL agent into the discriminator component to enhance its discrimination capability and adapt its behavior to optimize the training process. The proposed ASRGAN signifies that the SRGAN has been expanded with reinforcement learning, which can have implications for the training stability, discrimination ability, convergence, and fine-grained control of the super resolution process. The generalization capabilities of ASRGAN across diverse image types, highlighting its adaptability has also been brought through this work.

Keywords: GAN · Super Resolution · Reinforcement Learning

1 Introduction

Generative Adversarial Networks (GAN) [1] are Generative models that belong to Neural networks used for unsupervised learning. This model consists of two neural units, namely Generator and a Discriminator, which compete together in a zero-sum game framework. The GAN is broken down into three parts: Generative, Adversarial, and Network, where Generative unit is responsible of representing how data is generated, Adversarial block undertakes the process of training of a model. The networks are preferably deep neural networks for better training [2]. The Generator always generates faux data samples and renders to mislead the discriminator, whereas the discriminator will be trained to tell apart the real and fake samples. The generator and discriminator models are skilled in a complementary fashion where only one model will be trained at an instant. GANs

D. Pastor-Escuredo et al. (Eds.): ICDLAIR 2023, LNNS 1001, pp. 109–118, 2024.
https://doi.org/10.1007/978-3-031-60935-0_11

are used in several applications such as image synthesis [3,4], text generation [5], data augmentation [6], image-to-image translation [7], video synthesis [8], drug discovery [9], surveillance of videos [14] and anomaly detection [12].

High-quality images and videos, which is out current standard for view/store operations, can either be defined at the time of data collection or upgraded using computational techniques [13]. Visual technologies have a high demand on maintaining the quality of image and video data still reducing the memory requirements [11,14]. The images of low or poor quality can be improved based on the requirement through GAN models incorporated with Super Resolution [15–17]. Super-resolution has produced better images with applications ranging from Standard image analysis [14] to medical image enhancement [18]. The minimum standard resolution for the majority of viewing formats is 480p. The 640 × 480 pixel quality is available at this resolution, which was the norm in the early days of computers [19]. High Definitions (HD), starting with 720p, which typically has a pixel resolution of around 1280 × 720, move further along the scale. Then there is Full High Definition (FHD), whose short form is 1080p and stands for pixels that are 1920 × 1080, and Quad High Definition (QHD), whose short form is 1440p and for pixels that are 2560 × 1440. These three scales, which all have a 16:9 aspect ratio, are among those that are more frequently employed by conventional computing engines. The goal is to progressively enhance these visual and video quality to give viewers and audiences the best experience possible as technology advances by overcoming predominant challenges of super-resolution such as preserving image details, enhancing image quality, and improving image perception [20]. Input image can be lower quality due to variety of reasons such as type of camera lens, scaling capabilities, lack of advanced technology, poor editing, blurry backdrop capture, or other comparable ones [21]. Deploying deep learning neural networks for resolution, specifically the Super Resolution Generative Adversarial Networks (SRGANs) architecture, to transform the low-resolution images (or videos) into more high-quality content is more preferable for several advantages stated in this section.

In this work, better video and image quality through Super Resolution GANs have been proposed with an addition of Residual blocks where the augmentation is done to improve the noise improvement and the image enhancement through Reinforcement learning applied at discriminator.

2 Super Resolution

Super-Resolution is the technique to better the quality of a low-resolution photos by boosting the combined resolution creating a better view of the represented vision domain [22]. Super resolution is well known for its efficiency in improving the finer details of the image which can be improved from low quality images.

Bicubic interpolation is a mathematical method used for increasing the resolution of digital images or resizing them [23]. This method is one of the preferred methods for upscaling an image which minimal disruption to the smoothness and still simple to execute. This technique has significant loss of information while upscaling the image. Ankur Singh and Piyush Rai [24] has come with a semi-supervised approach for super-resolution that can engender high-resolution images utilizing as limited as 500 paired examples, which can be used beside supervised GAN-based super-resolution methods to better their performance. Ryan Dahl and et al. [25] has overcome the disadvantage of standard SR methods which produce blurry images, that was caused due to the averaging effects of individuals pixels of the images contributing to define a better image. It offers better resolution image through modelling of statistical dependencies of the low-resolution images. SRCNN stands for Super-Resolution Convolutional Neural Network [26]. It is a specific class of deep learning model applied to the goal of super-resolution of images. SRCNN displayed outstanding performance especially for the demand of super-resolution in a single image which could be fuzzy or pixelated and it was destined to create a better high-resolution image.

Fast Super-Resolution Convolutional Neural Network also termed as FSRCNN is similar to SRCNN with a special focus on the speed and effectiveness [27]. Efficiency, quick processing, interpolation-free, many stages, upscaling, feature extraction, and complete learning are some of the benefits of FSRCNN. Comparable to other super-resolution models, FSRCNN aims to enhance or preserve significant information swiftly while resulting high-resolution versions of low-resolution images. In contrast to some other models, FSRCNN is intended to be faster and more effective. LAPSRN stands for Deep Laplacian Pyramid Super-Resolution Network [28]. LapSRN produces produce high-quality super-resolution while being computationally effective which makes it suitable for real-time or near-real-time applications. Progressive super resolution, Pyramid Fusion, Deep Architecture, Skip connections, Learnable Upscaling, Complete Learning, efficient capturing of both global and local image characteristics and Multi scale Super Resolution are some of LAPSRN's features.

EDSR stands for Enhanced Deep Super-Resolution [29] is designed specially to improve the resolution through deep learning architecture. Deep Residual Network, Deep Architecture, Skip Connections, Efficient Training, Loss Function, No Interpolation, Multi Scale Super Resolution, High Quality Results, and Application are characteristics of EDSR. In order to make it more useful for real-time applications, efficient variations and optimizations have been suggested in literature. This technique requires large amount of training data and the applicability of the algorithm is limited.

The related work suggests that the loss function is not given much importance in the previous methods. Mode collapse is a serious problem in the above GAN-based SR methods resulting similar outputs for different inputs. The proposed method addresses these concerns to produce a better SR image.

3 Proposed Method

ASRGAN stands for "Augmented Super-Resolution Generative Adversarial Network". ASRGAN can generate high-resolution images from low-resolution inputs with the intention of maintaining or increasing image details. ASRGAN combines the power of GANs and perceptual loss through Reinforcement Learning. The properties of ASRGAN include Adversarial Training, Residual Blocks, Perceptual Loss, Multi Scale Super Resolution, Fine Tuning, and Practical Applications.

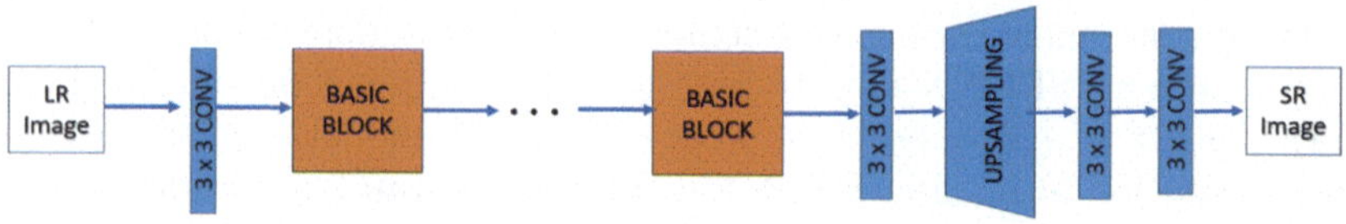

Fig. 1. Block Diagram of GAN including the process of upsampling used for Super Resolution

The fundamental goal of ASRGAN is to preserve and improve image quality by restoring complex textures and minute details during upscaling. The pixel-wise difference between the created high-resolution image and the ground truth is commonly measured using Mean Squared Error (MSE) loss. The PSNR (Peak Signal-to-Noise Ratio) is maximized by MSE, while perceptual qualities other than pixel-level accuracy is also emphasized by ASRGAN. Figure 1 illustrates the standard GAN architecture used for image resolution improvement. The integration of RL in ASRGAN advantages such as enhancing discrimination ability of discriminator enabling proper feedback to generator, stabilizing overall training of SRGAN by avoiding mode collapse and fine-grained control over super resolution process.

The block diagram shown in Fig. 2 describes the working of the proposed method which starts from the input image given to generator and then followed by discriminator with adversarial RL training. This technique updates the discriminator and loss calculation is performed. Another iteration of updating the generator produces the high resolution output.

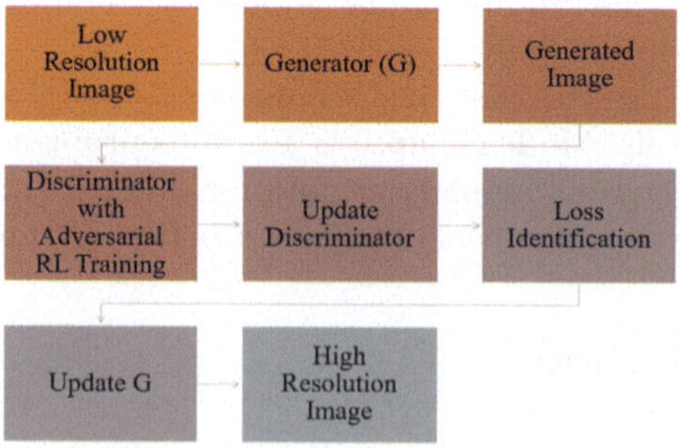

Fig. 2. Block Diagram of the Proposed ASRGAN

A modified form of Residual blocks in Super resolution GAN has been proposed and illustrated in Fig. 3. Receiving pairs of genuine high-resolution images from the generator (G), the discriminator engages with the environment. The discriminator divides the incoming photos into created and real categories. The RL agent's (discriminator's) ability to properly distinguish between real and produced images is maximized through policy learning. A feature extraction module, shuffle blocks, residual blocks, and a Global Residual Learning (GRL) module are some of the parts of ASRGAN architecture. The discriminator module is been augmented with Reinforcement Learning which can enable better learning compared to Relativistic GAN (RaGAN) and monitors the weight clipping [30].

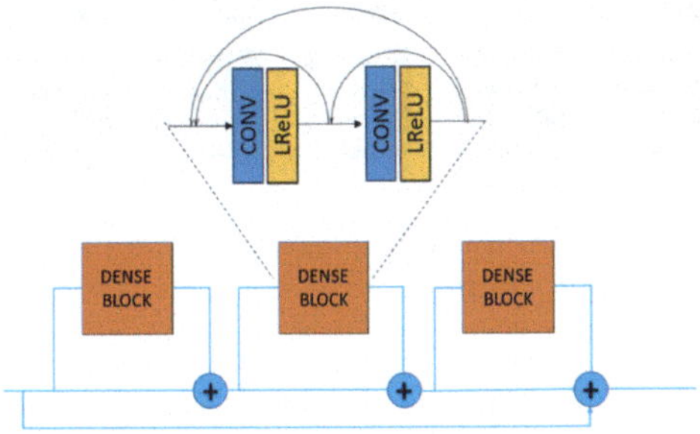

Fig. 3. Residual blocks of Proposed ASRGAN

The batch normalization units of a standard SRGAN has been removed to provide the facility of better training and higher capacity. The perceptual loss is the principle to access the possibility of reaching the required smoothness level of the image by assessing difference in the high level features of the improved image compared to the input image. Together, these components enable the creation of high-quality super-resolved images with improved details by extracting both low-level and high-level features from input images.

4 Results and Findings

The dataset used is Lytro dataset [31], which has been the benchmark for algorithms that work on depth understanding and image refocusing. It has been taken as the dataset in this work to justify the importance of proposed method in improving image resolution. The dataset contains 30 photographs which includes indoor, outdoor and motion blur. 5 sample images have been taken for testing the proposed method: Car, Flower, Road, Room and Toy.

The qualitative results have been portrayed in Fig. 4 indicating an improvement in the proposed method in terms of smoothness and clarity in the image. The quality metrics used are PSNR, MSE, SSIM, Correlation (r), Mutual Information, Image Entropy. PSNR is to understand the extent of interference in

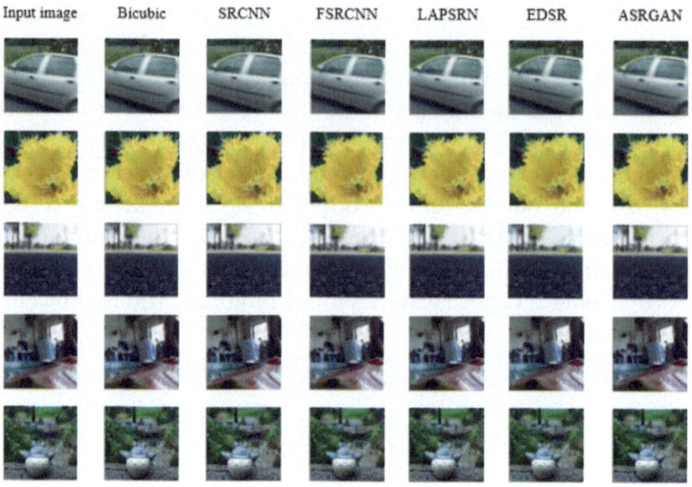

Fig. 4. The result of resolution improved images and the proposed method ASRGAN has been mentioned in the last column

image through noise signals. The maximum pixel value and average of mean square value of pixels are used in the calculation of PSNR.

Structural Similarity Index (SSIM) is to compare the structure, brightness, and contrast of two images in order to determine how similar they are and also provides a more thorough evaluation of image quality. The three main factors that go into calculating the SSIM index are luminance (brightness), contrast, and structure.

$$SSIM(X,Y) = \frac{(2.l.c + C_1).(2.s + C_2)}{(l^2 + c^2 + C_1).(s^2 + C_2)))} \tag{1}$$

In the Eq. (1), X and Y refers to the original image and processed image respectively. The variables l, c, s refer to local Mean of luminance, contrast and structure, C1, C2 are the Constants to prevent division by zero. The Pearson correlation coefficient, commonly abbreviated as r, calculates the linear relationship between two sets of data viz., the pixel values of two photographs. It has a range of −1 to 1. Mutual information (MI (X, Y)) is used to express how dependent or mutually informative two random variables are. Mutual information can be utilized in the context of images to determine how similar or dependent two images are, which is useful for image registration, image fusion, and other image analysis activities.

$$H(X) = -\sum_{i=1}^{N} p(x_i)log_2(p(x_i)) \tag{2}$$

Entropy of an image is a measure to quantify the amount of information or randomness present in a digital image. Image Entropy H(X) uses the $p(x_i)$ which is the probability of pixel in image having value x_i and it is given by Eq. (2).

Table 1. Table containing the values of PSNR, MSE, SSIM, r, MI and H for the images: Car and Flower from Lytro dataset [31]

	Car						Flower					
	PSNR	MSE	SSIM	r	MI	H	PSNR	MSE	SSIM	r	MI	H
Bicubic [23]	8.354	28495.78	0.277	0.217	0.438	6.592	7.749	32755.88	0.261	0.433	0.605	6.344
SRCNN [26]	4.267	73029.84	0.356	–	0.0	0.0	4.042	76908.36	0.378	–	0.0	0.0
FSRCNN [27]	4.884	63343.61	0.285	−0.22	0.343	3.131	5.152	59557.48	0.297	0.211	0.355	3.02
LAPSRN [28]	4.267	73029.84	0.356	–	0.0	0.0	4.042	76908.36	0.378	–	0.0	0.0
EDSR [23]	4.879	63429.23	0.285	−0.22	0.342	3.132	5.153	59551.68	0.297	0.211	0.352	2.994
ASRGAN	38.86	8.447	0.988	0.0	0.0	**0.0**	44.78	2.158	0.989	0.0	0.0	**0.0**

The results in Table 1 indicate that there is significant improvement in PSNR, MSE and SSIM value in the proposed method compared to exiting methods. The same trend has been seen in other selected images as elaborated in Table 2 and Table 3. The proposed ASRGAN has been found perform better compared to all the methods listed in Table 1, 2 and 3 by providing visual quality in images, perfect matching of the images with the real image. The method provides better generalization and robustness through the application of reinforcement learning. ASRGAN, being a deep learning-based approach, typically requires higher computational resources compared to traditional interpolation-based methods. However, the computational cost of ASRGAN is often justified by its ability to generate significantly improved results in terms of visual quality and perceptual fidelity.

Table 2. Table containing the values of PSNR, MSE, SSIM, r, MI and H for the images: Road from Lytro dataset [31]

H	Road					
	PSNR	MSE	SSIM	r	MI	
Bicubic [23]	8.443	27914.67	0.276	0.483	0.753	6.379
SRCNN [26]	3.711	83001.73	0.320	–	0.0	0.0
FSRCNN [27]	5.145	59656.45	0.298	0.272	0.429	3.037
LAPSRN [28]	3.711	83001.73	0.320	–	0.0	0.0
EDSR [29]	5.149	59598.99	0.299	0.272	0.432	3.047
ASRGAN	**44.96**	**2.071**	**0.995**	**0.0**	**0.0**	**0.0**

Augmented Super-resolution Generative Adversarial Networks (ASRGAN) with reinforcement learning techniques have several potential applications in computer vision. It can be useful in the applications where the model is required to Enhance Image Super resolution, video super resolution, medical imaging, remote sensing, forensic image analysis and art restoration.

Table 3. Table containing the values of PSNR, MSE, SSIM, r, MI and H for the images: Room and Toy from Lytro dataset [31]

	Room						Toy					
	PSNR	MSE	SSIM	r	MI	H	PSNR	MSE	SSIM	r	MI	H
Bicubic [23]	6.625	42428.16	0.187	−0.11	0.405	6.575	7.931	31412.18	0.135	0.212	0.284	6.477
SRCNN [26]	3.732	82600.65	0.296	–	0.0	0.0	3.282	91608.88	0.202	–	0.0	0.0
FSRCNN [27]	4.836	64048.18	0.242	0.048	0.221	3.122	4.551	68403.97	0.169	0.069	0.208	3.071
LAPSRN [28]	3.732	82600.65	0.296	–	0.0	0.0	3.282	91608.88	0.202	–	0.0	0.0
EDSR [29]	4.837	64033.29	0.242	0.048	0.218	3.118	4.548	68441.93	0.169	0.068	0.207	3.069
ASRGAN	**43.43**	**2.946**	**0.991**	**0.0**	**0.0**	**0.0**	**39.78**	**6.829**	**0.987**	**0.0**	**0.0**	**0.0**

5 Conclusion

In conclusion, our proposed ASRGAN, which combines the power of generative adversarial networks with reinforcement learning techniques, advances the field of image super resolution. The integration of RL in the discriminator component enhances training stability, discrimination ability, convergence, and fine-grained control. ASRGAN sets new benchmarks in generating high-quality super resolved images and opens up avenues for future research in the intersection of deep learning, adversarial training, and reinforcement learning for computer vision tasks. Experimental results demonstrate the effectiveness of the RL-augmented approach, showcasing significant improvements in image quality, convergence speed, and fine-grained control. As a future work, the work can be extended to multi-modal image restoration through adaptive reinforcement strategies which improve the feedback mechanism dynamically.

References

1. Goodfellow, I., et al.: Generative adversarial nets. In: Advances in Neural Information Processing Systems, vol. 27 (2014)
2. Salakhutdinov, R.: Learning deep generative models. Ann. Rev. Stat. Appl. **2**, 361–385 (2015)
3. Saradagi, A.B.: Evolutionary algorithm based encoder decoder network design for semantic inpainting and noise vector mapping in generative adversarial network. In: 6th International Conference on Inventive Computation Technologies (ICICT), pp. 108–113. IEEE (2021)
4. Nirmal, S., Sowmya, V., Soman, K.P.: Open set domain adaptation for hyperspectral image classification using generative adversarial network. In: Ranganathan, G., Chen, J., Rocha, Á. (eds.) ICICCT 2019. LNNS, vol. 89, pp. 819–827. Springer, Singapore (2019). https://doi.org/10.1007/978-981-15-0146-3_78
5. de Rosa, G.H., Papa, J.P.: A survey on text generation using generative adversarial networks. Pattern Recogn. **119**, 108098 (2021)
6. Shao, S., Wang, P., Yan, R.: Generative adversarial networks for data augmentation in machine fault diagnosis. Comput. Ind. **106**, 85–93 (2019)

7. Emami, H., Aliabadi, M.M., Dong, M., Chinnam, R.B.: SPA-GAN: spatial attention GAN for image-to-image translation. IEEE Trans. Multimedia **23**, 391–401 (2020)
8. Wei, D., Huang, K., Ma, L., Hua, J., Lai, B., Shen, H.: OAW-GAN: occlusion-aware warping GAN for unified human video synthesis. Appl. Intell. **53**(1), 616–633 (2023)
9. Ramesh, A., Rao, A.S., Moudgalya, S., Srinivas, K.S.: GAN based approach for drug design. In: 20th IEEE International Conference on Machine Learning and Applications (ICMLA), pp. 825–828. IEEE (2021)
10. Sree, K.V., Jeyakumar, G.: An evolutionary computing approach to solve object identification problem for fall detection in computer vision-based video surveillance applications. In: Hemanth, D., Kumar, B., Manavalan, G. (eds.) Recent Advances on Memetic Algorithms and Its Applications in Image Processing. SCI, vol. 873, pp. 1–18. Springer, Singapore (2020). https://doi.org/10.1007/978-981-15-1362-6_1
11. Bagavathi C., Saraniya O.: Enhanced texture classification through feature compaction using dihybrid bio-inspired computation techniques. Concurr. Comput. Pract. Exp. **33**(24) (2021). https://doi.org/10.1002/cpe.6453
12. Sabuhi, M., Zhou, M., Bezemer, C.P., Musilek, P.: Applications of generative adversarial networks in anomaly detection: a systematic literature review. IEEE Access **9**, 161003–161029 (2021)
13. Ma, S., Zhang, X., Jia, C., Zhao, Z., Wang, S., Wang, S.: Image and video compression with neural networks: a review. IEEE Trans. Circ. Syst. Video Technol. **30**(6), 1683–1698 (2019)
14. Aishwarya, G., Krishnan, K.R.: Generative adversarial networks for facial image inpainting and super-resolution. In: Journal of Physics: Conference Series, vol. 2070, no. 1, p. 012103. IOP Publishing (2021)
15. Ledig, C., et al.: Photo-realistic single image super-resolution using a generative adversarial network. In: Proceedings of the IEEE Conference on Computer Vision and Pattern Recognition, pp. 4681–4690 (2017)
16. Vamsi Kiran Reddy, P., Sajith Variyar, V.V.: Image enhancement using GAN (a re-modeling of SR-GAN for noise reduction). In: Kaiser, M.S., Xie, J., Rathore, V.S. (eds.) ICTCS 2020. LNNS, vol. 190, pp. 721–729. Springer, Singapore (2021). https://doi.org/10.1007/978-981-16-0882-7_64
17. Choudhary, R.R., Jisnu, K.K., Meena, G.: Image dehazing using deep learning techniques. Procedia Comput. Sci. **167**, 1110–1119 (2020)
18. Swapna, T.R., Indu, D., Chakraborty, C.: Macular region enhancement of Fundus Fluorescein Angiogram images using super resolution via sparse representation and quality analysis. Procedia Comput. Sci. **58**, 586–592 (2015)
19. Maheswari, B.U., Ramesh, T.K.: An improved delay-resistant and reliable hybrid overlay for peer-to-peer video streaming in wired and wireless networks. IEEE Access **6**, 56539–56550 (2018). https://doi.org/10.1109/ACCESS.2018.2871932
20. Daithankar, M.V., Ruikar, S.D.: Video super resolution by neural network: a theoretical aspect. J. Comput. Theor. Nanosci. **17**(9–10), 4202–4206 (2020)
21. Choudhary, R.R., Jangid, A., Meena, G.: A novel approach for edge detection for blurry images by using digital image processing. In: 2017 International Conference on Current Trends in Computer, Electrical, Electronics and Communication (CTCEEC), pp. 1029–1034. IEEE (2017)
22. Tang, Y., Yuan, Y., Yan, P., Li, X., Pan, X., Li, L.: Single-image super-resolution based on semi-supervised learning. In: The First Asian Conference on Pattern Recognition, pp. 52–56. IEEE (2011)

23. Patel, V., Mistree, K.: A review on different image interpolation techniques for image enhancement. Int. J. Emerg. Technol. Adv. Eng. **3**(12), 129–133 (2013)
24. Singh, A., Rai, P.: Semi-supervised super-resolution. arXiv preprint arXiv:2204.08192 (2022)
25. Dahl, R., Norouzi, M., Shlens, J.: Pixel recursive super resolution. In: Proceedings of the IEEE International Conference on Computer Vision, pp. 5439–5448 (2017)
26. Dong, C., Loy, C.C., He, K., Tang, X.: Image super-resolution using deep convolutional networks. IEEE Trans. Pattern Anal. Mach. Intell. **38**(2), 295–307 (2015)
27. Dong, C., Loy, C.C., Tang, X.: Accelerating the super-resolution convolutional neural network. In: Leibe, B., Matas, J., Sebe, N., Welling, M. (eds.) ECCV 2016. LNCS, vol. 9906, pp. 391–407. Springer, Cham (2016). https://doi.org/10.1007/978-3-319-46475-6_25
28. Lai, W.S., Huang, J.B., Ahuja, N., Yang, M.H.: Deep Laplacian pyramid networks for fast and accurate super-resolution. In: Proceedings of the IEEE Conference on Computer Vision and Pattern Recognition, pp. 624–632 (2017)
29. Lim, B., Son, S., Kim, H., Nah, S., Mu Lee, K.: Enhanced deep residual networks for single image super-resolution. In: Proceedings of the IEEE Conference on Computer Vision and Pattern Recognition Workshops, pp. 136–144 (2017)
30. Jolicoeur-Martineau, A.: The relativistic discriminator: a key element missing from standard GAN. arXiv preprint arXiv:1807.00734 (2018)
31. Mousnier, A., Vural, E., Guillemot, C.: Partial light field tomographic reconstruction from a fixed-camera focal stack. arXiv preprint arXiv:1503.01903 (2015)

Convolutional Block Attention Assisted Dense Stacked Bi-LSTM for the Generation of RDF Statements

Rubaya Khatun(✉) and Arup Sarkar

Department of Computer and Information Science, Raiganj University, University Road, College Para, Raiganj, West Bengal 733134, India
rkhatun18@gmail.com, arup.sarkar@raiganjuniversity.ac.in

Abstract. The rapid enhancement of the World Wide Web has attained an immense evaluation of information sources over the Internet. Due to heterogeneity and improper structure of web information sources, accessing significant information has resulted in redundant data. Hence, it is highly necessary to process the unstructured textual data for the better generation of the Resource Description Framework (RDF). Initially, the collected text data was pre-processed using one-hot encoding, z-score normalization, tokenization, stop word removal, and POS tagging. From the pre-processed data, effective features are extracted using the BERT-LSTM model (BLM), whereas similar features can be clustered using the Adaptive density K-means clustering (ADK-MC) mechanism to minimize the model complexity. The RDF triples can be classified using the convolutional block attention-assisted dense stacked Bi-LSTM (CBD-SBiL) model. The proposed method achieved better performance with 98.09% accuracy, 99.06% precision, 98.36% recall, and 98.51% F-measure.

Keywords: Normalization · One-hot encoding · unstructured text · RDF statement generation · K-means clustering · Stacked Bi-LSTM · BERT-LSTM

1 Introduction

The era of big data has arrived in human society thanks to the advancement of information technology. While the most recent technological revolution has made lives and work more convenient, many security concerns are now more obvious [1]. Different digital media, including images, audio, and texts, have grown in importance as means of information transmission in the age of big data [2]. They also make excellent cover for technical researchers who want to conceal information. The WWW includes more data sources due to the advancement of technology and population [3]. Managing this data consists of more complexity, such as security, data availability, etc.

Ontologies are designed to assist in marking up data with useful metadata that will make it easier to interpret. RDF (Resource Description Framework), a language where instances are represented by URIs (Uniform Resource Identifiers), expresses data as

D. Pastor-Escuredo et al. (Eds.): ICDLAIR 2023, LNNS 1001, pp. 119–130, 2024.
https://doi.org/10.1007/978-3-031-60935-0_12

triples subject-predicate-object, has been developed to formalize this new way of structuring data [4]. Due to the increase in complexity, the creation of RDF statements does not provide better results. Most existing works only provide a brief overview of fundamental architectures for producing an explanation of RDF statements [5]. Based on these issues, this research presents novel approaches to achieve improved performance. *The main contribution to this proposed work is described below.*

- To present the pre-processing approaches such as one-hot encoding, z-score normalization, tokenization, stop word removal, and POS tagging for transforming the text into the needed format and removing unwanted noises in the text.
- To improve the statement generation performance, the most particular features are extracted from the pre-processed textual data using the BERT-LSTM (BLM) model.
- To minimize the model complexity, similar features can be clustered using the Adaptive density K-means clustering (ADK-MC) mechanism.
- To classify the RDF statement generation using a hybrid deep learning model from the given unstructured textual data.

The rest of the paper is organized as follows; Sect. 1 provides Introduction of the RDF statement generation. Section 2 describes Various related works of RDF statement generation. Section 3 describes the proposed method in detail. Section 4 defines the result and discussion of the proposed methodology and Sect. 5 describes the Overall conclusion and future work.

2 Related Works

Some recent research works on RDF statement generation for processing unstructured text data techniques are described as follows.

Devi et al. [6] developed a method for extracting semantics (triplets) from medical reports and dengue patient case sheets and converting them into an RDF model. For extracting relations from text, a Text2Ontology framework is used, but it has been found to have some limitations.

Hari et al. [7] developed a deep learning model to create an unstructured text RDF-based ontology. This paper evaluates the method by constructing a general knowledge ontology from the corpora of newspaper articles. The developed model achieved an accuracy of 99.16% by using the NYT dataset and 93.52% accuracy by using the WebNLG dataset.

Moussallem et al. [8] presented a multilingual RDF verbalizer that uses a reification strategy and graph attention NN. This paper is based on the idea that since knowledge graphs are language-neutral, they are used to produce multilingual text. Results show that NABU attained reliable outcomes across all tongues in the multilingual consequence with 56.04 BiLingual Evaluation Understudy (BLEU), outperforming advanced English with 66.21 BLEU.

Soliman et al. [9] developed a Deep Learning Framework for RDF and knowledge graphs to maintain medical decision-making. This paper discusses the FCM Design and Big Data Tools Supporting eHealth Applications and Contradiction Algorithm.

Runumi et al. [10] developed a hybrid knowledge resource that allows for structure variations according to RDF triples extracted from the unstructured corpus. Only when an ontology is linked to textual units is it actually possible to interpret concepts and their properties.

3 Proposed Methodology

Initially, the collected text data was pre-processed using one-hot encoding, z-score normalization, tokenization, stop word removal, and POS tagging. From the pre-processed data, effective features are extracted using the BERT-LSTM model (BLM) whereas similar features can be clustered using the Adaptive density K-means clustering (ADK-MC) mechanism to minimize the model complexity. The diagram of the proposed model is shown in Fig. 1.

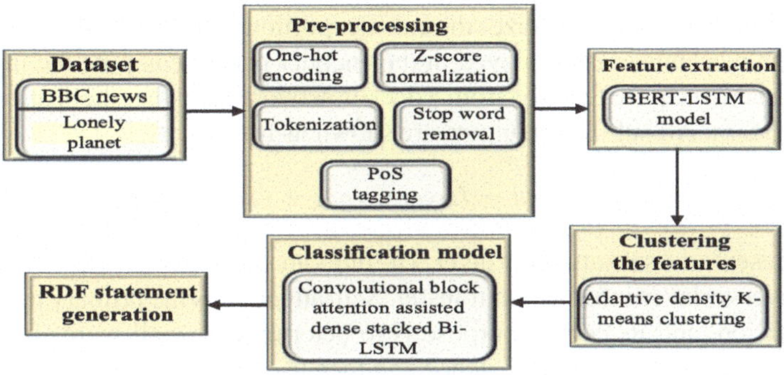

Fig. 1. Architecture of the proposed model

The RDF triples can be classified using the convolutional block attention-assisted dense stacked Bi-LSTM (CBD-SBiL) model. Finally, RDF statements generated by merging attained triples.

3.1 Pre-processing

One-hot encoding: One common method for managing categorical data is one-hot encoding. To prevent the correlations between independent variables from being interpreted incorrectly, it only draws attention to the features' variables.

Z-score normalization: Z-score normalization is a method for normalizing data that is based on the mean and standard deviation of the data (mean value and standard deviation). The mathematical expression for z-score normalization is denoted below.

$$z_{new} = \frac{z - \mu}{\sigma} = \frac{z - mean}{stddev(z)} \tag{1}$$

where z represents old value, z_{new} denotes new value, μ denotes the population mean, and σ implies standard deviation value.

Tokenization: Tokenization is the procedure of separating a string of characters into a series of fundamental units, such as words, phrases, symbols, and other so-called tokens.

Stop word removal: Elimination of the most prevalent words in a language, such as prepositions, numbers, and words that don't contain relevant information for the study, that are generally useless and ineffective for text mining.

POS tagging: Based on the word's placement in the sentence, the POS tagging determines the word class.

3.2 Feature Extraction Using the BERT-LSTM Model

Effective features are extracted from the pre-processed data using the BLM model. The primary purpose of the input layer is to transform the pre-processed text into the model-recognizable format. The input of point vectors, term vectors, and division vectors are basted and collected in the input layer. The twelve-layer Transformer network that makes up the BERT model primarily utilizes the encoder portion of the transformer system [11]. The attention mechanism is employed in the encoder part to compute the input-output connection and to acquire features from a deeper network level that the shallow network is unable to learn. All sub-layer outputs are shown as follows.

$$Sub\_layer\ output = LayerNorm(z + (Sub\_layer(z))) \tag{2}$$

where z represents the self-attention layer. The units are created with residual connections to enable layer-to-layer information transfer. Self-attention is the primary module in the encoder division, and its central purpose is to determine how all words in a sentence interact with other words in that sentence and then use that information to modify each word's weight.

3.3 Feature Clustering Using Adaptive Density K-means Clustering

Similar features can be clustered using the ADK-MC mechanism [12] to minimize the model complexity. Using an outlier detection method based on (Local Outlier Factor) LOF, perform a pretreatment on the data that will be clustered to get rid of outliers. The adaptive density K-means clustering is denoted as follows.

Input: Dataset $D = \{s_1, ...s_i, ...s_n\}$ to be clustered, the number of clusters c.

Output: The total distinction between all substances and their adjacent cluster centers is minimal when there are clusters.

Step 1: Calculate traversing the data set D,

$$LOF(q) = \frac{\sum_{x=1}^{c} \frac{LRD(x)}{LRD(q)}}{c} \tag{3}$$

where $LRD(x)$ denotes the density of the k-nearest neighbor of q, $LRD(q)$ represents the local density of q. Remove the isolated point $LOF(q)$ if it is significantly larger than 1; otherwise, leave $LOF(q)$; obtaining the most recent data set R.

Step 2: The first cluster center should be determined using the mean of the data set R.

$$m_1 = \frac{1}{R}\sum_{X=1}^{R} s_x \tag{4}$$

where m_1 denotes the mean of the dataset R and s_X denotes the data of the xth cluster.

Step 3: Determine the space between the cluster center and the remaining points.

$$L_n = \sum_{y}^{R} \max\left(d_{c-1}^{y} - ||s_n - s_y||^2, 0\right) \tag{5}$$

where s_n is the sample point who's L_n largest is the following cluster's initial center. Determine the space between the center of all clusters and each object s_x, then place it in the closest cluster. L_n represents the next cluster center point, d_{e-1}^{y} implies the distance between the remaining points, cluster objects represented as s_y. The average of the objects in each cluster is calculated to create a new cluster center.

Step 4: Return $m_1, m_2, ...m_c$ after repeating steps 3 and 4 until the criterion function is converged, then the algorithm stops.

3.4 RDF Classification Using Convolutional Block Attention Assisted Dense Stacked Bi-LSTM Model

The RDF statement generation can be performed using the proposed model. Every layer of the network is connected to every layer that follows it, and these connections are known as Dense Nets. They mention the layers among dense blocks, which consist of convolution and pooling, as transition layers, which are also defined as Transition blocks in this method. This is done specifically to facilitate down sampling in their architecture. A simplification of the logic function, the soft-max function can convert a $d-$ dimensional vector z covering any real number to additional $d-$ dimensional vector $\sigma(x)$, but each entry's range must be between [0, 1], and the total number of entries must be equal to 1. The definition of the Soft-max function is

$$\sigma(x)_k = \frac{e^{x_k}}{\sum_{d=1}^{d} e^{x_k}} \tag{6}$$

where $k = 1, ..., d$. In probability-based multi-classification issues, Soft-max is frequently used. The LSTM cell involves three gates: the input gate $\left(I_g\right)$, the forget gate $\left(F_g\right)$ and the output gate $\left(0_g\right)$. These gates regulate a vector that activates memory cells. The architecture of the CBD-SBiL is depicted in Fig. 2. In the Forgot gate, the quantity of material from an earlier state p_{g-1} is calculated using this gate. That, given the input I_g at hand, should either be remembered or ignored, and the hidden state is denoted as h_{g-1}. The forget gate's output ranges from 0 to 1.

The mathematical expression for the forgot gate is denoted as below.

$$F_g = Sigmoid\left(wm_{IF}I_g + wm_{hF}h_{g-1} + b_g\right) \tag{7}$$

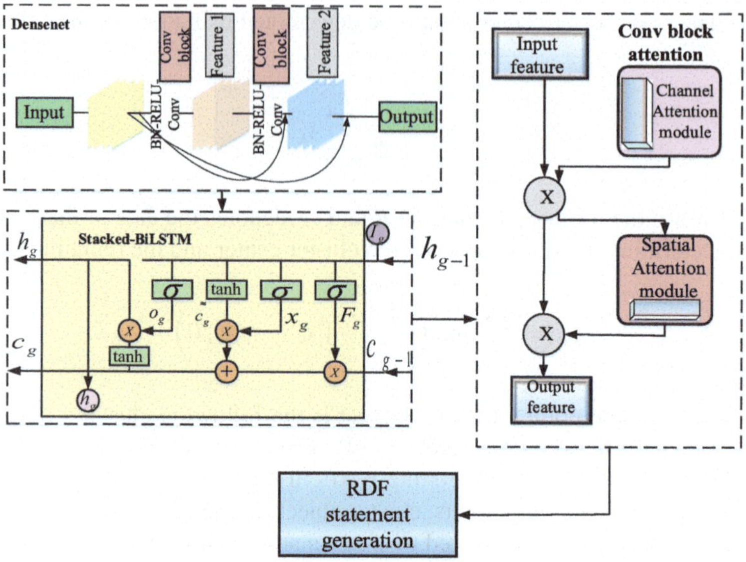

Fig. 2. Architecture of CBD-SBiL

where b_g represents bias vector, wm_{IF} denotes the weight matrix among I_g and F_g, and wm_{hF} denotes the weighted matrix between h_{g-1} and F_g. The input gate is used to estimate the capacity of the network's input I_g that needs to be kept for the state of the current cell c_g. The mathematical expression for the input gate is stated below.

$$I_g = Sigmoid\left(wm_{Ix}I_g + wm_{hx}h_{g-1} + b_x\right) \tag{8}$$

where b_x represents bias vector, wm_{hx} denotes the weight matrix among I_g and x_g, x_g denotes the $x - th$ gate and wm_{hF} denotes the weighted matrix between h_{g-1} and x_g. Output gate using the LSTM network's information transmission rate is calculated from the state of the cell c_g using the output value of this time h_g. h_g denotes the $h - th$ gate. The mathematical expression for the output gate is denoted as follows.

$$o_g = Sigmoid\left(wm_{Io} + I_g + w_{ho}h_{g-1} + b_o\right) \tag{9}$$

where b_o represents bias vector, wm_{Io} denotes the weight matrix among I_g and o_g, and wm_{h0} denotes the weighted matrix between h_{g-1} and o_g. The ending result of the LSTM cell is layer output h_g, and cell outcome state c_g is calculated using the below formulas.

$$h_g = o_g \otimes \tanh\left(c_g\right) \tag{10}$$

$$c_g = F_g \otimes c_{g-1} + x_g \otimes \overset{\approx}{c}_g \tag{11}$$

$\overset{\approx}{c}_g$ is denoted as the intermediate of the input state; this is calculated using the below equation. c_{g-1} input of the previous input.

$$\overset{\approx}{c}_g = \tanh\left(wm_{Ic}I_t + wm_{hc}h_{g-1} + b_c\right) \tag{12}$$

where b_c represents bias vector, wm_{Ic} denotes the weight matrix among I_g and $\overset{\approx}{c}_g$, and wm_{hc} denotes the weighted matrix between h_{g-1} and $\overset{\approx}{c}_g$. Since LSTM is unable to use the contextual data from upcoming tokens of websites, the bidirectional LSTM (BiLSTM) layer contains dual unknown LSTM layers that are oriented in opposition to one another and output the same result. A bidirectional LSTM (BiLSTM) combines data from both instructions of a phrase to produce word explanations. The forward LSTM computes the hidden state $\overrightarrow{h}_g$ by processing the input from left to right based on I_g and h_{g-1}, while the backward LSTM calculates the hidden state $\overrightarrow{h}_g$ and processes the input from right to left based on I_g and h_{g-1}. Although word embedding is shared, the forward and backward parameters in the BiLSTM network are independent of one and another. Lastly, BiLSTM is defined, which combines the headfirst and reverse direction vectors at the time step h_{ts}.

$$h_{ts} = \left[\overrightarrow{h}_{ts}, \overrightarrow{h}_{ts}\right] \tag{13}$$

where $\overrightarrow{h}_{ts}$ denotes the final state of BiLSTM combines the forward and reverse direction vectors at the time step ts. The mathematical expression for the channel attention and spatial attention modules is denoted as follows.

$$A_c(f) = \sigma(MLP(Avg\,pool(f) + MLP : (Max\,pool(f)))) \tag{14}$$

$$A_s(f) = \sigma\left(F^{7\times 7}(Concat(Avg\,pool(f), Max\,pool(f)))\right) \tag{15}$$

where f denotes input feature map, $A_c(f)$ represents channel attention module, sigmoid function is denoted as σ, fully connected multi-layer perception is denoted as MLP, and the activation function is Rectified Linear Unit (ReLU). The average pooling function is denoted as $Avg\,pool$, and the maximum pooling function is denoted as $Max\,pool$ and F implies 7×7 convolution operation. The Eq. (13) and (14) is replaced as soon as the residual network is introduced. It is denoted as follows.

$$f_c = A_c(f) \otimes f \tag{16}$$

$$f_s = A_s(f) \otimes f \tag{17}$$

where f_c is denoted as output feature map for the channel attention, A_c denotes the operation of channel attention mechanism, input feature map represents f, $\otimes$ implies the multiplication between the element and A_s denotes the operation of the spatial attention module. Finally, using the block attention module, the RDF triples classification is devised. Thus, using the proposed CBD-SBiL, the RDF triples are obtained through the Softmax activation function. RDF statement can be generated by merging attained triples.

4 Results and Discussion

In this section, various performance metrics are analyzed and compared with other existing systems for RDF statement generation using unstructured textual data.

The proposed method system configuration details; Intel(R) Core (TM) i5-3470 CPU @ 3.10GHz 3.10 GHz Processor, 16.0 GB installed RAM, 64-bit operating system, and x64-based processor system type. These details are used to evaluate the performance of the proposed method.

4.1 Dataset Description

The proposed model used two unstructured textual datasets, namely BBC News and Lonely Planet datasets. The BBC dataset includes 2225 text forms obtained by the BBC news website. The Lonely Planet datasets contain 96 classes of unstructured documents related to tourism spread across 1801 files. On 300 web pages, the textual description describes popular tourist destinations worldwide.

4.2 Performance Measures

The proposed method maintains various performance metrics that are analyzed, such as accuracy, precision, recall, and F-measure [13] for RDF statement generation using unstructured text data.

Accuracy: The total number of correct predictions in the total amount of predictions is defined as accuracy.

$$ACC = \frac{TN + TP}{TP + FP + FN + TN} \tag{18}$$

Here TP, FP, FN, TN denotes the number of true positives, false positives, false negatives, and true negatives, respectively.

Precision: Precision is referred to as the positive predictive value. The precision is calculated using the below equation,

$$PRE = \frac{TP}{TP + FP} \tag{19}$$

Recall: The recall parameter is used to correctly detect true positive instances by the Machine Learning (ML) model. It also measures the accuracy and how to identify relevant data.

$$recall = \frac{TP}{(TP + FN)} \tag{20}$$

F-Measure: F-Measure trade-off values between Recall and Precision which considers FP and FN. It measures the overall accuracy of an ML model, and it can be expressed in the below Equation;

$$F - Measure = \frac{2 * \Pr ecision * \text{Re}call}{\Pr ecision + \text{Re}call} \tag{21}$$

4.3 Performance Analysis

The performance is analyzed for various existing and proposed methods based on the RDF statement generation technique. Figure 3(a), (b), (c) and (d) shows the performance of accuracy, precision, Recall and F-Measure parameters. The proposed method compared with existing approaches such as Random Forest (RF), Support Vector Machine (SVM), Logistic Regression (LR), Long Short-Term Memory (LSTM), Convolutional Neural Network (CNN) and Multinomial Naïve Bayes (MNB).

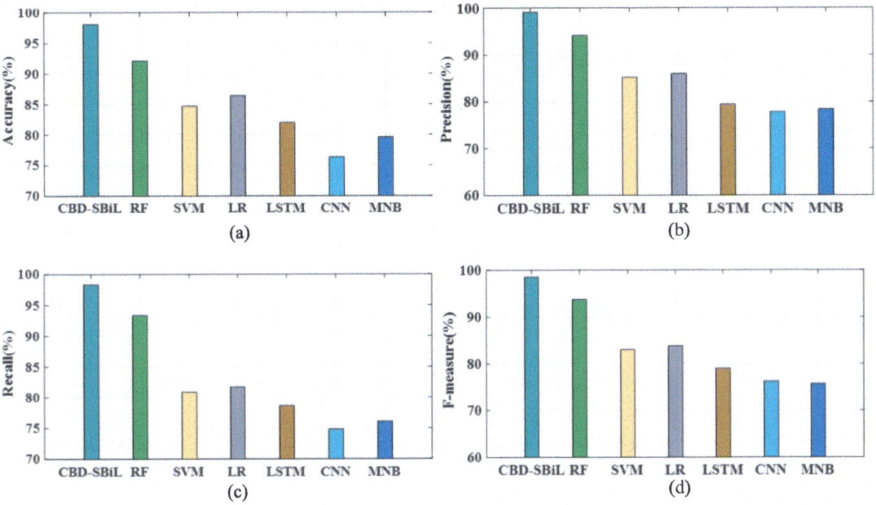

Fig. 3. Performance comparison of (a) Accuracy, (b) Precision, (c) Recall and (d) F-Measure

The proposed method can obtain better performance than other existing methods. The proposed model achieved an Accuracy of 98.09%, 99.06% in precision, 98.36% in Recall and 98.51% in the F-Measure parameter. The proposed model achieved high performance due to using novel approaches for generating RDF statements. Figure 4 shows the ROC analysis of the proposed method.

The ROC curve plotted between the false positive rate (FPR) and the true positive rate (TPR). In comparison to previous current technologies, the proposed method has produced improved ROC results. The proposed method achieved 0.99 AUC score in RDF statement generation process's analysis.

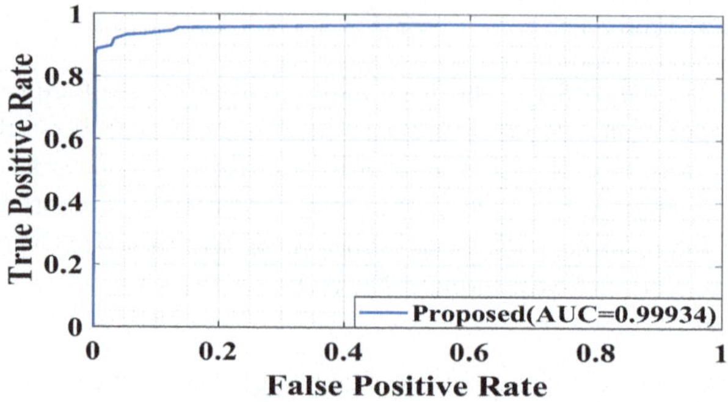

Fig. 4. ROC analysis

4.4 Discussion

The proposed CBD-SBiL RDF statement generation technique is compared with the existing approaches like KG Embedding framework [14], linear-chain CRF [15], OntoBestFit [16], AdaBoost [17], KHO–RDF [18]. Performance analysis of these existing and proposed method is shown in Table 1.

Table 1. Performance analysis

Model	Performance		
	Precision	Recall	Accuracy
KG Embedding framework [20]	0.74	0.83	0.77
linear-chain CRF [21]	0.32	0.30	0.67
OntoBestFit [22]	0.94	0.96	0.95
AdaBoost [23]	0.777	0.775	0.895
KHO–RDF [24]	0.72	4.92	0.77
Proposed	**0.99**	**0.98**	**0.98**

The above table illustrates the comparative examination of proposed technique and various existing technique. From the above table, it is clear that the proposed method attains better performance in terms of 98.09% of accuracy, 99.06% of precision and 98.36% of recall compared with existing techniques. The existing technique shows poor performance when huge number of datasets are trained. But the proposed method attains better outcome due to accurate RDF statement generation.

5 Conclusion

A novel model for RDF statement generation using unstructured textual data is presented in this research. One-hot encoding, z-score normalization, tokenization, stop word removal and POS tagging approaches were used in pre-processing the unstructured text data to enhance noise suppression. The appropriate features are then extracted using BLM. The extracted features are then clustered using ADK-MC to minimize the model complexity. After that, the triple classification is performed using CBD-SBiL model. The performance of the proposed method is compared to existing methods; the result shows that the proposed method achieved 98.09% accuracy, 99.06% precision, 98.36% Recall and 98.51% F-Score in RDF statement generation. In the future, the optimized deep learning method will be devised to optimize the loss function to enhance the statement generation accuracy; more datasets will be considered.

References

1. Duan, Y., Edwards, J.S., Dwivedi, Y.K.: Artificial intelligence for decision making in the era of Big Data–evolution, challenges and research agenda. Int. J. Inf. Manag. **48**, 63–71 (2019)
2. Rezaei, Z., Eslami, B., Amini, M.A., Eslami, M.: Hierarchical three-module method of text classification in web big data. In: 2020 6th International Conference on Web Research (ICWR), pp. 58–65. IEEE (2020)
3. Sun, H., et al.: IDF Diabetes Atlas: global, regional and country-level diabetes prevalence estimates for 2021 and projections for 2045. Diabetes Res. Clin. Pract. **183**, 109119 (2022)
4. Kiselev, B., Yakutenko, V.: An overview of massive open online course platforms: personalization and semantic web technologies and standards. Proc. Comput. Sci. **169**, 373–379 (2020)
5. Orlandi, F., Graux, D., O'Sullivan, D.: Benchmarking RDF metadata representations: reification, singleton property and RDF. In: 2021 IEEE 15th International Conference on Semantic Computing (ICSC), pp. 233–240 (2021)
6. Devi, R., Mehrotra, D., Baazaoui-Zghal, H.: RDF model generation for unstructured dengue patients' clinical and pathological data. Int. J. Inf. Syst. Model. Des. (IJISMD) **10**(4), 71–89 (2019)
7. Hari, A., Kumar, P.: WSD based ontology learning from unstructured text using transformer. Proc. Comput. Sci. **218**, 367–374 (2023)
8. Moussallem, D., Gnaneshwar, D., Castro Ferreira, T., Ngonga Ngomo, A.-C.: NABU – multilingual graph-based neural RDF verbalizer. In: Pan, J.Z., Tamma, V., d'Amato, C., Janowicz, K., Fu, Bo., Polleres, A., Seneviratne, O., Kagal, L. (eds.) ISWC 2020. LNCS, vol. 12506, pp. 420–437. Springer, Cham (2020). https://doi.org/10.1007/978-3-030-62419-4_24
9. Soliman, H.A.S.A., Tabak, F.: Deep learning framework for RDF and knowledge graphs using fuzzy maps to support medical decision. J. Int. Res. Med. Pharm. Sci. **14**(3), 92–97 (2020)
10. Devi, R., Mehrotra, D., Lamine, S.B.A.B., Baazaoui-Zghal, H.: A RDF model-based predictive model on dengue patients' diagnostic PDF reports. Int. J. Electron. Healthc. **13**(2), 113–133 (2023)
11. Devi, R., Mehrotra, D.: Enhancing ontology by integrating facts from unstructured data and mapping with linguistic knowledge. Int. J. Comput. Digit. Syst. **8**(6) (2019)
12. Pham-Hong, B.T., Chokshi, S.: PGSG at SemEval-2020 task 12: BERT-LSTM with tweets' pretrained model and noisy student training method. In: Proceedings of the Fourteenth Workshop on Semantic Evaluation, pp. 2111–2116 (2020)
13. Fränti, P., Sieranoja, S.: How much can k-means be improved by using better initialization and repeats? Pattern Recogn. **93**, 95–112 (2019)
14. Abu-Salih, B., et al.: Relational learning analysis of social politics using knowledge graph embedding. Data Min. Knowl. Disc. **35**(4), 1497–1536 (2021)
15. Cai, W., Liu, B., Wei, Z., Li, M., Kan, J.: TARDB-Net: triple-attention guided residual dense and BiLSTM networks for hyperspectral image classification. Multimed. Tools Appl. **80**, 11291–11312 (2021)

16. Deepak, G., Santhanavijayan, A.: OntoBestFit: a best-fit occurrence estimation strategy for RDF driven faceted semantic search. Comput. Commun. **160**, 284–298 (2020)
17. Cui, Q., Chen, Q., Liu, P., Liu, D., Wen, Z.: Clinical decision support model for tooth extraction therapy derived from electronic dental records. J. Prosthet. Dent. **126**(1), 83–90 (2021)
18. Ponnupandian, A., Veeramani, V.P.: Optimal design for SCAP/battery power management applied in electric vehicle (EV) applications: a KHO–RDF technique. Soft. Comput. **24**, 17247–17263 (2020)

Real-Time Permanent Change Proposals for Abandoned Object Detection

Aafaq Inamdar(✉) and Mukesh A. Zaveri

Department of Computer Science and Engineering, SVNIT, Surat, Surat, Gujarat, India
aafaq.in@gmail.com, mazaveri@coed.svnit.ac.in

Abstract. In this work, we address the significant challenge of detecting permanent changes in video scenes, a foundational element for the advancement of efficient surveillance systems. We introduce a unique two-step algorithm. The first step involves temporal smoothing by utilizing methods such as median operation and Mixture of Gaussians (MOG) to effectively model the background frame, thereby eliminating temporary objects within a dynamic window. The subsequent phase leverages a MOG-based strategy to identify alterations over the derived short-term background frames. Specifically tailored for real-time operations, our approach is adaptable to edge surveillance devices, even those equipped with minimal GPU capabilities. The algorithm adeptly identifies static foregrounds or lasting scene modifications, setting the stage for possible detection of abandoned objects in later processes. Tested on widely recognized datasets, our method demonstrates almost impeccable recall rates. Despite a lesser emphasis on precision, the produced false positives can serve as valuable data in subsequent surveillance stages. Optimized to operate near real-time, our system is compatible with compact embedded GPUs, such as the Nvidia Jetson. This approach offers a pragmatic and robust solution to a growing need in the automated video surveillance sector, aiming to bolster public safety. Notably, our method approaches state-of-the-art outcomes on standard benchmarks while significantly reducing computation time, especially when deployed on suitable modern edge computing devices.

Keywords: Video Scene Change Detection · Temporal Smoothing · Video Surveillance · Background Subtraction · GPU Computing · Mixture of Gaussians (MOG)

1 Introduction

With the widespread installation of Closed-Circuit Television (CCTV) cameras in most public and private spaces, these surveillance systems hold enormous

M. A. Zaveri—Contributing author.

D. Pastor-Escuredo et al. (Eds.): ICDLAIR 2023, LNNS 1001, pp. 131–146, 2024.
https://doi.org/10.1007/978-3-031-60935-0_13

potential as tools for ensuring public safety. Consequently, developing robust and efficient methods for processing and understanding the vast amounts of video data generated by these cameras has become an increasingly critical research topic. One of the distinctive aspects of this problem is the stationary nature of CCTV cameras. These cameras typically capture scenes with static backgrounds alongside dynamic entities such as people, cars, and changes in lighting conditions. This unique blend of static and dynamic elements poses specific challenges that necessitate careful attention. Compounding these complexities is the unpredictability of changes in scenes, such as the appearance of objects that remain in one place for an extended period. Such changes may occur in various shapes, sizes, and colors, which emphasizes the need for a versatile solution that requires minimal tuning for different cameras. The applications of detecting such prolonged changes, which we refer to as "abandoned objects" in this paper, are wide and impactful. To clarify, the term "abandoned object" denotes an item that becomes a constant feature in the scene over a substantial period. It's important to note that our algorithm does not identify whether an abandoned object is indeed an unattended item of concern; instead, it merely proposes potential items, thus generating alerts for further investigation by other systems which classify the alert like [1–6] or human operators. This preemptive detection mechanism can be instrumental in various scenarios. For instance, our algorithm can help detect instances of theft in both public and private spaces. By highlighting prolonged changes in the scene, unauthorized removal or addition of objects can be promptly identified, enabling quick response and potentially aiding in crime prevention. In the context of transportation and logistics, our solution can assist in identifying vehicles that have been stationary for too long or debris on highways, thereby contributing to improved road safety. Moreover, in crowded public areas such as airports, railway stations, and shopping malls, our system could alert security personnel to investigate objects that have remained stationary for a substantial time, thus providing a valuable tool for proactive security measures.

Existing deep learning models, despite their powerful capabilities, are not entirely suitable for this task. These models are often seen as black-boxes due to their complex internal structure and operations, leading to a lack of transparency in their decision-making processes. Furthermore, these models heavily rely on the availability of well-labelled data, which can be a significant challenge in real-world deployments where such data is often scarce or nonexistent. Another crucial challenge lies in the hardware limitations, especially considering the vast amount of data contained in high-resolution video frames. Scaling up these models to handle larger images, as suggested in papers like Google's EfficientDet [7], requires considerable computational resources, which poses challenges for deployments on edge devices, such as mobile and IoT devices. In response to these challenges, we propose an algorithm that primarily relies on non-deep learning methods. This approach significantly reduces the need for training data, making our algorithm more practical for real-world applications where collecting and annotating vast amounts of such data is often prohibitive.

Furthermore, by avoiding the use of deep learning models, we sidestep the issues associated with their lack of interpretability and the computational constraints of deploying them on edge devices.

Our proposed solution capitalizes on the rapid parallel processing capabilities of GPUs, employing Single Instruction, Multiple Data (SIMD) operations using CUDA to enhance its performance. This innovation distinguishes our algorithm from existing methods, as it offers a faster and more efficient solution for detecting abandoned objects, specifically designed to leverage the power of edge devices such as the Jetson Xavier AGX. We have tested our algorithm on several widely-used datasets, mainly PETS 2006 [8], AVSS2007iLIDS [9], CAVIAR [10], VISOR [11], and ABODA [12], to evaluate its accuracy. Additionally, we compared our algorithm's runtime performance against popular Stationary Foreground Detection (SFD) algorithms on the Jetson Xavier AGX edge device. Our solution demonstrated superior performance in these evaluations, further underscoring its practical applicability. The following sections of this paper provide a detailed overview of the related literature, our proposed method, and an extensive evaluation of its performance. The limitations of our current approach and avenues for future work are also discussed.

The focus of this work is to leverage the stationary nature of CCTV cameras and the static-dynamic interplay in their captured scenes. By harnessing these characteristic features, we aim to develop a robust and efficient solution for detecting abandoned objects in video scenes. In conclusion, our non-deep learning-based algorithm presents a practical solution to an increasingly critical problem in video surveillance. Its effective performance across multiple datasets and edge devices suggests a promising step forward in automated video surveillance. Through our work, we hope to stimulate further research into developing efficient and interpretable algorithms for understanding complex video scenes, thereby contributing to public safety.

2 Related Work

In the literature, the challenge of detecting abandoned objects or permanent scene changes is canonically approached through a multistage pipeline, as thoroughly outlined in [13]. This pipeline can be broadly segmented into the following stages: Foreground Segmentation, Stationary Foreground Detection, Candidate Generation, Candidate Validation, and finally, Abandoned Object Detection. This delineation provides a systematic framework for applying various specialized algorithms at each stage, culminating in an effective overall solution. Foreground Segmentation and Stationary Foreground Detection often rely on background subtraction methods. Following this initial stage, subsequent stages classify detected objects based on their relevance [Classification papers][2]. For background subtraction, the bgslibrary [14] provides a wealth of well-implemented algorithms. Our comparative analysis leveraged these algorithms, conducting comprehensive speed tests across our test-bench devices. Although we limited our evaluation to computational performance due to the unavailability of tuning parameters and specific dataset configurations, the speed test provides useful insights into the computational aspects of these algorithms. In the realm of Abandoned Object Detection, we utilized the results presented in [13] for comparison. There is, however, a noticeable gap in the literature when it comes to comprehensive evaluations that cover both accuracy across all datasets and computational efficiency. This lack of data is an area for future work and could significantly contribute to the field's advancement.

For an exhaustive and rigorous evaluation of our proposed method, we employed several standard benchmark datasets widely acknowledged in the field of video surveillance. These datasets are publicly available, and their widespread use in the community ensures a comprehensive comparison of our algorithm's performance with the state of the art. PETS 2006: The PETS 2006 dataset, provided by the International Workshop on Performance Evaluation of Tracking and Surveillance (PETS), is an extensive and diverse dataset focusing on multi-object tracking and surveillance, including scenarios with varying crowd densities, environmental conditions, and object interactions. AVSS 2007: The Advanced Video and Signal-based Surveillance (AVSS) 2007 dataset, created for the IEEE International Conference on Advanced Video and Signal-Based Surveillance, offers high-resolution videos encompassing various surveillance scenarios with different levels of complexity. It is often used to assess the robustness of algorithms under various challenging conditions. CAVIAR: The Context Aware Vision using Image-based Active Recognition (CAVIAR) dataset, provided by the EC Funded CAVIAR project, contains videos of people walking in an indoor environment with different viewpoints, including wide-angle scenes. This dataset is particularly suitable for evaluating the performance of surveillance algorithms in complex indoor environments. VISOR: The Video Surveillance Online Repository (VISOR) is a comprehensive and publicly accessible collection of video sequences specifically tailored for the evaluation of surveillance systems. The VISOR dataset provides a broad spectrum of real-world scenarios including varying light conditions, occlusions, and different types of activities. ABODA: The Abandoned Object Dataset (ABODA) is a specialized dataset for testing algorithms designed to detect abandoned objects. It contains a variety of scenes, both indoor and outdoor, with varying levels of pedestrian traffic and environmental conditions.

The Nvidia's Jetson series, was our platform of choice due to its widespread adoption in numerous real-world projects. This device strikes a remarkable balance between compactness, cost-effectiveness, and computational power, being equipped with an impressive built-in GPU tailored for the exigencies of real-time video processing. Its unified memory architecture, streamlining memory transfers between the CPU and GPU, further underscores its operational efficiency. Selecting this device as our testbed not only mirrors contemporary real-world applications but also ensures that the evaluated algorithms can be feasibly deployed to address practical challenges, demonstrating the tangible utility of our research.

3 Methodology

Applying a median operation on a sequence of frames captured by a static camera fundamentally aids in constructing a more accurate representation of the scene's unchanging background. This operation sorts the pixel intensity values at each location across the sequence and selects the middle value. Given enough frames, this process effectively filters out transient elements, such as moving objects or lighting changes, that do not consistently occupy a particular position. The resulting output is a robust estimation of the scene's static background, devoid of temporary changes, thus setting the stage for effective foreground detection.

Let's say we have a sequence of N frames, where each frame has a pixel intensity value X at a specific location (i, j) at time t.

$X_{i,j}(t) : i = 1, 2, ..., M; j = 1, 2, ..., N; t = 1, 2, ..., T$

Here, $M \times N$ is the resolution of the frames and T is the total number of frames in the sequence.

Let us denote the Short Term Background Belief as STGB and represent it using the Median operation. We use the placeholder variable $STGB_{i,j}$ to denote the background model at each pixel location (i, j):

$STGB_{i,j} = \text{Median}[X_{i,j}(t)]$ for $t = 1$ to T

Here, $STGB_{i,j}$ represents the background model at pixel location (i, j). The Median operation sorts the pixel intensity values $\{X_{i,j}(t)\}$ over time and selects the middle value.

The Mixture of Gaussians (MOG2) algorithm [15], as implemented in the OpenCV library [16], is an effective technique for foreground object detection, particularly useful in our context of identifying moving changes in a sequence of images.

MOG2 models each pixel as a mixture of K Gaussian distributions, where K is a predetermined number. The weights of the mixture represent the time proportions that those colors stay in the scene. The probable background colors are the ones which stay longer and more static.

The mathematical model for the MOG2 algorithm can be formulated as:

$$P(X_t) = \sum_{i=1}^{K} \omega_{t,i} \cdot \eta(X_t, \mu_{t,i}, \Sigma_{t,i})$$

Here, X_t is the pixel at time t, and the parameters of the model are:

- $\omega_{t,i}$: The weight of the i-th Gaussian in the mixture at time t, indicating the proportion of time that a particular color stays in the scene.
- $\mu_{t,i}$: The mean of the i-th Gaussian. This represents the color of the pixel.
- $\Sigma_{t,i}$: The covariance matrix of the i-th Gaussian, reflecting the variance of the color distribution.
- $\eta(., ., .)$: The Gaussian probability density function.

MOG2 iteratively updates these parameters and evaluates the likelihood of each pixel belonging to each Gaussian distribution, determining if a pixel belongs to the background or foreground.

In the context of permanent change detection, MOG2 serves as a foundational step, enabling the differentiation between background (stationary objects and the scene itself) and foreground (moving entities). By separating these, our algorithm can effectively detect the presence of new stationary objects (i.e., abandoned objects), which were once considered part of the foreground, and then became part of the static scene.

Our proposed method operates on the principle of constructing two distinct models of a video scene - a short-term model and a long-term model. These models, each capturing changes over different time spans, facilitate a more nuanced

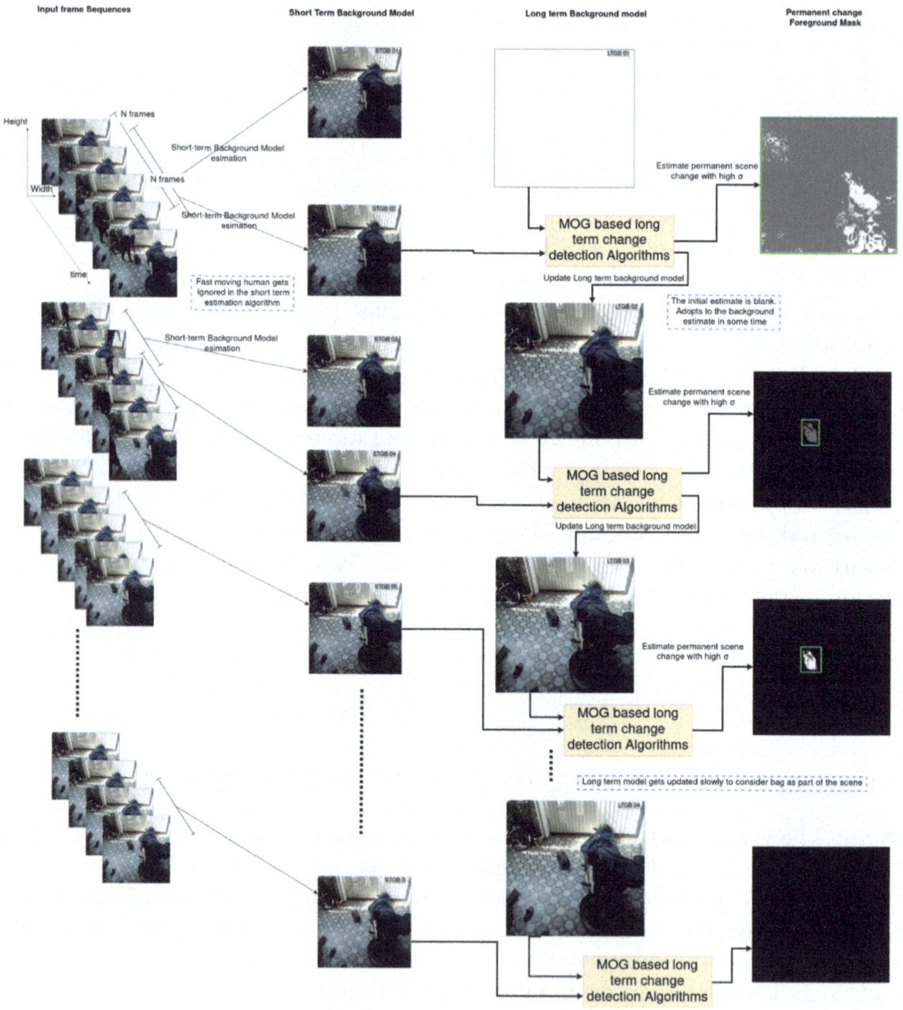

Fig. 1. Comprehensive overview of the proposed algorithm.

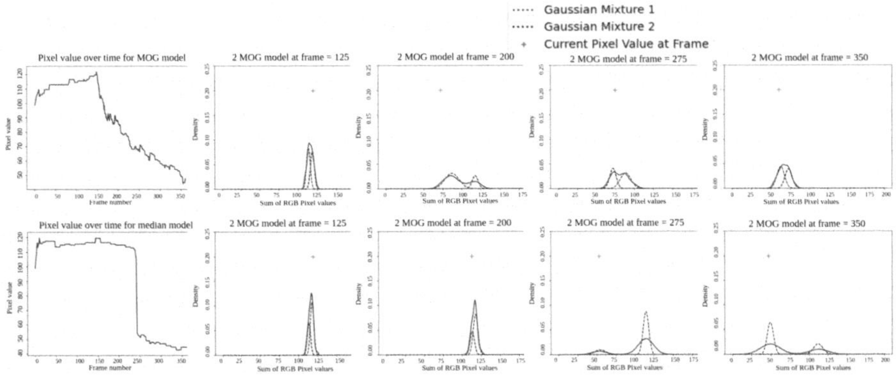

Fig. 2. Comparison of MOG based short term update versus median based.

understanding of the scene dynamics, enhancing our ability to discern and detect permanent changes.

The short-term model operates on a buffer queue of video frames. After every predetermined interval (denoted by *short_term_update*), a frame from the video is loaded into this buffer queue, maintaining a history of *median_history* number of frames. The queue follows a First-In-First-Out (FIFO) mechanism, thereby ensuring that the most recent *median_history* frames are always present in the queue. The short-term model, at any given point, is defined as the median frame across this queue along the time axis.

In parallel, the long-term model is maintained using the MOG2 algorithm. This model undergoes an update after every *long_term_update* seconds. However, during each update step, the latest frame from the short-term model (i.e., the median frame from the buffer queue) is taken into account. This mechanism ensures that the long-term model integrates the steady changes occurring in the scene, rather than the momentary or transient fluctuations.

The core of our permanent change detection lies in the comparison of the short-term and long-term models. Significant deviations between these models indicate substantial changes in the short-term scene, which, in turn, are flagged as potential permanent changes. Alongside, this comparison also serves to update the long-term model, ensuring its continual alignment with the evolving scene dynamics.

The frequency of updates for the short-term model is set to be higher than that of the long-term model ($short\_term\_update < long\_term\_update$). In our evaluations across various datasets, we found that $short\_term\_update = 2$, $long\_term\_update = 3$, and $median\_history = 20$ provided optimal results in most settings. Through this innovative approach, we effectively leverage the temporal dimensions of video data, exploiting the interplay of short-term and long-term dynamics to detect and highlight permanent changes within the scenes.

The rationale behind our choice of a median-based short-term model finds its roots in a comparative analysis demonstrated in Fig. 2. A Gaussian-based model was initially considered; however, it was observed that such models are slower to adapt to scene changes, or alterations in pixel values. Consequently, the resultant long-term model, given that it employs the short-term model for its updates, tends to exhibit slow adjustments to its scene estimation in response to gradual changes. Contrarily, the median-based short-term model demonstrates a rapid response to scene changes, especially when an object has been present in the scene for a period exceeding half of the *median_history* length. This sudden and drastic alteration in pixel values is easily captured by the long-term Gaussian model, enhancing its ability to recognize and adapt to enduring scene changes. To visualize this phenomenon, we refer to the diagram 2. The top row, representing the Gaussian-based short-term model, exhibits a slower rate of change in comparison to the bottom row, which depicts the Median-based model. The images on the leftmost side of both rows show the pixel values of the respective short-term models. Notice the sharp decline in the pixel values in the Median row. Adjacent to these images are plots illustrating the adaptation of a MOG-based model, assuming two Gaussian values (for simplicity), to the changes informed by the respective short-term models. Here, the green '+' signifies the current pixel value obtained from the short-term model (median or Gaussian). As we move from left to right, the plots portray the evolution of the MOG model's Gaussian estimates as the green '+' moves. The green '+' shifts slowly in the case of the Gaussian-based short-term model, but undergoes a single, drastic movement for the Median-based model. The red and blue Gaussians represent the background and foreground, respectively, with the background having the highest confidence. The farther the green '+' from the background Gaussian, the more probable it is to signify a foreground pixel. Therefore, when the long-term model is based on the MOG and the short-term model on the median operation, the marked difference in pixel values readily highlights the higher standard deviation, thereby facilitating the easier detection of permanent scene changes. This analysis underscores the effectiveness of the median operation in capturing substantial and rapid shifts in pixel values, providing a robust foundation for the proposed algorithm.

Figure 1 provides a comprehensive architectural overview of our proposed algorithm. The depicted scenario involves a human subject entering the scene, depositing a bag, and subsequently departing. The initial column represents the sequence of input frames, from which the second column – the Short-Term Background Model – is derived. This model is generated via a moving window approach, taking the median of the current frame buffer window. The buffer, designed as a queue, retains only the most recent frames up to a fixed size. Furthermore, we introduce a skipping mechanism here, whereby a single frame is sampled and added to the buffer for every set interval of *short_term_update* seconds. In the diagram the time is flowing in the positive direction from top to bottom. The third column portrays the Long-Term Model estimation. Like its short-term counterpart, this model incorporates a skipping mechanism, though it operates differently. It considers the backgrounds supplied by the short-term model only after each interval of *long_term_update* seconds. This column also enlightens the deviation of the current frame from the model's background estimation, with the white pixels indicating the foreground as determined by the MOG2 model. This indicates long-term scene changes after all steps are executed. Additionally, during this phase, the current frame is recursively fed back into the process to update the estimation of the long-term background. Finally, the algorithm incorporates several morphological operations and contour identification processes to enhance the estimation further. These mechanisms help in eliminating smaller, noise-induced pixel groups, retaining only those objects that meet specific height criteria. This carefully crafted architectural design, therefore, offers a robust solution to detect permanent scene changes efficiently and accurately.

4 Experiment Results

4.1 Accuracy Results

PETS 2006: Our method was evaluated using the PETS 2006 dataset, a collection of seven sequences varying in scene complexity. For the S1-T1-C sequence, our approach correctly detected the scene change involving a lone individual leaving an unattended bag (TP = 1), with no false positives or negatives. In the S2-T3-C sequence, the model correctly identified a scene change (TP = 1), but also flagged the movements of a garbage collection vehicle as scene changes, yielding four false positives (FP = 4). The S3-T7-A sequence, involving a brief placement and retrieval of a bag, did not trigger any detection, thus corroborating the algorithm's efficacy. Sequences S4-T5-A, S5-T1-G, S6-T3-H, and S7-T6-B, with unattended objects, were all correctly detected (TP = 1 for each sequence), with no false positives or negatives. The overall results of these sequences rendered a precision of 0.6 and a perfect recall of 1, yielding an F1 score of 0.75, thus demonstrating the robust potential of our approach. However, improvements are needed for complex scenes involving similar foreground and background movements.

AVSS2007: The performance of our algorithm was further validated using the AVSS2007 dataset, which comprises three sequences of varying complexity. In the AVSSS07_EASY sequence, the model achieved a true positive rate of 1, indicating robust detection capabilities in simpler scenarios. For the AVSSS07_MEDIUM sequence, despite increased complexity, our model maintained a true positive rate of 1 by adjusting the alert time. However, the AVSSS07_HARD sequence introduced increased complexity and occlusions, yielding two false positives due to the similarities in the color distribution between the humans and the object. This resulted in a precision drop to approximately 0.33, culminating in an F1 score of 0.5. These findings indicate the model's proficiency across various complexities while underscoring the need for further improvements in more challenging scenarios.

CAVIAR: Our model was also evaluated using the CAVIAR dataset, where it correctly detected abandoned objects in four out of five scenarios (TP = 1 for each successful detection), yielding a perfect precision of 1 and a recall of 0.8. However, the algorithm failed to detect an obscured object behind a chair, indicating some limitations when objects are not clearly visible.

VISOR: The algorithm exhibited exceptional performance on the VISOR dataset, identifying significant scene changes with perfect precision, recall, and an F1 score of 1. These results suggest our algorithm's outstanding performance as a first-stage alert system for detecting stopped vehicles and abandoned objects.

ABODA: Finally, the performance was tested on the ABODA dataset, with indoor and outdoor videos of abandoned objects. The algorithm demonstrated excellent performance in most scenarios, correctly detecting the abandoned objects with no false detections. However, some issues arose due to dynamic lighting conditions, resulting in multiple false positives. The algorithm also struggled with a chaotic scene, yielding multiple false positives and one false negative. In total, the algorithm achieved a precision of approximately 0.26, a recall of approximately 0.83, and an F1 score of 0.4 on this dataset. These results highlight the model's potential while emphasizing the need for further improvements in dynamic and challenging conditions. Table 1 briefly presents the results of the results.

Table 1. Performance Metrics of Proposed Algorithm for Popular Datasets

Dataset Name	TP	Precision	F1-Score
PETS 2006	6	0.6	0.75
AVSS2007iLIDS	3	0.6	0.75
CAVIAR	4	1	0.89
VISOR	15	1	1
ABODA	10	0.26	0.4

4.2 Accuracy Comparison

For a comprehensive and balanced evaluation, we referenced two influential surveys that have disclosed results on similar datasets. Survey [17] meticulously explores the datasets PETS2006 and AVSS2007. Notably, the survey [13] also utilized a custom dataset that aligns closely with our own, with both being an amalgamation of several widely recognized datasets. When interpreting our results, it is essential to note a key distinction. The results presented in surveys like [13] also encompass multiple facets of the abandoned object detection pipeline, which can help eliminate false positives and perform classifications. Our study, however, focuses primarily on the detection of permanent change, a difference that should be considered when drawing comparisons. Despite this, given the absence of complex false positive examples in our videos, we assert that our results offer valid insights into the performance of the full system pipeline on these datasets. The survey [13] presented a range of precision (P), recall (R), and F1-score results based on the KNN and PAWCS algorithms, which we used as our benchmark. Survey [17] disclosed results for Approaches 1–5, which are included in our comparative analysis in Table 2. The table includes he P, R, F1 score of proposed algorithm on the same dataset as well. We observed a spectrum of P, R, and F1 scores with different pipeline configurations, including MOG2, KNN, and PAWCS. These results were obtained by tuning the algorithms to prolong object detection. The metrics include Stationary Foreground Detection (SFD) using a foreground accumulation approach (ACC), Pedestrian Detection (PD) using Aggregated Channel Feature (ACF), and Abandoned Discrimination (AD) using High Gradients (HG). The MOG2-based pipeline demonstrated Precision of 0.38, Recall of 0.61, and F1-score of 0.47. In contrast, the KNN-based pipeline exhibited Precision of 0.75, Recall of 0.67, and F1-score of 0.71. Meanwhile, the PAWCS-based pipeline yielded Precision of 0.81, Recall of 0.72, and F1-score of 0.76. After less than optimal performance from the MOG2-based algorithms, the authors of the surveys focused on the KNN and PAWCS for the Foreground segmentation component. Consequently, KNN and PAWCS offered a range of P, R, and F1 scores with different algorithms used on the later stages after foreground segmentation the following was the observed range. The range for KNN was a minimum of (P 0.61, R 0.61, F1 0.61) and a maximum of (P 0.86, R 0.67, F1 0.75). The range for PAWCS was a minimum of (P 0.81, R 0.72, F1 0.76) and a maximum of (P 0.93, R 0.72, F1 0.81). Our dataset, although not identical, is closely aligned with the ones used in the survey [13]. Our average precision was 0.69, recall was 0.91, and F1 score was 0.75. These results affirm the effectiveness of our approach when compared with the benchmarks in the existing literature.

Table 2. Comparison of Approaches on PETS2006 and AVSS2007 Sequences

Approach	PETS2006 Sequences		AVSS2007 Sequences	
	P	R	P	R
Approach 1 (Stauffer and Grimson) [18]	0.05	1	0.01	1
Approach 2 (Guler and Farrow) [19]	0.6	1	0.1	1
Approach 3 (Mathew et al.) [20]	0.5	1	0.03	1
Approach 4 (Liao et al.) [21]	0.75	1	0.33	1
Approach 5 (Porikli et al.) [22]	0.37	1	0.05	1
Our proposed method	**0.6**	**1**	**0.6**	**1**

4.3 Runtime Speed Analysis

Our proposed method is purposefully architected for real-time operation on contemporary edge devices. To substantiate this claim, we conducted a comparative analysis of several foreground segmentation algorithms. These algorithms were evaluated on a 1449 frame video using the leading-edge Jetson Xavier AGX and all the implementation of the popular Background subtraction were adopted from the implementations given by [14]. The Jetson AGX, with its unified memory architecture for GPU memory and RAM, serves as an ideal testing environment due to its efficient handling of computational processes, particularly crucial as our algorithm necessitates CPU memory and GPU memory transfers.

Table 3 delineates the time each algorithm required to process the entire video. Of noteworthy mention is our proposed method, which clocked a mere 1.661 s, markedly outperforming all other examined algorithms.

However, a crucial distinction must be made in evaluating our algorithm. Unlike traditional methods that operate on each frame, our algorithm, customizable, executes between 10 to 30 times per second. Importantly, the previously disclosed precision-recall values are achieved with the same tuned parameters.

Our method fuses the tasks of foreground segmentation and stationary foreground segmentation, harnessing the computational prowess of the GPU via SIMD operations with CUDA. Conventional approaches in the abandoned object surveillance literature generally involve four steps: Stationary Foreground Detection (SFD), Pedestrian Detection (PD), and Abandoned Discrimination (AD). Table 3 distills the results of the foreground segmentation step executed by various popular methods.

The exceptional performance of our method can be attributed to its GPU-centric design, coupled with the device's unified memory architecture efficiency. This design circumvents the time-intensive transfers between GPU memory and RAM, frequently experienced at different processing stages.

These results underscore our method's accelerated performance on modern devices, underlining its prospective application in the surveillance of abandoned objects. The combination of precision, recall, and processing speed propels our proposed algorithm as a practical solution for real-world challenges.

Table 3. Time to Process One Frame for Various Algorithms on Jetson Xavier AGX in same environment we setup for benchmarking all algorithms

Algorithm Name	Time (seconds)
FrameDifference	7.54
StaticFrameDifference	5.74
WeightedMovingMean	32.07
WeightedMovingVariance	79.48
AdaptiveBackgroundLearning	26.81
AdaptiveSelectiveBackgroundLearning	2.84
MixtureOfGaussianV2	14.22
PixelBasedAdaptiveSegmenter [23]	80.41
SigmaDelta [24]	12.44
PAWCS [25]	787.58
ViBe [26]	4.57
CodeBook [27]	29.36
FuzzyChoquetIntegral [28]	170.71
LBSimpleGaussian [29]	11.54
LBFuzzyGaussian [30]	20.07
LBMixtureOfGaussians [31]	30.21
LBAdaptiveSOM [32]	54.75
VuMeter [33]	11.13
KDE [34]	90.90
KNN [35]	11.61
Our proposed method	**1.66**

5 Conclusion

In this research, we presented a novel approach for abandoned object detection. The comprehensive evaluation of our proposed method, using several well-known datasets, underscores its robustness and high efficiency in detecting scene changes indicative of unattended objects. Notably, our approach demonstrated commendable precision and recall values across varied complexities of video sequences.

One of the standout features of our method is its swift runtime speed, specifically engineered for deployment on modern edge devices. The utilization of GPU-centric design, harmonized with the device's unified memory architecture, resulted in a method that markedly outperformed a plethora of established algorithms in the realm of foreground segmentation.

Yet, as with any research, our method is not without its limitations. In scenarios replete with similar foreground and background movements or dynamic lighting conditions, the algorithm did encounter challenges. These areas of improvement are not setbacks but rather opportunities, leading us to anticipate avenues of further research.

6 Future Work

Enhanced Scene Analysis: One potential improvement lies in implementing a more profound scene analysis. Leveraging deep learning techniques, particularly Convolutional Neural Networks (CNNs), can help the model discern intricate details and distinguish between similar foreground and background movements more effectively.

Adaptive Lighting Compensation: To tackle the challenges arising from dynamic lighting conditions, future iterations can explore the integration of adaptive lighting compensation mechanisms. This would further bolster the algorithm's performance in both indoor and outdoor environments, irrespective of the prevailing lighting conditions.

Object Occlusion Management: Addressing the limitation encountered in the CAVIAR dataset, where the algorithm failed to detect an obscured object, we envisage an approach that can better handle object occlusions. Advanced object tracking and re-identification techniques can be employed to keep track of partially hidden or occluded objects.

Real-Time Feedback Loops: To enhance real-time responsiveness, the integration of a feedback loop that can dynamically adjust algorithm parameters based on the scene's complexity would be invaluable. This would enable the system to maintain high accuracy even in rapidly changing environments.

Expansion to Other Use-Cases: While our primary focus was on abandoned object detection, the foundational principles of our method can be tailored to other surveillance-related tasks, such as intrusion detection or anomalous behavior recognition. Future endeavors can explore these avenues, broadening the application scope of our proposed methodology.

In summary, the journey embarked upon in this research has laid a robust groundwork for advanced surveillance systems. With the advancements projected in future iterations, we are optimistic about presenting a holistic solution that addresses the multifaceted challenges inherent in real-world surveillance scenarios.

References

1. Nam, Y.: Real-time abandoned and stolen object detection based on spatiotemporal features in crowded scenes. Multimedia Tools Appl. **75**(12), 7003–7028 (2016)
2. Jo, K.-H., et al.: Cumulative dual foreground differences for illegally parked vehicles detection. IEEE Trans. Ind. Inf. **13**(5), 2464–2473 (2017)

3. Lin, C.-Y., Muchtar, K., Yeh, C.-H.: Robust techniques for abandoned and removed object detection based on Markov random field. J. Vis. Commun. Image Represent. **39**, 181–195 (2016)
4. Ferrando, S., Gera, G., Regazzoni, C.: Classification of unattended and stolen objects in video-surveillance system. In: 2006 IEEE International Conference on Video and Signal Based Surveillance, p. 21. IEEE (2006)
5. Ilias, D., El Mezouar, M.C., Taleb, N., Elbahri, M.: An edge-based method for effective abandoned luggage detection in complex surveillance videos. Comput. Vis. Image Underst. **158**, 141–151 (2017)
6. Lin, K., Chen, S.-C., Chen, C.-S., Lin, D.-T., Hung, Y.-P.: Abandoned object detection via temporal consistency modeling and back-tracing verification for visual surveillance. IEEE Trans. Inf. Forensics Secur. **10**(7), 1359–1370 (2015)
7. Tan, M., Pang, R., Le, Q.V.: EfficientDet: scalable and efficient object detection. CoRR abs/1911.09070 (2019). arXiv:1911.09070
8. PETS 2006 Benchmark Data (2006). http://www.cvg.reading.ac.uk/PETS2006/data.html
9. AVSS2007 i-LIDS Dataset (2007). http://www.elec.qmul.ac.uk/staffinfo/andrea/avss2007_d.html
10. CAVIAR Test Case Scenarios (2005). http://homepages.inf.ed.ac.uk/rbf/CAVIARDATA1/
11. VISOR Dataset. https://aimagelab.ing.unimore.it/visor/video_videosInCategory.asp?idcategory=12
12. Lin, K.: ABODA: abandoned object detection algorithm. https://github.com/kevinlin311tw/ABODA
13. Luna, E., San Miguel, J.C., Ortego, D., Martínez, J.M.: Abandoned object detection in video-surveillance: survey and comparison. Sensors **18**(12), 4290 (2018)
14. Sobral, A.: BGSLibrary: an OpenCV C++ background subtraction library. In: IX Workshop de Visão Computacional (WVC'2013), Rio de Janeiro, Brazil (2013). https://github.com/andrewssobral/bgslibrary
15. Zivkovic, Z.: Improved adaptive Gaussian mixture model for background subtraction. In: Proceedings of the 17th International Conference on Pattern Recognition 2004, ICPR 2004, vol. 2, pp. 28–31. IEEE (2004)
16. Bradski, G.: The OpenCV library. Dr. Dobb's J. Softw. Tools **25**(11), 120–123 (2000)
17. Bayona, Á., SanMiguel, J.C., Martínez, J.M.: Comparative evaluation of stationary foreground object detection algorithms based on background subtraction techniques. In: 2009 Sixth IEEE International Conference on Advanced Video and Signal Based Surveillance, pp. 25–30. IEEE (2009)
18. Stauffer, C., Grimson, W.E.L.: Adaptive background mixture models for real-time tracking. In: Proceedings of the 1999 IEEE Computer Society Conference on Computer Vision and Pattern Recognition (Cat. No. PR00149), vol. 2, pp. 246–252. IEEE (1999)
19. Guler, S., Farrow, M.K.: Abandoned object detection in crowded places. In: Proceedings of PETS, pp. 18–23. Citeseer (2006)
20. Mathew, R., Yu, Z., Zhang, J.: Detecting new stable objects in surveillance video. In: 2005 IEEE 7th Workshop on Multimedia Signal Processing, pp. 1–4. IEEE (2005)
21. Liao, H.-H., Chang, J.-Y., Chen, L.-G.: A localized approach to abandoned luggage detection with foreground-mask sampling. In: 2008 IEEE Fifth International Conference on Advanced Video and Signal Based Surveillance, pp. 132–139. IEEE (2008)

22. Porikli, F., Ivanov, Y., Haga, T.: Robust abandoned object detection using dual foregrounds. EURASIP J. Adv. Sig. Process. **2008**, 1–11 (2007)
23. Hofmann, M., Tiefenbacher, P., Rigoll, G.: Background segmentation with feedback: the pixel-based adaptive segmenter. In: 2012 IEEE Computer Society Conference on Computer Vision and Pattern Recognition Workshops, pp. 38–43 (2012). https://doi.org/10.1109/CVPRW.2012.6238925
24. Manzanera, A., Richefeu, J.C.: A new motion detection algorithm based on - background estimation. Pattern Recogn. Lett. **28**(3), 320–328 (2007). https://doi.org/10.1016/j.patrec.2006.04.007. Advances in Visual information Processing
25. St-Charles, P.-L., Bilodeau, G.-A., Bergevin, R.: Universal background subtraction using word consensus models. IEEE Trans. Image Process. **25**(10), 4768–4781 (2016). https://doi.org/10.1109/TIP.2016.2598691
26. Barnich, O., Van Droogenbroeck, M.: ViBe: a universal background subtraction algorithm for video sequences. IEEE Trans. Image Process. **20**(6), 1709–1724 (2011). https://doi.org/10.1109/TIP.2010.2101613
27. Kim, K., Chalidabhongse, T.H., Harwood, D., Davis, L.: Background modeling and subtraction by codebook construction. In: 2004 International Conference on Image Processing 2004, ICIP 2004, vol. 5, pp. 3061–30645 (2004). https://doi.org/10.1109/ICIP.2004.1421759
28. El Baf, F., Bouwmans, T., Vachon, B.: A fuzzy approach for background subtraction. In: 2008 15th IEEE International Conference on Image Processing, pp. 2648–2651. IEEE (2008)
29. Benezeth, Y., Jodoin, P.M., Emile, B., Laurent, H., Rosenberger, C.: Review and evaluation of commonly-implemented background subtraction algorithms. In: 2008 19th International Conference on Pattern Recognition, pp. 1–4 (2008). https://doi.org/10.1109/ICPR.2008.4760998
30. Wren, C.R., Azarbayejani, A., Darrell, T., Pentland, A.P.: Pfinder: real-time tracking of the human body. IEEE Trans. Pattern Anal. Mach. Intell. **19**(7), 780–785 (1997). https://doi.org/10.1109/34.598236
31. Bouwmans, T., El Baf, F., Vachon, B.: Background modeling using mixture of Gaussians for foreground detection - a survey. Recent Patents Comput. Sci. **1**(3), 219–237 (2008)
32. Maddalena, L., Petrosino, A.: A self-organizing approach to background subtraction for visual surveillance applications. Trans. Image Process. **17**(7), 1168–1177 (2008). https://doi.org/10.1109/TIP.2008.924285
33. Goyat, Y., Chateau, T., Malaterre, L., Trassoudaine, L.: Vehicle trajectories evaluation by static video sensors. In: 2006 IEEE Intelligent Transportation Systems Conference, pp. 864–869 (2006). https://doi.org/10.1109/ITSC.2006.1706852
34. Elgammal, A., Harwood, D., Davis, L.: Non-parametric model for background subtraction. In: Vernon, D. (ed.) ECCV 2000. LNCS, vol. 1843, pp. 751–767. Springer, Heidelberg (2000). https://doi.org/10.1007/3-540-45053-X_48
35. Zivkovic, Z., van der Heijden, F.: Efficient adaptive density estimation per image pixel for the task of background subtraction. Pattern Recogn. Lett. **27**(7), 773–780 (2006). https://doi.org/10.1016/j.patrec.2005.11.005

An Excursion to Ontology-Based Non-functional Requirements Specification

Krupa Patel[1] and Unnati Shah[2(✉)]

[1] Bhagwan Mahavir University, Surat 395007, India
[2] Utica University, Utica, NY 13495, USA
unshah@utica.edu

Abstract. It is challenging to ask stakeholders to explicitly provide Non-Functional Requirements (NFRs) for the software due to the domain-specific and context-sensitive nature. A formal and unambiguous specification of NFRs is provided by creating an NFRs ontology based on domain knowledge and quality models. Ontology concepts make NFRs specification definite, complete, consistent, and convenient to share and reuse. In this paper, we intend to examine the approaches for specifying NFRs. The purpose of this study is to disseminate, analyze, and deconstruct the published research in the field, and select metrics for a comparative assessment.

Keywords: Non-Functional Requirements · Quality Model · Ontology

1 Introduction

Functional Requirements (FRs) and Non-Functional Requirements (NFRs) are essential to the success of any software [1]. It is easy to identify FRs from the Requirements Specification Document (RSD). While capturing NFRs in RSD is difficult due to hidden and vague nature. The NFRs from the RSD are often manually identified by users and requirements engineers utilizing their experience and skills [2]. However, as NFRs are constantly associated with certain domains and influenced by context, it is challenging to request users to submit NFRs directly. Additionally, it is challenging to adapt to the environment's changing requirements and to explain them uniformly and consistently. Therefore, the requirements engineer faces a hurdle when trying to specify NFRs without having a thorough understanding of them. Specification of NFRs and its common interpretation among different stakeholders are promising issues, among others. It is difficult to state NFRs explicitly due to lack of domain knowledge. The domain knowledge helps requirements analyst to ask relevant questions to specify all possible NFRs at early stage by means of ontology [3, 4].

Ontology is a formal description of the concept of sharing, stressing the link between real entities [3]. Because of interoperability and machine reasoning, ontologies have become more widely used in computing in recent years. An ontology is a taxonomy of domain terms. The term ontology implies the modeling of domain rules, which provide an extra level of machine understanding. A formal, explicit specification of a shared

D. Pastor-Escuredo et al. (Eds.): ICDLAIR 2023, LNNS 1001, pp. 147–157, 2024.
https://doi.org/10.1007/978-3-031-60935-0_14

conceptualization is provided by the ontology's construction, which is based on domain knowledge and quality models [4]. It assists domain users in properly expressing their requirements, and it aids requirements analysts in accurately understanding and modeling the requirements. Additionally, RSD is definite, comprehensive, consistent, and easy to share and reuse because of the ontology concepts. There are several ontology-based NFRs specification methodologies in the literature [5–17]. The ontology can help developers have a shared understanding of NFRs and can serve as the foundation for NFRs standard.

2 Non-functional Requirements

"The hardest part of building a software system is deciding precisely what to build. No other part of the conceptual work is as difficult as establishing the detailed technical requirements, including all of the interfaces to people, to machines, and to other software systems. No other part of the work so cripples the resulting system if done wrong. No other part is more difficult to rectify later". [18].

Eliciting, analyzing, and writing precise RSD are the most difficult parts of software engineering. There are two types of requirements viz. FRs and NFRs. FRs depend on end users, system actions and business needs. Contrary, NFRs depends on properties that the system should have and the way in which the customers want the product to act. (viz. performance, security, usability, etc.) [19]. The main issue is when all FRs are met but the overall software still fails to meet some requirements because NFRs are not managed. Consider an ATM, for instance, whose primary purpose is to allow users to withdraw cash from banks. What if a transaction takes the system two hours to complete? What if the account information is not safe? Today, it is impossible to imagine living without smartphones. The phones save all data, including personal, professional, banking, and other details. What if the cell phone data is not secure? The NFRs are therefore crucial to the success of any software. In the literature, there exist various definitions of NFRs as shown in Table 1.

Each NFR has different quality characteristics (QCs) and roles during the software development process. Consider security NFRs describe numerous system security features including authentication, authorization, and privacy, whereas usability NFRs discuss numerous facets of the system's usability and usability. There are 248 different types of NFRs in the literature, including QCs. Out of 248 types of NFRs, researchers deal with 106 NFRs, out of which 28 NFRs are defined with detailed QCs, 29 NFRs are defined without QCs, and 54 NFRs are introduced without definition. This implies that there is still no definition for 50.94% of NFRs in the literature.

Table 1. Definitions of NFRs

Source	Definition
[20]	*"The term is not defined. The standard distinguishes design requirements, implementation requirements, interface requirements, performance requirements, and physical requirements."*
[21]	*"The required overall attributes of the system, including portability, reliability, efficiency, human engineering, testability, understandability, and modifiability."*
[19]	*"These are constraints on the services or functions offered by the system. They include timing constraints, constraints on the development process and standards. They often apply to the system as a whole. They do not usually just apply to individual system features or services."*
[22]	*"Describe the non-behavioral aspects of a system, capturing the properties and constraints under which a system must operate."*
[1]	*"The standard defines the categories functionality, external interfaces, performance, attributes (portability, security, etc.), and design constraints."*
[23]	*"A requirement that specifies system properties, such as environmental and implementation constraints, performance, platform dependencies, maintainability, extensibility, and reliability. A requirement that specifies physical constraints on a functional requirement."*
[24]	*"A property, or quality, that the product must have, such as an appearance, or a speed or accuracy property."*
[25]	*"The behavioral properties that the specified functions must have, such as performance, usability."*
[26]	*"A description of a property or characteristic that a software system must exhibit or a constraint that it must respect, other than an observable system behavior."*
[27]	*"There is not a formal definition or a complete list of nonfunctional requirements. They are global requirements on its development or operational cost, performance, reliability, maintainability, portability, robustness, etc. Informally stated, often contradictory, difficult to enforce during development and evaluate for the customer prior to delivery."*

3 Quality Models

A quality model is defined as *"the set of characteristics, and the relationships between them, that provides the basis for specifying quality requirements and evaluation"* by [29]. Models for evaluating software quality have been created, specifying the fundamental elements (also known as characteristics) and the sub factors within each of them. Each sub factor has a set of metrics for the actual evaluation. Quality model evolution is seen in [28]. This evolution has divided the models into two categories: 1. Basic Models, whose goal is to evaluate products holistically; and 2. Tailored Quality Models, which are focused on component evaluations. According to [1, 20] the degree to which a system, component, or process satisfies a user's wants or expectations is the same as the degree to which the system, component, or process satisfies specified requirements. The first published quality models for software date back to the late 1970s, when Boehm et al. [30] described quality characteristics and their decomposition.

These models have come in a number of modifications over time. The FURPS model, which divides quality into Functionality, Usability, Reliability, Performance, and Supportability, is one of the more well-known ones. The basic concept of these models, aside from this hierarchical decomposition, is that you deconstruct quality down to a level where you can measure and, hence, evaluate quality. The international standard ISO/IEC 9126 was based on these types of quality models. The standard proposes a few metrics for measuring them and establishes a standard decomposition into quality characteristics. However, these criteria do not account for all facets of quality. As a result, the standard does not guarantee full operational quality. The new standard ISO/IEC 25010, which replaces ISO/IEC 9126, modifies a few classifications but maintains the overall hierarchical breakdown.

These quality models are static and unreadable by machines. As a result, we discovered a number of methodologies in the literature that employ ontology to implement the quality models. The ontological quality models can be processed by a machine.

4 Ontology

The Greek words ontos (being) and logos (word) are the source of the word ontology. It disparages the study of being and the descriptions of how existence is arranged, identified, and categorized [31]. With the help of ontologies, individuals and application systems can communicate about a topic with a common understanding. There is an obvious link between the ontological modeling of a domain and the modeling that a requirements engineer would undertake during the requirements process since Requirements Engineering (RE) entails knowledge collecting and analysis. Nearly all RE formalisms are reduced to first-order logic, and they all require a specific conception. They thus share many characteristics with ontologies created using formal language. The following are some potential applications for building ontologies in RE: (i) Capture domain knowledge and offer a generally shared understanding of a subject. (ii). Allow interoperability and the reuse of domain knowledge by offering effective communication channels. (iii). Distinguish between operational and domain knowledge. (iv). Analyze subject-matter expertise. (v). Organize changeable and dynamic requirements.

In requirements engineering, authors [5] propose conceptual dependency of ontology, which illustrates the relationships between several ontologies, including requirements ontology, dependability ontology, and NFRs ontology. For capturing the fundamental stakeholder concerns during requirements engineering, including beliefs, desires, intentions, and evaluations, writers in [32] suggest a core ontology for requirements (dubbed CORE).

On the other hand, authors in [31] distinguish between three types of ontologies in RE: (i). Application domain ontology: It describes the business data needed to design the system, as well as the application domain knowledge. Understanding the domain aids in identifying dynamic and changing requirements. (ii). It captures the fundamental criteria, as well as their interdependencies and linkages. This ontology can be used to eliminate inadequate requirements definitions and reduce unclear requirements during requirements elicitation. An ontology for requirements can then be used for validation and verification. (iii). It describes the organization of requirements specification documents to help reduce the number of incomplete requirements specifications.

5 NFRs Specification

5.1 NFRs: A Hard or Soft Goal?

NFRs are considered as *quality* that take value in quality region either crisp (viz. 0–7 s) or vague (e.g., fast, slow). The crisp quality regions are defined as Quality Constraints (QCs). There is a need to identify which NFRs belongs in which quality hierarchy. The quality is not always stated explicitly; rather they are hidden in the requirements. Identifying those NFRs is an issue. In literature [27], NFRs are considered as a soft-goal. A soft goal means the requirements that have no clear-cut criteria of success. On the other hand, the hard goals (FRs) are considered as requirements having clear cut criteria of success. It is difficult to identify all situations that provide solutions to the soft-goal (vague requirements-can't decide the boundary) at an early stage of RE. In this case, we can identify the existence of the quality but cannot identify the scope of the quality. For example, *"design the system's main menu"* is a vague requirement and thus considered as a soft goal. After refining to NFRs goal, *"The menu buttons must have standard length and width."* yet the requirement is vague. The vague NFRs are continuously refined and operationalized, and hence such vagueness disappears [17].

On the other hand, some researchers claim that the NFRs/FRs could be hard-goal as well as soft goal, rather strictly NFRs as a soft goal and FRs as hard goal [32]. Furthermore, there exists no clear-cut boundary between NFRs and FRs and it is difficult to integrate NFRs with the FRs. In support to the claim, authors [14] provide precise definitions of soft-goal as a goal without a clear-cut definition of satisfaction, but an independent existence; Quality Goal as a quality requirement, which consists of a quality type, a desired quality region and a subject, with a different representation; Quality Constraints-precisely defines a desired quality region in the quality space of a quality type and is achieved or denied by tasks. They mention the incomplete ontology can be extended on demand. Furthermore, they present their ontology and axioms for satisfaction of quality goal.

The basic question is how to refine the NFRs to identify all possible solutions that satisfy them by means of measurement or remove the vagueness. The process of refinement conceptually deconstructs the NFR's qualities by means of identifying the qualities that are inherent in the quality associated with an NFR. For example, a *security* quality can be refined in its sub-qualities viz. *confidentiality*, *integrity* and *availability*. Another way to refine the NFRs is to identify whether the quality is a resultant quality, which can be conceptually reduced to the qualities of the parts of the bearer of the original quality. For example, consider a requirement, *"The user interface must have standard format"*. The requirement contains *"standard format"* as a quality that in terms parts of the bearer *"the interface"*. The interface is generally composed of buttons, fields, icons, etc., "*The menu buttons must have standard format*" illustrates a possible refinement of the requirement. The requirement can further refine because the button may decompose into "*size*", "*shape*", and "*color*". Considering "*size*", it may be further decomposed into "*height*" and "*width*". Hence, a further refinement is *"The menu buttons must have standard height and width"*.

Ultimately, we need to provide some values to make NFRs measurable that make NFRs in a quality constraint region known as operationalization NFRs. For example, the previous refined requirement can be rewritten with constraint as "*The menu buttons must have height 0.75 cm and width 1.75 cm*". It is not mandated that the qualities are constrained to have specific quality values, rather it may concern a region. For example, requirement "*The search functionality must be efficient*", operationalized by "*The search results shall be returned within 30 s after the user enters search criteria*". For instance, take the requirement *"Learning to operate the login functionality must be efficient"*. This NFR may be operationalized by "*The user should learn how to operate the login functionality within 3 min*". Thus "*efficient*" for learning the login functionality and for returning search results (previous example) may map to different regions in the time quality dimension.

5.2 NFRs Extraction: Interview-Based Approach to Ontology-Based Approach

At first stage requirements analyst extract NFRs by means of interview. At this stage NFRs are unclear and unstructured. Furthermore, requirements analyst may not have complete knowledge regarding the domain specific NFRs. Therefore, the requirements analyst must be supported by means of knowledge that will help them to identify pertinent NFRs and to ask right questions while eliciting NFRs at the time of interview. In literature, there exists a handful of proposals that give conceptual representation of NFRs. Much of the awareness about NFRs comes from the original work present in [27]. The NFR Framework is a goal-oriented approach, used to represent NFRs graphically by means of soft-goal interdependency graph. It provides methods of refinement, to further refine a goal into one or more goals.

To reduce the problem of integration, in [2], NFRs are integrated by means of extending UML models. However, this approach models certain NFRs due to limited knowledge regarding quality attributes. Instead, UML models, natural language processing helps to extract NFRs from requirements documents. Lee et al. [11] apply natural language processing to provide a common understanding of security requirements and to facilitate analysis at various decision points by making the required information readily available with appropriate context and format. However, they focus only on security requirements. Furthermore, natural language processing cannot help to provide additional information regarding application domain. NFRs are difficult to describe completely and precisely due to vague, conceptual, and subjective nature [11].

To identify NFRs there is a need to provide the domain knowledge support at the time of interview. Different NFRs has different meaning so ontology provides common understanding of the terms [10]. The authors [6] developed two ontology, one is to extract FRs using domain knowledge named functional domain ontology and another is to extract NFRs using standard quality model-ISO/IEC 9126 named quality ontology. To support the ontology the authors developed an interactive tool-ElicitO. They gather knowledge from textbooks and experts in the domain to build the functional ontology [6]. Furthermore, they present a scenario where the user and analyst seat together and use the tool to extract requirements, meanwhile using ontology appropriate NFRs are also considered that provide standardization compared to traditional approaches. To build the ontology authors follow the process presented in [7].

The generalized approach can be appropriate for any domain, but the quality level and the order of importance may vary from one domain to another. The aim of developed ontology is to provide the definition of the general concepts relevant to NFRs without reference to any domain. These general concepts can then act as a common foundation for describing NFRs attributes, metrics, etc. Furthermore, it provides a conceptual model for NFRs that contains rules which define the semantics of the defined concepts.

To provide the more generic solution, the concept of ontology used in [5, 8–10] in order to develop software rapidly. That is, common requirements in a specific domain should be treated as a special kind of reusable resources. In [8] authors explain the need of machine understandable Quality of Service (QoS) vocabulary and importance of sharing it for both static and dynamic systems architectures. The authors developed a tool named QoSOnt that provides an extension to OWL-S (non-Qos service provider). They focus on developing ontology for quality parameters viz. dependability, reliability and performance supporting metric representation. They build the complete ontology by means of interconnecting "base ontology" - separate ontology for all concept attributes of quality. The reason behind the separation is to refer the particular metrics of that quality attribute. In addition, they create a tool named SQRM to evaluate the validity of generated ontology.

One NFR can help or hinder the achievement of other NFRs at certain functionality. To identify such interaction manually is difficult. To make the process automate authors [3] focus not only extraction of NFRs, they provide relation of NFRs by means of identifying NFRs from different view perspective. They present three NFRs ontology views: 1. Intra-Model Dependency - NFRs relation with the other entities of the software system being developed. 2. Inter Model Dependency - contains the classes and properties intended to structure NFRs in terms of interdependent entities. 3. Measurable NFRs - measurement process and contains the concepts used to produce measures to measurable NFRs.

However, NFRs are not independent goals, rather they are related to other parts of software. If the requirements are classified as the NFRs it is necessary, that it belongs to at least one FR as it represents a set of attributes that bear on the existence of a set of functions and their specified properties according to the ISO9126 definition. For example, *"The interaction between the user and the software system while reading email messages must be secured"*. Here, reading email message is FR, while providing a secure environment while reading a message is NFRs.

5.3 Comparative Analysis and Issues

Table 2 presents a comparative study of existing approaches for specifying NFRs, considering parameters including elicitation, categorization, input, output, quality model, and dataset. Majority tools [5–17] uses portege as a ontology development tool. We observe that there exists no tool that provides validation of extracted NFRs except efforts given by [11]. Generally, the input is natural language. Most of the tools use input as a natural language except [3] uses XML representation of work flow as an input. A Service-Centric System (SCA) may have a static architecture decided at design-time.

While Service-Oriented Architectures (SOA) also open up the possibility of a dynamic architecture where certain components are only discovered and bound at runtime. Both approaches ideally require a service marketplace, in which functionally equivalent services compete for the same custom. In such a situation non-functional characteristic, i.e. QoS, become more important in distinguishing between the competing services. In either case, it is also desirable to have a machine understandable QoS vocabulary. In the static case this would prove useful in implementing design-time tools, in the dynamic case it is necessary to perform service selection at runtime based upon service QoS.

Table 2. Comparative Analysis of Approaches to Specify NFRs[1]

Citation	Elicit	Classify	Input	Output	Quality Model	Dataset
[5]	NFRs	Y	NL	NL	Generic	AutomotiveIndustry
[6]	FRs+NFRs	Y	NL	NL	ISO/IEC 9126	UniversityHelpdesk
[7]	FRs+NFRs	N	NL	Graph	Generic	AutomotiveIndustry
[9]	FRs+NFRs	Y	NL	NL	ISO/IEC 9126	UniversityHelpdesk
[10]	NFRs	N	XML	Graph	NFRs	Web Services
[3]	NFRs	Y	NL	Ontology	Generic	Email Application
[11]	Security	Y	NL	Ontology	Generic	Website
[13]	FRs+NFRs	Y	NL	NL	ISO9126	Concordia RE Corpus
[14]	FRs+NFRs	N	NL	Ontology	Generic	Urban Transportation Management
[15]	NFRs	Y	NL	Ontology	ISO/IEC25010	Promise
[16]	FRs+NFRs	Y	NL	Modeling language	ISO/IEC 9126	Promise
[17]	NFRs	Y	NL	Ontology	ISO/IEC25010	Promise

Ontologies are regarded as a valuable component of RE, although there are several challenges when creating ontology-based software development projects, including maintenance, merging, mapping, and versioning of ontologies. It is difficult to determine the "correct" level of expressiveness for the ontology language. Furthermore, there are numerous distinct ontology languages that are used for various reasons, and it is not immediately obvious how to map between ontologies that are expressed using various languages. In Table 3, we identify issues specifying NFRs using ontology.

[1] Y – Yes; N – No; NL – Natural Language; XML – Extensible Markup Language.

Table 3. Issues in Specifying NFRs using Ontology

Issue	Description
NFRs Conflict	It can happen that one NFR has a favorable or unfavorable impact on subsequent NFRs. Finding the conflict between the specified NFRs is necessary to provide the conceptualization
NFRs Traceability	One NFR may affect many functional requirements. While changing requirements affect both the functional and NFRs. There is a need to trace the relationship exist among them
NFRs Integration	How NFRs integrate with functional requirements?
NFRs Ambiguity	The representation of ontology gives a common interpretation among requirements engineers. The problem of ambiguity is still an issue from the customer's perspective
NFRs Modeling	How to represent NFRs effectively at requirements design stage?

6 Conclusions

In this paper, the concepts and quality models of the NFRs are thoroughly examined. The research also discusses a comparative examination of alternative ontology-based approaches to specify NFRs. The ontology has become more prominent in RE study areas, with a focus on defining NFRs. However, there are issues with ontology that are related to their application, such as merging, mapping, integrating, etc. Future research will focus on looking at potential solutions to the issues related to specifying NFRs such as conflict, traceability, ambiguity, and modeling.

References

1. Software Engineering Standards Committee: IEEE Recommended Practice for Software Requirements Specifications (1998)
2. Mijanur Rahman, M., Ripon, S.: Elicitation and modeling non-functional requirements – a POS case study. Int. J. Future Comput. Commun. **2**(5), 485–489 (2013)
3. Kassab, M., Ormandjieva, O., Daneva, M.: An ontology based approach to non-functional requirements conceptualization. In: 2009 Fourth International Conference on Software Engineering Advances, pp. 299–308 (2009)
4. Gruber, T.R.: A translation approach to portable ontology specifications. Knowl. Acquis. **5**(2), 199–220 (1993)
5. Dobson, G., Hall, S., Kotonya, G.: A domain-independent ontology for non-functional requirements. In: IEEE International Conference on E-business Engineering, pp. 563–566 (2007)
6. Al Balushi, T.H., Sampaio, P.R.F., Dabhi, D., Loucopoulos, P.: ElicitO: a quality ontology-guided NFR elicitation tool. In: Sawyer, P., Paech, B., Heymans, P. (eds.) REFSQ 2007. LNCS, vol. 4542, pp. 306–319. Springer, Heidelberg (2007).https://doi.org/10.1007/978-3-540-73031-6_23
7. Falbo, Rd.A., de Menezes, C.S., da Rocha, A.R.C.: A systematic approach for building ontologies. In: Coelho, H. (ed.) IBERAMIA 1998. LNCS (LNAI), vol. 1484, pp. 349–360. Springer, Heidelberg (1998). https://doi.org/10.1007/3-540-49795-1_31

8. Dobson, G., Lock, R., Sommerville, I.: Quality of service requirements specification using an ontology. In: SOCCER Workshop, Requirements Engineering, vol. 5 (2005)
9. Al Balushi, T.H., Sampaio, P.R.F., Dabhi, D., Loucopoulos, P.: Performing requirements elicitation activities supported by quality ontologies. In: 18th International Conference on Software Engineering and Knowledge Engineering, SEKE 2006, no. August 2015, pp. 343–348 (2006)
10. Hughes, C., Hillman, J.: QoS explorer: a tool for exploring QoS in composed services. In: International Conference on Web Services, ICWS 2006, pp. 797–806 (2006)
11. Lee, S.-W., Muthurajan, D., Gandhi, R.A., Yavagal, D., Ahn, G.-J.: Building decision support problem domain ontology from natural language requirements for software assurance. Int. J. Softw. Eng. Knowl. Eng. **16**(6), 851–884 (2006)
12. Li, F.-L., Horkoff, J., Mylopoulos, J., Liu, L., Borgida, A.: Non-functional requirements revisited. In: Proceedings of the 6th International i* Workshop, vol. 978, no. iStar, pp. 109–114 (2013)
13. Rashwan, A., Ormandjieva, O., Witte, R.: Ontology-based classification of non-functional requirements in software specifications: a new corpus and SVM-based classifier. In: COMPSAC 2013 Proceedings of the 2013 IEEE 37th Annual Computer Software and Applications Conference, no. ii, pp. 381–386 (2013)
14. Jingbai, T., Keqing, H., Chong, W., Wei, L.: A context awareness non-functional requirements metamodel based on domain ontology. In: Proceedings - 1st IEEE International Workshop on Semantic Computing and Systems, WSCS 2008, pp. 1–7 (2008)
15. Li, F.-L., et al.: Non-functional requirements as qualities, with a spice of ontology. In: Proceedings of the 22nd International Requirements Engineering Conference, pp. 293–302 (2014)
16. Li, F.-L., Horkoff, J., Borgida, A., Guizzardi, G., Liu, L., Mylopoulos, J.: From stakeholder requirements to formal specifications through refinement. In: Fricker, S.A., Schneider, K. (eds.) REFSQ 2015. LNCS, vol. 9013, pp. 164–180. Springer, Cham (2015). https://doi.org/10.1007/978-3-319-16101-3_11
17. Guizzardi, R., Li, F., Borgida, A., Guizzardi, G.: An ontological interpretation of non-functional requirements. In: Formal Ontology in Information Systems, pp. 344–357. IOS Press (2014)
18. Brooks, F.P., Bullet, N.S.: Essence and accidents of software engineering. Computer **20**, 10–19 (1987)
19. Sommerville, I.: Software Engineering, 5th edn (1995)
20. IEEE Standard: IEEE Standard Glossary of Software Engineering Terminology (1990)
21. Davis, A.M.: Software Requirements: Objects, Functions, and States. Prentice-Hall Inc., Upper Saddle River (1993)
22. Anton, A.I.: Goal Identification and Refinement in the Specification of Software-based Information Systems. Georgia Institute of Technology, Atlanta, GA, USA (1997)
23. G. Jacobson, I., Booch, G., Rumbaugh, J., Rumbaugh, J., & Booch, I. Jacobson, G. Booch, and J. Rumbaugh. (1999)
24. Robertson, S., Robertson, J.: Mastering the Requirements Process (1999)
25. Ncube, C.: A requirements engineering method for COTS-based systems development (2000)
26. Karl, W.: Software requirements (2003)
27. Chung, L., Nixon, B.A., Yu, E., Mylopoulos, J.: Non-Functional Requirements in Software Engineering. Springer, New York (2012)
28. Miguel, J.P., Mauricio, D., Rodríguez, G.: A review of software quality models for the evaluation of software products. Int. J. Softw. Eng. Appl. **5**(6), 31–53 (2014)
29. IEC 9126-1: Software Engineering-Product Quality-Part 1: Quality Model (2001)

30. Boehm, B.W., Brown, J.R., Lipow, M.: Quantitative evaluation of software quality. In: Proceedings of the 2nd International Conference on Software Engineering, pp. 592–605 (1976)
31. Castaneda, V., Ballejos, L., Caliusco, L., Galli, R.: The use of ontologies in requirements. Glob. J. Res. Eng. **10**(6), 2–8 (2010)
32. Jureta, I.J., Mylopoulos, J., Faulkner, S.: Revisiting the core ontology and problem in requirements engineering. In: Proceedings of the 16th IEEE International Requirements Engineering Conference, RE 2008, no. November 2015, pp. 71–80 (2008)

A Review of Traditional and Neural Network Methods for Protecting Privacy in Big Data Analytics

C. A. Thasna(✉), Meenu Chawla, and Namita Tiwari

Department of Computer Science and Engineering, MANIT Bhopal, Bhopal, India
thasnaansari2000@gmail.com, chawlam@manit.ac.in,
namita_tiwari21@rediffmail.com

Abstract. Computing technology suffused the human environment and engenders copious amounts of data. An individual's right to privacy means that they can choose what information is shared. The area of data analytics has changed as a result of big data's introduction. If private information is not managed properly on the web, it becomes public. Data collection has become quicker because of modern technology. However, breaching people's privacy poses an extreme concern. This study evaluates current privacy-preserving methods including traditional and neural network methods and describes the challenges of using them for heterogeneous data and how fuzzy neural networks are used in big data analytics to protect privacy.

Keywords: big data · anonymization · privacy techniques · neural networks · Heterogeneous Data

1 Introduction

Big data is the term used to describe datasets that are excessively big or intricate for software used for traditional data processing to manage. Regarding the use of big data technologies for analytics, privacy is a crucial concern. As a greater amount of data is collected, data aggregation and data analytics could result in breaches of user privacy. Organizations aim to improve customer fulfillment through big data analytics, but they must preserve user privacy in order to do so.

Big Data is growing quite quickly. Today, most data are produced in an unstructured format [15]. Unstructured data exists in a variety of formats and is heterogeneous and varied in nature. Big data architecture contains 3 components. Multisource big data collection, distributed big data storing, and big data processing [4]. The life cycle of big data is depicted in Fig. 1. During extensive data collection eavesdropping is feasible and data may be accidentally leaked. So physical protection methods are needed. In big data storage disclosure of more individual personal information can occur. As part of big data analytics, the processing of big data is essential. If the data analytics is outsourced, third-party

D. Pastor-Escuredo et al. (Eds.): ICDLAIR 2023, LNNS 1001, pp. 158–167, 2024.
https://doi.org/10.1007/978-3-031-60935-0_15

entities can extract useful information and they can use it maliciously. The volume and complexity of big data frequently exceeds the capabilities of conventional data processing methods [26]. Big data analytics offers the ability to predict epidemic breakouts and avert diseases that can be prevented. It has the greatest application in big data analytics. Surveillance, disclosure, discrimination, personal embracement, and discrimination are the key privacy threats [1].

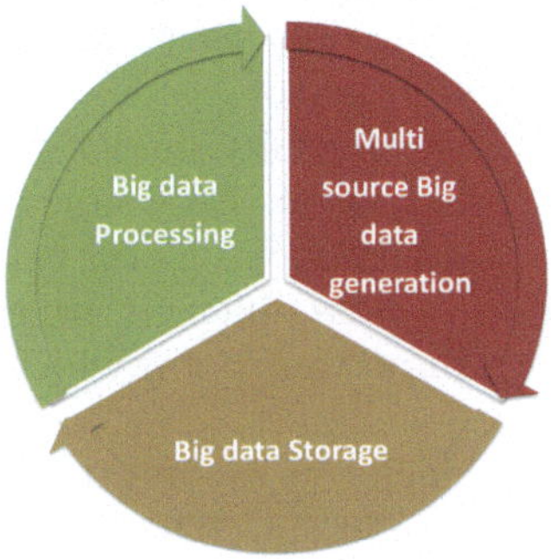

Fig. 1. Life Cycle of Big Data [5]

The five parameters of big data features include volume, velocity, variety, veracity, and value [5]. The authors of survey paper [2] reviewed privacy preservation approaches from the perspective of communication, and it explains in detail how social networking services and location-based services violate an individual's privacy. In earlier times privacy preservation methods for knowledge discovery were categorized into two groups: Data Perturbation (Adding uniform noise, Hiding private data) and Cryptographic Methods. The goal is to enable the extraction of accurate and meaningful information from a database while protecting the privacy of individuals whose data is included.

This paper examines various methods for protecting the privacy of large data analysis, as well as the difficulties involved in handling heterogeneous data. The rest of the paper is structured as follows. Section 2 presents the different Anonymization strategies. The data distribution technique is described in Sect. 3. Section 4 describes The randomization technique. Section 5 delves into privacy-preserving neural networks. Section 6 of this study points out how fuzzy neural networks address these issues and guarantee the confidentiality of big data research on different datasets. Finally, the conclusion is provided in Sect. 7.

2 Anonymization Techniques

The technique of changing data before it is used is called anonymization. In order to protect people's privacy, data anonymization is a technique of stripping or altering identifying information from a dataset [25]. A minimal set of characteristics known as a Quasi-identifier (QI) is used to identify people specifically. To satisfy k-anonymity, a table must contain at least k instances of each sequence of quasi-identifier values in the dataset [16]. Using suppression and generalization operations to change the value of the quasi-identifier characteristic, K anonymity aims to produce an equivalence class of records with identical quasi-identifier values. By doing this, the attacker's ability to ascertain the identity of the people in a particular data collection will be restricted however the disclosed data will continue to be as helpful as is practical.

Sensitive attributes refer to the data that has to stay a secret from others who don't have the right to obtain person-specific information [17]. Besides the k-anonymity model, there is the l-diversity model which is intended to prevent the publication of personal information. Each equivalence class is required to have L prominent values for the sensitive property by l-diversity. When sensitive attribute values in an equivalence class differ yet are conceptually equivalent, an attacker may be able to learn important data, which is the main disadvantage of l-diversity.

The t-closeness technique, which gets around the k-anonymity and l-diversity restrictions, is proposed in the study [18]. The equivalence class is regarded to possess t-closeness if there is a difference of less than a threshold t with the distribution of a sensitive property inside the class and the distribution of the attribute for the entire table. A table is referred to as t-close if all equivalence classes have t-closeness.

3 Data Distribution Technique

The data in this method is dispersed among numerous places. The following methods of data distribution are available.

3.1 Horizontal Distribution

The term "horizontal distribution" refers to the dispersion of data over many locations with similar features. It is only appropriate to distribute data horizontally when specific collective actions or procedures must be performed on the information without directly distributing the data. Even though, while analyzing data, the owner of the data might be required to share the information with an outside analyst, which could result in a privacy infringement. Despite the fact that techniques for grouping and classification can be utilized to distribute data, privacy is not always guaranteed. If the data is dispersed among numerous websites run by different organizations, the results of the combined functions could enable one party to determine which party owns the data. In these circumstances, authors presume that all participating websites will be sincere with one another [20].

3.2 Vertical Distribution

The spread of individual user information across numerous websites managed by multiple organizations is referred to in this phrase. When information from all of these websites has to be combined for analytics, there is a chance that privacy laws may be broken. Executing data analytics on vertically distributed data is highly challenging when the characteristics are dispersed over several places and are managed by multiple parties.

If datasets are shared, it is difficult to guarantee privacy. Anonymization is ineffective when full records are not required for analysis. Data distribution closely mimics encryption techniques, however it does not guarantee privacy preservation.

4 Randomization Technique

Before using the dataset, the randomization approach entails adding noise to it. However, because to time complexity and data utility, randomization cannot be applied to a big dataset. The fundamental issue is that some outlier records continue to be exposed to adversary attacks and are unaffected by randomization [1].

5 Privacy Preserving Neural Networks

Big data analysis requires massive data collecting, which inevitably raises privacy concerns. The businesses that gather users private, extremely sensitive data, like images and voice recordings, keep it forever. Users are unable to delete it or limit its use. Due to privacy and confidentiality concerns, many data owners are prevented from sharing valuable data and enjoying the advantages of large-scale deep learning, including medical organizations that might want to use these techniques on clinical records [12].

In the paper, [11] The authors suggested a deep learning method that protects anonymity and allows each participant to train the dataset locally. The approach includes a parameter communication mechanism that enables users to receive the most recent parameter values at each local stochastic gradient descent epoch and upload the gradients of a few chosen neural network parameters to the parameter server, which stores the global parameters. Each person can individually and privately assess the network on fresh data after it has been trained, without interacting with other participants. Privacy-preserving neural networks (PPNN) have gained importance recently owing to the rise of big data and the rise in concerns about data security and privacy. Several privacy-preserving learning techniques were created as a result of this.

The authors add secure sum and secure matrix addition to the neural network classification algorithm in [3]. Here the data is distributed among the semi-honest parties by the horizontally distributed dataset with the same data schema. In the proposed algorithm Each party finds a local weight change then aggregate

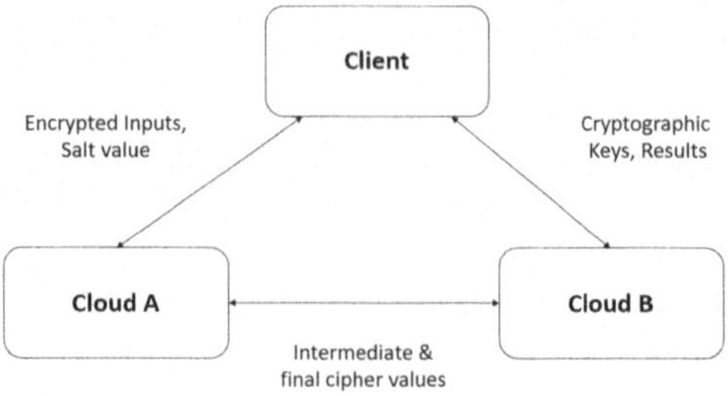

Fig. 2. System Component View of Dual Cloud System [6]

the results and finds the global weight change. Then they update each party's weight locally based on the backpropagation algorithm. It is working on the assumption that all parties are semi-honest and do not collude.

5.1 Cloud-Based Neural Networks

There are multiple techniques for implementing homomorphic encryption in neural network models [14]. It is employed to safeguard both data and neural network models from the cloud service provider. These methods are hampered by the significant computational burden of applying homomorphic encryption to the neural network model's hidden layers.

The paper [7] proposed a privacy-preserving technique for training neural networks in a decentralized setting such as a cloud computing platform. This allows multiple parties to collaboratively train a neural network while keeping their input data private. This method uses homomorphic encryption to encrypt the sensitive data and neural network model and they distribute the computation of the backpropagation algorithm across multiple cloud nodes. Homomorphic encryption enables computations on encrypted material without having first to decode it [13]. However, this method takes so much time for the encryption of the data sets. Also, all the computation overhead is on a single cloud system.

In paper [6] the author presented a non-colluding dual cloud computing system that eliminates the computational overhead on the single cloud. Each neuron's input is handled by one cloud, while its output is handled by another cloud. The data belongs to the client. When a client requests the use of a neural network to do classification it works with Cloud A on behalf of the client to oversee the neural network-based categorization. Cloud A is aware of how many neurons there are as well as any hidden layers. Before transferring data to Cloud B, the Client and Cloud B use a common secret key to encrypt the data. Keys are unique to each customer. The weighted sum arriving from cloud A is decrypted by cloud B, and the activation function is then used. The activation function results are then encrypted and transmitted back to cloud A using the secret key. These encrypted outputs serve as a fresh source of input for the following neuron operating in hidden layers. To Cloud B first, then to Client, are given the final categorization results. Untrustworthy Cloud A could try to conduct weighted sums and activations independently without sharing with Cloud B. This can cause Cloud A to start making too much noise. The system component view of dual cloud system is depicted in Fig. 2.

5.2 Clustering Based Algorithms

Paper [8] described a Clustering-Based neural network for Privacy-Preserving Data Mining to ensure privacy while still allowing meaningful data mining. The data is first preprocessed to eliminate any identifiable or sensitive information. The modified data is then used to train a neural network, which discovers the underlying patterns and structure. The data is then clustered into categories based on similarity using the trained neural network.

The individual records are then further protected by the clusters being subjected to a privacy-preserving procedure. The "privacy cluster center," a brand-new idea presented by the authors, stands for each cluster's privacy-preserving data. In order to maintain privacy while still maintaining the key traits of the clusters, the privacy cluster centers are generated using the neural network's learned representation of the input.

Fuzzy clustering and neural networks are combined in [9] to increase the classification accuracy of ECG arrhythmias. After applying fuzzy clustering to preprocess the ECG data, the algorithm will assign a membership degree to each feature vector. The neural network that classifies the signals into different arrhythmias using the backpropagation technique will use this membership level as an input.

Database sizes are growing as a result of the increase in digital information. This vast volume of data [21] includes both structured and semi-structured data. Any data that exhibits a substantial degree of heterogeneity in data kinds and formats is referred to as heterogeneous data. They may be ambiguous and of low quality due to missing values, significant data redundancy, and untruthfulness. Heterogeneous data must be carefully combined in order to fulfill the organization's expectations [22].

Neural network techniques that use clustering and cloud networks provide data privacy for each individual. However, these techniques struggle with heterogeneous data.

Table 1 enumerates the several techniques used to protect privacy when analysing data, along with the difficulties in applying each technique when dealing with heterogeneous data.

6 Privacy Preservation Using Fuzzy Neural Network

Big data that is diverse often contain sensitive personal data. Another unfavorable feature of heterogeneous large data is uncertainty. Fuzzy systems that can handle ambiguous input are combined with neural networks' learning capabilities in fuzzy neural networks [11]. Uncertainty may be handled with fuzzy neural networks. The authors introduced a PP-HFNN (Privacy Preserving-Hierarchical

Table 1. Comparison Study of Different Privacy Preservation Techniques

Author	Year	Name of Method	Technique	Implementation	Features	Challenges to apply Heterogeneous data
Ajun, C. [24]	2015	Generalization	Anonymization-(K-anonymity)	Use generalized values to modify specific attribute values	QI-values in each class is semantically similar	Only valid in Low dimensional data
Chen, R. et al. [23]	2013	Suppression	Anonymization-(K-anonymity)	Uses special symbols to replace attribute values	Special form of generalization	Same problem with generalization
Li, N. et al. [18]	2013	l-diversity	Anonymization	l-well represented values for SA in each equivalence class	Secure individual identity from disclosure	Difficult and insufficient to prevent attribute disclosure
Liang, H. et al. [19]	2007	t-closeness	Anonymization	Based on the difference between the distribution of SA in an equivalence class and entire table	Overcomes the limitation of other anonymization techniques	No efficient algorithm that can find t-close anonymization of a given dataset
Aggarwal, C. C. et al. [20]	2008	Horizontal and vertical distribution	Data Distribution	Data is dispersed among numerical places	Aggregate functions can be performed on the data without sharing it	Assumption that all the participating websites are trustful
Ram Mohan Rao, P. et al. [1]	2018	Randomization	Differential Privacy	Add random noise	Low computational complexity	Exposed to adversary attacks
Yuan, J. and Yu, S. [7]	2014	PPNN with single cloud	Cloud based Neural Networks	Uses Homomorphic encryption to encrypt the data	Allows collaboratively training a neural network while keeping their input data private	Encryption time is high for simple dataset
Baryalai, M. et al. [6]	2016	PPNN with Dual cloud	Cloud based Neural Networks	One cloud handles input to the neurons another cloud handles the output of each neuron	Overcomes the computation overhead on a single cloud system	For large dataset set size of the cipher text is high

Fuzzy Neural Network) [10]. It effectively addresses the issues of data heterogeneity and uncertainty using fuzzy if-then rules and hierarchical structure.

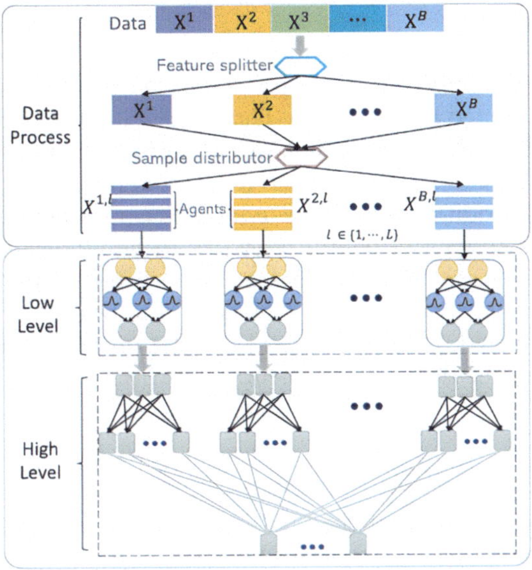

Fig. 3. PP-HFNN structure [10]

A dual-level hierarchical structure underlies the PP-HFNN. The upper level receives the outputs from the FNNs that make up the lower level. Heterogeneous data is split at the feature level and sample levels during the data pre-processing stage. For feature-level segmentation, a feature splitter separates the heterogeneous features into a number of subsets, making sure that each group has only homogenous features. Only one subset is under the control of a low-level FNN. The samples of every low-level FNN are randomly allocated to different agents.

Then, in order to safeguard data privacy and control the enormous volumes of data samples, a distributed computing framework is made available to a low-level FNN's agent. Data preprocessing and PP-HFNN are shown in Fig. 3 as ways to protect the privacy of heterogeneous large data.

Low-level FNN has four levels overall. Each element of the input vector is used in Layer 1 to create the fuzzified outputs using a Gaussian membership function that makes use of the center and width of the associated fuzzy set. Layer 2 implements fuzzy rules, while layer 3 fuzzified the results. The output layer is layer 4. Each FNN's antecedent layer's parameters were identified by the authors using a distributed k-means approach. The generalized HRSS (High storage system data) and EEG-DD (EEG signals from the distracted driving experiment) datasets are being used by the authors to assess the performance of the PP-HFNN. Additionally, it makes the multiagents out to be reliable parties. HRSS data is highly heterogeneous. On this HRSS dataset, this PP-HFNN achieves high performance while safeguarding privacy by using the distributed clustering approach.

7 Conclusion

An extensive survey of different methods for preserving privacy in big data analytics has been conducted in this paper. There are many methods for protecting a dataset's privacy. In Table 1, the difficulties of various approaches are displayed. However, the traditional privacy-preserving techniques are inadequate as the dataset size grows. Additionally, this survey aids researchers in comprehending the critical functions of fuzzy neural networks in handling heterogeneous data during the analysis stage. On the dataset, the current fuzzy neural network-based techniques are providing good accuracy. However, the authors only take into account generalized datasets. A good architecture to address the problems posed by heterogeneous data is fuzzy neural networks.

References

1. Ram Mohan Rao, P., Murali Krishna, S., Siva Kumar, A.P.: Privacy preservation techniques in big data analytics: a survey. J. Big Data **5**, 1–12 (2018)
2. Wang, T., Zheng, Z., Rehmani, M.H., Yao, S., Huo, Z.: Privacy preservation in big data from the communication perspective-a survey. IEEE Commun. Surv. Tutor. **21**(1), 753–778 (2018)
3. Schlitter, N.: A protocol for privacy preserving neural network learning on horizontal partitioned data. PSD (2008)
4. Lu, R., Zhu, H., Liu, X., Liu, J.K., Shao, J.: Toward efficient and privacy-preserving computing in big data era. IEEE Netw. **28**(4), 46–50 (2014)
5. Devi, A.S., Chinnasamy, A.: Privacy preservation of sensitive data in big data analytics-a survey. In: 2021 10th International Conference on Internet of Everything, Microwave Engineering, Communication and Networks (IEMECON), pp. 01–05. IEEE (2021)
6. Baryalai, M., Jang-Jaccard, J., Liu, D.: Towards privacy-preserving classification in neural networks. In: 2016 14th Annual Conference on Privacy, Security and Trust (PST), pp. 392–399. IEEE (2016)
7. Yuan, J., Yu, S.: Privacy preserving back-propagation neural network learning made practical with cloud computing. IEEE Trans. Parallel Distrib. Syst. **25**(1), 212–221 (2013)
8. Tsiafoulis, S., Zorkadis, V.C., Karras, D.A.: A neural-network clustering-based algorithm for privacy preserving data mining. In: Kim, T., Yau, S.S., Gervasi, O., Kang, B.H., Stoica, A., Ślęzak, D. (eds.) GDC CA 2010. CCIS, vol. 121, pp. 269–276. Springer, Heidelberg (2010). https://doi.org/10.1007/978-3-642-17625-8_27
9. Özbay, Y., Ceylan, R., Karlik, B.: A fuzzy clustering neural network architecture for classification of ECG arrhythmias. Comput. Biol. Med. **36**(4), 376–388 (2006)
10. Zhang, L., Shi, Y., Chang, Y.C., Lin, C.T.: Hierarchical fuzzy neural networks with privacy preservation for heterogeneous big data. IEEE Trans. Fuzzy Syst. **29**(1), 46–58 (2020)

11. de Campos Souza, P.V.: Fuzzy neural networks and neuro-fuzzy networks: a review the main techniques and applications used in the literature. Appl. Soft Comput. **92**, 106275 (2020)
12. Shokri, R., Shmatikov, V.: Privacy-preserving deep learning. In: Proceedings of the 22nd ACM SIGSAC Conference on Computer and Communications Security, pp. 1310–1321 (2015)
13. Lauter, K., Naehrig, M., Vaikuntanathan, V.: Can homomorphic encryption be practical? (2011)
14. Pulido-Gaytan, B., et al.: Privacy-preserving neural networks with homomorphic encryption: challenges and opportunities. Peer-to-Peer Netw. Appl. **14**(3), 1666–1691 (2021)
15. Das, T.K., Kumar, P.M.: Big data analytics: a framework for unstructured data analysis. Int. J. Eng. Sci. Technol. **5**(1), 153 (2013)
16. Sweeney, L.: k-anonymity: a model for protecting privacy. Int. J. Uncertainty Fuzziness Knowl.-Based Syst. **10**(05), 557–570 (2002)
17. Machanavajjhala, A., Kifer, D., Gehrke, J., Venkitasubramaniam, M.: l-diversity: privacy beyond k-anonymity. ACM Trans. Knowl. Discov. Data (TKDD) **1**(1), 3-es (2007)
18. Li, N., Li, T., Venkatasubramanian, S.: t-closeness: privacy beyond k-anonymity and l-diversity. In: 2007 IEEE 23rd International Conference on Data Engineering, pp. 106–115. IEEE (2006)
19. Liang, H., Yuan, H.: On the complexity of t-closeness anonymization and related problems. In: Meng, W., Feng, L., Bressan, S., Winiwarter, W., Song, W. (eds.) DASFAA 2013. LNCS, vol. 7825, pp. 331–345. Springer, Heidelberg (2013). https://doi.org/10.1007/978-3-642-37487-6_26
20. Aggarwal, C.C., Philip, S.Y. (eds.): Privacy-Preserving Data Mining: Models and Algorithms. Springer, New York (2008)
21. Kalra, M., Lal, N.: Data mining of heterogeneous data with research challenges. In: 2016 Symposium on Colossal Data Analysis and Networking (CDAN), pp. 1–6. IEEE (2016)
22. Wang, L.: Heterogeneous data and big data analytics. Autom. Control Inf. Sci. **3**(1), 8–15 (2017)
23. Chen, R., Fung, B.C., Mohammed, N., Desai, B.C., Wang, K.: Privacy-preserving trajectory data publishing by local suppression. Inf. Sci. **231**, 83–97 (2013)
24. Ajun, C.: Slicing: a new approach to privacy preserving data publishing. IJESC (2015)
25. Cuzzocrea, A.: Privacy-preserving big data stream mining: opportunities, challenges, directions. In: 2017 IEEE International Conference on Data Mining Workshops (ICDMW), pp. 992–994. IEEE (2017)
26. Gupta, M., Rai, A., Pandey, B., Gupta, A., Verma, V.: Big data privacy: a survey paper. In: 2023 International Conference on Intelligent Computing and Advanced Technologies (ICICAT), pp. 1–6. IEEE (2023)

A Long Short-Term Memory Learning Based Malicious Node Detection for Clustering in Wireless Sensor Networks

Rajpreet Kaur[1] and Surjit Singh[2](✉)

[1] School of Mathematics, Thapar Institute of Engineering and Technology, Bhadson Road, Patiala 147004, Punjab, India
rkaur_msc21@thapar.edu

[2] Computer Science and Engineering Department, Thapar Institute of Engineering and Technology, Bhadson Road, Patiala 147004, Punjab, India
surjit.singh@thapar.edu

Abstract. WSNs (wireless sensor networks) are essential for everyday tasks such as security, medical care, and industrial automation, making them susceptible to many kinds of Denial of service (DoS) attacks. This work explores the various models such as Artificial neural-networks (ANN), Recurrent neural-networks (RNN), and Long Short-Term Memory (LSTM) networks for attack detection. The models' performance is analyzed across multiple evaluation metrics, including RMSE, R2 score, and accuracy. The results show that LSTM outperform other models with overall RMSE 0.17, R2 score 0.9033 and accuracy 99.08%. The findings showcase the potential of various models to improve WSN security through accurate attack identification and offer insights into model selection for specific applications.

Keywords: Wireless Sensor Networks · Machine Learning · Clustering · Attack Detection

1 Introduction

Wireless Sensor Networks (WSNs) are the foundation, for emerging applications such as surveillance, healthcare management and industrial automation [1]. These networks consist of small energy efficient sensor nodes that work together to collect and send data to a system [2]. The dynamic and distributed nature of these networks introduces security challenges [3].

WSN data streams faces security threats, including attacks, greyhole attacks, flooding attacks, Time Division Multiple Access (TDMA) manipulations and compromised normal nodes [4]. A blackhole attack occurs when a malicious node falsely claims to have the path to a sink node in order to divert data [5]. Greyhole attacks involve dropping packets to cause data loss and disrupt communication. Flooding attacks overwhelm the network with packets leading

D. Pastor-Escuredo et al. (Eds.): ICDLAIR 2023, LNNS 1001, pp. 168–177, 2024.
https://doi.org/10.1007/978-3-031-60935-0_16

to congestion and resource depletion. Manipulating TDMA schedules disrupts synchronized communication among nodes while compromised nodes give access, to adversaries [6].

The factors for clustering in WSNs include scalability, load balancing, increased connectivity, etc. The network is efficiently organized into a connected hierarchy using load-balanced clustering [3]. Thus, we design a machine learning based prediction framework applying for attack prediction for clustering in WSNs. The paper organizes research findings systematically, presenting a structured exploration of the chosen topic to facilitate comprehension and analysis.

The remaining work is arranged as: Related work is included in Sect. 2. Section 3 outlines the suggested architecture. Results and analysis are given in Sect. 4, and the work presented is concluded in Sect. 5.

2 Related Work

WSNs have become widely used in critical applications like military surveillance, but are vulnerable to security threats. Intrusion Detection Systems (IDS) are needed to ensure WSN security. However, there is a lack of specialized techniques for training and evaluating IDS in WSNs. The LEACH protocol, a recognized clustering mechanism has proven its effectiveness in Wireless Sensor Networks (WSNs) [2,7–9]. This protocol functions by dividing the network into clusters managed by cluster heads. These cluster heads play a role, in collecting and transmitting data to a base station. The research [10] utilizes the data gathered using the Low Energy Adaptive Clustering Hierarchy (LEACH) protocol on Wireless Sensor Networks. Using the LEACH protocol [10] , their work assemble a robust data streams that captures the challenges of dynamic spectrum-based WSNs. This sets the platform to try clever machine-learning algorithms to solve network problems. These algorithms are essential for identifying and mitigating potential attacks across the network landscape. The authors in [11] presents an approach for detecting intrusions and classifying Denial of Service (DoS) attacks in Wireless Sensor Networks (WSNs). They first provide background on WSNs and the LEACH hierarchical routing protocol. LEACH forms clusters and uses cluster heads to aggregate and transmit data to conserve energy. The authors in [12,13] & [14], presents the related work on secure LEACH-based protocols, such as, DoS attack detection and cryptographic solutions for clustering in WSNs. Most resolve only 1–2 attacks and cannot handle multiple concurrent attacks without high overhead.

The study [15], offers a thorough analysis of the function of machine learning (ML) and deep learning (DL) techniques in addressing security and privacy issues in Internet of Things (IoT) networks. The authors first discuss the motivation for using ML in IoT security due to the limitations of traditional cryptographic solutions in coping with dynamic and heterogeneous IoT environments. Key security challenges and threat models in IoT networks are then outlined, spanning different network protocol stack layers. An overview of basic ML algorithms, including supervised and unsupervised reinforcement learning and their applications, is provided. ML techniques like support vector machines, random forests, neural networks, deep learning, and others for IoT security problems, including authentication, access control, attack detection, malware analysis, and intrusion detection systems, are reviewed. ML and DL-based solutions from existing literature are surveyed in detail for each security domain. The work in [15], highlights several limitations of current ML approaches, such as lack of interpretability in DL models, instability in reinforcement learning, susceptibility to adversarial attacks, and overfitting issues. Emerging techniques like federated learning and generative adversarial networks are discussed as promising future directions. Key challenges related to the availability of training datasets, data heterogeneity, model efficiency and complexity, real-time learning requirements, and legislative issues are also analyzed.

In summary, the related literature provides a comprehensive analysis of the role of ML in addressing IoT security issues and identifies open challenges and future opportunities in this domain. The detailed review of existing literature and discussion of limitations sets the context for developing more robust, efficient, and tailored ML solutions for securing dynamic and decentralized IoT networks.

Therefore, the main goal of this research is to improve the security of WSN data streams against the attacks by using machine learning algorithms. Literature reveals that traditional security measures may not be sufficient to handle the ever-changing nature of attacks. As a result, this research propose integrating machine learning techniques to create a powerful defense mechanism that can adapt to new attack patterns.

Our approach involves the application of deep learning algorithms to identify and classify malicious activities within the WSN environment. Specifically, we explore the effectiveness of Artificial Neural Networks (ANN), Recurrent Neural Networks (RNN), and Long Short-Term Memory (LSTM) networks in detecting and mitigating attacks. These algorithms have demonstrated success in various domains and offer the potential to enhance the security of WSN by learning patterns from data and adapting to new attack scenarios.

3 Proposed System Architecture

Machine learning is a concept that allows systems to learn from training data and automate the process of building analytical models to solve tasks. It relies on problem-specific training data to make predictions or classifications [16]. A subfield of machine learning referred as "deep learning" uses artificial neural networks. Deep neural networks with numerous layers must first be trained to extract intricate patterns and features from data. In many applications, deep learning models have shown superior performance to shallow machine learning models and traditional data analysis approaches [16]. The models used in this work are as follows:

ANN [17]**:** The human brain's biological neural networks were used as a model for artificial neural networks (ANNs), which are computer programs. An artificial neural network (ANN) comprises synthetic neurons or nodes connected in layers that transmit signals from input to output nodes. Each connection has an associated weight that can be adjusted through a learning process to improve performance. ANNs are trained using algorithms like backpropagation to modify weights and minimize error iteratively. Key advantages of ANNs include the ability to model complex nonlinear relationships, learn from examples, and generalize well on new data. ANNs have found wide application in pattern recognition, prediction, optimization, and data clustering problems across domains like finance, medicine, engineering, and science. Popular ANN architectures include multi-layer perceptrons, recurrent neural networks, convolutional neural networks, and self-organizing maps.

RNN [18]**:** An artificial neural network called a recurrent neural network (RNN) is made for processing sequential data, such as time series, audio, and text. Unlike traditional feedforward neural networks, RNNs can keep track of earlier inputs thanks to connections that loop back on themselves. This enables them to capture temporal dependencies and patterns in sequential data, making them suitable for tasks like language modeling, machine translation, and speech recognition. However, standard RNNs can struggle with capturing long-range dependencies due to vanishing or exploding gradient problems. To overcome these drawbacks and increase their efficiency in acquiring and using long-term contextual information, a number of improved RNN architectures, including Long Short-Term Memory (LSTM) and Gated Recurrent Unit (GRU), have been created.

LSTM [19]**:** The vanishing gradient problem is solved by the recurrent neural network design known as long short-term memory (LSTM), which enables learning long-term dependencies. The LSTM architecture contains special units called memory cells with an internal recurrence (a self-loop) to produce paths where gradients can flow over many steps. This allows the network to learn correlations between events separated by significant time lags. Key components of the LSTM cells are the input, output, and forget gates that modulate the flow of information. LSTMs have proven successful in time series forecasting, speech recognition, and natural language processing tasks like translation and text generation. Advantages of LSTMs include the ability to remember the long-term context, handle varied length sequences, and model temporal dynamics. Variants of LSTMs include bidirectional LSTMs and LSTMs combined with convolutional neural networks. Overall, LSTMs offer a powerful mechanism for sequential data modeling.

The Fig. 1 illustrates the step-by-step structure of the proposed system architecture. It is the model to secure the attacks on the wireless sensor networks using various deep learning techniques such as ANN, RNN and LSTM. Firstly, cluster heads receive input from sensor nodes that sense their surroundings. The data is fed into the DL models during clustering, and the models successfully classify the type of attacks.

3.1 Parameter Tuning, and Evaluation Metrics

In this work, the performance metrics used are as follows:

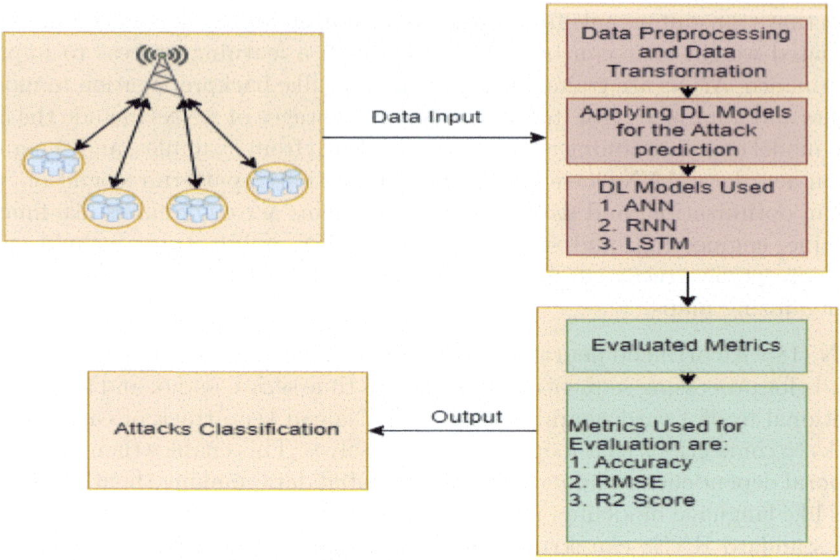

Fig. 1. System Architecture

- Accuracy: Accuracy is the sum of correctly classified positive class and correctly classified negative class divided by total number of instances
- RMSE [20]:

$$RMSE = \sqrt{(\sum((truevalues - predictedvalues)^2)/n)} \tag{1}$$

- R2-score
- Hyper-parameters used: The batch size used for ANN is 64 and; RNN and LSTM is 10. Number of hidden layers are 3.

4 Results and Discussions

We compare our models based on the various comparison metrics such as RMSE, Accuracy and R2-score. It measures how closely the actual values match the expected values. The model is closer to the actual values when the RMSE is lower. Table 1 shows the RMSE for the models as it decreases as the number of steps increases. This means the LSTM model is better at predicting the actual values as it sees more data and the RMSE score is 0.170. The RNN and ANN models also show a similar trend with the RMSE score of 0.1767 and 0.1900 respectively, but the LSTM model has the lowest RMSE for all steps. Figure 2 shows the graphical representation of RMSE trends.

It is the percentage of the variance in the actual values that the model explains. A higher R2 score indicates a better model. The Table 2 shows that the R2 score for all three models increases as the number of steps increases.

Table 1. RMSE Analysis

RMSE	RMSE			
Model	Steps			
	5	10	25	Overall
ANN	0.21	0.19	0.17	0.1900
RNN	0.22	0.13	0.18	0.1767
LSTM	0.21	0.19	0.11	**0.1700**

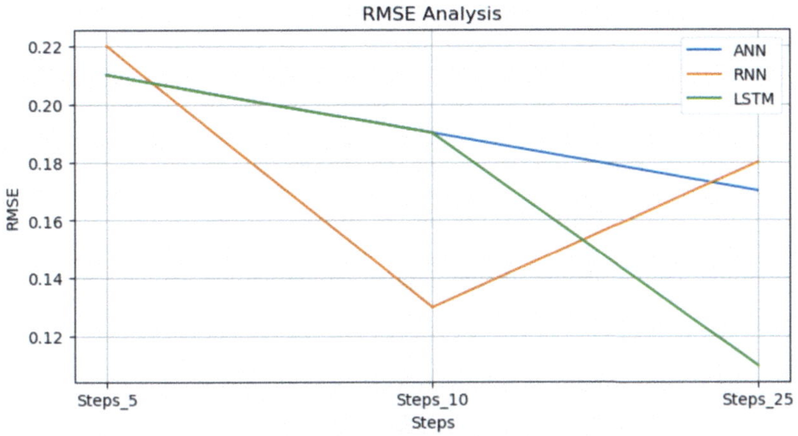

Fig. 2. RMSE Analysis Graphical Representation

Table 2. R2 Score

R2-Score	R2-Score			
Model	Steps			
	5	10	25	Overall
ANN	0.86	0.88	0.91	0.8833
RNN	0.85	0.95	0.90	0.9000
LSTM	0.86	0.89	0.96	**0.9033**

This means all three models better explain the variance in the actual values as they see more data. The LSTM model has the highest R2 score for all number of steps. Figure 3 shows that the R2 for the LSTM is highest as compare with others. LSTM has the lowest RMSE of 0.170 and the highest R2 score of 0.9033 for all steps.

Table 3 and Table 4 show the values for Training and Test Accuracy. Figure 4 and Fig. 5 show the graphical representation of Training and Test Accuracy. For training accuracy, as the number of training steps increases from 5 to 25, the ANN show an upward trend from 0.9871 to 0.9922 with an average accuracy of

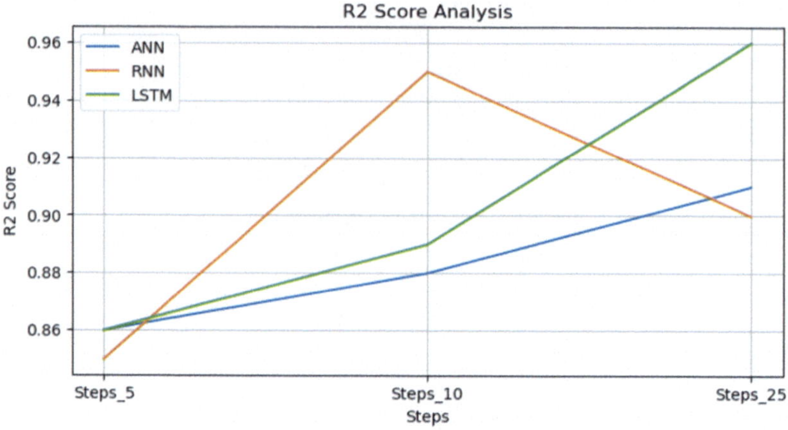

Fig. 3. R2 Score Graphical Representation

Table 3. Training Accuracy Analysis

Accuracy	Training Data Accuracy			
Model	Steps			
	5	10	25	Overall
ANN	0.9871	0.9891	0.9922	0.9895
RNN	0.9884	0.9941	0.9949	0.9925
LSTM	0.9879	0.9903	0.9945	0.9909

0.9895, for RNN the upward consistency from 0.9884 to 0.9949 with an average of 0.9925 with the results of the LSTM consistently improving their accuracy from 0.9879 to 0.9945 with the average of 0.9909. This indicates their capacity to learn and adapt to the training data. For test accuracy, similar to training accuracy, the models generally exhibit improved performance as the number of training steps increases with the overall accuracy of 0.9908. This demonstrates that the models effectively generalize what they've learned from the training data to unseen test data. Therefore, results show that the LSTM model is the best-

Table 4. Test Accuracy Analysis

Accuracy	Test Data Accuracy			
Model	Steps			
	5	10	25	Overall
ANN	0.9875	0.9894	0.9913	0.9894
RNN	0.9870	0.9948	0.9903	0.9907
LSTM	0.9870	0.9897	**0.9958**	**0.9908**

Fig. 4. Training Accuracy Analysis

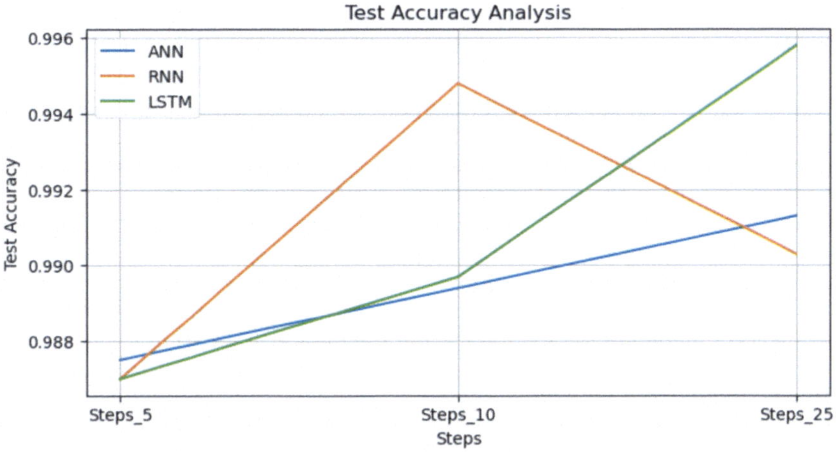

Fig. 5. Test Accuracy Analysis

performing model for this task. The marginal differences in accuracy between different models across training steps underscore the competitiveness of all three models (ANN, RNN, LSTM). The similarity between training and test data accuracy concludes that overfitting is successfully controlled.

5 Conclusion

In this work, we conducted a comprehensive analysis of different machine learning models applied to finding the Attack type of the WSN network. We aimed to identify the most effective model for accurately classifying and predicting the Attack type of the WSN network. The models evaluated in this work include Artificial Neural-Networks (ANN), Recurrent Neural-Networks (RNN), and Long Short-Term Memory (LSTM). Our findings revealed distinct performance trends across various evaluation metrics and model configurations. Furthermore, our analysis demonstrated that ANN, RNN, and LSTM models could achieve high training and testing data accuracy. Additionally, the LSTM model consistently exhibited superior RMSE values of 0.17 and R2 scores of 0.9033 for regression tasks, indicating its proficiency in capturing underlying data patterns with also the high test data accuracy of 99.08%. The execution time of LSTM was significantly higher than ANN but less than RNN, which could be attributed to its inherent complexity.

Acknowledgments. This work is supported by Thapar Institute of Engineering and Technology, Patiala, India under SEED research grant.

References

1. Gungor, V.C., Bin, L., Hancke, G.P.: Opportunities and challenges of wireless sensor networks in smart grid. IEEE Trans. Industr. Electron. **57**(10), 3557–3564 (2010)
2. Abbasi, A.A., Younis, M.: A survey on clustering algorithms for wireless sensor networks. Comput. Commun. **30**(14–15), 2826–2841 (2007)
3. Singh, S., Sharma, R.M.: HSCA: a novel harmony search based efficient clustering in heterogeneous WSNs. Telecommun. Syst. **67**, 651–667 (2018)
4. Rassam, M.A., Maarof, M.A., Zainal, A.: A survey of intrusion detection schemes in wireless sensor networks. Am. J. Appl. Sci. **9**(10), 1636 (2012)
5. Modares, H., Salleh, R., Moravejosharieh, A.: Overview of security issues in wireless sensor networks. In: 2011 Third International Conference on Computational Intelligence, Modelling & Simulation, pp. 308–311. IEEE (2011)
6. Sen, J.: Security in wireless sensor networks. Wireless Sens. Netw.: Curr. Status Future Trends **407**, 408 (2012)
7. Heinzelman, W.R., Chandrakasan, A., Balakrishnan, H.: Energy-efficient communication protocol for wireless microsensor networks. In: Proceedings of the 33rd Annual Hawaii International Conference on System Sciences, p. 10 (2000)
8. Daanoune, I., Abdennaceur, B., Ballouk, A.: A comprehensive survey on LEACH-based clustering routing protocols in Wireless Sensor Networks. Ad Hoc Netw. **114**, 102409 (2021)
9. Del-Valle-Soto, C., Rodríguez, A., Ascencio-Piña, C.R.: A survey of energy-efficient clustering routing protocols for wireless sensor networks based on metaheuristic approaches. Artif. Intell. Rev. 1–72 (2023)
10. Begum, B.A., Nandury, S.V.: Data aggregation protocols for WSN and IoT applications-a comprehensive survey. J. King Saud Univ.-Comput. Inf. Sci. (2023)
11. Almomani, I., Al-Kasasbeh, B., Al-Akhras, M.: WSN-DS: a dataset for intrusion detection systems in wireless sensor networks. J. Sens. **2016** (2016)
12. Masdari, M., Bazarchi, S.M., Bidaki, M.: Analysis of secure LEACH-based clustering protocols in wireless sensor networks. J. Netw. Comput. Appl. **36**(4), 1243–1260 (2013)
13. Tandel, R.I.: Leach protocol in wireless sensor network: a survey. Int. J. Comput. Sci. Inf. Technol. **7**(4), 1894–1896 (2016)

14. Urooj, S., Lata, S., Ahmad, S., Mehfuz, S., Kalathil, S.: Cryptographic data security for reliable wireless sensor network. Alex. Eng. J. **72**, 37–50 (2023)
15. Janiesch, C., Zschech, P., Heinrich, K.: Machine learning and deep learning. Electron. Mark. **31**(3), 685–695 (2021)
16. Hussain, F., Hussain, R., Hassan, S.A., Hossain, E.: Machine learning in IoT security: current solutions and future challenges. IEEE Commun. Surv. Tutor. **22**(3), 1686–1721 (2020). https://doi.org/10.1109/COMST.2020.2986444
17. Agatonovic-Kustrin, S., Beresford, R.: Basic concepts of artificial neural network (ANN) modeling and its application in pharmaceutical research. J. Pharmac. Biomed. Anal. **22**(5), 717–727 (2000)
18. Sherstinsky, A.: Fundamentals of recurrent neural network (RNN) and long short-term memory (LSTM) network. Phys. D **404**, 132306 (2020)
19. Huang, Z., Xu, W., Yu, K.: Bidirectional LSTM-CRF models for sequence tagging. arXiv preprint arXiv:1508.01991 (2015)
20. Willmott, C.J., Matsuura, K.: Advantages of the mean absolute error (MAE) over the root mean square error (RMSE) in assessing average model performance. Clim. Res. **30**(1), 79–82 (2005)

Experimental Analysis for Sensor Reduction to Depict Real-Time Applications Through Regression Techniques

K. Vinodha(✉), E. S. Gopi, and Bapeswar Vinnakota

National Institute of Technology Tiruchirappalli,
Tiruchirappalli 620015, Tamil Nadu, India
vinodhakamaraj@gmail.com, esgopi@nitt.edu

Abstract. The expenses and maintenance requirements rise along with the number of sensors. To solve this problem, strategically placing sensors in real-time applications is essential. This work presents a novel approach that leverages regression analysis to identify representative and non-representative sensors, leading to the removal of redundant sensors without compromising accuracy. The proof-of-concept study employed a fan as the test object to collect sensor data. However, the proposed method can be extended to real-time scenarios such as vehicles, engines, turbines, and gym or sports equipment. Establishing an approximate correlation between representative sensors (close to the item) and non-representative sensors (placed at a distance) can eliminate the need for a larger number of sensor placements. It utilizes different regression models and analyzes the gathered data to determine the use of representative and non-representative sensors. The evaluation process involves techniques such as correlation matrix and eigenvalues to identify usual sensors. A machine learning regression method is used to predict non-representative sensor outputs, including linear regressions (LR) and Gaussian regressions (GR). A mean square error (MSE) is calculated to compare the accuracy of predicted values obtained from both regression techniques. The proposed method offers several benefits, including simplified maintenance, reduced costs, and improved performance with fewer sensors. Accurately identifying representative sensors provides a practical solution for optimizing sensor deployment in real-time applications.

Keywords: Arduino Mega 2650 · Piezo Electric Vibrational Sensor · Regression Model · Correlation Matrix · Mean Square Error

1 Introduction

The number of sensors required in real-time applications can lead to increased expenses and maintenance complexities. However, by strategically placing sensors and leveraging regression analysis, it is possible to reduce the sensor count

D. Pastor-Escuredo et al. (Eds.): ICDLAIR 2023, LNNS 1001, pp. 178–194, 2024.
https://doi.org/10.1007/978-3-031-60935-0_17

without compromising accuracy. The novel approach to identifying representative and non-representative sensors using regression analysis enables cost-effective sensor deployment and improved performance.

In nature, vibration is a common occurrence. Vibration detection is one of the most significant physical measuring parameters in scientific research and engineering applications. Vibration sensors include optical fiber, piezoelectric, etc. The use of different sensors to detect vibration signals depends on the vibration measurement parameters, such as displacement, velocity, and acceleration. The piezoelectric sensor belongs to the inertial acceleration sensor [1]. Over the last few years, medical and sports researchers have focused on human activity recognition (HAR), which enables the identification of daily life activities. Inertial sensors, such as accelerometers, are wearable sensors that have proven very useful in monitoring human activity. Accelerometers calculate the sum of inertial linear and gravitational accelerations along their sensitive axis [2]. Vibration sensors are essential in industrial and medical research. Many studies have been conducted on defect diagnosis, prediction, and sensor utilization.

2 Related Work

A case study on engine failure prediction uses Deep learning algorithms for predictive maintenance. It also discussed the usage of sensors in the industry and electrochemical sensor usage in predictive maintenance in the future. The optimal sensor sites can be updated using an online regulation technique that offers essential information on the evolution of the sensitivity distributions-displaying the primary findings on a complex thermal scenario in 2D geometry. It demonstrates which sensors should be used over time when determining an unknown mobile heating flux online [4]. Compared to previous approaches, a methodology for combining domain expert knowledge and temporal data to design an IoT system was established, resulting in a highly resilient and accurate method with lower cost due to sensor reduction, size, and complexity. This was demonstrated in a case study of a mass-produced electrophotographic printer in a system meant to classify media kinds. Compared to a more traditional strategy that did not use domain expert knowledge to enhance the dataset, the proposed methodology increased classifier accuracy by 16% and classifier acceptability by 6.5% [5]. The Multi-objective Simulated Annealing algorithm performs the optimization for sensor reduction in shell structure[6]. Using the Systematic Sensor Selection Strategy (S4), developed at NASA, they reduced the sensors present at various locations for a rocket engine from 58 to 31. Representative results from S4 analyses of a boost-stage rocket engine previously under development as part of NASA's Next Generation Launch Technology (NGLT) program [7].

The existing research focuses on leveraging advanced techniques in sensor data analysis. Specifically, they emphasize the importance of efficient data processing and analysis in various domains, such as federated learning, human activity recognition, and wireless sensor networks [1–3, 6]. These studies demonstrate the effectiveness of incorporating AI-based approaches, including unsupervised representation learning, communication-efficient methods, and data reduction techniques, to enhance accuracy, scalability, and resource utilization in complex sensor-based systems. System health monitoring in real-time involves the use of more sensors. Sensor placement at appropriate locations also helps to reduce sensor count. As a result, the number of sensors on the market steadily declined.

We attempted to accomplish this by developing a regression model using Machine Learning techniques. With a modest amount of study on sensor count reduction, sensor data is now mainly used for noise reduction and problem detection. As a result, utilizing sensors to monitor the object's health and reducing the number of sensors used helps us save money on the number of sensors used and the amount of time it takes to monitor the object [8–11].

3 Proposed Methodology

This research aims to optimize sensor deployment in real-time applications by leveraging Piezo Electric Vibrational Sensors. By utilizing machine learning regression techniques, we seek to determine the representative and non-representative sensors for a proof-of-concept fan as an object. Our objective is to identify scenarios where a sensor can be removed if an approximate close value can be obtained between the sensor positioned close to the object and a sensor placed at a short distance away (Fig. 1).

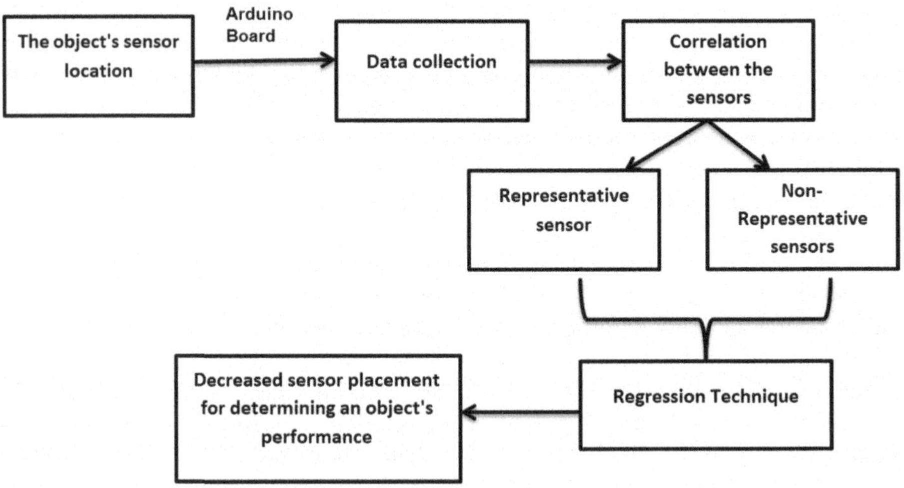

Fig. 1. Block Diagram of proposed work

The first step involved placing the Piezoelectric vibration sensors on the targeted object. These sensors were strategically positioned to capture data on the fan's behavior and performance in various modes of operation. Data was collected using the sensors, and the corresponding sensor readings were recorded using Arduino Mega 2650.

A correlation analysis was conducted to assess the relationship between the sensors. The correlation matrix was calculated using the collected dataset, revealing the link between each unique reading and other readings. This analysis allowed us to identify the sensors that exhibited strong correlations with each other.

Based on the correlation matrix, the representative sensors were identified. The values of the correlation matrix vary from 0 to 1. Sensors with high correlations (close to 1) were considered representative, while those with lower correlations were classified as non-representative (Refer Table 1). The eigenvalues for a given dataset are calculated and sorted in ascending order. They are then normalized to forecast how many representative sensors can be used for user convenience, using the most significant result as a baseline. This step helped determine which sensors could potentially be removed or replaced without significantly affecting the accuracy of the predictions. Regression techniques, such as linear regression and Gaussian regression, were applied to the dataset. These techniques allowed us to predict the target variable (non-representative sensor outputs) based on the inputs from the representative sensors. By comparing the predicted values with the actual values, the performance of each regression technique was evaluated.

Table 1. The value represents the correlation strength between two sensors.

Coefficient's	Interpretation
0.00–0.10	Negligible Correlation
0.10–0.39	Weak Correlation
0.40–0.69	Moderate Correlation
0.70–0.89	Strong Correlation
0.90–1.00	Very Strong Correlation

The Mean Square Error (MSE) was calculated to compare the accuracy of the predicted values from both the linear regression and Gaussian regression models. The MSE quantifies the average squared difference between the predicted and actual values, providing a measure of the model's performance.

3.1 Regression Technique

Regression is a statistical strategy used in machine learning to predict a continuous output based on the value of one or more input variables [12]. The collection of input vectors $x_1, x_2, \ldots, x_N$ (random vector outcome) influences the system to produce the equivalent set of output vectors $t_1, t_2, \ldots, t_N$. E(t/x) is the Minimum Mean Square Estimation (MMSE) solution. Typically, a parametric method with a linear model is used to handle this problem.

$$t_n = y(x_n, w) + \epsilon, \text{where } y\ (x_n) = w^T x_n$$

Table 2. Eigenvalues for various operating modes

Max. speed		Max. speed with osc.		Min. speed		Min. speed with osc.		Min. to max. speed		All five mode	
Desc.	Norm.	Desc.	Norm.	Desc.	Norm.	Desc.	Norm.	Desc.	Norm.	Desc.	Norm.
1.47E−05	**1.00E+00**	**1.80E−06**	**1.00E+00**	**1.58E−05**	**1.00E+00**	**2.25E−05**	**1.00E+00**	**1.04E−05**	**1.00E+00**	**3.04E−06**	**1.00E+0**
5.04E−07	3.42E−02	**5.86E−06**	**3.26E−01**	**1.79E−06**	**1.13E−01**	**2.15E−05**	**9.57E−01**	**1.07E−06**	**1.03E−01**	**1.39E−06**	**4.56E−01**
5.72E−08	3.88E−03	1.82E−08	1.01E−03	**8.76E−06**	**5.54E−01**	**6.27E−06**	**2.79E−01**	**2.59E−07**	**2.49E−02**	**9.83E−08**	**3.23E−02**
1.60E−08	1.09E−03	6.49E−08	3.61E−03	4.86E−08	3.07E−03	7.36E−08	3.28E−03	**1.41E−07**	**1.36E−02**	**4.81E−08**	**1.58E−02**
1.32E−08	8.94E−04	3.16E−08	1.75E−03	2.38E−08	1.50E−03	5.42E−08	2.41E−03	2.45E−08	2.36E−03	1.93E−09	6.35E−04
9.17E−09	6.22E−04	2.43E−08	1.35E−03	1.34E−08	8.50E−04	4.00E−08	1.78E−03	1.68E−08	1.62E−03	1.67E−09	5.48E−04
6.17E−09	4.18E−04	1.27E−08	7.04E−04	7.84E−09	4.96E−04	1.64E−08	7.32E−04	1.28E−08	1.23E−03	8.69E−11	2.86E−05
5.07E−09	3.43E−04	1.17E−08	6.53E−04	6.84E−09	4.33E−04	1.49E−08	6.64E−04	8.93E−09	8.59E−04	4.63E−11	1.52E−05
4.03E−09	2.73E−04	6.26E−09	3.48E−04	4.93E−09	3.12E−04	8.10E−09	3.61E−04	5.72E−09	5.50E−04	1.71E−11	5.60E
2.53E−09	1,71E−04	3.32E−09	1.84E−04	3.08E−09	1.94E−04	5.96E−09	2.65E−04	3.65E−09	3.52E−04	1.32E−11	4.33E−06

In our case, the dataset is divided into 80% for training and 20% for testing. There are two main types: Simple regression and Multivariable regression. Simple linear regression uses the usual slope-intercept form, with "m" and "b" as the variables, and our algorithm will attempt to "learn" to make the most accurate predictions. Our input data is represented by "x", and our target is meant by "Y".

$$Y = mx + b$$

Multi-variable linear equations are more sophisticated, where "w" represents the coefficients or weights that our model will learn.

$$F(x, y, z) = w_1x + w_2y + w_3z$$

Each observation's properties, or unique pieces of information, are represented by the variables x, y, and z. Gaussian Processes (GP) is a supervised learning method that can address problems like regression and probabilistic categorization. The prediction is probabilistic (Gaussian), so empirical confidence intervals can be computed and used to determine if the forecast should be refitted (online fitting, adaptive fitting) in a particular region of interest. Different kernels can be provided, making it versatile.

Standard kernels are included; however, custom kernels can also be specified. They are not sparse because they employ all the information from the samples/features to make the prediction. When elements exceed a few hundred, they lose efficiency in high-dimensional spaces.

Table 2 represents the experimental calculation of Eigenvalues for sensors placed at various operating modes, such as (a) Maximum Speed, (b) Maximum Speed with Oscillation, (c)Minimum Speed, (d)Minimum Speed with Oscillation, (e) Minimum to Maximum Speed (Acceleration) and (f) whole dataset, all five modes of operations.

3.2 Mathematical Model

The data set is separated into three sections: 70% for training, 20% for testing, and 10% for validation.

$$Y_{pred}(x) = \sum_{K=1}^{n} K(x, x_K)t_K$$

x is the data belonging training set of inputs (representative sensors), x_k is the data belonging validation set of inputs (representative sensors) t_k the data belonging validation set of target or outputs (non-representative sensors) K is the Gaussian Kernel, which is given by

$$K = E^{-\alpha(x-x_k)^T(x-x_k)}$$

α is changed from −1000 to 1000 in the training state of the dataset until the Mean Square Error (MSE) is the least. After finding the fair value of α, narrow the range to the exact acceptable value. Repeat the technique until the MSE falls inside the −1000 range. The result is predicted by replacing the training data with testing data and using the α gained during training. Finally, the mean square error compares the actual and expected output values (MSE). The difference between the expected and actual values is first calculated and squared to find the mean square error. The Mean Squared Error is calculated by taking the mean of the obtained values.

4 Experiments and Results

This section provides an overview of the data collection process conducted to optimize sensor deployment in a fan object. The data collection methodology involved the use of an Arduino Mega 2650 microcontroller board and a Piezoelectric vibration sensor. Additionally, the Tera Term software was employed for data acquisition and analysis (Refer Fig. 2).

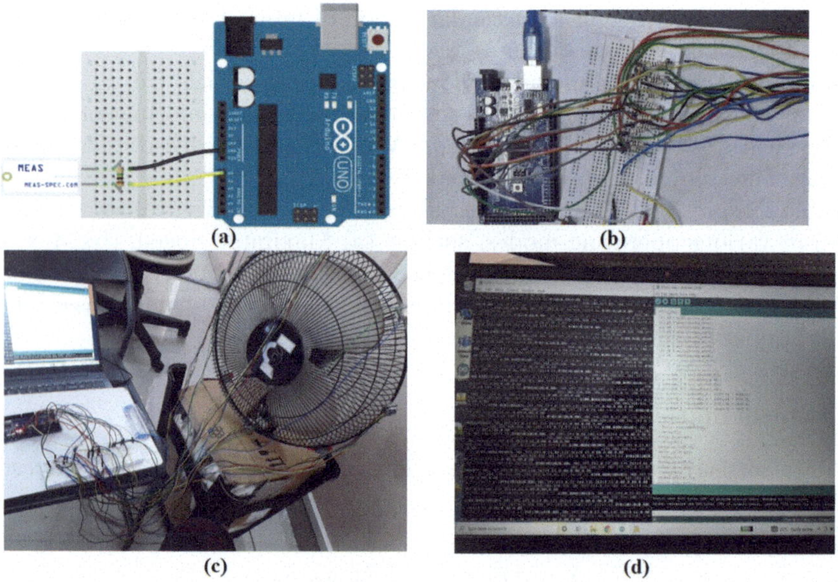

Fig. 2. (a) Circuit with 1 Piezo sensor. (b) Circuit Connections with 10 Piezo Sensors. (c) Fan and Arduino Mega 2650. (d) Tera Term Software for data collection

The fan object was selected as the target item for sensor placement. To gather comprehensive data, various modes of operation were considered, including the maximum speed, maximum speed with oscillation, minimum speed, minimum speed with oscillation, minimum to maximum speed, and all five modes. Each mode provided unique insights into the behavior and characteristics of the fan. The Piezoelectric vibration sensor was strategically positioned on top of the fan to capture relevant data. These sensors were chosen due to their capability to measure vibration and acceleration, providing valuable information about the fan's performance and operation. The Arduino Mega 2650 microcontroller board was used to collect data from the sensors. The Tera Term software facilitated real-time communication between the microcontroller and the computer, enabling the recording and storage of sensor data. Figure 3 depicts the visualization of sensor placement in various locations throughout the targeted object. The Arduino board was used to collect data on multiple modes of operation. To identify the representative sensors for each mode of operation, we conducted a thorough analysis. The correlation matrix and eigenvalues were generated to assess the relationship between the sensor outputs and the target variable. This statistical analysis provided insights into the sensors that exhibited strong correlations and influences on the fan's behavior. To validate the initially estimated representative sensors, the collected datasets were divided into training, testing, and validation subsets for each mode of operation. Regression techniques, such as linear regression and Gaussian regression, were applied to predict the target variable based on the representative sensors. The Mean Square Error (MSE) was calculated to evaluate the accuracy of the predictions and determine the optimal regression technique.

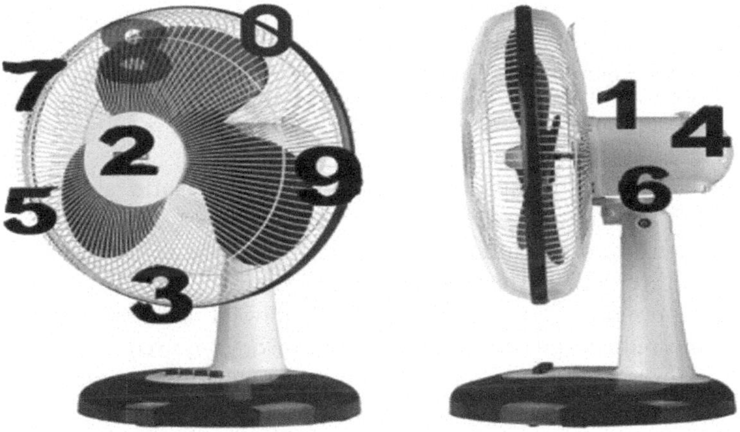

Fig. 3. Sensors placed at different positions

The data collection process enabled us to gather comprehensive sensor data for different modes of operation in the fan object. The data was collected at a rate of approximately 300 samples/sec from 10 piezoelectric vibrational sensors placed at different locations at various operating modes. The operational modes are: (a) Maximum Speed: Number of samples obtained is n = 41,905; (b) Maximum Speed with Oscillation: Number of samples obtained is n = 34,823; (c) Minimum Speed: Number of samples obtained is n = 38,631; (d) Minimum Speed with Oscillation: Number of samples obtained is n = 27,325; (e) Minimum to Maximum Speed (Acceleration): Number of samples obtained is n = 55,393 and (f) whole dataset, all five modes of operations: Number of samples obtained is n =1,98,073. This data forms the foundation for subsequent analyses and sensor optimization. By identifying the representative sensors and assessing their performance through regression analysis, we can effectively reduce the number of sensors required while ensuring accurate predictions.

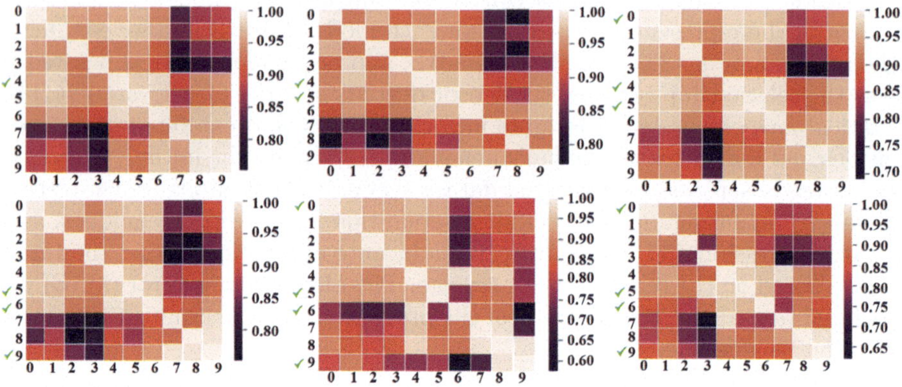

Fig. 4. The Correlation Matrix is visualized as a heat map. (a) Maximum Speed, (b) Maximum Speed with Oscillation, (c) Minimum Speed, (d) Minimum Speed with Oscillation (e) Minimum to Maximum Speed (Acceleration), and (f) whole dataset, all modes of operations

The correlation analysis provided insights into the relationships between the sensors used in the targeted fan object. By examining the correlation matrix, we identified which sensors exhibited strong correlations and could be considered representative, as well as those with weaker correlations that could be classified as non-representative (Refer Fig. 4). This allowed us to optimize the sensor placement and reduce the overall number of sensors required.

Figure 5 shows the correlation between each sensor and the findings of representative sensors at various modes of operation.

Figures 6, 7, 8, 9, 10, and 11 shows the plot between original and predicted values of the non-representative sensor outputs using representative sensors at various mode of operation. In the first case with Maximum speed, it is observed that V4 is the representative sensor, and V0, V1, V2, V3, V5, V6, V7, V8, and V9 are the non-representative sensors. In case two, Maximum speed with oscillation, V4 and V5 are the representative sensors, and V0, V1, V2, V3, V6, V7, V8, and V9 are the non-representative sensors. Third case with Minimum speed mode: V0, V4, and V5 are the representative sensors, and V1, V2, V3, V6, V7, V8, and V9 are the non-representative sensors. Fourth case Minimum to maximum speed mode: V0, V4, V5, and V9 are the representative sensors, and V1, V2, V3, V6, V7, and V8 are the non-representative sensors. Fifth case Minimum speed with oscillation modes V4, V5, and V9 are the representative sensors, and V0,

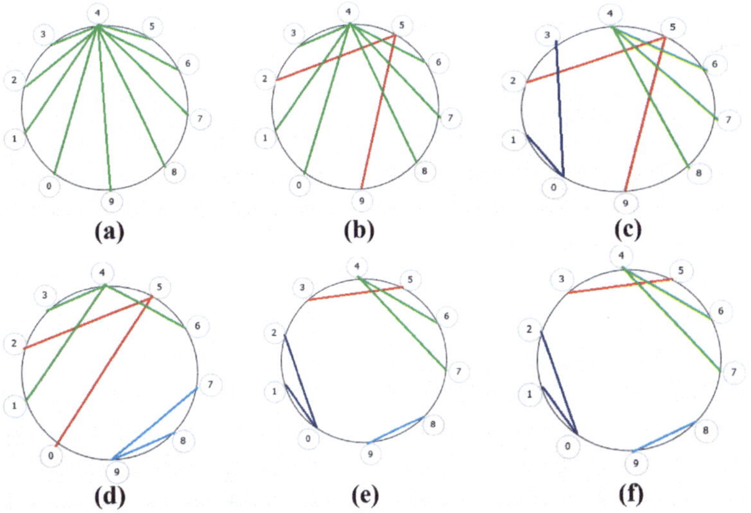

Fig. 5. A model of a representative sensor. (a) Maximum Speed, (b) Maximum Speed with Oscillation, (c) Minimum Speed, (d) Minimum Speed with Oscillation (e) Minimum to Maximum Speed (Acceleration), and (f) whole dataset, all modes of operations

V1, V2, V3, V6, V7, and V8 are the non-representative sensors. Final sixth case with all five modes together(combining all operational mode dataset together) it is found that V0, V4, V5, and V9 are the representative sensor, and V1, V2, V3, V6, V7, and V8 are the non-representative sensors. In all cases, the predicted values are closer to the original values. Figure 12 represents the findings of the total number of representative sensors that can replace all ten sensors.

The regression techniques, specifically linear regression, and Gaussian regression, were applied to predict the values of the non-representative sensor outputs using the readings from the representative sensors. The MSE values obtained for both regression models were compared to evaluate their performance. Lower MSE values indicate more accurate predictions. The results showed that both linear regression and Gaussian regression models provided reasonable predictions for the non-representative sensor outputs. However, the MSE analysis indicated that the linear regression model outperformed the Gaussian regression model, resulting in lower prediction errors. Therefore, the linear regression model was identified as the optimal approach for predicting the non-representative sensor outputs in the fan object.

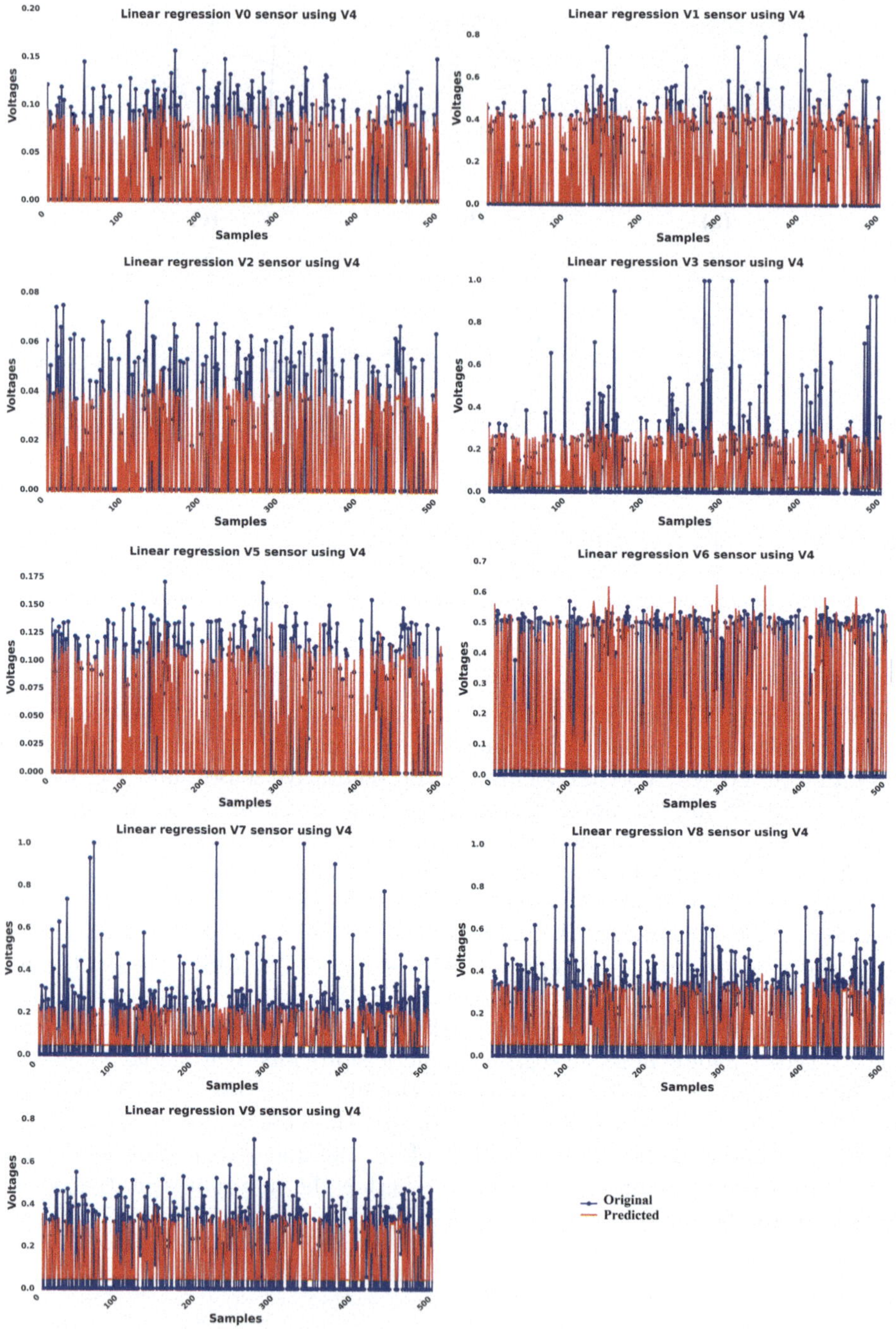

Fig. 6. Maximum speed mode: Original and predicted values of non-representative sensors using representative sensors.

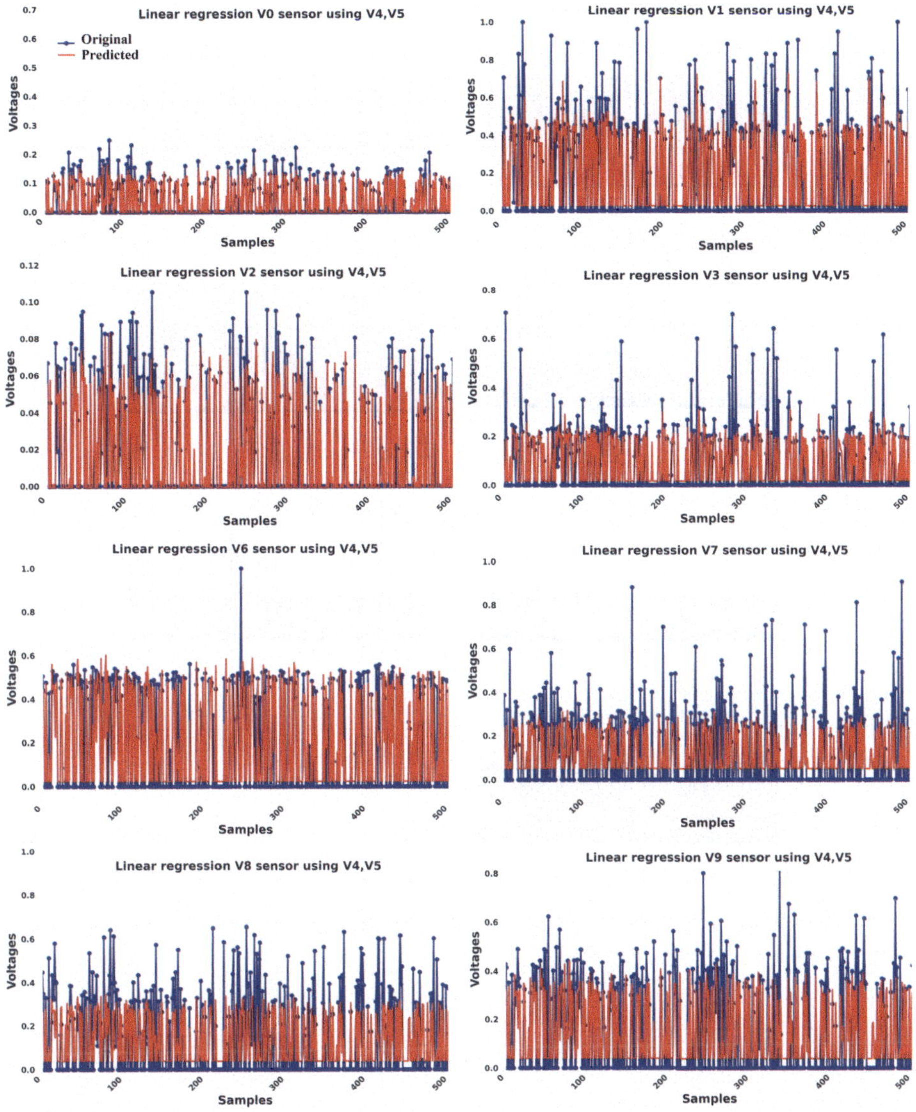

Fig. 7. Maximum speed with oscillation mode: Original and predicted values of non-representative sensors using representative sensors.

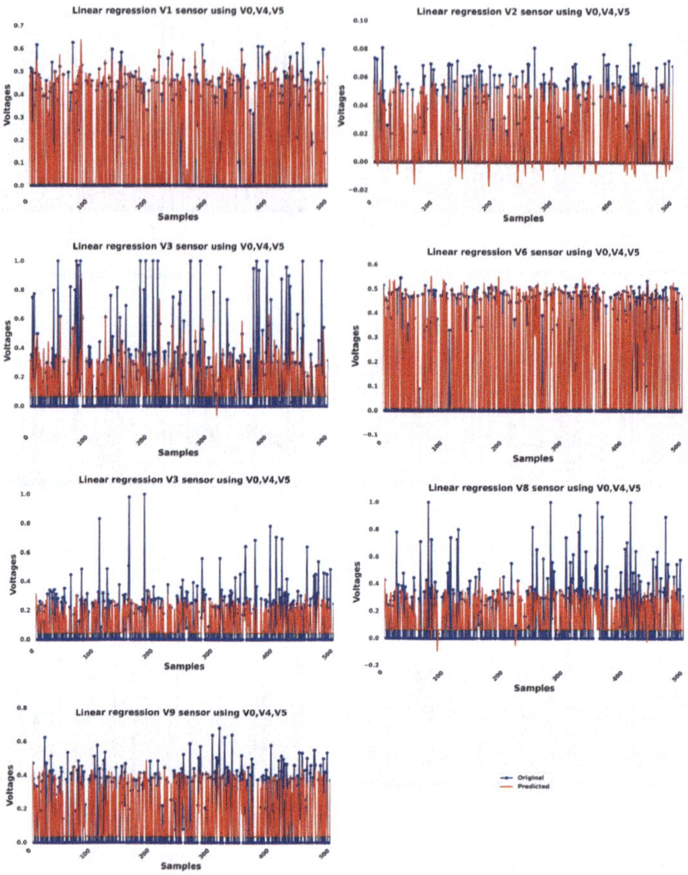

Fig. 8. Minimum speed mode: Original and predicted values of non-representative sensors using representative sensors.

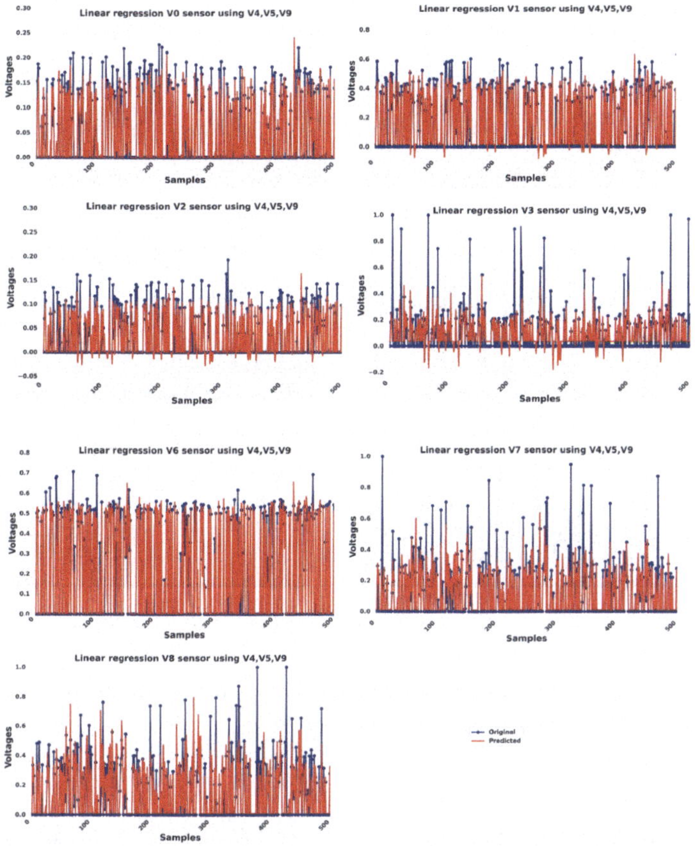

Fig. 9. Minimum speed with oscillation mode: Original and predicted values of non-representative sensors using representative sensors.

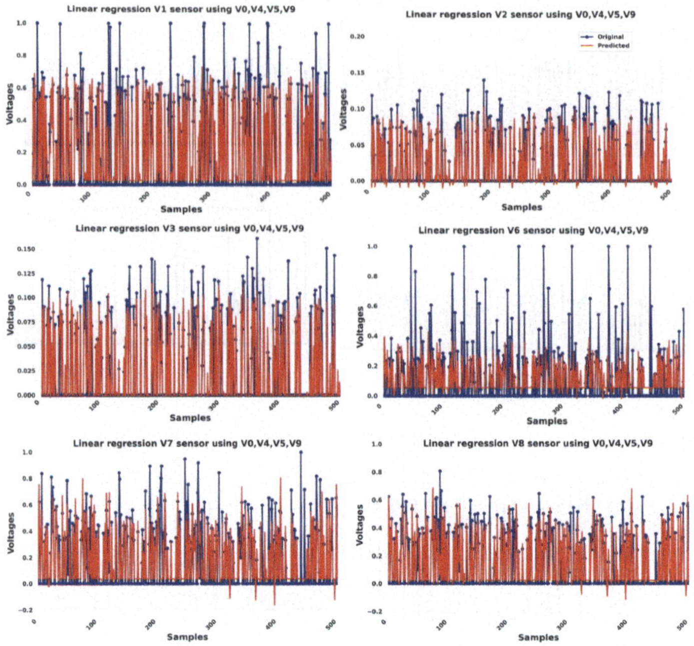

Fig. 10. Minimum to maximum speed mode: Original and predicted values of non-representative sensors using representative sensors.

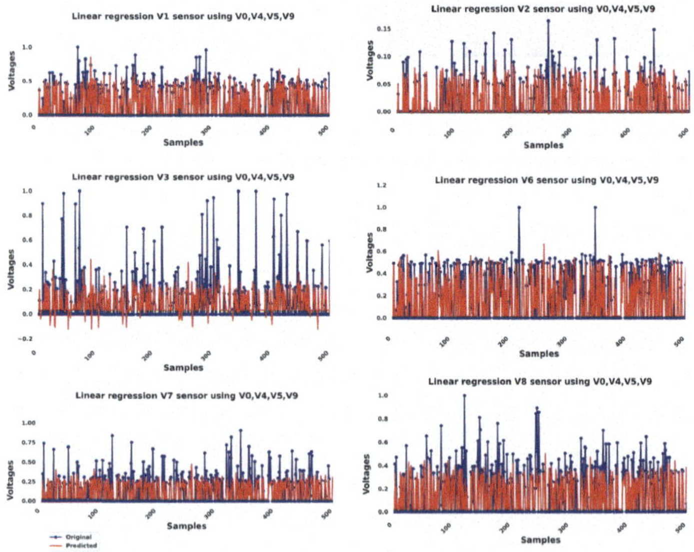

Fig. 11. All five modes together (maximum speed, maximum speed with oscillation, minimum speed, minimum speed with oscillation, and minimum to maximum Speed): Original and predicted values of non-representative sensors using representative sensors.

Fig. 12. Final four representative sensors that can replace all 10 sensors.

5 Conclusion

In this study, we successfully employed a methodology for optimizing sensor placement at different locations on the target object based on regression and correlation analyses. We reduced the required sensors by identifying representative and non-representative sensors using the correlation matrix. The corresponding mean square error (MSE) is calculated while maintaining accurate predictions of the non-representative sensor outputs by applying linear and Gaussian regression techniques. It was observed that the Linear regression technique gives less MSE **0.01805**. Sensor number **V0, V4, V5 and V6 can replace all the 10** sensors to get minimum MSE. The regression techniques employed in our work revealed that they provide superior predictions. This finding suggests that the regression model is better suited for predicting the non-representative sensor outputs in the fan object.

Overall, this methodology offers a practical approach to sensor optimization, enabling cost reduction and improved performance in real-time applications. By selecting representative sensors and utilizing regression techniques, accurate predictions can be achieved with a smaller number of sensors, leading to easier maintenance, cost reduction, and enhanced performance.

References

1. Arora, G.D., Chopra, N.K., Gopinath, N., Ravichandran, S.K., Chakravarthi, M.K., Gangodkar, D.: Data Reduction techniques in wireless sensor networks with AI. In: IEEE 5th International Conference on Contemporary Computing and Informatics (IC3I), Uttar Pradesh, India (2022)

2. Shrestha, S., Fu, X.: Communication-efficient federated linear and deep generalized canonical correlation analysis. IEEE Trans. Signal Process. **71**, 1379–1394 (2023)
3. Takenaka, K., Hasegawa, T.: Unsupervised representation learning method in sensor based human activity recognition. In: IEEE International Conference on Machine Learning and Cybernetics (ICMLC), Japan (2022)
4. Chao, J., Ruihong, J., Wen, H., Jiani, H.: Comparative experiments of optical fiber sensor and piezoelectric sensor based on vibration detection. In: IEEE 4th International Conference on Frontiers of Sensors Technologies (ICFST), Shanghai, China (2020)
5. Burgos, C.P., et al.: In-ear accelerometer-based sensor for gait classification. IEEE Sens. J. **20**(21), 12895–12902 (2020)
6. Vergnaud, A., Perez, L., Autrique, L.: Adaptive selection of relevant sensors in a network for unknown mobile heating flux estimation. IEEE Sens. J. **20**(24), 15133–15142 (2020)
7. Namuduri, S., Narayanan, B.N., Davuluru, V.S.P., Burton, L., Bhansali, S.: Review-deep learning methods for sensor based predictive maintenance and future perspectives for electrochemical sensors. J. Electrochem. Soc. **167**(3) (2020)
8. Murrell, N., Bradley, R., Bajaj, N., Whitney, J.G., Chiu, G.T.-C.: A method for sensor reduction in a supervised machine learning classification system. IEEE/ASME Trans. Mechatron. **24**(1), 197–206 (2019)
9. Weickgenannt, M., Neuhaeuser, S., Henke, B., Sobek, W., Sawodny, O.: Optimal sensor placement for state estimation of a thin double-curved shell structure. Mechatronic **23**(1), 346–354 (2013)
10. Santi, L.M., Sowers, T.S., Aguilar, R.B.: Optimal sensor selection for health monitoring systems. In: 41st AIAA/ASME/SAE/ASEE Joint Propulsion Conference and Exhibit. National Aeronautics and Space Administration, Glenn Research Center, Cleveland (2005). https://doi.org/10.2514/6.2005-4485
11. Schober, P., Boer, C., Schwarte, L.A.: Correlation coefficients: appropriate use and interpretation. Anesth. Analg. **126**(5), 1763–1768 (2018). https://doi.org/10.1213/ANE.0000000000002864
12. Gopi, E.S.: Pattern Recognition and Computational Intelligence Techniques Using Matlab, pp. 69–120. Springer, Cham (2020). https://doi.org/10.1007/978-3-030-22273-4

Multi-resolution Neural Network for Road Scene Segmentation

I. Dheivya(✉) and Gurunathan Saravana Kumar

Indian Institute of Technology Madras, Chennai 600 036, India
idheivya@gmail.com, ed16d010@smail.iitm.ac.in, gsaravana@iitm.ac.in
https://ed.iitm.ac.in/~gsaravana/

Abstract. This study presents a deep neural network for semantic segmentation in road scene images. We have used the convolution neural network in the encoder-decoder format. The architecture consists of convolution layers, max-pooling layers, batch normalization layers, and ReLU as nonlinear layers in contracting and expanding paths. This study uses the Cambridge-driving Labeled Video Database (CamVid) dataset for implementation. The multi-resolution convolution blocks in the proposed neural network enable hierarchical feature learning and contextual understanding. In scenarios like road scenes, where the objects are of different sizes, the multi-resolution convolution blocks provide the adaptability to label the pixels efficiently. We trained the networks with controlled hyperparameters. Also, we compare the performance of our proposed neural network with the existing state-of-the-art models. Hausdorff distance is noted as 199.69 ± 31.83, 204.82 ± 28.78, 202.21 ± 30.92, 203.57 ± 32.39, 195.46 ± 30.12, 200.37 ± 31.48 and 135.36 ± 23.34 for Pretrained SegNet, SegNet, VGG16, MobileNetV2, Pretrained DeepLabv3+, Pretrained FCN and the proposed model, respectively. Hausdorff's distance of the proposed model emphasizes efficient delineation of object boundaries, and the overall quality of the segmented images is analyzed using mean Intersection of Union (mIoU), mean accuracy, and mean BF score. Observation shows that the proposed network is robust in segmenting the less frequent classes and class imbalance objects.

Keywords: road scene segmentation · multi-resolution CNN · encoder-decoder · CamVid dataset

1 Introduction

Autonomous driving has made significant progress in recent years, owing primarily to developments in artificial intelligence and deep learning technology. Road scene segmentation, a crucial task in autonomous driving, involves identifying and delineating various objects and regions within a scene, focusing on distinguishing the road and its surroundings. This task is pivotal for self-driving vehicles to accurately assess their surroundings, navigate safely, and make intelligent decisions. Because of differences in lighting, weather, road conditions, and

D. Pastor-Escuredo et al. (Eds.): ICDLAIR 2023, LNNS 1001, pp. 195–205, 2024.
https://doi.org/10.1007/978-3-031-60935-0_18

occlusions, simple image processing techniques cannot produce consistent results in complicated road situations. With the advent of deep learning, particularly convolutional neural networks (CNNs), road scene segmentation has undergone a paradigm shift, achieving state-of-the-art performance and offering numerous benefits for automatic driving. The advantages include precision, real-time processing that adapts to diverse road conditions, generalization across different environments that may require understanding occlusion and semantics, and continuous improvement through learning.

This study proposes a novel CNN architecture to obtain better accuracy in road scene segmentation problems. Low-level and high-level feature extraction can be effectively done at the same level of the deep neural architecture using multi-resolution feature extraction. The encoder and decoder layers can deploy multiple kernels of different dimensions to capture low-level features like edges, textures, color gradients, and the high-level semantic segmentation map. This multi-resolution feature extraction will help for a comprehensive understanding of the road scene and improve overall segmentation accuracy.

This paper is organized as follows: Sect. 2 discusses the related works in road scene segmentation; Sect. 3 describes the proposed neural network model, dataset, and experimental setup; Sect. 4 discusses the results and the performance analysis; and Sect. 5 provides a conclusion and future direction.

2 Related Works

Memory-efficient models have proven with higher scores in categorizing the objects in the images from ImageNet and COCO datasets [7]. Multi-resolution images used in training enabled us to solve the bottleneck between the number of samples available when it is insufficient to train the deep neural network [10]. Radial Basis function (RBF) kernel is used in the convolutional neural layers as the activation for a stable performance [2]. Due to the stability issues with RBF, DeepLABNet requires additional training time as it is a deep model and because of its higher number of learnable. This results in a trade-off between overfitting and larger variance. Other than model selection for an application, another challenge exists in initializing the model weights. Instead of starting with zero weights for the convolution layers, weight initialization methods are used for faster minimization of loss function. Xavier method helps to avoid the vanishing or exploding gradient problem [1], and the He method addresses the dead neuron. The dynamics of the model behave differently when the non-linearity, weight initialization, and the features of interest in the input change [5]. The Intermediate deconvolution layers learn the weights through bilinear Upsampling. Final deconvolutional layers are fixed with linear interpolation weights [10].

Numerous architectures utilize intricate methods and multi-stage training procedures to attain impressive accuracy on road scene segmentation. To tackle the complications in the evaluation, controlled benchmarking by batch normalization for uniform training is equipped [13]. DeepLabV3 [6] utilized a modified atrous convolution pyramidal pooling strategy with the incorporation of

Image Pooling to circumvent the loss of contextual information. These architectural choices facilitate the concurrent learning of both low-level and high-level intricacies through the utilization of dilation and encoder-decoder techniques. Lightweight bilateral asymmetric residual network (LBARNet) extracts features across multiple network stages that leverage shallow, dense texture and location data to capture spatial details that achieved 67.2% segmentation accuracy on the CamVid dataset [15]. BiSeNet V3 uses the Sobel operator for conventional edge detection that would require fine annotated images [12]. Generalizing the Sobel operator parameter for a wide range of training images may affect the network's performance. Default filters in the U-Net architecture are replaced with the Gabor filters, which gives efficient spatial locality detection with increased directional selectivity of 65.47% meanIoU [9]. The boundary-aware semantic segmentation model (BASeg) with ResNet101 architecture as the backbone, pretrained on the Cityscapes dataset, excels in the CamVid road scene segmentation with 77.3% meanIoU [14].

In road-scene understanding, numerous networks evolved post the introduction of FCN [10], leveraging ImageNet backbone networks [4,8,11] designed for ILSVRC [3]. Despite adopting dilated convolutions for a broader field of view, their deep structure and dense convolutions posed challenges in real-time deployment on devices with limited computational resources. It is critical to bridge this gap for neural networks to generalize effectively in diverse tasks using the CamVid dataset.

The contributions presented in this work encompass:

1. Proposed a novel neural network model for road scene segmentation on the Cambridge-driving Labeled Video Database (CAMVID) dataset
2. Conducted a comprehensive comparative analysis with existing models, highlighting performance using metrics: mean accuracy, mean IoU, mean BF score, and Hausdorff distance, demonstrating the model's effectiveness

3 Proposed Work

This section explains the proposed neural model for segmentation, dataset, and benchmarking models.

3.1 Proposed Neural Network

This study introduces a CNN for semantic segmentation in the road scene frames. The proposed model employs various kernel dimensions in the specific layers to improve feature learning. A multi-resolution neural network captures fine and coarse details, enabling effective feature extraction across different scales and improving the model's ability to delineate objects with varying sizes and complexities. The proposed model is illustrated in Fig. 1. We used 3 * 3 and 5 * 5 kernels for the convolution layers, 2 * 2 kernels, and stride 1 for the max pooling layers. Different kernel dimensions in the convolution layers enhance the segmentation accuracy, fostering robust performance across diverse spatial scales and contributing to the model's adaptability in road scene segmentation. To effectively prevent the vanishing gradient problem, we used the Rectified Linear Unit (ReLu). The loss function used is cross-entropy. The model uses batch normalization layers and is trained for 100 epochs to prevent overfitting issues. The proposed network has weights initialized by the Gloret algorithm. We used mini-batch size 8 and a learning rate of 0.001 without dropout.

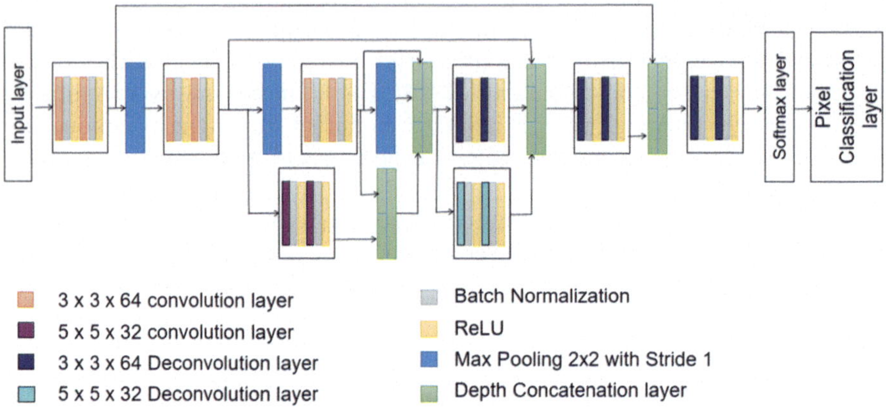

Fig. 1. Illustration of the proposed neural network model

3.2 Dataset

The CamVid dataset is pivotal for road scene segmentation research, providing high-quality annotations and real-world scenarios. Its benchmark status fosters algorithm development, enabling robust model evaluation. The dataset's size and diversity cater to autonomous vehicle applications, advancing computer vision solutions for road safety. From the CamVid dataset, 701 labeled road scene images are used. We divided the samples into 561 images for training, 70 for validation, and 70 images for testing the network's performance. Samples are RGB images of dimension $960 * 720 * 3$. Road scene frames have objects to be labeled with one of the 11 classes: sky, building, tree, road, pavement, fence, pedestrian, column pole, sign symbol, car, and bicyclist. Few classes, like column poles and sign symbols, have fewer pixels than road, sky, building, etc. We used the representation of the classes in the images with the existing class imbalance to train and test.

3.3 Benchmarking

The effectiveness of our suggested model in labeling the objects in the road scene is evaluated in comparison to other state-of-the-art models for the segmentation technique. We implemented SegNet, VGG16, DeepLabv3+, MobileNetV2, and Fully connected network (FCN) for the comparison study. DeepLabv3+ uses ResNet18 weights, while SegNet and the decoder of pretrained SegNet weights are initialized using the He technique. FCN and the encoder network of pretrained SegNet used the weights from VGG16 trained on the ImageNet dataset. VGG16 and MobileNetV2 have weights initialized using the Gloret algorithm. We chose a learning rate of 0.001 and a mini-batch size of 8 for all the models to create a controlled environment for training the networks. When the network is trained, processing the samples in a batch helps to generalize the weights for a collection of samples. As a loss function for all the models, we train them using cross-entropy. On a Titan Xp GPU, training and testing are conducted using MATLAB 2021a.

4 Results and Discussion

Mean intersection over union (mIoU), mean accuracy, global accuracy, and mean F1 score are used to evaluate the performance of the proposed model with state-of-the-art methods. These overlap-based metrics are generally more sensitive to class imbalance than distance-based metrics like Dice score and Hausdorff distance. The comparison of networks at different iterations is given in the Table 1. Pretrained models perform better in the initial phase (less than 300 iterations). The proposed model quickly gains better performance with increased iterations and shows marked improvement in global accuracy. The performance of the networks is illustrated with example segmentation in Fig. 2. Samples here correspond to the highest mean IoU from the test set. The proposed model shows better performance in demarcating the boundaries. Under controlled training parameters, proposed model exhibits a seamless segmentation of all classes compared to the pretrained networks used in the benchmarking. Without any pre- and post-processing of the input samples and outcomes, proposed model shows a better and smoother delineation of the small classes.

CamVid dataset has a class imbalance problem, which we use with inherent different labels as such. Therefore, to have a more balanced assessment of segmentation performance, the Dice score and Hausdorff distance are used to evaluate the segmentation performance. Table 2 gives the classwise Dice score of all the networks. Amongst the object classes, the road and the sky are best segmented by all the networks. The object classes pole and cyclist are segmented with the least performance metric. This is mainly due to the class imbalance inherently present in the data. Amongst the networks, the performance of the proposed model is better for all classes as compared with the networks used for the study. Hausdorff distance-based performance evaluation is presented in Fig. 3. The box plot shows the outliers using red plus symbols. The average performance of the various networks tested in the study is similar, but the standard deviation is the least with SegNet. The proposed model shows good performance compared to the other networks in the mean performance and has no outlier performance. The results demonstrate that the proposed network has better segmentation performance than the other networks considered for the study.

4.1 Discussion

The Table 1 at each iteration, global accuracy is higher than the mean accuracy for all network models. This is due to the class imbalance in multiple classes and the diverse complexities of different classes. While global accuracy provides an overall view of the model's performance, mean accuracy offers a deeper understanding of how well the model performs for each class. All the models demonstrate good learning with their increasing values for metrics over the iterations. DeepLabv3+ shows better values at the initial iterations because of the efficient weights initialized from the training on a larger dataset. At iteration 14000, we could see the DeepLabv3+ is progressing to better metrics values, but the FCN outperforms in global accuracy, Mean IoU, and Mean BF score. The proposed model has the highest mean accuracy at iteration 14000. At the end of the training, i.e., at iteration 28000, proposed model's progressively increasing scores suggest a more comprehensive learning process, potentially capturing intricate and context-dependent features of the classes. At the end of 100 epochs (28000 iterations), the mean IoU of the proposed model is 71.82% and is comparable and better to the values 70.8%, 60.10%, and 65.47% in the related studies [15],[13] and [9], respectively. The proposed model has marginally better global accuracy, 92.46%, than the value of 91.44% mentioned in one related study [9].

Table 2 shows performance metrics of segmentation class-wise. The highest Dice score in each of the classes is given in bold. The dice score of the proposed model signifies a robust alignment between the network's prediction and the label, indicative of a high-performing model. The imbalanced class distribution influences the Dice score in the images. This can be clearly seen in the Dice score of the small-scale objects like poles, sign symbols, fences, and bicyclists. In the road scene images, the small class objects present along the large classes, like roads, pavement, trees, and buildings, are affected by occlusion in their background. For example, in the sample images given in Fig. 2, all models have difficulty identifying the sign board present just before the tree.

As an outlier-sensitive metric, Hausdorff distance effectively captures extreme deviations between the boundaries in the prediction and the labels. Figure 3 gives the central tendencies, dispersion, and potential outliers for each model's performance on the test set. The medians of the boxes are shown in the red line, and the dotted lines represent whiskers. The negative skew in our proposed neural network model indicates that the predictions have lower boundary deviations from the ground truth. This suggests that the model captures intricate road scene details and maintains accuracy. The larger positive whiskers of other models display challenges in accurately capturing complex structures or handling variations in the position of the objects or boundary determination due to occlusions.

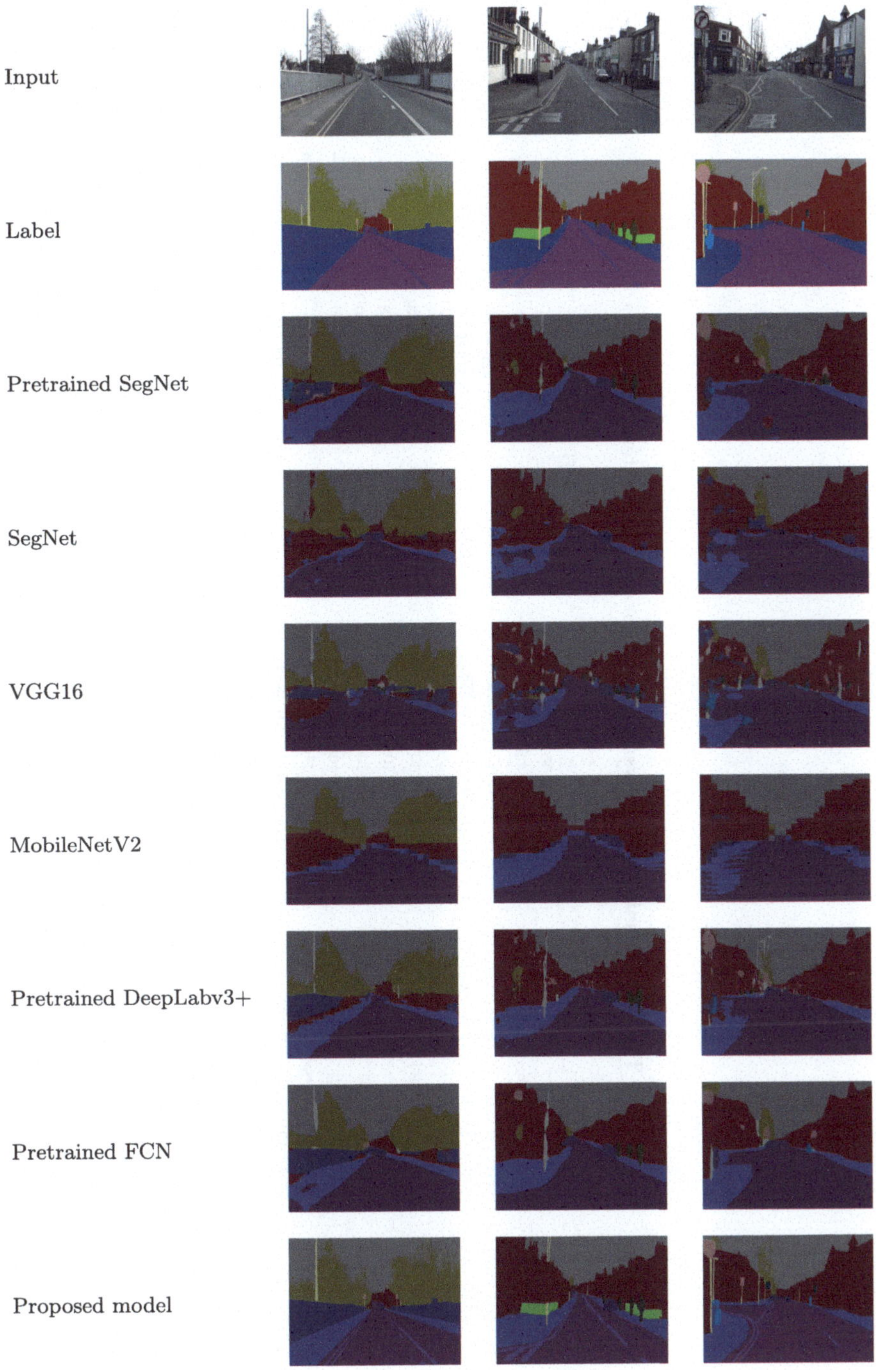

Fig. 2. Results of segmentation by different models

Table 1. Quantitative analysis of networks at different iterations

Iteration Number	Metrics	Pretrained Segnet	Segnet	VGG16	MobileNetV2	Pretrained DeepLabv3+	Pretrained FCN	Proposed model
280	Mean Accuracy	0.1889	0.2121	0.1553	0.0915	**0.5977**	0.5586	0.3747
	Global Accuracy	0.5403	0.4905	0.4246	0.2998	**0.8941**	0.8910	0.3968
	Mean IoU	0.1243	0.1290	0.0876	0.0278	**0.5189**	0.4914	0.2343
	Mean BF score	0.1518	0.1529	0.0871	0.1581	**0.5503**	0.5859	0.2157
14000	Mean Accuracy	0.5376	0.4530	0.5071	0.3407	0.6576	0.6712	**0.6891**
	Global Accuracy	0.9055	0.8518	0.8507	0.7839	0.9113	**0.9163**	0.8001
	Mean IoU	0.4703	0.3696	0.4109	0.2795	0.5866	**0.5908**	0.5184
	Mean BF score	0.5611	0.5448	0.4601	0.4490	0.6270	**0.6360**	0.5086
28000	Mean Accuracy	0.6236	0.4612	0.5462	0.4012	0.6976	0.6992	**0.7097**
	Global Accuracy	0.9084	0.8551	0.8276	0.8078	0.9149	0.9175	**0.9246**
	Mean IoU	0.5267	0.3808	0.4101	0.3239	0.6103	0.6110	**0.7182**
	Mean BF score	0.5731	0.5054	0.4493	0.4897	0.6475	0.6497	**0.6499**

Table 2. Performance analysis using Dice score

Region of Interest	Pretrained Segnet	Segnet	VGG16	MobileNetV2	Pretrained DeepLabv3+	Pretrained FCN	Proposed model
Sky	0.91 ± 0.02	0.94 ± 0.02	0.94 ± 0.03	0.92 ± 0.03	0.96 ± 0.02	0.95 ± 0.03	**0.97 ± 0.02**
Building	0.85 ± 0.14	0.78 ± 0.18	0.74 ± 0.18	0.75 ± 0.20	0.83 ± 0.17	0.85 ± 0.16	**0.93 ± 0.06**
Pole	0.14 ± 0.11	0.0 ± 0.0	0.17 ± 0.12	0.0 ± 0.0	0.30 ± 0.16	0.18 ± 0.16	**0.37 ± 0.09**
Road	**0.97 ± 0.02**	0.94 ± 0.04	0.93 ± 0.05	0.92 ± 0.05	**0.97 ± 0.02**	**0.97 ± 0.02**	**0.97 ± 0.02**
Pavement	0.88 ± 0.06	0.75 ± 0.12	0.75 ± 0.17	0.60 ± 0.17	0.86 ± 0.08	0.87 ± 0.07	**0.94 ± 0.08**
Tree	0.69 ± 0.18	0.63 ± 0.21	0.59 ± 0.23	0.43 ± 0.32	0.64 ± 0.23	0.67 ± 0.20	**0.87 ± 0.12**
Sign Symbol	0.33 ± 0.19	0.0 ± 0.0	0.19 ± 0.17	0.0 ± 0.0	0.34 ± 0.24	0.38 ± 0.23	**0.51 ± 0.14**
Fence	0.39 ± 0.24	0.14 ± 0.14	0.28 ± 0.18	0.0 ± 0.01	0.48 ± 0.31	0.48 ± 0.35	**0.51 ± 0.02**
Car	0.64 ± 0.24	0.43 ± 0.25	0.46 ± 0.26	0.25 ± 0.27	0.65 ± 0.24	0.65 ± 0.22	**0.84 ± 0.16**
Pedestrian	0.27 ± 0.24	0.0 ± 0.0	0.19 ± 0.18	0.0 ± 0.0	0.37 ± 0.27	0.39 ± 0.30	**0.64 ± 0.12**
Bicyclist	0.13 ± 0.18	0.01 ± 0.01	0.31 ± 0.24	0.0 ± 0.0	0.30 ± 0.37	0.33 ± 0.33	**0.35 ± 0.22**

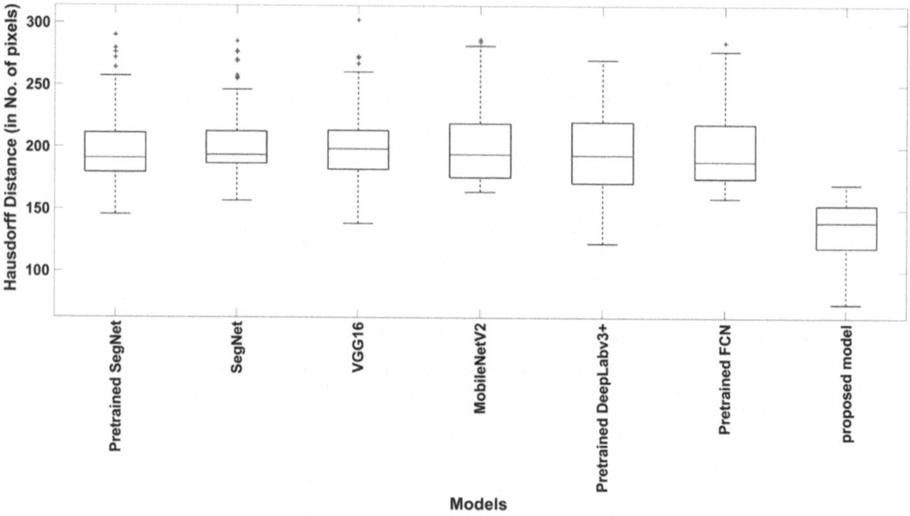

Fig. 3. Comparison of Hausdorff distance of various neural network models

5 Conclusion

A neural network that can extract multi-scale features simultaneously is proposed for road scene segmentation on the CamVid dataset. We evaluated the proposed CNN-based model with state-of-the-art neural network models. The proposed model showcases better accuracy and robustness over the iterations. Quantitative comparison at different iterations indicates the model's better pixel-level accuracy and adaptability in the given dataset. This could probably be improved further with real-world scenarios having more representation from the small class objects. Dice score and Hausdorff distance-based comparison prove the overall effectiveness of the proposed network in segmenting each class and the boundary delineation. Increasing the dataset size may empower the identification of potential shortcomings in boundary delineation, aiding in the iterative refinement of neural network architectures to enhance road scene segmentation precision for the small class objects. The future work includes conducting ablation studies to analyze the impact of model components on performance, providing insights for future refinement and optimization. The prospective direction of this research finds scope in self-supervised learning. Model generalization for the real-world deployment shall be focused on using transfer learning across diverse scenarios using attention mechanisms.

References

1. Glorot, X., Bengio, Y.: Understanding the difficulty of training deep feedforward neural networks. In: Proceedings of the Thirteenth International Conference on Artificial Intelligence and Statistics, pp. 249–256. JMLR Workshop and Conference Proceedings (2010)
2. Hryniowski, A., Wong, A.: DeepLABNet: end-to-end learning of deep radial basis networks. J. Comput. Vision Imaging Syst. **5**(1), 1 (2020). https://openjournals.uwaterloo.ca/index.php/vsl/article/view/1663
3. Deng, J., Dong, W., Socher, R., Li, L.-J., Li, K., Fei-Fei, L.: ImageNet: a large-scale hierarchical image database. In: IEEE Conference on Computer Vision and Pattern Recognition, pp. 248–255 (2009). https://doi.org/10.1109/CVPR.2009.5206848
4. He, K., Zhang, X., Ren, S., Sun, J.: Deep residual learning for image recognition. In: IEEE Conference on Computer Vision and Pattern Recognition (CVPR), Las Vegas, NV, USA, pp. 770–778 (2016). https://doi.org/10.1109/CVPR.2016.90
5. Kumar, S.K.: On weight initialization in deep neural networks. arXiv (2017). http://arxiv.org/abs/1704.08863
6. Chen, L.C., Papandreou, G., Schroff, F., Adam, H.: Rethinking atrous convolution for semantic image segmentation. arXiv (2017). https://doi.org/10.48550/arXiv.1706.05587
7. Sandler, M., Howard, A., Zhu, M., Zhmoginov, A., Chen, L.-C.: MobileNetV2: inverted residuals and linear bottlenecks. In: IEEE/CVF Conference on Computer Vision and Pattern Recognition (2018). https://doi.org/10.1109/CVPR.2018.00474
8. Liu, S., Deng, W.: Very deep convolutional neural network based image classification using small training sample size. In: 3rd IAPR Asian Conference on Pattern Recognition (ACPR), Kuala Lumpur, Malaysia, pp. 730–734 (2015). https://doi.org/10.1109/ACPR.2015.7486599
9. Nemade, S.B., Sonavane, S.P.: Semantic segmentation using GSAUNet. ICT Express (2023). https://doi.org/10.1016/j.icte.2022.09.007
10. Shelhamer, E., Long, J., Darrell, T.: Fully Convolutional networks for semantic segmentation. IEEE Trans. Pattern Anal. Mach. Intell. (2017). https://doi.org/10.1109/TPAMI.2016.2572683
11. Szegedy C., et al.: Going deeper with convolutions. In: IEEE Conference on Computer Vision and Pattern Recognition (CVPR), Boston, MA, USA, pp. 1–9 (2015). https://doi.org/10.1109/CVPR.2015.7298594
12. Tsai, T.-H., Tseng, Y.-W.: BiSeNet V3: bilateral segmentation network with coordinate attention for real-time semantic segmentation. Neurocomputing (2023). https://doi.org/10.1016/j.neucom.2023.02.025
13. Badrinarayanan, V., Kendall, A., Cipolla, R.: SegNet: a deep convolutional encoder-decoder architecture for image segmentation. IEEE Trans. Pattern Anal. Mach. Intell. **39**(12), 2481–2495 (2017). https://doi.org/10.1109/TPAMI.2016.2644615
14. Xiao, X., et al.: BASeg: boundary aware semantic segmentation for autonomous driving. Neural Netw. (2023). https://doi.org/10.1016/j.neunet.2022.10.034
15. Hu, X., Zhou, B.: LBARNet: lightweight bilateral asymmetric residual network for real-time semantic segmentation. Comput. Graph. (2023). https://doi.org/10.1016/j.cag.2023.07.039

A CNN-Based Road Accident Detection and Comparison of Classification Techniques

Jasmeet Singh Gosal, Lopamudra Hota, and Arun Kumar(✉)

National Institute of Technology, Rourkela, Rourkela, Odisha, India
{119cs0527,519cs1015,kumararun}@nitrkl.ac.in

Abstract. The likelihood of accidents rises due to an increase in the number of vehicles on the road. Traffic collision detection can help to lessen fatalities and catastrophic injuries. This paper considers the different frames in road accident videos, creates a deep learning feature extraction algorithm, and classifies images as accident or non-accident. The last pooling layer output from the Convolutional Neural Network (CNN) is used to extract features that should be used for detecting an accident. Different layers of CNN with filters are used to extract features and also padding is applied to avoid any information loss in the convolution process. The fully connected layer classifies after receiving the extracted features. CNN overcomes the different shortcomings of conventional approaches for accident analysis. Also, different machine learning classification algorithms are used with the feature extraction by CNN to improve the accuracy of the classification of images as accident or non-accident.

Keywords: Traffic Accident Detection · Convolutional Neural Networks · Accident · Images Classification · Feature Extraction

1 Introduction

Consequences of motor vehicle accidents have frequently been severe, including loss of life and property. Traffic accidents have an impact on traffic operations and flow, and they can result in fatalities or very serious injuries. According to the World Health Organization (WHO), 1.35 million individuals worldwide pass away every year as a result of road accidents [3]. Despite significant advancements in automobile technology, traffic engineering strategies, and crash data analytics, the number of crashes and fatalities is still too high. These technologies allow for the collection of data from many sources to assist those who have been hurt. Image sources are among the most significant data sources. All major cities across the world have already placed a sizable number of cameras for traffic surveillance reasons due to numerous security concerns [7]. Nevertheless, present systems heavily rely on human observation. A feasible approach would be to use these

D. Pastor-Escuredo et al. (Eds.): ICDLAIR 2023, LNNS 1001, pp. 206–219, 2024.
https://doi.org/10.1007/978-3-031-60935-0_19

already-existing surveillance camera networks. It is practically impossible for human observers to follow and recognize strange happenings in such a significant number of camera scenes instantaneously without missing them [6]. As a result, it highlights the necessity for automated accident detection technologies.

Applications for traffic accident/abnormality detection include traffic flow management, pedestrian protection, autonomous driving, and road safety alert. Many studies have been done on accident analysis [5]. Several image-processing techniques have been developed so that accidents can be detected and support can be provided. The development of automatic detection systems utilizing computer vision and pattern recognition techniques has been the focus of research from both industry and academia, but the current state of technology does not allow for their practical application. It is quite difficult to create a vision-based algorithm for this task. Many factors can have an impact on how well a vision-based algorithm performs. These variables include the imaging environment (variable lighting and varied weather conditions, urban, highways). Various feature extraction techniques have been proposed but these techniques usually extract non-useful information which reduces the system accuracy [2].

Several computer vision tasks, including picture classification, semantic segmentation, object detection, human action/activity recognition, visual tracking, etc., have been tackled using deep learning approaches. Deep learning techniques' remarkable success is largely attributable to how well they represent visual data. In this paper, CNN is used for the particular task of classification of accident images. We use accident images in this paper but video data can also be used it is also required to get frame-by-frame images for that particular video and then act as an input to any computer vision technique and be used for accident detection. For classification purposes, a feature extractor is required and CNN is used for that particular purpose. Accident analysis is prioritized in order to identify accident occurrences and categorize images as accidents or non-accidents. The CNN max-pooling layer extracts the features, which it then feeds to the fully connected layer. This forms the basic baseline CNN structure.

In this research work the features are extracted using the CNN algorithm and then fed into different machine-learning classifiers. Machine learning algorithms like Logistic Regression, Support Vector Machine(SVM), K-Nearest Neighbour(KNN), Naive Bayes and Random Forest(RF) are implemented on the dataset to detect accident images and distinguish them from normal cases. The performance of the RF algorithm, in terms of different evaluation metrics used, was found superior compared to all the other classification algorithms that are used.

1.1 Motivation

The number of vehicles increasing running on the road, also lead to more chances of being met with accidents. Also, cities around the world have installed various cameras for traffic monitoring. Use of these rely on human observers but human observers may miss events due to a large number of cameras. So an automated system for detection is required.

1.2 Objective

The main objective of this paper is to classify video frames into accident or non-accident based on the feature extracted from the frame and using classification techniques.

1.3 Contribution

The main contribution includes the creation of the dataset by videos from YouTube. Then the CNN-based learning network is implemented to extract the features and ML algorithms to detect accidents.

1.4 Paper Organization

The organization of the rest of the paper is; Sect. 2 presents the related works, followed by the proposed work in Sect. 3. Section 4 demonstrates the results and discussion, and finally, Sect. 5 presents the conclusion and future work.

2 Related Works

Earlier, techniques like Histogram of Gradient (HOG), Local Binary Pattern (LBP), Speeded-Up Robust Features (SURF), and Maximally Stable Extremal Regions (MSER) were employed to extract features from accident photos.

For the analysis of 2D textures, LBP [4] has been created. This technique applies a local structure to the images and is used to basically compare nearby pixels to the pixel in consideration. LBP is a texture descriptor that can be used to describe a feature with a strong discrimination power as compared to others. A value of unity is assigned to the neighbours with a grey level greater than the centre pixel in a preset patch else a value of zero. The central pixel is then given a number in terms of binary.

Each pixel's gradient in the x and y axes is calculated by HOG [11], which also provides the size and direction of each gradient. It serves the aim of object detection in image processing and computer vision. In the confined area of an image, the technique counts instances of gradient orientation. The output that HOG gives out mainly focuses on the object's shape and structure. HOG uses the angle of the gradient and the magnitude to compute the features so it turns out to be better than edge descriptors. Using the gradient's magnitude and orientations, it generates histograms for the various areas of the image.

MSER [10] is one of the image processing techniques for detecting bubbles. It finds stable zones by using the intensity range of the supplied image. Different zones are considered as defined when a certain threshold is defined and colour intensity variances are below that. The various thresholds are examined by this attribute. These alterations must be smaller than the input value for the region to be recognized as stable. The wide-baseline matching is aided by this technique of extracting a substantial number of related image elements, and it has improved stereo matching and object recognition algorithms.

For applications like object recognition, picture registration, categorization, or 3D reconstruction, SURF [8] can be employed. The points are chosen using a square filter (Hessian matrix). A point is taken as a potential candidate for SURF features when the norm of the Hessian matrix is maximum. The Hessian matrix measures the local changes around the pixels. SURF descriptors have been applied to extract points of interest, recreate 3D scenes, track objects, and find and identify objects, people, or faces. A descriptor's purpose is to offer a distinctive and thorough description of an image feature, for instance by characterizing the intensity distribution of the pixels close to the place of interest. Since the majority of descriptors are therefore computed locally, a description is obtained for each point of interest that has been found.

All these feature extractors are to be run on a single input image. These traits are extracted, and then they are combined into a matrix. The matrix becomes redundant as a result. It is a good idea to have a backup plan in case something goes wrong. So there is a requirement for the identification of appropriate features for classification. The matrix is very vast, and the majority of the entries are not profitable, which can reduce accuracy.

Classification accuracy was not acceptable by using these feature extractors [14]. We can probably not examine the incidents based on the typical aspects because the automobiles engaged in the accident scene collapsed and the colours were disturbed. In reality, the corner, form, and colour of the automobiles in such photographs are wholly atypical. Also, the accident photographs were gathered from various sources, and frequently, different image properties like the light intensity and the angle of the image are not the same for each image. In these situations, extracting certain deep and hidden features may increase the classification accuracy and help to extract meaningful characteristics.

2.1 Machine Learning Methods

Machine learning techniques make an effort to classify new samples using the methodology that was discovered from categorized examples. Since the classification algorithm is given access to the training sample and its associated results, the term "supervised learning" is also used in literature to refer to classification. Each result represents that sample's class.

1. **Support Vector Machine (SVM)**
   SVM is a machine learning algorithm used for the purpose of classification and regression tasks. It works by diving various classes in high-dimensional space using a hyperplane. The first step of the algorithm is to convert low-dimension data points to high-dimension data points using different functions called kernel functions. Finding a hyperplane that can linearly separate the data points in such a manner is done even when the original data cannot be linearly separated. The algorithm searches for the hyperplane that maximizes the margin between the two classes after the conversion of data points is done to a higher dimension. The margin is the separation between the closest data points from each class and the hyperplane. The algorithm finds a hyperplane

that generalizes effectively to fresh, untested data by maximizing the margin. The algorithm can use a soft margin approach when the data cannot be separated linearly. To find a hyperplane that still maximizes the margin as much as possible means allowing some misclassifications.

2. **Random Forest (RF)**
   A particular kind of machine learning algorithm called a random forest classifier categorizes data using a group of decision trees. Each of the decision trees in a random forest is made by using only some randomly selected features of the dataset, and the final prediction is based on the collective decision of the individual trees. To begin, a random subset of features from the dataset is chosen by the algorithm. The outcome is then predicted using a decision tree that is constructed using these features. This procedure is repeated several times, with a different subset of features being used to construct each decision tree. The majority vote of all the decision trees serves as the foundation for the final prediction. A random forest classifier has several advantages over a single decision tree, one of which is that it is less prone to overfitting. The algorithm generalizes better to new, untested data by combining the predictions of multiple decision trees.
3. **Logistic Regression (LR)**
   For binary classification issues, a well-liked machine learning algorithm is logistic regression [1]. The logistic function, on which it is based, converts any input value into a probability between 0 and 1. To maximize the likelihood of the observed data, the logistic regression algorithm seeks out the best parameters. Based on one or more predictor variables, it calculates the likelihood of the binary outcome. The algorithm converts the linear equation into a probability between 0 and 1 by using a sigmoid function. In order to reduce the discrepancy between the predicted probability and the actual probability of the observed data the algorithm uses an optimization algorithm, such as gradient descent to modify the weights of predictor variables. Once trained, the model can be used to forecast the binary result for fresh data points. The algorithm uses the logistic function to determine the outcome's probability before applying a threshold (typically 0.5) to categorize the result as positive or negative.
4. **Naive Bayes (NB)**
   An algorithm for machine learning called a Naive Bayes classifier is based on the Bayes theorem. It is a straightforward but efficient algorithm for classification issues. This algorithm estimates a particular event probability based on some kind of prior knowledge about the event. The algorithm operates by first figuring out each class' prior probability based on the training data. This is achieved by measuring the total number of occurrences of each class and dividing by the total number of training samples. Next, for each feature in the data, the algorithm calculates the likelihood of that feature belonging to each class. The number of times the feature appears in each class divided by the total number of samples in that class is done for the particular purpose. In order to determine the posterior probability of each class given the observed features, the algorithm uses Bayes' theorem as its final step. The

predicted class is determined to be the one with the highest posterior probability. The "naive" assumption behind Naive Bayes is that, given the class, the features are independent of one another. By making this assumption, the calculations are made easier and the algorithm is able to scale effectively to high-dimensional data.

5. **K-Nearest Neighbour (KNN)**
An instance-based learning algorithm for classification problems is known as a KNN [13] classifier. It is predicated on the notion that similar data points frequently fall into the same class. The number of nearest neighbours to take into account when making a prediction, k, must first be set in the algorithm. The algorithm is used to determine the separation between a new data point and a different data point in the training set given a new data point. Different metrics, such as the Manhattan distance or the Euclidean distance can be used to measure the distance that is used by the algorithm. The algorithm chooses the k data points from the training set that are nearest to the data point to be classified by calculating the distances. The k-nearest neighbours are these k data points. The new data point is then given a class based on the classes of its closest neighbours by the algorithm. A majority vote of the classes of the k-nearest neighbours is used to decide this. The algorithm can use a number of tie-breaking techniques, such as choosing the class of the closest neighbour or choosing a random class if there is a tie. Several methods, such as weighting the neighbours according to how far away they are from the new data point or switching to distance-weighted voting instead of simple majority voting, can be used to further improve the performance of the KNN classifier.

3 Proposed Work

This section presents the proposed model, the architecture used, feature extraction via CNN and prediction by ML algorithms.

3.1 Feature Extraction by CNN

One of the techniques for processing images is CNN [12]. By pooling layers, the CNN becomes capable of extracting the hidden features which can be helpful for different purposes. The final pooling layer's outputs are typically used for classification and regression tasks. CNN can also be used for real-time image processing and can be very helpful in solving different challenging situations. Many convolutional, nonlinear, and max-pooling layers are used by CNN to capture the image. It gives a decent description of the image together with helpful details that may be used to classify it. By removing hidden features from distorted images, CNN might significantly increase detection accuracy. The capability of CNN to extract the appropriate features from massive data dispersion is one of its key advantages.

3.2 Architecture Used

1. **Dataset Preparation**
   Different accident videos were obtained through youtube and online websites. The videos consisted of several clips in which accidents happened and were visible in the video. The obtained videos were then converted to frames using software for converting videos into jpg format. The interval used for capturing frames was set to be 1 sec meaning after 1 sec for each video a frame was captured. The dataset has been updated in our github [https://github.com/Research-cloud/Accident-Detection-Dataset]. Each frame obtained by applying the process was then analyzed individually and any redundant frame or unclear frames were removed. A total of 1509 frames were obtained and then each frame was classified as "accident" or "non-accident" and placed in the respective folders for it to be used to train the CNN model. Also, the dataset is divided into training and testing data so that analysis can be performed (Fig. 1).

Fig. 1. Sample accident images used in the dataset

2. **CNN Architecture Used**
   The input layer, convolutional layer, maximum pooling layer, and fully connected layer are all included in the CNN architecture used (Fig. 2).

Fig. 2. Sample non-accident images used in the dataset.

The CNN architecture is adapted from [9]. The architecture consists of:

- **Input layer**
  The input layer gets an input image with the specified size and the 3 colour layers of RGB.
- **Convolution layer**
  Then the convolutional layers are used. The specified filters in each of the tiers of the convolutional layer are used accordingly. Padding is done with the specified size. Padding is done on the matrix so that when the process of convolution occurs there would be no loss of information. For the project zero padding has been used.
  The first convolution layer receives input of size $102 \times 102 \times 3$ after the application of zero padding and gives an output of $100 \times 100 \times 16$ due to 16 filters for the first tier and all the other convolution layers give results accordingly after the implication of padding before each layer. Also for each convolution layer, the activation function of relu is applied.
- **Max-pooling layer**
  The recommended model makes use of four max-pooling layers. This layer's pool size is as specified. The size of the stride which is what the pooling filter has to jump on the feature extracted matrix is [2, 2]. The pooling layer extracts the maximum value in its given pool size on the

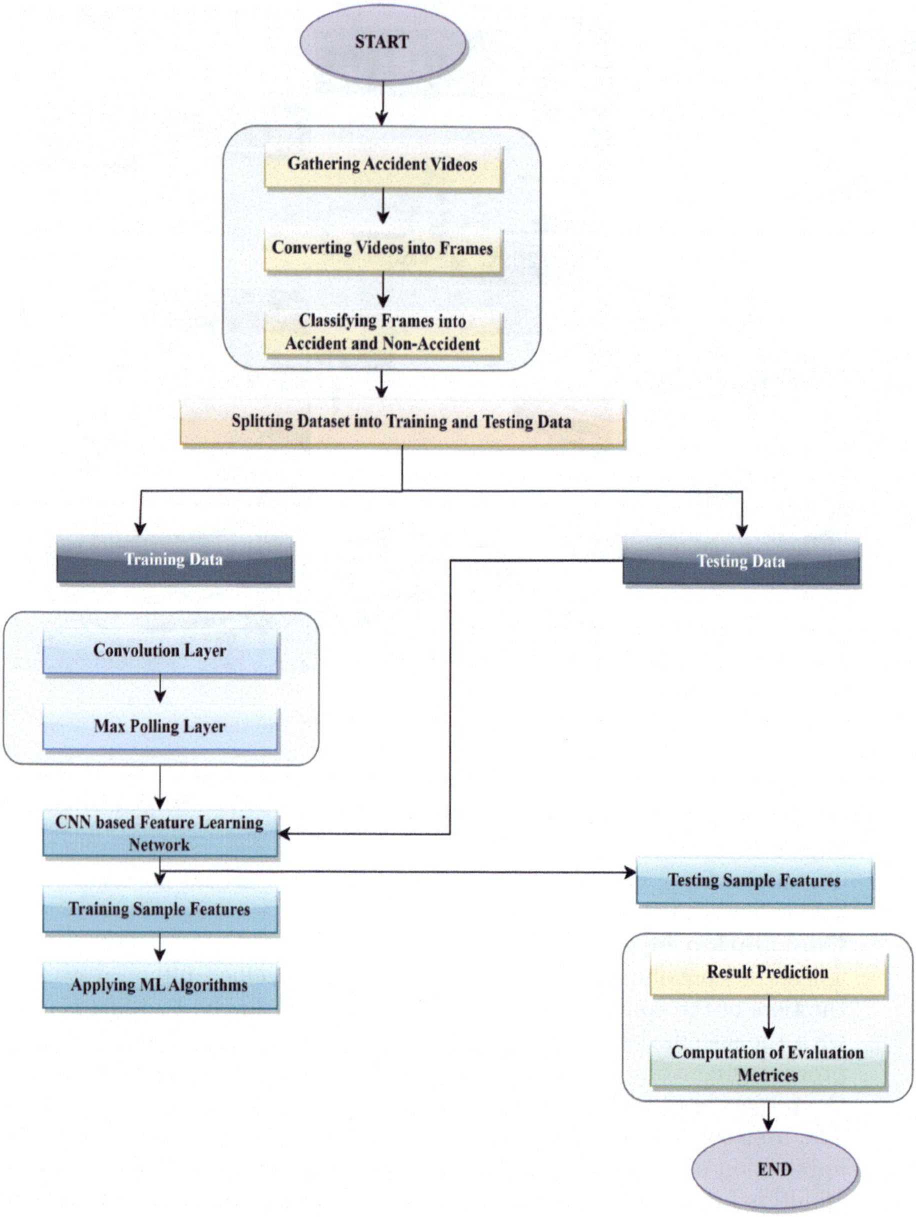

Fig. 3. Flowchart of classification of frames

input matrix and moves according to the given stride size. The pooling layer is essential for extracting all the important information on the given image (Fig. 3).

- **Flatten**
  After the fourth max-pooling layer the final matrix is obtained. Now the matrix is flattened meaning converted into a 1-D array for being able to feed further in the architecture to be used for classification purposes.
- **Dense layer**
  The output from the flattening process is fed into the dense or fully connected layer. This layer works basically like a classification model. The categorization technique uses a Multi-layer perceptron (MLP). Fully linked outputs are converted to a probability distribution on the classes, in this case, accident, and non-accident, using the activation function softmax.

Three tasks are carried out by CNN: fully connected layer classification, Values of point of interest by pooling layer from the convolution layer, and feature extraction by convolution layer. The primary features that a CNN extracts have to do with visual texture and colour. CNN extracts these features and tries to classify them.

3. **Machine Learning Algorithms Used**
   After the training samples are fed through the CNN architecture, this CNN can now be used to either classify the given testing samples on its own or it can be used to extract features that can be used for classification purposes from a particular sample frame. The CNN architecture was now used to extract features for both the training and the testing samples. Then different supervised machine learning algorithms were applied to these extracted features and different metrics for evaluation were calculated. Machine Learning Algorithms that were used were RF, Naive Bayes, SVM, Logistic Regression, and KNN.

4 Results and Analysis

This section presents the results and analysis for the prediction analysis. The parameters used for experimental analysis are presented in Table 1. The parameter values are taken based on extensive execution results with different values.

While evaluating different classification algorithms and comparing them, four different terms are obtained. True Positive (TP) is a term used in binary classification, which refers to the number of instances where the model correctly predicts the positive class (i.e., the class of interest). True Negative (TN) is a term used in binary classification, which refers to the number of instances where the model correctly predicts the negative class (i.e., the class that is not of interest). False Positive (FP) is a term used which refers to a number of instances where the model incorrectly predicts a positive class. False Negative (FN) is a term that refers to a number of instances where the model incorrectly predicts a negative class.

Table 1. Parameters used and their corresponding values

Parameter	Value
Batch size	100
Image height	100
Image width	100
Training data size	1205
Testing data size	304
Convolution layers	4
Number of Filters	16, 32, 64, 128
Filters size	3×3
Padding size	1
Pooling layers	4
Pooling layer size	2×2

Different evaluation metrics used in the analysis are defined below:

- **Accuracy**
  It is defined as the ratio of TP added with TN to the sum of all four terms0. Accuracy is used to measure the correct predictions of the model.

$$Accuracy = \frac{TP + TN}{TP + TN + FP + FN} \tag{1}$$

- **Precision**
  It is defined as the ratio of TP to the total number of positive predictions made by the model. Precision metrics are used to measure the correct prediction of the target class by the model among all the instances it predicts as the target class. It is a measure of the accuracy of the positive predictions made by the model.

$$Precision = \frac{TP}{TP + FP} \tag{2}$$

- **Recall**
  It is defined as the ratio of TP to the total number of actual positive samples in the dataset. Recall metrics are used to measure the identification of the target class among all the target class instances. It is a measure of the completeness of the positive predictions made by the model.

$$Recall = \frac{TP}{TP + FN} \tag{3}$$

- **F1-score**
  The F-measure is defined as the harmonic mean between precision and recall obtained in the evaluation metrics. A higher value obtained for F-measure indicates a better performance of the machine learning algorithm.

$$F1 = \frac{2 * (Precision) * (Recall)}{Precision + Recall} \tag{4}$$

The "Accident-detection" dataset goal is to use image processing to find accidents when they happen. It has two subfolders, "accident" and "non-accident", respectively. The dataset consisted of training data and testing data. The above-discussed model was implemented on Google Colab and results were obtained. The training data consisted of 1205 frames and the testing data consisted of 304 frames.

First, the CNN model discussed earlier was provided with training data, and the model was trained. After that, the model was evaluated using testing data, and the accuracy of testing data was calculated. These results were obtained with the default baseline CNN and no other machine-learning classification algorithm was used. The number of epochs was set to 20 and the values of the testing accuracy at each epoch were plotted as shown in Fig. 4.

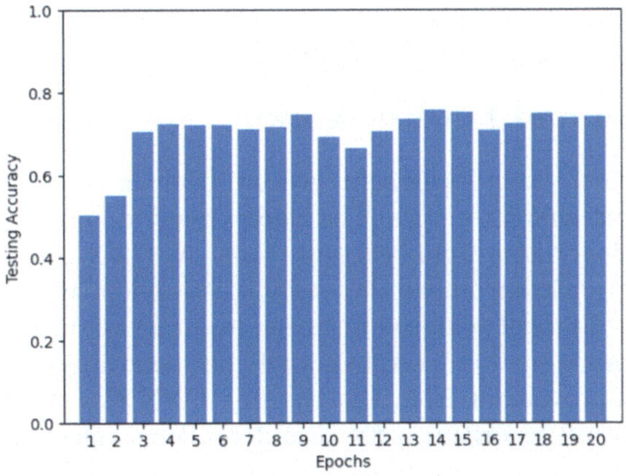

Fig. 4. Testing Accuracy at different epochs

At each epoch, the model adjusted the value of the filter that was used so as to give minimum loss function values. The maximum testing accuracy reached was 0.757. As for the training accuracy, it kept on improving with more epochs.

As clearly observed that the testing accuracy for baseline CNN is not very good, hence there is a requirement of using better machine learning algorithms for the classification of data. The CNN architecture in the previous case which was used now was used only up to the part of feature extraction. The feature extracted for a particular frame was stored as numpy array. An array consisting of labels for particular frames was also generated. These labels and the feature-extracted numpy arrays were then used for training in each of the classification models.

The best values for each of the evaluation metrics are shown in bold numbers. It can be easily observed from Table 2, RF classification algorithm provides the best value for each of the evaluation metrics.

Table 2. Feature Extraction with CNN and classification using different classifiers

Feature Extractor	Classifier	Precision	Recall	F1-score	Accuracy
CNN	SVM	87.688	93.168	90.345	88.630
	Random Forest	**98.674**	**97.383**	**98.024**	**97.759**
	Logistic Regression	87.568	93.168	90.281	88.547
	KNN	90.603	93.895	92.219	90.954
	Naive Bayes	88.049	93.168	90.536	88.879
Baseline CNN		81.016	79.340	80.169	75.760

Even though all the machine learning models have improved the accuracy for testing samples significantly as compared to that of the baseline CNN model architecture, the RF algorithm comes out on the very top with reaching an amazing accuracy of 0.97, while other algorithms have it around 0.90.

5 Conclusion and Future Work

Accident management centers benefit greatly from having a thorough understanding of accident characteristics. The frames generated from videos are useful informational tools for identifying certain traits. As CNN is a well-suited model for feature extraction from image datasets, it is implemented for this research work. The proposed model achieved a testing accuracy of 0.75. Various ML classification algorithms were then used to classify accident images and they resulted in a significant improvement from the baseline CNN classification. Among the different classification algorithms, the Random Forest algorithm proved to be the best with the highest accuracy of 0.977. It also has the highest precision and recall values which indicate that it predicts the positive class very well which will be very helpful as it reduces the false alarm rate which is a huge problem in automatic accident detection scenarios.

As a part of future work, we will implement some more Neural Networks frameworks to improve the accuracy of prediction and detect accidents in real-time scenarios rather than using datasets.

References

1. Agarwal, S., Kachroo, P., Regentova, E.: A hybrid model using logistic regression and wavelet transformation to detect traffic incidents. Iatss Res. **40**(1), 56–63 (2016)

2. Ali, F., Ali, A., Imran, M., Naqvi, R.A., Siddiqi, M.H., Kwak, K.S.: Traffic accident detection and condition analysis based on social networking data. Accid. Anal. Prevent. **151**, 105973 (2021)
3. Aqib, M., Mehmood, R., Alzahrani, A., Katib, I.: In-memory deep learning computations on GPUs for prediction of road traffic incidents using big data fusion. In: Mehmood, R., See, S., Katib, I., Chlamtac, I. (eds.) Smart Infrastructure and Applications. EAI/Springer Innovations in Communication and Computing, pp. 79–114. Springer, Cham (2020). https://doi.org/10.1007/978-3-030-13705-2_4
4. Chen, J., Kellokumpu, V., Zhao, G., Pietikäinen, M.: RLBP: robust local binary pattern. In: BMVC (2013)
5. Chen, Y., Boukerche, A.: A novel lane departure warning system for improving road safety. In: ICC 2020-2020 IEEE International Conference on Communications (ICC), pp. 1–6. IEEE (2020)
6. Djahel, S., Doolan, R., Muntean, G.M., Murphy, J.: A communications-oriented perspective on traffic management systems for smart cities: challenges and innovative approaches. IEEE Commun. Surv. Tutor. **17**(1), 125–151 (2014)
7. Mateen, A., Hanif, M.Z., Khatri, N., Lee, S., Nam, S.Y.: Smart roads for autonomous accident detection and warnings. Sensors **22**(6), 2077 (2022)
8. Oyallon, E., Rabin, J.: An analysis of the surf method. Image Process. On Line **5**, 176–218 (2015)
9. Pashaei, A., Ghatee, M., Sajedi, H.: Convolution neural network joint with mixture of extreme learning machines for feature extraction and classification of accident images. J. Real-Time Image Proc. **17**, 1051–1066 (2020)
10. Samat, A., Li, E., Wang, W., Liu, S., Lin, C., Abuduwaili, J.: Meta-XGBoost for hyperspectral image classification using extended MSER-guided morphological profiles. Remote Sens. **12**(12), 1973 (2020)
11. Seemanthini, K., Manjunath, S.: Human detection and tracking using hog for action recognition. Procedia Comput. Sci. **132**, 1317–1326 (2018)
12. Yang, D., Wu, Y., Sun, F., Chen, J., Zhai, D., Fu, C.: Freeway accident detection and classification based on the multi-vehicle trajectory data and deep learning model. Transp. Res. Part C: Emerg. Technol. **130**, 103303 (2021)
13. Zhang, S., Cheng, D., Deng, Z., Zong, M., Deng, X.: A novel kNN algorithm with data-driven k parameter computation. Pattern Recogn. Lett. **109**, 44–54 (2018)
14. Zhou, Z., Dong, X., Li, Z., Yu, K., Ding, C., Yang, Y.: Spatio-temporal feature encoding for traffic accident detection in VANET environment. IEEE Trans. Intell. Transp. Syst. **23**(10), 19772–19781 (2022)

Football Match Result Prediction Using Twitter Statistical/Historical Data

Dipesh Jangir, Lopamudra Hota, Biraja Prasad Nayak, and Arun Kumar(✉)

National Institute of Technology, Rourkela, Rourkela, Odisha, India
{119cs0503,519cs1015,519cs1005,kumararun}@nitrkl.ac.in

Abstract. The results of a football game provide a fascinating test because football is one of the most popular and widely played games. Forecasting can also assist clubs and administrators in making the right decisions to win associations and competitions. Many studies have been conducted on football match prediction using statistical or historical data and various models for prediction. However, in this paper, historical data and Twitter data were extracted from tweets related to the premier league season 2021/22. Various classification models are employed, including Random Forest, SVM, Naïve Bayes, and KNN algorithm to see the outcomes of these models and their accuracy based on historical and Twitter datasets.

Keywords: Football Match · Result Prediction · Twitter · Statistical · A Random Forest · SVM · Naive Bayes · KNN

1 Introduction

One of the most exciting sports is football. Many people, including fans and punters, find predicting the outcomes of football games entertaining. Due to the fact that the result of a football game depends on a variety of variables, including a team's morale, talents, current score, etc., it is also intriguing as a study subject, in part because of how challenging it is [1]. Therefore, it is quite difficult to forecast the precise outcomes of football matches, even for football specialists. One of the leagues that the public supports the most is the English Premier League. According to ESPN statistics for the 2017–2018 football season, 14,56 million people watch these games. Due to the widespread interest in football in society, both social media and the public frequently attempt to analyze the results of soccer matches [2]. Twenty clubs make up the Premier League, a professional football league. According to official statistics, there were 4.7 billion viewers during the 2010–2011 football season across 212 nations and territories and 80 distinct broadcasters. Each team participates in a total of thirty-eight regular season games from August through May of each year. The participating teams in this tournament play at least one game per week throughout the season, and more significantly, every single one of these games is aired nationally and internationally. This gives many people the chance to leave comments, voice their

D. Pastor-Escuredo et al. (Eds.): ICDLAIR 2023, LNNS 1001, pp. 220–236, 2024.
https://doi.org/10.1007/978-3-031-60935-0_20

ideas, or start new dialogues on social media platforms like Twitter [3]. As betting involves financial resources, determining which side is likely to win is crucial. For this reason, bookmakers, fans, and prospective bids are all interested in knowing the approximate odds of a game ahead of time [4]. Due to the volatile nature of sporting events, particularly football tournaments, this task, which serves as the foundation of the betting industry, is quite challenging. The quality of the offence and defence, home-field advantage, and other game-related elements are some of the many determinants of goal scoring. Additionally, unforeseen elements like a player's dismissal (red card, penalty, rigidity of the referee, and many others) might have an impact on the result [5]. Many studies used only historical or statistical data for forecasting like players' attributes, managers' attributes, home ground advantage, number of goals etc. [6]. This paper has taken advantage of the sentiments of fans and tweets about the match from Twitter for the premier league season 2021/22 and compared the results of models based on the historical dataset and Twitter dataset.

1.1 Motivation

Sports are a widespread and well-liked way to pass the time, stay fit and healthy, and make money. The management of sports teams is spending a lot of money on updating their players' facilities and technical equipment. Data collection on players and games enables analysts to examine past team performance as a whole and determine the effects of various factors on the team's performance. Football (soccer), one of the most popular sports, has a multi-billion-dollar international market and a massive infrastructure. The finest players are supposed to have transfer values in the tens of millions of dollars range. To make the best investment selection, an investor requires professional assistance. However, a football coach must provide direction to his players as they formulate a plan of attack against the opposition. The conventional option is to enlist an expert, but that does not mean it is perfect. Making decisions with human support is not always dependable. Some psychological factors may affect how the outcome is analyzed and lead to improper conclusions being made. In spite of several studies in this area, there is room for improvement in terms of developing a system with higher prediction accuracy without much complexity.

1.2 Objective

The main objective of this paper is to predict outcomes of the premier league season 2021/22 using teams' historical data, players' attributes, and Twitter data using various existing machine learning classification algorithms.

1.3 Contribution

The primary contribution of this research work includes sentiment analysis of Twitter data to predict the match outcome. The dataset has been created by scraping data from the tweets.

2 Literature Survey

This literature has presented various models used for prediction of football match prediction and various factors affecting the prediction. The following literature has been studied in order to build the most effective model for football game predictions. One literature used the C5.0, Random Forest, and XGBoost algorithms from the decision tree method to make predictions about the outcome and the backward wrapper technique was used for feature selection. The prediction accuracy of C5.0 was 64.87% and Prediction accuracy of the random forest was 68.55% and the prediction accuracy of XGBoost algo was 67.89% [2]. In [3] used three different approaches. First, the authors assessed the model's performance which was based solely on Twitter features. Then, They built a model that was based on past match history and other statistical data about the football teams and they compared this model with the Twitter model. Finally, they combined both model approaches in order to analyse whether, in addition to the information that was already available through historical data, Twitter can offer any extra information. Igiri has conducted additional research for predicting football matches using the SVM machine learning approach with fifteen features, with a prediction accuracy of 53.3%. But research has proven that it is not an effective way to predict soccer matches [7]. A paper showed a variety of experiments based on a dataset including information on player traits and match history. The traits were obtained from the computer game FIFA, which was integrated with information gathered from actual football games. This study Produced 3 main model structures overall - Classification model, Regression model and Dense model. The best accurate model was the LSTM regression model, with an average forecasting accuracy of 52.479% [5]. In one research, Prasetio and Harlili used the logistic regression method with the model features home offence, home defence, away offence, and away defence, with prediction percentages reaching 69.5%. A drawback in using logistic regression to forecast soccer match outcomes is that the results only have two values(win and lose), whereas, in fact, a match's outcomes can produce three possible final scores: win, draw, or lose [8]. In the research, author used A deep multilayer neural network based on elastic net regularization. The reason for which this NN structure is adopted is that it has the capacity to produce good forecasting results using limited data, LeakyReLU activation functions used for input data and softmax activation function for neurons of the output layer was used. The prediction accuracy of the model was 61.14% [9].

Summary of Literature Review: The summary of the literature studied is given in Table 1. For the purpose of visualization and explanation, information relating to football match prediction some existing literature had been examined and summarised in a table.

3 Methodology

After reviewing existing literature, two models are designed based on historical data of the matches combined with players' attributes and Twitter data retrieved

Table 1. Summary of literature survey

Author	*Year*	*Dataset Used*	*Model Used*	*Accuracy*	*Limitations*
Kampakis et al. [3]	2014	English Premier League (2014)	Random forest, Naive Bayes	69.4%	Used some matches of the league not the entire league for twitter data so every club did not have the same number of matches
Igiri et al. [7]	2015	English Premier league (2014–2015)	SVM	53.3%	Findings showed that SVM was not an appropriate technique for feature sets used for this analysis
Prasetio et al. [8]	2016	Barclay's Premier league (2015/2016)	Logistic regression	69.5%	Classified only 2 classes - home victory and away victory not considered draw class
Razali et al. [10]	2017	English premier league (2010/2011 - 2012/2013)	Bayesian networks	75.09%	Very few matches used for training and testing, divided 380 matches into 10 groups and then 90% of each group used for training and 10% used for testing individually
Norbert et al. [5]	2018	Players stats - FIFA video game Match history-Leagues from Europe	LSTM neural networks	52.479%	Used only limited attributes for players which were gained from video game not from real match data and few attributes for teams
Alfredo et al. [2]	2019	English Premier League (2007/2008 - 2016/2017)	i) Random forest, ii) C5.0, iii) Gradient Boosting	i) 68.55% ii) 64.87% ii) 67.89%	Used limited no. of features, more significant features could be used. Players' attributes not considered.
Anfilets et al. [9]	2020	English premier league	Deep multilayer neural network	61.14%	Random attributes were used and players' attributes not taken into consideration and social media sentiments also not used for prediction
Rahman et al. [11]	2020	International Football results (1872–2018), FIFA world cup 2018 Dataset.	Deep neural network	63.3%	Accurate dataset was not used, quite good in group stage matches, but quarter final, semi final and final prediction were not up to the mark
Rodrigues et al. [6]	2022	English premier League (2013/2014 - 2018/2019)	SVM, KNN, Naïve bayes,C5.0. Random forest	Best Acc - 65.26%	For RF Used teams statistical data, players' attributes not taken into consideration and social media sentiments also not used

from Twitter API. Flowcharts of models are given below. Figure 1 shows a flowchart for prediction using statistical data and Fig. 2 shows a flowchart for prediction using Twitter data.

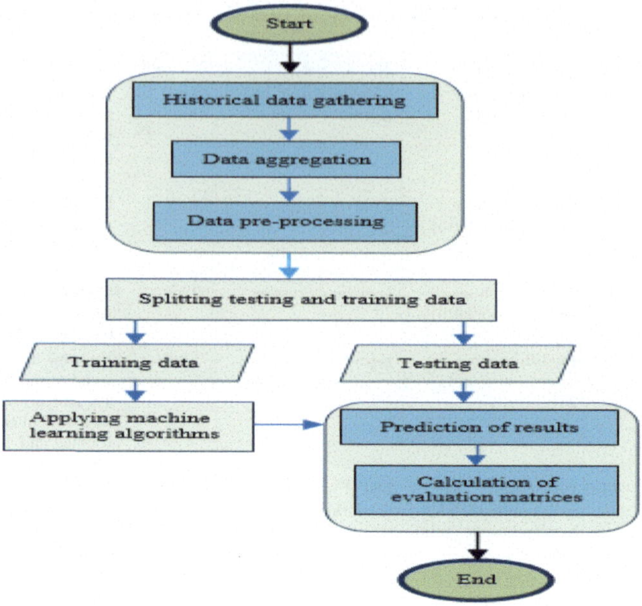

Fig. 1. Flowchart for prediction using historical dataset

3.1 Datasets

The description of the prepared dataset in our GitHub [https://github.com/Research-cloud/Football-match-Dataset] is presented in this section.

Historical Dataset. Historical data has been scraped from the website

[https://fbref.com/en/comps/9/2021-2022/2021]. Premier league season 2021/22 which consists of 380 matches and 20 teams. Each team plays 38 matches. Data was scraped from the start of the league to the end of the league. Figure 3 shows the first five rows of the dataset. Players' attributes and the history of their past game were used to make predictions which are also scraped from the website https://fbref.com/en/comps/9/2021-2022/2021. Players' details were scraped based on match date, which means on match date which players are playing and then scraped their past history. 537 players' details had been scraped, which are played in season 2021/22. Figure 4 shows the first five players' data which consist of name, a match played (MP), Goal against (GA), Goal against per 90 min (GA90), Shots on Target Against (SoTA), Save a percentage of goals (Saves%), Percentage of matches that results in a Clean Sheet (CS%), Goals scored (Gls), Assists (Ast), Penalty Kicks made (PK), Penalty Kicks attempted (Pkatt), Yellow cards (CrdY), Red cards (CrdR), Shots total

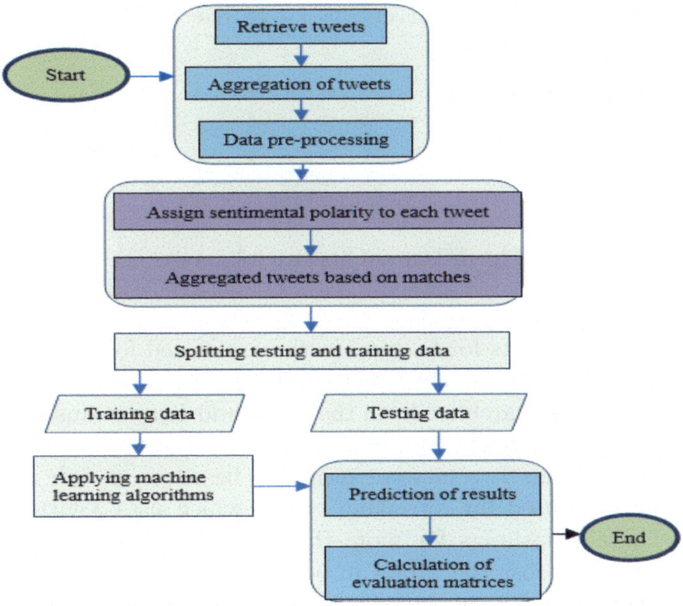

Fig. 2. Flowchart for prediction using Twitter dataset

	Date	Home_Team	Away_Team	Home_Goal_L5	Away_Goal_L5	Home_Win%_L5	Away_Win%_L5	Result
0	2021-08-15	Tottenham Hotspur	Manchester City	0	0	0.000000	0.000000	W
1	2021-08-21	Manchester City	Norwich City	0	0	0.000000	0.000000	W
2	2021-08-28	Manchester City	Arsenal	5	0	50.000000	0.000000	W
3	2021-09-11	Leicester City	Manchester City	4	10	66.666667	66.666667	L
4	2021-09-18	Manchester City	Southampton	11	4	75.000000	0.000000	D

Fig. 3. First five rows of teams' dataset

(Sh), Percentage of Shots that are on Target (SoT%) and Goals per Shot (G/Sh). The final dataset was made by combining these two datasets and making predictions. Each team has 11 players for a match and use substitutes in the ongoing match, so all squad had been used instead of only 11 players for each team some teams had only 13 players and some had 15 players in their squad, so 15 players for both teams had been taken for a particular match and 15 attributes of each player. Total number of features was 454.

name	MP	GA	GA90	SoTA	Saves%	CS%	Gls	Ast	PK	PKatt	CrdY	CrdR	Sh	SoT%	G/Sh
Joe Willock	178	0	0.0	0.0	0.0	0.0	30	14	0	1	10	0	162	21.418519	2.75
Francisco Trincao	136	0	0.0	0.0	0.0	0.0	20	12	0	0	6	0	179	28.933333	1.10
Pontus Jansson	351	0	0.0	0.0	0.0	0.0	19	4	0	1	66	4	101	7.268000	0.38
Sergio Reguilon	134	0	0.0	0.0	0.0	0.0	5	16	0	0	27	1	98	14.605556	0.34
Douglas Luiz	205	0	0.0	0.0	0.0	0.0	11	12	0	0	47	2	174	28.540000	1.32

Fig. 4. Data of first five players'

Twitter Dataset. Tweets for each team were retrieved for each match. Team hashtags were used for retrieving tweets and retrieved tweets for a particular match 48 h prior to the match for both the home and away teams. Table 2 shows hashtags for some of the teams gathered from different online sources.

Using these hashtags tweets were retrieved based on match date. Python library snscrape was used for retrieving tweets which uses Twitter API. As a tweet might contain hashtags for more than one team which would result in massive duplication across the tables in our dataset and the focus was on the tweets which were comments on a particular match from the perspective of one of the two teams. Those tweets were discarded which contained hashtags for more than one team. So, if a tweet contained any number of hashtags corresponding to exactly one team, the tweet was assigned to that team.
For pre-processing of tweet content NLP library of python was used for the removal of stop words, and punctuation and removed links, and emojis in the tweets. 6.03 lacs tweets had been retrieved and used for prediction.

Assigned Sentimental Polarity: After retrieving all the tweets and pre-processing of tweets sentimental analysis had been done using a transformer tool for each of the tweets and assigning polarity to the tweet whether the tweet was positive, negative or neutral. There was no use of neutral tweets so discarded those tweets. Figure 5 shows the first five rows of the Twitter dataset. which consist of Date, Home team, Away team, Positive tweets for the home team(Home_Pos), Negative tweets for the home team(Home_Neg), Positive tweets for the away team(Away_Pos), negative tweets for the away team(Away_Neg) and result.

	Date	Home_Team	Away_Team	Home_Pos	Away_Pos	Home_Neg	Away_Neg	Result
0	15-08-2021	Tottenham Hotspur	Manchester City	471	99	207	56	W
1	21-08-2021	Manchester City	Norwich City	96	91	29	42	W
2	28-08-2021	Manchester City	Arsenal	224	363	270	209	W
3	11-09-2021	Leicester City	Manchester City	74	66	31	21	L
4	18-09-2021	Manchester City	Southampton	140	90	150	102	D

Fig. 5. First five rows of Twitter data

Figure 6 shows the number of tweets used for some of the teams, As no. of tweets for some of the teams was huge and for some of the teams was less, so needed to normalize the data for prediction, a normalized formula was used : Positive tweets = positive tweets / (positive tweets + negative tweets) [12] Fig. 7 shows the updated normalized dataset.

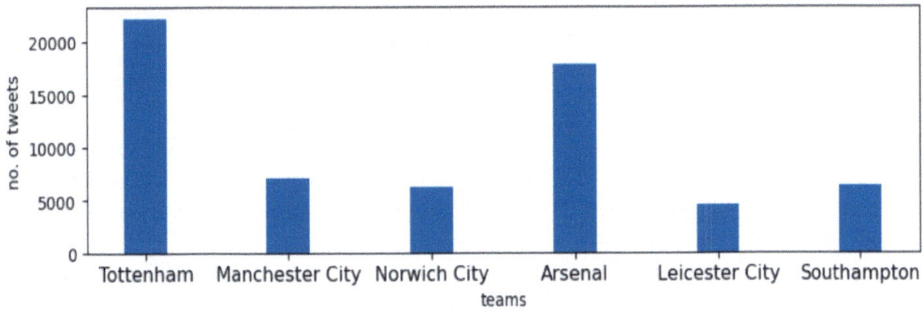

Fig. 6. Number of tweets for some of the teams

	Date	Home_Team	Away_Team	Home_Pos	Away_Pos	Result
0	15-08-2021	Tottenham Hotspur	Manchester City	0.694690	0.638710	W
1	21-08-2021	Manchester City	Norwich City	0.768000	0.684211	W
2	28-08-2021	Manchester City	Arsenal	0.453441	0.634615	W
3	11-09-2021	Leicester City	Manchester City	0.704762	0.758621	L

Fig. 7. Normalized twitter dataset

3.2 Training and Testing Splitting

Dataset had 380 matches, it was split into 80:20 ratio for training and testing dataset to train our model and measure accuracy based on training and testing dataset respectively. Dataset was split in this ratio because it is commonly used and there is no clear guidance on what ratio is best or optimal for a given dataset. Thus, training dataset size was 304 matches and the testing dataset size was 76 matches.

3.3 Models Used

As prediction was done for the home team of a particular match given some features, possible outcomes are win, draw or lost. So basically it was a classification problem, existing classification machine learning models were used and their accuracy was measured. The models used were random forest classifier,

Table 2. Hashtags of some of the teams

Team	Hashtags
Liverpool	#liverpoolfc, #YNWA, #LFC
Everton	#evertonFC, #EFC, #efc
Arsenal	#AFC, #ArsenalFC, #coyg
Manchester city	#MCFC, #ManCityFC, #mancityFC
Chelsea	#CFC, #ChelseaFC, #CFCXMAS

SVM classifier, naïve bayes algorithm and KNN classification algorithm. Iteration was done over various values of random forest parameter max_depth for better accuracy. For SVM all types of kernels were used for prediction for better accuracy.

3.4 Evaluation Matrices

Confusion matrix: A confusion matrix is an N * N matrix where N is the number of outcome classes, since in our problem there were three possible outcome classes win, draw and lost, thus 3 * 3 matrix was obtained. There were nine distinct predicted and actual value combinations in the table. We can measure Recall, Precision, F1-score along with Accuracy using a confusion matrix, which are the metrics used to measure the performance of ML models.

Precision: For a class, it is the proportion of values that were correctly predicted.

$$Precision = TP/(TP + FP) \tag{1}$$

Recall: For a class, it is a portion of values that are correctly predicted.

$$Recall = TP/(TP + FN) \tag{2}$$

F1-Score: It is used when we are trying to get the best precision and recall at the same time. It is a harmonic mean of recall and precision.

$$F1 - score = 2 * Precision * Recall/(Precision + Recall) \tag{3}$$

Accuracy: It is the proportion of correct prediction and sample size.

$$Accuracy = (TP + TN)/(TP + TN + FP + FN) \tag{4}$$

4 Results and Discussion

Four machine learning classification models were used for both datasets and measured their accuracy.

4.1 Random Forest Classifier

For Historical Dataset. Figure 8 shows occurrences of different possible outcomes and Fig. 9 shows the confusion matrix for the model. Table 3 shows the values of different evaluation matrices. The accuracy for this model was 61%.

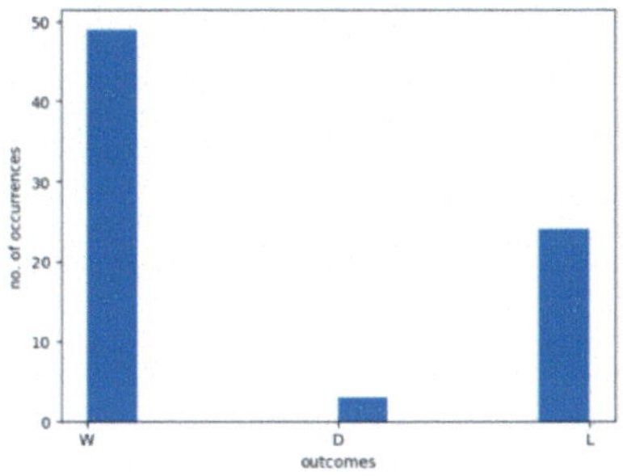

Fig. 8. Occurrences of different possible outcomes for random forest for historical data

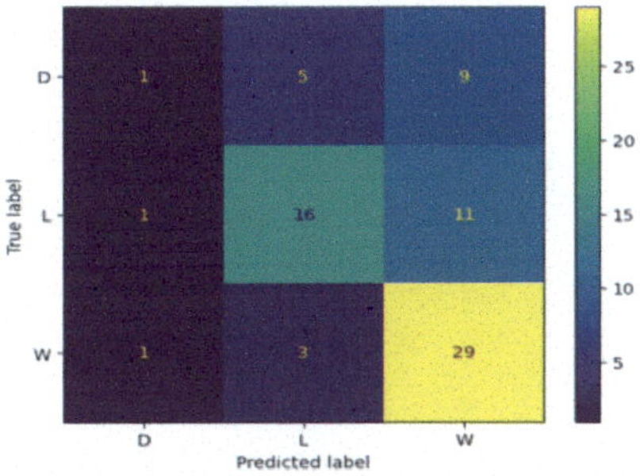

Fig. 9. Confusion matrix for random forest for historical dataset

Table 3. Evaluation matrices for random forest for historical dataset

Class	*Precision*	*Recall*	*F1-score*	*Support*
D	0.33	0.07	0.11	15
L	0.67	0.57	0.62	28
W	0.59	0.88	0.71	33

For Twitter Dataset. Figure 10 shows occurrences of different possible outcomes and Fig. 11 shows the confusion matrix for the model for Twitter data. Table 4 shows the values of different evaluation matrices. The accuracy of the Twitter dataset for this model was 68%.

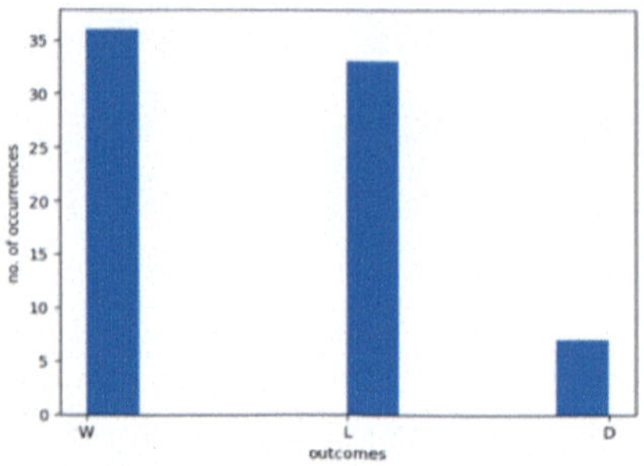

Fig. 10. Occurrences of different possible outcomes for random forest for Twitter data

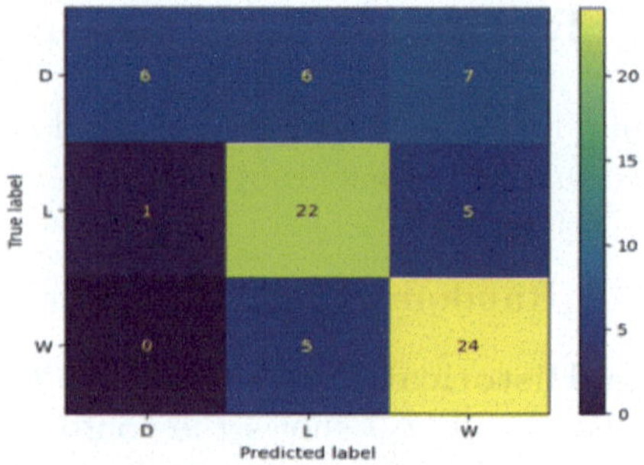

Fig. 11. Confusion matrix for random forest for Twitter dataset

Table 4. Evaluation matrices for random forest for Twitter dataset

Class	*Precision*	*Recall*	*F1-score*	*Support*
D	0.86	0.32	0.46	19
L	0.67	0.79	0.72	28
W	0.67	0.83	0.74	29

4.2 SVM Classifier

For Historical Dataset Table 5, Table 6, Table 7 and Table 8 shows evaluation matrices for kernel type linear, rbf, sigmoid and polynomial respectively. Table 9 shows the accuracy for different kernel types.

Table 5. Evaluation matrices for linear kernel for historical dataset

Class	*Precision*	*Recall*	*F1-score*	*Support*
D	0.13	0.20	0.16	15
L	0.48	0.36	0.41	28
W	0.53	0.52	0.52	33

Table 6. Evaluation matrices for rbf kernel for historical dataset

Class	*Precision*	*Recall*	*F1-score*	*Support*
D	1.00	0.07	0.12	15
L	0.68	0.54	0.60	28
W	0.55	0.88	0.67	33

Table 7. Evaluation matrices for sigmoid kernel for historical dataset

Class	*Precision*	*Recall*	*F1-score*	*Support*
D	0.00	0.00	0.00	15
L	0.64	0.32	0.43	28
W	0.50	0.94	0.65	33

Table 8. Evaluation matrices for polynomial kernel for historical dataset

Class	*Precision*	*Recall*	*F1-score*	*Support*
D	0.00	0.00	0.00	15
L	0.71	0.43	0.53	28
W	0.53	0.94	0.68	33

Table 9. Accuracy of different kernels for historical dataset

Type of kernel	*Linear*	*Rbf*	*Sigmoid*	*Polynomial*
Accuracy	39%	59%	53%	57%

For Twitter Dataset. Table 10, Table 11, Table 12 and Table 13 show evaluation matrices for kernel type linear, rbf, sigmoid and polynomial respectively for the Twitter dataset. Table 14 shows the accuracy for different kernel types for the Twitter dataset.

Table 10. Evaluation matrices for linear kernel for Twitter dataset

Class	*Precision*	*Recall*	*F1-score*	*Support*
D	0.00	0.00	0.00	19
L	0.62	0.46	0.53	28
W	0.42	0.79	0.55	29

Table 11. Evaluation matrices for rbf kernel for Twitter dataset

Class	*Precision*	*Recall*	*F1-score*	*Support*
D	0.00	0.00	0.00	19
L	0.65	0.79	0.71	28
W	0.53	0.72	0.61	29

Table 12. Evaluation matrices for sigmoid kernel for Twitter dataset

Class	*Precision*	*Recall*	*F1-score*	*Support*
D	0.00	0.00	0.00	19
L	0.00	0.00	0.00	28
W	0.38	1.00	0.55	29

Table 13. Evaluation matrices for Polynomial kernel for Twitter dataset

Class	*Precision*	*Recall*	*F1-score*	*Support*
D	0.00	0.00	0.00	19
L	0.66	0.68	0.67	28
W	0.49	0.79	0.61	29

Table 14. Accuracy of different kernels for Twitter dataset

Type of kernel	*Linear*	*Rbf*	*Sigmoid*	*Polynomial*
Accuracy	47%	57%	38%	55%

4.3 Naive Bayes Classifier

For Historical Dataset. Figure 12 shows the number of different possible outcomes, Fig. 13 shows the confusion matrix and Table 15 shows values of evaluation matrices for the Naïve Bayes model for the historical dataset. The accuracy for this model was 54%.

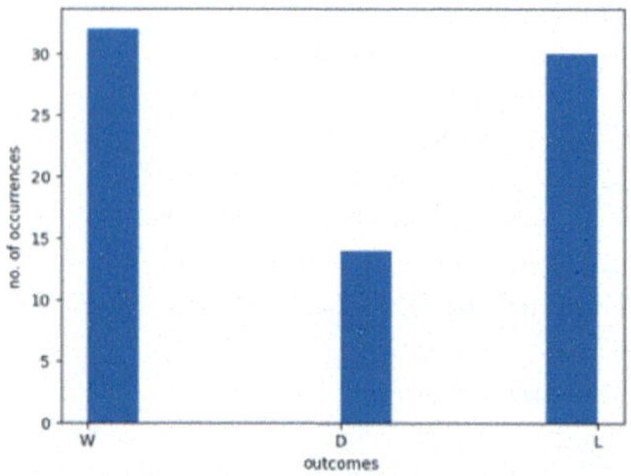

Fig. 12. Occurrences of different possible outcomes for naive bayes for historical dataset

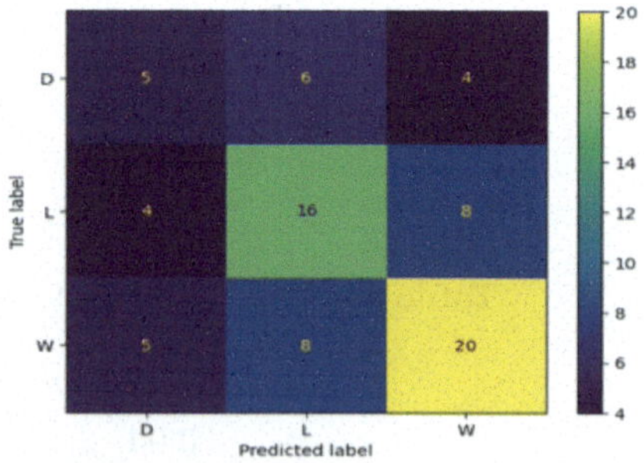

Fig. 13. Confusion matrix for naive bayes for historical dataset

Table 15. Evaluation matrices for naive bayes for historical dataset

Class	*Precision*	*Recall*	*F1-score*	*Support*
D	0.36	0.33	0.34	15
L	0.53	0.57	0.55	28
W	0.62	0.61	0.62	33

For Twitter Dataset. Figure 14 shows the number of different possible outcomes, Fig. 15 shows the confusion matrix and Table 16 shows the values of evaluation matrices for the Naïve Bayes model for the Twitter dataset. This model had obtained only win as outcome and accuracy was 38% which seemed it was very bad model for this dataset.

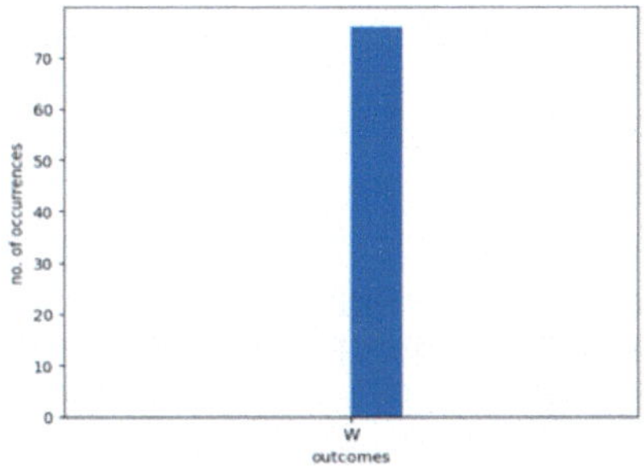

Fig. 14. Occurrences of different possible outcomes for naive bayes for Twitter data

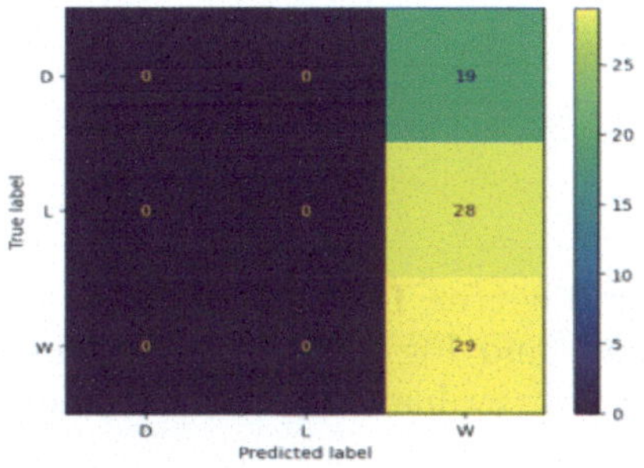

Fig. 15. Confusion matrix for naive bayes for Twitter dataset

Table 16. Evaluation matrices for naive bayes for twitter dataset

Class	*Precision*	*Recall*	*F1-score*	*Support*
D	0.00	0.00	0.00	19
L	0.00	0.00	0.00	28
W	0.38	1.00	0.55	29

4.4 KNN Classifier

For Historical Dataset. Figure 16 shows the number of different possible outcomes and Fig. 17 shows the confusion matrix. Table 17 shows the values of different evaluation matrices. The accuracy of this model for the historical dataset was 73%.

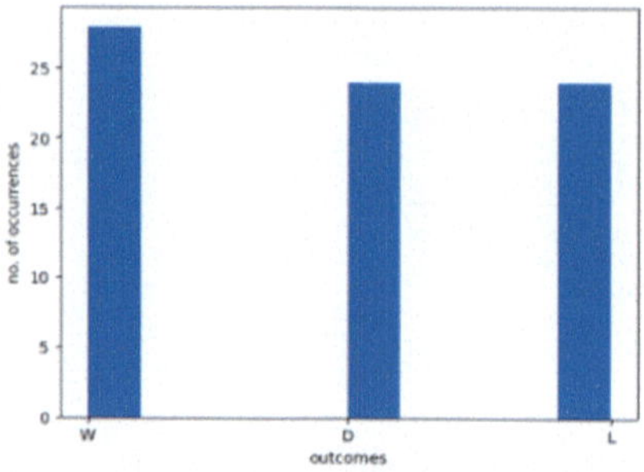

Fig. 16. Occurrences of different possible outcomes for KNN for historical data

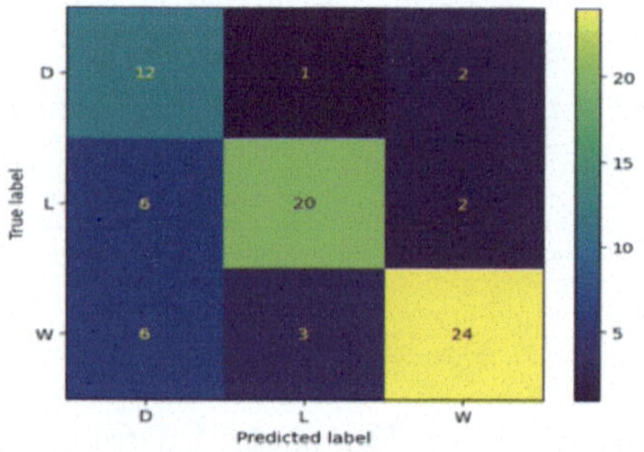

Fig. 17. Confusion matrix for KNN for historical dataset

Table 17. Evaluation matrices for KNN for historical dataset

Class	*Precision*	*Recall*	*F1-score*	*Support*
D	0.50	0.80	0.62	15
L	0.83	0.71	0.77	28
W	0.86	0.73	0.79	33

For Twitter Dataset. Figure 18 shows the number of different possible outcomes and Fig. 19 shows the confusion matrix. Table 18 shows the values of different evaluation matrices. The accuracy of this model for Twitter dataset was 76%.

Table 18. Evaluation matrices for KNN for Twitter dataset

Class	*Precision*	*Recall*	*F1-score*	*Support*
D	0.67	0.84	0.74	19
L	0.74	0.71	0.73	28
W	0.88	0.76	0.81	29

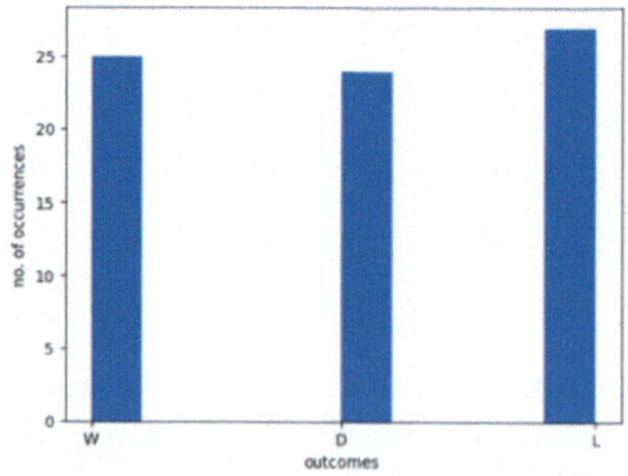

Fig. 18. Occurrences of different possible outcomes for KNN for Twitter data

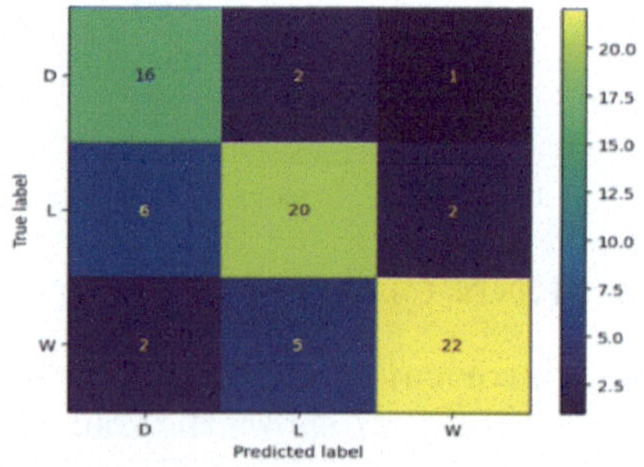

Fig. 19. Confusion matrix for KNN for Twitter dataset

4.5 Comparison of all Models Used

Table 19 shows a comparison of accuracy for all models used. Regarding the historical dataset, KNN algorithm was the best classifier it had the best recall and precision scores for all classes and regarding the Twitter dataset KNN algorithm was the best classifier, it had the best recall and precision scores for all outcomes. It was observed that using Naïve Bayes model had obtained only one outcome for Twitter dataset which was the worst classifier for the dataset. It was observed that using random forest and KNN accuracy was improved for prediction using Twitter dataset in comparison to the historical dataset because social media might contain important information rather than historical data. Random forest and KNN performs better than SVM because a number of attributes are large and SVM perform better on a small set of attributes and for two class predictions. By analyzing the final results we can observe that information extracted from Twitter can indeed be a valuable resource for making predictions about Premier League games.

Table 19. Accuracy of all models used

Model	*Random forest*	*SVM*	*Naive bayes*	*KNN*
Historical dataset	61%	59%	54%	74%
Twitter dataset	68%	57%	38%	76%

4.6 Comparison with Existing Models

Comparative analysis with other academic research projects is not straightforward, because each one was trained on a distinct dataset with different traits and accuracy measures. Considering the same type of dataset prediction using players' attributes, there was an LSTM model implementation with an accuracy of 52.479% [5], which was trained on data from the Europe league. Our models' performance was better. One of the papers used the EPL dataset and SVM classifier and obtained an accuracy of 53.3% [7] and our model with SVM classifier performed better with an accuracy of 59%. One paper used a similar dataset and used KNN with an accuracy of 57.63% [6] and our models performed best for the KNN classifier with an accuracy of 74%.

5 Conclusion and Future Work

In this research, a study is done on various models used for football match prediction and features set used for prediction, and the importance of social media in prediction. Most of the existing papers are based on statistical or historical data of the teams and not taking into account various factors which can significantly affect the prediction like weather, injury of key players and social media (tweets about match). In this paper, different existing machine learning algorithms were used for both historical datasets which consist of teams' data and players' attributes, and the Twitter dataset and compared, among all models KNN model performed better.

A comparative study of the historical dataset model and Twitter dataset model had been done using various classification models, the combination of both datasets can improve the accuracy of our model. For the historical dataset players' attributes had been used, so effective attributes can be selected instead of all attributes, which can improve accuracy significantly. For teams, head-to-head matches record can be used as one of the features for better accuracy.

References

1. Wunderlich, F., Memmert, D.: Forecasting the outcomes of sports events: a review. Eur. J. Sport Sci. **21**(7), 944–957 (2021)
2. Alfredo, Y.F., Isa, S.M.: Football match prediction with tree based model classification. Int. J. Intell. Syst. Appl. **11**(7), 20–28 (2019)
3. Kampakis, S., Adamides, A.: Using twitter to predict football outcomes. arXiv preprint arXiv:1411.1243 (2014)
4. Stübinger, J., Mangold, B., Knoll, J.: Machine learning in football betting: prediction of match results based on player characteristics. Appl. Sci. **10**(1), 46 (2019)
5. Danisik, N., Lacko, P., Farkas, M.: Football match prediction using players attributes. In: World Symposium on Digital Intelligence for Systems and Machines (DISA), pp. 201–206 (2018)
6. Rodrigues, F., Pinto, Â.: Prediction of football match results with machine learning. Procedia Comput. Sci. **204**, 463–470 (2022)
7. Chinwe Peace Igiri: Support vector machine-based prediction system for a football match result. IOSR J. Comput. Eng. **17**(3), 21–26 (2015)
8. Prasetio, D., et al.: Predicting football match results with logistic regression. In: International Conference on Advanced Informatics: Concepts, Theory And Application (ICAICTA), pp. 1–5 (2016)
9. Anfilets, S., et al.: Deep multilayer neural network for predicting the winner of football matches. Int. J. Comput. **19**(1), 70–77 (2020)
10. Razali, N., Mustapha, A., Yatim, F.A., Ab Aziz, R.: Predicting football matches results using bayesian networks for English premier league (EPL). In: IOP Conference Series: Materials Science and Engineering, vol. 226, p. 012099. IOP Publishing (2017)
11. Rahman, M.A.: A deep learning framework for football match prediction. SN Appl. Sci. **2**(2), 165 (2020)
12. Schumaker, R.P., Jarmoszko, A.T., Labedz Jr., C.S.: Predicting wins and spread in the premier league using a sentiment analysis of Twitter. Decis. Support Syst. **88**, 76–84 (2016)

Safeguarding Ecosystems and Efficiency in Peer-to-Peer File Sharing Systems: An IoT-Inspired Approach to Pollution Mitigation

Ritu(✉)

Chandigarh College of Engineering, Chandigarh Group of Colleges, Jhanjeri 140307, India
ratheeritu@yahoo.in

Abstract. The disruption of the environment's equilibrium has occurred as a result of mishandling and contamination of vital components of the ecosystem, which are essential for supporting human life. This situation poses a significant risk to the harmonious coexistence of humans and nature. To effectively address this issue, it is crucial to assess the extent of the damage and implement measures to prevent further harm. The rapidly advancing field of the Internet of Things (IoT) offers numerous promising models that can contribute to achieving these goals. This paper aims to provide an overview of selected models surveyed and evaluate their effectiveness. Peer-to-peer (P2P) file sharing systems face a significant challenge when a large number of fraudulent files are introduced, causing pollution. Since P2P users serve each other, their actions directly influence pollution levels. Consequently, we conducted a comprehensive study to analyse user behaviour and identified cooperative behaviour and pollution awareness as critical factors. Our research findings indicate that users generally exhibit apathy towards pollution in peer-to-peer (P2P) file sharing systems. Their behavior shows a distinct distribution regarding the time interval between downloading a file and verifying its quality. In order to assess the impact of pollution on the popularity of files, we propose a mathematical model. Through our analysis, we reveal that user awareness plays a crucial role in the dynamics of pollution. Additionally, we evaluate the effects of pollution on P2P traffic loads and find that it can result in a four-fold increase in loads in the worst-case scenario. These findings underscore the importance of addressing pollution-related issues in P2P systems to ensure their efficiency and long-term sustainability.

Keywords: Environmental Degradation · reliability · nature · P2P · pollutants

1 Introduction

Environmental pollution and contamination have emerged as significant challenges, especially in developing nations. While distinguishing between pollution and contamination may not always be straightforward, pollution typically involves the presence of harmful substances in concentrated amounts, while contamination can occur at lower levels yet

The original version of the chapter has been revised. The author Ritu's affiliation has been updated. A correction to this chapter can be found at
https://doi.org/10.1007/978-3-031-60935-0_71

D. Pastor-Escuredo et al. (Eds.): ICDLAIR 2023, LNNS 1001, pp. 237–245, 2024.
https://doi.org/10.1007/978-3-031-60935-0_21

render the affected environment unusable or pose minimal risks. Both urban and rural regions in industrialized countries have experienced substantial losses attributable to pollution and contamination, resulting in significant human and financial consequences. Developed nations like the United States, Europe, and Asia allocate substantial financial resources each year to combat these problems. Effectively controlling and addressing the extensive generation of pollutants and contaminants, which can originate naturally or from human activities like industrial operations, agriculture, and waste disposal systems, is of utmost importance. These pollutants can assume diverse forms and exhibit varying persistence in the environment, causing harm to distinct components such as the atmosphere, soil, water bodies, land, and living organisms. This article focuses specifically on pollution in the hydrosphere, particularly within the groundwater domain, and its repercussions in developing nations. It explores the origins and categories of pollution, the processes contributing to their emergence, and their implications for geology and hydrogeology. Additionally, the paper outlines potential health hazards associated with groundwater pollution in developing countries, proposes improved methods to combat pollution, and provides recommendations to enhance pollution management techniques in countries where they are lacking. The paper highlights the need for a comprehensive management strategy to mitigate the degradation of potable water supplies and protect the environment. It emphasizes the importance of prioritizing and demonstrating concern through governmental authorities, allocating necessary resources and personnel for basic research, and enforcing environmental. In recent research, several P2P user models have been proposed to investigate general pollution dynamics in P2P systems. However, recent empirical investigations have uncovered that the degree of pollution in existing P2P networks surpasses the predictions made by conventional models. For example, a study [7] analyzing the renowned P2P system KaZaA discovered that an astounding 79.8% of the 1,816,663 versions of the song "My Band" were tainted within the network, exceeding the expectations derived from these models using reasonable parameters. The main aim of this paper is to develop a straightforward yet realistic extension to the prevailing P2P user models, aiming to gain a more profound understanding of pollution dynamics in P2P file sharing systems. To achieve this objective, the paper initially presents findings derived from a user survey, which compellingly suggest that even well-informed P2P users unwittingly contribute to the proliferation of polluted files due to their lack of awareness regarding the downloaded files' contamination. Building upon these insights, the paper proposes a fresh P2P user model that incorporates user consciousness of pollution, specifically capturing the proportion of users who identify pollution in downloaded files and subsequently remove them. The analysis reveals the significant impact of this awareness factor on the final pollution level, and integrating it into the model yields predictions that closely align with the observed pollution levels. To the best of our knowledge, this study is the pioneering effort to consider user awareness and investigate its influence on the overall pollution level in the network. The research brings to light several noteworthy findings, including:

A considerable portion of users demonstrate a lack of concern towards pollution. Despite a significant number of users conducting immediate quality checks on downloaded files (approximately 65% in the study), a notable proportion of users postpone these checks for an extended duration, often exceeding 12 hours. As a result, this leads to a two-pronged distribution in the time interval. Moreover, even when users do verify the quality of a downloaded file, a significant percentage fails to recognize the presence of pollution. The research reveals that for specific types of pollution, more than 70% of users are unable to detect it, inadvertently spreading the contaminated file within the network. The analysis emphasizes that awareness of pollution is a key factor influencing the overall level of contamination.

2 Related Work

Polymeric encapsulation play a vital role in the safe and reliable operation of PV modules by electrically insulating the solar cells and protecting them from weathering and mechanical stresses (Boyd et al., 2018; Cheacharoen et al., 2018; Gagliardi and Paggi, 2019; Kopecek and Libal, 2018; Snaith and Hacke, 2018). Many new encapsulation materials and backsheet products have been developed in the past decade to compete with the more prevalent technologies (Geretschläger et al., 2016; Shimpi et al., 2019). Some of these new backsheet products are based on lower cost non-fluoropolymer materials, such as polyethylene terephthalate (PET) and polyamide (PA) (Liu et al., 2014; Omazic et al., 2019; Reese et al., 2018; Yuen et al., 2019). However, their long-term performance is unknown, and their lower cost may come with reduced weatherability and lifetime (Fairbrother et al., 2017; Haegel et al., 2019; Kurtz, 2018). This increased sensitivity to environmental stresses means that a greater understanding on the impact of various weathering variables is necessary to design more accurate accelerated laboratory weathering tests and reliability assessment methodology (Haegel et al., 2019; Sharma and Chandel, 2016; Snaith and Hacke, 2018). Backsheets are typically composed of three primary layers: a weathering resistant outer layer, an electrically insulating core layer, and an adhesion promoting inner layer. The PA-based backsheet of this study consists of three primary layers of PA with different additives (Lyu et al., 2018). They have emerged as an alternative to the laminated backsheet structure because they are formed by a coextrusion process. Potential advantages of such a process are reduced manufacturing costs (by elimination of the lamination step) and improved the inter-layer adhesion strength. However, some PA-based backsheets have been found to exhibit catastrophic failure by cracking after only a few years of field exposure (Eder et al., 2019; Fairbrother et al., 2017; Han et al., 2018). Very recently, Eder et al. (Eder et al., 2019) reported various types of cracking such as square cracks along the intercellular spacings and longitudinal cracks beneath the busbar of the cells for polyamide based backsheet deployed in the field for 5 to 6 years. That has been troubling because it is far short of the >25-year expected lifetime of PV modules. Of even greater concern is the fact that these particular backsheet products had passed all relevant standardized tests for design and safety, and were not identified as "bad" products, and that the mechanisms behind cracking are not fully understood. The main environmental stresses on polymeric materials include ultraviolet (UV) light, heat, and moisture, which act to degrade backsheets and cause failures such as cracking or delamination (Fairbrother et al., 2018; Jordan et al., 2017). Polyamide in particular has been shown to be susceptible to UV degradation and dimensional changes due to water uptake (Chaupart et al., 1998). These and other stressors are present in varying levels in different climates around the globe, and the degree and type of degradation can vary by location. For example, field surveys across multiple climates have found higher defect rates for backsheets in hot arid climates than tropical or temperate ones (Felder et al., 2017; Gambogi, W.J. et al., 2014). For PA-based backsheets, cracking has been observed in many, but not all arrays (Eder et al., 2019). Although the difference in manufacturing the backsheet or lamination process could be a possible contribution for this observation, it is also suggested that the environmental stresses which induce cracking may not be similarly profound in every location, that a more complex relationship (antagonistic, synergistic) may exist between environmental stressors, or that it may take more time in some locations. A clearer understanding of such relationships would aid development of a new generation of accelerated laboratory tests that are better able to evaluate the reliability of new backsheet products.

3 Methodology

Our survey consisted of two main sections. The primary objective of the first section was to assess the participants' familiarity with P2P systems and determine how their level of knowledge might influence the survey results. The second section aimed to gain insights into users' practices when it comes to downloading and managing files in P2P networks, with the ultimate goal of identifying the key factors that impact pollution dynamics. To gauge familiarity with P2P systems, we included a set of five questions that explored participants' experience and understanding of P2P file sharing. The results indicated that a majority of participants (60%) provided affirmative responses to all five questions, which was expected given that most of them were graduate students in the Computer Science Department. In the second section of the survey, we delved into an examination of the three stages of P2P client usage: preparation for downloading, the actual download process, and post-download activities. Concerning the preparation for downloading, the primary consideration for a majority of participants (57%) was the file's quality, with file availability (20%) and file size (just below 20%) following as secondary factors. During the downloading stage, a significant majority of participants (83%) opt to initiate a new download process if the speed fails to meet their expectations, while 63% frequently engage in simultaneous downloads of multiple files. In the post-download stage, 70% of participants reported encountering polluted files in the past, of which 30% initially mistook them for genuine versions but later realized their contamination. Additionally, the survey revealed that not all users actively share the files they download, and users predominantly base their download decisions on their anticipated quality. Overall, the survey findings highlight that even experienced P2P users occasionally encounter challenges in identifying polluted files and do not consistently assess the quality and authenticity of their downloaded files.

In the provided figure that is in Fig. 1, it is depicted that the process of extracting knowledge involves several steps. Firstly, a large amount of data is collected, which is then stored in a data warehouse. The next step involves preprocessing the data, followed by mining and evaluating patterns within the data. Finally, the knowledge is extracted from this extensive dataset, resulting in reliable information.

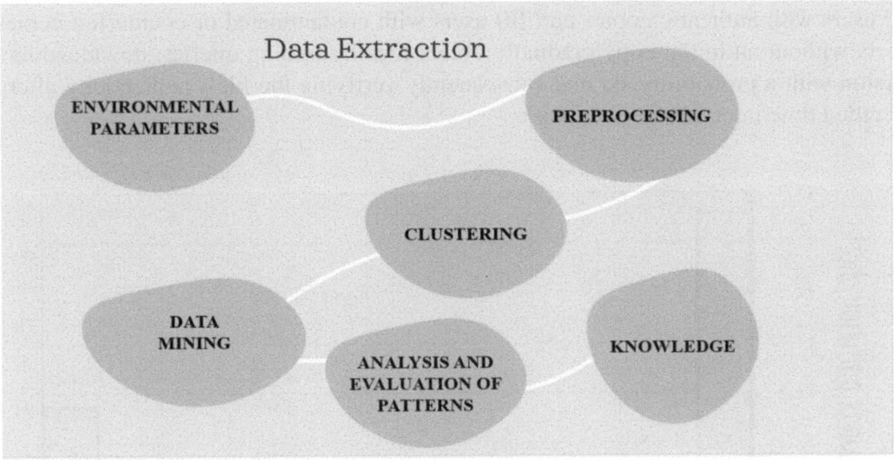

Fig. 1. Data Extraction Model Flowchart

Table 1. Crucial Parameters for Indexing and analysis of Pollution

Parameters	Original Sample(kg/acre)	Pit 1 (50 g of lead + 100 g of PCB) (Kg/acre)	Pit 2(100 g of lead) (Kg/acre)	Pit 3(150 g of lead) (Kg/acre)
Calcium Carbonate	Medium	High	High	High
Phosphorus	255	155	100	155
Nitrogen	72.4	66	50	35.6
Potassium	245	60	115	75
pH	8.1	9.3	9.3	9.3
Electrical Conductivity	0.21	1.2	1.4	0.3

In Table 1, an analysis has been conducted to determine the factors influencing pollution. These factors include the presence of nitrogen, calcium, potassium, pH value, and electrical conductivity. These parameters are considered when evaluating the original sample.

4 Experiments and Results

In this section, we present a novel analytical model that extends the previous research discussed in reference [2]. The model integrates the insights obtained from the survey and experiment conducted in the preceding section to examine the dynamics of pollution in peer-to-peer (P2P) networks. Our model assumes the presence of M users in the network, each maintaining a single version of a specific topic. Initially, the network consists of

G0 users with authentic copies and B0 users with contaminated or counterfeit copies. Users without an initial copy gradually obtain files by issuing queries, downloading a version with a probability sk, and subsequently verifying the file's authenticity after a specified time interval referred to as t.

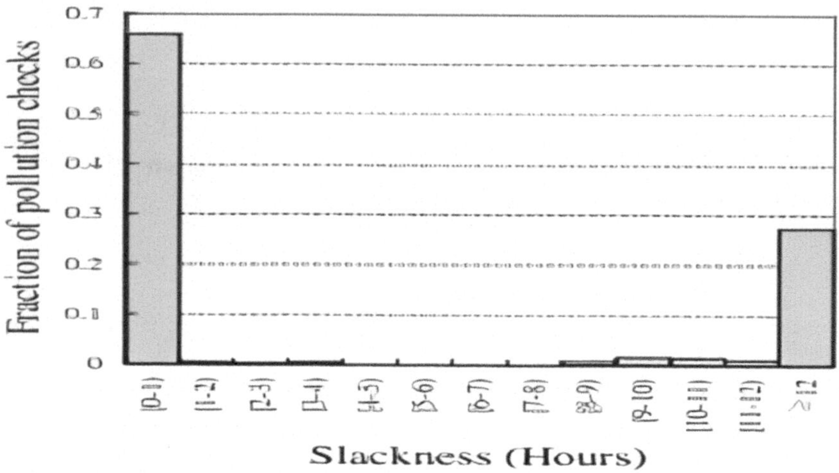

Fig. 2. Function based analysis for analysis of Pollution level

If a user detects pollution in the file, they promptly delete it. However, there is a chance that the user may not notice the pollution and mistakenly perceive the file as genuine, with a probability of 1−pa. After verifying the file's authenticity, users decide whether to continue sharing it or not, based on a cooperative behavior probability of pc. The number of users holding authentic and contaminated copies at a specific time step, denoted as k, is represented by Gk and Bk, respectively. Dk represents the total number of users who have downloaded a file by the previous time step, k−1. At time step k, a fraction sk of the remaining M−Dk users download a file, with gk and bk indicating the number of users downloading genuine and polluted versions, respectively. The time interval between downloading a file and verifying its authenticity is denoted by the random variable t, and the probability of experiencing delays in the verification process is represented as pSt. The number of genuine files at time step k + 1 can be calculated using a formula that considers the number of users downloading genuine files, the probability of uncooperative users leaving the system after j slots, and the slack probability. If a user downloads a polluted file and becomes aware of the pollution, they delete the file. Conversely, if the user fails to recognize the pollution, there is a probability of pc that they will share the file. These events are independent, and the probability of file deletion is determined by Eq. 1:

$$pD = pa + (1 - pa)(1 - pc) \quad (1)$$

The deletion occurs during time slot j with a probability pSj. Therefore, Eq. 2 is valid:

$$\boldsymbol{Bk+1 = Bk + bk - pD} \tag{2}$$

Moreover, it is worth highlighting that retrials take place only when users are aware of the presence of pollution (pa) and have the intention to download files again in the next time step (pr). As a result, the number of retrials at time step k+1 can be calculated using equation 3:

$$\boldsymbol{rk+1 = papr * \Sigma(j = 1\ to\ L)\,(bk+1-j * pSj)} \tag{3}$$

In order to determine the different level to determine pollution, the best way is to calculate it at various function level. The illustration of pollution level at function level is determined in Fig. 2.

4.1 Analysis

The researchers make an assumption that random users have a cooperation probability (pc) of 0.25. They incorporate the measured value of awareness (pa), which is determined to be 0.76 based on the analysis of the song "The Pussycat Dolls-Don't Cha". The pollution level is defined as the ratio between the number of polluted copies and the number of genuine copies during a specific time period. The study focuses on the initial pollution level, referred to as PL-k, where k represents the ratio. The research aims to examine the impact of increasing the initial pollution level set by the polluter and investigates the effects of both the retry probability (pr) and awareness (pa) on pollution dynamics. Specifically, the analysis explores two different pollution levels: PL-1 and PL-20. The findings indicate that higher pollution levels are observed when there is an increase in the retrial probability and a decrease in awareness.

Figure 2 presents the relationship between the reliability of degradation in environmental issues and the fraction of pollution checks. The figure illustrates how the reliability of addressing environmental problems changes as the fraction of pollution checks varies.

Peer-to-peer (P2P) networks don't automatically cause damage in the environment. P2P networks allow users to share files and resources directly with one another. They are a type of decentralized communication and file-sharing system. P2P networks' effects on the environment are mostly determined by the kind of material exchanged, how energy-efficient the devices are, and the network's general architecture. P2P networks may possibly exacerbate environmental issues in the following ways. Sharing of Files with High Resource Utilization:

Sharing huge files on a regular basis by users, particularly those that need a lot of processing power to develop or handle, may be indirectly responsible for pollution in the environment. For instance, distributing big software programmes or high-definition films may lead to higher energy usage during the upload and download phases.

Use of Data Centers: P2P networks may depend on centralized servers or data centers to help peers establish their first connection. Data centers have the potential to use large quantities of energy, which can pollute the environment if they are fueled by non-renewable resources. Device Energy Consumption: The energy efficiency of the devices participating in P2P networks is a significant factor. If users are running power-hungry devices or leaving them on for extended periods while participating in P2P activities, it could contribute to overall energy consumption and environmental impact. Network Infrastructure: The efficiency and environmental impact of the underlying internet infrastructure also play a role. If data transmission relies on energy-intensive networks or if the infrastructure is not optimized for energy efficiency, it could contribute to pollution indirectly. It's crucial to remember that P2P networks' environmental effects rely on their particular implementation, the kind of material being shared, and the energy sources used to power the infrastructure involved. These factors are not fundamental to the P2P concept itself. Furthermore, technological developments and a move towards renewable energy sources might lessen the possible negative environmental effects of digital activities, especially P2P network-related ones.

4.2 Impact

When a user downloads a polluted version, there is a potential for them to attempt downloading it again, resulting in additional network traffic. The magnitude of traffic generated by pollution can be estimated by considering the popularity of topics, which reflects user interest. In a study conducted on the KaZaA platform, it was discovered that only a small fraction of topics experience frequent queries, and popular files have short lifetimes, making them prime targets for polluters [8]. The repetitive nature of pollution attacks leads to a notable increase in unnecessary downloads, and this increase can be quantified using Eq. 6. As the system transitions to a steady state, the total number of retrials can be determined using Eq. 7. Figure 7 visualizes the relationship described by Eq. 7, considering awareness and retrial probability, with a pollution level set to PL-15. Remarkably, in the worst-case scenario, the number of retrials can surpass triple the number of initial trials, resulting in a quadrupling of the P2P traffic load [8]. Given that P2P activity constitutes a significant portion (60%) of Internet traffic, a pollution attack has the potential to exert a considerable impact on the overall load of Internet traffic [8].

5 Conclusion

According to a study involving human participants, it was observed that users had a limited level of awareness regarding different forms of pollution. The study also revealed that users tended to check their downloaded files immediately or after a significant delay, resulting in a bimodal distribution of slackness. Based on these findings, a mathematical model was developed to gain deeper insights into pollution dynamics. The analysis using this model emphasized the crucial role of user awareness in influencing pollution dynamics. It was further discovered that for large-scale attacks to be effective, polluters needed to decrease user awareness. Additionally, targeting popular files could result in a fourfold increase in P2P traffic loads. To advance research in this field, future investigations should explore user behavior when downloading files other than music, such as movies and software. Exploring other parameters used in the model, including cooperativeness and retrial probabilities, also holds potential for further insights. An interesting avenue for exploration would be the design of a reputation system that incorporates user awareness, considering the study's indication that even "honest" users may not accurately recognize the quality of files.

References

1. Christin, N., Weigend, A.S., Chuang, J.: Content availability, pollution and poisoning in peer-to-peer file sharing networks. In: ACM E-Commerce Conference (EC 2005), June 2005
2. Dumitriu, D., Knightly, E., Stoica, I., Zwaenepoel, W.: Denial-of-service resilience in peer-to-peer file sharing systems. In: Proceedings of the ACM SIGMETRICS 2005, June 2005
3. Good, N., Krekelberg, A.: Usability and privacy: a study of KaZaA P2P file-sharing. In: Proceedings of the SIGCHI 2003 (2003)
4. Gummadi, K.P., Dunn, R.J., Saroiu, S., Gribble, S.D., Levy, H.M., Zahorjan, J.: Measurement, modeling, and analysis of a peer-to-peer file-sharing workload. In: Proceedings of the SOSP 2003, October 2003
5. Kamvar, S.D., Schlosser, M.T., Garcia-Molina, H.: The EigenTrust algorithm for reputation management in P2P networks. In: Proceedings of the WWW 2003, May 2003
6. Lee, U., Choi, M., Sanadidi, M.Y., Gerla, M.: Understanding pollution dynamics in P2P file sharing, UCLA CSD Technical Report, October 2005
7. Liang, J., Kumar, R., Xi, Y., Ross, K.: Pollution in P2P file sharing systems. In: Proceedings of the INFOCOM 2005, May 2005
8. Walsh, K., Sirer, E.G.: Fighting peer-to-peer SPAM and decoys with object reputation. In: Proceedings of the P2PECON 2005, August 2005
9. Kurra, S.S., Naidu, S.G., Chowdala, S., Yellanki, S.C., Sunanda, D.B.E.: Water quality prediction using machine learning. Int. Res. J. Mod. Eng. Technol. Sci. (2022)
10. Haghiabi, A.H., Nasrolahi, A.H., Parsaie, A.: Water quality prediction using machine learning methods. Water Qual. Res. J. **53**(1), 3–13 (2018)
11. Chen, K., et al.: Comparative analysis of surface water quality prediction performance and identification of key water parameters using different machine learning models based on big data. Water Res. **171**, 115454 (2020)
12. Ahmed, U., Mumtaz, R., Anwar, H., Shah, A.A., Irfan, R., García-Nieto, J.: Efficient water quality prediction using supervised machine learning. Water **11**(11), 2210 (2019)
13. Du, S., Li, T., Yang, Y., Horng, S.J.: Deep air quality forecasting using hybrid deep learning framework. IEEE Trans. Knowl. Data Eng.Knowl. Data Eng. **33**(6), 2412–2424 (2019)
14. Bhambri, P.: A CAD system for software effort estimation. In: 2022 2nd International Conference on Technological Advancements in Computational Sciences (ICTACS), 10 October 2022, pp. 140–146. IEEE (2022)
15. Fathima, S.A., Hundewale, N.: Comparitive analysis of machine learning techniques for classification of arbovirus. In: Proceedings of 2012 IEEE-EMBS International Conference on Biomedical and Health Informatics, pp. 376–379. IEEE, January 2012
16. BaniMustafa, A.: Predicting software effort estimation using machine learning techniques. In: 2018 8th International Conference on Computer Science and Information Technology (CSIT), pp. 249–256. IEEE, July 2018

A Heuristic for Minimizing Resource Requirement for Quantum Graph Neural Networks

Srinath Devale(✉), Karthick Seshadri, and S. Nagesh Bhattu

Department of Computer Science and Engineering, National Institute of Technology Andhra Pradesh, Tadepalligudem 534101, India
mtcs2201@student.nitandhra.ac.in,
{karthick.seshadri,nageshbhattu}@nitandhra.ac.in

Abstract. Quantum graph neural networks (QGNNs) utilize qubit entanglement and superposition for better representation of the graph data, but they are limited by the number of qubits available on the quantum hardware. To offset this limitation, we propose an approach called hybrid quantum-classical GNNs (QCGNNs) involving graph clustering algorithms for partitioning the nodes of a graph into clusters and learning representation of clusters using QGNNs for subsequent graph classification. QCGNNs require lesser number of qubits and runs on quantum hardware in comparison to existing QGNNs without significantly impacting accuracy. Suitability of different clustering algorithms for implementing QCGNNs was also assessed in this research attempt. Experimental results show an 81.33% reduction in qubit requirement on the MUTAG dataset. All the experiments reported in this paper are performed on the SV1 quantum simulator provided by Amazon Braket environment.

Keywords: quantum computing · hybrid quantum-classical graph neural network · quantum machine learning · braket · graph classification · graph clustering

1 Introduction

Since the advent of quantum computers, the confluence of quantum computing and neural networks has started a new field at the intersection of machine learning and quantum computing. Quantum computers have the ability to solve complex problems exponentially faster than classical computers, on the other hand, neural networks have produced very high proficiency in learning patterns and representations from data. A justifiable combination of these two domains resulted in the conceptualization of Quantum Graph Neural Networks (QGNNs); this approach leverages the power of quantum computing for graph representation learning.

In specific, QGNNs are well-suited to domains where the model involves complex-structured data; this is facilitated by entanglement and superposition

D. Pastor-Escuredo et al. (Eds.): ICDLAIR 2023, LNNS 1001, pp. 246–255, 2024.
https://doi.org/10.1007/978-3-031-60935-0_22

properties of quantum systems, which in turn enables to learn hidden patterns, relationships, and features that were hard to extract otherwise, thus facilitating advancements in fields, such as social network analysis, cryptography, and artificial intelligence.

The approach used to construct a QGNN circuit for graph classification is referred to as Ego-Graph representation learning approach by Xing et al. [2], which is briefly explained in the methodology section. To overcome the restriction of the number of qubits and the runs required on a quantum hardware, in this research attempt, we have implemented a hybrid quantum graph classification approach, that requires lesser number of qubits and runs on quantum hardware. We have analyzed the performance benefits of the proposed approach on the State Vector Simulator (SV1) provided by the Amazon Braket environment. SV1 is a utility computing based state vector simulator provisioned with 34 qubits to simulate circuits and generates as output a full state vector or an array of amplitudes [1].

The objectives of this research attempt are as follows:

1. To design a hybrid quantum-classical graph neural network that requires lesser number of qubits to implement.
2. To assess and evaluate for which class of input graphs QCGNNs are preferable to QGNNs.

The rest of the paper is organized as follows. In Sect. 2 we introduce the research attempts on QGNNs; Some preliminaries regarding quantum circuits, and their implementations in the Braket environment are outlined in Sect. 3. Section 4 outlines the methodology for mapping graph features to quantum circuits and for subsequent quantum graph classification. In Sect. 5 we discuss the experimental results to validate and verify our QCGNN implementation on quantum computers and simulators. The final section has the concluding summary of the research work and a brief remark on the extensions that can be carried out on this research.

2 Related Work

There have been various research works carried out to represent graph neural networks in the form of a quantum circuit to take advantage of quantum resources. The first model for quantum graph neural networks was developed by Verdon et al. [11]. This model made use of quadratic Hamiltonian and quantum circuits to extract relevant representations of the graph or nodes of the graph in Hilbert space. Following this, Mernyei et al. [8] proposed a quantum circuit-based approach that captures the permutation-invariant topologies of the input graph. However, this model is only restricted to small datasets, due to linear scaling of qubits with the number of nodes.

Xing et al. [2] proposed a hybrid-quantum algorithm for graph classification, called the ego-graph-based representational learning approach. This algorithm works by decomposing a graph into a set of ego-graphs. These sub-graphs were used for feature extraction with the help of quantum circuits. The authors demonstrate that this algorithm could achieve an accurate at-par graph classification with 1.68% lesser parameters as compared to other models such as, QGNN [11], QGCN [14], and GraphQNTK [10], that have been previously used for classification. The quantum circuit proposed in their model takes a set of ego-graphs as inputs and applies rotatory gates based on the learned parameters for generating graph representations. Subsequently, a tensor product of all the quantum states is computed to obtain a graph embedding which was used for classification.

Chalumuri et al. [4] proposed a method for implementing multi-class classification of data items on IBM Qx hardware and evaluated the accuracy of the model. The authors made use of the properties of quantum mechanics such as entanglement and superposition for feature extraction. Gradient descent was used for learning the parameters, which were leveraged by the rotatory gates for feature extraction. The authors used separate qubits to encode classical data and ancilla qubits for getting measurements from the quantum circuit. Subsequently, the softmax function was applied to predict the class label.

This work is motivated by and based on the QGNN proposed by Xing et al. [2] wherein, the number of ego-graphs generated is as high as the number of vertices in the graph. However, for adopting the QGNN model for large graphs, we have proposed a heuristic based on clustering the graph into subgraphs and evaluated the merit of the heuristic.

3 Preliminaries

3.1 GNNs for Graph Classification

GNNs are a type of artificial neural networks that are designed for the processing and analysis of graph data. The idea of GNNs is to aggregate and propagate information across nodes in a graph, enabling GNNs to capture complex patterns and connectivities present in the data. Given a graph $G(V, E)$ where V is the set of vertices, E is the set of edges, and $v \in V$, the approach for graph classification is given by Xu et al. [13] as mentioned in Eq. 1.

$$h_k^v = \sigma\left(h_{(k-1)}^v W_{(k-1)}^{(1)} + \sum_{\mu \in N(v)} h_{(k-1)}^{\mu} W_{(k-1)}^{(2)}\right) \tag{1}$$

In Eq. 1, h_k^v is the feature vector of node v at the k-th iteration; $N(v)$ is the set of neighboring nodes of v; $W_{(k-1)}^1$ and $W_{(k-1)}^2$ are trainable weight matrices at the k-th iteration for the node v. For classification, the aggregation of all the node representations is performed to obtain the complete graph representation h_G, as mentioned in Eq. 2.

$$h_G = \sum_{v \in V} h_k^v \tag{2}$$

3.2 Quantum Circuits

A quantum circuit is a mathematical and visual representation of quantum operations performed on qubits. It is similar to gate operations performed on classical bits. In Braket environment, there are several quantum gates capable of manipulating the state of the qubits. They include fundamental gates like Pauli gate, Hadamard gate, and CNOT gate [9]. Quantum gates provide an abstraction for implementing quantum algorithms by making use of the unique properties of qubits.

Pauli Gate (X, Y, Z): The Pauli gate [9], rotates the quantum state around the X, Y, and Z axes of the Bloch sphere, resulting in a superposition of quantum states due to phase shifts.

Hadamard Gate: The Hadamard gate [9], creates a superposition by rotating qubit states to a combination of $|0\rangle$ and $|1\rangle$ states, which is a key step in many quantum algorithms.

CNOT Gate: The CNOT gate [9], operates on two qubits, a control qubit (q_c), and a target qubit (q_t). CNOT gate flips q_t if q_c is in state $|1\rangle$; otherwise q_t is not changed by the CNOT gate.

RY Gate: The RY gate [9], is a type of single-qubit gate that rotates the state around the Y-axis of the Bloch sphere by a specified angle. The angle of rotation is denoted by θ. When applied to a qubit in the $|0\rangle$ state, the RY gate transforms the state to a superposition of $|0\rangle$ and $|1\rangle$ with different amplitudes, determined by the angle θ.

3.3 Hybrid-QGNNs for Graph Classification

The approach for graph classification process using QGNNs was proposed by Xing et al. [2]. Their algorithm uses a mapping from the classical metric space to the Hibert space for parameter training. The algorithm uses a graph decomposition method to generate ego-graphs. The classifier circuit makes use of the ego-graphs to generate their respective features, using which the ego-graphs are classified based on Von Neuman entropy. The mapping and classification process is explained in detail in the methodology section.

3.4 Clustering Algorithms

Spectral clustering is a type of clustering algorithm that uses eigenvalues and eigenvectors to transform the data into lower dimensions followed by clustering of nodes [12]. To compute the similarity between the nodes a similarity matrix is constructed. The data is transformed into a lower-dimensional space by use of a Graph Laplacian matrix. The subsequent step requires clustering of the lower dimensional data points generally by using K-Means clustering algorithm.

K-Means clustering is an iterative algorithm for the clustering of nodes [6]. It involves randomly initializing cluster centroids and assigning all the nodes to their nearest cluster centroid. After each iteration cluster centroids are updated by taking the mean of nodes in respective clusters. This process is carried out until convergence. The optimal number of clusters K can be decided based on the elbow method.

Agglomerative clustering is a type of hierarchical clustering approach [6] in which, initially each node is considered to be a separate cluster and then iteratively pairs of clusters separated by the least distance are merged. This step is repeated until the required number of clusters are computed.

4 Methodology

The first step in our approach is to apply a clustering algorithm on the input graph to form subgraphs or clusters. Subsequently, the sub-graphs are input to a QGNN circuit to learn a quantum representation of the sub-graphs in the Hilbert space; this representation is used for classifying the sub-graphs. The parameters of the QGNN circuit are learned in such a way that the classification error is minimized. The step-by-step approach adopted is illustrated in Fig. 1 and explained in the following subsections.

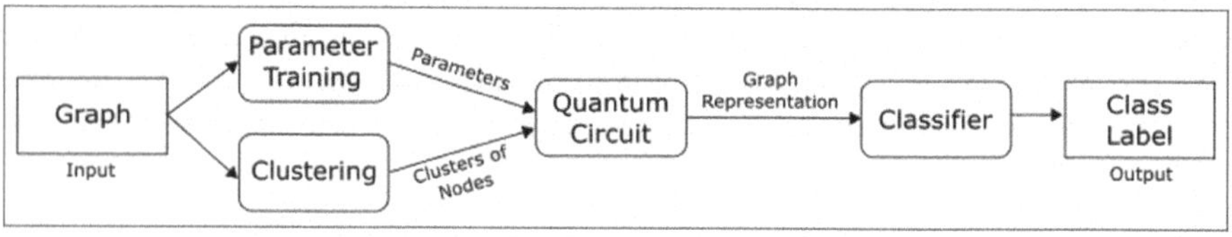

Fig. 1. Flow diagram for the proposed approach

4.1 Parameter Training

For a given graph $G(V, E)$, each node $v_i \in V$ is associated with a feature vector $x_i = \{x_{i_1}, x_{i_2}, ..x_{i_n}\}$. Here N and n are the number of nodes in the graph and the length of the feature vector of each node respectively. The following steps proposed by Xing et al. [2] can be adopted to learn the parameters of a QGNN corresponding to a node in the graph.

1. For each node, $v_i \in V$ we randomly assign a list of parameters $P_i = \{p_1, p_2, .., p_n\}$ in the range $[0, \pi/2]$. Distance between vertex pairs $v_i, v_j \in V$ are computed using Eq. 3.

$$D_{ij} = x_i \cdot x_j \forall i, j \in \{1, 2, ...N\} \tag{3}$$

2. For each node $v_i \in V$, the objective is to generate an equivalent quantum state $|\phi_i\rangle$ using the quantum circuit shown in Fig. 2 as stated in Eq. 4. The quantum mapper (QM) circuit takes the parameter set P_i and feature vector x_i as inputs, for generating the quantum state representing the node v_i using a single qubit.

$$|\phi_i\rangle \leftarrow QM(P_i, x_i) \tag{4}$$

3. For each node $v_i \in V$, we calculate its distance from the node v_j, where $i, j \in [1, N]$ and $j \neq i$ in Hilbert space, using the quantum state representations of $\{v_i, v_j\}$ i.e. $\{\phi_i, \phi_j\}$ using Eq. 5.

$$D'_{ij} = arccos(\langle\phi_i|\phi_j\rangle) \tag{5}$$

4. The loss is calculated as the aggregated difference between D_{ij} and D'_{ij} $\forall j \in \{1, 2, ..N\}\backslash\{i\}$ using Eq. 6.

$$Loss_i = Loss_i + \sum_{j \in \{1,2,..N\}\backslash\{i\}} |D_{ij} - D'_{ij}| \tag{6}$$

5. We select the parameter P_i, which minimizes the loss as shown in Eq. 7.

$$P_i = \arg\min_{P} Loss_i \tag{7}$$

```
T  : |0|1|2|3|4|5|6|

q0 : -U-U-U-U-U-U-U-

T  : |0|1|2|3|4|5|6|
```

Fig. 2. Quantum mapper circuit

4.2 Classification

The hybrid-QGNNs as discussed in Sect. 4.1, can be directly used to perform graph classification as shown by Xing et al. [2]. As the number of qubits required by this approach depends on the maximum number of adjacency nodes in the graph and the number of runs on quantum hardware is proportional to number of nodes in the graph, for smaller graphs, it requires a lesser number of qubits (due to limited number of maximum adjacency nodes), but for a larger graph, it may require a significant number of qubits and more runs on a quantum circuit. To overcome this limitation, we propose a hybrid QCGNN utilizing clustering methods to perform graph classification. As our approach involves creating a quantum circuit for each of K sub-graphs (clusters), the total runs required on the quantum hardware is K and the number of qubits required is $\theta(N/K)$ assuming balanced clustering. The following are the steps proposed to perform graph classification:

1. Segment the input graph $G(V, E)$ into a set of 'K' clusters $C = \{c_1, c_2, ..c_K\}$ by applying either spectral or k-means or agglomerative clustering algorithm.
2. For each c_i in the set C, we calculate the quantum state of that cluster $|\phi_{c_i}\rangle$ by calling the QC circuit as stated in Eq. 8. The QC circuit containing rotational gates R_x, R_y, and R_z for a cluster having three nodes is shown in Fig. 3.

$$|\phi_{c_i}\rangle \leftarrow QC(c_i) \tag{8}$$

3. The density matrix ρ_{c_i} for cluster c_i is computed by taking the outer product of the quantum state of c_i with itself by using Eq. 9.

$$\rho_i = |\phi_{c_i}\rangle\langle\phi_{c_i}| \tag{9}$$

4. The set of density matrices $\rho = \{\rho_1, \rho_2, .., \rho_K\}$ is obtained by performing steps 2 and 3 for each cluster $c_i \in C$. The representation of graph G (h_G) can be computed using Eq. 10.

$$h_G = \sum_{i \in \{1,2,..K\}} \rho_i \tag{10}$$

5. The classification of a given graph is done by comparing the difference of ρ_0 and ρ_1 with the graph representation matrix h_G as per Eq. 11, where $S(\rho_0)$ and $S(\rho_1)$ are the Von Neumann entropies corresponding to class 0 and 1 respectively, which are computed using the training graph dataset with the class labels 0 and 1 respectively.

$$y_G = \begin{cases} 0, & \text{if } |S(\rho_0) - h_G| \geq |S(\rho_1) - h_G| \\ 1, & \text{else} \end{cases} \tag{11}$$

The Von Neumann entropy [3] of a density matrix ρ_i can be computed using Eq. 12.

$$S(\rho_i) = -tr(\rho_i log(\rho_i)) \tag{12}$$

5 Experiments and Results

We performed graph classification on a truncated MUTAG dataset [5] consisting of 40 graphs having an average of 18 nodes and 20 edges, using the proposed QCGNN approach discussed in the previous section. The classification task on

```
T  : |0|     1     |     2     |     3     |4|5|6|Result Types|

q0 : -H-Rx(21.88)-Ry(21.88)-Rz(21.88)-X---C-Probability--
                                      |   |
q1 : -H-Rx(21.88)-Ry(21.88)-Rz(21.88)-C-X-|-------------
                                        | |
q2 : -H-Rx(21.88)-Ry(21.88)-Rz(21.88)---C-X-------------
```

Fig. 3. QC circuit

this dataset is performed on the graph as an extrinsic task to measure the performance of the proposed approach. Several metrics have been computed for each of the Spectral, K-Means, and Agglomerative QCGNNs as shown in Table 1. Evaluation of accuracy on different models is shown in Table 2.

Table 1. Performance on MUTAG

	Accuracy	Precision	Recall	F1-Score
Spectral QCGNN	0.65	0.69	0.75	0.71
K-Means QCGNN	0.61	0.61	0.75	0.67
Agglomerative QCGNN	0.60	0.60	0.75	0.66

Table 2. Evaluation for graph classification accuracy of different models on MUTAG

Model	Accuracy
Gra+QCNN [14]	81.04 ± 14.1
GraphQNTK [10]	88.4 ± 6.5
egoQGNN [2]	85.83 ± 3.7
Spectral QCGNN	0.65
K-Means QCGNN	0.61
Agglomerative QCGNN	0.60

We also performed classification task on a truncated OGBN-arxiv [7] graph consisting 40 nodes and 92 edges. The task is performed through the proposed approach as well as QGNN approach, for which the performance metrics observed are reported in Tables 3 and 4. QGNN has an average accuracy of 0.75.

From the performed experiments following observations can be made:

1. Spectral QCGNN performs relatively better than K-Means and Agglomerative QCGNNs on the MUTAG dataset. It could be attributed to spectral clustering algorithm capturing complex graph similarities relatively better.

Table 3. Performance on OGBN-arxiv graph

	Accuracy	Precision	Recall	F1-Score
Spectral QCGNN	0.70	0.63	0.78	0.69
K-Means QCGNN	0.65	0.70	0.63	0.66
Agglomerative QCGNN	0.65	0.66	0.60	0.62

Table 4. Quantum resources required

			Mean-Qubits	
	Qubits	Runs	MUTAG	OGBN
Proposed QCGNN	$\theta(N/K)^*$	$O(K)$	3.6	5.3
EgoQGNN	$\theta(N)$	$O(N)$	18	40

*assuming balanced clustering

2. The qubits required in the proposed approach are equal to the maximum number of nodes in the cluster, however in case of EgoQGNN depends on the maximum degree in the graph. For a dataset like MUTAG consisting of smaller graphs the EgoQGNN approach requires lesser number of qubits, while also requiring a higher number of runs on the quantum hardware compared to the proposed approach.
3. Experimental results show an 81.33% reduction in qubit requirements compared to egoQGNN model on MUTAG dataset. Calculated using $(N - K) * 100/N$, where N is number of nodes in graph and K is number of node clusters formed.
4. The performance of Spectral QCGNN on a large graph is almost at par with the EgoQGNN as can be seen from Table 3. For a large dense graph, the proposed approach requires lesser number of qubits and a lesser number of runs on the quantum hardware.

6 Conclusion

In this paper we proposed a graph classification approach which consists of implementing a clustering algorithm followed by an application of QGNN on the resultant clusters. We analyzed the performance metrics for assessing the suitability of different clustering algorithms on a MUTAG dataset as well as on a citation graph. Our approach requires lesser quantum resources compared to the previously implemented EgoQGNN approach while at the same time minimizing the impact on the accuracy of the classification task. The future work may involve extending the proposed approach for multi-class classification. Research on other alternative parameter training approaches for learning the quantum circuit of QGNN is also a potential future research direction.

Acknowledgements. The authors would like to express their sincere gratitude to the Ministry of Electronics and Information Technology, India (MEITY) and Amazon Web Services (AWS) for providing access to Quantum computers and Simulators under an R&D project with Ref. No. N-21/17/2020-Ne-GD sanctioned to the second and third authors of this paper. We would also like to express our sincere gratitude to Dr. Anjani Priyadarsini, Quantum Computing Evangelist - AWS India.

Funding Information. This research work is sponsored by the Ministry of Electronics and Information Technology, India and Amazon AWS under a project with the sanction number N-21/17/2020-Ne-GD.

References

1. Amazon services homepage (2023). https://docs.aws.amazon.com/braket/
2. Ai, X., Zhang, Z., Sun, L., Yan, J., Hancock, E.: Decompositional quantum graph neural network. arXiv preprint arXiv:2201.05158 (2022)
3. Boes, P., Eisert, J., Gallego, R., Müller, M.P., Wilming, H.: Von Neumann entropy from unitarity. Phys. Rev. Lett. **122**(21), 210402 (2019)
4. Chalumuri, A., Kune, R., Manoj, B.: A hybrid classical-quantum approach for multi-class classification. Quantum Inf. Process. **20**(3), 119 (2021)
5. Debnath, A.K., de Compadre, R.L.L., Debnath, G., Shusterman, A.J., Hansch, C.: Structure-activity relationship of mutagenic aromatic and heteroaromatic nitro compounds. Correlation with molecular orbital energies and hydrophobicity. J. Med. Chem. **34**(2), 786–797 (1991)
6. Han, J., Pei, J., Tong, H.: Data Mining: Concepts and Techniques. Morgan Kaufmann (2022)
7. Hu, W., et al.: Open graph benchmark: datasets for machine learning on graphs. Adv. Neural. Inf. Process. Syst. **33**, 22118–22133 (2020)
8. Mernyei, P., Meichanetzidis, K., Ceylan, İİ: Equivariant quantum graph circuits: constructions for universal approximation over graphs. Quantum Mach. Intell. **5**(1), 6 (2023)
9. Nielsen, M.A., Chuang, I.L.: Quantum computation and quantum information. Phys. Today **54**(2), 60 (2001)
10. Tang, Y., Yan, J.: GraphQNTK: quantum neural tangent kernel for graph data. Adv. Neural. Inf. Process. Syst. **35**, 6104–6118 (2022)
11. Verdon, G., McCourt, T., Luzhnica, E., Singh, V., Leichenauer, S., Hidary, J.: Quantum graph neural networks. arXiv preprint arXiv:1909.12264 (2019)
12. Von Luxburg, U.: A tutorial on spectral clustering. Stat. Comput. **17**, 395–416 (2007)
13. Xu, K., Hu, W., Leskovec, J., Jegelka, S.: How powerful are graph neural networks? arXiv preprint arXiv:1810.00826 (2018)
14. Zheng, J., Gao, Q., Lü, Y.: Quantum graph convolutional neural networks. In: 2021 40th Chinese Control Conference (CCC), pp. 6335–6340. IEEE (2021)

Light-Gated Recurrent Unit Based Acoustic Modeling for Improved Hindi ASR

Tripti Choudhary(✉), Vishal Goyal, and Atul Bansal

Department of Electronics and Communication Engineering, GLA University, Mathura, India
triptichoudhary06@gmail.com

Abstract. Speech Recognition field gets more advantage of deep learning in the last few years. The substantial improvement was reported by Recurrent Neural Network (RNN), which can model long term temporal context. The traditional RNN's have vanishing gradient problem. This problem was recently tackled by Long Short Term Memory (LSTM) with more promising results. However, the computational requirement of the Bi-directional LSTM model was quite high due to the use of three gating units. Later-on, this computational complexity was reduced by Gated Recurrent Unit (GRU) by using two gating mechanism instead of three. Despite these remarkable achievements, still, there is a need for improvement to beat human accuracy. In this series, we present a novel neural network architecture by combining the recently proposed Light-Gated Recurrent Unit (Li-GRU), Multi-layer perceptron (MLP), and softmax module. In this proposed architecture, MLP was used in the first and last layers as DNNs are good enough in feature mapping and class-based discrimination. In between DNN layers, we stacked Li-GRU layers to take advantage of RNN architecture. This architecture was tested on the Hindi Speech dataset, and noticeable improvement was recorded.

Keywords: Hindi Speech Recognition · Deep Neural Network · Light-Gated Recurrent Unit · Acoustic Modeling

1 Introduction

Deep learning techniques have significantly improved the performance of the Automatic Speech Recognition (ASR) system [1]. After Hinton et al. [2], Deep learning has been successfully applied in various areas like Natural Language Processing (NLP) [3], image recognition [4], machine translation [5], and speech recognition [6] with great success. In the last few years, a wide range of neural network architectures has been proposed. Each architecture has some advantages and some limitations. DNNs are good enough to map features at the initial layers and performing class-based discrimination very well at upper layers [7]. DNNs can't persist past information as they have no connection to previous layers,

D. Pastor-Escuredo et al. (Eds.): ICDLAIR 2023, LNNS 1001, pp. 256–265, 2024.
https://doi.org/10.1007/978-3-031-60935-0_23

which makes DNN less suitable choice for speech recognition. DNN can also process the fixed size acoustic frame window (9–13 frame window), which makes them unable to process temporal dependencies and different speaking rates.

RNN is a well-known architecture of deep learning. RNNs loop-like structure allows to retaining its previous information. RNN flows data in concealed layers which persevere past information and makes it good choice for speech recognition. It has vanishing and exploding gradient problem [8]. RNN varients like LSTM [9], GRU [10], and Li-GRU [11] resolve this issue by using gating units.

In this work, we proposed the multilayer network by combining DNN and li-GRU layers to take the merits of each model. Only a few attempts have been recorded by fusing DNNs plus gated RNNs [12–15] in different machine learning areas. These models outperform compared to other state-of-the-art models but suffered from the heavy computational load. In this work, we reduce the computational load significantly by li-GRU layers in which only one gating mechanism is used. The proposed model was implemented and tested on the standard Hindi speech dataset [16]. The proposed model reduces the training time by a factor of 1.2 compared to the DNN-LSTM model. The main contributions of this article are as follows:

1. A novel multilayer network based on a combination of DNN and li-GRU was proposed for Hindi speech recognition.
2. We investigate the role of various feature combinations and hyperparameters over ASR performance.
3. The proposed system significantly reduces the training time compared to other combinations and shows high accuracy.

The rest of the paper is organized as follows. Section 2 presents the details of the proposed architecture. Section 3 describes the corpus details and experimental setup. Section 4 presents the results, and Sect. 5 shows the conclusion.

2 DNN-Li-GRU Hybrid Architecture

DNN based acoustic modeling is a current trend of speech recognition. DNNs are good enough of mapping features at initial layers and class-based discrimination at the last layer of the neural network [12]. On the other side, DNNs have some serious limitations like they are not capable to handle the longer context-window of speech frames.

RNN is an appropriate choice for acoustic modeling. It has an ability to predict the future sequence with the help of previously observed speech features. GRU model uses fewer parameters compared to the LSTM model, which significantly reduces the computational load of the system. As the hybrid architecture of DNN-GRU became more complex and requires more computational power, we replace the GRU with Li-GRU, which further reduces the computational load of the system with better results. The hybrid architecture for low resource Hindi ASR is shown in Fig. 1. The input features for the hybrid acoustic model were the combination of 39-MFCC features, 40-FBANK features, and 40-fMLLR

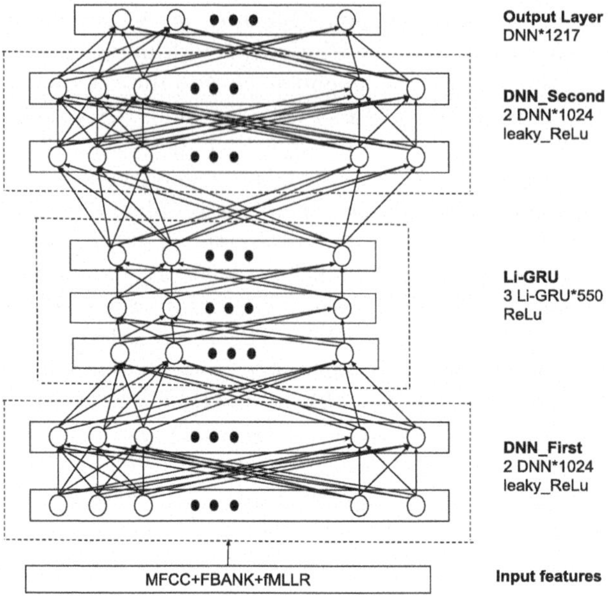

Fig. 1. Proposed hybrid DNN-LiGRU architecture.

features. This combined feature vector was feed into DNN layers. The initial two-layer DNN consists of 1024 units per layer. We choose a leaky-ReLu activation function.

The Li-GRU layers were appended over DNN layers. We used 3 Li-GRU layers with 550 units per layer. On top of that two DNN layers were put with 1024 units per layer. Finally, one output layer with a softmax activation function with 1217 units was used to calculate the posterior probabilities. All the details can be found in Table 1 related to the proposed novel hybrid architecture. We used Kaldi decoder with the beam set to 13, and the lattice beam was set to 8. The acoustic weight was set to 0.2. This configuration was used to test all the experiments. For Hindi ASR, the evaluation of ASR performance was done on Word Error Rate (WER). These results were compared with previously published results on the same dataset.

Table 1. Hybrid Architecture Details

Architecture	Layers	#Neurons	Activation Function	Dropout	Learning rate	Batch-Normmalization
DNN_First	2	1024	Leaky_ReLu	0.15	0.0004	True
Li-GRU	3	550	ReLu	0.2	0.0004	True
DNN_Second	2	1024	Leaky_ReLu	0.15	0.0004	False

3 Corpus Details and Experimental Setup

3.1 Corpus Details

Hindi speech dataset developed by TIFR, Mumbai was used in this work [16]. This dataset is the collection of 100 speakers with total size of 2.5 h. It was recorded on 16 kHz sampling rate. We used ratio of 80-10-10 for train, validation and testing. The total output classes are 1217 in this dataset. To increase the size of this dataset, we used speed perturbation technique (Table 2).

Table 2. Hindi Corpus Details

Dataset	#Sentences	Duration	Duration after 5 fold speed perturbation
Training	800	2.1 h	10.5 h
Development	100	10 Min	50 Min
Test	100	10 Min.	–

3.2 Experimental Setup

In this work, various acoustic features were used to train the acoustic models. These Features are MFCC (Mel Frequency cepstral coefficient) with dimension of 39 (13 static+ Δ + $\Delta\Delta$), 39 PLP (13 static + Δ + $\Delta\Delta$), 40 log-Mel filterbank features (FBANKS) and 40 FMLLR Features. Except for these Features, various combinations of acoustic Features were also combined in this work. These acoustic Features combinations are MFCC + PLP, MFCC + FBANK, MFCC + FBANK + FMLLR. All these features were extracted as reported in the TIMIT recipe of Kaldi [17]. The concentration of different acoustic Features was done by the PyTorch-Kaldi toolkit [18].

The feed-forward connections in the deep supervised neural network are not initialized as classical feed-forward neural networks. We follow scheme describe in [19] to initialize feed-forward connections in the architecture. We used orthogonal matrices to initialize the recurrent weights. It has been shown that the acoustic models trained with dropout have better generalization capabilities and significant performance gain. In this work, recurrent dropout has been used as a regularization technique. The dropout technique works well for the feed-forward neural network but not in case of recurrent connection. To solve this issue, we used the same dropout mask across all time steps as suggested by [20]. Dropout with value 0.2 has been used in this work. Batch Normalization addresses the internal covariate shift problem. Batch-Normalization speed-up the training time and helps to improve system performance. In this work, Batch-Normalization was also used only for feed-forward connections as suggested in [21]. As mentioned in [22], we initialized the gain factor gama of batch-normalization and shift parameter beta with a value of 0.1 and 0, respectively.

All training sentences were sorted into ascending order of their length before the training. We used the minibatch size of 8 sentences. Start from the short utterances, this minibatch size was progressively processed by the training algorithm. As the training of the recurrent model on very long sentences requires more memory and most of the times out-of-memory issues arise. Here, we truncate the long sentences having value more than 1000 to mitigate the out-of-memory issue. There are three main reasons to apply the sorting approach.

1. It minimize the need of zero-padding to form minibatches which help to avoid biases on batch-normalization
2. helps to slightly improve the performance
3. helps to ensure the numerical stability of gradients.

Adaptive Moment Estimation algorithm was used for optimization. The hyperparameters of the acoustic models are learning rate, number of hidden layers, hidden neurons per layer, dropout factor, etc. These hyperparameters were optimized on the development set. We used a simple utility provided by PyTorch-Kaldi [18] for hyperparameters tuning. This utility implements a random search algorithm. A total of 25 random configuration files were generated by changing the above-mentioned hyperparameters. The system was trained and examined for all the 25 configurations and the best one was chosen. This utility takes the input configuration with some initial values according to our experience. As a result, an initial learning rate of 0.0125, batch-size 8, ReLu activation function, and dropout factor of 0.2 were chosen for all experiments. The optimal number of hidden layers and hidden neurons were chosen from the range of 3 to 5 hidden layers with 300 to 500 neurons on each layer. The speech recognition labels were driven from triphone based acoustic model trained by the Kaldi toolkit [17]. These labels were generated by applying forced alignment procedure on test, dev and train set. We used the Kaldi WSJ recipe to train the baseline model. PyTorch-Kaldi [18] uses the Kaldi baseline features and labels to train the complex state of the art neural network acoustic models.

4 Experiments

4.1 Experiment with Different Acoustic Features

Table 3 shows the performance of Hindi ASR with neural network based acoustic modeling using a different combination of acoustic features set. All the neural network architecture in this experiment contains three hidden layers with 550 hidden neurons in each layer. The first column of Table 3 reports the WER achieved by RNN with ReLu activation function. This architecture doesn't include any gating mechanism. The best WER 12.7% was achieved via Li-GRU acoustic model. Due to the single gating mechanism, it saves 33% computational power over GRU acoustic model in this work. The combination of MFCC, FBANK, and fMLLR feature set gives the lowest WER in all neural network architecture. The SRI language model (SRILM) toolkit [23] was used to train the tri-gram language model on the available text transcriptions.

Table 3. Performance evaluation on different feature-set with different acoustic models

Features/Architecture	Language Mode	WER%			
		ReLu-RNN	LSTM	GRU	Li-GRU
MFCC	3G	20.4	19.2	17.5	16.9
FBANK	3G	19.9	18.6	16.9	16.2
fMLLR	3G	18.0	16.7	15.2	14.4
MFCC+FBANK	3G	17.4	16.1	14.9	13.6
MFCC+FBANK+ fMLLR	3G	16.8	15.9	14.2	12.7

4.2 Role of Batch-Normalization

Table 4 takes the gated and non-gated model from the previous experiment with the combined feature set (MFCC + FBANK + FMLLR). This experiment shows the importance of batch normalization. Table 4 clearly shows that batch normalization improves performance. The Relative improvement of 1–1.5% was observed in all neural network models. Batch Normalization did not work for a high learning rate [21]. According to this, we used a small learning rate in Li-GRU which gave better results. The best WER of 11.6% was achieved by the Li-GRU model. This experiment strongly recommends to use the batch normalization along with these acoustic models.

Table 4. Role of Batch-Normalization

Parameters	WER%			
	ReLu-RNN	LSTM	GRU	Li-GRU
without batch-normalization	16.8	15.9	14.2	12.7
with batch-normalization	15.6	14.6	13.3	11.6

4.3 Impact of Diffent Techniques

In this experiment, we investigate the role of other techniques to improve ASR performance. The baseline models were chosen from experiment 1 with the combined feature set (MFCC + FBANK + FMLLR). In the second row, we progressively increase the sequence length during training (Table 5). The starting sequence length for training was chosen as 100 (i.e. 1-s duration). We double this value at every epoch. The maximum sequence length was allowed during training was 1000. By using this strategy, our model learns the short-term dependencies in the initial stage of training and forced to learn long-term dependencies in the later stage of training.

Next, we add recurrent dropouts and measured performance improvement. As mention in section (Experimental setup), we use the same dropout valve of

Table 5. Role of other techniques

Parameters	WER%			
	ReLu-RNN	LSTM	GRU	Li-GRU
Baseline	16.8	15.9	14.2	12.7
+Increase Sequence length	16.5	14.7	13.8	12.2
+Recurrent droupout	16.1	14.5	13.2	11.3
+Batch-Normalization	15.5	14.1	12.8	11.1
+Monophone Regularization	15.2	13.8	12.5	10.8

0.2 to avoid gradient vanishing problem. In the fourth row, we applied the batch-normalization technique. In the earlier section, we already investigate the role of batch normalization without apply increase sequence length and recurrent dropout. Finally, the impact of mono-phone regularization was investigated. In this, two softmax layers were used, one for context-dependent state and other for mono-phone states. This technique helps to improve ASR performance. The best WER of 10.8% was achieved by progressively and applying these techniques.

4.4 Computational Efficiency of Different Models

In this experiment, we compare the computational efficiency of each acoustic model reported in earlier sections. Table 6 shows per epoch training time of RNN and other gated architectures. For the Hindi dataset approx 18% reduction in training time was achieved by Li- GRU model. Due to single gated recurrent architecture, Li- GRU model trained with less number of parameters compared to GRU and LSTM model. In this experiment, we consider only 5, and 7 layer architecture with 550 hidden neurons in each hidden layer.

Table 6. Training time comparision

Architecture	#Layers	#Neurons	#Parameters	Training time
ReLu-RNN	5	550	5.5M	500 s
LSTM	5	550	20.1M	1369 s
GRU	5	550	15.2M	1207 s
Li-GRU	5	550	10.4M	997 s
MLP+Li-GRU+MLP	7	550	8.7M	870 s

4.5 Hybrid Acoustic Modeling

In this experiment, we integrate the feed-forward neural network with recurrent models. We tested Hindi ASR on four hybrid acoustic models with a different set of acoustic features. The hybrid acoustic models are comparatively heavy compared to the standard recurrent models. The integrated feature set of MFCC+FBANK+fMLLR outperforms to other feature combinations. The hybrid acoustic model with Li-GRU also perform well compared to other models. The best WER 9.8% was reported by DNN+Li-GRU+DNN acoustic model with the MFCC+FBANK+fMLLR feature set. All values remained the same as discussed in the earlier Section (Table 7).

Table 7. Performance comparision on different hybrid acoustic models

Features	WER%			
	MLP+RNN +MLP	MLP+LSTM +MLP	MLP+GRU +MLP	MLP+Li-GRU +MLP
MFCC	17.6	15.5	15.8	13.8
FBANK	17.1	16.1	14.6	13.1
fMLLR	15.2	14.2	12.9	11.2
MFCC+FBANK	14.8	13.8	12.5	10.6
MFCC+FBANK+fMLLR	14.1	12.9	11.7	9.8

5 Conclusion

This paper describes the impact of the proposed hybrid DNN-Li-GRU acoustic model in Hindi ASR. Our hybrid acoustic model has shown significant WER reduction compared to other acoustic models only based on RNN's. We also investigate the role of different techniques like batch-normalization, integrated feature-set, recurrent-dropout, and mono-phone regularization on Hindi ASR. We obtained the lowest WER of 9.8% using the proposed model. The findings of this work can be summarized as follow:

1. The combined feature-set of (MFCC+FBANK+fMLLR) reports 4% relative improvement over MFCC features.
2. Batch-Normalization also helps to improve the performance of Hindi ASR with 1% WER reduction
3. The proposed model is having more layers with less number of parameters. This significantly reduces the computational load of the ASR system and increase the generalization capability.

The experimental results also support the fact that hybrid of DNN-Li-GRU is computationally efficient and performing better compared to other acoustic models.

References

1. Deng, L., Yu, D., et al.: Deep learning: methods and applications. Found. Trends® Signal Process. **7**(3–4), 197–387 (2014)
2. Hinton, G.E., Osindero, S., Teh, Y.W.: A fast learning algorithm for deep belief nets. Neural Comput. **18**(7), 1527–1554 (2006)
3. Collobert, R., Weston, J.: A unified architecture for natural language processing: Deep neural networks with multitask learning. In: Proceedings of the 25th International Conference on Machine Learning, pp. 160–167 (2008)
4. Cireşan, D.C., Meier, U., Gambardella, L.M., Schmidhuber, J.: Deep, big, simple neural nets for handwritten digit recognition. Neural Comput. **22**(12), 3207–3220 (2010)
5. Deselaers, T., Hasan, S., Bender, O., Ney, H.: A deep learning approach to machine transliteration. In: Proceedings of the Fourth Workshop on Statistical Machine Translation, pp. 233–241 (2009)
6. Mohamed, A.R., Dahl, G., Hinton, G., et al.: Deep belief networks for phone recognition. In: Nips Workshop on Deep Learning for Speech Recognition and Related Applications, Vancouver, Canada, vol. 1, p. 39 (2009)
7. Sainath, T., Weiss, R.J., Wilson, K., Senior, A.W., Vinyals, O.: Learning the speech front-end with raw waveform CLDNNs (2015)
8. Bengio, Y., Simard, P., Frasconi, P.: Learning long-term dependencies with gradient descent is difficult. IEEE Trans. Neural Netw. **5**(2), 157–166 (1994)
9. Hochreiter, S., Schmidhuber, J.: Long short-term memory. Neural Comput. **9**(8), 1735–1780 (1997)
10. Cho, K., Van Merriënboer, B., Bahdanau, D., Bengio, Y.: On the properties of neural machine translation: encoder-decoder approaches. arXiv preprint arXiv:1409.1259 (2014)
11. Ravanelli, M., Brakel, P., Omologo, M., Bengio, Y.: Light gated recurrent units for speech recognition. IEEE Trans. Emerg. Topics Comput. Intell. **2**(2), 92–102 (2018)
12. Jung, G., Cho, N., Kim, H., Cho, H.: DNN-GRU multiple layers for VAD in PC game cafe. In: 2018 IEEE International Conference on Consumer Electronics-Asia (ICCE-Asia), pp. 206–212. IEEE (2018)
13. Chen, Y., Wang, Y., Ma, J., Jin, Q.: Brim: an accurate electricity spot price prediction scheme-based bidirectional recurrent neural network and integrated market. Energies **12**(12), 2241 (2019)
14. Shashidhar, R., Patilkulkarni, S., Puneeth, S.: Combining audio and visual speech recognition using LSTM and deep convolutional neural network. Int. J. Inf. Technol. **14**(7), 3425–3436 (2022)
15. Ye, F., Yang, J.: A deep neural network model for speaker identification. Appl. Sci. **11**(8), 3603 (2021)
16. Samudravijaya, K., Rao, P., Agrawal, S.: Hindi speech database. In: Sixth International Conference on Spoken Language Processing (2000)
17. Povey, D., et al.: The kaldi speech recognition toolkit. In: IEEE 2011 workshop on automatic speech recognition and understanding, CONF. IEEE Signal Processing Society (2011)
18. Ravanelli, M., Parcollet, T., Bengio, Y.: The pytorch-kaldi speech recognition toolkit. In: ICASSP 2019-2019 IEEE International Conference on Acoustics, Speech and Signal Processing (ICASSP), pp. 6465–6469. IEEE (2019)

19. Glorot, X., Bengio, Y.: Understanding the difficulty of training deep feedforward neural networks. In: Proceedings of the Thirteenth International Conference on Artificial Intelligence and Statistics, pp. 249–256. JMLR Workshop and Conference Proceedings (2010)
20. Moon, T., Choi, H., Lee, H., Song, I.: Rnndrop: a novel dropout for RNNs in ASR. In: 2015 IEEE Workshop on Automatic Speech Recognition and Understanding (ASRU), pp. 65–70. IEEE (2015)
21. Laurent, C., Pereyra, G., Brakel, P., Zhang, Y., Bengio, Y.: Batch normalized recurrent neural networks. In: 2016 IEEE International Conference on Acoustics, Speech and Signal Processing (ICASSP), pp. 2657–2661. IEEE (2016)
22. Ravanelli, M., Brakel, P., Omologo, M., Bengio, Y.: Batch-normalized joint training for DNN-based distant speech recognition. In: 2016 IEEE Spoken Language Technology Workshop (SLT), pp. 28–34. IEEE (2016)
23. Stolcke, A.: Srilm-an extensible language modeling toolkit. In: Seventh International Conference on Spoken Language Processing (2002)

Detecting Phishing URLs Using Machine Learning: A Review

Kritika Kapse(✉), Meenu Chawla, Namita Tiwari, and Richa Goenka

Department of Computer Science and Engineering, MANIT, Bhopal, India
kritika10kapse@gmail.com

Abstract. The Internet's explosive expansion has led many people to switch from conventional banking to online banking. Unfortunately, this transition has resulted in cybercrimes such as domain fraud, ransomware, hacking, social engineering, and phishing. Attackers easily use phishing websites to steal delicate personal data because of the Internet's inherent anonymity, such as passwords and user identities. Deterring these phishing scams is critical for protecting online organisations and individual users. Therefore, the purpose of this study is to investigate and evaluate the current state of the art in detecting phishing URLs using machine learning. Various methods, including feature extraction methods and classification algorithms, were studied. The review discusses limitations and effectiveness evaluation metrics while providing possible improvements. This comprehensive review aims to provide insight into URL-based phishing detection techniques through machine-learning approaches, which promise more robust ways of enhancing online security against fraudulent activities.

Keywords: Malicious URL detection · Phishing Detection · Phishing Attacks · Cybersecurity · Machine Learning

1 Introduction

In today's hyper-connected world, when cutting-edge technology and ubiquitous Internet access are the norm, phishing attacks have evolved into a big problem. Phishing is essentially when malicious individuals appear to be genuine sources in order to deceive unsuspecting users into providing critical information such as credentials or personal information. These naive consumers may not even be aware that they are being tricked [1]. This fraudulent activity can take many forms, including deceiving websites or persuasive emails that trick individuals by causing significant risks related to privacy and security threats, including financial losses. Figure 1 demonstrates the steps and how a typical phishing attack takes place [2].

(a) Planning Phase: During the planning phase, the first step is identifying the victim or victims. Then, the specific information that needs to be collected

D. Pastor-Escuredo et al. (Eds.): ICDLAIR 2023, LNNS 1001, pp. 266–275, 2024.
https://doi.org/10.1007/978-3-031-60935-0_24

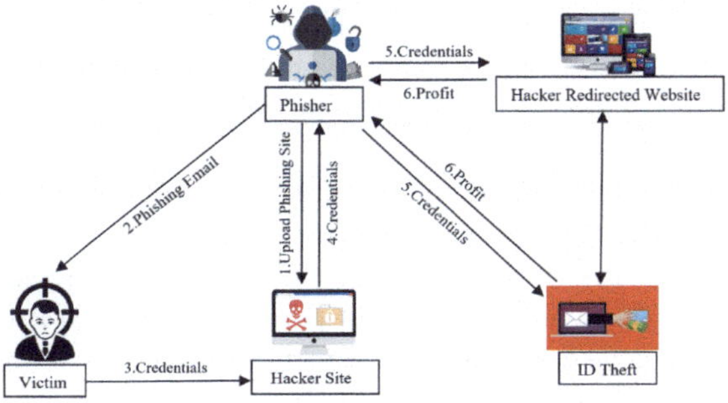

Fig. 1. Life Cycle of Phishing Attack

is determined, and a phishing website is created. To host the phishing website, a web hosting service is used.

(b) Spreading Phase: In the spreading phase, attackers send phishing links to chosen victims via email, SMS, social media, or messaging. These malicious links exploit various communication channels to deceive and compromise unsuspecting individuals.

(c) Striking Phase: The striking phase of a phishing scam takes place when a victim clicks on a fake link. This procedure launches a phoney webpage developed by scammers. Con artists can use victims' personal information or login credentials illegally if they accidentally share it on a fake site. Additionally, malware may be installed on the victim's computer at this phase.

(d) Ending Phase: After achieving the objective, victims are redirected to the authentic site. Steps are then taken to erase phishing evidence, involving the removal of the fake website and the deletion of pertinent log files [1].

According to a report by the Anti-Phishing Working Group (APWG) in 2023, phishing attacks increased significantly from April to June, with 1,286,208 incidents, the third-highest quarterly total on record. There were a total of 597,789 unique URLs identified as phishing sites. The financial sector, particularly banks, was the most heavily impacted industry, accounting for 23.5% of all targeted phishing attacks [3]. Phishing detection methods are diverse in nature, employing distinct approaches and techniques for each purpose.

(a) List-based approaches: A list-based method for phishing detection involves using a database of trusted and potentially malicious websites to identify the difference between the two. This technique falls into two categories: blacklisting and whitelisting [4]. However, when it comes to zero-day assaults, blacklist-based detection in phishing isn't as effective. Its low false positives and ease of use are notable, but it can't keep up with evolving threats because it can't classify new, unlisted harmful sites [5,6].

(b) Heuristic-based approaches: Phishing detection employs heuristic methods using webpage elements to discern real from fake sites. These features power a classifier, creating an effective model for detecting phishing, a vital cyber defense [7]. URL elements, web page contents, and third-party components help create distinct fraudulent website detection characteristics [8]. Although effective in quickly identifying risks upon page load and recognising new attacks, heuristic-based systems have difficulty with novel patterns beyond usual phishing threats [9].

(c) Visual similarity-based approach: Identifying possible phishing websites can be done with the use of a method called visual similarity. In place of looking at the websites' source codes, this technique evaluates their visual appearance in terms of layout and design [10]. Using image processing to compare snapshots of target sites against a database demands more computational resources than saving URLs due to photo storage space [11] (Table 1).

Table 1. Comparison of Various Phishing Detection Techniques.

Techniques & Parameters	List-Based	Heuristics with URL-based Features	Heuristics with Third party-based features	Heuristics with source code-based features	Visual similarity based
Zero-day Detection	No	Yes	Yes	Yes	No
Database Independent	No	Yes	Yes	Yes	No
Third-party Independent	Yes	Yes	No	Yes	Yes
Language Independent	Yes	Yes	No	No	Yes

Examining existing literature reveals that no single strategy effectively identifies all sorts of phishing attacks. Blacklist/whitelist or visual similarity-based methods are unable to identify zero-day attacks, and heuristic algorithms may overlook attacks involving embedded objects, resulting in a high rate of false positives. Thus, cybersecurity experts use machine learning to spot phishing scams by identifying suspicious URLs. A Uniform Resource Locator (URL) is a standardised web address that enables internet users to access specified resources. It identifies the communication protocol by including a protocol identifier and a resource name, and it distinguishes between the domain name and the IP address by using a colon and double forward slashes as separators [9]. Figure 2 depicts this setup.

Fig. 2. URL Components

Machine learning algorithms are particularly effective in identifying phishing URLs because they are efficient and accurate. The features of a URL play a key role in determining if a website or page is legitimate or fraudulent. Choosing the right URL characteristics can improve the detection results. Figure 3 shows the key performance evaluation metrics for calculating phishing URL detection [22]. This paper's main objective is to present an extensive summary of workable tactics for immediately repelling phishing attacks. It examines the fundamental life cycle of a phishing attack, which begins when a user clicks on a phishing link and employs technical tactics to alert the user. In addition to various tools like visual similarity analysis, blacklists, whitelists, and image processing, this study explains phishing URL detection based on machine learning. This study examines state-of-the-art solutions, analyses their challenges and limitations, and suggests future research directions.

SNo.	Name	Definition	Formula
1.	False Positive Rate (FPR)	It is the proportion of legitimate URLs incorrectly identified as phishing.	$\frac{FPR}{FPR + TNR}$
2.	False Negative Rate (FNR)	It signifies the ratio of Phishing URLs incorrectly classified as safe.	$\frac{FNR}{FNR + TPR}$
3.	True Positive Rate (TPR) or Recall	It signifies the proportion of actual Phishing URLs correctly identified as phishing.	$\frac{TPR}{TPR + FNR}$
4.	True Negative Rate (TNR)	It is the proportion of Legitimate URLs correctly identified as safe.	$\frac{TNR}{TNR + FPR}$
5.	Accuracy	It is a measure of correctly identifying Phishing URLs among all URLs tested.	$\frac{TPR + TNR}{TPR + TNR + FNR + FPR}$
6.	Precision	It is the proportion of correctly identified phishing URLs among all phishing URLs.	$\frac{TPR}{TPR + FPR}$
7.	F1-Score	It provides a harmonic mean to assess a model's classification accuracy, balancing precision and recall.	$\frac{2*TPR}{2*TPR + FPR + FNR}$

Fig. 3. Performance Metrics

This paper consists of multiple sections. Section 2 encompasses the literature review, which incorporates viewpoints from various studies in this area. In contrast, Sects. 3 and 4 focus on specific challenges documented among the papers analysed, offering prospective directions for future research to propel and expand this field.

2 Literature Review

In Jain, A. K. et al. [12], the authors suggested a novel phishing prevention method that uses machine learning to extract information from the client side directly. Through a thorough analysis of both genuine and phishing URLs, they identified 19 standout features that could distinguish between legitimate and deceptive web pages with high accuracy (99.09%). These independent features are derived from URLs as well as source code and have outperformed other approaches in identifying phishing websites.

Mahajan et al. [13] have suggested that the machine learning technology employed in this study extracts 16 features from both genuine and phishing URLs to identify the latter. The dataset comprises a total of 36,711 URLs, including 17,058 legitimate and 19,653 fraudulent URLs. To gauge the classifier's performance, it has been split into three parts. Results show that using the Random Forest technique on a training dataset of 90% data yields an impressive detection accuracy of 97.14%.

Alswailem et al. [14] developed a supervised learning method to identify phishing websites using machine-learning algorithms. They first analysed 6116 dataset entries, eliminating duplicates and missing data. Then, they collected features such as URLs, page content, and page rank to train the model with 26 significant features. The Random Forest algorithm was chosen for its excellent efficacy, resulting in an overall accuracy of 98.8% for identifying phishing websites.

In Sahingoz et al. [15], researchers suggested a stand-alone anti-phishing solution in real-time based on NLP-based features and seven classification algorithms. It can detect new phishing attacks and works regardless of language. They worked with a large dataset of 73,575 URLs and incorporated various machine learning techniques such as word vectors, hybrid features, and NLP-based characteristics to train the model. Out of all these approaches, Random Forest using solely NLP-based attributes achieved an outstanding accuracy rate of 97.98%.

In Gandotra and Gupta [16], authors presented a system for detecting phishing URLs that uses six distinct classification methods, and to divide the features effectively, they are classified into website-related, URL-based, and HTML-based groups. The dataset includes 5223 URLs with 26 distinct features to train the machine learning system, among the other algorithms of classification presented in this research paper. Random forest has the best accuracy rate of 99.5%, outperforming others significantly based on FPR and FNR criteria.

In Alam et al. [17], researchers utilised machine-learning techniques, especially decision trees and random forests, to create a model that can distinguish phishing attacks. They trained the ML systems using standard datasets of phishing attacks from Kaggle and applied principal component analysis to recognise and categorise the components of these datasets. The website was mainly classified with the help of decision trees and random forest algorithms, which used 32 features altogether. The random forest technique scored a high accuracy rate of 97%.

In Kumar, Jitendra, et al. [18], the authors used five machine-learning algorithms to differentiate between phishing and genuine URLs. They utilised lexical analysis for feature extraction, statistical analysis for classification, and dataset randomization techniques. Additionally, a regular expression technique was used during the feature engineering process. The team utilised a training set of 100k URLs, with Naive Bayes producing the highest accuracy of 98% compared to other classifiers in this approach.

In Sánchez-Paniagua et al. [19], the authors found that genuine login URLs were more effective than previous approaches in detecting phishing attempts on the login page. To address accuracy issues caused by outdated training datasets, they created a comprehensive dataset called PILU-90K, containing 30K phishing URLs and 60K valid URLs with both index and login pages. The logistic regression model performed exceptionally well with TF-IDF feature extraction using the newly given dataset of login URLs, attaining an excellent 96.50% accuracy. Various research works in the field of detecting phishing attacks on malicious URLs through machine-learning techniques have been summarised in Table 2.

This research identified a highly efficient technique known as Random Forest (RF), which has been implemented in numerous other investigations. RF is an ML method that helps build decision trees and can be used for both binary and multiclass classification. RF doesn't overfit data, works well with huge numbers, and is very reliable. It can function effectively on multiple features and suits both classification and regression. RF has some good points, but it also has some problems. There is no way to understand it, and if the data set is noisy, it may overfit. A large dataset can make training memory-intensive and slow [20].

Table 2. Compilation of Literature Review

Cite	ML Algorithms Used	Datasets	Advantages	Limitations	Performance Metrics
[12]	SVM Naive-Bayes Logistic Regression Neural Network Random Forest	1. Legitimate URLs: 1,918 URLs from Alexa, Payment Gateway 2. Phishing URLs: 2,141 URLs from Phishtank and Openphish	1. A Feature Selection Technique is used (Pearson Correlation) 2. No third-party Features are used	1. Used a small-sized Dataset (4,059 URLs)	1. Accuracy: 99.09% 2. Precision: 99.39% 3. F-Measure: 98.75% 4. FPR: 1.25% 5. FNR - 0.61%
[13]	SVM Decision Tree Random Forest	1. Legitimate URLs: 17,058 URLs from Alexa 2. Phishing URLs: 19,653 URLs from Phishtank	1. Used a Large Dataset 2. High Detection Accuracy	1. No applicable Feature Selection method is used	1. Accuracy: 97.14% 2. FPR: 2.61% 3. FNR: 3.14%
[14]	Random Forest	1. Legitimate URLs: 4,000 URLs collected daily by 10 users 2. Phishing URLs: 12,000 URLs from Phishtank The final dataset contained 2,456 URLs.	1. Higher Accuracy Rate 2. A Feature Selection Technique is used (Combinatory (min-max) approach)	1. Asymmetric Dataset used 2. No Comparison with other ML Algorithms were conducted	1. Accuracy: 98.8%
[15]	Naive Bayes Random Forest KNN (n = 3) AdaBoost K-star SMO Decision Tree	1. Legitimate URLs: 36,400 URLs 2. Phishing URLs: 37,175 URLs (Ebbu2017 Phishing Dataset, 2017)	1. Usage of Feature-rich Classifiers 2. A real-time execution 3. Can detect Zero-day Phishing websites 4. Used a large volume of both genuine and Phishing datasets	1. Could not classify shorter domain URLs	1. Accuracy: 97.98% 2. Precision: 97% 3. F-Measure: 98%
[16]	KNN Random Forest Decision Tree SVM Naive-Bayes AdaBoost	1. Legitimate URLs: 2,723 URLs from Alexa and Payment Gateway 2. Phishing URLs: 2,500 URLs from Phishtank and Openphish	1. The ratio of fraudulent to Legitimate Datasets is almost equal 2. 20 URL features are used.	1. No applicable Feature Selection methods were used 2. Small dataset size (5,233 URLs)	1. Accuracy: 99.5% 2. Precision: 99.4% 3. F-Measure: 99.4% 4. FPR: 0.6% 5. FNR: 0.5%
[17]	Random Forest Decision Tree	1. 2,211 Legitimate and Phishing URLs from Kaggle	1. Different Feature Selection Techniques are used 2. 32 sets of Features are used for Classification	1. A single source dataset is used 2. The size of dataset is small (2,211 URLs)	1. Accuracy: 97% 2. Precision: 96.89% 3. F1-Score: 58.74% 4. Recall: 42.16%
[18]	Random Forest Naive-Bayes KNN Logistic Regression Decision Tree	1. Dataset of 100K URLs which has 57,000 to 60,000 Phishing URLs and rest were Legitimate URLs	1. Used a Huge dataset 2. 26 URL features are used for Classification	1. A single source of dataset is used 2. There are no applicable Feature Selection strategies	1. Accuracy: 98% 2. Precision: 100% 3. F1-Score: 97% 4. Recall: 95%
[19]	Light GBM XGBoost AdaBoost Random Forest KNN SVM Logistic Regression Naive-Bayes	1. 30K Legitimate URLs of Homepages 2. 30K Legitimate Login URLs 3. 30K Phishing URLs (Phishing Indexed Login URLs PILU-90K)	1. Used the Popular TF-IDF Technique for Feature Extraction 2. Used a Large Dataset	1. Accuracy slightly reduces due to utilising login websites for training	1. Accuracy: 96.50% 2. Precision: 96.54% 3. F1-Score: 96.48% 4. Recall: 96.51%

3 Issues and Challenges

Utilising machine learning to distinguish phishing from noxious URLs includes several challenges. Among these issues are:

3.1 Datasets and Insufficient Training Data

The efficacy of machine learning algorithms is significantly influenced by the quality and quantity of the training data they are provided with. However, overfitting issues can arise if the data isn't distributed uniformly or if the samples are repeated or unbalanced [21]. It is similar to outdated datasets, does not accurately reflect real-world attack situations and behaviours, and lacks real-time features. Since dataset size influences classifier performance, each strategy uses a different dataset size. They intended to utilise a larger dataset because a larger dataset would enable the classifier to detect fraudulent URLs more precisely and efficiently [22].

3.2 Dynamic and Changing Nature of Phishing Techniques

Phishing techniques are constantly changing and evolving, making it difficult for machine-learning algorithms trained on previous data to detect new tactics. Even classical ML methods struggle with zero-day phishing attacks due to a lack of efficiency [23]. To efficiently combat these kinds of attacks, the detection model must dynamically adapt and explore new behaviours in newly evolving phishing patterns. Unfortunately, existing classification techniques are not capable of this level of adaptability [24].

3.3 Feature Extraction Challenges

Feature extraction is crucial to machine learning phishing detection. The procedure is difficult and time-consuming, requiring experimentation and human knowledge, leading to errors [25]. Some researchers choose features only from personal experience, while others use statistical methods to narrow down the optimum set [8]. The time and expertise needed to manually craft features limits phishing detection. These hurdles hinder identifying specific URL elements that indicate phishing attacks [6]. URL attributes, such as the use of the HTTPS protocol, can affect performance. The process of feature extraction is difficult; careful selection improves performance, while bad selection decreases it [22].

3.4 Class Imbalance

A class imbalance issue occurs when a significantly higher number of legitimate URLs than phishing ones are present. This imbalance affects machine-learning algorithms' ability to detect phishing URLs, resulting in biassed and inaccurate models. This limitation compromises their ability to detect potential cyberattacks.

4 Future Research Directions

Machine learning algorithms offer substantial potential for improving phishing detection from malicious URLs in the future. Areas of inquiry and development moving forward might include:

4.1 Improved Feature Extraction and Selection

Enhancing the extraction and selection of features is vital to improving phishing detection through machine learning. However, identifying and extracting relevant features from a URL can be challenging. In order to achieve top accuracy, it's crucial to carefully select the URL features for the classifier. Utilising URL features is most effective in stopping users from accessing fake websites. This approach swiftly identifies various phishing links, relying solely on host-based and lexical features. Machine-learning algorithms can achieve higher accuracy by advancing feature selection and extraction processes, leading to more reliable and robust phishing detection systems.

4.2 Better Training Data

Machine learning algorithms need lots of high-quality training data to identify phishing URLs. Future studies should focus on gathering and labelling large datasets that are specific to phishing URLs with this end in mind. Many phishing databases have duplicate data. Thus, eliminating it optimises performance. Most training methods use 70% training and 30% testing data. Randomised dataset splitting with 10-fold cross-validation improves outcomes [22]. With additional training data, machine-learning models can learn to recognise complicated phishing URL patterns and attributes.

4.3 Hybrid Approaches

Hybrid approaches that combine multiple machine-learning algorithms with rule-based techniques can enhance phishing detection accuracy. By leveraging the unique strengths of each process, such as the flexibility of machine learning and the explicit rules of rule-based systems, this comprehensive technique offers a more reliable method for identifying phishing URLs.

4.4 Integration with User Education

By including user education alongside machine-learning algorithms, cybersecurity against phishing attacks can be significantly enhanced. While machine learning effectively detects phishing URLs, educating users about recognising and preventing such attacks is critical to maintaining security. Future research should investigate the integration of machine-learning algorithms with user education programs for a holistic cybersecurity approach. Empowering users with knowledge can complement the capabilities of machine learning in combating phishing threats and improving overall resilience.

5 Conclusion

In conclusion, phishing attacks are among the most common and dangerous cyberattacks that steal sensitive data from individuals, businesses, and organisations. Such assaults must be prevented to avoid financial losses and data theft. Currently, machine learning algorithms offer an encouraging approach to detecting phishing URLs effectively and quickly. This review paper presents an insight into the ongoing research in this field. The significance of detecting malicious URL-based phishing and the available detection approaches have been discussed. Furthermore, we conducted a study on the benefits and drawbacks of different machine learning techniques used for detecting phishing attempts. We also provided an extensive analysis of the latest research findings in this field.

Despite advances in machine learning algorithms for phishing URL detection, insufficient training data, dynamic phishing methods, and feature extraction remain. Additionally, future studies can explore novel ways to improve machine-learning models and detect new phishing methods. User education and these insights may form a comprehensive cybersecurity strategy. Detecting phishing attacks from malicious URLs using machine learning is quickly increasing and could help battle this. However, accuracy and efficacy are crucial metrics that necessitate continuing research and the creation of new strategies to remain effective against the ever-changing nature of phishing attacks.

References

1. Jain, A.K., Gupta, B.B.: A survey of phishing attack techniques, defence mechanisms and open research challenges. Enterp. Inf. Syst. **16**(4), 527–565 (2022)
2. Basit, A., Zafar, M., Liu, X., Javed, A.R., Jalil, Z., Kifayat, K.: A comprehensive survey of AI-enabled phishing attacks detection techniques. Telecommun. Syst. **76**, 139–154 (2021)
3. Anti-Phishing Working Group, Phishing Attack Trends Report - 2nd Quarter (2023). https://docs.apwg.org/reports/apwg_trends_report_q2_2023.pdf
4. Das, A., Baki, S., El Aassal, A., Verma, R., Dunbar, A.: SoK: a comprehensive reexamination of phishing research from the security perspective. IEEE Commun. Surv. Tutor. **22**(1), 671–708 (2019)
5. Prakash, P., Kumar, M., Kompella, R.R., Gupta, M.: Phishnet: predictive blacklisting to detect phishing attacks. In: 2010 Proceedings IEEE INFOCOM, pp. 1–5. IEEE (2010)
6. Do, N.Q., Selamat, A., Krejcar, O., Herrera-Viedma, E., Fujita, H.: Deep learning for phishing detection: taxonomy, current challenges and future directions. IEEE Access **10**, 36429–36463 (2022)
7. Zuraiq, A.A., Alkasassbeh, M.: Phishing detection approaches. In: 2019 2nd International Conference on new Trends in Computing Sciences (ICTCS), pp. 1–6. IEEE (2019)
8. Aljofey, A., Jiang, Q., Qu, Q., Huang, M., Niyigena, J.P.: An effective phishing detection model based on character level convolutional neural network from URL. Electronics **9**(9), 1514 (2020)
9. Sahoo, D., Liu, C., Hoi, S. C.: Malicious URL detection using machine learning: a survey. arXiv preprint arXiv:1701.07179 (2017)

10. Gupta, B.B., Tewari, A., Jain, A.K., Agrawal, D.P.: Fighting against phishing attacks: state of the art and future challenges. Neural Comput. Appl. **28**, 3629–3654 (2017)
11. Feng, J., Zou, L., Nan, T.: A phishing webpage detection method based on stacked autoencoder and correlation coefficients. J. Comput. Inf. Technol. **27**(2), 41–54 (2019)
12. Jain, A.K., Gupta, B.B.: Towards detection of phishing websites on client-side using machine learning based approach. Telecommun. Syst. **68**, 687–700 (2018)
13. Mahajan, R., Siddavatam, I.: Phishing website detection using machine learning algorithms. Int. J. Comput. Appl. **181**(23), 45–47 (2018)
14. Alswailem, A., Alabdullah, B., Alrumayh, N., Alsedrani, A.: Detecting phishing websites using machine learning. In: 2019 2nd International Conference on Computer Applications & Information Security (ICCAIS), pp. 1–6. IEEE (2019)
15. Sahingoz, O.K., Buber, E., Demir, O., Diri, B.: Machine learning based phishing detection from URLs. Expert Syst. Appl. **117**, 345–357 (2019)
16. Gandotra, E., Gupta, D.: Improving spoofed website detection using machine learning. Cybern. Syst. **52**(2), 169–190 (2021)
17. Alam, M.N., Sarma, D., Lima, F.F., Saha, I., Hossain, S.: Phishing attacks detection using machine learning approach. In: 2020 Third International Conference on Smart Systems and Inventive Technology (ICSSIT), pp. 1173–1179. IEEE (2020)
18. Kumar, J., Santhanavijayan, A., Janet, B., Rajendran, B., Bindhumadhava, B.S.: Phishing website classification and detection using machine learning. In: 2020 International Conference on Computer Communication and Informatics (ICCCI), pp. 1–6. IEEE (2020)
19. Sánchez-Paniagua, M., Fernández, E.F., Alegre, E., Al-Nabki, W., Gonzalez-Castro, V.: Phishing URL detection: a real-case scenario through login URLs. IEEE Access **10**, 42949–42960 (2022)
20. Jagadeesan, S., Chaturvedi, A., Kumar, S.: URL phishing analysis using random forest. Int. J. Pure Appl. Math. **118**(20), 4159–4163 (2018)
21. Mahdavifar, S., Ghorbani, A.A.: Application of deep learning to cybersecurity: a survey. Neurocomputing **347**, 149–176 (2019)
22. Jalil, S., Usman, M.: A review of phishing URL detection using machine learning classifiers. In: Intelligent Systems and Applications: Proceedings of the 2020 Intelligent Systems Conference (IntelliSys), vol. 2, pp. 646–665. Springer, Heidelberg (2021). https://doi.org/10.1007/978-3-030-55187-2_47
23. Abbasi, A., Dobolyi, D., Vance, A., Zahedi, F.M.: The phishing funnel model: a design artifact to predict user susceptibility to phishing websites. Inf. Syst. Res. **32**(2), 410–436 (2021)
24. Robic-Butez, P., Win, T.Y.: Detection of phishing websites using generative adversarial network. In: 2019 IEEE International Conference on Big Data (Big Data), pp. 3216–3221. IEEE (2019)
25. Ahmad, R., Alsmadi, I.: Machine learning approaches to IoT security: a systematic literature review. Internet of Things **14**, 100365 (2021)

Comparative Analysis of Pneumonia Detection from Chest X-ray Using Deep Learning

Sharda Rani(✉) and Navdeep Kaur

Sri Guru Granth Sahib World University, Fatehgarh Sahib, Punjab, India
duhan.sharda@gmail.com

Abstract. In health sector, various automated system has been proposed for detection of pneumonia. This study represents survey on pneumonia detection from chest X-ray. In this study usability, seamless, goodness factor, computational complexities of the algorithms, various methodologies, and their limitation is to be analyzed. This study provide help to select the most effective and efficient methods from a real perspective. The purpose of this study is analysis and comparison of various methodologies related to the detection of pneumonia. This study also discusses the quality, usability and size off available X-ray datasets. After comparisons of studies, mostly available datasets are highly imbalanced and limited. Advanced pre-processing techniques are required; quality plays a major role to increase the performance, so that significant features are to be extracted in deep learning. Some parameters which are more clinically applicable like Positive likelihood ratio (PLR), Negative likelihood ratio (NLR), Positive predictive value (PPV) and Negative predictive value (NPV) are considered to reflect clinical efficiency.

Keywords: CNN · CXR · DEEP LEARNING · NEURAL NETWORK

1 Introduction

Pneumonia is a severe lung infection, which may be caused by bacteria, viruses and fungi, in which small air sacs filled with pus that leads to high fever and breathing problem. Due to lack of early diagnosis, the world Health Organization (WHO) estimates that, more than four million deaths are caused by pneumonia. And UNICEF claims the lives of 700,000 children every year [1]. Mainly children or senior citizen are infected with pneumonia. One common method to diagnose pneumonia is chest X-ray. Resolution of CXRs is lower than those of MRI and CT, so can't easily interpret even by experienced radiologists. It becomes challenging for researchers to determine whether the spot is due to pneumonia fluid or some other opacity. Severe condition like high fever, cough in chest, similar symptoms of other overlapping respiratory disease. So a novel approach needed to improve early diagnosis of pneumonia with accurate prediction value.

Study also describes publicly available CXR datasets used for pneumonia detection and issue related with these datasets. Author used recent deep learning model to detect

D. Pastor-Escuredo et al. (Eds.): ICDLAIR 2023, LNNS 1001, pp. 276–287, 2024.
https://doi.org/10.1007/978-3-031-60935-0_25

disease such as ResNet, DenseNet, VGG, RetinaNet and ensemble learning models with various pre-processing techniques to improve imbalance and image quality issue.

Over the past decade, Convolution neural network (CNN), is type of deep neural network is working on medical imaging are giving good result. Digital image processing is efficient and reliable method in the classification of the disease. DL based algorithms such as SegNet, U-Net, Cardiac Net, Chexnet have becomes the default choice for medical imaging application.

1.1 Transfer Learning

Transfer learning is a technique in deep learning that stores knowledge of one problem and applies it to another problem. It improves learning by learned from one domain can be reused on a different domain. A model which is trained on a dataset and is used to solve a similar task is called pre-trained model. For e.g. VGG, Inception Net, DenseNet, ResNet, MobileNet and U-Net. These models have many versions of their own. Due to such a vast dataset, the models have high accuracy in image processing. The objective of training a deep learning model is to find the correct weight for the network by numerous forward and backward iterations. One of the advantages of a pre-trained model is the reduced cost of training for the new model. This is because pre-trained weight was used, and the model only has to learn the weight of the last few layers. Transfer learning can be used in two ways: (a) fine-tuning, in fine tuning, the weight of the pre-trained CNN model is preserved on some of the layer and tuned in the other. Weight of initial layers in a model are frozen while top layer of the pre-trained models is retained so that the model learned high level features specific to the new dataset. (b) Using CNN as feature extractor, for tasks where small dataset is available, by removing last fully-connected layer and then using the network as a fixed feature extractor for new dataset.

1.2 Ensemble Learning

When more than one classifier is combined to make a prediction, this is known as ensemble learning. The ensemble method can be applied to improve model accuracy. The ensemble method is computational expensive because it requires multiple architecture to be combined. Ensemble learning technique in deep learning works on the basis of majority voting. In which every model makes a prediction for each test instance, votes for a class label, makes a prediction for each test instance, votes for a class label, and the final prediction is the label that received the most votes. In other means weighted majority voting. Sometimes it gives promising result. Assigning correct weightage to each model in ensemble learning is a topic of further research, but once the weights are given, the result depend on the ratio of the weight assigned. Many researchers prove that ensemble learning model perform better than other models.

2 Available Datasets

1. **The chest X-Ray 14(c)**: Wang et al. [2] Presented a chest X-ray database, which contains 112,120 images of 14 diseases obtained from 30,805 patients. The mostly used by researchers. The CXR are collected at the (US) National Institute of Health (NIH).

The common disease i.e., atelectasis, effusion, infiltration, cardiomegaly, mass, nodule, pneumonia and pneumothorax. And the associated images were extracted. Less than 1500 pneumonia images were found in the dataset. The images are distributed as 8-bit grayscale images scaled to 1024 × 1024 pixels.

2. **Pediatric CXRs (PP):** Kermany et al. [3] collected and labeled pediatric CXRs from the Guangzhou Woman and Children Medical Centre (Guangzhou, China), which contains 5,856 CXR images from pediatric patients aged 1–5 years. 4,273 pneumonia images (2,780 bacterial and 1,493 viral) and 1,583 normal images. The images are distributed in 8-bit grayscale images scaled in various resolutions such as 1040 × 664, 1224 × 1000 and 1848 × 1632 pixels.
3. **RSNA-Pneumonia (RP):** The Radiological Society of North America (RSNA) is the organizer of the competition. Recently, this dataset has been used in a competition of the Kaggle platform called RSNA Pneumonia Detection Challenge [4]. RSNA dataset is a subset of NIH dataset; there is numerous differences that make it an improved dataset focused on pneumonia cases. It consists 30,287 CXRs images, out of which 9555 are pneumonia images and rest 20672 are normal images. These images are acquired from chest X-Ray 14 and 8-bit grayscale with 1024 × 1024 resolution. Annotation is added by radiologists using bounding boxes indicating normal, lung opacity, not normal.
4. **MIMIC CXR:** currently MIMIC CXR [5] is largest available dataset. It contains 371,920 CXRs from 64,588 patients from the Beth Israel Deaconess Medical Centre (Massachusetts, USA) from 2011 to 2016. The CXRs are publicly available in full resolution (2544 × 3056) with 14 chest diseases. MIMIC-CXR was later released including the anonymized radiology reports and DICOM files.
5. **JSRT Dataset**: The Japanese Society of Radiology Technology collected 247 images, among which 154 with pulmonary nodules and 93 were nodule free, with resolution of 2048 × 2048, 0.175 mm pixel size and 12-bit depth [6]. The reference standard for heart and lung segmentation of these images is provided by SCR dataset.
6. **Open-I INDIANA dataset**: Demner-Fushman et al. [7] collected the Indiana (USA) dataset from multiple hospital associated with the Indiana University school of Medicine. It is a dataset consisting of 7910 CXRs from 3955 patient, with resolution 512 × 512 pixels. The images are distributed as anonymized with disease view (frontal or lateral) in DICOM. This dataset is publicly available for users but contains only 40 pneumonia sample.
7. **BIMCV (B) dataset**: it is a large dataset from Valencia Region Medical Image Bank (BIMCV) in 2020 [8]. It containing chest X-ray image and computed tomography (CT) images of Covid 19 + patient. It includes 3293 CXRs image from which 1305 covid-19 positive. All images are in 16-bit PNG format with high resolution and entities are localized with anatomical labels in a medical imaging Data Structure (MIDS) format.

3 Issues with Datasets

A Chest X-ray is common and cost-effective medical imaging technique for clinical diagnosis. There is a scarcity of publicly available dataset. So, it is difficult to deal in clinical application with following problem.

1. **DICOM Format**: Mostly dataset of CXRs in Digital Imaging and Communication in Medicine format. DICOM format is a special format widely used in the medical fields that include the radiographic images and patients' data. First, the information of the concerned patient is de-identified to satisfy the privacy standards. After de-identification, DICOM images changes into JPEG or PNG format.
2. **Dimension**: dimension of normal X-ray images is difficult to process due to high computational cost. So, X-ray images reduced to optimal size while conserve information in it.
3. **Class Imbalance**: Mostly dataset suffer with class-imbalance problem. Most of them are under sampling. One way to increase images in dataset without obtain new images is image augmentation.
4. **Quality**: Mostly publicly available dataset has images with lower quality with respect to image size or bit depth. So, sometimes it becomes cause of poor performance in deep learning system.
5. **Labeling**: sometime labels defined in the public dataset are given confusion to researcher for e.g., consolidation and pneumonia
6. **View position**: Posterior-Anterior (PA), it is standard position for find chest radiograph. PA film is useful if lung disease patient in fixed position. Anterior-Posterior (AP), it is conducted when patient is incapable with PA process. It seems more extended than PA position.

4 Comparative Analysis of Pneumonia Detection Approaches

Most of deep learning method makes "image level prediction" in which classification is based on labels, that is implemented by analysis of entire X-image. But some method works on small patches or a specific region within the image, to identify abnormalities and make prediction. This section describes different method used by different researchers, their advantages and disadvantages (Table 1).

Table 1. Comparison of Pneumonia detection approaches

S.No	Authors/Year	Method	Dataset	Advantages	Limitations
1	Rajpurkar P. et al. [9] 2017	CheXNet 121 layers	Chest X-rays 14	Acc = 76.8%, F1-score = 0.43, heatmaps to visualize the disease area,	Only frontal X-rays available, needs to be improvement
2	Wang et al. [10] 2017	Transfer learning of pre-trained AlexNet, GoogleNet, VGG16, ResNet	Chest X-ray8	Acc = 0.75 for pneumonia	Accuracy not promising, the dataset contains eight diseases

(*continued*)

Table 1. (*continued*)

S.No	Authors/Year	Method	Dataset	Advantages	Limitations
3	Kermany et al. [11] 2018	Transfer Learning	Pediatric chest X-ray	Acc = 92.8, Sensitivity = 93.2 Specificity = 90.1	Accuracy needs improvement
4	Rajarman et al. [12] 2018	Customized VGG 16, grad-CAM	Pediatric chest X-ray	Accuracy = 96%, AUC = 0.99 Precision = 0.97, Specificity = 0.96 b/w normal vs pneumonia	Specificity 0.85, when detecting viral vs bacterial pneumonia
5	Gu et al. [13] 2018	FCN, DCNN, SVM	JSRT, M.C Chest X-ray	Acc = 0.8048 ± 0.020 Sensitivity = 0.7755 ± 0.029, Segment the lungs	Low sensitivity and specificity of ensemble Learning
6	Kumar et al., [14] 2019	Mask-RCNN based method	RSNA pneumonia dataset	Image segmentation combined with image augmentation along with dropout and L2 regularization	Performance can be further improved
7	Liang et al. [15] 2019	49-layer CNN with residual DL architecture	Pediatric CXRs	Accuracy = 90.05, Recall = 0.967, F1-score = 0.927, AUC = 0.95, Precision = 0.891	Low Precision, Needs improvement
8	Sirazitdinov et al. [16] 2019	RetinaNet and Mask R-CNN	RSNA pneumonia dataset	Precision = 0.75, recall = 0.79, F1 score = 0.77, meta information	Needs to improvement

(*continued*)

Table 1. (*continued*)

S.No	Authors/Year	Method	Dataset	Advantages	Limitations
9	Ayan et al. [17] 2019	Transfer learning of Xception and VGG16	Pediatric chest X-ray	Accuracy, sensitivity, and specificity were 0.87,0.82,0.91 for VGG and 0.82, 0.85, 0.76 for Xception	Performance is not promising; needs improvements
10	Togacar et al. [18] 2019	AlexNet, VGG-16, VGG-19, (mRMR,SGD)	Chest X-ray Dataset	Acc = 99.41, Sensitivity = 99.61, Specificity = 99.22	No standard data division
11	Stephan et al. [19] 2019	Use CNN with four layers from scratch	Pediatric Dataset Kaggle	T. Acc = 95%, V. Acc = 93%, less complex model	Overfitting, Nonstandard division of dataset, no other metrics
12	Elshennawy et al. [20] 2020	ResNet152 V2 and MobileNet V2, CNN and LSTM	Pneumonia dataset Kaggle	Acc = 99.2 Precision = 99.4 F1 score = 99.4 Recall = 99.4	Preexisting model (ResNet152 V2) accuracy more than custom CNN, Overfitting, no explanation of result
13	Acharya et al. [21] 2020	Deep Siamese based NN	Kaggle dataset	Improved generalization	Only 2 metrics are considered (cross validation & confusion matrix)
14	Yu et al. [22] 2020	CGNet	Pediatric chest X-rays	Acc = 0.98, Sensitivity of 1, Specificity = 0.97	Size of dataset very small
15	Tobias et al., [23] 2020	MobileNet V2	Kaggle dataset	Acc = 90%, low diagnostics tool	Accuracy needs improvement

(*continued*)

Table 1. (*continued*)

S.No	Authors/Year	Method	Dataset	Advantages	Limitations
16	Rahman et al. [24] 2020	Transfer Learning (AlexNet, ResNet18, DenseNet 201, SqueezeNet)	Pneumonia dataset Kaggle	Acc = 98%, Activation map for different models	Need some improvements, no result justification, limited dataset
17	Chouhan et al. [25] 2020	Ensemble (AlexNet, InceptionV3, ResNet, GoogleNet, and DenseNet 121	Pediatric CXRs dataset	Accuracy = 96.4%, Sensitivity = 99.0%	Low Precision (93.2%), Computationally complex, overfitting
18	Hashmi et al. [26] 2020	Weighted Classifier (ResNet18, Xception, InceptionV3, DenseNet121, MobileNetV3)	Pediatric chest X-ray	Acc = 98.43, Precision = 98.26, AUC score = 99.7	Weighted classifiers time consuming and complex, scarcity of data, no result explanation
19	Sarkar et al. [27] 2020	Deep residual learning, (CLAHE, Heatmaps)	Pediatric chest X-ray	Acc = 98.288, AUC = 0.99726, Sensitivity = 0.9927, Specificity = 0.9555	No standard data division and comparison
20	Khatri et al. [28] 2020	Preprocessing and EMD (Earth Mover's Distance)	36 chest X-ray	Acc = 83.5%, Precision = 80%, Recall = 89%	Accuracy needs improvements, EMD is sensitive to rotation and scaling
21	Kundu et al. [29] 2021	GoogLeNet, ResNet-18, DenseNet-121	Chest X-ray, RSNA	Acc = 98.81, 86.85 Sensitivity = 98.80, 87.02 Precision = 98.82, 86.89 F1-Score = 98.79, 86.95	In some instances, framework classify incorrectly

5 Role of Pre-processing in Deep Learning

Pre-processing techniques is used to improve image quality, by applying it, changed to original to more informative form. Most of datasets have issue like different image size, low quality, and imbalanced ratio, low resolution and unwanted noise. So pre-processing helps to remove the irrelevant data and improve the quality of images to make region of interest (ROI) clearer. Widely used pre-processing techniques on CXR images are:

Augmentation: Stephen et al. [19] used several data augmentation methods to increase size and quality of the dataset like rescale, rotation, width shift, height shift, shear range, Zoom range and horizontal flip. Rahman et al. [24] used three techniques (rotation, scaling and translation) to increase size of dataset. And then X-ray images resized to 227×227 pixel for AlexNet and squeezeNet and 224×224 pixel for ResNet 18 and DenseNet 201. To avoid over-fitted problem Chouhan et al. [25] firstly resized image to $224 \times 224 \times 3$, then three augmentation techniques were used Random horizontal flip, Random resized crop and finally augmenting image with a varying intensity of images.

Segmentation: Rajaraman et al. [12] used a method in which lung boundaries are cropped to the size of a bounding box to include all the lungs pixels that formed ROI for current application. And then cropped bounding box resampled to 1024×1024 pixel dimension and normalized for farther operation. Sirazitdinov et al. [16] Considered that a small region of chest X-ray where pneumonia is found, and it became challenging for modern detector. So author represent the RSNA dataset with median height equals to 304 pixels(29.6% of image height) and median width equals to 219 pixels (21.3% of image width). Acharya et al. [21] worked to improve poor quality, by dividing each X-ray into two segment left and right region, and then segmented image is normalized to process it. In bacterial pneumonia case, alveoli filled with white inflammatory fluid and in viral pneumonia left part is infected with white spot.

Enhancement: Kumar et al. [14] Also performed augmentation on lung opacities and image data with random scaling including shifting in coordinate space ((x1, y1), (x2,y2)) and increasing /decreasing brightness and contrast including blurring with Gaussian blur under batches. Researchers focused on different techniques such as pixel-difference, edge detection, correlation for image enhancement, Histogram Equalization(HE), Contrast-Limited Adaptive Histogram Equalization (CLAHE), Noise Filter, Discrete Wavelets Transform algorithm improve contrast and preserve brightness and widely used by researchers for medical images. Sharma et al. [30] also proposed a study In which various techniques (HE, CLAHE, Gamma, TVF, AMF) is compared with different neural network, and gave advice single method is not enough to get good result, so author focus on advance pre-processing technique. M. Sharma et al. [31] Proposed a study in which CLAHE method is used and compare with combination like morphological operation, median filter, and discrete cosine transform (DCT), discrete wavelet transform (DWT) and find that CLAHE + DWT is better than other and obtained 29.62 value of PSNR.

6 Discussion and Comparative Analysis

This section discusses the approaches and dataset used for pneumonia detection and their limitations. Chest X-ray images are used for pneumonia detection due to availability and cost-effective. But currently it becomes complex task, because physicians observed there is overlapping with other lung abnormalities.

It can be seen from table, multiple datasets are used by researchers, but Chest X-ray 14, Pediatric dataset and Pneumonia dataset Kaggle are considered common most relevant datasets. In most of studies dataset suffer with data imbalance, in which sample of one class is higher than the other class. For example chest X-ray 14 has few labeled images of pneumonia (1500) which became cause of unbalanced classification in deep learning. Many researchers use Augmentation techniques, which helps to increase data in dataset by including rotation, flipping and transformation. Size of X-ray images are very large, so many researchers apply resizing, cropping, dividing and segmentation techniques to reduce computational time. Pre-processing is beneficial technique in deep learning, but sometime it create problem of over-enhancing and noise amplification. So researchers need advanced pre-processing techniques based on deep learning to increase resolution of X-ray images.

In most of studies performance checking tool is Accuracy, Sensitivity (SE), Specificity (SP), Precision, F1-Score and Recall. We know that a perfect classifier has all non-zero value on the diagonal, and zero everywhere else. But there is need to explore the confusion matrix to explain what kind of error made on what class. Some parameters which are clinically applicable are also considered. Fagan monogram should be applied to interpret clinical utility of deep learning algorithms. Positive Predictive Value (PPV), Negative Predictive value (NPV), Positive likelihood Ratio (PLR), Negative Likelihood Ratio (NLR), Diagnostic Odds Ratio (DOR) reflect more clinical practice [32].

A wide variety of architectures are considered in studies, in some studies work on image-level prediction task, compares the performance of various approaches. But it's very difficult to suggest which architecture perform better then another for predication. The choice of model (AlexNet, ResNet, VGG) and depth level (ResNet-18, ResNet-50, ResNet-121) also varies and no standard design. These are some factors that affect the performance of dep neural network. In some studies Saliency map (Grad-Cam) is considered to find region of interest, but still lack of transparency, explainability and understandability of result, because Blackbox solutions are not acceptable in real application.

Challenge to implement ensemble learning, the base classifiers should have a very low correlation. It is expected that base classifiers will complement each other to produce better classification result. The use of transfer learning (combination of NN) gives remarkable result, due to complex structure sometime it suffer with negative transfer, because pre-trained model mismatch with particular task.

It is extremely computationally expensive to train large size (dataset), time consuming to train a deep neural network model with GPU software. Cloud computing provide helps to handle huge size of images and datasets. Multiple GPU can be available in reasonable cost.

Future Works: After a remarkable performance of Deep algorithms, there are some limitations and methodological deficiencies in studies, on which researchers have work on. Studies should reflect clinical practice and compared with health care professionals. Limited, poor quality, poor labeling datasets which is barrier b/w performance and clinical environment must be improved.

Some more statistical analysis like

Positive predictive value (PPV) = 100 * TP/ (TP + FP)
Negative predictive value (NPV) = 100 * TN/ (FN + TN)
Positive likelihood ratio (PLR) = [sensitivity/(1- specificity)]
Negative likelihood ratio (NLR) = [(1- sensitivity)/specificity]

Above parameter should be reported in each study. Consider Fagan Monogram that are widely understand by health professional.. A diagnostic test related with reproducibility (result remains same when different testing is applied). Lack of transparency, Model interpretation is important for disease detection approaches. Several technical challenges remain to be solved.

7 Conclusion

After review, deep learning approaches for pneumonia detection is highly significant, future research will focus on right way to improve performance of disease detection system. Now it is necessary to perform model interpretation because it improves clinical trust in deep neural network. Because bacterial pneumonia need antibiotics and viral pneumonia need different treatment.

Due to limitation of publically available datasets and quality of X-ray images, use of very deep neural network increase complexity. Researchers focus on to increase the quality of X-ray images (high PSNR) with super resolution convolution neural network, then high resolute images is used for classification to achieve more valuable result. Mostly researchers focus on overall accuracy, but that is not sufficient, so positive predictive value (PPV) and negative predictive value (NPV) are consider to specify the limit of specificity and sensitivity, which are parameter relative to the clinical truth. Positive and negative predictive value is new key piece in the diagnostic pathway. PPV of 99% shows that positive result 99% chance of being correct. It provides high level of accuracy with confidence.

This study provides a direction to improve the efficiency and accuracy in pneumonia detection application. And avoid common mistake when dealing with this application.

References

1. UNICEF: Pneumonia, UNICEF Data, June 2018. https://data.unicef.org/topic/child-health/pneumonia/
2. Wang, X., Peng, Y., Lu, L., Lu, Z., Bagheri, M., Summers, R.M.: Chestx-ray8: hospital-scale chest x-ray database and benchmarks on weakly-supervised classification and localization of common thorax diseases. In: Proceedings of the IEEE conference on computer vision and pattern recognition, pp. 2097–2106 (2017). https://doi.org/10.1109/CVPR.2017.369

3. Kermany, D.S., Goldbaum, M., Cai, W., et al.: Identifying medical diagnoses and treatable diseases by image-based deep learning. Cell. **172**(5), 1122–1131 (2018). https://doi.org/10.1016/j.cell.2018.02.010
4. RSNA 2018, RSNA pneumonia detection challenge, Library Catalog. https://www.kaggle.com
5. Johnson, A.E., et al.: MIMIC-CXR-JPG, a large publicly available database of labeled chest radiographs. arXiv preprint arXiv:1901.07042 (2019). https://doi.org/10.48550/arXiv.1901.07042
6. Shiraishi, J.: Development of a digital image database for chest radiographs with and without a lung nodule: receiver operating characteristic analysis of radiologists' detection of pulmonary nodules. AJR Am. J. Roentgenol. **174**(1), 71–74 (2000). https://doi.org/10.2214/ajr.174.1.1740071. PMID: 10628457
7. Demner-Fushman, D., et al.: Preparing a collection of radiology examinations for distribution and retrieval. J. Am. Med. Inf. Assoc. **23**(2), 304–310 (2016). https://doi.org/10.1093/jamia/ocv080. Epub 2015 Jul 1. PMID: 26133894; PMCID: PMC5009925
8. de la Iglesia Vayá, M., et al.: BIMCV COVID-19-: a large annotated dataset of RX and CT images from COVID-19 patients. IEEE Dataport (2021). https://doi.org/10.21227/m4j2-ap59
9. Rajpurkar, P., et al.: CheXNet: radiologist-level pneumonia detection on chest x-rays with deep learning. arXiv preprint arXiv:1711.05225 (2017). https://doi.org/10.48550/arXiv.1711.05225
10. Wang, X., Peng, Y., Lu, L., Lu, Z., Bagheri, M., Summers, R.M.: ChestX-Ray8: hospital-scale chest X-Ray database and benchmarks on weakly-supervised classification and localization of common thorax diseases. In: 2017 IEEE Conference on Computer Vision and Pattern Recognition (CVPR), Honolulu, HI, USA, pp. 3462–3471 (2017). https://doi.org/10.1109/CVPR.2017.369
11. Kermany, D.S., et al.: Identifying medical diagnoses and treatable diseases by image-based deep learning. Cell. **172**(5), 1122–1131 (2018). https://doi.org/10.1016/j.cell.2018.02.010.e9. PMID: 29474911
12. Rajaraman, S., Candemir, S., Kim, I., Thoma, G., Antani, S.: Visualization and interpretation of convolutional neural network predictions in detecting pneumonia in pediatric chest radiographs. Appl. Sci. **8**(10), 1715 (2018). https://doi.org/10.3390/app8101715
13. Gu, X., Pan, L., Liang, H.-Y., Yang, R.: Classification of bacterial and viral childhood pneumonia using deep learning in chest radiography. In: ICMIP 2018: Proceedings of the 3rd International Conference on Multimedia and Image Processing, pp. 88–93 (2018). https://doi.org/10.1145/3195588.3195597
14. Jaiswal, A.K., Tiwari, P., Kumar, S., Gupta, D., Khanna, A., Rodrigues, J.J.: Identifying pneumonia in chest X-rays: a deep learning approach. Measurement **145**, 511–518 (2019). https://doi.org/10.1016/j.measurement.2019.05.076
15. Liang, G., Zheng, L.: A transfer learning method with deep residual network for pediatric pneumonia diagnosis. Comput Methods Programs Biomed. **187**, 104964 (2020). https://doi.org/10.1016/j.cmpb.2019.06.023. Epub 2019 Jun 26. PMID: 31262537
16. Sirazitdinov, I., Kholiavchenko, M., Mustafaev, T., Yuan, Y., Kuleev, R., Ibragimov, B.: Deep neural network ensemble for pneumonia localization from a large-scale chest x-ray database. Comput. Electr. Eng. **78**, 388–399 (2019). https://doi.org/10.1016/j.compeleceng.2019.08.004
17. Ayan, E., Ünver, H.: Diagnosis of pneumonia from chest X-Ray images using deep learning. 1–5 (2019). https://doi.org/10.1109/EBBT.2019.8741582
18. Toğaçar, M., Ergen, B., Cömert, Z.: A deep feature learning model for pneumonia detection applying a combination of mRMR feature selection and machine learning models. IRBM (2019). https://doi.org/10.1016/j.irbm.2019.10.006

19. Stephen, O., Sain, M., Maduh, U.J., Jeong, D.U.: An efficient deep learning approach to pneumonia classification in healthcare. J. Healthc. Eng. **27**(2019), 4180949 (2019). https://doi.org/10.1155/2019/4180949. PMID:31049186;PMCID:PMC6458916
20. Elshennawy, N.M., Ibrahim, D.M.: Deep-pneumonia framework using deep learning models based on chest X-Ray images. Diagnostics. **10**(9), 649 (2020). https://doi.org/10.3390/diagnostics10090649
21. Acharya, A.K, Satapathy, R.: A deep learning based approach towards the automatic diagnosis of pneumonia from chest radio-graphs. Biomed. Pharmacol. J. **13**(1) (2020). https://doi.org/10.13005/bpj/1905
22. Yu, X., Wang, S.H., Zhang, Y.D.: CGNet: a graph-knowledge embedded convolutional neural network for detection of pneumonia. Inf. Process. Manag. **58**(1), 102411 (2021). https://doi.org/10.1016/j.ipm.2020.102411. Epub 2020 Oct 19. PMID: 33100482; PMCID: PMC7569413
23. Tobias, R.R.: CNN-based deep learning model for chest X-ray health classification using TensorFlow (2020). https://doi.org/10.1109/RIVF48685.2020.9140733
24. Rahman, T., et al.: Transfer learning with deep convolutional neural network (CNN) for pneumonia detection using chest X-ray. Appl. Sci. **10**(9), 3233 (2020). https://doi.org/10.3390/app10093233
25. Chouhan, V., et al.: A novel transfer learning based approach for pneumonia detection in chest X-ray images. Appl. Sci. **10**(2), 559 (2020). https://doi.org/10.3390/app10020559
26. Hashmi, M.F., Katiyar, S., Keskar, A.G., Bokde, N.D., Geem, Z.W.: Efficient pneumonia detection in chest Xray images using deep transfer learning. diagnostics (Basel). **10**(6), 417 (2020). https://doi.org/10.3390/diagnostics10060417. PMID: 32575475; PMCID: PMC7345724
27. Sarkar, R., Hazra, A., Sadhu, K., Ghosh, P.: A novel method for pneumonia diagnosis from chest X-ray images using deep residual learning with separable convolutional networks. In: Gupta, M., Konar, D., Bhattacharyya, S., Biswas, S. (eds.) Computer Vision and Machine Intelligence in Medical Image Analysis. AISC, vol. 992, pp. 1–12. Springer, Singapore (2020). https://doi.org/10.1007/978-981-13-8798-2_1
28. Khatri, A., Jain, R., Vashista, H., Mittal, N., Ranjan, P., Janardhanan, R.: Pneumonia identification in chest X-Ray images using EMD. In: Sarma, H.K.D., Bhuyan, B., Borah, S., Dutta, N. (eds.) Trends in Communication, Cloud, and Big Data. LNNS, vol. 99, pp. 87–98. Springer, Singapore (2020). https://doi.org/10.1007/978-981-15-1624-5_9
29. Kundu, R., Das, R., Geem, Z.W., Han, G.-T., Sarkar, R.: Pneumonia detection in chest X-ray images using an ensemble of deep learning models. PLoS ONE **16**(9), e0256630 (2021). https://doi.org/10.1371/journal.pone.0256630
30. Sharma, A., Mishra, P.K.: Image enhancement techniques on deep learning approaches for automated diagnosis of COVID-19 features using CXR images. Multimed Tools Appl **81**, 42649–42690 (2022). https://doi.org/10.1007/s11042-022-13486-8
31. Sharma, M., Kumar, D.: Analysis of image enhancement techniques for chest X-ray images. In: International Conference on Computational Intelligence and Sustainable Engineering Solutions (CISES), Greater Noida, India, pp. 130–135. 9844364 (2022). https://doi.org/10.1109/CISES54857.2022
32. Trevethan, R:. Sensitivity, specificity, and predictive values: foundations, pliabilities, and pitfalls in research and practice. Front. Public Health. **5**, 307 (2017). https://doi.org/10.3389/fpubh.2017.0037. PMID: 29209603; PMCID: PMC5701930

Forging Futures: Empowering Decentralised AI Marketplaces Through Blockchain Innovation

Shriya Singh[1(✉)] and Piyush Gupta[2]

[1] Birla Institute of Technology, Mesra, Ranchi, India
shriyasingh2111@gmail.com
[2] Department of CSE, Birla Institute of Technology, Mesra, Ranchi, India

Abstract. Artificial intelligence (AI) is growing rapidly into a pervasive technology with applications in a wide range of sectors. In response to these issues, the concept of a decentralised AI marketplace has developed as a potential solution, combining the concepts of decentralised technologies such as blockchain with the world of artificial intelligence. It is therefore perfectly suited to the AI marketplace. Automated transactions, which can shorten time and money, could also be carried out with the use of Blockchain. The objectives of this paper are to examine the complex nature of an autonomous AI marketplace. This innovative concept promises to tackle key issues like data sovereignty, privacy breaches and unequal access to AI developments. We've discovered significant findings from our research on decentralised AI platforms that help us to understand the potential of such innovation fusion. Our study has shown that this marketplace is capable of democratising access to AI, empowering smaller players and individuals with the ability to use advanced algorithms. Despite smart contracts simplifying transactions, issues such as scalability continue to be a challenge. However, before a widespread adoption of Blockchain on the Artificial Intelligence market can take place, it is necessary to address certain challenges such as scale and regulatory compliance.

Keywords: AI marketplace · Blockchain · Decentralised · Pervasive and data sovereignty

1 Introduction: Decentralized Artificial Intelligence Marketplace Bridging Innovation and Trust

A new era of possibilities has emerged as a result of the combination of AI and decentralised technologies, in which both sectors are synergistically exploited. In the face of this, a revolutionary solution that not only addresses these constraints but also opens up entirely new avenues for collaboration, innovation and democratisation emerges: Decentralised Artificial Intelligence Market. A paradigm shift is being created with the integration of principles into artificial intelligence, allowing stakeholders to engage in decentralised AI activities that are more transparent, secure and fair.

D. Pastor-Escuredo et al. (Eds.): ICDLAIR 2023, LNNS 1001, pp. 288–297, 2024.
https://doi.org/10.1007/978-3-031-60935-0_26

This research report looks at the dynamic situation of decentralised AI market places, exploring their multidisciplinary implications, challenges and potential to influence future innovation in artificial intelligence. These platforms promise to transform the way we share information, develop models and collaborate with each other in a range of different industries by breaking old barriers and adopting principles of decentralisation. The aspects of an AI marketplace are as varied as they are revolutionary, from data monetization to decentralised learning, governance models to crosschain interoperability.

2 What is Decentralized AI Marketplace?

Decentralized Artificial Intelligence marketplace refers to an ecosystem or platform where artificial intelligence resources, e.g. AI models, algorithms, information and computing power can be exchanged, shared, sold by participants in a decentralised manner. In order to address some of the challenges and limitations inherent in traditional centralised artificial intelligence models and platforms, the concept of decentralised AI markets has been proposed.

2.1 The Decentralised Artificial Intelligence Market Seeks to Address These Issues by

Data Privacy and Ownership: Participants can retain ownership of their data and control who accesses it, promoting privacy and security.

1. **Access and democratization:** It is possible to provide greater access to AI resources for individuals, researchers or organisations with the aim of democratising innovation in this area.
2. **Trust and transparency**: Using blockchain technology, transactions and interactions on the market can be recorded in a way that is unobservable but secure enough to enhance trust.
3. **Incentives:** Tokenisation has been widely used to attract contributions and participation on decentralised marketplaces, fostering a cooperative ecosystem.
4. **Collaboration:** By working together more effectively, artificial intelligence developers, data providers and users can benefit from better AI models and solutions.
5. **Innovation:** Different forms of artificial intelligence development can be encouraged through the Decentralized Artificial Intelligence.

3 Blockchain and Smart Contracts

3.1 Immutable Record Keeping

The decentralised management of the Blockchain provides participants with a collective tool for shaping rules and decisions, promoting fairness and transparency. This immutability will make it possible to establish a clear and lasting record of data exchange, model acquisition and licensing agreements in relation to artificial intelligence transactions. In addition, participants can easily trace the provenance of AI models, monitor ownership transfers and verify that transactions are legitimate.

3.2 Smart Contract Automation

Smart Contracts and self executing AI programs running on blockchain can help to streamline all the processes of a decentralised artificial intelligence marketplace Smart contracts are a crucial element of the efficiency of all aspects of the market, as far as AI transactions are concerned.

- **Licensing:** The licensing of AI models and data can be automated by smart contracts. The associated license terms will be entered into a smart contract once an AI model is acquired by the user.
- **Payments:** Automated and transparent payment mechanisms shall be facilitated by intelligent contracts. The payment process can be triggered by the contract when a user has access to an AI model or dataset.
- **Data Sharing:** Smart contracts may enable data sharing to be secured and controlled when it comes to data driven AI models. Conditions for access to data can be laid down by the data owners.

4 Data Monetization and Ownership

4.1 Data Sovereignty and Control

Data sovereignty refers to the right of individuals to control their personal data. This issue is addressed through decentralised artificial intelligence markets, where users are able to keep ownership and control of their data. Users can choose to share specific data attributes while keeping sensitive information private in a decentralised AI market-place.

4.2 Tokenized Data Exchange

Tokenization is the use of a digital token on a blockchain to represent realworld assets. With regard to data, the exchange of tokenized data enables users to transform their input into intrinsicly valuable tokens that are part of the marketplace ecosystem. Creating a direct link between data contribution and monetary rewards.

4.3 Fair Compensation

Informed consent is the provision of full information to users as to how their data will be exploited before they are shared. That contrasts with the centralised platforms, where terms may be routinely unnegotiable.

5 Federated Learning and Privacy in a Decentralized AI Marketplace

The incorporation of federated learning is an innovative approach to the critical issue of data privacy and security in a decentralised AI marketplace.

5.1 Federated Learning: A Decentralized Approach to Learning

Federated learning has become a powerful solution, as part of the decentralised artificial intelligence marketplace. There are a range of key advantages to being integrated into federation learning, which include:

1. Privacy Preservation

Concentrating on localized data storage, federated learning promotes a seamless integration with the principles of decentralization. The groundbreaking work done by Google Gboard, whereby personalised language models are trained jointly between user computers without sharing individual inputs in order to protect their privacy, is a prime example.

2. Privacy-Preserving AI Services:

The philosophy of privacy protection is at the core of AI services in a decentralised marketplace. By ensuring that services built on it respect data protection, federation learning contributes to this proposition.

6 Dynamic Pricing and Allocation of Resources in a Decentralised Artificial Intelligence Marketplace

This section aims at exploring the complex functioning of dynamic pricing, with a focus on its benefits, challenges and potential future developments.

6.1 Dynamic Pricing Mechanisms

Dynamic pricing is a mechanism to adapt the cost of AI services, taking into consideration actual market conditions. Dynamic pricing relies upon the dynamics of supply and demand, allowing for a fluid pricing model to adapt to changes in circumstances (Fig. 1).

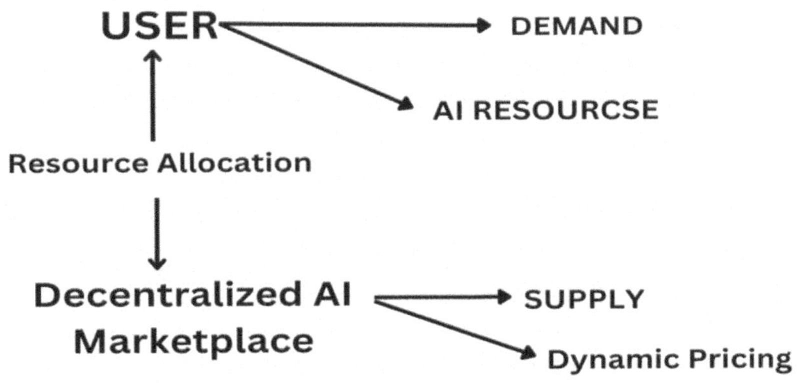

Fig. 1. This flowchart represents the dynamic and decentralized nature of the AI marketplace

6.2 User-Centric Pricing

An innovative extension of dynamic pricing is the implementation of personalized pricing models. These models take into account the user's profile, preferences and historic interactions in order to determine pricing for individual users.

The advantages of dynamic pricing are as follows:

- *Accessibility*: Dynamic pricing makes AI services available to users with various budget constraints.
- *Efficient allocation of resources:* Dynamic pricing provides incentives for users to use AI services when demand is low.
- *Incentivizing Contributors:* Dynamic pricing provides an incentive to contribute resources during the peak demand period to ensure a continuous supply of high quality AI models.
- *Resource Allocation Algorithms:* In addition to dynamic pricing algorithms, the allocation of computational resources among diverse AI tasks and users in a market is determined by resource allocation algorithms.

6.3 Obstacles and Considerations

A series of challenges need to be thoughtfully dealt with in the context of Dynamic Pricing and Resource Allocation, despite its benefits:

- **Fairness**: The key challenge is to ensure equal access to AI services.
- **Transparency**: It is vital that the pricing mechanism be transparent. The way in which prices are established, the factors that influence changes and historical price developments should be clearly understood by users.
- **Algorithmic Bias:** Rigorous monitoring and design of resource allocation algorithms are essential to mitigate bias, promoting fairness in resource distribution and preventing favoritism.

7 Decentralized Governance Models AI

7.1 Decentralized Autonomous Organizations (DAOs)

Decentralized autonomous organizations, through smart contracts, are offering the self executing model of governance.
Implications:

- *Transparency:* Transparent management and allocation of resources are ensured by smart contracts.
- *Autonomy:* DAOs bring down reliance on intermediaries, and allow the community to play an active role in shaping market direction.

7.2 Consensus Mechanisms for Governance

Advanced consensus mechanisms introduce a variety of ways of influencing decisions, in addition to simple voting, which encourages more thoughtful participation.

- **Mechanisms:** Proportional voting, assigning a price to votes. Thus discouraging the monopolization of power.
- **Quadratic funding:** to encourage wider support the contribution shall be matched by contributions for proposals.
- **Futarchy:** the use of market forecasts to inform decision making.

7.3 Compliance and Governance in the Field of Regulation

In order to ensure compliance with legal requirements, it is necessary to integrate regulatory compliance into governance.

Approach:

- *Regulator Sandbox:* controlled experiments with governance models within legal frameworks.
- *Collaboration:* Collaboration between developers, regulators, and industry stakeholders for responsible innovation.

8 Cross-Chain Interoperability

The ability of various blockchains to exchange information and cooperate with one another is referred to as cross chain interoperability. Cross chain interoperability plays an important role as regards the decentralised AI marketplace (Fig. 2).

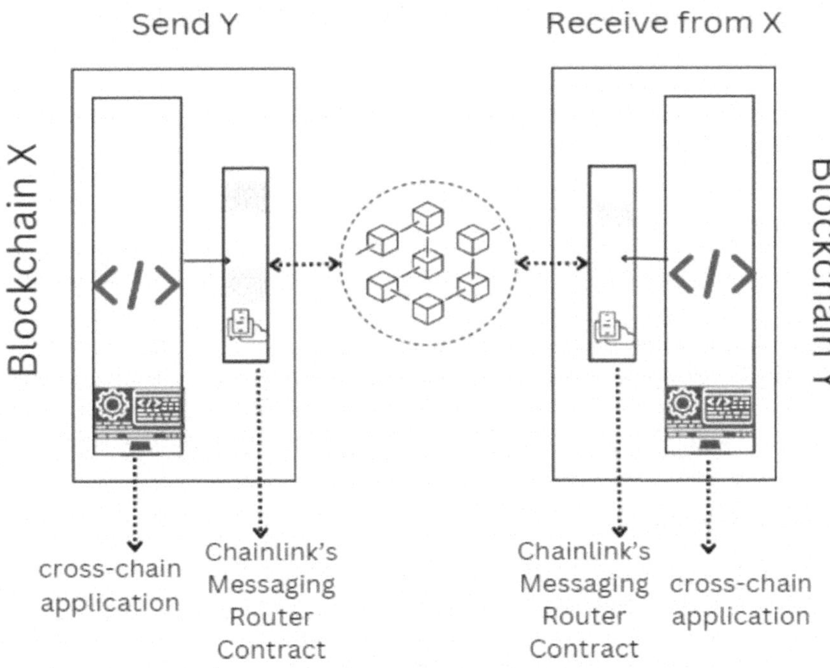

Fig. 2. A figure sending message between blockchain X to Y

8.1 Key Aspects

- **Interconnected Ecosystem:**
- Cross-chain interoperability aims to break down these silos and create an interconnected ecosystem where assets and data can flow freely across different chains. By eliminating the obstacles imposed by the various limitations of the blockchain, this interconnectedness can increase the reach and impact of AI resources.
- **Interoperability Protocols shall be used as follows:**
- The main role of interoperability protocols, such as Polkadot, Cosmos or Wanchain, in facilitating CrossChain Communication is crucial. The common standards and mechanisms for data exchange set out in these protocols will ensure that the various blockchains are able to communicate effectively.
- **Reducing Fragmentation:**
- Defragmenting of the network will occur as many different projects create their own separate chains. This fragmentation is dealt with through crosschain interoperability, which enables the free movement of assets and data between chains.
- **MultiChain Deployment:**
- Strategically distributing components of a Decentralized AI marketplace across various Blockchain networks is an essential part of the multichain deployment concept. This approach aims to improve the efficiency and balance of the online marketplace ecosystem by increasing the strength of different blockchains.

9 Challenges and Solutions in Decentralized AI Marketplaces

The intersection of AI and Blockchain creates a set of unique challenges with regard to establishing an autonomous marketplace for artificial intelligence. This section addresses key issues and proposes possible solutions.

9.1 Scalability

There is a significant scaling challenge in running AI intensive algorithms on a decentralised network. The processing of many transactions in a parallel manner may be challenging for old Blockchains, which could have an impact on the efficiency of AI related tasks. Consideration should be given to implementation of sharding in order to mitigate this challenge. The sharding is a process of splitting the blockchain between small, separate segments which allow for transactions to be performed independently.

9.2 Quality Assurance

It is essential to ensure that participants can trust each other, by ensuring the accuracy and reliability of AI models and data in a decentralised marketplace. The establishment of a reputation assessment system which assesses participants based on their contributions and the quality of models submitted would be an appropriate solution. It can leverage smart contracts in enforcing certain criteria that models need to fulfil prior to their inclusion on the market.

9.3 Data Privacy and Ownership

It is possible to protect privacy through the use of advanced techniques, such as zero knowledge proofs and homomorphic encryption. These methods provide participants with the ability to use artificial intelligence models and train them, without prejudice to data confidentiality.

10 Case Study - Ocean Protocol

A decentralised marketplace for the trading of artificial intelligence models and services can be established through use of the Ocean Protocol's architecture and features, making it safe, transparent as well as effective.

10.1 Data Assets Representation

Data assets in the Ocean Protocol are represented as non-fungible tokens (NFTs) on the blockchain. The NFTs shall provide a single identifier for each data asset to enable seamless tracking, ownership verification and access control. Data assets are easy to trade and interoperable in the ocean ecosystem by means of NFTs and datatokens.

10.2 Smart Contracts

For the purpose of automation and enforcement of data exchange rules, this protocol uses clever contracts that have already been implemented on the Ethereum Mainnet as well as any compatible network. Smart contracts are contractual arrangements that facilitate the transfers of data assets between parties based on predetermined conditions.

10.3 Compute-To-Data (C2D)

The Ocean Protocol ecosystem's paradigm for computing and analysing data revolutionizes their way of processing and analyzing it. It provides a secure transport of algorithms into data sources, allowing computation to take place locally and not requiring the disclosure of sensitive information.

10.4 Data Monetisation

Through the creation of ERC20 Tokens called DataTokens, which allow data providers to buy access to information that can be purchased with OCEAN Tokens and other exchange formats used to offer rewards such as staking, the Ocean Protocol facilitates the use of data assets by data providers.

10.5 Decentralised Governance Mechanisms

The Ocean Protocol allows decentralised management schemes to be implemented, enabling participants to participate in decisions and contribute to development of the Platform.

11 Conclusion

In conclusion, a key advance has been the shift to democratisation of artificial intelligence by means of decentralised markets. These platforms are enabling individuals and organisations from all walks of life to gain access to resources that bridge the Digital Divide, while promoting Global Innovation. Crossdisciplinary cooperation and the development of novel business models are encouraged by collaboration in these markets. However, there is a need to deal with issues such as fairness in AI results, establishment of adequate mechanisms for accountability and the legal difficulties that must be overcome. It is crucial to overcome these challenges so as to fully exploit the potential of decentralised artificial intelligence markets.

Looking ahead, the impact of these marketplaces is profound. They have the potential for supporting relief projects, addressing global challenges and accelerating research and innovation which will lead to a meaningful impact on society.

Essentially, the decentralized AI marketplace signals a paradigm shift that extends beyond technical concepts. They are actively contributing to the creation of an innovation landscape, defined by accessibility, fairness and progress, through their involvement in this concept. Our shared commitment to a future where AI will contribute positively to global change is reflected in the dynamic path that has taken us towards decentralized artificial intelligence markets.

References

1. Singh, A.: Decentralized AI marketplaces: a new frontier for innovation. J. Emerg. Technol. **8**(2), 76–89 (2020)
2. Smith, J.K.: The role of blockchain in decentralized AI marketplaces. In: International Conference on Artificial Intelligence and Blockchain, pp. 123–135 (2021)
3. Ocean Protocol. How It Works: Decentralized Data Exchange Protocol. https://oceanprotocol.com/
4. United Nations Development Programme. Sustainable Development Goals (2015)
5. Garcia, M., Kim, M.: Addressing algorithmic bias: a collaborative approach in decentralized AI marketplaces. In: Ethical AI Symposium Proceedings, pp. 45–56 (2019)
6. Johnson, L.M.: Tokenomics in decentralized AI ecosystems: a comprehensive analysis. J. Blockchain Econ. **12**(3), 210–227 (2023)
7. National Institutes of Health. AI-enabled disease detection: leveraging decentralized AI marketplaces. J. Med. Technol. **56**(4), 278–291 (2022)
8. Bélanger, J., Mélançon, B.: Decentralized governance in AI and robotics. In: Ethical and Social Issues in AI and Robotics, pp. 255–270. Springer (2020)
9. Hardjono, T., Pentland, A.: Towards a decentralized AI. IEEE Comput. Soc. **52**(6), 60–67 (2019)
10. McMahan, H.B., Moore, E., Ramage, D., Hampson, S., Arcas, B.A.: Communication - efficient learning of deep networks from decentralized data. In: Proceedings of the 20th International Conference on Artificial Intelligence and Statistics (AISTATS) (2017)
11. Chen, Y., Weinberg, S.M.: Dynamic pricing in two-sided markets: when to charge listing fees. Mark. Sci. **38**(3), 403–418 (2019)
12. Ocean Protocol. White Paper. https://oceanprotocol.com/techwhitepaper.pdf
13. Lee, J.H., Kim, K.H., Kim, J.H.: Decentralized artificial intelligence: overview and perspectives. IEEE Access **8**, 102837–102850 (2020)

14. McMahan, H.B.: Federated learning: Strategies for improving communication effi ciency. 1610.05492 (2016)
15. Kairouz, P.: Advances and open problems in federated learning. 1912.04977 (2019)
16. Bonawitz, K.: Towards federated learning at scale: system design. 1902.01046 (2019)
17. https://wazirx.com/blog/why-is-cross-chain-technology-crucial-in-block-chaininteroperability/
18. https://blog.chain.link/introducing-the-cross-chain-interoperability-protocolccip/
19. Wood, G.: Polkadot Whitepaper (2016). https://polkadot.network/PolkaDotPaper.pdf
20. Buchman, E.: Cosmos Whitepaper (2016). https://cosmos.network/resources/whitepaper

Citrus Leaf Disease Prediction: Deep Feature Fusion Perspective

Shantilata Palei[1(✉)], Rakesh Kumar Lenka[1], Santi Kumari Behera[2], Prabira Kumar Sethy[3], Sandeep Nanda[4], and Rabindra Kumar Barik[5]

[1] Computer Science and Engineering, International Institute of Information Technology, Bhubaneswar, Bhubaneswar 751003, Odisha, India
c122011@iiit-bh.ac.in

[2] Computer Science and Engineering, Veer Surendra Sai University of Technology, Sambalpur 768018, Odisha, India

[3] Electronics, Sambalpur University, Sambalpur 768019, Odisha, India

[4] Computer Science and Engineering, Sanskriti University, Mathura 281401, Uttar Pradesh, India

[5] School of Computer Applications, Kalinga Institute of Industrial Technology, Bhubaneswar 751024, Odisha, India

Abstract. Citrus disease is a major problem of economic and productive loss in agriculture worldwide. Crops are affected by uneven climatic conditions, which leads to lower agricultural yields. This will affect the world's agricultural economy. This paper proposes an image recognition method based on machine learning (ML) and deep learning (DL) feature fusion techniques for citrus leaf disease prediction. The three most popular CNN models were considered. The CNN models were AlexNet, vgg16, and vgg19. The features of these CNN models were extracted and fed into the support vector machine (SVM) to classify citrus leaf diseases. These classification models were evaluated in terms of performance metrics calculation. Subsequently, the multilayer feature fusion technique was adapted to the best-performing CNN models. Here, vgg16 performed the best among AlexNet, vgg16, and vgg19. Then, the different combinations of feature fusion of vgg16 are adapted. These fused features were fed into the SVM for citrus leaf disease classification. The feature fusion of 'fc6 plus fc7' of vgg16 with SVM yielded the best results; that is, the maximum values of accuracy, sensitivity, specificity, precision, FPR, F1 score, MCC, and Kappa coefficient were 97%, 97%, 98.09%, 97.06%, 1.58%, 95.34%, 84.4%, and 86.79% respectively. This method is very much helpful to the farmer as well as to the customers.

Keywords: Citrus diseases · Deep learning · Feature fusion · Leaf diseases · Diseases diagnosis

1 Introduction

Citrus is India's third largest fruit industry, followed by mangoes and bananas. India ranks tenth in orange production, accounting for 3% of the world's total. It exports 1.72% of its production. Therefore, it is necessary to increase citrus production in India [1]. Bacterial

D. Pastor-Escuredo et al. (Eds.): ICDLAIR 2023, LNNS 1001, pp. 298–306, 2024.
https://doi.org/10.1007/978-3-031-60935-0_27

and fungal diseases constantly threaten orange trees [2]. The increased usage of cell phones and deep learning-powered computer vision has allowed automated agricultural disease detection. Proper implementation of ML impacts predictions and subsequent on-farm decisions. For traditional methods, accurate parameterization is critical. In DL, defining architecture and parameters is crucial. DL designs are problem-dependent; diverse architectures must be considered when addressing novel agronomic issues [3]. DL-based CNN is mainly used for feature extraction, and various MLs like SVM, random forest, decision tree, and xgboost, etc., are used for classifying the extracted features [4–6]. This study proposes a deep feature fusion approach for identifying citrus leaf diseases. Early detection of citrus disease plays a significant role in the sustainability and production of citrus fruits [7]. Diseases such as black spots, cankers, greening, and melanoses affect the growth and production of citrus plants. To avoid the impact of these diseases, proper identification is required to implement the necessary treatment. Exact treatment, such as applying pesticides and fertilizers, is possible with the actual identification of diseases. Using human specialists to identify citrus diseases is a common practice in agriculture. Image processing has enabled the automatic prediction of diseases from raw images.

1.1 Contributions

This contributions is summed up as follows:

- CNN Models hybrid with linear SVM for citrus leaf disease prediction.
- The multilayer feature fusion technique was adapted to the best-performing CNN model.
- These fused features were fed into the linear SVM for classification.

1.2 Organisations

The rest of this paper is dispose as follows: Sect. 2 reviews the related articles, Sect. 3 presents the material and methodology, details of the dataset and the adapted methodology. The results and discussion are shown in Sect. 4. Finally, Sect. 5 concludes the study.

2 Related Work

The use of ML & DL technologies for crop disease prediction is a popular trend. Many research articles on citrus disease diagnosis based on CNN have been published. Iqbal et al. [8] Proposed infection detection using segmentation, feature extraction and classification. Sharif et al. [9] proposed a hybrid method for diseases classification. Zhang et al. [10] identified fruit using optimization and variable selection. Partel et al. [11] proposed a technique to recognize infections by the CNN model. Liu, et al. [12] introduced a Method to detect diseases using MobileNetV2. Lan et al. [13] presented a method to recognize disease using ML algorithms. Singh et al. [14] given their review on plant disease detection. Using CNN and SVM, Argu¨eso et al. [15] presented a methodology to detect huanglongbing (HLB) disease in plants. Barman et al. [16] proposed a method to detect the diseases of the citrus leaves using KNN and a deep neural network classifier. Table 1 shows an analysis of the citrus disease detection articles.

Table 1. Analysis of Citrus Disease Detection Articles

Ref.	Author	Name & No. of Data	Methods	Accuracy (%)
[17]	Cruz et al.	HLB (2680)	AlexNet, Inception v3, ResNet-101, & SqueezeNet	99.33
[18]	Ferentinos et al.	HLB (5507)	AlexNet & VGG	99.48
[19]	Badnakhe et al.	Gummosis	Multilinear regression	NA
[20]	Deng et al.	HLB (108)	DT, SVM, & KNN	96
[21]	Rauf et al.	Black spot (171), Canker(163), Scab (15), Greening (204), and Melanose(13)	Gaussian function & PCA	NA
[22]	Barman et al.	Greening (1004), CTV (1021)	MobileNet	99
[23]	Atila et al.	HLB	ResNet50, VGG16 and InceptionV3	99.97
[24]	Sujatha et al.	Black spot (171), Melanose (13)	SVM, RF, VGG	89.5
[25]	Ashwin et al.	Early Blight (1000), Late Blight (1909), Leaf Mold (952), Target Spot (60)	VGG-16 & AlexNet	98.7
[26]	Tiwari et al.	Black Spot (171), Canker (163), HLB (204)	MobileNet-v2 & DenseNet 201	99.25

3 Material and Methodology

The details of the dataset distribution and adapted methodology are described in this section.

3.1 About Dataset

The dataset was collected from the Kaggle repository [21], and the samples of diseased citrus leaves and healthy leaves are displayed in Fig. 1. Table 2 shows the different citrus leaf diseases & the number of samples collected for each category.

Table 2. Dataset Description

Name of Data	Number of Data
Black spot	171
Canker	163
Greening	204
Melanose	13
Healthy	58

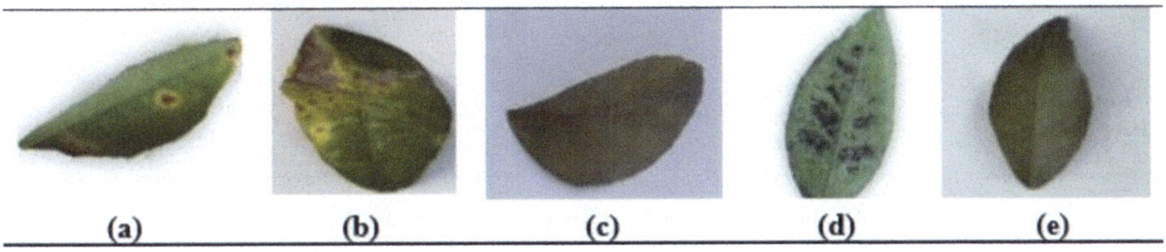

Fig. 1. Citrus Leaf Images (a) Black spot, (b) Canker, (c) Greening, (d) Melanose, (e) Healthy.

3.2 Methodology

A disease prediction method based on DL is proposed. Here, the three most popular CNN models are taken into consideration. The CNN models are alexnet, vgg16, and vgg19. The features of these CNN models are extracted and fed to SVM for the classification of citrus leaf diseases. These classification models are evaluated using a performance matrix. Then, the multilayer feature fusion technique is adapted to best-performed CNN models. The block diagram of this methodology is shown in Fig. 2.

The methodology consisted of two stages. In the first stage, AlexNet, vgg16, and vgg19 with an SVM were evaluated to classify four typical citrus and healthy leaf diseases. First, the deep features of these three CNN models were individually fed to the SVM to classify citrus leaves. The deep features were extracted from the fully connected layer of the CNN models, that is, fc8. The evaluation method is illustrated in Fig. 3.

After evaluating each model, the top-performing CNN model implemented the feature fusion approach. Here, the aim is to improve the performance of classification models with a limited available dataset. In the DL approach, performance depends on the feature vector's dimensions. CNN models, such as AlexNet, vgg16, and vgg19 have 4096, 4096, and 1000 features in their fc6, fc7, and fc8 layers, respectively. Therefore, multilayer feature fusion was adopted to enhance the feature vector dimension. Here, which model performed better than the other two CNN models, feature fusion is implemented in that best-performed model. The adapted feature-fusion approach for classifying citrus leaves is shown in Fig. 4.

Here, four-way feature fusion was implemented. In these four methods, different features are combined, and an enhanced feature vector is fed to the SVM for the classification of citrus leaves. The combinations of features are; 'fc6 (fully connected6) plus fc7', 'fc6 plus fc8', 'fc7 plus fc8' & 'fc6 plus fc7 plus fc8' and enhanced features are 8192, 5096, 5096 and 9192 respectively.

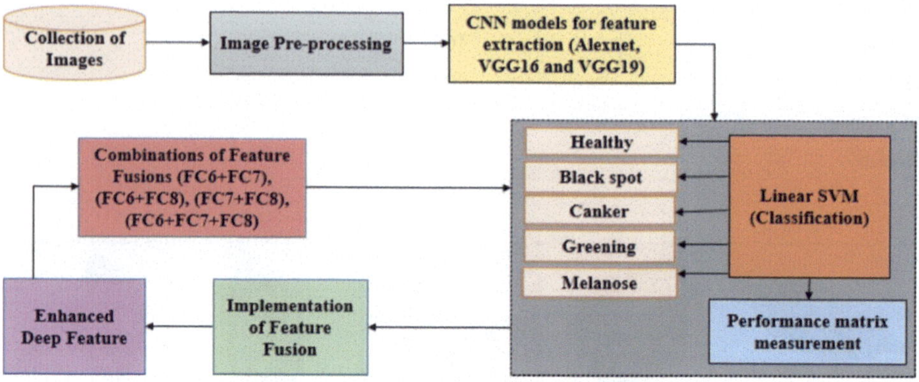

Fig. 2. Proposed Model for Citrus Leaf diseases Prediction.

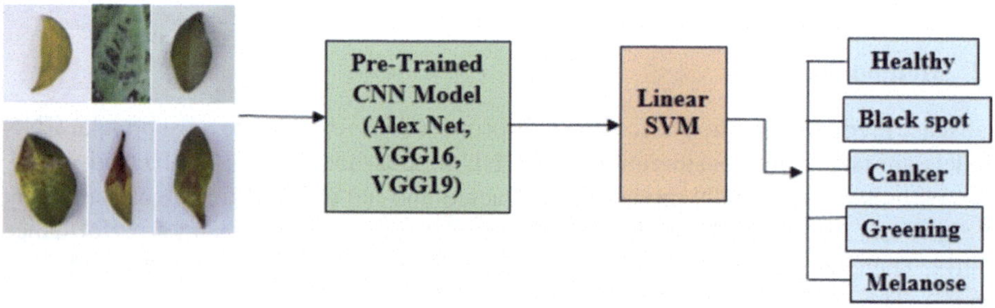

Fig. 3. DL Approach for Classification of Leaf Diseases using Alex Net, VGG16, and VGG19 with SVM.

4 Result and Discussion

Our study required 609 citrus leaves. The models were run using MATLAB 2020a on a Windows 10 laptop and an NVIDIA GeForce GPU. Stochastic gradient descent with momentum was used for training, using 32 mini-batches, 0.001 beginning learning rate, 10 maximum epochs, and five validation frequencies. Mathematical expressions of accuracy, sensitivity, specificity, precision, FPR, F1 Score, Matthew's correlation coefficient (MCC) and Kappa coefficients are mentioned in Table 3 in terms of True Positives (TP), True Negatives (TN), False Positives (FP), False Negatives (FN). In the kappa coefficient, i is the number of classes, N is the total number of class values, $m_{i,i}$ are several values belonging to the truth class i that have also been classified as class i, C_i is the total value of prediction and G_i is the total truth values. Table 3 shows all the mathematical formulas of performance matrix parameters.

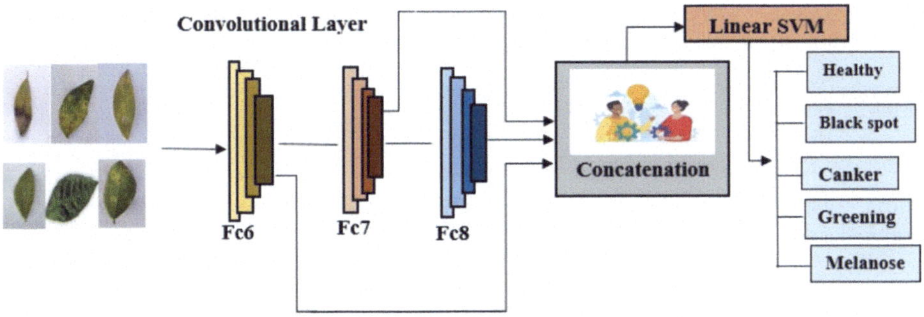

Fig. 4. Feature Fusion Approach for Classification of Citrus Leaf Disease Prediction.

Table 4 and Table 5 show three CNN model's results. From that, we observed that vgg16 performs better than AlexNet and vgg19. The combinations of different features are fed to the SVM to classify citrus leaves, and the results are presented in Tables 6 and Table 7. Table 8 compares this paper with existing papers.

Table 3. Performance matrix parameters

Parameters	Mathematical Formula
Accuracy	$\frac{TP+TN}{TP+TN+FP+FN}$ (1)
Sensitivity or Recall	$\frac{TP}{TP+FN}$ (2)
Specificity	$\frac{TN}{TN+FP}$ (3)
Precision	$\frac{TP}{TP+FP}$ (4)
FPR	$\frac{FP}{TN+FP}$ (5)
F1 Score	$\frac{2TP}{2TP+FP+FN}$ (6)
MCC	$\frac{TP.TN-FP.FN}{\sqrt{(TP+FP)(TP+FN)(TN+FP)(TN+FN)}}$ (7)
Kappa coefficient	$K = \frac{N\sum_{i=1}^{n} m_{i,i}-\sum_{i=1}^{n}(G_i,C_i)}{N^2-\sum_{i=1}^{n}(G_i,C_i)}$ (8)

Table 4. AlexNet, VGG16, and VGG19 performance for Citrus leaf Disease Classification in terms of Accuracy, Sensitivity, Specificity, and Precision in %

Models	Accuracy			Sensitivity			Specificity			Precision		
	Mean	Min	Max	Mean	Min	Max	Mean	Min	Max	Mean	Min	Max
AlexNet	80.78	64.54	89.26	80.78	64.54	92.04	95.19	91.13	98.01	84.32	70.83	91.93
VGG16	85.21	75.99	94.97	85.21	75.91	94.94	96.3	93.39	98.74	87.94	81.57	95.12
VGG19	84.72	65.28	93.52	84.72	65.28	93.52	96.18	91.32	98.38	88.2	80.17	93.54

Table 5. AlexNet, VGG16, and VGG19 performance for Citrus leaf Disease Classification in terms of FPR, F1 Score, MCC and Kappa in %.

Models	FPR			F1 Score			MCC			kappa		
	Mean	Min	Max	Mean	Min	Max	Mean	Min	Max	Mean	Min	Max
AlexNet	0.48	0.198	0.886	80.61	62.72	91.94	77.26	57.61	89.99	40.75	00.02	75.13
VGG16	0.369	0.125	0.602	85.03	74.24	94.99	82.51	72.03	93.77	53.78	29.94	84.28
VGG19	0.381	0.161	0.867	84.8	67.07	93.5	82.26	63.07	91.18	52.6	07.12	79.76

Table 6. Evaluation of VGG16 in different Feature Fusion Approach (Accuracy, Sensitivity, Specificity, and Precision in %) for citrus leaf disease predictions

Feature Fusion	Accuracy			Sensitivity			Specificity			Precision		
	Mean	Min	Max	Mean	Min	Max	Mean	Min	Max	Mean	Min	Max
FC6+FC7	95.01	91.02	97.00	95.01	91.02	97.00	95.75	95.03	98.09	95.21	93.05	97.06
FC7+FC8	79.97	76.00	80.05	79.97	76.01	80.05	94.99	95.75	98.99	83.53	81.55	85.50
FC6+FC8	94.50	92.45	95.67	94.50	92.45	96.87	98.62	96.05	98.99	94.6	92.03	96.66
FC6+FC7+FC8	85.36	83.06	87.22	85.36	83.45	87.65	96.34	94.56	98.56	87.03	85.65	88.99

Table 7. Evaluation of VGG16 in different Feature Fusion Approach (FPR, F1Score, MCC, Kappa in %) for citrus leaf disease predictions

Feature Fusion	FPR			F1 Score			MCC			kappa		
	Mean	Min	Max	Mean	Min	Max	Mean	Min	Max	Mean	Min	Max
FC6+FC7	1.25	1.03	1.58	94.98	92.45	96.50	93.83	91.25	95.44	84.40	82.34	86.79
FC7+FC8	0.501	0.301	0.721	77.71	75.43	79.54	75.59	73.28	77.60	37.42	35.68	39.78
FC6+FC8	1.38	1.10	1.56	94.53	92.45	96.50	93.17	91.13	95.40	82.81	80.22	84.59
FC6+FC7+FC8	3.66	1.66	5.99	85.49	83.27	87.36	82.36	80.11	85.45	54.26	52.64	56.45

Table 8. Comparison Table

References	Accuracy	Sensitivity	Specificity	Precision	FPR	F1 Score	MCC	Kappa
[4]	Yes	Yes	No	Yes	No	Yes	No	No
[10]	Yes	No	No	No	No	No	No	No
[11]	No	Yes	No	Yes	No	No	No	No
[12]	Yes	No	No	No	No	No	No	No

(continued)

Table 8. (*continued*)

References	Accuracy	Sensitivity	Specificity	Precision	FPR	F1 Score	MCC	Kappa
[13]	Yes	Yes	Yes	Yes	No	Yes	No	No
[15]	Yes	Yes	No	Yes	No	Yes	No	No
[25]	Yes	Yes	No	Yes	No	Yes	No	Yes
[24]	Yes	Yes	No	Yes	No	Yes	No	No
This paper	Yes	Yes	Yes	Yes	Yes	Yes	Yes	Yes

5 Conclusion

This study proposes a hybrid approach to predict four types of diseases and healthy leaves of citrus. The hybrid method is a combination of ML & DL. First, the CNN models perform feature extraction. Those features are extracted from the fully connected layer. The classification task is then performed using the ML approach, SVM. Individual features such as fc6, fc7, and fc8 of AlexNet, VGG16, and VGG19 with SVM were evaluated to classify the five types of citrus leaves. It was found that VGG16 with SVM using features extracted from fc8 resulted in the maximum values. Furthermore, after adapting the feature fusion technique, the fc6 and fc7 of vgg16 combined features performed well with the SVM. This method is helpful for farmers to identify citrus diseases. This research may be extended by implementing optimization techniques and autoencoder methods.

Acknowledgements. The authors gratefully credit the IoT & Cloud Computing Lab at IIIT Bhubaneswar for providing computational resources.

References

1. Rai, N., et al.: Applications of deep learning in precision weed management: a review. Comput. Electron. Agric. **206**, 107698 (2023)
2. Palei, S., Behera, S.K., Sethy, P.K.: A systematic review of citrus disease perceptions and fruit grading using machine vision. Procedia Comput. Sci. **218**, 2504–2519 (2023)
3. Richetti, J., Diakogianis, F.I., Bender, A., Colaço, A.F., Lawes, R.A.: A methods guideline for deep learning for tabular data in agriculture with a case study to forecast cereal yield. Comput. Electron. Agric. **205**, 107642 (2023)
4. Ullah, Z., Alsubaie, N., Jamjoom, M., Alajmani, S.H., Saleem, F.: Effimob-net: A deep learning-based hybrid model for detection and identification of tomato diseases using leaf images. Agriculture **13**(3), 737 (2023)
5. Mahmood, A., Tiwari, A.K., Singh, S.K., Udmale, S.S.: Contemporary machine learning applications in agriculture: Quo vadis? Concurr. Comput.: Pract. Exp. **34**(15), 6940 (2022)
6. Javeed, D., Gao, T., Saeed, M.S., Kumar, P.: An intrusion detection system for edge-envisioned smart agriculture in extreme environment. IEEE Internet Things J. (2023)
7. Palei, S., Lenka, R.K., Sahoo, T.K., Mallick, S.R., Patra, D., Sahoo, A.: Citrus fruit disease prediction using ml on a decentralized and secure smart platform. In: 2023 International Conference on Computer, Electronics & Electrical Engineering & Their Applications (IC2E3), pp. 1–6 (2023). IEEE

8. Iqbal, Z., Khan, M.A., Sharif, M., Shah, J.H., Rehman, M.H., Javed, K.: An automated detection and classification of citrus plant diseases using image processing techniques: a review. Comput. Electron. Agric. **153**, 12–32 (2018)
9. Sharif, M., Khan, M.A., Iqbal, Z., Azam, M.F., Lali, M.I.U., Javed, M.Y.: Detection and classification of citrus diseases in agriculture based on optimized weighted segmentation and feature selection. Comput. Electron. Agric. **150**, 220–234 (2018)
10. Zhang, Y., Lee, W.S., Li, M., Zheng, L., Ritenour, M.A.: Non-destructive recognition and classification of citrus fruit blemishes based on ant colony optimized spectral information. Postharvest Biol. Technol. **143**, 119–128 (2018)
11. Partel, V., Nunes, L., Stansly, P., Ampatzidis, Y.: Automated vision-based system for monitoring asian citrus psyllid in orchards utilizing artificial intelligence. Comput. Electron. Agric. **162**, 328–336 (2019)
12. Liu, Z., Xiang, X., Qin, J., Ma, Y., Zhang, Q., Xiong, N.N.: Image recognition of citrus diseases based on deep learning. CMC-Comput. Mater. Continua **66**(1), 457–466 (2021)
13. Lan, Y., et al.: Comparison of machine learning methods for citrus greening detection on uav multispectral images. Comput. Electron. Agric. **171**, 105234 (2020)
14. Singh, V., Sharma, N., Singh, S.: A review of imaging techniques for plant disease detection. Artif. Intell. Agric. **4**, 229–242 (2020)
15. Argu¨eso, D., Picon, A., Irusta, U., Medela, A., San-Emeterio, M.G., Bereciartua, A., Alvarez-Gila, A.: Few-shot learning approach for plant disease classification using images taken in the field. Comput. Electron. Agric. **175**, 105542 (2020)
16. Barman, U., Choudhury, R.D.: Smartphone image based digital chlorophyll meter to estimate the value of citrus leaves chlorophyll using linear regression, LMBP-ANN and SCGBP-ANN. J. King Saud Univ.-Comput. Inf. Sci. **34**(6), 2938–2950 (2022)
17. Cruz, A., et al.: Detection of grapevine yellows symptoms in vitis vinifera L. with artificial intelligence. Comput. Electron. Agric. **157**, 63–76 (2019)
18. Ferentinos, K.P.: Deep learning models for plant disease detection and diagnosis. Comput. Electron. Agric. **145**, 311–318 (2018)
19. Badnakhe, M.R., Durbha, S.S., Jagarlapudi, A., Gade, R.M.: Evaluation of citrus gummosis disease dynamics and predictions with weather and inversion based leaf optical model. Comput. Electron. Agric. **155**, 130–141 (2018)
20. Deng, X., Huang, Z., Zheng, Z., Lan, Y., Dai, F.: Field detection and classification of citrus huanglongbing based on hyperspectral reflectance. Comput. Electron. Agric. **167**, 105006 (2019)
21. Rauf, H.T., Saleem, B.A., Lali, M.I.U., Khan, M.A., Sharif, M., Bukhari, S.A.C.: A citrus fruits and leaves dataset for detection and classification of citrus diseases through machine learning. Data Brief **26**, 104340 (2019)
22. Barman, U., Choudhury, R.D., Sahu, D., Barman, G.G.: Comparison of convolution neural networks for smartphone image based real time classification of citrus leaf disease. Comput. Electron. Agric. **177**, 105661 (2020)
23. Atila, Ü., U¸car, M., Akyol, K., Ucar, E.: Plant leaf disease classification using efficientnet deep learning model. Ecol. Inform. **61**, 101182 (2021)
24. Sujatha, R., Chatterjee, J.M., Jhanjhi, N., Brohi, S.N.: Performance of deep learning vs machine learning in plant leaf disease detection. Microprocess. Microsyst. **80**, 103615 (2021)
25. Ashwinkumar, S., Rajagopal, S., Manimaran, V., Jegajothi, B.: Automated plant leaf disease detection and classification using optimal mobilenet based convolutional neural networks. Mater. Today: Proc. **51**, 480–487 (2022)
26. Tiwari, V., Joshi, R.C., Dutta, M.K.: Dense convolutional neural networks based multiclass plant disease detection and classification using leaf images. Eco. Inform. **63**, 101289 (2021)

Badminton Shot Recognition with LSTM Network

Avinav Jain, Shubham Agrawal, Gaurav Singh Chauhan, I. Sai Shruti, and Preety Singh(✉)

The LNM Institute of Information Technology, Jaipur 302031, Rajasthan, India
{avinavjain.y19,shubhamagrawal.y19,gauravchauhan.y19, saishrutii.y19,preety}@lnmiit.ac.in

Abstract. Deep learning has gained widespread application in action recognition and classification in sports. This paper focuses on the classification of three fundamental shots—*Clear*, *Serve*, and *Smash*—played in the game of badminton. For our experiments, we curated a dedicated dataset. Our approach involves the implementation of a Long Short-Term Memory (LSTM) deep learning model, trained to discern the movements of key points on a player's body during the execution of each specific shot. The outcome of our experiments is noteworthy, demonstrating an impressive 89.49% accuracy in the classification of badminton shots.

Keywords: badminton · action recognition · Long Short Term Memory · cross-validation

1 Introduction

The statistical aspects of a game and performances of players hold significant appeal for sports enthusiasts. Automated action recognition in sports can be very useful for the purpose of analyzing the game and identifying the types of shots played [18]. A player's posture is important to elevate the skill levels in the game. Determining the correctness of the player's posture while playing various shots can help coaches to detect inadequacies and to plan training schedules accordingly [11,14]. Action recognition can be utilized to identify the weaknesses and strengths of a player facilitating targeted improvements [19].

The use of deep learning for action recognition and classification in sports has gained popularity in recent years [6,13,16]. Recurrent Neural Networks (RNNs) and Long Short Term Memory (LSTM) have been explored extensively in this field [2,4,9]. In games like cricket, computer vision and deep learning play a pivotal role in predicting outcomes, such as determining if a batsman is *out* by tracing the probable trajectory of the ball or predicting a cricket shot [21]. Similarly, in tennis, trained machine models are used to tell if the ball played is *fair* or *foul* [14]. In football, it can be used to predict the outcome of a match [10]. Match prediction is also one of the major applications of machine learning in the field of sports [15]. By utilizing data from past matches, a trained system can

D. Pastor-Escuredo et al. (Eds.): ICDLAIR 2023, LNNS 1001, pp. 307–315, 2024.
https://doi.org/10.1007/978-3-031-60935-0_28

predict the winning probability of each team on the basis of the current match scenario.

Badminton is one of the fastest racket sport and a major sport at global level. However, despite its popularity, there is limited research in this game due to non-availability of a freely available dataset. In this paper, we have used our own dataset for recognition of three different kinds of shots played in the game of badminton. For the purpose of this study, we have taken three classes of shots and used a sequential model to predict the shot played on basis of body movements of the player. Most of the research related to the game of badminton focuses on the posture of the player and or classifying the game of the player as *offensive* or *defensive* [3]. Our main focus in this paper is to classify the shot played by the sportsperson. The main contributions of our research are as follows:

- Created a new badminton dataset comprising of three classes of shots, namely *Serve*, *Smash* and *Clear*. The recorded dataset consists of 984 videos (328 videos per class), which has been augmented to a total of 5516 videos.
- Trained a time-distributed sequential model to classify badminton shots based on movements of the player in a sequence of frames.

The rest of the paper is organized as follows: Sect. 2 presents related literature in the field of sports. Section 3 discusses our proposed methodology. Section 4 presents our experiments and results while Sect. 5 concludes the paper.

2 Related Work

In this section, we present few research papers in the domain of action recognition in the field of sports [1,7,15]. In [5], Ghosh et al. proposed a generic pipeline for the analysis of racket games. In this work, they have provided an analysis of the broadcast videos of badminton matches. Having detected players and shots for each frame in a match, they have computed the player's reaction time, dominance, positioning and footwork around the court, etc. Their player detection model uses color histogram features to cluster the image into two groups using the Gaussian mixture model, each representing a player. This model has a mAP@0.5 value of 97.85% for the bottom player, and 96.90% for the top player. (The bottom player refers to the player closer to the camera, whose back is visible, and the top player refers to the player who is far away from the camera, whose front is visible).

In [3], Chu et al. proposed a framework to classify the strategy of an individual player in the game of badminton as *offensive* or *defensive*. They have detected court lines and players by background subtraction, and classify players' postures into six categories of shot types. Taking the sequence of observations of shot type and the movements of the player, they predict the strategy of the gameplay. They have reported an average strategy classification accuracy of 70%.

In [19], Fok et al. have presented a framework to recognize various styles of swimming, such as *butterfly*, *freestyle*, etc. by feeding the posture key-points to a neural network. They have shown 92.9% test accuracy. In [12], Mora et al.

have classified strokes in the game of tennis into four action groups and have trained a 3-layer LSTM model on an independent dataset. They have reported an accuracy of 84.10% when tested on a mixed set of matches of amateurs and professionals.

In [8], Kumar et al. have worked on the game of cricket. They have used the optical flow technique to identify whether the ball was hit by a batsman. Because of the small dataset, they have used the k-fold cross-validation spitting technique and obtained an accuracy of 80%. In [17], Tsunoda et al. have used multi-person centered features and temporal dynamics of features to recognize various futsal plays in the game of soccer recorded from multiple cameras in multi-view. They have integrated one more state in the general LSTM, called *Keeping state*, which is externally controlled. Their proposed method with RGB image and meta-information inputs yielded an accuracy of 70.90%.

3 Proposed Methodology

The block diagram of our proposed methodology is given in Fig. 1. We have created a dataset for our experiments. The dataset was augmented using various techniques. From each frame of a shot video, we have extracted key-points of the player to track his/her movements. The dataset is split into two: the training dataset is used to train a deep neural network which is then evaluated with the test dataset.

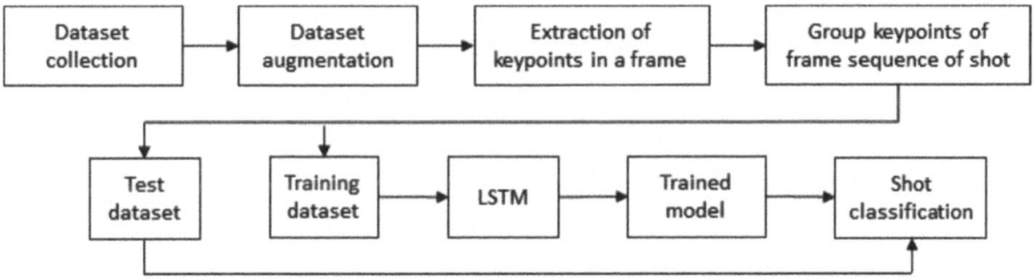

Fig. 1. Block diagram of proposed methodology.

3.1 Dataset Collection

We have created our dataset by recording shots of three classes, namely, *Serve*, *Smash* and *Clear* in the badminton game. We have collected 328 videos for each class of shot. We used the Microsoft Xbox Kinect One Sensor for recording. The videos have been recorded at 30fps. To simulate behind the baseline view, we placed the sensor around 80cm above the ground, behind the player, and parallel to the net. The *bottom* player (whose back is visible) is more in view as compared to the *top* player across the net. To record the front view, we placed the sensor directly under the net at a similar height. Five files are available for each sample of a shot - a color video, a depth video, a raw depth video, a skeleton points video, and a JSON file with information about the skeleton joints coordinates. Few samples are shown in Fig. 2.

Fig. 2. Snapshots of the colour, depth and skeleton views obtained for a *Smash* shot.

The Kinect contains two cameras with different resolutions: a color camera and a depth camera. The color camera has a resolution of 1920×1080. The depth camera has a resolution of 512×424. While the Kinect sensor provides both color and depth videos, we have used only the color videos so that the experimental results can be extended to videos recorded from regular video cameras without depending on any special camera sensor. Moreover, as usually the whole game is shown from behind the baseline in a professional match, we have considered the recordings of the bottom player only for our experiments.

3.2 Video Augmentation and Labelling

Table 1. Details of augmented dataset.

Shot	No. of Videos	Average frames per video
Serve	1859	11.94
Smash	2109	11.49
Clear	1710	10.77

To increase the number of videos in our dataset, we have applied augmentation on videos by applying contrast changes, rotations, flipping the frames horizontally, and smoothening the frames. The total number of videos after augmentation is 5516. The total number of videos for each shot and average number of frames per video are shown in Table 1.

3.3 Key-Point Detection

To predict the type of shot played, we have extracted 33 key-points from the body posture of the bottom player in each frame (refer Fig. 3) by applying a key-point detection model on the frames of the videos using MediaPipe pose detector. Each key-point comprises of the three-dimensional coordinates of the human joint. This results in a total of 33×3 feature points per frame.

Fig. 3. Skeleton of the player for key-point extraction.

3.4 Classification Model

We have used the Long Short Term Memory (LSTM) architecture as our deep learning classification model to capture the important features of the shot play and to classify the shots. Our dataset is divided into two parts: approximately

90% for training, and 10% for testing, using a stratified split. During training of the model we have used the 5-fold cross-validation method. The number of videos used for training and testing are given in Table 2.

Table 2. Number of videos in training and test datasets.

	Serve	Smash	Clear
Training	1668	1895	1401
Test	192	215	145

4 Experiments and Results

Our badminton dataset consists of three classes of shots, *Clear*, *Serve*, and *Smash*. From the colour video of each shot video, we have extracted key-points from the skeleton of the bottom player for each shot played using the MediaPipe Key-points Holistics Model [20]. As the number of frames vary in different videos, we have fixed the number of frames to 15, after carefully observing that in almost all the shots are captured in 15 frames. Those videos which exceeded 15 frames have been trimmed as they generally contained some redundant frames post playing the shot. For videos which had frames less than 15, we have applied post-padding with zero value frames.

Table 3. Architecture of LSTM model.

Layer Type	Activation function	Output Shape	No. of Parameters
Time Distributed	–	(None, 15, 225)	0
LSTM	tanh	(None, 15, 512)	1511424
LSTM	relu	(None, 15, 256)	787456
LSTM	relu	(None, 128)	197120
Dense	relu	(None, 64)	8256
Dense	relu	(None, 32)	2080
Dense	softmax	(None, 3)	99

The key-point sequences extracted from each of the 15 frames are fed to an LSTM model to train it to classify the shots. The architecture of the model is shown in Table 3. The total number of parameters are 256435. We have used the Adam optimizer and the categorical cross-entropy loss function. To evaluate the performance of our model on the test set, we have used Precision, Recall, Accuracy, $F1$-score and Support as our evaluation metrics. On evaluating our

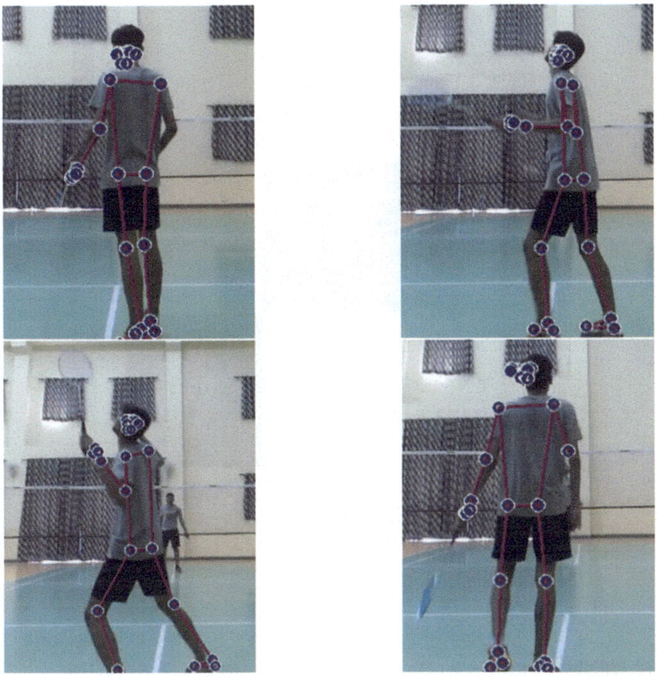

Fig. 4. Player key-points on frames from a video of *Clear* shot

test set on the trained model, we obtained an overall accuracy of 89.49% and support of 552. The other evaluation metrics for the different shots are shown in Table 4. The confusion matrix is shown in Fig. 5.

Table 4. Evaluation metrics for different shots of the game.

	Precision	Recall	F1-score	Support
Serve	0.95	0.94	0.94	194
Smash	0.82	0.95	0.88	186
Clear	0.93	0.78	0.85	172

As seen in the confusion matrix, our model works best for the *serve* shot, followed by the *clear* shot. Some of the *smash* shots are getting classified as *clear*. This could be due to the close resemblance of the posture of the player while playing these shots. This accuracy can be improved by incorporating the trajectory of the shuttle cock along with the player's body movements as input to the model.

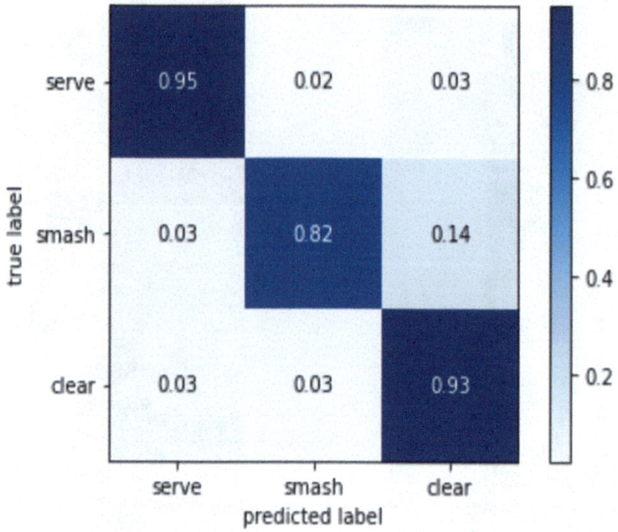

Fig. 5. Confusion matrix.

5 Conclusion

In this paper, we have introduced a labeled dataset containing three types of shots played in the game of badminton. We have used a time-distributed deep learning model to classify the shots into three classes, *Clear*, *Serve*, and *Smash*. We have achieved 89.49% accuracy on the test set. In future, this study can be extended by introducing more classes of shots in the dataset. Classification of shots over frames of a given game video can help in segmenting videos for the various shots for analysis of the game.

References

1. Ban, K.W., See, J. Abdullah, J., Loh, Y.P.: BadmintonDB: a badminton dataset for player-specific match analysis and prediction. In: Proceedings of the 5th International ACM Workshop on Multimedia Content Analysis in Sports, pp. 47–54. ACM (2022)
2. Bhatnagar, B.L., Singh, S., Arora, C., Jawahar, C.V.: Unsupervised learning of deep feature representation for clustering egocentric actions. In: Proceedings of the 26th International Joint Conference on Artificial Intelligence, pp. 1447–1453 (2017)
3. Chu, W.T., Situmeang, S.: Badminton video analysis based on spatiotemporal and stroke features. In: Proceedings of the 2017 ACM on International Conference on Multimedia Retrieval, pp. 448–451. ACM (2017)
4. Fathi, A., Rehg, J.M.: Modeling actions through state changes. In: Conference on Computer Vision and Pattern Recognition, pp. 2579–2586. IEEE (2013)

5. Ghosh, A., Singh, S., Jawahar, C.V.: Towards structured analysis of broadcast badminton videos. In: IEEE Winter Conference on Applications of Computer Vision, pp. 296–304 (2018)
6. He, K., Zhang, X., Ren, S., Sun, J.: Deep residual learning for image recognition. In: Conference on Computer Vision and Pattern Recognition, pp. 770–778. IEEE (2016)
7. Kulkarni, K.M., Shenoy, S.: Table tennis stroke recognition using two-dimensional human pose estimation. In: IEEE/CVF Conference on Computer Vision and Pattern Recognition Workshops, pp. 4571–4579 (2021)
8. Kumar, A., Garg, J., Mukerjee, A.: Cricket activity detection. In: International Image Processing, Applications and Systems Conference, pp. 1–6 (2014)
9. Lea, C., Reiter, A., Vidal, R., Hager, G.D.: Segmental spatiotemporal CNNs for fine-grained action segmentation. In: Leibe, B., Matas, J., Sebe, N., Welling, M. (eds.) Computer Vision – ECCV 2016. ECCV 2016. LNCS, vol. 9907, pp. 36–52. Springer, Cham (2016). https://doi.org/10.1007/978-3-319-46487-9_3
10. Rudrapal, D., Boro, S., Srivastava, J., Singh, S.: A deep learning approach to predict football match result. In: Behera, H., Nayak, J., Naik, B., Pelusi, D. (eds.) Computational Intelligence in Data Mining. AISC, vol. 990, pp. 93–99. Springer, Singapore (2020). https://doi.org/10.1007/978-981-13-8676-3_9
11. Mlakar, M., Luštrek, M.: Analyzing tennis game through sensor data with machine learning and multi-objective optimization. In: Proceedings of the International Joint Conference on Pervasive and Ubiquitous Computing and Proceedings of the 2017 ACM International Symposium on Wearable Computers, pp. 153–156. ACM (2017)
12. Mora, S.V., Knottenbelt, W.J.: Deep learning for domain-specific action recognition in tennis. In: Conference on Computer Vision and Pattern Recognition Workshops, pp. 170–178. IEEE (2017)
13. Ren, S., He, K., Girshick, R., Sun, J.: Faster R-CNN: towards real-time object detection with region proposal networks. IEEE Trans. Pattern Anal. Mach. Intell. **39**(6), 1137–1149 (2017)
14. Renò, V., et al.: A technology platform for automatic high-level tennis game analysis. Comput. Vis. Image Underst. **159**, 164–175 (2017)
15. Sharma, M., Lamba, M., Kumar, N., Kumar, P.: Badminton match outcome prediction model using Naïve Bayes and feature weighting technique. J. Ambient Intell. Humaniz. Comput. 8441–8455 (2021)
16. Simonyan, K., Zisserman, A.: Very deep convolutional networks for large-scale image recognition. In: ICLR. arXiv (2014)
17. Tsunoda, T., Komori, Y., Matsugu, M., Harada, T.: Football action recognition using hierarchical LSTM. In: Conference on Computer Vision and Pattern Recognition Workshops, pp. 155–163 (2017)
18. Wang, L., Qiao, Y., Tang, X.: Action recognition with trajectory-pooled deep-convolutional descriptors. In: Conference on Computer Vision and Pattern Recognition, pp. 4305–4314. IEEE (2015)
19. Fok, W.W.T., Chan, L.C.W., Chen, C.: Artificial intelligence for sport actions and performance analysis using recurrent neural network (RNN) with long short-term memory (LSTM). In: Proceedings of the 2018 4th International Conference on Robotics and Artificial Intelligence, pp. 40–44. ACM (2018)
20. Lugaresi, C., et al.: MediaPipe: A Framework for Building Perception Pipelines. ArXiv, abs/1906.08172 (2019)
21. Siddiqui, H.U.R., et al.: Enhancing cricket performance analysis with human pose estimation and machine learning. Sensors **23**(15), 6839 (2023)

An Ensemble Learning Model for Automatic Detection of Cyberbullying on Instagram Platform

S. Abarna[1(✉)], J. I. Sheeba[1], and S. Pradeep Devaneyan[2]

[1] Puducherry Technological University, Pondicherry, India
abarna@pec.edu
[2] Pondicherry University, Pondicherry, India

Abstract. Social media usage has increased dramatically in recent years with the expansion of the Internet and also has emerged as the most powerful networking tool of the twenty-first century. However, increased social networking often has a detrimental influence on society, contributing to a number of unpleasant behaviors including online abuse, harassment, cyber-bullying, cyber-crime, as well as online trolling. Cyberbullying commonly causes severe emotional and physical anguish, especially among women as well as children, and may even lead to suicide attempts. Because of its significant detrimental societal effect, online harassment garners attention. Many occurrences have lately happened throughout the globe as a result of online bullying, such as the publishing of private conversations, rumors, and sexual comments. As a result, experts are increasingly interested in detecting bullying on social media via text or communication. The main purpose of this research is to create and refine a strategy for identifying cyber harassment and bullying comments by Machine Learning (ML) algorithms. In this paper, multi classifier system is termed an Ensemble-based learning model. The suggested model contains recommended ML algorithms including Support Vector Machine (SVM), Naive Bayes (NB), as well as Random Forest (RF) method as the classifier for this persistence. Finally Boosting Ensemble classifier has combined the best performance in detecting the abusing messages. The proposed model will be evaluated by various performance metrics.

Keywords: Cyberbullying · Multi Classifier System · Boosting Ensemble Classifier · Supervised Classifiers · Online harassment

1 Introduction

The use of social networking sites is one of the most effective ways to get to know new people. However, as the use of social media has increased in popularity, people have found illegal and unethical ways to abuse these networks [1]. People, particularly teenagers and early adults, are inventing new methods to attack each other through the Internet. In Symantec research, over 25% of parents indicated that their kid had been engaged in a cyberbullying event to their knowledge [2]. Though social media allows

D. Pastor-Escuredo et al. (Eds.): ICDLAIR 2023, LNNS 1001, pp. 316–325, 2024.
https://doi.org/10.1007/978-3-031-60935-0_29

individuals to interact, it also introduces them to potentially dangerous situations such as violent cyberbullying, which may lead to despair and suicide ideation [3]. Cyberbullying may occur via text messaging, remarks, and other forms of online communication [4]. The social lifestyle transcends the protection layer of human connection and includes unsupervised communication with strangers, the circumstances of cyberbullying must be examined and studied [5]. Because online is simply a click away, cyberbullying gives the sufferer the impression that they are being assaulted everywhere [6]. There is a need to develop a tailored system that can effectively and efficiently retrieve such text [7].

Instagram also has its own set of features, such as a base of supporters, usernames, profile images, locations, biographies, hashtags, and so on. It is believed that over 80% of Instagram messages are sent using cell phones [8]. The social networking site is already turning into a haven for cyberbullies. Using a machine-learning technique, this attempts to recognize and detect bullying in worldwide social media platforms by using behavioral characteristics inherent in the material [9]. This necessitates research methodologies that are more intimate with participants. One such method is the constructivist grounded theory (CGT) method, which collects data and conducts analysis while simultaneously attempting to be attentive to the fundamental concerns and social processes of the participants [10].

The emphasis of our research is on an ensemble learning technique. It aids in the identification of linguistic trends and the development of cyberbullying detection algorithms. We used datasets from sites like Instagram to train the algorithm. The remaining 20% of both sets of data are retained as test data, while 80% of both sets of data are preserved as training data. Finally, we analyzed each model's precision and performance by applying it to the same dataset and scoring it. The section on relevant works is discussed in Sect. 2, the section on our suggested methodology is discussed in Sect. 3, the section on experimental analysis is discussed in Sect. 4, and the section on conclusion as well as future work is discussed in Sect. 5.

2 Related Works

The qualitative investigation of bullying & cyberbullying offers a study approach that can bring details such as motives as well as difficulties that quantitative studies may not be able to reach. The current paper's [11] objective is to highlight how qualitative interviews contextualize quantitative results and to give a new method using longitudinal & multi-perspective mixed methods research. This research [12] explores the relationship between victimization fear, forms of victimization, and advocacy on 834 students across 16 Iranian state middle schools, controlling for grade and gender. Multilevel linear modelling was used to investigate the relationships between different forms of victimization and victimization fear. The purpose of this study [13] was to investigate the long-term connections that may exist between bullying participation character behavior as well as the sense of guidance they received from peers and instructors. In addition, the perception of support from friends and a healthy balance within the family unit are both protective factors against the act of cyberbullying [14].

Cyberbullying on social media among children and adults, as well as the link between such conduct and well-being. It examines numerous potential indicators of cyberbullying on social media, such as inappropriate sharing, moments spent on social networking sites, and personality factors. This study [15] addresses possible harmful repercussions on well-being that are related to cyberbullying on social media, such as psychological distress, reduced levels of life satisfaction, and thoughts of suicide. It finishes the research with recommendations for prevention and intervention, such as employing detection technologies of cyberbullying when posting such bullying comments [16]. One of the famous natural language processing problems is the identification of cyberbullying on social media utilizing synthetic or real-world information. In order to do an analysis of a certain text, one must first identify the semantic, structural, and geographical linkages that are present [17].

In recent years, online social networks (OSN) including social media have grown in popularity. On the contrary side, OSN and other platforms for social media put great emphasis on protecting users' personal information and data privacy. Cyberbullying (CB), on the other hand, is a major issue that must be handled on social media networks [18]. Victimization and distress have been researched in connection to bad sentiments on social media activities as well as the degree to which students experience peer hostility at school [19]. Both of these factors have been shown to be related to negative emotions.

The detection of harassment within textual information using ML was the primary focus of this paper [20], which was written on the topic. The issue of automatic detection of cyberbullying is increasingly receiving attention, particularly within the field of Machine Learning. This work [21] presents a comprehensive evaluation of contemporary approaches, machine learning algorithms, and technological tools used for the identification of harassment including abusive conduct of individuals inside the online information domain. This study [22] encompasses a total of twenty-three low-resource dialects and languages. It offers recommendations for enhancing the overall research practices in the field of online harassment detection, including a specific emphasis on low-resource linguistics.

The phenomenon of cyberbullying is widely prevalent within the realm of online social media platforms, when an individual assumes the role of a bully and engages in the mistreatment of a targeted victim via the medium of an online social networking session. This survey [23] study presents a framework that encompasses four distinct stages that session-based surveillance for cyberbullying should undergo. The proliferation of technological advancements has given rise to a new kind of bullying, often resulting in the development of adverse social stigmatization. The objective of this study [24] is to use methods based on machine learning within the language of Arabic to facilitate the automated identification of harassment.

Social media sites provide users with several chances to engage in information sharing, collaboration, and constructive communication. This study [25] introduces a resilient hybrid machine learning model designed for the purpose of detecting instances of harassment in the possible Bengali culture as seen on social media platforms.

3 Proposed System

In this paper, multi classifier system is termed an Ensemble-based learning model. For this persistence, the proposed model has preferred supervised machine learning algorithms such as SVM, RF, and NB classifiers. Finally, XG Boosting Ensemble classifier has combined the best performance in detecting the abusing messages. The proposed model will be evaluated by various performance metrics. An overall architecture diagram for the ensemble model for cyberbullying detection is shown in Fig. 1.

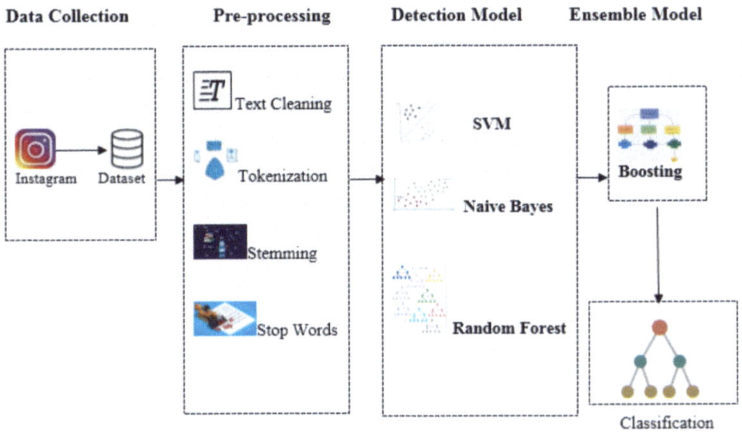

Fig. 1. Ensemble Model for Cyber Bullying Detection

3.1 Pre-processing of Data

Data cleansing begins with pre-processing, which is a critical functional step. The bulk of the attempts in the reviewed literature used the NLTK library14 to tokenization, remove blank rows, remove extraneous characters, fix misspellings, lemmatize, and/or stem raw data. Additional methods were often used in the context of Instagram datasets, such as modifying user mentions Url, incorporating hashtags with odd characters, and eliminating duplicates. The processed data are of great assistance to the models in their analysis of the raw data. The processed data makes it easier for computers to interpret human language, which in turn leads to improved categorization results. The following is a list of techniques employed by this model:

- Tokenization: Break down comments into words or subword tokens.
- Text cleaning: Remove special characters, URLs, and irrelevant information.
- Stopword removal: Eliminate common words that do not carry significant meaning.
- Text normalization: Convert text to lowercase to ensure uniformity.

3.2 Detection Model Using Supervised Machine Learning Algorithm

Cyberbullying Detection uses our programmed, machine-learning algorithms to detect harmful comments in communications received from a user. The algorithm assigns a value to the message and then determines whether or not the remark is severe enough to be changed based on using pre-trained data.

3.2.1 Support Vector Machine

A type of data analysis known as "supervised learning" is one in which the intended result is either predetermined or labelled. SVM is a method of machine learning that goes beyond traditional X/Y prediction by using techniques to train as well as classify data across levels of polarity. These algorithms may be found in the SVM software. After that, the Support Vector Machine will choose the optimal hyperplane for dividing the tags. Simply put, this is a line that exists in two dimensions. In terms of sentiment analysis, for instance, it would be considered both good and negative.

SVM constructs a set of hyperplanes in an infinitely space and therefore is performed using a module that turns a given input range into the proper shape. Kernel Function, for instance, uses the standard dot product of any two items, as seen below:

$$x(n, ni) = \Sigma(\mathrm{n} * \mathrm{ni}) \quad (1)$$

3.2.2 Random Forest Algorithm

This algorithm is a trustworthy combination of many Decision Trees (or CARTs), albeit it is more often used in classification applications than in regression ones. The individual trees in this case are divided and created using boosting and fewer characteristics. The random forest with uncorrelated trees that results from this has less variation, making it more resistant to data change and retaining its prediction accuracy for new data.

Perhaps by using a single tree structure, the random forest evaluates each created tree's prognosis and determines the final outcome based just on projections. For example, if there are 2 classes, X and y, and the overwhelming tree structure predicts target class Y of any case, RF will decide on class label Y as follows:

$$X(n) = majority\ of\ tree\ as\ Y \quad (2)$$

3.2.3 Naive Bayes Algorithm

Even though we might not realize it, this is the methodology that is most often used to sort through spam on Instagram. It uses Bayes's theorem and something called "posterior probability" to classify the unstructured data. To do so, it makes the naive assumption that perhaps the predictions are autonomous, which may or may not be correct. The model performs effectively with a little training dataset if all of the categorical predictor's classes are present.

In accordance with Bayes' Theorem, the following formula is used to compute the probability that an event will take place given the possibility that another occurrence will take place:

$$\mathrm{P(X|Y)} = \mathrm{P(Y|X)} * \mathrm{P(Y)} / \mathrm{X} \quad (3)$$

where $X = x1, x2, x3...xn$, Y is the variable that represents the class.

3.2.4 Ensemble Learning Model

Ensemble techniques are a kind of machine learning that mixes numerous basic models. In order to create a single best prediction model. The greatest performance in recognizing abusive communications has been combined by the XG Boosting Ensemble classifier. Extreme Gradient Boosting, often known as XG Boost, is a technique of machine learning that makes use of the gradient boosting ensemble approach, which is a methodology that focuses on decision trees. ML is much better than any other algorithms or frameworks when it comes to the problem-solving ability required to make accurate predictions from unstructured data. XG Boost is utilized for supervised learning issues in which we predict a target variable using training data. Especially nowadays, the supervised learning method known as XG Boost is widely used for performing substantial regression and classification. It does this by using sequentially built shallow decision trees in conjunction with a scalability training technique that strives to prevent overfitting. The end result is dependable results.

Our model may be expressed mathematically as given in Eq. 4,

$$E = \sum_{t=1}^{n} f_t(x_i), f_t \in C \tag{4}$$

where, the number of trees is denoted as t, functional space is stated as f of a given Cart and C is the possibilities of CARTs.

Algorithm Steps for Ensemble Model

Input: Dataset **Traning** $T = \{(A1),, (A2), \ldots, (An)\}$
Classifiers Model $\boldsymbol{M}$. // SVM, RF and NB
Output: Trained Model of Supervised ML Classifier E
Step 1: BEGIN
Step 2: Train fundamental models by using three different classifiers, M *to dataset* T.
Step 3: for $M = m1, m2, m3$ **do**
Step 3: $E_i = M(T)$
Step 4: end for
Step 5: Predicting Trained Data P(T)
Step 6: for i = 1,,.., n **do**
Step 7: Classify P(T) Predicted Trained Data
Step 8: end for
Step 9: $P(T) = (M)$, where $M =$ *Classifier algorithm* $(m1, m2, m3)$ // SVM, RF, NB
Step 10: Ei = P(T) // **Ensembling Classifiers**
Step 11: Return E_i
Step 12: end

The multi-classifier system is referred to as an Ensemble-based learning model in the proposed study. For this purpose, the proposed model includes preferred DL unsupervised classifiers such as SVM, Random Forest technique, and Naive Bayes classifier. Finally, the Boosting Ensemble classifier was the most effective in detecting abusive messages. The XG Boosting Ensemble classifier has combined the best performance in recognizing abusive messages.

4 Experimental Results

4.1 Description of Dataset

As the digital domain has increased and technology has evolved, cyberbullying, also referred to as bullying, has grown increasingly frequent, particularly among youngsters. There are many datasets available that include abusive text and harsh language used by individuals of all ages on various social networking networks. The Instagram datasets are available at https://www.kaggle.com/datasets/marklvl/sentiment-labelled-sentences-data-set. These data are saved as CSV files, with two columns: 'comments' and 'class'. Each comment's class indicates its categorization. Non-offensive Instagram comments are tagged as '0,' while offensive ones are marked as '1'.

The data distribution is shown in Fig. 2 below. The dataset's null value rows were eliminated. The final dataset for the whole project has 14998 rows as well as 4 columns. The data labelled Class '0' reflects non-cyberbullying material. The collection included 10807 records of Class '0' data. The cyberbullying material was classified as Class '1'. There were 4191 entries containing cyberbullying material in total.

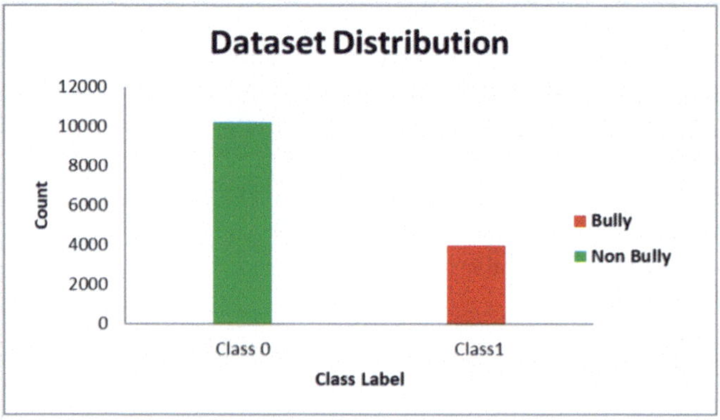

Fig. 2. Dataset Distribution

4.2 Performance Evaluation Metrics

This paper focuses on three models, SVM, Random Forest method, and Naive Bayes algorithm utilizing XG Boost, which was combined to form the Ensemble model. Examining the evaluation measures is essential for analyzing model performance. Models were evaluated using parameters including Accuracy, Precision, Recall, and F1-score.

A summary of the precision, recall and F1-score of the four models that were evaluated is given in Table 1.

Table 1. Performance of Different Algorithms

Measures	SVM			Random Forest			Naive Bayes			Ensemble XG Boosting		
	P	R	F	P	R	F	P	R	F	P	R	F
Without Cyberbullying	0.84	0.87	0.86	0.85	0.89	0.88	0.83	0.86	0.84	0.91	0.98	0.95
With Cyberbullying	0.86	0.85	0.87	0.88	0.86	0.89	0.83	0.85	0.86	0.91	0.93	0.94

*P-Precision; R-Recall; F- F1-Score;

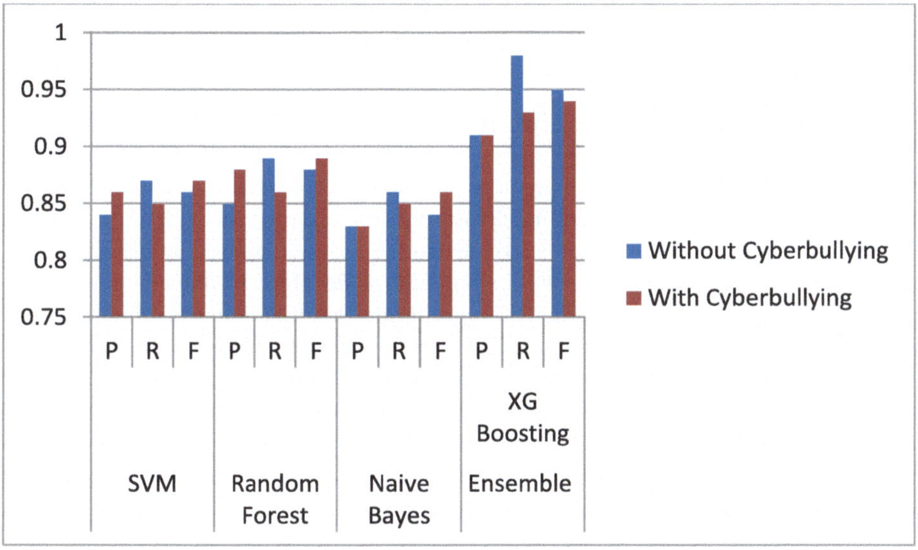

Fig. 3. Performance Evaluation of different algorithms

From Fig. 3 shows the performance of different algorithms using the parameters of average and its metrices.

Finally, Fig. 4 compares the ensemble method to the other three algorithms in terms of accuracy for detecting the cyberbullying in the text commands. As a consequence, we find that the Ensemble XG Boosting method outperforms the other algorithms in terms of accuracy.

The multi-classifier system will be referred to as an Ensemble-based learning model in the work that is being presented. The recommended model incorporates favored unsupervised deep learning classifiers for this persistence. These classifiers include the SVM, Random Forest technique, and Naive Bayes classifier. At the end, the Boosting Ensemble classifier proved to be the most effective in identifying harassing or threatening messages. The XG Boosting Ensemble classifier has been coupled to get the highest performance in recognizing abusive messages.

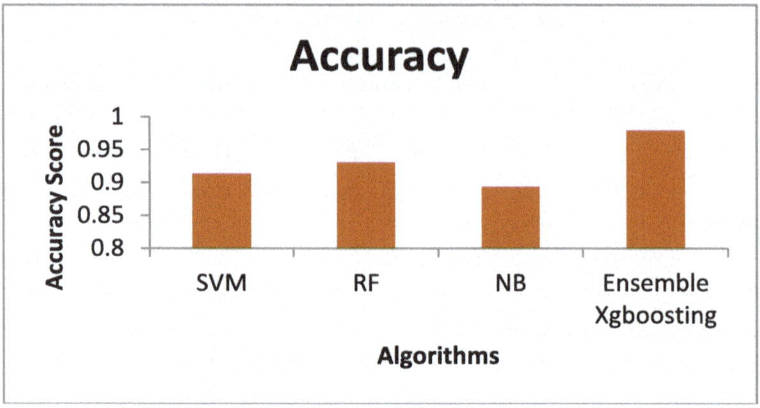

Fig. 4. Accuracy of the Model

5 Conclusion and Future Enhancement

Particularly, cyberbullying has grown more prevalent and began to raise important societal difficulties as a proliferation result of platforms in social media as well as the growing usage of social media by adolescents. To prevent the negative impacts of cyberbullying, an automated cyberbullying detection system must be developed. Given the importance of cyberbullying detection, we studied the algorithmic identification of postings on social media linked to cyberbullying using the XG Boosting features in this research. To recognize bullying language, three machine learning techniques are used: SVM, RF, and NB. In the future, machine learning unsupervised classifiers will be used to create a framework for automated detection and categorization of bullying from Instagram comments. Finally, the Boosting Ensemble classifier performed the best in recognizing abusive communications.

References

1. Wang, S.: Standing up or standing by: Bystander intervention in cyberbullying on social media. New Media Soc. **23**(6), 1379–1397 (2021)
2. López-Vizcaíno, M.F., Nóvoa, F.J., Carneiro, V., Cacheda, F.: Early detection of cyberbullying on social media networks. Futur. Gener. Comput. Syst. **118**, 219–229 (2021)
3. Kumari, K., Singh, J.P.: Identification of cyberbullying on multi-modal social media posts using genetic algorithm. Trans. Emerg. Telecommun. Technol. **32**(2), e3907 (2021)
4. Huang, J., Zhong, Z., Zhang, H., Li, L.: Cyberbullying in social media and online games among Chinese college students and its associated factors. Int. J. Environ. Res. Public Health **18**(9), 4819 (2021)
5. Cheng, L., Mosallanezhad, A., Silva, Y., Hall, D., Liu, H.: Mitigating bias in session-based cyberbullying detection: a non-compromising approach. In: Proceedings of the 59th Annual Meeting of the Association for Computational Linguistics and the 11th International Joint Conference on Natural Language Processing (Volume 1: Long Papers), pp. 2158–2168 (2021)
6. Chelmis, C., Zois, D.S.: Dynamic, incremental, and continuous detection of cyberbullying in online social media. ACM Trans. Web (TWEB) **15**(3), 1–33 (2021)

7. Perera, A., Fernando, P.: Accurate cyberbullying detection and prevention on social media. Procedia Comput. Sci. **181**, 605–611 (2021)
8. Bozyiğit, A., Utku, S., Nasibov, E.: Cyberbullying detection: utilizing social media features. Expert Syst. Appl. **179**, 115001 (2021)
9. Alotaibi, M., Alotaibi, B., Razaque, A.: A multichannel deep learning framework for cyberbullying detection on social media. Electronics **10**(21), 2664 (2021)
10. Forsberg, C.: The importance of being attentive to social processes in school bullying research: adopting a constructivist grounded theory approach. Int. J. Bullying Prev. **4**(3), 180–189 (2022). https://doi.org/10.1007/s42380-022-00132-y
11. Mishna, F., Birze, A., Greenblatt, A.: Understanding bullying and cyberbullying through an ecological systems framework: the value of qualitative interviewing in a mixed methods approach. Int. J. Bullying Prev. **4**, 1–10 (2022). https://doi.org/10.1007/s42380-022-00126-w
12. Rezapour, M., Nezam Tabar, A., Khanjani, N.: Fear of bullying victimization in middle school: types of victimization and advocacy considerations. Int. J. Bullying Prev. **4**(2), 89–98 (2022). https://doi.org/10.1007/s42380-021-00089-4
13. Riffle, L.N., Demaray, M.L.: Bully participant role behavior and social support from teachers and classmates: a mediation analysis. Int. J. Bullying Prev. **4**(2), 130–143 (2022). https://doi.org/10.1007/s42380-021-00092-9
14. Arató, N., Zsidó, A.N., Rivnyák, A., Péley, B., Lábadi, B.: Risk and protective factors in cyberbullying: the role of family, social support and emotion regulation. Int. J. Bullying Prev. **4**(2), 160–173 (2022). https://doi.org/10.1007/s42380-021-00097-4
15. Triantafyllopoulou, P., Clark-Hughes, C., Langdon, P.E.: Social media and cyber-bullying in autistic adults. J. Autism Dev. Disorders **52**, 4966–4974 (2022). https://doi.org/10.1007/s10803-021-05361-6
16. Giumetti, G.W., Kowalski, R.M.: Cyberbullying via social media and well-being. Current Opin. Psychol. **45**, 101314 (2022)
17. Kumar, A., Sachdeva, N.: A Bi-GRU with attention and CapsNet hybrid model for cyberbullying detection on social media. World Wide Web **25**(4), 1537–1550 (2022). https://doi.org/10.1007/s11280-021-00920-4
18. Chandrasekaran, S., Singh Pundir, A.K., Lingaiah, T.B.: Deep learning approaches for cyberbullying detection and classification on social media. Comput. Intell. Neurosci. (2022)
19. Parris, L., Lannin, D.G., Hynes, K., Yazedjian, A.: Exploring social media rumination: associations with bullying, cyberbullying, and distress. J. Interpers. Violence, **37**(5–6), NP3041-NP3061 (2022)
20. Jain, V., Kumar, V., Pal, V., Vishwakarma, D.K.: Detection of cyberbullying on social media using machine learning. In: 2021 5th International Conference on Computing Methodologies and Communication (ICCMC), pp. 1091–1096. IEEE (2021)
21. Sultan, D., et al.: A review of machine learning techniques in cyberbullying detection. Comput. Mater. Continua, **74**(3) (2023)
22. Mahmud, T., Ptaszynski, M., Eronen, J., Masui, F.: Cyberbullying detection for low-resource languages and dialects: review of the state of the art. Inf. Process. Manage **60**(5), 103454 (2023)
23. Yi, P., Zubiaga, A.: Session-based cyberbullying detection in social media: a survey. Online Soc. Netw. Media **36**, 100250 (2023)
24. Alduailaj, A.M., Belghith, A.: Detecting Arabic cyberbullying tweets using machine learning. Mach. Learn. Knowl. Extr. **5**(1), 29–42 (2023)
25. Akhter, A., Acharjee, U.K., Talukder, M.A., Islam, M.M., Uddin, M.A.: A robust hybrid machine learning model for Bengali cyber bullying detection in social media. Nat. Lang. Process. J. **4**, 100027 (2023)

Enhancing Plant Pathology Detection and Stratification with a Tailored Mask R-CNN Framework

Miriyala Sai Manikanta(✉), G. Kalyani, and G. Prasanna kumar

Department of Information Technology, Velagapudi Ramakrishna Siddhartha Engineering College, Vijayawada, India
saimanikantamiriyala5756@gmail.com

Abstract. Our everyday lives are significantly impacted by the agriculture sector, which is crucial to the global economy. Despite its importance, a lack of knowledge in recognizing and controlling leaf diseases causes many farmers to have lower yields. It is crucial to quickly and accurately identify plant diseases since they have a direct impact on agricultural profit and loss margins. In order to identify and categorize plant leaf diseases, the Mask-RCNN model is used in this work to propose a novel solution to the problem. The first step in our technique is to eliminate noise from plant photos before feature extraction using the ResNet-50 model. The Regional Proposal Network (RPN) processes these properties after which they are divided and categorized using the Region-of-Interest (ROI) Align procedure. This study shows potential in changing disease detection in huge agricultural areas with a remarkable accuracy of 92.45%.By offering early warnings and interventions, these developments can be of considerable help to farmers and agronomists in assuring healthier crops and more effective agricultural methods. It is impossible to overstate the importance of identifying plant leaf diseases since it affects not only the agricultural industry but also related fields like biological research and agricultural institutes.

Keywords: Plant leaf diseases · Deep Learning · Mask R-CNN · Object detection · segmentation · accuracy · and time to lesion identification

1 Introduction

It's pivotal for effective agrarian operation to use aerial view of crop data to estimate crop stress. By using aerial view, this system enables early discovery of stress factors similar. Due to its broad content, it's possible to cover husbandry on a large scale, which facilitates the identification of patterns and trends. To distribute water, toxin, and fungicides as efficiently as possible, perfection resource operation may be employed with the use of this data. Growers are more suited to lower stress and boost crop productivity because real- time data enables quick decision- timber.

Agriculture can profit from using aerial vehicles to identify crop stress in a number of ways [1]. Prompt mitigation of the goods of stressors similar nutrient scarcities, water

D. Pastor-Escuredo et al. (Eds.): ICDLAIR 2023, LNNS 1001, pp. 326–339, 2024.
https://doi.org/10.1007/978-3-031-60935-0_30

stress, pests, and pathogens is made possible by early opinion of crop stress. The wide-ranging content of aerial vehicles makes it possible to cover crops across enormous distances, making it simpler to spot patterns and trends. This information assists in the effective use of coffers by enabling precise input operation, reducing waste, and limiting environmental goods [2]. Quick decision- making made possible by real- time data enables growers to increase productivity by changing their styles. Also, crop soothsaying and threat analysis using data might give pivotal knowledge for preparing and mollifying unborn issues. Several typical cost-saving methods Pest and plant disease detection methods include: Visual inspection, Spectral imaging, Machine learning analysis.

Plant yield and quality can be dramatically reduced by a variety of ailments and pests. For the implementation of efficient management measures and for the prevention of the problem from spreading, early discovery of these difficulties is essential. Visual inspection [3], automated monitoring systems, and the application of genetic techniques are only a few of the methods used to diagnose plant diseases and pests. Which approach is most successful depends on the kind of pest or disease, the kinds of plants, and the available resources. Each method has advantages and disadvantages. The article's major arguments focus on early detection, increased accuracy, economics, enhanced productivity, and sustainable farming [4].

Digital technology allows for early detection. This project's major goal is to use

[5] aerial data of crops to determine the crop's level of stress. Similar to that, this project's many goals include tracking crop development and growth through time, evaluating the crop's general health, estimating prospective yield. Although everyone agrees that agriculture is important to the world's food security, plant diseases continue to pose a serious threat to it. Plant diseases can lower global food supply by 10–16%, according to studies like the one published in "Nature Ecology & Evolution" by Savary et al. (2019). This highlights the significant influence that these diseases have on agricultural productivity. Oerke (2006), underscoring the economic ramifications, projected that plant diseases could cause losses of up to 30% in important crops globally in the Journal of Agricultural Science. According to the Food and Agriculture Organization (FAO), smallholder farmers in underdeveloped nations are most affected by this problem because they frequently do not have access to the necessary tools and information for efficient disease management. These difficulties are exemplified by historical events such as the significant production losses resulting from Wheat Rust across Africa, Asia, and the Middle East, and the up to 70% yield decline in East and Central Africa from Cassava Mosaic Disease. A study published in the 'Annual Review of Phytopathology' by Garrett et al. (2006) also raises the possibility that plant diseases may become more severe and spread more widely as a result of climate change, which highlights the need for improved methods and expertise in plant disease prevention. Together, these results highlight how crucial it is to enhance agricultural leaf disease detection and control in order to protect food production and financial stability.

2 Literature Survey

According to Prasanna Mohanty et al.'s study "Deep learning for based on pictures plant identification" [6], convolutional neural network training can be used to detect plant illnesses. In order to identify both healthy and unhealthy plants from 14 different species, the CNN model has been trained. The model's accuracy on the test set of data is 99.35%. When used using images taken from trustworthy internet sources, this is superior to a straightforward model of random selection; nevertheless, the accuracy can be increased by using a larger variety of training data. The model achieves a 31.4% validity. Furthermore, various model or neural network training strategies might result in improved accuracy, paving the way for widespread accessibility of plant disease detection.

In their article, "Detection and Classification of Leaf Disease Using Artificial Neural Network," Malvika Ranjan et al. made a suggestion for a technique to identify plant ailments using the image of the ill leaf. Artificial neural networks, also known as ANN, must be trained by carefully choosing feature values in order to distinguish between healthy samples and diseased plants. The ANN model has an error rate of 80%.

According to the study "Detecting unhealthy leaf regions and categorizing diseases of plant leaves using textural features," [8], the process of diagnosing a disease entails four fundamental steps: The green pixels are first recognized and excluded using a specific threshold value, after which the input RGB image is given a color transformation structure. Following segmentation, texture statistics are generated to create useable segments. Finally, the disease is classified using the obtained attributes by a classifier.

Since the suggested method depends upon an ANN classifier for classification and a Gabor filter for feature extraction, it yields better results with a recognition rate of up to 91%.

Plant disease detection using CNN and GAN has been proposed as a method by Emaneul Cortes in his article "Plant illness diagnosis using GAN" [10].Background segmentation is carried out to ensure accurate feature extraction and output mapping. Even though it is obvious that using Gans could be helpful for classifying plant diseases, background-based segmentation had no appreciable impact on accuracy. In their research titled "Convolutional neural network using Inception 3 Model for Animal Classification," Jyotsna Bankar et al. recommended utilizing the Inception v3 model to classify animals into different species. Because Inception v3 can classify and categorize objects, it is advantageous for a number of distinct image classifiers.

Overall, these studies demonstrate the promise of deep learning and machine learning techniques for detecting pests and illnesses of plants. They underline the importance of accurate and timely identification of plant diseases and pests for better crop management and output.

3 Proposed Methodology

1. Data Gathering and Preprocessing: To develop a comprehensive dataset that included pictures of both healthy and damaged plant leaves, the first step was to gather photographs from various sources. A vast and diverse dataset was produced by combining the data from these several sources. The photos underwent a number of preprocessing processes after being captured to maintain consistency and improve data quality. In order to maintain a consistent color format throughout all of the photographs, they were first converted to grayscale. Then noise-reduction techniques were used to assure clarity and remove any stray artifacts. Each photograph's pixels connected to leaves

and those connected to the environment were carefully identified and noted. In order for the model to eventually provide flawless segmentation, this accurate labeling is essential.

2. The Mask R-CNN model receives the images before they are passed to the ResNet-50 architecture, which is utilized to extract prominent features from the preprocessed images. For the steps of detection and classification that follow, this feature extraction is essential.
3. A Regional Proposal Network (RPN) based on the ResNet-50 architecture processes the features after feature extraction. Mask R-CNN and RPN Model Training. This RPN offers ideas for possible rectangular enclosures for locations that are covered with vegetation. The Mask R-CNN detector is then used to polish these concepts. In order to precisely detect disease, it computes pixel-level segmentation for each proposal and categorizes the type of disease present on the leaves.
4. Model Evaluation: A second test set of images that the Mask R-CNN model has never seen previously is used in order to fully evaluate the performance of the model. Among the evaluation metrics are memory, precision, and MAP (mean average precision). A weighted average of memory and precision across many class criteria is provided by MAP. Recall determines the percentage of legitimately used sick days (Fig. 1).

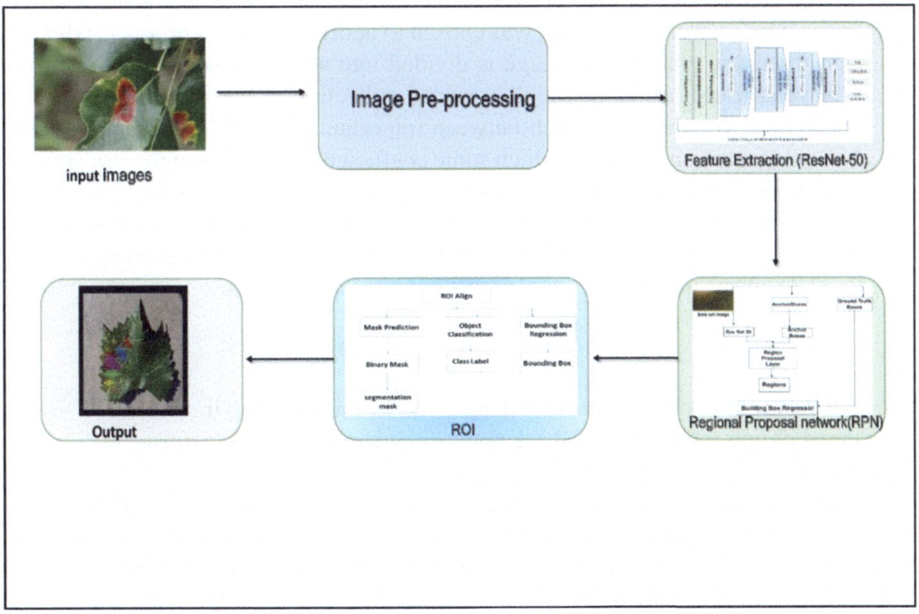

Fig. 1. Proposed Methodology Architecture

A. Image Pre-processing: The readiness and quality of the input data are crucial in the effort to develop a reliable plant disease detection model. The systematic preprocessing procedures described in this part optimize the dataset for the upcoming stages of model training and evaluation by ensuring its integrity and consistency.
   1) Fusion and Data Gathering: The initial step was gathering different datasets. We searched a number of archives and databases and found images of both healthy and damaged plant leaves. Data from these many sources were combined in order to create a comprehensive and heterogeneous dataset that addressed the complete range of plant diseases.
   2) Grayscale Conversion: The next step was to convert color images, which are naturally identifiable by their Red, Green, and Blue (RGB) components, to grayscale. This conversion makes use of the formula:

$$\text{Gray} = 0.299 * \boldsymbol{Red} + \mathbf{0.587} * \boldsymbol{Green} + \mathbf{0.114} * \boldsymbol{Blue}$$

   This kind of adjustment is crucial since it uniformizes the dataset and ensures uniformity in color representation. The photos' intensity distribution is preserved while the color nuances are simplified, allowing the model to focus more on structural patterns than color variations.
   3) Wavelet Denoising for Noise Removal: Noise problems existed since the images were taken from diverse sources. Because it has the ability to handle noise at various scales, wavelet denoising was chosen to address this. The process is divided into three key phases: An image is divided into several frequency bands using a wavelet transformation, which creates a collection of wavelet coefficients for each band. We can distinguish between important visual information and noise by setting a threshold above which some coefficients are kept and the rest are cancelled out. After thresholding, the image is translated into its spatial form using the inverse wavelet transformation, creating a noise-free, accurate representation. This denoising procedure can be summed up by the following equation:

   The encapsulating formula for this denoising process is:

$$\boldsymbol{I'} = \mathbf{W'}(T(W(I)))$$

   Where $\boldsymbol{I}'$ is the denoised image, **W** is the wavelet transform, **T** is the thresholding function, and $\boldsymbol{W}'$ denotes the inverse wavlet transform

B. RESNET-50 model's feature extraction technique:

   The extraction of intrinsic features from the images, which comes after the initial preprocessing procedures, is the crucial stage in our methodology for precisely detecting flaws. Figure 2 illustrates the modified ResNet-50 architecture we use for this. This model is skilled at spotting details like texture, color moment us, and hue saturation. ResNet-50, with its deep convolutional structure, is renowned for its proficiency in feature extraction because it blends convolutional and pooling layers. ResNet-50 stands out for its residual connections that omit a few tiers. The accompanying graphic shows how these short cuts successfully solve the vanishing gradient problem and enable efficient training in deeper networks. Our curated plant disease photographs are processed by this ResNet-50 to yield a wealth of insights.

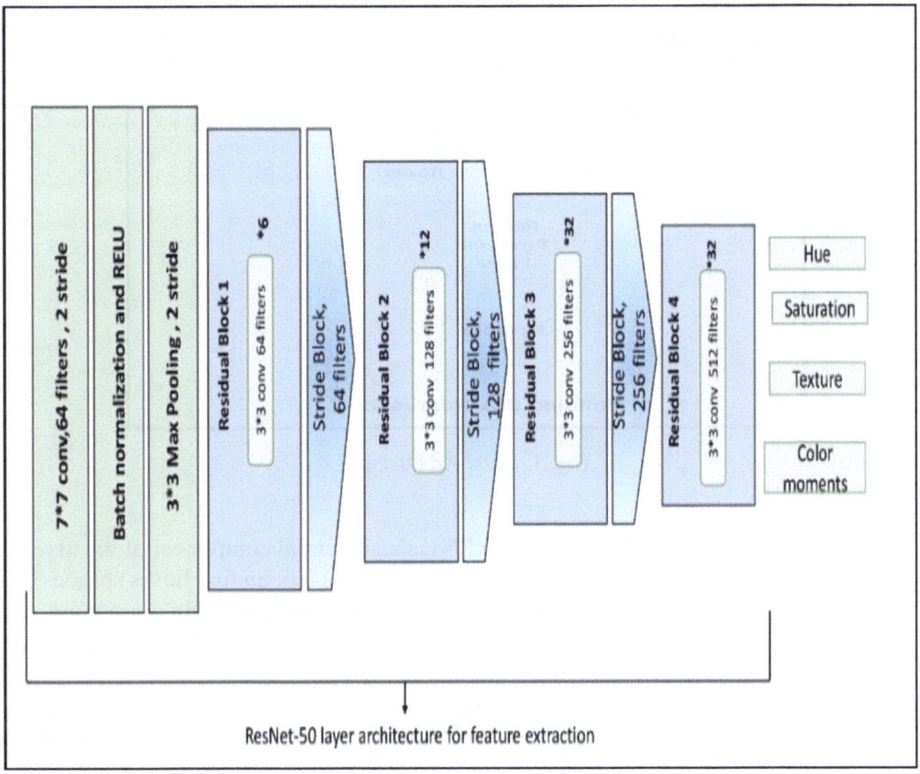

Fig. 2. ResNet50 layer architecture for feature extraction

A 7x7 convolutional layer with a stride of two is first applied to the input image to aid in the identification of the fundamental visual elements. After that, a ReLU activation function and a batch normalization function stabilize the features while adding nonlinearity to the network. The spatial dimensions are then reduced by a second 3x3 max pooling layer with a stride of 2, while maintaining crucial features. This design's core consists of the following four distinct residual block groupings:

Residual Block 2:Six 3x3 convolutional layers with 64 filters each make up Residual Block 1, which is designed to distinguish spatial characteristics.

Residual Block 2: This block enlarges the architecture with twelve 3x3 convolutional layers, each with 128 filters, to capture more complicated patterns.

Residual Block 3: To expose more intricate visual details, it goes deeper by utilizing 32 3x3 convolutional layers with 256 filters in each one.

Residual Block 4: Using 32 3x3 convolutional layers once more to design the network for extensive feature extraction while maintaining the depth constant.

Each of these blocks is fundamentally composed of the shortcut or skip connection, a ResNets distinguishing feature. This link immediately merges the input and output of the block, ensuring gradient flow during training.

C. Regional Proposal Network:

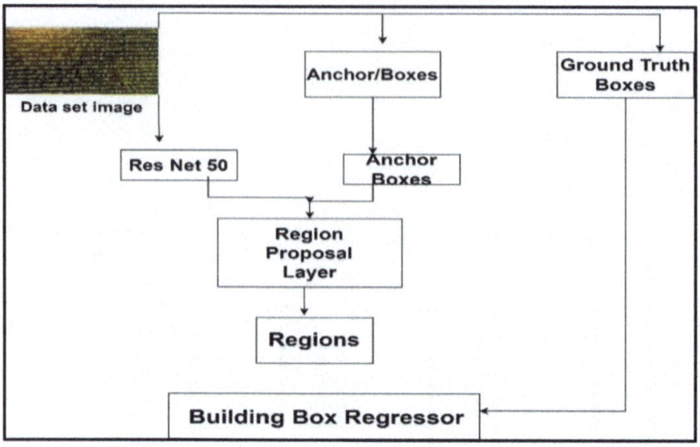

Fig. 3. RPN Architecture

The Regional Proposal Network (RPN) is an essential component of the disease detection tool Mask R-CNN. It offers region proposals (bounding boxes) based on a convolutional feature map. Anchor boxes and prediction object ratings are improved by RPN. Binary classification and bounding box regression loss functions are used in its instruction. RPN creates potential bounding boxes using scores. Redundancy is decreased when non- maximum suppression is used. These recommendations are added to the ROI Align layer of the Mask R-CNN for classification and mask prediction. RPN improves region-of- interest efficiency and accuracy, two elements necessary for correctly classifying diseases in images (Fig. 3).

D. ROI align:

The ROI (Region-of-Interest) Align operation, a crucial part of our method, guarantees precise segmentation and classification of detected portions inside an image. This intricate process is divided into several interconnected processes. Beginning with mask prediction, the strategy aims to create a binary mask for each identified region of interest. The result is the creation of a segmentation mask that successfully separates the regions of interest from the background and other unimportant parts of the image. After this segmentation, the emphasis shifts to object classification. Here, each segmented region is analyzed, which results in the designation of a specific class label that corresponds to its special characteristics. This categorization is followed by a vital phase known as bounding box regression (Fig. 4).

Algorithm: An Improved Mask R-CNN for Plant Disease Detection
Input:- Plant Image *I*
Output:- Classification of diseases, bounding boxes, and segmented masks
Method:

1. Preprocessing:
   1.1 Use the formula Gray = 0.299 x Red + 0.587 x Green + 0.114 Blue to turn I into grayscale.

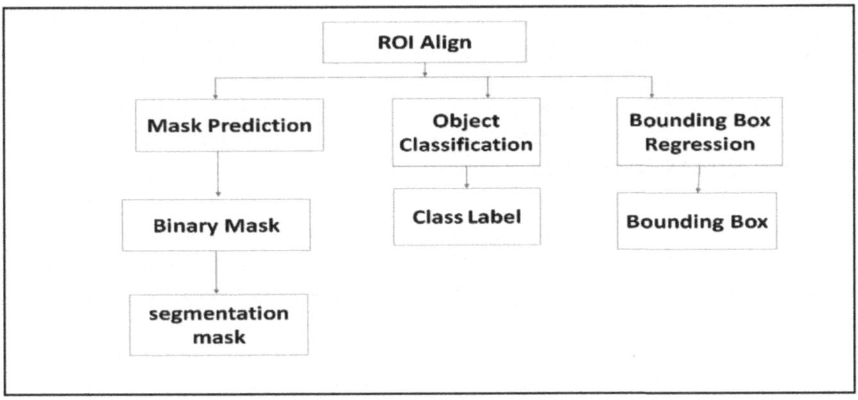

Fig. 4. ROI Architecture

1.2 Apply wavelet denoising to *I* by transforming it into *I' = W'(T(W(I)))*, where Wi is the wavelet transform, *T* is the thresholding function, and W' is the inverse wavelet transform.

2. Feature Extraction using ResNet-50:
   2.1 Transmit I' using the ResNet-50 model.
   2.2 It is necessary to extract the high-level properties from I that are represented by F.
3. Region Proposal using RPN with ResNet-50:
   3.1 To create region proposals, use the ResNet-50 model.
   3.2 Consider Fusing the RPN, which is denoted by the formula R = RPN (F to acquire suggested regions R).
4. Detection of diseases and rol alignment:
   4.1 For each region r in R:
      4.1.1 To get features that are aligned, apply ROI Align to r. Ay.
      4.1.2 To determine whether a disease is present, classify A using classifier function C so that L, = C (Ar).
      4.1.3 Disease is indicated if L.:
         4.1.3.1 To obtain a binary mask M that highlights the sick area, use segment A.
         4.1.3.2 Create a bounding box B that contains the sick area, denoted by B, = Box (Ar).
      4.1.4 For the area, store Lr, Mr., and B.
5. Output:
   5.1 Send back a list of all the regions' classifying labels, L.
   5.2 Get a list of bounding boxes B for all the sick areas that were found.
   5.3 For each divided sick region, provide a list of masks M.

Considering these drawbacks, our research adds something special to the field. One noteworthy improvement is the Mask R-CNN model's modification for the particular goal of plant disease detection in agriculture. By applying customized preprocessing approaches that are specifically made to manage the complexity of agricultural photos, our solution improves the detection accuracy. Furthermore, our method improves the field by offering disease identification that is more precise and effective than existing methods—a critical component for early intervention and control in crop production.

4 Results and Discussions

This section compares our methodology to "state-of-the-art" techniques by categorizing plant illnesses using our dataset analysis and our method for obtaining disease area in a leaf. The results of our strategy will be assessed in the following section using experimental data.

4.1 Dataset Description

There are 250 different types of plant leaves and background photos can be found among the 91,486 images in the Crop diseases of leaves dataset with augmentation data-set. To boost the volume of data gathered, we employed six distinct augmentation methods. Image flipping, scaling, rotation, noise via injection, PCA color augmentation, noise injection, and noise injection are some of these methods. A total of 30052 photos with 24 labels are used in the Phyto Leaf Diseases collection.

4.2 Quantitative Metrics

We'll go over the quantitative tests performed to validate the Mask-RCNN model's precision in identifying and categorizing plant ailments in this part. Figures depicting additional visual substantiations of the model's ability will be provided with these evaluations. These visuals will act as supporting evidence in addition to quantitative performance metrics like accuracy, precision, recall, and the F1-score. Our objective is to present a comprehensive evaluation that demonstrates the model's suitability for practical applications in Mask- RCNN plant disease detection.

In order to recognize and explain the existence and severity of leaf diseases, specific locations on plant leaves are recognized in photographs. Automated disease detection and monitoring is made possible by this technology using deep learning models (Figs. 5 and 6).

4.3 Performance Evaluation Metrics

1. **Accuracy:** A model's accuracy is measured as a proportion of correctly predicted outcomes. It is calculated by splitting the model's overall prediction output by the proportion of accurate forecasts.

$$\textbf{Accuracy} = \frac{\boldsymbol{TP + TN}}{\boldsymbol{TP + FP + FN + TN}}$$

Fig. 5. Annotation of Disease Spotted area a) Apple_scrab, b) corn_gray_leaf_spot, c) potato_early_blight, d) Tomato_bacterial_spot

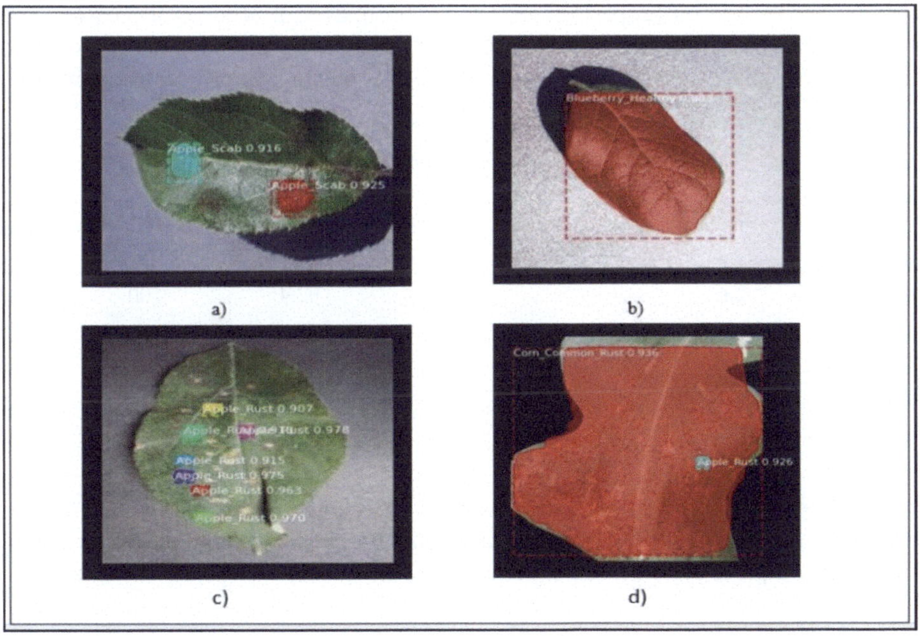

Fig. 6. Emphasizing imperfections imposed on by the disease a) Apple_rust, b) Corn_CommonRust, c) Blueberry_healthy, d) Apple_scab

2. **Precision:** Precision is the percentage of positive right predictions across all the optimistic claims the model makes. It is calculated by splitting the total number of true positives by the sum of true positives and false positives.

$$\textbf{Precision} = \frac{\boldsymbol{TP}}{\boldsymbol{TP} + \boldsymbol{FP}}$$

3. **Recall:** The proportion of true positive cases that are truly true positives is known as recall. It is calculated by dividing the number of successful outcomes by the sum of successful results and unsuccessful results.

$$\textbf{Recall} = \frac{\boldsymbol{TP}}{\boldsymbol{TP} + \boldsymbol{FN}}$$

4. **F1-score:** The F1-score is the average of recall and precision. It is a fair measurement that takes memory and precision into account.

4.4 Performance Evaluation

When evaluated side by side with traditional approaches, our recently invented Mask RCNN methodology exhibits great performance in the realm of plant disease categorization. The comparison between faster-RCNN, CNN and our MASK-RCNN is displayed in Table1 and focuses on key performance indicators like recall, accuracy, f-measure, and precision. This assessment is especially relevant given the state of our dataset. Includes, such as Apple_rust , Corn_CommonRust, Blueberry_healthy, Apple_scab and other

Table 1. Comparison of Performance Evaluation Metrics

Model	Precision (%)	Recall (%)	F1-Score (%)	Accuracy (%)	Inference Time (ms)
Mask R-CNN	82.6	89.5	90.3	92.45	100
YOLOv6	85.0	87.0	86.0	91.50	50
Faster R-CNN	80.0	85.0	82.5	90.00	130
SSD	75.0	80.0	77.5	88.00	40

The empirical data highlight our network model's higher accuracy, and Fig. 7 elaborates on the accuracy comparisons. Additionally, Fig. 8 shows in graphical form how our model compares other models in terms of accuracy, recollection, and f-measure.

4.5 Computation Time

The efficiency of computational time stands as another metric of interest. Our approach strives for computational simplicity, a feature that is evident when compared to other state- of-the-art techniques. Table 2 presents a head-to-head comparison of the computational times for our Mask-RCNN model and other existing methods.

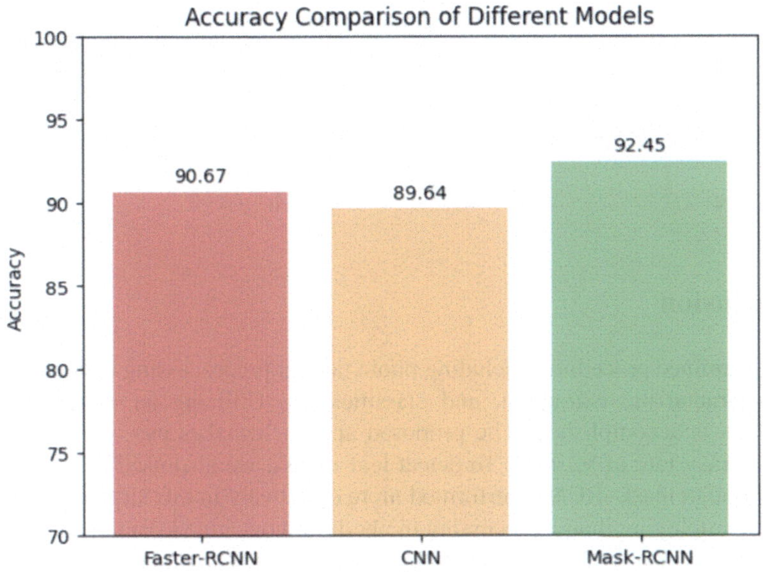

Fig. 7. Analysis of Accuracy based on different techniques

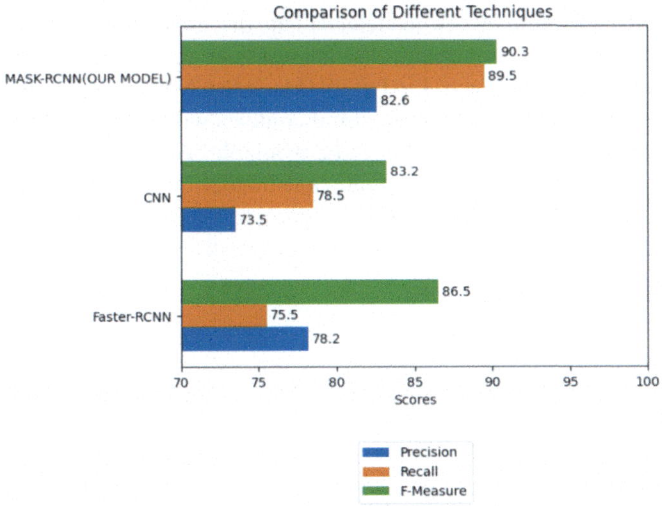

Fig. 8. Metric-based Comparative Analysis

Table 2. Computational Time (in seconds)

Method	Time
Mask-RCNN(Our Model)	0.16
CNN	0.23
Faster-RCNN	0.19

5 Conclusion

We have examined procedures including photo taking, preprocessing of pictures, image slicing, characteristic extraction, and classification. Utilizing an image processing method, this is accomplished. The proposed approach makes use of R-CNN, which has an accuracy rate of 92.45%. To detect leaf illness, we also used the faster-RCNN model, although mask- RCNN performed more effectively in this situation. It has been shown that employing image processing methods and machine learning algorithms can help diagnose agricultural illnesses with high accuracy. Last but not least, the suggested technology has the potential to greatly assist farmers by making it possible for them to promptly detect and treat agricultural problems. This will eventually result in better crop management, greater yields, and fewer losses. In fact, our study's method—which uses the Mask R-CNN framework to detect plant diseases—represents a substantial breakthrough in agricultural technology. We do, however, recognize the restrictions and difficulties related to this approach. First of all, Mask R-CNN's computational intensity could be problematic, especially in agricultural contexts with low resources where such computational resources might not always be available. Additionally, the caliber and variety of the training dataset have a significant impact on how well our model performs. It is still very difficult to compile a comprehensive dataset that fairly depicts the vast array of plant diseases under various circumstances.

References

1. Jung, et al.: Deep learning-based tomato leaf disease recognition using single leaf image (2021)
2. Kalamkar et al.: Automated detection of brown spot disease in rice using deep learning techniques (2021)
3. Shah, et al.: Real-time detection and classification of plant diseases and pests using convolutional neural networks (2022)
4. Salawu, et al.: Deep learning-based leaf disease detection and classification for cassava plantation (2021)
5. Mahajan, et al.: Machine learning-based detection of banana leaf diseases using image processing techniques (2022)
6. Mohanty, S.P., Hughes, D.P., Salathé, M.: Using deep learning for image based plant disease detection. Front. Plant Sci. **7**, 1419 (2016)
7. Wallelign, S., Polceanu, M., Buche, C.: Soybean plant disease identification using convolutional neural network. In: FLAIRS-31, May 2018, Melbourne, United States (2018)

8. Arivazhagan, S., Shebiah, R.N., Ananthi, S., Varthini, S.V.: Detection of unhealthy region of plant leaves and classification of plant leaf diseases using texture features. Agric. Eng. Int. CIGR J. **15**, 211–217 (2013)
9. Huu Cap, H.Q., Suwa, K., Fujita, E., Kagiwada, S., Uga, H., Iyatomi, H.: A deep learning approach for on-site plant leaf detection. In: 2018 IEEE 14th International Colloquium on Signal Processing & Its Applications (CSPA) (2018)
10. Santhosh Kumar, S., Raghavendra, B.K.: Diseases detection of various plant leaf using image processing techniques: a review. In: 2019 5th International Conference on Advanced Computing & Communication Systems (ICACCS) (2019)
11. Thakur, et al.: Transfer learning for automated plant disease detection using deep convolutional neural networks (2020)
12. Kumar, et al.: A novel automated deep learning-based leaf disease detection model for mango plants (2021)
13. Ozturk, et al.: Early detection of tomato plant diseases using machine learning techniques (2020)
14. Sharma, et al.: A review on deep learning techniques for plant disease detection (2020)

Deep Neural Network Regression Based Device Free Localization Technique in Changing Indoor Environment

K. S. Anusha(✉) and R. Ramanathan

Department of Electronics and Communication Engineering, Amrita School of Engineering, Amrita Vishwa Vidyapeetham, Coimbatore 641112, India
{ks_anusha,r_ramanathan}@cb.amrita.edu

Abstract. Conventional device free localization techniques are more of a classification approach in which the target location is identified as a probable region. For small room setups and datasets, classification-based device free localization can be used conveniently to find the target's location area. However, many application scenarios demand device free localization algorithm that can accurately find the location coordinates of the target rather than the probable region. Also, finding the exact location is another tedious task for large spaces and limited datasets. For such scenarios, regression based device free localization is an obvious choice. Deep learning in device free localization is becoming popular due to its potential merits in finding the location coordinates accurately in real time scenarios. The proposed link distance based device free localization using a deep neural network is found to be low complex and highly accurate. The highest mean localization error is 21cm, with a mean execution time of 0.0567 ms.

Keywords: Device free localization · Deep neural network · Regression

1 Introduction

Classification based fingerprinting treats the localization problem as discrete, whereas target localization is by nature a continuous problem that estimates the coordinates by regression. Recent research has proved experimentally that SVM regression (SVR) could perform better than its classification counterpart and Bayesian classifier concerning localization [1]. There is an upcoming trend of deep neural networks addressing classification and regression problems due to their implicit feature extraction quality, which has increased applications in target localization and artificial intelligence in recent years. In real-time application scenarios like intruder detection and elderly person tracking indoors, the target's location is estimated in a single room and within a single floor of a house or a building with multiple rooms [2]. In the case of a single floor, the multipath effect will be higher because of multiple walls in every room and other objects in the building. Moreover, with the increase in complexity of indoor, shadow effect, fading, multipath propagation, blockage, etc., localization performance with RSSI parameter also gets affected [3].

D. Pastor-Escuredo et al. (Eds.): ICDLAIR 2023, LNNS 1001, pp. 340–351, 2024.
https://doi.org/10.1007/978-3-031-60935-0_31

Thus, when it comes to a cluttered floor setup of a building or a house, the accuracy degrades significantly with traditional machine learning approaches, and centimeter-level accuracy may not be achievable, which is one of the upcoming challenges in device free localization techniques [4]. Further, if the environment changes frequently, effective models are to be introduced to estimate the target's location accurately. Thus, to achieve centimeter-level accuracy even in a cluttered indoor setup with a high multipath effect, the dataset is vast, but with missing values, a regression based deep learning model is a probable solution. Handcrafted machine learning models widely address the indoor localization problem [5]. There are many recent works with deep learning based DFL due to its sensitivity to the input data format, whereas RSS fingerprint data is vectorized data for each location coordinate. In that, deep neural networks (DNN) are popularly chosen due to some potential advantages. It is easier for DNN to achieve global optimization by backpropagation and fine tuning network parameters [6]. Also, deep neural networks provide efficient generalization performance in an indoor environment and swiftness in estimating the target location during the testing phase [7].

The significant outcomes of the work are:

- This work proposes a link distance-based deep neural network regression model that works well in a changing environment and utilizes only a single model.
- Proposing a regression approach to develop a highly accurate link distance-based device-free localization technique that is validated through experimental data.
- The dependency of RSS fingerprints with target location is modeled using DNN regression to achieve centimeter-grade accuracy.

The organization of the paper is as follows: Sect. 2 depicts the proposed link distance deep neural network (LD-DNNR) Regression based DFL technique, methodology, and architecture details. The results and discussion are explained in Sect. 3. Section 4 gives conclusive remarks and the scope of the work.

2 Proposed Link Distance Deep Neural Network (LD-DNN) Regression Based DFL Technique

In this section, an LD-DNNR model is proposed in which a deep model using link distance estimates the target's location, similar to LD-SVR [8]. As in LD-SVR, since the target location estimation uses the link distance concept, DNNR is denoted as LD-DNNR. Each link in the RSS matrix will estimate the target location. It is to be noted that the LD-SVR model used in our previous research work [8] is multiple modeled in which different SVR models are used for different link distances in estimating the target location. Through this work, the researchers have attempted to eliminate this multiple-model requirement through the LD-DNNR DFL technique, which employs only a single estimation model. The LD-DNNR is modeled by employing the dependency of location coordinates with RSS fingerprints from the network, thereby finding the exact target location. Deep neural networks are one of the accurate estimation deep models employed for regression compared to deep belief networks and their variants. The system architecture of the LD-DNNR based DFL technique proposed for a changing indoor environment is explained in the forthcoming section.

2.1 A System Architecture of Proposed LD-DNNR Model in Real Time Scenario

The architecture of the proposed LD-DNNR model validated through experimental data is shown in Fig. 1. In real-time, the RSS values are affected not only by the presence or absence of a target but also due to various multipath effects. So, in a real-time environment, each target-affected link contributes to target location estimation by taking entire RSS values as input to the proposed LD-DNNR model. The estimated output is the local target coordinate. This is to account for the impact of the multipath effect on RSS values in real time scenarios. In the LD-SVR model with simulated data [8], each target-affected link estimates the target position with the RSS value of a particular affected link alone as input and the target's distance from a transmitter as output.

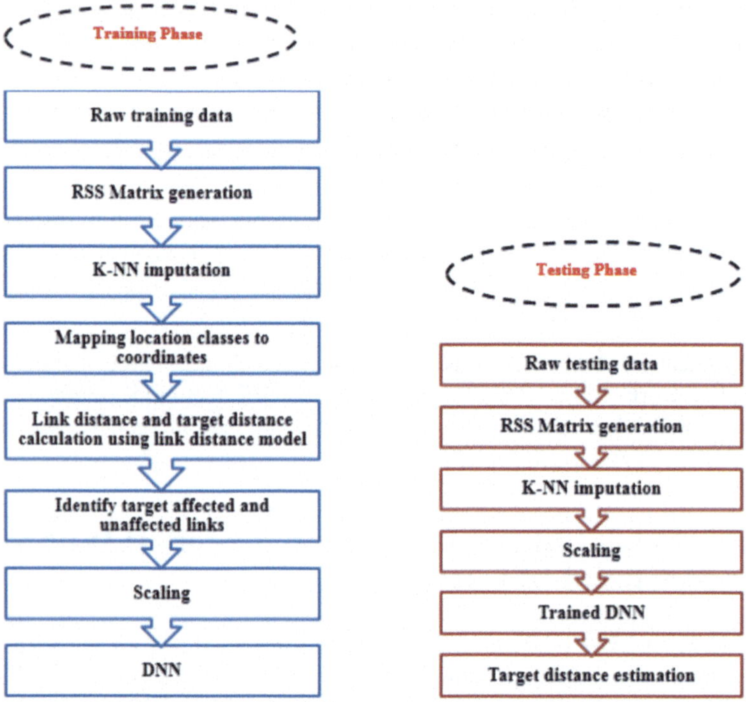

Fig. 1. System architecture of the proposed LD-DNNR based DFL technique in real data applications

In this research, the raw input data obtained from the Span Lab, University of Utah database is converted to a location based RSS matrix, which is further fixed for missing values using the K-NN imputation technique [9, 10]. The output is the corresponding location coordinates of the location classes used from the dataset since DNN is modeled as a regression model. During the training phase, for each link in the RSS matrix, the link and target distance from the transmitter are calculated as in the LD-SVR model in [8]. The target-affected links are identified by comparing the target's distance from a transmitter with the link distance. For each target-affected link, the local target coordinate is calculated; for each unaffected link, the local target coordinate is marked as zero. The RSS values are scaled within 0 and 1 for faster convergence. The corresponding output is the local target coordinate. With this input-output pair, the proposed LD-DNNR model is trained during training. In the testing phase, location based RSS matrices are obtained after preprocessing the raw data. The missing rows are fixed using K-NN imputation as in training. This RSS test data, after scaling, is given to the trained LD-DNNR for estimating the local target coordinate.

3 Deep Neural Network Regression Model

This section describes the deep neural network regression model proposed in this research for finding the target location in a real-time environment.

3.1 MLP Based Deep Neural Network Regression

The first and foremost choice of a regression model is a multilayer perceptron based deep neural network due to its superior capabilities. The MLP network outperforms other networks in terms of localization accuracy due to its tuning via backpropagation. This helps to reduce over-fitting problems. Several features of MLP can be updated per the features of the data. MLP is also known for its excellence in pattern recognition in complex and deep studies related to classification and regression problems [11].

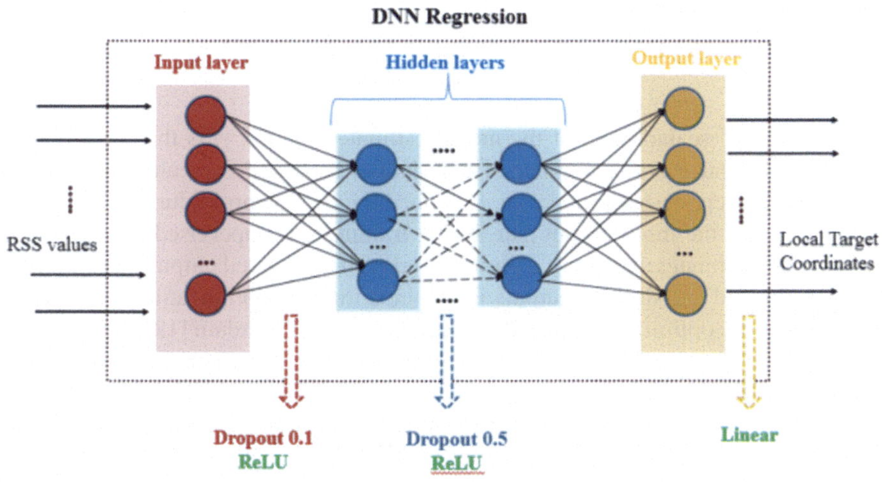

Fig. 2. MLP based DNNR architecture

The DNN utilizes a multilayer perceptron (MLP) network with an input, output, and multiple hidden layers. The proposed model is denoted as LD-DNNR, being a regression model. The proposed MLP based deep neural network architecture is shown in Fig. 2. For the data obtained from a real-time environment, every link with RSS values alone may not contribute to target location estimation. The RSS values in each link get affected due to many other objects present and multipath effects other than the target position. So, it is ideal to take the RSS values of all the links for the link distance based DFL technique, and all the links will estimate the target location from the transmitter or local target coordinate d_1. Thus LD-DNNR based DFL technique is an ideal choice in such scenarios and is validated using experimental data. The MLP is normally represented as fully connected layers. It is a class of feed forward networks in which the mapping function is defined as:

$$y = \varphi\left(w^T x + b\right) \tag{1}$$

The activation function used in hidden layers is ReLU and Linear in the output layer. A ReLU function is defined as:

$$s(x) = \begin{cases} 0, x \leq 0 \\ x, x > 0 \end{cases} \tag{2}$$

The derivative of ReLU when $x < 0$ is zero when $x > 0$ is unity, and at $x = 0$, the derivative is undefined. The range and order of continuity of the ReLU function are from $[0, \infty)$ and have a C^0, respectively. Apart from the vanishing gradient problem, the ReLU implementation is very simple, unlike other activation functions like tanh and sigmoid, which require an exponential function. Being a regression problem, the activation function in the output layer is Linear. The backpropagation algorithm appropriately updates the hidden layer weights via loss function L in (3). It is calculated by taking the average of the sum of the squared difference between the estimated target location and the actual target location.

$$L(\hat{y_j}, y) = \frac{1}{n} \sum_{j=1}^{n} (\hat{y_j} - y_j)^2 \tag{3}$$

where $\hat{y_j}$ is the jth scalar value of the model estimated output, y_j is the corresponding target location, and n is the number of test dataset samples. Typically, the objective of training the deep learning models is to reduce the loss incurred during the training. A model is given the input-output pair during training in a supervised learning mode. When a model estimates an output, it is compared with the desired output. The difference between estimated and desired outputs is obtained through optimization algorithms to reduce errors. The optimization algorithm used in our work is Adam [12]. Adam utilizes momentum and learning rates, which are adaptive to faster convergence even with noisy, incomplete, or sparse datasets.

For input layers, the dropout is 0.1; for hidden layers, the dropout is 0.5 for better convergence. For hidden layers, the dropout rate of 0.5 is an ideal choice for large networks. In the case of an input layer, it is ideally kept at 0.2 or lower. The reason for this limit is the training gets affected by increased dropout in the input layer. This is the only regularization parameter used in this work. Some links are affected by a target position, and others are unaffected within the given network. If the link distance is d and the target distance from the transmitter of that link is d_1, then d_1 can be calculated from the link distance model proposed by the researchers in (4), as explained in [8].

$$P_r = \frac{P_t L_1 k^2}{(d - d_1)^2 {d_1}^2} \tag{4}$$

$$\text{where } k = G_t G_r \left(\frac{\lambda}{4\pi}\right)^2 \tag{5}$$

Here, G_t is the gain of the transmitter, G_r is the receiver's gain, and λ is the transmitted signal wavelength expressed in meters. Since experimental data from a real-time environment is used in this work, L_1 need not be separately modeled. From the dataset, RSS values corresponding to each location class are available. The location class is mapped to the corresponding location coordinate. Once all the link distances and target distances from each link are calculated, the links are identified as target-affected and unaffected links as per (6).

$$\hat{y_j} = \begin{cases} d_1, d_1 < d \\ 0, d_1 > d \end{cases} \tag{6}$$

The affected and unaffected links are identified in real time scenarios by checking the target position between the transmitter and the receiver for each link. Thus, if the calculated target distance from both the transmitter and receiver of a given link is less than the link distance, the target's position is assumed to be between the transmitter and receiver. That link is considered to be affected. For affected links, the distance of the target from the transmitter d_1 is calculated, and for unaffected links distance of the target from the transmitter is marked as zero.

Thus, in a given setup, RSS values from every link are given as input to the single DNNR model, and the same number of outputs is also obtained for a target position as every link is estimating. The mean of these values is the estimated target distance/local target coordinate. The difference between the actual and the estimated target distance by each link is localization error, and the mean gives mean localization error given in (7):

$$Mean\,Localization\,error = \frac{1}{n}\sum\nolimits_{j=1}^{n}\sqrt{(predicted\,(d_1)_j - actual\,value(d_1)_j)^2} \tag{7}$$

4 Results and Discussion

This section assesses the experimental setup used for target localization, the simulation parameters used, and the performance analysis of the proposed LD-DNNR model using relevant performance metrics.

4.1 Experimental Setup

The experimental data for the proposed LD-DNNR based DFL technique was adopted from the database of SPAN Lab, University of Utah [9]. The parameters like transmitted power and gain are calculated for Texas Instruments CC2531 dongles, which operate in the 2.4 GHz range. The data is obtained from the Set 2 dataset comprising eight channels with 19 test configurations denoting different environments. The simulation based experiments are done in Python. The hyperparameters used by the proposed LD-DNNR models in real-time setup are tabulated in Table 1. For iterative optimizer, loss function and regularization parameters are selected based on the performance of proposed DNN regression models using the trial and error method. And so the choice of dropout rate, hidden layers and activation function made with respect to the performance of the proposed LD-DNNR model.

Table 1. Hyperparameters of LD-DNNR

Hyperparameters	LD-DNNR
Dropout/Batch normalization	0.5/nil
Learning Rate	0.001
Learning Optimizer	Adam
Activation Function (hidden layers/Output Layer)	ReLU/Linear
No. of hidden layers No. of neurons in the output layer	4 Same as the number of inputs

After preprocessing of the data, the location based RSS value matrix is generated. This preprocessed data is handled for missing values using the KNN imputation technique. Most of the missing samples from all 8 channels were recovered with K-NN imputation. The location classes are mapped to respective coordinates. The training samples from all eight channel is 11412 in number, and the testing samples are 123402. The affected and unaffected links are identified for each target location, and the corresponding outputs are marked as d1 and zero, respectively. This input-output pair is given to the proposed LD-DNNR model for learning their dependency in the training phase. Thus, all the links in the network estimate the output. The mean of this target distance for a given target position is the estimated target distance from the transmitter or local target coordinate. The estimated and actual target distance difference gives the mean localization error.

4.2 Performance Metrics

Following performance metrics are used to assess the LD based regression model with real data, which are the popular metrics used by the researchers [13].

$$Mean\,Localization\,error = \frac{1}{n}\sum\nolimits_{j=1}^{n}\sqrt{\left(\widehat{y_j} - y_j\right)^2} \tag{8}$$

$$StandardDeviation = \sqrt{\frac{\sum_{i=j}^{n}\left(y_j - \hat{y}\right)^2}{n}} \tag{9}$$

4.3 Performance Comparison of the Proposed LD-DNNR Based DFL Technique with Benchmark Models

The performance of the proposed LD-DNNR model is compared with benchmark models of different literature to assess the effectiveness of the proposed model in a changing indoor environment. This is assessed by mean localization error and standard deviation. The mean localization error helps to understand how much mean target distance error is developed using the proposed LD-DNNR model. Similarly, the standard deviation metric is used to assess the extent of deviation from the mean value of target distance estimation by each link.

Overall Performance of the Proposed LD-DNNR Model

The overall performance of the proposed LD-DNNR model is mainly assessed by the mean of localization error between the actual and estimated target distance from the transmitter of each link and the standard deviation of target distance calculated by each link and mean target distance.

Table 2. Overall performance of the Proposed LD-DNNR based model

Classifier	Mean Localization Error (m)	Standard Deviation (m)	Maximum Error (m)	Minimum Error (m)
Proposed LD-DNNR	0.21	0.61	3.70	0.10
PCA-SVM [14]	2.8	1.08	5.74	0.34
RF [9]	1.25	0.83	4.14	0.08
AugRF [13]	0.89	0.58	3.6	0.06
CAE [15]	1.17	0.79	3.87	0.06
CNN [16]	0.23	–	–	–

The maximum and minimum error obtained by the proposed LD-DNNR model is also an important parameter metric in assessing the range of errors. It is evident from Table 2 that the proposed LD-DNNR based DFL technique is performing better than the RF model used in the benchmark literature [9] with a mean localization error of 1.25m and a standard deviation of 0.83m using the same dataset. The performance of the PCA-SVM model proposed in the literature [14] is also underperforming, with a mean error of 2.8m and a standard deviation of 1.08m.

Apart from conventional machine learning architectures, the proposed LD-DNNR model is compared with a deep convolutional autoencoder in literature [15], which exhibited a mean localization error of 1.17m, far higher than the proposed LD-DNNR model. The mean localization error using the deep learning based model Augmented RF (AugRF) is 0.89m, which is much higher than the proposed LD-DNNR model [13]. Standard deviation measured using the proposed LD-DNNR is found to excel/comparable compared to the same benchmark models. Recent work in literature [16] also proposed a CNN model that could estimate the target location with a mean localization error of 0.23m.

Complexity of the Proposed LD-DNNR Model

The complexity of the proposed LD-DNNR model is analyzed to understand its effectiveness in a changing indoor or real-time environment. In certain critical real-time application scenarios, the swiftness of the machine learning model is important. The mean execution time taken for estimating one target location is estimated for the proposed LD-DNNR model, and comparisons with benchmark models are tabulated in Table 3. Table 3 shows that the proposed LD-DNNR model outperforms the benchmark models AugRF [13] and CAE [15]. It exhibited a comparable performance with other benchmark models.

Table 3. Mean execution time Proposed LD-DNNR based Model

Classifier	Mean execution time (ms)
Proposed LD-DNNR	0.0567
PCA-SVM [14]	0.020
RF [9]	0.017
AugRF [13]	1.065
CAE [15]	0.935
CNN [16]	0.0291

4.4 Performance of Proposed LD-DNNR Based DFL Technique in Changing Indoor Environment

Regarding the experimental data obtained from a real-time environment, the localization setup is not confined to a room but a house floor with many obstructions. Thus, considering the realistic scenario and dataset quality, it is ideal to go with estimations made by the proposed LD-DNNR model.

Channel-wise Mean Localization Error

The effectiveness of the proposed LD-DNNR model is assessed by mean localization error with different channel-wise data. As mentioned in experimental setup, eight channel data is used in local target coordinate estimation to understand how effectively the proposed LD-DNNR model handles different channel data and associated noise. It is clear from Table 4 that the mean localization error is comparable with all the eight channel data, which shows the supreme performance of the proposed LD-DNNR model with data with different levels of noise in a real-time environment. The effectiveness of using proposed LD-DNNR model with different channel data is highlighted in this section.

Table 4. Channel-wise Mean Localization Error for the Proposed LD-DNNR model

Channels	Mean Localization error (m)
12	21.29
14	21.33
16	21.30
18	21.33
20	21.19
22	21.40
24	21.24
26	21.27

Proposed LD-DNNR Model Performance with Different Hidden Layers. The performance of the proposed LD-DNNR model is analyzed with several hidden layers to understand the suitability of hidden layers in the proposed LD-DNNR model. The performance comparison is depicted in Fig. 3. This is assessed by understanding its impact on mean localization error (MLE), standard deviation (SD), and maximum error (ME). Figure 3 shows that the lowest mean localization error of 16cm is obtained with two hidden layers.

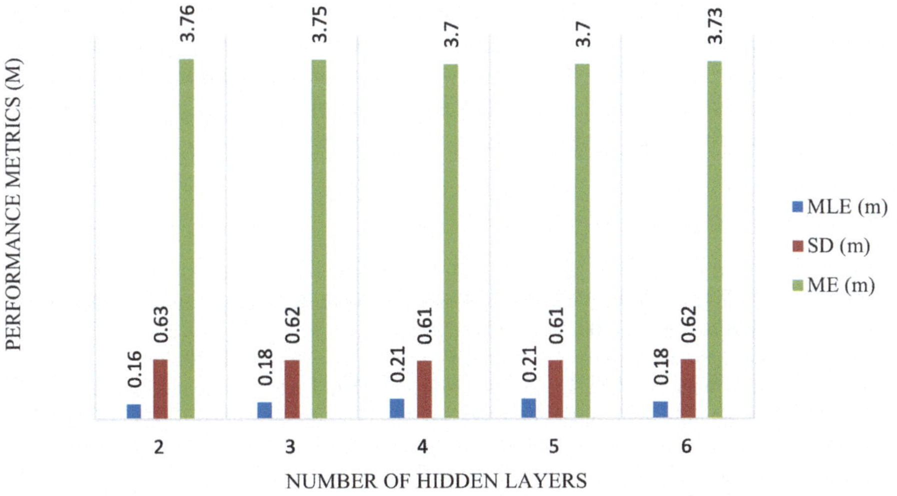

Fig. 3. Proposed LD-DNNR Model with Different Hidden Layers

However, the standard deviation and the maximum error obtained are comparatively higher in two hidden layers when compared to the higher number of hidden layers. With the proposed LD-DNNR based DFL technique, local target coordinate estimation is made by taking the mean of estimations made by all the links in the network. Thus, the standard deviation is a decisive parameter for choosing the number of hidden layers for the proposed LD-DNNR model apart from accuracy. The lowest standard deviation is attained with four hidden layers. Also, it is observed that the variation in mean localization error across different layers is less than 4cm in the proposed LD-DNNR model.

5 Conclusion

A new regression based DFL technique using a deep learning model is proposed in this study, which can accurately localize the target in the monitoring area by implementing the concept of link distance based device free localization regression models with a single model. The proposed LD-DNNR framework could achieve the centimeter grade average localization error, which performs better than the state of the art framework. This work introduces a single model to reduce the memory footprint associated with multiple models. The proposed LD-DNNR model in target location estimation utilized multi-channel real data in which each link estimates the target distance from the transmitter and respective localization error. The average localization error of all the links is the mean localization error. The proposed LD-DNNR model with experimental data outperforms the benchmark models with the same and similar dataset. The mean localization error of 21 cm is obtained, thus achieving the centimeter-level accuracy. The effectiveness of the proposed LD based DNNR model is assessed by analyzing the same with different channel-wise data, and it was found that the performance of the LD-DNNR model is comparable for different channel data. In this work, all three evaluation metrics, namely mean localization error, standard deviation, and maximum localization error, are utilized to assess the choice of hidden layers, unlike the many literature that stick only to the mean localization error metric. The lower value of mean localization error concerning benchmark models emphasizes the effectiveness and uniqueness of using link distance based device free localization techniques in changing environments.

References

1. Hemalatha, T., Ramesh, M.V., Rangan, V.P.: Effective and accelerated forewarning of landslides using wireless sensor networks and machine learning. IEEE Sens. J. **19**, 9964–9975 (2019). https://doi.org/10.1109/JSEN.2019.2928358
2. Vivek Sidhaarthan, S., Mukul, A., Ragul, P., Gokul Krishna, R., Bharathi, D.: Offline 3D indoor navigation using RSSI. In: Lecture Notes on Data Engineering and Communications Technologies, pp. 831–846. Springer Singapore (2021). https://doi.org/10.1007/978-981-15-9509-7_67
3. Nessa, A., Adhikari, B., Hussain, F., Fernando, X.N.: A survey of machine learning for indoor positioning. IEEE Access **8**, 214945–214965 (2020). https://doi.org/10.1109/ACCESS.2020.3039271
4. Shit, R.C., et al.: Ubiquitous localization (UbiLoc): a survey and taxonomy on device free localization for smart world. IEEE Commun. Surv. Tutorials. **21**, 3532–3564 (2019). https://doi.org/10.1109/COMST.2019.2915923
5. Altaf Khattak, S.B., Fawad, Nasralla, M.M., Esmail, M.A., Mostafa, H., Jia, M.: WLAN RSS-based fingerprinting for indoor localization: a machine learning inspired bag-of-features approach. Sensors **22**, 1–16 (2022). https://doi.org/10.3390/s22145236

6. Yang, Y., Zheng, K., Wu, C., Yang, Y.: Improving the classification effectiveness of intrusion detection by using improved conditional variational autoencoder and deep neural network. Sens. (Switz.) **19** (2019). https://doi.org/10.3390/s19112528
7. Félix, G., Siller, M., Álvarez, E.N.: A fingerprinting indoor localization algorithm based deep learning. In: Internatinaol Conference Ubiquitous Future Networks, ICUFN. 2016-Augus, pp.1006–1011(2016). https://doi.org/10.1109/ICUFN.2016.7536949
8. Anusha, K.S., Ramanathan, R., Jayakumar, M.: Link distance-support vector regression (LD-SVR) based device free localization technique in indoor environment. Eng. Sci. Technol. an Int. J. **23**, 483–493 (2020). https://doi.org/10.1016/j.jestch.2019.09.004
9. Mager, B., Lundrigan, P., Patwari, N.: Fingerprint-based device-free localization performance in changing environments. IEEE J. Sel. Areas Commun. **33**, 2429–2438 (2015). https://doi.org/10.1109/JSAC.2015.2430515
10. Anusha, K.S., Ramanathan, R., Jayakumar, M.: Impact of K-NN imputation technique on performance of deep learning based DFL Algorithm. In: 2021 International Conference Wireless Communication Signal Process. Networking, WiSPNET 2021, pp. 153–157 (2021). https://doi.org/10.1109/WiSPNET51692.2021.9419414
11. Adege, A.B., Lin, H.P., Tarekegn, G.B., Jeng, S.S.: Applying deep neural network (DNN) for robust indoor localization in multi-building environment. Appl. Sci. **8**, 1–14 (2018). https://doi.org/10.3390/app8071062
12. Zaheer, R., Shaziya, H.: A study of the optimization algorithms in deep learning. In: Proceedings 3rd International Conference Inventive System Control ICISC 2019, pp. 536–539 (2019). https://doi.org/10.1109/ICISC44355.2019.9036442
13. Ma, Y., Ning, W., Wang, B., Liang, X.: A data augmentation-based method for robust device-free localization in changing environments of passive radio frequency identification system. IEEE Trans. Instrum. Meas. **70**, 1–13 (2021). https://doi.org/10.1109/TIM.2021.3065426
14. Zhou, R., Lu, X., Zhao, P., Chen, J.: Device-free presence detection and localization with SVM and CSI fingerprinting. IEEE Sens. J. **17**, 7990–7999 (2017). https://doi.org/10.1109/JSEN.2017.2762428
15. Zhao, L., Huang, H., Li, X., Ding, S., Zhao, H., Han, Z.: An accurate and robust approach of device-free localization with convolutional autoencoder. IEEE Internet Things J. **6**, 5825–5840 (2019). https://doi.org/10.1109/JIOT.2019.2907580
16. Lee, J., Park, K., Kim, Y.: Deep learning-based device-free localization scheme for simultaneous estimation of indoor location and posture using FMCW radars. Sensors. **22**, 4447 (2022). https://doi.org/10.3390/s22124447

An Ensemble-Based Approach for Cyber Attack Detection in Financial Systems

Dhanisht Kumar, Harshit Srivastava, Kartik S. Bhamare, and Malay Kumar(✉)

Department of Computer Science and Engineering, Indian Institute of Information Technology, Dharwad, Karnataka, India
malay.kumar@iiitdwd.ac.in

Abstract. Digital commerce is the process of exchanging money or assets using electronic or digital channels such as the internet or mobile devices. Instead of physical cash or checks, individuals or businesses can use digital platforms to transfer money, make payments or conduct financial transactions securely and efficiently. As more and more people conduct financial transactions online, the risk of cyberattacks targeting financial institutions will increase. This article presents an integrated approach to detecting cyber-attacks in the financial sector. The proposed method combines learning models including decision trees, random forests, and gradient boosting to increase the accuracy and robustness of the model.

Keywords: Cyber Attack · Digital commerce · Financial Systems · Fraud · Machine Learning

1 Introduction

In the past, cash was often used to exchange goods and services due to the lack of network resources and smart device support [1]. However, with the advancement in software, networks, and hardware (such as 5G networks) and the increasing use of mobile phones, we now have the necessary tools to help the businesses to digitize. As a result, business technology has evolved into a digital model [2]. While the transition to digital commerce has brought benefits such as convenience and speed, it has also created new opportunities for cyber attackers to exploit security vulnerabilities in financial systems [3]. Therefore, there is an urgent need to develop effective methods to detect and prevent cyber-attacks on financial markets. As the technology advances, more and more payment methods are added to financial transactions, this article proposes a novel algorithm for detecting cyber-attacks in all types of transactions such as ***Unified Payment Interface (UPI)***, ***Online Banking***, ***Mobile Wallets***, ***Credit and Debit Cards***, ***Aadhar-Enabled Payment System*** (AEPS), ***USSD*** etc [1, 2]. The experiment has been performed on credit card data only due to unavailability of fraudulent transaction data on other payment systems. In recent years, machine learning techniques have shown promise results in detecting and preventing cyberattacks. Hybrid methods combining multiple models to improve performance are appreciated for their ability to solve problems such as overfitting and bias [3].

D. Pastor-Escuredo et al. (Eds.): ICDLAIR 2023, LNNS 1001, pp. 352–363, 2024.
https://doi.org/10.1007/978-3-031-60935-0_32

2 Background Study

2.1 Digital Transaction and Payment Methods

In 1994, Stanford Federal Credit Union introduced its online banking service [3] and other banks did the same, finally allowing their customers to log in to online payments in 1995. In the year 1996 the Credit and Investment Corporation of India (ICICI Bank Ltd.) become the first bank in India to offer online banking services to its customers [3]. However, it wasn't the year 2008 that the National Payments Corporation of India (NPCI) established a better payment system, paving the way for many new products and services in India for digital payment systems. More importantly, this period also saw the 2010 launch of Aadhaar, India's unique digital number system, which has become an integral part of the country's digital ecosystem [4]. Now there are various payments are in use for digital transaction **such as** ***Banking Cards (Credit/Debit), Unstructured Supplementary Service Data (USSD), Aadhar Enabled Payment System (AEPS), Unified Payment Interface (UPI), Mobile Wallets, Internet Banking, Mobile Banking***.

2.2 Online Transaction System

The general framework for online transaction systems is presented in Fig. 1. The process begins with a user accessing a payment gateway to initiate a transaction. The payment processor transmits the transaction details to the payment processing network (PPN) for authorization. Upon receiving authorization, the issuing bank assesses the availability of funds and approves or declines the transaction accordingly. Subsequently, the transaction flows through the electronic or mobile network to the processor, and an approval code is transmitted to the sales device at the merchant's location. The issuing bank then transfers the funds to the processing company as reimbursement for the completed purchase. Finally, the processor deposits the merchant's funds into their bank account.

2.3 Digital Payment Fraud

According to the Federal Trade Commission, the total economic loss in the United States for 2022 is $8.8 billion [5]. India lacks reporting from any authentic financial institution. However, the media reports many cases of financial fraud. According to the Economic Times report, a total of 1.4 corer fraud cases were reported in the 2022 fiscal year, including credit card and UPI fraud. However, in our research, we did not find any publicly available report or data on credit card or UPI fraud specific from Indian financial institutions. There are various types are fraud reported in the literature such as Identity theft, Skimming, Phishing, Account Takeover, Man-in-the-Middle Attacks, Denial of Service (DoS) Attacks, Log-Based Attack.

Currently, many machine learning (ML) algorithms are used to identify vulnerabilities in financial systems. Figure 1 illustrates the latest developments in this field, showing machine learning-based methods for identifying vulnerabilities. The tree structure (Fig. 1) shows the various algorithms used and their relative accuracy levels and provides an overview of the current state of vision security. The classification of the research attack is shown in Fig. 1. Interested reader may refer to the article in [4–14].

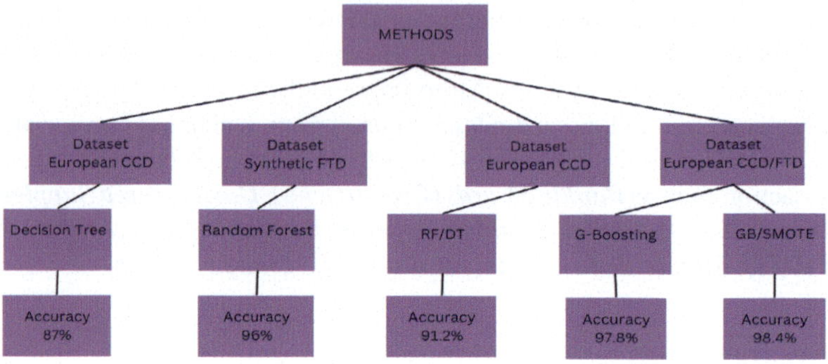

Fig. 1. Taxonomy of Attack Detection Method

3 Methodology

The general framework in Fig 2, representing the flow of transaction and the interaction between the user, merchant, payment gateway and the bank.

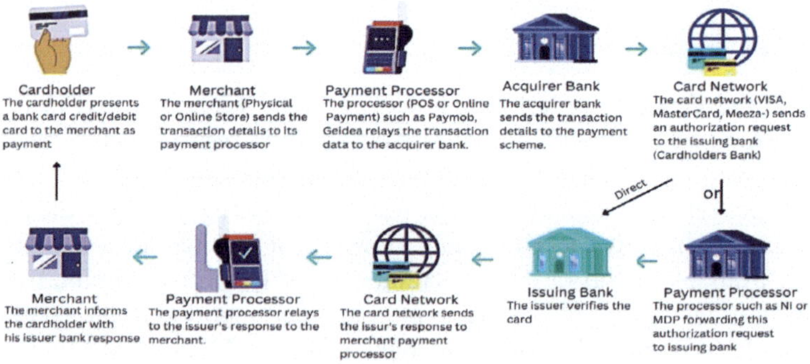

Fig. 2. General Framework for Transaction Systems

3.1 Operational Flow

The work includes data collection, preprocessing and modeling, which includes collecting and pre-processing relevant data and transforming the data into research-appropriate data. The entire process flow is shown in Fig. 3 below. **Dataset:** Pay-Sim synthetic dataset has been included for study, features including transaction type, amount, customer names, balances, cash-in, cash-out, and attack indicator class [15] (Fig. 4).

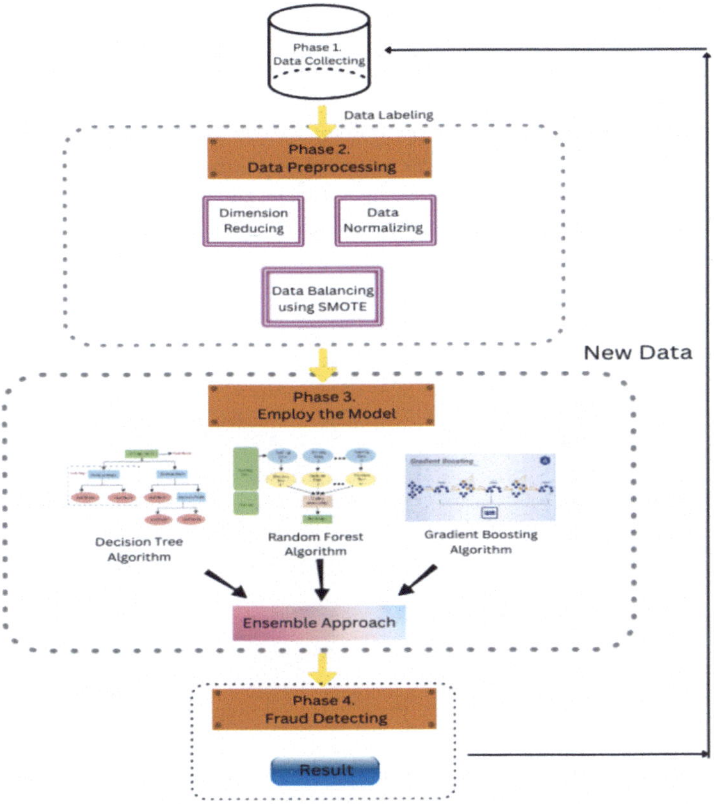

Fig. 3. Process flow diagram

Algorithm 1: SMOTE

Input: *Imbalanced data*
- Minority class examples X_min
- Number of synthetic examples to generate N
- Number of nearest neighbors k
Intermediate result:
- New synthetic examples X_syn
Initialize *X_syn as an empty set*
For *each x in X_min:*
| *Find the k nearest neighbors of x from X_min (excluding x itself)*
| ***For*** *i from 1 to N:*
| | *Choose one of the k nearest neighbors randomly and call it x'*
| | *Generate a new synthetic example as follows:*
| | $X\_new = x + rand(0,1) * (x' - x)$
| | ***Add*** *X_new to X_syn*
Return *X_syn*
Output: *Balanced data*

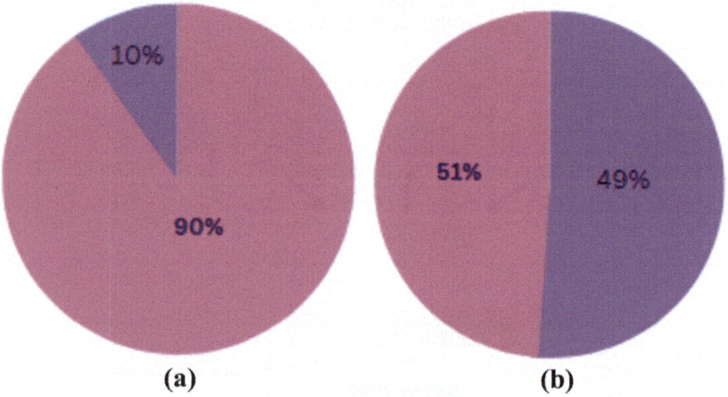

Fig. 4. (a) Unbalance Dataset (b) balance Dataset

Algorithm 2: Decision Tree

Input: *Training dataset D*
Features F
Maximum depth max_depth
Output: *Decision tree model*

```
function build_decision_tree(D, F, depth):
| if depth > max_depth or all samples in D belong to the same class:
| | return LeafNode(majority_class(D))
| else:
| | Select the best feature F_best to split D based on some criterion
| | Split D into subsets D1, D2, ..., Dk based on the values of F_best
| | node = InternalNode(feature=F_best)
| | for each subset Di:
| | | if Di is not empty:
| | | | child_node = build_decision_tree(Di, F, depth + 1)
| | | | node.add_child(child_node, value=Fi)
| | return node
root_node = build_decision_tree(D, F, depth=1)
return DecisionTree(model=root_node)
```

Algorithm 3: Random Forest model

```
Input: Training dataset D, Number of trees num_trees, Maximum depth max_depth,
Number of features k_features
Output: Random Forest model
function calculate_gini(D):
| Calculate class distribution frequencies
| Gini = 1 − Σ (p_i)^2 for each class i
| return Gini
function calculate_split_gini(D, F):
| Split D into subsets D1 and D2 by feature F
| Split_gini = (|D1| / |D|) * calculate_gini(D1) + (|D2| / |D|) *
calculate_gini(D2)
| return Split_gini
function select_best_feature(D, features):
| best_feature, best_split_gini = None, ∞
| for each feature F in features:
| | split_gini = calculate_split_gini(D, F)
| | if split_gini < best_split_gini:
| | | best_split_gini, best_feature = split_gini, F
| return best_feature
function build_tree(D, features, depth):
| if depth > max_depth or samples in D are same class:
|    return LeafNode(majority_class(D))
| else:
| | best_feature = select_best_feature(D, features)
| | node = InternalNode(feature=best_feature)
| | for each subset Di:
| | | if Di is not empty:
| | | | child_node = build_tree(Di, features, depth + 1)
| | | | node.add_child(child_node, value=Fi)
| return node
function build_forest(D, num_trees, max_depth, k_features):
| forest = []
| for i = 1 to num_trees:
| | D_subset = Randomly sample subset of D
| | tree = build_tree(D_subset, k_features, max_depth)
| | forest.append(tree)
| return RandomForest(model=forest)
training_data = LoadTrainingData()
num_trees = 10, max_depth = 5, k_features = √(number_of_features)
random_forest_model = build_forest(training_data, num_trees, max_depth, k_fea-
tures)
return random_forest_model
```

Algorithm 4: Gradient Boosting

Input: *Training dataset D, Number of boosting iterations num_iterations, Learning rate eta*
Output: *Gradient Boosting model*
function *initialize_predictions(D):*
| *Initialize predictions as zeros: F_0(x) = 0*
| ***return*** *predictions*
function *calculate_residuals(y_true, predictions):*
| Calculate residuals: $r\_i = y\_i - F\_\{i-1\}(x\_i)$
| ***return*** *residuals*
function *build_weak_learner(D, residuals):*
| *Fit a weak learner (e.g., decision tree) to minimize residuals*
| ***return*** *fitted_weak_learner*
function *update_predictions(predictions, fitted_weak_learner, eta):*
| Update predictions: $F\_i(x) = F\_\{i-1\}(x) + eta * h\_i(x)$
| ***return*** *updated_predictions*
function *build_gradient_boosting(D, num_iterations, eta):*
| *predictions = initialize_predictions(D)*
| ***for*** *i = 1 to num_iterations:*
| | *residuals = calculate_residuals(y_true, predictions)*
| | *fitted_weak_learner = build_weak_learner(D, residuals)*
| | *updated_predictions = update_predictions(predictions, fitted_weak_learner, eta)*
| ***return*** *GradientBoosting(model=updated_predictions)*
training_data *= LoadTrainingData()*
num_iterations = 100, learning_rate = 0.1
gradient_boosting_model = build_gradient_boosting(training_data, num_iterations, learning_rate)
return *gradient_boosting_model*

Algorithm 5: ENSEMBLE Algorithm

Input:
- Training data X_train and y_train
- Number of decision trees T
- Maximum tree depth d
- Number of features to consider at each split f
- Learning rate for gradient boosting eta
- Random seed for reproducibility
Output of Decision Tree:
- Ensemble model consisting of T decision trees
Initialize *an empty list of decision trees*
Pick *from 1 to T:*
| *Randomly select a subset of features of size f*
| *Fit a decision tree of maximum depth d on X_train and y_train, using the selected subset of features*
| *Add the decision tree to the list of trees*
Define *a function predict(x) that takes an input feature vector x and returns the ensemble prediction:*
| ***Initialize*** *an empty list of predictions preds*
| ***For*** *each decision tree in the list of trees:*
| | ***Compute*** *the predicted value for x using the decision tree*
| | *Add the predicted value to the list of predictions preds*
| ***Compute*** *the ensemble prediction by taking the weighted average of the predictions:*
| *y_pred = sum(preds) / T*
Define a function *train(X, y) that takes a training set X and corresponding labels y, and trains the ensemble model:*
| *Fit a gradient boosting model on X_train and y_train, with learning rate eta and the ensemble prediction function as the base learner*
| ***Return*** *the trained ensemble model*

The proposed ensemble method integrates decision trees, random forests, and gradient boosting, from different implemented phases. Decision trees establish a hierarchical structure for making decisions based on input features, while random forests aggregate multiple trees to improve accuracy. Gradient boosting iteratively adjusts the model by minimizing residual errors from previous iterations. This combined approach enables the ensemble model to effectively detect fraudulent activity in financial transactions. One of the pivotal components of this ensemble method is gradient boosting, a technique that fine-tunes the model iteratively by minimizing residual errors from previous iterations. In the context of fraud detection, gradient boosting starts with a base model and then sequentially adds subsequent models, each focusing on correcting the errors made by the previous ones. This is achieved by training each new model on the residuals (the differences between predicted and actual values) of the previous model. By gradually improving predictions with each iteration, gradient boosting effectively enhances the model accuracy.

4 Result and Analysis

Evaluating a model's performance based solely on accuracy may leave some information about model effectiveness. Metrics such as ROC score, TP/FP rates, and F1-score should also be considered for better interpretability. Assessing multiple performance factors enables a comprehensive evaluation of the model's ability to accurately classify target outcomes. The ROC score serves as an evaluation metric for determining a model's effectiveness in distinguishing between positive and negative classes. A higher ROC score indicates superior performance of the model in accurately classifying positive and negative instances. In Table 2, the ROC scores for both the separate and ensemble approaches are summarized, providing an overview of their respective performance in terms of classification accuracy.

4.1 Formulation

Out-of-Bag Error Formula (OOB error): Eq. (3) defines the out-of-bag error for a predictive model. It calculates the average squared difference between the actual target values (y) and the model's predictions (f(x)) for the instances not included in the bootstrap sample. This error metric helps assess how well the model generalizes to new, unseen data (Table 1 and Fig. 5).

Table 1. Notation & Parameters

Notions	Identification
$f(x)$	Predicted label by corresponding Tree
y	Actual label
n	Sample size

- **Precision:** Precision of the machine learning model describes how sensible our model is against false positive case.

$$Precision = TP/(TP + FP) \tag{1}$$

- **Recall:** Recall of the machine learning model describes how sensible our model is against false negative case. As there is always tradeoff between precision and recall, In this system we want to maximize the recall value as our model aims to avoid any instances where it incorrectly predicts a positive case as negative (false negatives). The focus is on ensuring that all actual positive cases are accurately identified and that no positive cases are overlooked. So consequently we have to minimize the FN value.

$$Recall = TP/(TP + FN) \tag{2}$$

$$F1\,Score = 2 * (Precision * Recall)/(Precision + Recall) \quad (3)$$

$$Accuracy = (TP + TN)/(TP + TN + FP + FN) \quad (4)$$

$$Weight\ of\ ModelI = (1 - error\ rate\ of\ model\ i)/(Sum\ of\ weight\ of\ all\ models) \quad (5)$$

Here, the error rate of model i is calculated on the validation set, and the weights of all models are summed up to normalize the weights to add up to one. The higher the performance of a model, the higher its weight in the ensemble.

Table 2. Score Matrix

Method	Precision	F1 Score	Sensitivity	Accuracy
Decision Tree	97.22	97.90	98.59	98.16
Random Forest	96.77	97.56	98.36	97.90
Gradient Boosting	98.04	98.52	99.01	98.79
Ensemble	99.24	99.24	99.62	99.65

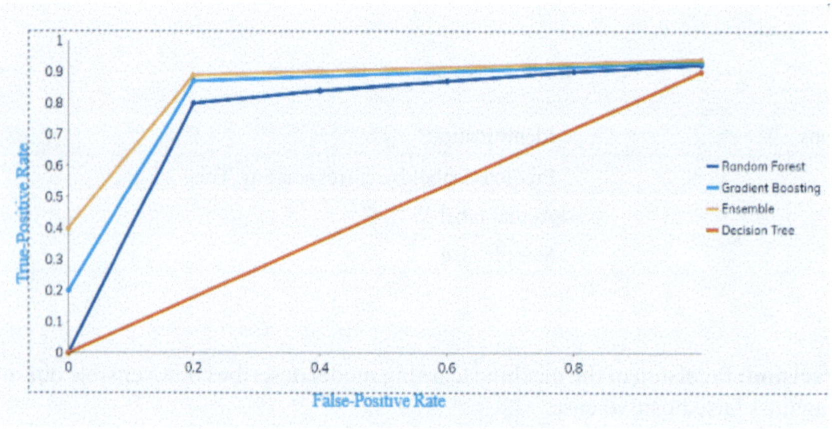

Fig. 5. True-Positive ROC Curve

5 Conclusion

The paper proposes an ensemble-based approach using machine learning models to detect cyber-attacks in financial transactions, improving accuracy and robustness by combining diverse models. The research offers a comprehensive overview of digital transactions and the need for effective detection and prevention of cyber-attacks, providing a general framework for online transaction systems. The proposed approach offers a reliable and effective solution to detect cyber-attacks on financial systems, emphasizing the importance of developing effective methods in the current digital era.

References

1. Digital Payment Methods. Ministry of Electronics and Information Technology, Government of India. http://cashlessindia.gov.in/digital_payment_methods.html. Accessed on 13 Sep 2023
2. Federal Trade Commission. New FTC Data Show Consumers Reported Losing Nearly $88 Billion to Scams in 2022. Press release (2023). https://www.ftc.gov/news-events/news/press-releases/2023/02/new-ftc-data-show-consumers-reported-losing-nearly-88-billion-scams-2022
3. Alarfaj, F.K., Malik, I., Khan, H.U., Almusallam, N., Ramzan, M., Ahmed, M.: Credit card fraud detection using state-of-the-art machine learning and deep learning algorithms. IEEE Access **9**, 50090–50100 (2021). https://doi.org/10.1109/ACCESS.2021.3061626.(GB)
4. Zhang, X., Han, Y., Xu, W., Wang, Q.: HOBA: a novel feature engineering methodology for credit card fraud detection with a deep learning architecture. Knowl.-Based Syst. **161**, 145–154 (2019). https://doi.org/10.1016/j.knosys.2018.09.017
5. Taha, A.A., Malebary, S.J.: An intelligent approach to credit card fraud detection using an optimized light gradient boosting machine. IEEE Access **9**, 51989–51999 (2021). https://doi.org/10.1109/ACCESS.2021.3072879
6. Han, S., Zhu, K., Zhou, M., Cai, X.: Competition-driven multimodal multiobjective optimization and its application to feature selection for credit card fraud detection. IEEE Trans. Syst. Man Cybern.: Syst. **50**(9), 3369–3381 (2020). https://doi.org/10.1109/TSMC.2019.2926782
7. Zhang, X., Han, Y., Xu, W., Wang, Q.: HOBA: a novel feature engineering methodology for credit card fraud detection with a deep learning architecture. Expert Syst. Appl. **142**, 113068 (2020). https://doi.org/10.1016/j.eswa.2019.113068
8. Carcillo, F., Le Borgne, Y.A., Caelen, O., Kessaci, Y., Oblé, F., Bontempi, G.: Combining unsupervised and supervised learning in credit card fraud detection. Neurocomputing **275**, 237–245 (2018). https://doi.org/10.1016/j.neucom.2017.08.063
9. Cherif, A., Badhib, A., Ammar, H., Alshehri, S., Kalkatawi, M., Imine, A.: Credit card fraud detection in the era of disruptive technologies: A systematic review. J. Retail. Consum. Serv. **61**, 102570 (2021). https://doi.org/10.1016/j.jretconser.2021.102570
10. Forough, J., Momtazi, S.: Ensemble of deep sequential models for credit card fraud detection. Expert Syst. Appl. **132**, 67–80 (2019). https://doi.org/10.1016/j.eswa.2019.04.038
11. Fathy, M., Elhoseny, M., Alkinani, H.H., Hassanien, A.E.: A proposed model for fraud detection based on CatBoost and deep neural network. J. King Saud Univ. – Comput. Inf. Sci. **33**(2), 185–195 (2021)
12. Li, L., Li, S., Li, Y., Li, W.: An intelligent approach to financial fraud detection using an optimized light gradient boosting machine. J. Comput. Sci. **51**, 101331 (2021)
13. Shen, J., Chen, Y., He, X., Wang, L.: Deep representation learning with full center loss for fraud detection. IEEE Access **7**, 177342–177352 (2019)
14. Osegi, E.N., Jumbo, E.F.: Comparative analysis of credit card fraud detection in simulated annealing trained artificial neural network and hierarchical temporal memory. Heliyon **7**(4), e06822 (2021). https://doi.org/10.1016/j.heliyon.2021.e06822
15. Pay-Sim Dataset. https://www.kaggle.com/datasets/harinip/paysim-data-set-with-network-features. Accessed on September-2023

LoHCEO (Low Overhead and High Congestion Reduction in Epidemic Mode for Opportunistic Networks)

Tanvi Gautam[1,2(✉)] and Rahul Johari[3]

[1] Jaypee Institute of Information and Technology, Sector 62, Noida 201301, Uttar Pradesh, India
tanvi.gautam@mail.jiit.ac.in, tanvi.gautam@jiit.ac.in
[2] GGS Indraprastha University, Dwarka 110078, New Delhi, India
tanvigautam99@gmail.com
[3] USAR, GGS Indraprastha University, East Campus Surajmal Vihar 110092, New Delhi, India

Abstract. In an environment where nodes possess characteristics features such as: intermittent connectivity, dynamic network topology, mobility, store—carry—forward approach, high error rate et al. Opportunistic Networks (OppNets) plays a vital role. To ensure ~ 100 percent packet delivery ratio, Epidemic routing is usually followed by OppNet wherein a multiple copies of message packet is flooded to all neighboring nodes but it incurs high network resource consumption such as network link, bandwidth, buffer space, power, energy, bottleneck ratio et al. In this research work, we proposed a routing scheme Low Overhead and High Congestion reduction in Epidemic mode for Opportunistic Networks (LoHCEO) which is an enhanced version of SERPO (Selective Enhanced Routing protocol for OppNet). In SERPO, the packet is routed only to the shortest path selected from all available neighboring nodes using Dijkstra's algorithm and First – in – First – Out (FIFO) for packet queuing that causes high network resources consumption. To overcome such issue LoHCEO has been proposed by implementing WFQ packet queuing mechanism for packet forwarding among source and destination nodes. The simulation of proposed work has been done in ONE simulator on three network performance metrics namely packet delivery ratio, packet overhead ratio and average latency ratio respectively. Additionally, a comparison study also presented with the SERPO for showing the efficacy of proposed scheme. We also presented the scope of improvement and conclusion based on the simulation.

Keywords: Low Overhead and High Congestion Enhancement in Epidemic for OppNets (LoHCEO) · Opportunistic Networks (OppNets) · Selective Epidemic Routing Protocol (SERPO) · WFQ (Weighted Fair Queuing)

1 Introduction

A network that possesses characteristics such as variable packet size, intermittent connectivity, dynamic network topology, varying congestion and high latency are referred as a Delay Tolerant Network (DTN) [2]. The broad classification of network has been

D. Pastor-Escuredo et al. (Eds.): ICDLAIR 2023, LNNS 1001, pp. 364–373, 2024.
https://doi.org/10.1007/978-3-031-60935-0_33

done into: Wired Network and Wireless Network which are further categorized on the basis of infrastructure as: Infrastructure and Infra—structure less (also known as Adhoc Network) respectively [2]. Several sub classes of Adhoc Network that have evolved over period of time are; Opportunistic Networks (OppNets), Wireless Sensor Network (WSN), Vehicular Adhoc Network (VAN) [3]. OppNet is nothing but an improvised and derived class of DTN that follows as store—carry—forward [4] packet forwarding fashion and intermittent nodes. In DTN, a traditional approach used for making the routing mechanism feasible throughout the network is Epidemic. Epidemic is a type of flooding routing protocol [5] based on the principle of "packet replication" in which a multiple copy of packet has been generated and it would be broadcasted to all connected neighboring nodes until it encounters the destination node. Epidemic routing scheme ensures the message packet delivery ratio would be as high as possible but it suffers with the low performance in terms of high overhead ratio, high network resource consumption and high bottleneck probability ratio. During Epidemic routing in an OppNet factor's that it has been observed by many researchers [6–11] are as:

- **High bottleneck probability ratio and congestion in a network:** One of a main reason behind it, is that, during flooding multiple copies of a packet are generated and flooded to all neighboring nodes resulting in more network traffic causing high ratio of bottleneck, congestion and packet overhead respectively.
- **High network resource consumption:** Due to the flooding mechanism high network resource consumption (such as buffer space, energy, power and numerous constraints) are being required by a network which in turn increases high demand of more resource availability.
- **Store—carry—forward packet routing mechanism:** OppNet inherited a feature of store—carry—forward approach for packet routing to its destined node. Each node is required to store a message packet till it encounters a neighboring node which is a most suitable node for packet routing to its destination node.

The majority of OppNet routing protocols and their variations are [6–11] supported in improving the network performance metrics: message delivery ratio, message relaying (packet transferring) time and message access delay ratio that ensure the network resource optimization usage (buffer) and reduction of congestion. A promising approach had been proposed called "SERPO (Selective Epidemic routing scheme) [1]" which primarily works on the concept of message forwarding only on a selected routing path using Dijkstra's algorithm [12] and FIFO (First – In – First – Out) as a packet queuing mechanism implementing the concept of Epidemic. But SERPO causes the following mentioned issues:

a. Loss of message packet resulting in poor Packet Delivery Ratio (PDR).
b. High usage of available network resources which reduces the network performance.
c. High network congestion of message packets causing low overhead ratio.

The routing scheme proposed in this research work called Low overhead and High Congestion reduction in Epidemic mode for Opportunistic Networks (LoHCEO) is an enhanced/optimized routing scheme of SERPO. LoHCEO improves the issue of high congestion, low overhead by using the Weighted Fair Queuing (WFQ). LoHCEO not only improves the SERPO network performance metrics but also improvises the packet transmission in initial phase. For the sake of comfort and convenience in reading, rest of the paper is organized as: Sect. 2 presents the literature survey of paper that explains the work that has been proposed by researchers in the field of OppNet Epidemic routing protocol. Section 2.4 explains the experimental work that has been conducted on enhancement of SERPO by presenting its extended work LoHCEO in which route node cost metric improvement has been done through the implementation of WFQ packet queuing mechanism. Section 2.6 explains the observations and result analysis that has been presented for the three network performance metrics as: average packet delivery ratio, average latency ratio, and average packet overhead ratio, respectively for LoHCEO. Section 4 presents the future work and the conclusion section which presents the research areas on which the LoHCEO routing protocol performance can be improved further. Figure 1 outline the organization of the paper.

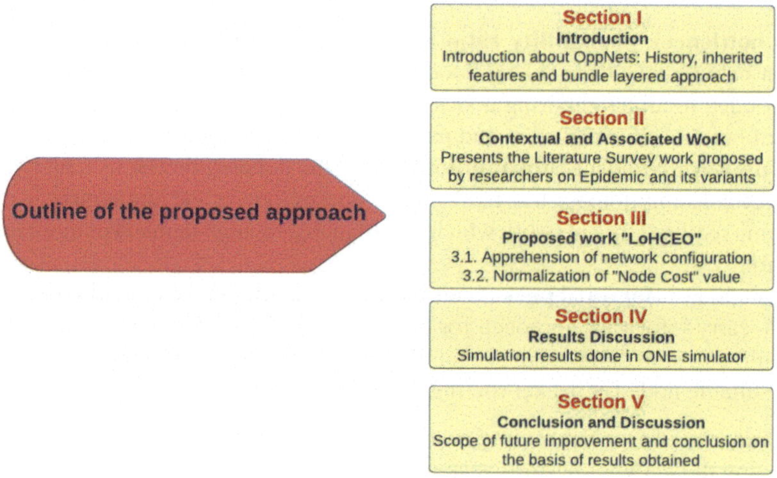

Fig. 1. Outline of the Proposed approach "LOHCEO"

2 Contextual and Associated Work

In this section, we present an outline of several variants of epidemic work that has been proposed by several researchers in OppNets: Epsoc: Social-based epidemic-based routing protocol in opportunistic mobile social network [6], Enhanced epidemic routing protocol in delay tolerant networks [7], Analysis of Epidemic, PROPHET and Spray and Wait Routing Protocols in the Mobile Opportunistic Networks [8], Congestion control for Epidemic routing in opportunistic networks [9], Buffer Adaptive Epidemic Protocols for DTN [10], Prioritized Epidemic Routing for Opportunistic Networks [11].

2.1 Epsoc: Social-Based Epidemic-Based Routing Protocol in Opportunistic Mobile Social Network

Epsoc [6] is a hybrid routing protocol scheme that has been proposed by combining scheme of Epidemic flooding routing scheme and exploitation of social features such as degree of centrality during packet forwarding in an OppNet. It is a two-step routing scheme: primarily adjustment of message centrality on the basis of TTL (Time to live) value of each node and secondly for controlling the rate of message packet replication "message blocking mechanism" is used that help in maintaining the level of flooding throughout the network. This scheme helps in improving the network performance metrics: packet delivery ratio, average number of hop counts and average latency ratio as compared to Epidemic.

2.2 Enhanced Epidemic Routing Protocol in Delay Tolerant Networks

Enhanced epidemic routing protocol in delay tolerant networks [7] improves the packet delivery ratio and tries to ensure that packet has been delivered to the intended destination as soon as possible. But, as well-known Epidemic protocol consumes network resources such as buffer space, power, node energy level and high traffic through the network resulting in high packet overhead and latency ratio respectively. Therefore, in routing scheme [5] an enhancement of Epidemic scheme has been done on the basis of node energy level (battery) and buffer space while forwarding the message packet during routing.

2.3 Analysis of Epidemic, PROPHET and Spray and Wait Routing Protocols in the Mobile Opportunistic Networks

Analysis of Epidemic, PROPHET and Spray and Wait Routing Protocols in the Mobile Opportunistic Networks [8] presents a thorough investigation on existing Opportunistic Mobile Networks (OMN) focusing on Epidemic, PROPHET and Spray and Wait routing protocols. As we know that OppNet is a type of MANET that possess all inherited feature of DTN in which the path existence between sender and receiver is an intermittent path that too in rigorous and challenging environments. The main focus of this work is to explore all characteristic features of three routing protocols with the help of two characteristic features namely mobility of nodes and store—carry—forward packet forwarding scheme among source and destination.

2.4 Congestion Control for Epidemic Routing in Opportunistic Networks

Congestion control for Epidemic Routing in Opportunistic Networks [9] proposed a routing scheme based on the concept of flooding—based routing scheme like Epidemic in which a multiple replicas of message packet has been flooded to all neighboring nodes until it encounters the destination node. As a network traffic increases, due to the creation of packet replica's, which in turn increases the packet overhead ratio and average latency ratio of network. In the proposed scheme, a strategy has been proposed for controlling the level of congestion by exploiting an additional node information to make coordination possible for packet transmission. The proposed solution has been formulated in a manner which is using a feasible node central or global information through probability—based decision with the help of node ephemeral information.

2.5 Buffer Adaptive Epidemic Protocols for DTN

Buffer adaptive epidemic protocols for DTN [10] proposed a routing scheme in which a node floods message packet to all neighboring nodes until it reaches to an intended destination through all neighboring and connected nodes. As well-known with the fact that Epidemic works on the principle of FIFO (First in First out) packet queuing mechanism. Here protocol follows three steps: primarily deletion of those bundle nodes whose message TTL has been expired, secondly selection of node has been done for those who possess high probability in process of encountering the destination node fast and thirdly node bundles farthest from the source node.

2.6 Prioritized Epidemic Routing for Opportunistic Networks

Prioritized Epidemic Routing Protocol [11] can be called as an extended version/variant of Epidemic [5]. In such a scheme, the message packet has been called as “fractional message bundled” surveyed with the help of DTNRG (Delay Tolerant Network Research Group) [13] envisioned. A message precedence depends on the following inputs: up -to—date packet, node engaging rate, from source to intended end point, data packet engaging rate cost from source to intended node, packet finish time, and packet access delay time respectively. Out of all of them, a novel metric has been used for the computation of inter-node cost and priority called as “Average Availability”.

3 Proposed Work

For the sake of better understanding, the proposed work has been neatly explained in following sections and sub-sections: Sect. 2.4, explains the working of LoHCEO followed by agenda points of issues in SERPO [1]. Section 2.3 describe the apprehension of network configuration, an initial phase of "*LA*" link availability and "*RC*" route cost metric computation respectively, by calculating the shortest routing path using Dijkstra’s Algorithm.

3.1 Apprehension of Network Configuration

This phase is an initial phase of SERPO wherein we did the computation of "*LA*" metric on the basis of concept of the node availability whose value is obtained using Eq. 1. The parameter "*LA*" plays a role in cost determination among the nodes through calculating a link availability metric using link discovery algorithm [14] as detailed in Eq. (1):

$$LA = \frac{L_u(i,j)}{L_T(i,j)} \tag{1}$$

where:

L_u: Total time for which link were active in network in past.

L_T: Total time for which link is active at the present.

i, j: Denotes the source (i) and destination node (j) respectively.

With the help of "*LA*" value, we compute the shortest path between nodes called the lowest route cost using Dijkstra's algorithm known as "*RC*" is done using Eq. 2.

$$RC = 0.99 + 0.01 \frac{\max[RC(i,j)]}{total[RC(i,j)]}(0 < RC \leq 1) \quad (2)$$

It can be referred from the Eq. 2 that value of "*RC*" must lies between 0 and 1 only. So, for reducing the possibility of both congestion as well as overhead ratio values must be lies between $0 < RC \leq 1$. The working of the proposed work can be referred from Fig. 2 followed by the normalization of route cost metric.

4 Result Discussion

In our improved approach LoHCEO with the normalization process, we performed the simulation in ONE simulator along with the comparative examination with existing OppNet routing protocol SERPO. The simulation environment of ONE is known as "community model" wherein a generation of nodes occurs randomly in an area, packet payload size is "1000 bytes", TTL (Time to Live) of packet is "300 min". Table 1, refers the simulation setting for the proposed work. In the current research work, pre – defined and pre – configured dataset which is available as default dataset has been used for carrying out simulation in ONE.

Table 1. Simulation parameter

Parameter	Value
Area dimension	3500 * 4000 m
Number of nodes	{n = 20, 40, 60, 80, 100}
Speed (in m/s)	{10, 20, 30, 40, 50}
Routing protocol	SERPO, LoHCEO
Simulation time	86,400 s
Type of Interface	Bluetooth
Message TTL	300 min
Buffer size	5 – 100 MB
Message packet size	50 KB – 1 MB
Packet payload length	1000 bytes
Transmission speed	250 KB/sec (2 Mbps)
Type of traffic	2 Base stations
	One set of mobile nodes

4.1 Packet Transfer Ratio (PDR)

A fraction of the number of messages forwarded to entire number of messages.

$$PDR = \frac{Number\ of\ message\ packets\ transferred}{Total\ number\ of\ messages} \tag{3}$$

In LoHCEO, it can be seen that at n = 60 the performance seems to improve by 32.4% when nodes utilize a buffer space in an efficient manner. It can also be seen that for the bottleneck issue, which is occurring at n = 20 and n = 40 is high seems to improve which can be inferred from Fig. 2. The observations can be concluded are less buffer consumption and packet dropping issue seems to be improve by 32.4%.

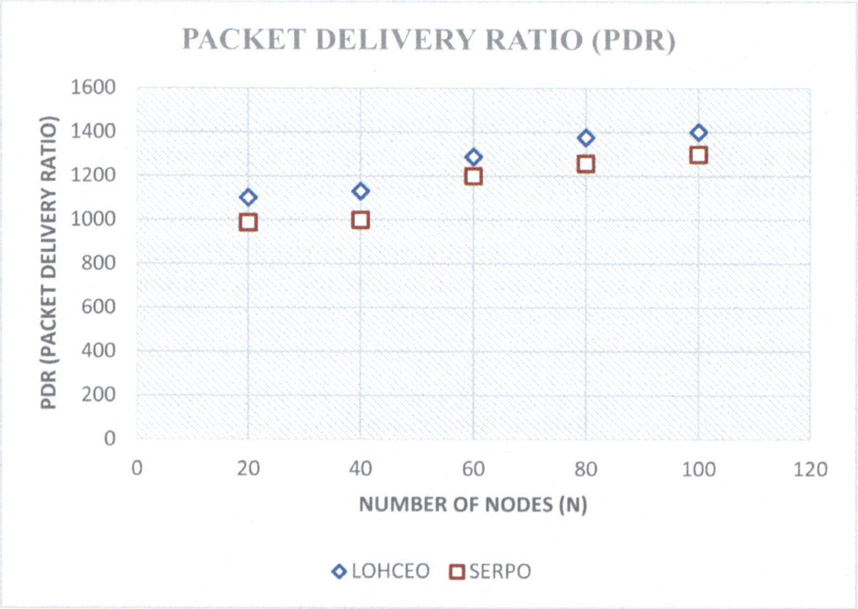

Fig. 2. Study of Packet Delivery Ratio (PDR).

4.2 Packet Overhead Ratio

A fraction of total message packet transferred by the size of a message packet.

$$POR = \frac{Number\ of\ message\ packets\ transferred}{Size\ of\ a\ packet} \tag{4}$$

The overhead ratio in a network which states that for a best performance of a network, the overhead ratio value should be less. Figure 3, represents the overhead ratio obtained using LoHCEO. The observations can be concluded are less buffer consumption and issue of bottleneck has been improved by 2.35% and 4.56% respectively, in comparison with Epidemic at n = 80 and n = 100.

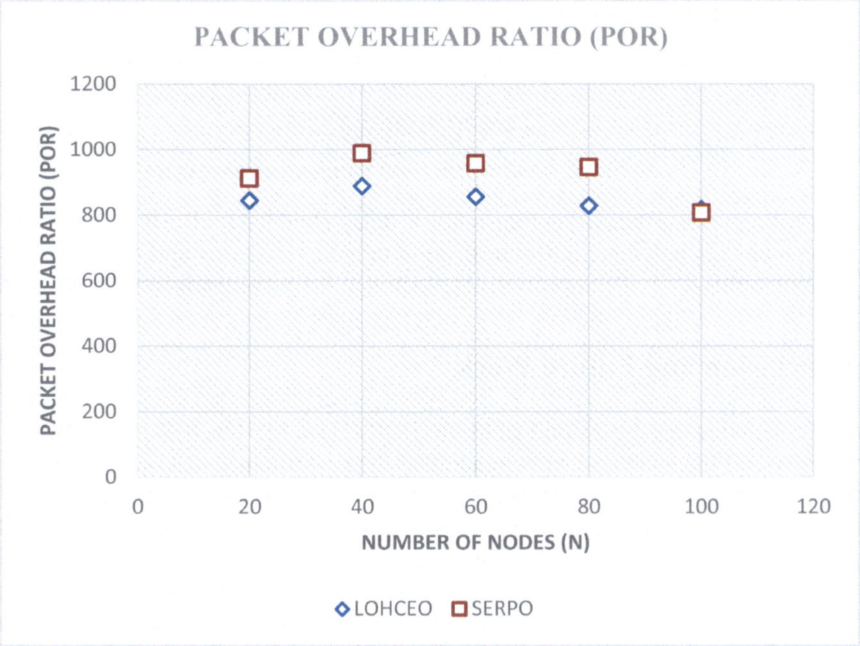

Fig. 3. Study of Packet Overhead Ratio (POR)

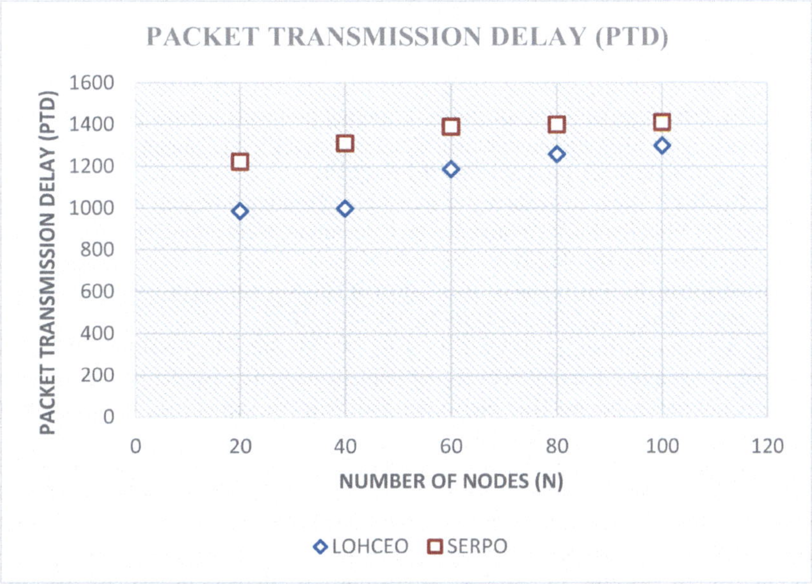

Fig. 4. Study of Packet Transmission Delay (PTD)

4.3 Packet Transmission Delay

A fraction of time is taken to put the data packet on the transmission link.

$$TD = \frac{\frac{Packet\ queue\ length(Buffer\ Size)}{Size\ of\ a\ packet}}{Network\ Bandwidth(Total\ number\ of\ messages)} \tag{5}$$

Figure 4 represents the packet transmission delay obtained for LoHCEO. The observations can be concluded are buffer utilization in an optimal way and improvement of congestion/bottleneck issue improve by 9.9% at n = 60 in comparison with SERPO.

5 Conclusion and Future Work

It can be concluded that the average of MDR, POR, TD is 13.46%, 8.472%, and 11.07%, respectively and the analysis can be made that the performance of LoHCEO is best in comparison with SERPO. For MDR, it can be noted that its performance reduces by 0.01% at n = 60 shows that there is a possibility of congestion occurrence. For POR, it is being observed that at n = 40, the performance of LoHCEO decreases by 4.4%, which shows that a high network resource consumption is being done. For future work, the further improvement can be done by following the process of normalization of route cost metric so as to reduce the congestion possibility. Additionally, further comparison work can be done with existing routing protocols Epidemic on the network performance metrics namely PDR, POR and PTD respectively.

References

1. Gautam, T., Dev, A.: Opportunistic network routing protocols: challenges, implementation and evaluation. In: 2019 9th International Conference on Cloud Computing, Data Science & Engineering (Confluence), pp. 100–106. IEEE (2019)
2. Hom, J., Good, L., Yang, S.: A survey of social-based routing protocols in delay tolerant networks. In: 2017 International Conference on Computing, Networking and Communications (ICNC), pp. 788–792. IEEE (2017)
3. Gautam, T., Johari, R.: Performance assessment of OppNet routing protocols with real world mobility traces. Intell. Dec. Technol. **16**(2), 337–355 (2022)
4. Penning, A., Baumgärtner, L., Höchst, J., Sterz, A., Mezini, M., Freisleben, B.: DTN7: an open-source disruption-tolerant networking implementation of bundle protocol 7. In: International Conference on Ad-Hoc Networks and Wireless, pp. 196–209. Springer, Cham (2019)
5. Vahdat, A., David, B.: Epidemic routing for partially connected ad hoc networks **2019** (2000)
6. Lenando, H., Alrfaay, M.: Epsoc: social-based epidemic-based routing protocol in opportunistic mobile social network. Mobile Inf. Syst. **2018** (2018)
7. Garg, P., Hemang, K., Rahul, J., Prabhanshu, G., Riya, B.: Enhanced epidemic routing protocol in delay tolerant networks. In: 2018 5th International Conference on Signal Processing and Integrated Networks (SPIN), pp. 396–401. IEEE (2018)
8. Singh, J., Raman, M.: Analysis of epidemic, PROPHET and spray and wait routing protocols in the mobile opportunistic networks. Recent Adv. Comput. Sci. Commun. (Formerly: Recent Patents Comput. Sci.) **14**(6), 1974–1983 (2021)

9. Bialon, R., Kalman, G.: Congestion control for epidemic routing in opportunistic networks. In: 2019 International Conference on Internet of Things (iThings) and IEEE Green Computing and Communications (GreenCom) and IEEE Cyber, Physical and Social Computing (CPSCom) and IEEE Smart Data (SmartData), pp. 102–109. IEEE (2019)
10. Chauhan, A., Alka, A.: Buffer adaptive epidemic protocols for DTN. In: 2019 Global Conference for Advancement in Technology (GCAT), pp. 1–4. IEEE (2019)
11. Srivastav, S.K., Tripathi, M.M.: Priority-based buffer management technique against dropping attack in opportunistic networks. In: Proceedings of International Conference on Computing and Communication Networks, pp. 153–163. Springer, Singapore (2022)
12. Rosita, Y.D., Rosyida, E.E., Rudiyanto, M.A.: Implementation of dijkstra algorithm and multi-criteria decision-making for optimal route distribution. Procedia Comput. Sci. **161**, 378–385 (2019)
13. Fall, K., Ott. J.: Delay-tolerant networking research group-DTNRG **34** (2002)
14. Xiu-Yang, C., Yu-Bin, S.: Link prediction in opportunistic networks based on network representation learning. J. Beijing Univ. Posts Telecommun.: 0

MetaMis: A Study of Identifying Missed Labels or Mislabels of Chest Radiographic Images Using Meta Learning

D. Suganya(✉) and R. Kalpana

Department of Computer Science and Engineering, Puducherry Technological University, Puducherry 605014, India
{suganyajagathguru,rkalpana}@pec.edu

Abstract. Chest radiographic images of lungs were useful for assessing the compilations, prognostication of lungs conditions, monitor treatment for various diseases like COVID-19, Lung cancer, Pneumonia etc. Manually mislabeling and missed labeling occurs which fails to find the appropriate diseases and patient's lung conditions. Artificial intelligence helps to investigate various medical reports and find solutions without human interventions. To diagnose the missed labeling and mislabeling of the chest computerized tomography (CCT) reports, AI based Meta learning algorithms were used. In this study, nearly 15 million reports from 14 different organization databases were combined; it covers the years from 1999 to 2023 and looked for any chest radiographic reports that included appendices. The 3469 CCT reports that had addenda removed because of typos, the error during writing reports, losses of some important section were removed by a thoracic radiologist. Additionally missed or error in diagnoses such as pneumothorax, fibrosis, lung nodules, consolidations, fractures in rib etc. were present in the remaining 279 patient reports. The misdiagnosis and missed diagnosis for these CCT were analyzed with various metrics. Specific findings (sensitivity, specificity, accuracy) associated with the AI based Meta learning included nodules (96.32%, 98.13%, 96.42%), pneumothorax (84.91%, 93.41%, 85.89%), pleural effusion (96.87%, 53.92%, 71.09%), consolidation (97.52%, 91.25%, 96.42%), and rib fractures (87.65%, 95.67%, 94.83%) respectively**.** When compared to various AI algorithms, Meta Learning can accurately identify mislabelled and overlooked findings and reduces the number of inaccuracies that occur when radiographic findings are discovered and labeled.

Keywords: Artificial intelligence · chest CT scan · CCT Meta Learning · pleural effusion · pneumothorax · consolidation · lung nodules

1 Introduction

The most common imaging treatment, a chest computerized tomography (CCT) makes approximately 20% of all procedures on images [1]. The studies show that up to 30% of all diagnostic imaging procedures involve CCTs [2]. This equates to an annual average

D. Pastor-Escuredo et al. (Eds.): ICDLAIR 2023, LNNS 1001, pp. 374–386, 2024.
https://doi.org/10.1007/978-3-031-60935-0_34

of 236 CCTs per 1000 patients. In the United States in 2010, 15,900 radiologic units completed 183 million radiography operations. CCTs made for 44% of all radiographs taken [3]. As a result of its accessibility, portability, widespread recognition, and low cost (in comparison to other imaging tests), it is routinely employed in medical practise to treat a wide spectrum of heart and lung diseases [4]. Despite their widespread use, CCTs are notoriously challenging to interpret, with substantial inter- and intra-observer variation. To the best of our knowledge, no articles have investigated the effect of a Meta learning algorithm on the number of mislabelled or misread CCT findings in daily clinical praxis. Therefore, we looked into whether an artificial intelligence (AI) CCT programme might detect CCT abnormalities that were missed or mislabelled in radiology reports [5]. The report's sections 2–5 provide in-depth descriptions of the study's resources and procedures, findings and analyses, and final thoughts.

1.1 Missed diagnosis

Several diagnoses may miss diagnosis which leads to sever healthy complications. Examples of some missed diagnosis or lesions are lung nodules, pneumonia, rib fracture, tracheal stenosis, fibrosis, consolidations, pneumoperitoneum etc.

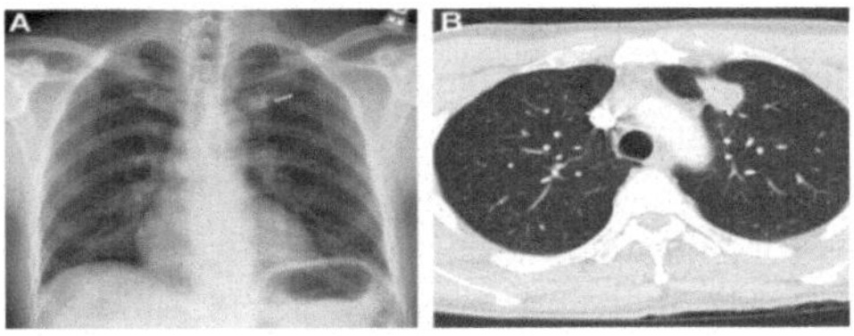

Fig. 1. Images of missed diagnosis of small cell lung cancer.

The image is from 51 years old women whose right upper lobe is affected by small cell lung cancer. The posteroanterior CXR misdiagnosed it as asymmetric opacity, the CCT detect with the super im-positional structure of the given image which leads to small cell lung cancer.

Table 1. Various types of reporting error

Pre-reporting errors	Reporting error made by clinicians	Post-Reporting error
Missing of some important sequences	Don't have enough knowledge	Not understand the communication
Study with no contrast	Cant able to find the exact location properly	Reporter made mistake while reporting
Modality choices were inappropriate	Error related to examining priory	
Some technical issues	Misunderstanding and false reasoning	Recommendations and medicines are not properly communicated
The study of patient is incomplete	Incomplete report generation	

Several types of errors may occur during diagnosis. Table 1 mentioned some of the common errors occur before, after and during disease detection by the radiologist.

AI algorithms help to identify the exact location of the abnormality and reduce errors. Meta learning helps us to diagnose everything easily by learning the data by fine tuning process.

The main contribution of this work is to study and analyze the most important chest radiological principle concerned with missed pulmonary lesions for various pulmonary disorders. Chest CT is helpful for early diagnosis and to avoid mistakes when it is identified using Meta learning algorithms. The summary of various AI, Machine learning and Meta learning algorithms were analyzed to avoid misdiagnosis. The structure of this paper organized as follows, related works and summary of CCT and its algorithms were mentioned in Sect. 2. Several methods of AI and Meta learning algorithms using CCT were analyzed in Sect. 3. The statistical analyses of missed finding and mislabeling were analyzed and discussed in Sects. 4 and 5. Conclusion section comes at the end of this paper.

2 Related Work

A retrospective research found that radiologists and physicians previously agreed on 78% of chest x-rays (CXRs) [6]. Researchers have discovered that radiologists can have different opinions on how to interpret a chest x-ray [7]. Some estimates put the number of incorrectly interpreted CCTs at 30%. Negligence in radiology can have fatal effects. Research from 1999 found that only a little over 20% of lung tumours that appear as nodules on the lungs are picked up by CCTs [8]. Such missed findings may have unfavourable outcomes for both the patient and the reporting doctor. According to the Institute of Medicine (IOM), there are between 43,985 and 97,987 avoidable deaths in the United States each year. Artificial intelligence may help reduce the number of misclassified or misread CCTs, in our opinion. In addition to making CCTs simpler to read, multiple studies [9] have demonstrated that their findings can now be interpreted more accurately. Algorithms trained on medical imaging data are becoming increasingly useful for triage and diagnosis of CCT abnormalities such as pneumonia, pneumothorax, pleural effusion, and pulmonary nodules.

3 Methods

3.1 Chest Radiographies

In order to compile all CCT reports with accompanying materials from 1999 through 2023, we searched over 13 million records from 13 radiology search databases. The terms "chest radiograph" and "addendum" were searched for in the files. Posterio-anterior CCTs were reported being taken both while lying down and while standing. We used the mPower search engine and the Render search engine to compile our data. When there were multiple reports for the same patient with the same medical record and accession codes, they were discarded. Between January 2015 and March 2021, a total of 3,469 unique CCT reports were discovered. We did not use electronic health records for our patients until after 2015. Massachusetts General Hospital (MGH), Brigham and Women's Hospital (BWH), Faulkner Health Centre (FH), Martha's Vineyard Hospital

(MVH), Salem Hospital (NSMC), Newton-Wellesley Hospital (NWH), and Spaulding Rehabilitation Hospital (SRH) all met the incorporation criteria of having DICOM CCT images available [10]. Before performing the analysis, we masked the domain names to protect the sites' anonymity. More addenda were linked to more reported CCTs, but not to more accurate reporting; therefore this was done to address that concern. If a radiographic finding was initially overlooked on a CCT report but later added in an update, the record is considered to have missing findings. Since the first CCT reports included inaccurate side labels for the CCT findings, corrections had to be included to some of them.

CCT reports with typographical errors (e = 746), inaccurate report form (e = 490), misplaced signoff sections (e = 195), and lack of context (e = 1264) were extracted for investigation by a thoracic radiologist (MKK) and a post-doctoral radiology researcher. In addition, n = 302 of examination duplicates and patient reports were not included. Since the AI algorithm could not evaluate the data, details about 32 cases of cardiac calcification, 79 cases of diaphragmatic hernia, 31 cases of clavicle or humeral fractures or dislocations, 89 cases of lines, and 41 cases of devices were omitted. After accounting for exclusions, the final sample size was 311 CCTs and supplementary materials from 311 patients. The demographics of the patients are briefly discussed here. The total number of missed or wrongly labelled CCT findings is summarised in Figure 1. Addendum co-researchers recorded the organisation and the name of the finding errors on CCT reports, which may have included pulmonary nodules, rib fractures, fibrosis and pleural-effusion. A chest radiologist (MKK) reviewed all of the CCTs with unrecognised or incorrectly labelled nodules and assigned a severity rating from 1 (not significant, the nodule is definitely benign such as granuloma) to 3 (definitely of clinical importance). According to the findings, more imaging or a follow-up test is required.

3.2 AI and Meta Learning Algorithms

Before being processed by an artificial intelligence programme on a computer inside our institution's firewall, 279 anterior Chest CT was anonym zed and exported as DICOM images. The Meta Learning algorithm used in our research has been granted regulatory permission in a number of Asian, African, and European nations, but the United States is not one of them. Over 37 lakhs of CCTs and radiography findings from hospitals and clinics around the world were used to train the algorithm. In order to detect anomalies in frontal CCTs, the method is built from many convolutional neural networks (CNNs). The technique initially resizes and normalises the CCTs in an effort to reduce variability during acquisition. In order to identify chest CTs from other types of radiographs, the system modifies the DenseNet/ResNet network architecture. Then, many networks, such as DenseNet and ResNet, are used to process the various CCT results. More in-depth information on the algorithm's implementation can be found in the prior art [11].

Over 95,432 CCTs were used to test the algorithm, and they came from all over India. No CCTs from the testing sites were used in either the training or validation data sets. The imaging algorithms (qXR v3) were trained and fine-tuned using 37 lakhs Chest CTs and the records that accompanied them. Youden's Index and an algorithm created by Qure.ai were employed with other standard approaches to establish the optimal cutoffs. The cutoffs were verified using data from over 93,000 CCTs that were not used during training. Over fifty distinct models make up the algorithm, each dedicated to scouring CCTs for a different set of problems or characteristics.

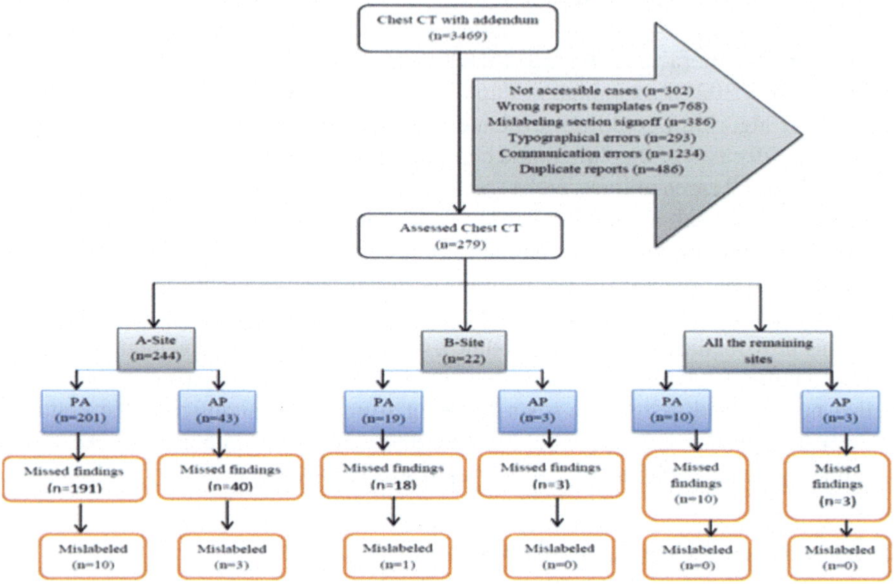

Fig. 2. Flow chart for summarizing the methodologies of the study and the number of errors or wrongly labeled location at various sites and chest CT types (PA-posterior-anterior view & portable views).

During the training process, various architectures are chosen and used, each with a unique combination of layer counts and tuning settings. EfficientNet b 5, Resnet 50, EfficientNet B6, ResNet101, EfficientNet7 and ResNeXt-101 are just few of the network architectures used by the algorithms. The typical number of parameters in a model is between 20,000,000,000 and 50,000. Common optimizers, such as SGD and ADAM, are used to train models for approximately 200 epochs. The timetables for how rapidly a model is anticipated to pick up new information differ from one model to the next. The process is described in full at https://arxiv.org/abs/1807.07455 (accessed on 19 July 2018) and was developed by Qure.ai in collaboration with qCT. Using the paid version of qCT, the cutoff values were calculated in advance. Each CCT is analyzed by the algorithm in less than 10 seconds.

3.3 Statistical analysis

Excel (Microsoft Corporation; Redmond, Washington) was used for the statistical analysis. To gauge the precision of the Meta learning system, we catalogued examples we deemed to be correct (positive), incorrect (negative), or both (false positive and false negative). Findings that are missing in both the supplement and the Meta learning result are promising. For instance, both the artificial intelligence and the supporting materials agree that a pneumothorax is not present. When both the addendum and the AI conclude that no pneumothorax is present, we say that there is agreement between the two sources (the addendum described a missed result that did not correlate to that discovered by the Meta learning; for example, the addendum documented the presence of pneumothorax which the AI did not detect) as shown in Fig. 2. Using Microsoft Excel 16 (Microsoft Inc., Redmond, WA, USA) and SPSS version (IBM Inc., Chicago, IL, USA), we determined the sensitivity, specificity, accuracy, receiver operating characteristics (ROC), and area under the curve (AUC).

4 Statistical Analysis

The average age of the 279 patients who had CCTs with additional data was 59.20. Of the total 279 patients, 143 were men and 136 were women. There were 230 fixed CCTs from Pennsylvania and 49 mobile CCTs in the data collection. We reported pneumonia and atelectasis as consolidation because that's how the algorithm categorised them. The rates at which different CCT findings were missed or mislabeled (as posterior or portable, respectively) are listed in Table 2. The significant majority of erroneous or incompletely labelled addendum reports concerned posteroanterior CCTs as compared to portable CCTs (p 0.001). It was more likely that a discovery would be overlooked than mislabeled. The most common misdiagnoses included pneumothoraces (358/100), consolidation (62/279), pulmonary nodules (19.4%), rib fractures (17.2%), and pleural effusions (5.7%).

Table 2. Summary of missed (MS) and mislabelled (ML) findings in poster anterior (PA) and portable CCTs at different sites include in our study.

CT Site View	Site A				Site B			
	Posterior-Anterior		*Portable*		*Posterior-Anterior*		*Portable*	
Findings	Missed labels	Mislabels	Missed labels	Mislabels	Missed labels	Mislabels	Missed labels	Mislabels
Lung nodules	27	2	4	2	9	1	4	1
Consolidation	51	5	7	1	5	2	3	1
Pneumothorax	57	7	24	2	3	0	5	0
Pleural effusion	14	0	1	2	5	0	3	0
Rib-fracture	38	0	8	0	2	0	0	0

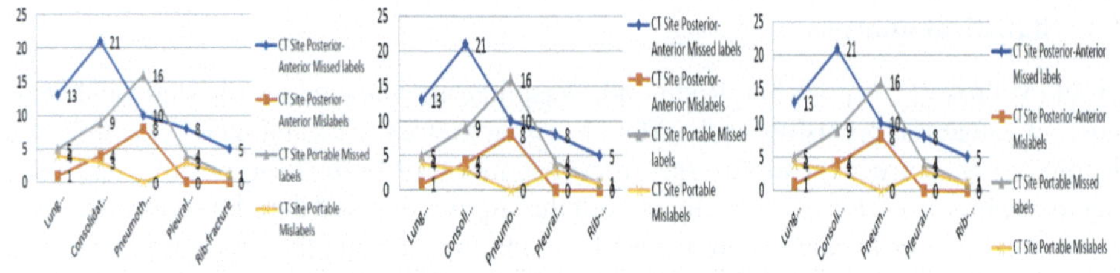

Fig. 3. Missed and Mislabels views from CT sites

The line chart of different CT sites with various finding mainly mentioned the missed labels and mislabels of portable and posterior-anterior models. Fig. 3: Shows the CT site –A, Cite-B and all the remaining sites excluding these two sites. Various metrics such as sensitivity, specificity, accuracy, and AUC of the Meta learning algorithm are summarized for each result in Table 3. When it came to detecting abnormalities, the AI was most effective at finding pulmonary nodules and consolidation and least effective at detecting pleural effusion (low specificity).

Table 3. Summary statistics of AI performance detecting missed findings on CCTs. The numbers in parentheses represent 95% confidence interval for the area under the curve (AUC).

Findings	Sensitivity (%)	Specificity (%)	Accuracy (%)	AUC
Lung nodule	96.32±0.33	98.13±0.12	96.42±0.33	0.9843 (0.94-1.00)
Consolidation	97.52±0.86	91.25±3.91	96.42±0.91	0.9901 (0.97 – 1.00)
Rib fracture	87.65±2.51	95.67±1.92	94.83±1.58	0.9465 (0.85 – 1.00)
Pleural effusion	96.87±1.42	53.93±9.53	71.06±5.82	0.8273 (0.54 – 1.00)
Pneumothorax	84.91±3.26	93.41±2.78	85.89±3.87	0.9281 (0.86 – 0.98)

There was no significant difference in Meta learning performance ($p > 0.5$) between site A and site B, or between portable and postero-anterior CCTs. Due to insufficient statistical power, data from the other sites were not compared. This is because some findings were missed or mislabeled. The AUC was moderate to high in every case of undetected failure. In contrast to the supplementary radiological reports, the Meta learning -annotated true positive heat map placed the lesion on the right side. Figure 4 shows the numerous "true positives" (findings that were missed but were discovered by the Meta learning algorithm) that were found.

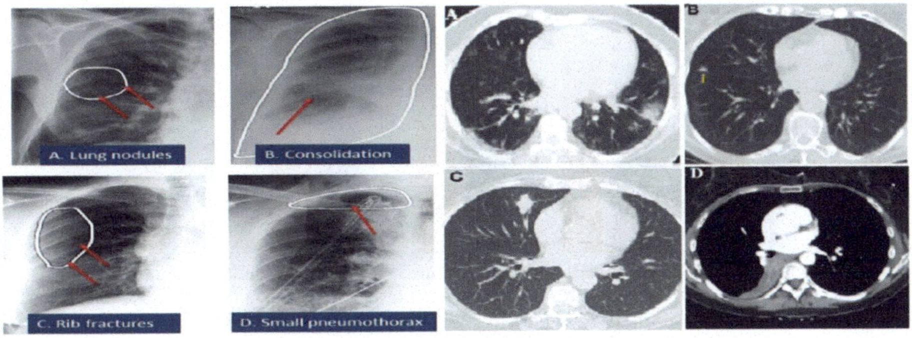

Fig. 4. Spectrum of missed CXT findings: pulmonary nodule (**A**), pneumonia (**B**), rib fracture (**C**) pneumothorax (**D**) for which radiologists issued addenda to their original radiology reports and Missed CCT findings: A. Lung nodules, B. Covid Pneumonia C. Pneumothorax D. Lung fibrosis

The AUCs for the various radiographic outcomes are summarized. There was no significant difference in Meta learning performance ($p > 0.5$) between site A and site B, or between portable and posteroanterior CCTs. Due to insufficient statistical power, data from the other sites were not compared. This is because some findings were missed or mislabeled. The AUC was moderate to high in every case of undetected failure. In contrast to the supplementary radiological reports, the Meta learning -annotated true positive heat map placed the lesion on the right side. Meta learning -generated annotations were shown as white ellipses. The discovery was denoted by red arrows. There were 41 requests for supplemental information on rib fractures, pneumothoraces, and pulmonary nodules out of a total of 279 CCTs. A chest tube was inserted for anyone with a missed pneumothorax. Three patients had lung nodule biopsies, and all 34 initial CCTs that were negative for nodules were followed up with CT scans (one benign nodule, two malignant nodules). CCT and/or CT scans performed as follow-ups revealed that the 31 pulmonary nodules that had persisted were stable. The size of each nodule is different. The chest CT helps to find pulmonary nodules, pneumothorax, pneumonia and fibrosis.

Table 4. Performance comparison of Various AI algorithms with Meta learning

Findings	AI based Accuracy (%)	MetaAI based Accuracy (%)
Lung nodule	96.42 ± 0.33	97.12 ± 0.12
Consolidation	96.42 ± 0.91	98.52 ± 0.54
Rib fracture	94.83 ± 1.58	96.65 ± 2.02
Pleural effusion	71.06 ± 5.82	96.87 ± 1.32
Pneumothorax	85.89 ± 3.87	93.91 ± 2.63

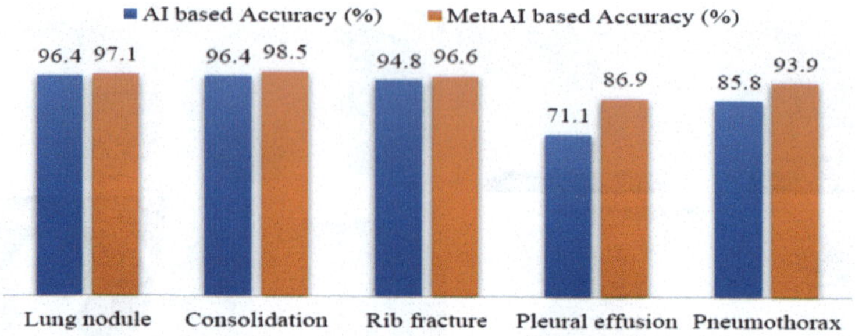

Fig. 5. Comparison chart of Meta learning over AI algorithms

The AI algorithms were used to find the mis-diagnosis and missed diagnosis, when compared to various AI methods Meta learning model produces the best results. The Table 4 and Fig. 5 shows the comparison Meta learning algorithms over other AI algorithms.

5 Discussion

In this paper, we demonstrate that specific findings on CCT are commonly missed or mislabeled before being renamed and corrected. The evaluated Meta learning technique can aid in discovering such errors and advancements (AUC 0.82–0.99 respectively). Even though previous studies have shown that AIs perform similarly whether used alone or with a 2nd reader. Many of these surveys have looked at continuous or few selective CCTs without much time spend at the clinical importance of radiography detections that were found or missed. Misdiagnosis from CCT results has been researched before. An artificial intelligence program has been recently described, that is as accurate as (or more accurate than) third-year radiology students at spotting anomalies in frontal Chest CTs. The average AUC for all Meta learning algorithms was 0.772. The use of these Meta learning algorithms has the potential to improve both productivity and quality. Addenda to radiological reports are always examined and authorized by the on-call doctor a completely qualified trainee such as a medicos before being included in the final report. Thus, the excellent accuracy (up to 0.99) of the tested META LEARNING system is in line with discoveries that doctors who did not complete residency training did not identify.

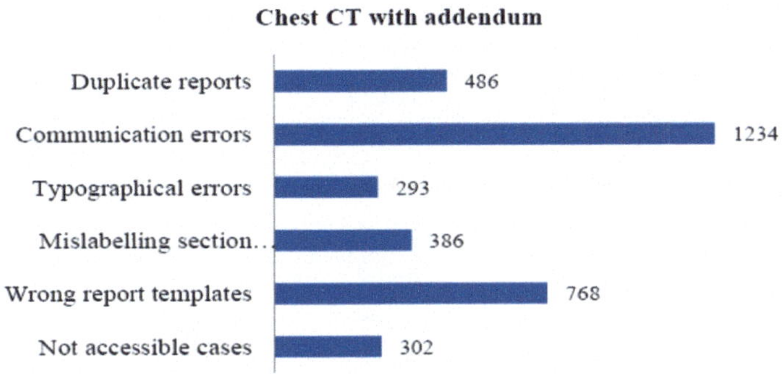

Fig. 6. Number of Chest CT with addendum analyzed

We employed a Meta learning system with great accuracy and AUCs equivalent to prior research to identify consolidation and lung nodules. [12] reported that a Meta learning model could detect consolidation on CCTs with an accuracy of 94.67%. Like previous AI algorithms able to detect lung nodules regardless of their aetiology. It is found that the sensitivity of nodules for all causes was 86.2% and the specificity was 85%. The accuracy of Meta learning in identifying malignant nodules in digital CCTs was likewise shown to be higher (up to 100%) than that of radiologists (94.1% accuracy). When compared to other multicenter studies utilising different AI algorithms, such as the one conducted on [14] whose AUC ranged from 0.91 to 0.97 for diagnosing pneumothorax, the Meta learning algorithm's sensitivity, specificity, accuracy, and AUC stand out as particularly impressive. We have analyzed various model for diagnosing errors in radiographic images, the comparison of various state of art methods is mentioned in Table 5.

Our findings show that Meta Learning is useful and might be used in CCT reporting as shoen in Figure 6. Misinterpretations of CCTs are common because of the huge volume of these tests performed in hospitals, the poor remuneration for interpreting them, the need to report fast and properly, and the very subjective nature of projectional radiography. Artificial intelligence algorithms can help reduce the risk of missing or incorrectly labelling typical CCT findings, as demonstrated by our own and other studies [13]. Also, the need to transmit supplemental information to previously reported tests is reducing since routine CCT Meta learning interpretation has made it possible to ignore common reporting inaccuracies. High false positive outputs from Meta learning systems may cause a shift in the dialogue surrounding radiography findings from non-records or over records or under records. Such problems can be addressed and resolved through the use of appropriately trained artificial intelligence models and the determination of cut-off values that keep a reasonable balancing for recall and specificity across various units of radiology and quality of radiography. The area under curves of our Meta Learning system were comparable to those who reports for testing open access CCT datasets, while we did not make a direct comparison between the two.

Table 5. Analysis of various States of art models

State of Art Models	F1-Score	Recall	Accuracy	Specificity
ResNetV2	92.43	94.67	95.32	94.81
DarkNet	89.25	90.43	91.23	91.09
Covid-CheXNet	93.52	94.34	94.57	96.82
InceptionV3	87.93	89.67	90.21	91.82
ResNeXt	91.23	91.98	92.83	93.17
LSTM-CNN	90.02	94.68	95.47	94.65
SVM, RF	88.23	89.69	93.45	92.91
Meta Learning	94.89	96.38	96.42	98.12

We used the Meta learning algorithm to categorise radiological findings during our study. Other research have looked at how may be used to prioritise the interpretation of Chest X-rays used to increase the speed for recording the abnormalities of CCTs and other specific detections. Prioritising tasks using Meta learning helped minimise the amount of time it required to submit critical CCT data [16]. Evidence suggests that employing Meta learning to prioritise work lists can reduce reporting times for head CTs for cerebral hemorrhage and other types of imaging procedures. Our findings show Meta learning has the potential to reduce radiology's high rate of missed diagnoses and inaccurate reporting. Before our findings can be extended to other sites and practises, however, larger investigations are necessary, ideally with progressively more balanced samples of each missed finding. CCTs showing lung nodules or pleural effusions should be interpreted with caution because these abnormalities are so infrequent in our data set.

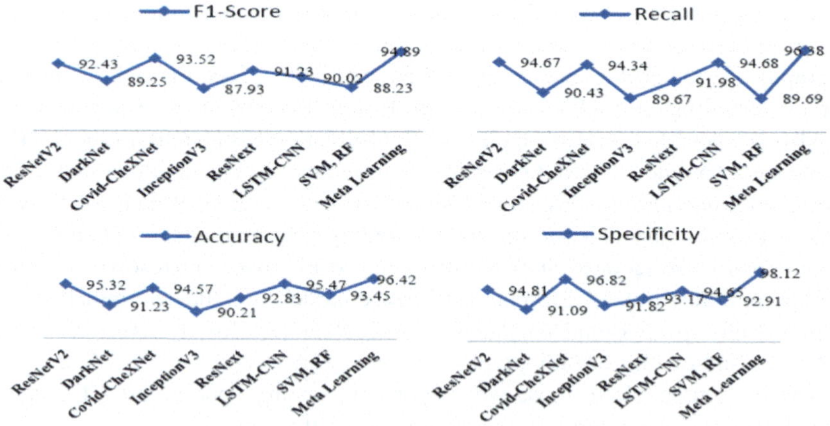

Fig. 7. Analysis of various states of art models

There are few holes in our research. The size of sample (e = 311 CCTs) was low because of the strict inclusion and exclusion criteria, however we did search through approximately 129 lakhs radiology records between 2000 and April 2022 and found 2984 CCT records with appendix. Our research's biggest shortcoming is in this particular area. Even though a bigger dataset would be desirable for a Meta learning program, our study paints a good picture of missed and wrongly tagged discoveries on consecutive eligible CCTs from different links [15]. Addendums to radiographic records may not accurately represent the number of such errors because most missing or mislabeled findings are not able to detect else addressed in following CXRs or other imaging tests like CCTs. The research work is not aimed to determine what percentages of CCT findings are overlooked. Instead, we wanted to see how well the Meta learning algorithm did at not missing or incorrectly labelling important results that were flagged by referring doctors and/or radiologists. We were unable to assess how often the AI failed to recognise anomalies in the mediastinum and hilum, cavities, or pulmonary fibrosis, due to the small sample size. Additionally, the effectiveness of Meta learning can change based on the characteristics of the data. Because of this, it's possible that our results won't apply in contexts where CCT results are uncovered in a different way. Figure 7 shows the comparative analysis of various states of art models. Our current Meta learning algorithm also fails to properly interpret some of our data, such as the positioning of lines and gadgets. This is a primary reason why supplementary data wasn't included in the portable CCT reports. We used real time CCTs with addendum records to include error findings, but didn't incorporate real time randomized CCT datasets without addendum records because the performance of the Meta learning algorithm we used in this study has already been reported on [17]. Unfortunately, we did not perform a thorough algorithm analysis to check for domain bias in the research sites. A larger diversity of study, radiography apparatus and radiologists would have helped us limit the possibility of influence in our study. Eventually, running the CCTs through any other Artificial Intelligence algorithms is not done; we are unable to remark on their efficacy.

6 Conclusion

Misdiagnosis and missed diagnosis of several wounds and other mild conditions of lungs makes a big trouble in future. This study shows that Meta learning can help reveal various discoveries on Chest CTs that were find either missing or mislabelled in the past. The examined Meta learning system can aid radiologists in the detection, description, and labelling of radiographic abnormalities such as consolidation, lung nodules, pneumothorax, rib fractures, and, to a lesser extent, pleural effusions when employed as a secondary reader. Future research on a grander scale will be necessary to fully understand the role that Meta learning algorithms play in improving the precision of radiology reporting for Chest CT images.

References

1. Ekpo, E.U., Egbe, N.O., Akpan, B.E.: Radiographers' performance in chest X-ray interpretation: the Nigerian experience. Br. J. Radiol. **88**, 20150023 (2015)

2. Speets, A.M., et al.: Chest radiography in general practice: indications, diagnostic yield and consequences for patient management.Br. J. Gen. Pract. **56**, 574–578 (2006)
3. Forrest,J.V., Friedman,P. J.: Radiologic errors in patients with lung cancer. West J. Med. **134**, 485–490 (1981)
4. Kelly, B.:Thechestradiograph.Ulster Med. J. **81**, 143–148 (2012)
5. Schaefer-Prokop, C., Neitzel, U., Venema, H.W., Uffmann, M., Prokop, M.: Digital chest radiography: an update on modern technology, dose containment and control of image quality. Eur. Radiol. **18**, 1818–1830 (2008)
6. Stia, I., Bashagha, S., Bibi, A., Ahmed, R., Mellor, S., Zaman, F.: AssessingtheaccuracyandcertaintyininterpretingchestX-raysinthemedicaldivision. Clin. Med. **13**, 349–352 (2013)
7. Fancourt, N., et al.: Standardized interpretation of chest radiographs in cases of pediatric pneumonia from the PERCH study. Clin. Infect. Dis. **64**(Suppl.3), S253–S261 (2017)
8. Berlin, L.: Reporting the "missed" radiologic diagnosis: medico legal and ethical considerations. Radiology **192**,183–187 (1994)
9. Quekel, L.G., Kessels, A.G., Goei, R., vanEngelshoven, J.M.: Missrateoflungcanceronthechestradiographinclinicalpractice. Chest **115**, 720–724 (1999)
10. Suganya, D., Kalpana, R.: Automated detection of Covid-19 waves with computerized tomography scan using deep learning. In: Awasthi, S., Sanyal, G., Travieso-Gonzalez, C.M., Kumar Srivastava, P., Singh, D.K., Kant, R. (eds) Sustainable Computing. Springer, Cham (2023). https://doi.org/10.1007/978-3-031-13577-4_3
11. Suganya D., Kalpana R.: Prognosticating various acute Covid lung disorders from COVID-19 patient using chest CT Images, Eng. Appl. Artificial Intell. 105820 (2023). ISSN 0952-1976. https://doi.org/10.1016/j.engappai.2023.105820
12. Suganya, D., Kalpana, R.: Automated COVID-19 diagnosis using Deep Multiple Instance Learning with CycleGAN. Int. Conf. Smart Technol. Syst. Next Gener. Comput. (ICSTSN) **232022**, 1–6 (2022). https://doi.org/10.1109/ICSTSN53084.2022.9761334
13. Wu, J.T., et al.: Comparison of chest radiograph interpretations by artificial intelligence algorithm vs radiology residents. JAMA Netw. Open **3**,e2022779 (2020)
14. Li, X., et al.: Multi-resolution convolutional networks for chest X-ray radiograph based lung nodule detection. Artif. Intell. Med. **103**,101744 (2020)
15. Lan, C.C., et al.: Deep learning-based artificial intelligence improves accuracy of error-prone lung nodules. Int. J. Med. Sci. **19**, 490 (2022)
16. Zhang, Y., et al.: Lung nodule detectability of artificial intelligence-assisted CT image reading in lung cancer screening. Curr. Med. Imaging **18**, 327–334 (2022)
17. Rudolph, J., et al.: Artificial intelligence in chest radiography reporting accuracy: added clinical value in the emergency unit setting without 24/7 radiology coverage. Investig. Radiol. **57**, 90–98 (2022)

An Automatic Process of Online Handwriting Recognition and Its Challenges

Mamta(✉) and Gurpreet Singh

CUIET, Chitkara University, Rajpura Punjab, India
{mamta.2325,gurpreet.1082}@chitkara.edu.in

Abstract. Online Handwriting Recognition (OHR) has been proven as best alternative against keystroke pattern input. During this process of OHR, input stroke patterns are captured with the help of stylus and digitizer environment. Pen up and Pen down events in continuation results in ink strokes. These ink strokes may have some uneven shapes. Sometime these uneven shapes cannot be compared further with the standards well known shapes. This leads to the biggest challenge while recognition of these stroke patterns. In this paper, the automatic process of OHR is explained by focusing on the challenging issues like identification of uneven shapes, script recognition, character segmentation, word segmentation etc. The main emphasis is given over writer independent and open vocabulary systems.

Keywords: Online Handwriting Recognition · Key strokes · script recognition · character segmentation · word segmentation

1 Introduction

The development of new fields such as artificial intelligence, machine learning, computer vision, wireless sensors, cloud computing, big data analytics, data science, deep learning, optimization techniques, etc. has resulted in a constant change in technical trends [1]. In contrast to conventional methods, all of these emerging innovations provide a significant computational environment that supports a variety of computer application software. With a very broad solution area, these recent advancements also offer solutions to some issues that are incredibly complicated. One of these challenging issues has been the computerized identification of handwriting. The uneven shapes of the characters used for writing have been cited as the primary cause of the complex structure of handwriting recognition software. The automatic handwriting recognition process is also impacted by other complexity issues, such as the presence of numerous languages, multiple writing scripts, distinct writing styles of various authors, or even the same author over time [2]. The execution of every computer system programme or application involves three primary steps in accordance with the Von Neumann architecture. The initial step is to take into account user input using different input methods, such as mouse clicks, keyboard keystrokes, scanner, joystick, etc. The second stage entails executing the appropriate functionality as specified by the user under evaluation on the data with the assistance of a processor [3]. The user is presented with the estimated output in the third and final

D. Pastor-Escuredo et al. (Eds.): ICDLAIR 2023, LNNS 1001, pp. 387–394, 2024.
https://doi.org/10.1007/978-3-031-60935-0_35

stage via external devices such a printer, monitor, etc. Second stage is very important as processing is involved into it.

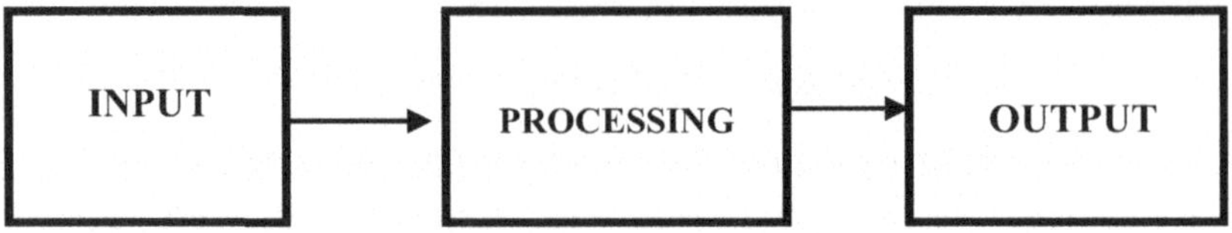

Fig. 1. Von Neumann Architecture [1]

Numerous optimization methods have been employed in conjunction with technologies like AI, ML, deep learning, etc. to increase the processing step's effectiveness. However, despite the existence of the fast-processing techniques, the operation's total performance can be impacted if the user input stage involves providing data at a slow rate for analysis [4]. As already said, keyboards, mouse, and other devices are being employed to collect input data. The majority of the time, key stroke data is used to collect the information. The user's typing pace has been taken into consideration when using a keyboard in order to give quick input. Users who are less proficient at typing on a keyboard might not be able to input data quickly into computer applications. The use of handwritten input as a substitute to keystrokes data has been acknowledged for software applications [5–7]. This method appears to be faster as comparing to keystroke inputs. In addition, the user's typing speed varies while using a keyboard, and there is always the possibility of human error. Therefore, various handwriting recognition technologies open the door for faster user input.

The emerging technologies in the area of computer science is to adapt HCI (Human Computer Interaction) systems to conduct same to corresponding HHI (Human-Human Interaction) models. In essence, HCI systems take into account user input in a digital information that the system can immediately understand and translate into the necessary output after processing. Examples of these HCI systems include chatbot apps, autonomous robotics, etc. The major objective of these kinds of systems is to see behaviour similar to that seen in human-to-human conversation. For instance, input for chatbot applications typically takes the form of written characters, words, or even questions as text strings but for automated robotic systems, input takes the form of expression, navigation system inputs, etc. to continue to process [8].These processing units' entire input categorization for all of these sources is that of input signal. In a chatbot application, the system can respond to user concerns in a way that mimics how a human would respond to them by utilising the capabilities of artificial intelligence (AI). Therefore, the behaviour of a chatbot programme resembles any human-human conversation [9].The robot responds similarly to any interpersonal interactions in autonomous robotic applications based on the digital input received. The HHI method includes both chatbot and robotic applications. Handwriting samples are a type of keystroke input substitute, as discussed earlier in this chapter. It is also among the quickest methods for giving input to any computer application [10]. The subset of non-digital input includes handwritten samples though. Therefore, it is challenging to anticipate an HHI behaviour from a computer program if these examples have been used as input. Since these technologies demand digital input, handwriting recognition systems are vital in transforming non-digital information into digital input in this situation. One of the most challenging topics in computer engineering is the area of automatic handwriting recognition. The handwriting recognition industry is challenging for the reasons listed below:

- There are around 6500 different languages spoken worldwide.
- The availability of several writing systems and character sets for all linguistic varieties.
- Each writer has a unique writing style.
- Changes in a writer's writing style throughout time.
- The intricate designs of the many characters in the character set for the relevant language.

The estimation of the layout or pattern of the letters or lines used to create handwriting samples may be influenced by all of the preceding complication factors [7, 11]. The computer programmes that use handwritten observations as input must work similarly to HHI systems, which means they must have existing experience about the regular expression of the handwritten script, the handwriting type of the authors, as well as other complexity aspects related to distinct scripts. They must also have prior understanding of the language used to construct the handwritten examples. In the case of handwriting recognition systems, the systems can be classified into the two groups based on the kind of input given and the demand for various applications, as below given in Fig. 2.

- Systems for offline-handwriting recognition.
- Systems for offline-handwriting recognition.

The input received as a digitized picture in these kinds of handwriting recognition systems. These pictures were actually text that has written in a particular language's script. Systems of this nature are included in the broad category of digital image processing (DIP) [11]. The information needed to identify and recognise the text included in these photos is extracted using the various intensities of each pixel.

The general technique used to transform the handwritten characters from these captured images to its digital information follows an identical pattern to that used when offline printed documents converting apps from their image files are converted. The asymmetrical layout and the size of the characters portrayed by the user while supplying input samples, however, are the main complication as opposed to printed text apps in offline handwriting identification apps. Figure 2 depicts the input for systems that operate offline.

The general technique used to transform the handwritten characters from these captured images to its digital information follows an identical pattern to that used when offline printed documents converting apps from their image files are converted. The asymmetrical layout and the size of the characters portrayed by the user while supplying input samples, however, are the main complication as opposed to printed text apps in offline handwriting identification apps. Figure 3 depicts the input for systems that operate offline.

In order to retrieve text, phrases, and symbols, two fundamental techniques—horizontally projection and vertically projection—are used. The feature extraction procedure was important after segmentation. The major characteristics taken into account

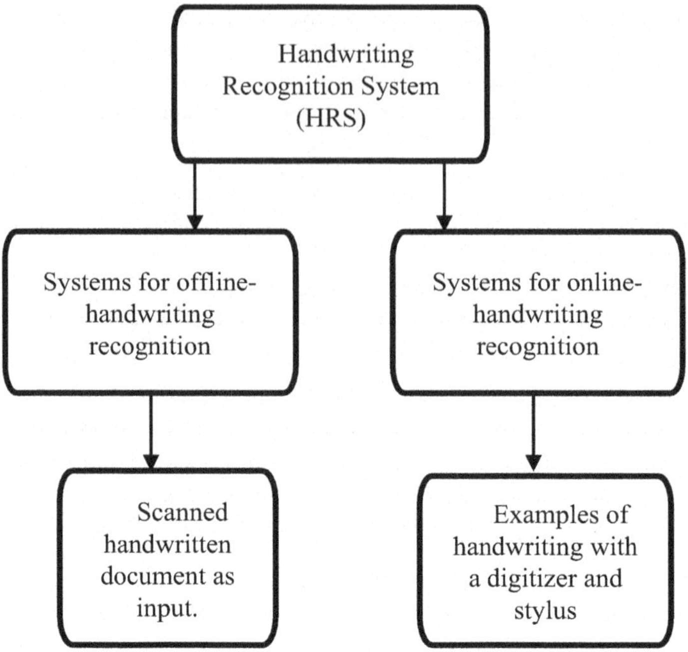

Fig. 2. Types of Handwriting Recognition Systems

Fig. 3. Digitized written content as offline input [13]

in an offline system include obtaining integrate correspondence points for compartmentalised characters, identifying each character's slope and framework, measuring the region taken up by various letters, determining the number of loops in segmentation is performed symbols, etc. The retrieved data is given to a few suitable classifiers based on the chosen features. The classifier is trained using these data [12]. This knowledge aids in the extraction of pertinent facts at the time of the examination. At the conclusion, the certain data is utilized to determine the required symbol. These sorts of offline technology are essential for activities like digitising detailed notes, automating historic scribbled textbooks, digitising some text materials without a scanned image already available, and more.

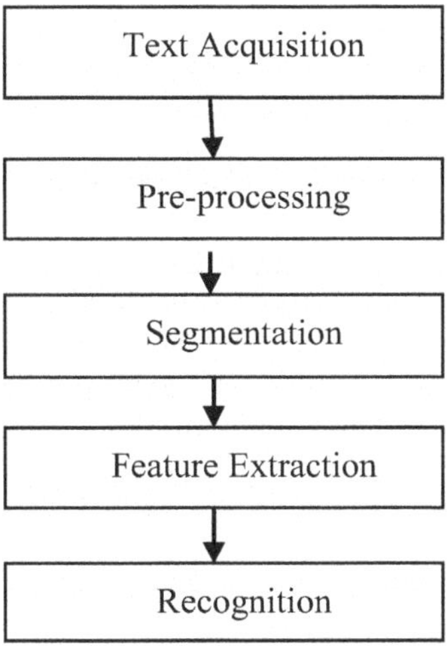

Fig. 4. Automatic Process of Online Handwriting Recognition [14]

2 Online Handwriting Recognition Systems

Special hardware like a digitizer and stylus are needed for these types of handwritten recognition systems. Each written interface, including touch screens, digitized pads, and other devices, can be a digitizer. It could be a computing device, a digitized tablet, etc. In reality, a stylus is a pen that is used to write on touch screen surfaces [13]. Stylus can be categorised into either an engaged stylus with a pointed end like a regular pen used to write in everyday lives or as a passive pen with touch capabilities equivalent to single touch. The records of each stroke is collected when collecting additional documents and entered into any computerized handwritten object identification. A stroke is a group of coordinate points that the stylus, when being used to type on a digital surface, contacts in a particular order while putting the pen down and taking it back up [14]. It can be viewed as a 2D array storing the data of each spot's X and Y positions that the stylus contacted on the digitizer's surface [16–20]. In other words, during the process of online handwritten character recognition, a single letter of any script or a whole word could be produced by using one or more strokes. Throughout this method of online handwritten character recognition, each stroke has been treated as a separate input unit. In subsequent consecutive stages of the handwritten identification system, a stroke serves as the starting point for additional processing [21–24]. The benefits of utilizing online handwritten identification systems include the following:

- There is no requirement to use vertical and horizontal projection techniques to locate the pixels involved in the creation of letters or sentences.
- The data about the pixels involved in the production of the stroke is contained in the stroke's information individually.
- It is not necessary to manage pictures and their enormous dimensions.
- Online identification systems can also record directional movement during stroke creation.

- Retrieval of additional knowledge, such as load created by the writer throughout stroke generation on the digitizer's surface, can aid in the forecasting of data for purposes like predicting character or social interactions.

3 Challenges in Case of Online Handwriting Recognition

- Writers use a variety of writing techniques.
- The sam*e author has* used various style of writing throughout decades.
- There is no predefined font style accessible for producing strokes.
- The technology used for the digitizer and stylus is of high class active or passive stylus). Various different languages are utilized in different dialects.

4 Challenges Based on Writer's Category and Vocabulary Type

The kind of handwriting recognition systems, based on the parameters of the dataset are used to train the algorithms, a variety of handwriting recognition methods have been developed. These are mostly divided into the four groups listed below:

- Independent writer
- Dependent Writer
- Restricted Vocabulary
- Open Vocabulary

Programs that depend on authors demand data for training taken from the exact group of writers that will be used to test the system. This implies that these technologies can only evaluate or recognise supporting documentation from individuals who have already participated to creating the training examples. These algorithms might not always be capable of producing the desired digital text if handwritten examples from unfamiliar or new writers are provided. However, with writer individual systems, the devices cannot rely solely on the given dataset that is given to them. The system must be capable to react to handwritten patterns from individuals who did not participate to the testing set. The idea behind these systems is to extract several similarities for each symbol existing in the regular expression of the relevant scripts [13]. The connected elements from the training examples are subsequently taken into consideration for the transformation of the textual content if comparable similarities are found in the samples submitted by anonymous authors during testing. Due to the need for a substantial and robust testing set, this author independent procedure is regarded as difficult and demanding. Many academics focused on the idea of closed vocabulary structures from the beginning of their studies in the field of computerized handwritten character recognition. These systems require handwritten patterns to be compared only to a predetermined collection of words or characters. The letters or lines chosen for the testing must be identical to those used to create the training set. It is possible for this restricted vocabulary system to be either author dependent or independent. Open vocabulary systems, on the other hand, never occur in the context of any limited set of words or characters. The testing step allows for the testing of any word in any format. There is no requirement that the word being tested be trained during the phase of training. The characteristics of open terminology and writer-independent techniques must be taken into account in a comprehensive dynamic solution for automated handwriting recognition systems.

5 Conclusion

In this paper, the main challenges related to automated process of online handwriting recognition have been explored. The challenges started from the familiarity level of writers with digitizer and stylus environment. The writers have been identified as per their familiarity level as moderate, perfect or new user. Then the next one involved with the problem of different writing styles of writer or the different writing style of same writer at two different times. Uneven structure of different character sets is also considered to be a big challenge Apart from these, the challenges related to the category of system as writer independent or writer dependent, where initial data is required in a bulky manner also affects the complexity of the overall system. Open vocabulary and closed vocabulary systems also required different algorithmic approaches to identify the strokes under consideration.

References

1. Chauhan, V.K., Singh, S., Sharma, A.: HCR-Net: a deep learning based script independent handwritten character recognition network. vol. 1, no. 1 (2021). http://arxiv.org/abs/2108.06663
2. Flores, C.A., Figueroa, R.L., Pezoa, J.E.: Active learning for biomedical text classification based on automatically generated regular expressions. IEEE Access **9**, 38767–38777 (2021). https://doi.org/10.1109/ACCESS.2021.3064000
3. "Von Neumann代数の生成," SUGAKU, vol. 22, no. 4, pp. 292–298 (1970). https://doi.org/10.11429/sugaku1947.22.292
4. Singh, H., Sharma, R.K., Singh, V.P.: Online handwriting recognition systems for Indic and non-Indic scripts: a review, vol. 54, no. 2. Springer Netherlands (2021). https://doi.org/10.1007/s10462-020-09886-7
5. Singh, H., Sharma, R.K., Singh, V.P., Kumar, M.: Recognition of online handwritten Gurmukhi characters using recurrent neural network classifier. Soft. Comput. **25**(8), 6329–6338 (2021). https://doi.org/10.1007/s00500-021-05620-9
6. Modha, S., Majumder, P., Mandl, T.: An empirical evaluation of text representation schemes to filter the social media stream. J. Exp. Theor. Artif. Intell. **34**(3), 499–525 (2022). https://doi.org/10.1080/0952813X.2021.1907792
7. Hamdan, Y.B., Sathesh, A.: Construction of statistical SVM based recognition model for handwritten character recognition. J. Inf. Technol. Digit. World **3**(2), 92–107 (2021). https://doi.org/10.36548/jitdw.2021.2.003
8. Liu, S., Ni'mah, I., Menkovski, V., Mocanu, D.C., Pechenizkiy, M.: Efficient and effective training of sparse recurrent neural networks. Neural Comput. Appl. **33**(15), 9625–9636 (2021). https://doi.org/10.1007/s00521-021-05727-y
9. Ali, A.A.A., Mallaiah, S.: Intelligent handwritten recognition using hybrid CNN architectures based-SVM classifier with dropout. J. King Saud Univ. - Comput. Inf. Sci. **34**(6), 3294–3300 (2022). https://doi.org/10.1016/j.jksuci.2021.01.012
10. Nguyen, H.T., Nakamura, T., Nguyen, C.T., Nakawaga, M.: Online trajectory recovery from offline handwritten Japanese kanji characters of multiple strokes. In: 2020 25th International Conference on Pattern Recognition (ICPR), pp. 8320–8327 (2020). https://doi.org/10.1109/ICPR48806.2021.9413294
11. A. Resende, D. Railsback, R. Dowsley, A. C. A. Nascimento, and D. F. Aranha, "Fast Privacy-Preserving Text Classification Based on Secure Multiparty Computation," vol. 17, pp. 428–442, 2022

12. Li, Y., et al.: Sentence-level Online Handwritten Chinese Character Recognition, vol. 1, no. 1. Association for Computing Machinery (2021). http://arxiv.org/abs/2108.02561
13. Scheidi, H.: towardsdatascience.com. https://towardsdatascience.com/build-a-handwritten-text-recognition-system-using-tensorflow-2326a3487cd5%0A%0A
14. Li, Y., et al.: Fast and robust online handwritten chinese character recognition with deep spatial & contextual information fusion network. IEEE Trans. Multimed. 9210(c), 1–13 (2022) https://doi.org/10.1109/TMM.2022.3143324
15. Dey, R., et al.: Sliding window based off-line handwritten text recognition using edit distance. IEEE Trans. Multimedia Tools Appl. **81**, 22761–22788 (2022).https://doi.org/10.1007/s11042-021-10988-9
16. Kaur, R., Singh, G., Gaur, P.K.: Hybrid Classification Method for the Human Activity Detection. In: 2021 2nd Global Conference for Advancement in Technology, GCAT 2021 (2021)
17. Singh, G., Sachan, K.: A Bilingual (Gurmukhi-Roman) Online Handwriting Identification and Recognition System (2019)
18. Singh, G., Sachan, M.: A Framework of Online Handwritten Gurmukhi Script Recognition. vol. 8491, pp. 52–56 (2015)
19. Singh, G., Sachan, M.K.: An unconstrained and effective approach of script identification for online bilingual handwritten text. Natl. Acad. Sci. Lett. **43**(5), 453–456 (2020)
20. Kanwar, S., Sachan, M., Singh, G.: N-grams solution for error detection and correction in Hindi language. Int. J. Adv. Res. Comput. Sci. **8**(7), 667–670 (2017)
21. Singh, G., Sachan, M.K.: Performance comparison of classifiers for bilingual gurmukhi-roman online handwriting recognition system. Int. J. Eng. Adv. Technol. **8**(5), 573–581 (2019)
22. Singh, G., Sachan, M.K.: Data capturing process for online Gurmukhi script recognition system. In: 2015 IEEE International Conference on Computational Intelligence and Computing Research, ICCIC 2015 (2016)
23. Kaur, R., Singh, G.: Performance comparison of AI models for digital image captioning. In: 2022 International Conference on Computational Modelling, Simulation and Optimization (ICCMSO), Pathum Thani, Thailand, pp. 257–261 (2022). https://doi.org/10.1109/ICCMSO58359.2022.00058
24. Nisha Singh, G.: Social network analysis of Twitter tweets. In: 2021 3rd International Conference on Advances in Computing, Communication Control and Networking (ICAC3N), Greater Noida, India, pp. 99–103 (2021). https://doi.org/10.1109/ICAC3N53548.2021.9725406

IoT-Enabled Smart Restaurant Management System

Charanpreet Singh[1], Yohjit Chopra[1], Ishaan Gupta[1], Harshdeep Singh[1], Amit Kumar[1(✉)], and Imali Dias[2]

[1] Chitkara University Institute of Engineering and Technology, Chitkara University, Rajpura 140401, Punjab, India
amit.pandey@chitkara.edu.in

[2] School of Information Technology, Deakin University, Geelong 3220, Australia

Abstract. The restaurants have several operational features such as taking online/offline customer orders, staff and resource management, fire prevention, and safety. Due to recurring costs, an exhaustive installation process, regular maintenance, and a lack of technically skilled personnel, many restaurants are not able to comply with these tasks completely, thus resulting in opting for only customer management tasks and overlooking other features. This article reviews existing systems designed for such purposes and presents the implementation of a more comprehensive and independent Internet of Things (IoT)-based smart restaurant management system. The proposed system integrates a personalized and efficient food ordering system with a reliable and responsive fire management system that functions even in the absence of Internet connectivity. The system has been successfully tested for responsiveness, with an average response time of 0.5 s for the latency-sensitive fire management system. The system is also fault-tolerant against hardware and network errors and has minimal set up and maintenance costs.

Keywords: Internet of Things · Embedded Systems · Smart Shopping · RFID · Arduino Nano 33 IoT

1 Introduction

Restaurants are an imperative part of the hospitality industry around the world. In India, restaurants of varying types and levels exists that attract and cater to a wide variety of customers, thereby contributing to significant revenue growth. In any restaurant, the owners prioritize the customer satisfaction and health and safety of their employees and customers. The inclusion of recent advancements such as the Internet of Things (IoT) [1,2] and Industry 4.0, specifically sensor-based systems, in modern restaurants, can help automate these tasks easily. Due to the recent developments in embedded systems and communication technologies such as wireless, wired, and optical networks, IoT-based solutions are extensively deployed in these areas. These IoT solutions depend on a variety of sensors to detect and collect data from their surroundings and process it

D. Pastor-Escuredo et al. (Eds.): ICDLAIR 2023, LNNS 1001, pp. 395–404, 2024.
https://doi.org/10.1007/978-3-031-60935-0_36

using microcontrollers to execute the desired actuation/action. Nowadays, such solutions are deployed in modern restaurants to provide a more personalized customer experience and improve the safety standard of the entire restaurant system [3].

There are many aspects to customer satisfaction in a restaurant starting from ordering the meals to enjoying them with an appealing ambiance [4]. For example, when ordering food, the use of an e-menu offers more details about the food and mitigates human error in taking orders. Using technology devices such as tablets placed at tables will help multiple customers place their orders simultaneously, thereby reducing the waiting time for the servers. Further, presenting the customers with food suggestions based on their previous orders can have a positive impact on customer satisfaction and also can make the ordering process more efficient and personalized. Further, ensuring the health and safety of the customers and employees due to reputation and regulatory considerations is very important for restaurant owners. In this regard, liquid petroleum gas (LPG) leakages are considered to pose a serious risk of fire when used in the vicinity of kitchens. To prevent accidents caused by LPG leakages, it is important to develop a reliable and responsive fire safety mechanisms. More specifically, IoT-based solutions have been designed for such purposes where gas sensors are placed to detect leakages [5]. Although standalone systems that address either customer satisfaction or health and safety have been proposed in existing literature, a system that integrates both of these systems is yet to be evaluated.

Taking these key requirements into account, this work presents a comprehensive restaurant management system that makes the ordering process more efficient and personalized while also ensuring the safety of the customers and staff. The proposed system (a) displays a dynamic, user-friendly, and up-to-date menu that is easy to navigate; (b) takes orders through the electronic display, (c) incorporates a recommendation system based on previous dining of the customers and (d) integrates a responsive safety system to ensure the health and safety of the patrons. The rest of the paper is structured as follows. In Sect. 2, we present similar work done in restaurant management systems to highlight the novelty and advantages of proposed system. Section 3 discusses the proposed system in detail. Testing and evaluation of the system are provided in Sect. 4.5 with concluding remarks in Sect. 5.

2 Existing Literature

This section presents an overview of different types of existing restaurant management systems and more specific advancements on their sub systems such as food ordering and fire safety.

2.1 Restaurant Management Systems

The growing demand for restaurant management systems has led to the development of various types of software and hardware solutions for such systems. In [6], a personal data assistant (PDA)-based wireless ordering system is presented. The proposed solution uses Wi-Fi and wireless local area network (WLAN) technologies to realize the functionalities of the system. Similarly in [7], the authors have used Bluetooth as the wireless technology and Peripheral Interface Controller (PIC) as the controlling hardware device. The work reported in [8] proposes a LAN-based solution for e-restaurants where the touch screens connected to a LAN can receive food orders from customers. The objective of this work, however, was to make the system more affordable and customized. The use of robots is also an avenue explored in restaurant management systems [9].

Integrating existing technologies to the restaurant system to enhance the user experience is also explored in current industry and literature. Sarkar et al. highlighted the use of Android to integrate touch technologies in eateries [10]. This system consists of a dynamic database tool that retrieved all of the data from a single database. The Android application with restaurant and menu information is contained in a tablet placed at the customer table. In the work reported in [11], the authors have introduced the near field communication (NFC) for restaurants for customers to place their orders before they enter the restaurant. In [12], a Bluetooth enabled Android phone is used where the details of food items in restaurants are included in an Android application and the database is maintained on a cloud-based server.

Based on existing research, it is evident that significant work has been done in integrating communication protocols, novel technologies, and even robotics into improving the customer experience at the restaurants. However, the use of prior user preferences are yet to be included into a smart restaurant architecture due to the challenges associated with storing different user data and also ensuring secure use of such data.

2.2 Fire Prevention Systems

Fire prevention systems are being investigated and designed, not just for restaurants but also for other premises such as smart homes and hospitals. The research on this area investigates different types of sensors, micro-controllers, communication methods, and technologies. Kurzekar et al. proposed a micro-controller-based LPG detection and monitoring system where the sensor data is sent to an Android platform using ZigBee-based Bluetooth [13]. Similarly, a LPG leak detection device has been created using NodeMCU, MQ2 gas sensor, and flame sensor in[14], with data collected by the sensors presented on an Android-based smartphone. In the system proposed in [15], a MQ6 sensor connected to a PIC16F877A detects LPG leakages. The system proposed in [16] is suitable for both residential and industrial environments and not only contains a gas detection module but also an exhaust fan to reduce the LPG concentration in the surrounding air.

Shrestha et al. proposed an IoT-enabled device that not only detects gas leakages in cylinders and sends an alert SMS to the user but also send a SMS to an automatic booking system when the cylinder is empty [17]. Kodali et al. proposed an inexpensive industrial plant safety system for detecting gas leaks using IoT technology [18]. Due to various types of gases that can be present in an industrial plant environment, different sensors such as MQ4, MQ6, and MQ135 are used in this solution to detect LPG, Methane, and Benzene Gas, respectively. The data collected by these sensors is then uploaded to cloud storage using an ESP-32 Wi-Fi module. Zualkernan et al. proposed an Emergency Evacuation System designed for indoor emergencies such as earthquakes or fires [19]. The system employs smoke and temperature sensors in order to identify areas of danger and offer guidance for evacuation. A similar system with A* algorithm to find the shortest path is proposed in [20].

As evident from these existing studies, the fire prevention systems are designed as standalone entities. However, in a restaurant environment, it is important the fire prevention/alarm system is used effectively to inform the patrons in the case of a fire. Also importantly, the fire management system should function even in situations with no network connectivity, which could be the case in a fire emergency. Taking all these into consideration, in this paper, we have designed and implemented a comprehensive restaurant management and fire management system that (a) uses LCD-based interactive user input system to order meals, (b) incorporate previous meal orders of the customers to make meal suggestions, (c) uses the meal ordering system to inform the patrons in case of a fire emergency, and (d) a reliable and responsive fire detection and management system that does not rely on Internet connectivity.

3 Proposed System

As shown in Fig. 1, the proposed system comprises of two sub-systems: food ordering and safety. The overall function of these sub-systems are as follows (Fig. 2).

3.1 Fire Safety

For safety purposes, the kitchen is designed using smart sensors such as smoke detectors, solenoid locks, wireless connectivity module (nRF) to send sensed data to the central database of the system and onto the Arduino IoT cloud as well. As shown in Fig. 4 (a), the MQ-2 gas sensor which is connected to the cloud

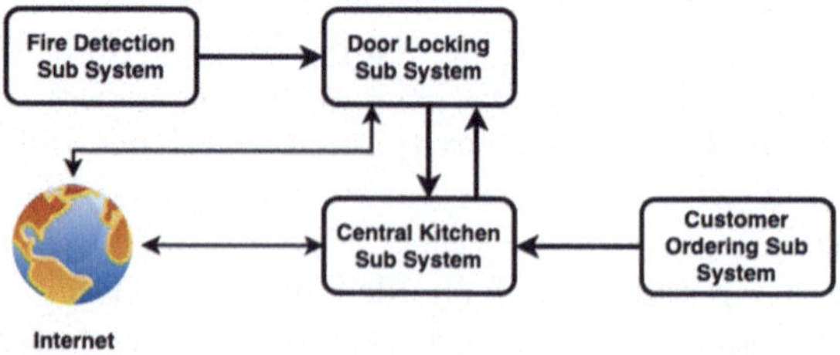

Fig. 1. Block diagram of the food ordering and fire prevention system

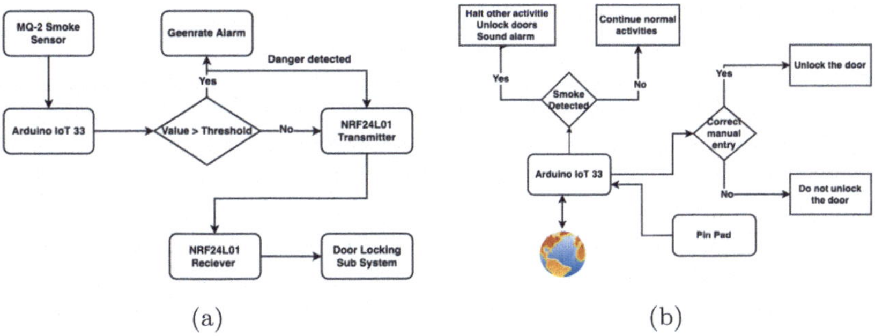

Fig. 2. Fire management systems with (a) door locking sub system and (b) door locking subsystem

through the Arduino IoT 33 board continuously sense the smoke level in the vicinity of the kitchen. This sensed data is transmitted to the Arduino IoT 33 board that controls the door locking system via a nRF transmitter-receiver pair. If the smoke level detected is above a pre-determined threshold, the patrons will be notified through the ordering system. The door locking system will pause all the other activities and initiate unlocking the doors for patrons to reach safety. If the smoke detected is below the threshold, normal operation will continue uninterrupted.

3.2 Food Ordering

The food ordering sub-system has features such as electronic tablet-based menus, customer registration through face recognition, online payment system, and a customer feedback system as shown in Fig. 3. The customer has to first get registered into the system using a facial recognition system and collect a unique customer ID. It will serve as a reference for future recommendations on food when a customer revisits the restaurant. Once the customer has completed the initial registration, he/she can proceed to order food. When an order is placed through, the information is conveyed to the kitchen-side web interface. This information is used by the chef and other staff members of the restaurant and is also stored in the central database of the proposed system. The customer has the option of making the payment online thereby reducing the stress and time involved in standing in the queue for the final billing and improving the overall customer experience. The central database also contains details of all restaurant staff, which can be used to perform an analysis of staff performance as and when required. The system also uses Arduino IoT cloud to display the real-time sensor data of the kitchen and the number of orders received and delivered on a particular day.

3.3 Face Recognition System

The facial recognition system utilizes the Local Binary Pattern Histogram (LBPH) algorithm from the OpenCV library. The fundamental principle of

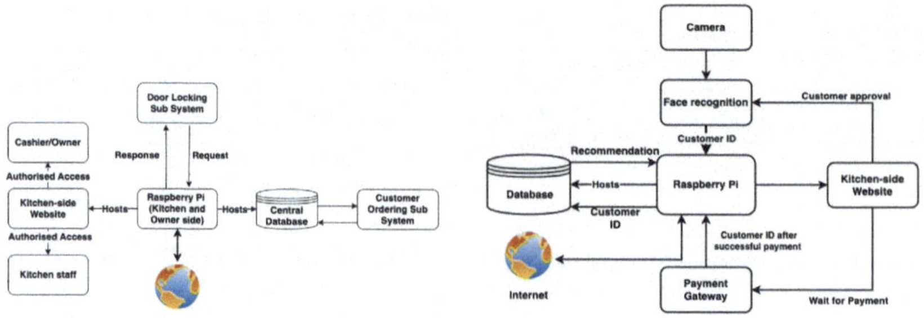

(a) (b)

Fig. 3. Food ordering system with (a) central kitchen side (b) customer side

LBPH involves using the Local Binary Pattern (LBP) texture descriptor to identify patterns within facial images. During the training phase, using a dataset of labeled facial images, serialized and stored using pickle, LBPH extracts local binary pattern histograms from segmented facial regions. The algorithm systematically forges associations between these extracted features and their corresponding identity labels. During the recognition phase, LBPH extracts local binary pattern histograms from the input facial image and compares against known faces within the training set and identifies the person under a certain confidence thresholds. The accuracy of the proposed model depends on multiple parameters such as lighting conditions, pose variations, and the size of the training dataset. It is important to note the confidence level intervals of the model can be adjusted based on the application.

4 Performance Analysis

The proposed system is designed to satisfy different embedded system design principles as discussed in the following subsections.

4.1 Design for Hardware Fault Tolerance

As shown in Fig. 1, the overall system consists of multiple hardware subsystems such as gas leakage detection, automatic and manual door control, and a network communication system. The gas leakage subsystem is responsible for accurately detecting the LPG in the kitchen and generating an alarm when it crosses a pre-determined threshold level. The MQ2 sensor (LPG detector) is connected to the Arduino nano33 IoT board, which uses a nRF transceiver module to send signals to the door locking system in the case of a fire emergency. The hardware design of this subsystem ensures hardware faults such as failure of nRF module initialization and failure of power and the microcontroller unit are dealt using audio and visual indicators. For this purpose, during the initial boot-up, the system will check if the nRF transceiver is operating as intended. If the transceiver has not responded after a certain time lapse, the system will sound a buzzer to indicate that there is a fault in the underlying hardware. Further, the system will also print "Hardware is not responding" onto the display screen. However, if the transceiver responds correctly and there is no error in either the fire or ordering systems, a green LED will glow, indicating a successful start of the system. Further, in case of a power or microcontroller failure, the green LED will not glow. These audio and visual alarms will therefore help notify the staff in case there are any hardware failures in the system.

4.2 Design for Network Fault Tolerance

As illustrated in Fig. 1, the overall system uses two types of wireless communication protocols: IEEE 802.15.4 for nRF modules and IEEE 802.11 for WiFi communication between the Arduino Nano 33 IoT board and the central database server deployed on the Raspberry Pi. To ensure a reliable connection to the Internet, in the case of a Wi-Fi failure, the system will repeatedly try to reconnect to Wi-Fi. Most importantly, for transmitting data collected by the sensors to the door locking system, nRF transceivers are used in the proposed system, which enable communication between two transceivers even in the absence of Wi-Fi connectivity to provide QoS(2) services. On the other hand, the central database server, food ordering system, and door locking subsystems are connected over the Internet to ensure QoS(0) services to the whole system. As a result, even during a fire or a hazardous situation where Wi-Fi connectivity is lost, the proposed system is still able to detect a fire and activate the door locking system. Furthermore, these radio transceivers (with a range of more than 60 m) outperform conventional Wi-Fi (with a range of less than 50 m).

4.3 Design for Performance

In order to maximize its performance, FreeRTOS, a real-time operating system for embedded devices, is incorporated into the proposed system. The door locking system specifically has three main tasks that run in parallel to improve the overall performance and can activate the system based on where the control signal is originated from, namely (1) update from database, (2) keypad unlock, and (3) nRFReciever.

The **Update from database** task is designed to get live updates from the central database to open/close the door remotely through the website. For this purpose, the task continuously checks the value that is present at the central database. If the value is set as "lock", it will lock the door or continue to keep the door locked. Once the value is set as "unlock" in the central database, either manually by a user or automatically in case of an emergency, the system will unlock the door.

The **Keypad Unlock task** checks if a user has entered a certain key combination through the keypad. If they enter the correct password, the door will be unlocked for only 4 s. However, the system will first suspend "Update from Server" task as the doors are unlocked manually and do not want interference

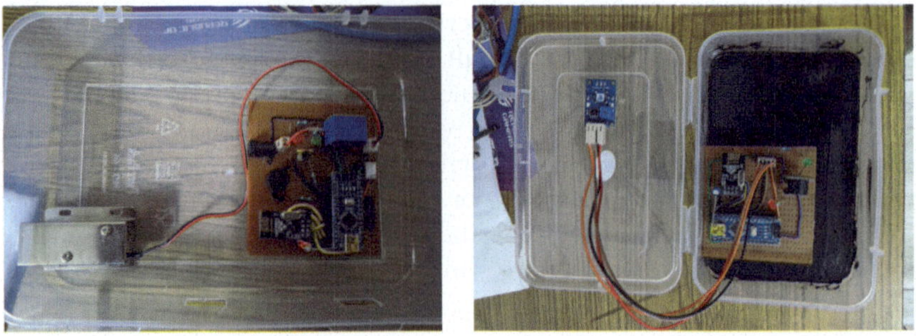

(a) (b)

Fig. 4. Door locking system (a) with all the components and (b) with major components removed from the PCB

from the server. Moreover, the door automatically locks itself and resumes the "Update from Server" tasks.

The **nRFReceive** task uses the data received from the sensors to control the door lock system. If it receives "danger" values from the sensor, it will automatically alarm the users by turning on a buzzer and unlocking the emergency exits. Moreover, this task will stop all the tasks running in the background so that there can be no interference from the server. This task can only be terminated by physically pressing the reset button.

4.4 Design for Maintenance

One of the key features of the proposed system is the ease of maintainability as discussed with respect to the door looking system. Figures 4a and 4b illustrate the system deployed in the door locking system with all the components and with major components removed, respectively. Most of the components have a plug-and-play functionality and can be replaced by a new part easily without needing to actually open and loosen all the connections. The replaceable components include buzzer, relay, nRF24L01, keypad, lock, and microcontroller, which is the Arduino Nano IoT 33 is shown in Fig. 4.

4.5 Unit Testing

The overall system relies on three main functions, namely, face recognition, door locking system, and transceiver functionality, and were tested separately under unit testing as follows. In order to test the face recognition system, each individual was assigned a number by the software. For example, if Person A is given the number 1, the software will print 1 in the terminal if it successfully recognizes the person. If the system misidentifies the person, it will print an unknown or a number associated with other person's face stored in the database. Using this iterative approach, the system was fine-tuned for better results and high accuracy. To test the response time of the data transmitted from the sensor to the door, it is important to first understand the data transmission between the transceivers. The data transmission of the radio transceiver works in such a way that once the data is sent, it will wait for 130 microseconds for an acknowledgment. If the data is successfully transmitted, it will be acknowledged. Using the Arduino's built-in function millies(), the time just before transmission and the time just after transmission are noted and then subtracted to get the final time, which is the total time that the proposed system took to transmit the data to the other system. The system reported an average response time of 0.5 s.

5 Conclusion

The proposed system offers a solution to two significant challenges faced by the restaurant industry: inefficient manual order-taking with no personalizing and the lack of a reliable and responsive fire management system that functions even in the absence of Internet connectivity. Further, the system is designed for performance, fault tolerance, both hardware and network, and reports an average response time of 0.5 s which is sufficient for the underlying application. In future, the proposed system will be upgraded with web-application through which food ordering can be made by the customer own smart device and they will receive future recommendations and offers for their next visit, which will help in increasing revenue growth for restaurant industry.

References

1. Rani, S., Ahmed, S. H., Rastogi, R.: Dynamic clustering approach based on wireless sensor networks genetic algorithm for IoT applications. Wirel. Netw. **26**(4), 2307–2316 (2020). https://doi.org/10.1007/s11276-019-02083-7
2. Kumar, A., Sharma, S., Goyal, N., Singh, A., Cheng, X., Singh, P.: Secure and energy-efficient smart building architecture with emerging technology IoT. Comput. Commun. **176**, 207–217 (2021). https://doi.org/10.1016/j.comcom.2021.06.003
3. Ranjan, V., Masiwal, N., Verma, N., Bharadi, V.: e-Restaurant: online restaurant management system for android. International Journal of Advanced Computer Science and Applications (IJACSA), Special Issue on Selected Papers from International Conference and Workshop On Advance Computing (2013). https://doi.org/10.14569/SpecialIssue.2013.030108
4. Qamar, T., Tariq, S., Aqeel, A., Jaffri, F., Hussain, T. (n.d.). SMARTABLE - Digital Ordering System for Restaurant Using Android
5. Yahaya, Z., Zailani, M.N.M., Soh, Z.H.C., Ahmad, K.A.: IoT based system for monitoring and control of gas leaking. In: 2020 1st International Conference on Information Technology, Advanced Mechanical and Electrical Engineering (ICITAMEE), pp. 122–127, Yogyakarta, Indonesia (2020). https://doi.org/10.1109/ICITAMEE50454.2020.9398384
6. Khairunnisa, K., Johari, A., Wahab, M., Erdi, M., Mustafa, Ayob, M., Ayob, A.: The application of wireless food ordering system. MASAUM J. Comput. **1**(2), 178–184 (2009)

7. Hashim, N.M.Z., Ali, N.A., Ja'afar, A.S., Mohamad, N.R., Salahuddin, L.B., Ishak, N.A.: Smart ordering system via Bluetooth. Int. J. Comput. Trends Technol. (IJCTT) **4**(7), 2253–2256 (2013)
8. Patil, S., Salunkhe, S., Kulkarni, N., Savant, P.: E-Converse: an affordable touch screen solution to intrigue dinning experience. Int. J. Comput. Commun. Technol. **7**(1), 21–25 (2016)
9. Wadile, K., Sushma, A.: E-Restaurant management system using robot. Int. J. Inf. Futuristic Res. **2**(6), 1840–1847 (2015)
10. Sarkar, S., Shinde, R., Thakare, P., Dhomne, N., Bhakare, K.: Integration of touch technology in restaurants using android. Int. J. Comput. Sci. Mob. Comput. **3**(2), 721–728 (2014). https://doi.org/10.47893/IJCCT.2016.1328
11. Sarmaputra, S.: The food pre-order system for restaurant using NFC based smartphone. In: International Conference on Advances in Information Technology (2012). https://api.semanticscholar.org/CorpusID:40005657
12. Bharadi, V.A., Ranjan, V., Masiwal, N., Varma, N.: Intelligent e-restaurant using android OS. In: IJAIS Proceedings on International Conference and Workshop on Advanced Computing, ICWAC(4), pp. 18–24 (2013)
13. Kurzekar, R., Arora, H., Shrestha, R.: Embedded hardware prototype for gas detection and monitoring system in android mobile platform. In: 2017 IEEE International Symposium on Nanoelectronic and Information Systems (iNIS), pp. 6–10, Bhopal, India (2017). https://doi.org/10.1109/iNIS.2017.11
14. Santiputri, M., Tio, M.: IoT-based gas leak detection device. In: 2018 International Conference on Applied Engineering (ICAE), pp. 1–4, Batam, Indonesia (2018). https://doi.org/10.1109/INCAE.2018.8579396
15. Unnikrishnan, S., Razil, M., Benny, J., Varghese, S., Hari, C.V.: LPG monitoring and leakage detection system. In: 2017 International Conference on Wireless Communications, Signal Processing and Networking (WiSPNET), pp. 1990–1993, Chennai, India (2017). https://doi.org/10.1109/wispnet.2017.8300109
16. Shingan, G.G., Sambhare, S.V., Bhokare, V.S., Nikam, A.L., Shinde, H.D.: Smart gas cylinder: leakage alert and automatic booking. In: 2017 International Conference on Energy, Communication, Data Analytics and Soft Computing (ICECDS), pp. 1127–1130, Chennai, India (2017). https://doi.org/10.1109/ICECDS.2017.8389616
17. Shrestha, S., Anne, V. P. K., Chaitanya, A. R.: IoT based smart gas management system. In: 2019 3rd International Conference on Trends in Electronics and Informatics (ICOEI), pp. 550–555, Tirunelveli, India (2019). https://doi.org/10.1109/ICOEI.2019.8862639
18. Kodali, R.K., Greeshma, R.N.V., Nimmanapalli, K.P., Borra, Y.K.Y.: IoT-based industrial plant safety gas leakage detection system. In: 2018 4th International Conference on Computing Communication and Automation (ICCCA), pp. 1–5, Greater Noida, India (2018). https://doi.org/10.1109/CCAA.2018.8777463
19. Zualkernan, I.A., Aloul, F.A., Sakkia, V., Noman, H.A., Sowdagar, S., Hammadi, O.A.: An IoT-based emergency evacuation system. In: 2019 IEEE International Conference on Internet of Things and Intelligence System (IoTaIS), pp. 62–66, Bali, Indonesia (2019). https://doi.org/10.1109/IoTaIS47347.2019.8980381
20. Khan, A., Aesha, A.A., Aka, J.S., Rahman, S.M.F., Rahman, M.J.U.: An IoT based intelligent fire evacuation system. In: 2018 21st International Conference of Computer and Information Technology (ICCIT), December 2018. https://doi.org/10.1109/iccitechn.2018.8631945

Design Framework for Building Smart Shopping Stores Using IoT

Shravel Sharma[1], Sohil Nagpal[1], Yuvraj Bansal[1], Amit Kumar[1(✉)], and Imali Dias[2]

[1] Chitkara University Institute of Engineering and Technology, Chitkara University, Rajpura 140401, Punjab, India
amit.pandey@chitkara.edu.in

[2] School of Information Technology, Deakin University, Geelong 3220, Australia

Abstract. The advancement in the electronics industry in conjunction with wired and wireless communication technologies, has resulted in an immense growth in electronic commerce (e-commerce) in recent years. These days, shopping and making purchases at various shopping premises have become a daily pastime. Improving the experience of shoppers during these shopping endeavors can have a positive impact on individual shopping patterns and in turn, can result in further revenue growth. To this end, this work focuses on the design and development of a low-cost, reliable, and standalone solution with a simple installation procedure that can be used for shopping, inventory management, and security needs. The system specifically assists diverse range of shoppers in making decisions about their spending when choosing the items, thereby dynamically changing their shopping cart while they shop. The overall automated system will also ensure an easy checkout process and security and theft management system at shopping venues.

Keywords: Internet of Things · Embedded Systems · Smart Shopping · RFID · Arduino Nano 33 IoT · Raspberry Pi

1 Introduction

In recent years, technological advancements paired with infrastructure developments have seen a shift in how we shop [1]. The accessibility, the availability of range of products at one place, and promotional offers have drawn customers to small to large retail establishments such as convenience stores, supermarkets, and shopping malls for their daily and specialized shopping needs, with the customer experience playing a vital role in choosing a product. The competitive market structure, large numbers of products, varying scale of usability, etc. open avenues for the latest technologies to provide better customer experience [2].

The increased popularity of these shopping venues on the other hand, has resulted in multiple challenges such as having to spend a significant amount of time locating the necessary items, calculate the cost of the items against the money available, waiting in long queues, manual scanning of the items, and

D. Pastor-Escuredo et al. (Eds.): ICDLAIR 2023, LNNS 1001, pp. 405–414, 2024.
https://doi.org/10.1007/978-3-031-60935-0_37

human errors in generating bills [3]. Further, there can be other challenges in a shopping environment such as inventory, security, and theft management.

The exponential growth in Internet of Things (IoT) in recent years has seen many manual activities been automated across multiple areas [4,5]. To overcome the aforementioned challenges, IoT-based automated systems combined with the latest technologies and tools have been identified as an efficient solution [6–10]. For this purpose, the current literature in e-commerce has investigated smart shopping cart systems [7], mobile and web applications [10], and machine learning and other artificial intelligence (AI)-based solutions [11]. The proposed systems in existing literature however, address one or more of the aforementioned challenges. A comprehensive framework that addresses customer satisfaction at different phases of their shopping experience is yet to be proposed. To this end, this paper proposes an intelligent shopping cart framework using an IoT architecture and smart sensors to enhance customer experiences, mainly at retail stores and/or shopping malls. The system provides an end-to-end solution for retail shop owners with an enhanced user experience and a complete automatic process for purchase, inventory management, and security. In the next section, we will present the existing work carried out in smart shopping systems to highlight the motivation behind our framework.

2 Existing Literature

The current literature in intelligent shopping systems have addressed strategies to improve user experience in both online [12] and on-site shopping. This sections presents such work done as hardware improvements or technology advancements integrated into such hardware solutions.

Hardware Improvements: The existing work on hardware-based solutions mainly focuses on integrating bar code readers, Radio Frequency Identification (RFID), different micro controllers such as Arduino, Raspberry Pi, and ESP32 into embedded systems. Among such work, extensive work has been carried out in designing and deploying smart shopping carts. In the study reported in [9], the authors have proposed a smart shopping cart equipped with a bar code scanner which is integrated with the customer mobile phone to instantly show the customers the products they have selected.

In [6], Ali *et al.* have reviewed the use of RFID readers to identify the items selected by the shoppers. The appropriateness of the RFID technology over bar codes is presented in [7]. Characteristics such as being able to read multiple tags simultaneously, being able to both read and write, no requirement for line of sight and improved security have been identified as favorable features over bar code scanners. The study also presents a system consisting of a RFID reader, RFID tag, LCD display, ESP8266 Wi-Fi module, and a database integrated with a website to maintain product and customer information. The smart shopping cart proposed in [8] performs similar functions, but uses ZigBee technology to communicate between the shopping cart and the central billing unit. Taking the RFID technology a step further, both cost and space management is addressed in [13]. The system allows both online and onsite shoppers to locate the items which will then be placed on a conveyor belt to reach the customer. Similarly in [14], Rouniyar *et al.* have proposed a RFID-based system which displays the product information including the price at the exit gate of the shopping center.

In addition to displaying the item information, some studies have also proposed systems that guides the shoppers to a certain product, thereby reducing the time spent in searching for the item. In the studies reported in [3] and [16], the authors have proposed a RFID based system paired with an mobile application to automatically pay the bills and guide the shopper to a certain product. The automated shopping cart proposed in [17] direct the cart towards the item, rather than the shoppers manually handling it. The proposed solution also ensures security and reliability of the system.

Integrating Supporting Technologies: In conjunction with advancements in hardware systems, algorithm-based approaches such as machine learning and AI have also been incorporated into smart shopping systems in recent literature. For example, in [18], Chi *et al.* have proposed an image recognition system to identify the items in a shopping cart. The study has used deep learning techniques such as Faster R-CNN, YOLOv2, and YOLOv2-Tiny in their work to identify content in the video feed. A linear regression model is proposed in [11] to predict the amount of food required based on time-series data. The RFID-enabled shopping cart in [19] uses Apriori algorithm to identify frequently bought items and make recommendations accordingly.

In summary, the studies reported in Subsects. 2 and 2 have incorporated many advancements such as latest microcontrollers, sensors, machine learning, AI, and context-aware communication into their solutions to overcome different challenges. However, majority of these solutions addresses a single challenge while some rely on availability of customer equipment such as mobile phones as a key component of the system. Such solutions may limit the overall usability of the system and most importantly, the range of customers who can be catered by the system.

The current literature therefore, warrants a system that improves the overall shopping experience across a wide range of customers using a simple but a reliable solution that also assists with inventory management, security and theft control at the shopping centers. For this purpose, in Sect. 1, we present our smart shopping cart system, which is a simple, energy-efficient, and cost effective solution also characterized by real-time data exchange, multi-layer authentication, proactive customer communication, and physical security measures for improved customer experience and store management.

3 Proposed Framework

The proposed embedded systems framework for the smart shopping center is presented in Fig. 1. The framework consists of three sub-sections: a smart shopping cart, an automated billing counter, and sensor-enabled exit gates.

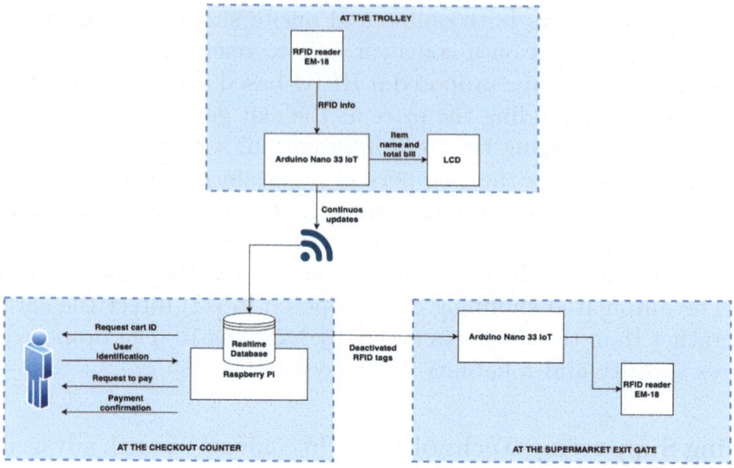

Fig. 1. Design Framework for Smart Shopping Store

Under the proposed solution, the existing shopping carts at retail stores will be upgraded with a device that consists of RFID receivers and LCD display unit connected to an Arduino nano 33 IoT. The carts are connected to an automated billing counter through wireless connectivity, and display real-time updates of products purchased by customers with their total amount at any given time. This helps customers keep track of their total bill, decide on products to be purchased, and avoid queuing and hassle during payments at the billing counters. Further, the automated billing counter only requires the shopping cart identification number and customer ID to complete the final billing and provides both online and offline payment options. This in turn reduces the time the customers have to stand up in the queue and wait for their turn to make payment. In the end, sensor-based security gates add an extra level of security to the entire retail shopping store, as the system automatically fetches the data from the billing counter and runs an RFID scan of all products simultaneously while the customer walks through the exit gate. If any item is left un-scanned, an alert will be generated for the store owner and the person at the billing counter. Thus, the overall embedded system framework provides an improved shopping experience for all customers and employees working in the store.

3.1 Smart Shopping Cart

The design architecture of the shopping cart and the prototype developed are shown in Fig. 2 and 3, respectively. The shopping cart has an IoT node that comprises a microcontroller device, Arduino Nano 33 IoT, an EM-18 RFID reader, a 16×2 LCD, a buzzer, and a battery-operated power unit.

The EM-18 RFID reader is interfaced with the Arduino Nano 33 IoT microcontroller using the Universal Asynchronous Receiver Transmitter (UART) com-

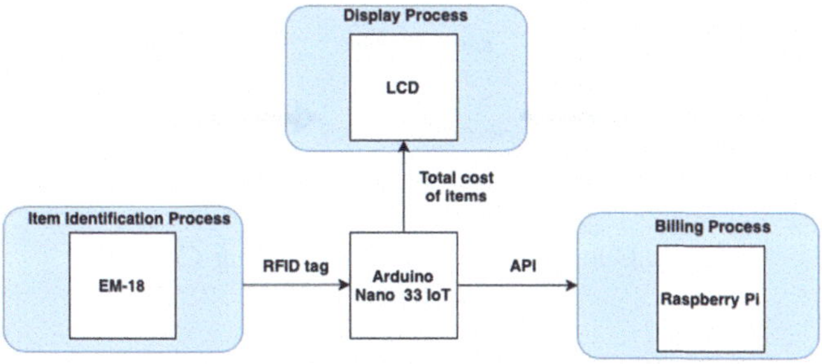

Fig. 2. Smart Shopping Cart System Design

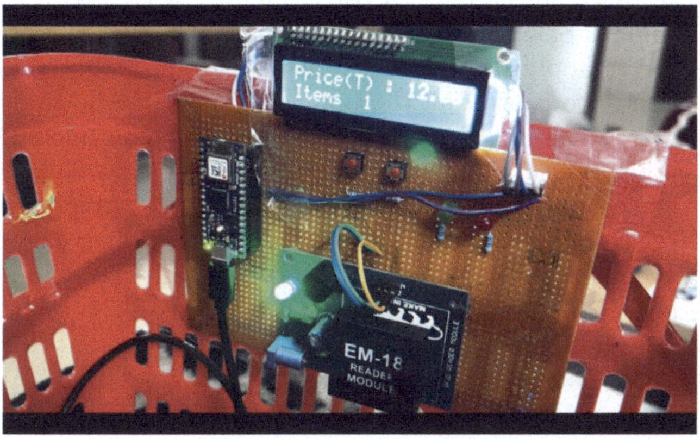

Fig. 3. Prototype Design for the Shopping Cart

munication protocol and scans passive RFID tags at a frequency of 125 kHz. It requires a 5-volt DC voltage and a current of less than 50 mA, ensuring low power consumption for the system. The RFID reader automatically identifies and tracks tags attached to objects using electromagnetic waves. Further, the display unit,16×2 LCD, of this node is interfaced using the Inter-Integrated Circuit (I2C) protocol with the microcontroller. The node also contains two push buttons, which help customers add or delete the respective scanned product from their cart during shopping at any point in time. This node is further connected to an automated billing counter using Wi-Fi for real-time data exchange.

The overall system uses Google Firebase as the central database to store information of all customers and store inventory as shown in Fig. 4. The control unit of the cart is also connected to the Firebase to exchange real-time data updates during shopping. Every time a new item is added/deleted to the cart, it will update the values in the central database. It also holds the data of all the smart shopping carts present in the store. The customers can view their total bill on this IoT node and proceed toward the billing counter after they finish scanning all the required products.

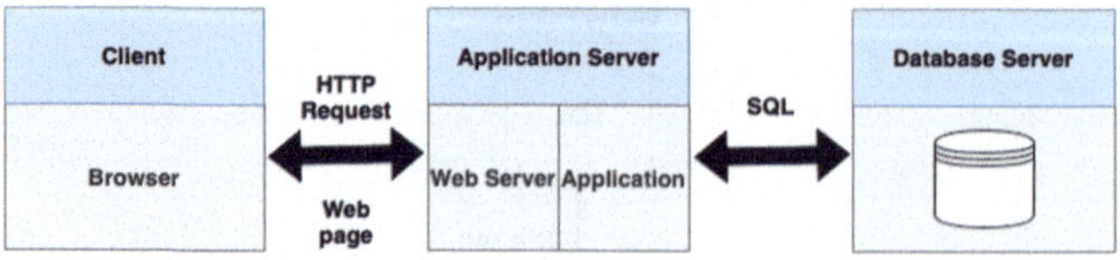

Fig. 4. Client-Server model for Billing Counter

3.2 Automated Billing Counter

The automated billing counter system comprises of Raspberry Pi, a display unit, either capacitive touch screen or TFT/LCD monitor screen with keyboard and mouse, and a Point of Sale (PoS) machine for payment. A customized graphical user interface (GUI) application designed using Python programming language and Tkinter Package to control all the billing processes is shown in Fig. 5. The UI is easy to navigate and use by customers or billing personnel. The customer can sign up for the application for future visits or log in as a Guest for one-time access. After the sign-up process, each user will be assigned a unique customer ID for future reference. After successful login, the customer has to enter the shopping cart ID as given on the IoT node of the cart. The billing system will automatically retrieve all information on products available in the shopping cart through the central database. Further, for making payments, the system offers four payment methods: cash, wallet payment, UPI payment, and credit or debit card swipe, through a PoS machine. Once the payment is made, the RFID tags of the items paid get deactivated and updated in the database.

3.3 Inventory Management

The store staff regularly creates and updates the product inventory using the smart shopping cart. Each product is attached to a passive RFID tag, and then it gets scanned through the IoT node of the cart, with information is sent to the central database. After that, all scanned items are put on different shelves for sale purposes. Now this inventory information is available at all automated billing counters.

3.4 Sensor-Enabled Exit Gate

The design framework offers sensor-enabled security gates to provide an extra layer of security to the overall system. The exit gate also contains an IoT node designed using an Arduino Nano 33 IoT board, EM18-RFID reader, and a buzzer. The security system is constantly retrieving data from the billing counter, updating all purchased items as well as their tags. The automatic exit gate will open only when the passive RFID tags of all the items in the cart get deactivated at the billing counter. As mentioned before, ss soon as the billing counter receives the payment, it deactivates passive RFID tags and updates this information in

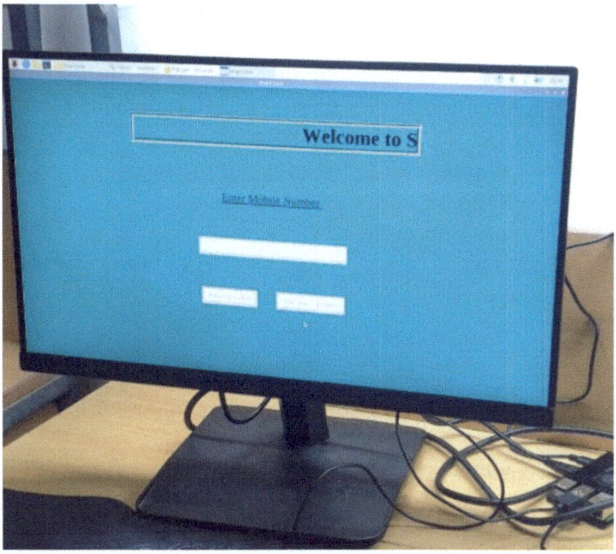

Fig. 5. Graphical User Interface of the Prototype

the Google Firebase. The control unit at the exit gate, Arduino Nano 33 IoT board, fetches this information from Firebase and runs real-time scanning of RFID product tags in one go. If any tag remains active, it generates an alarm to inform the person at the billing counter and automatically locks the exit gates. The proposed framework therefore, ensures the theft and inventory management of in the retail or shopping store.

4 Results Analysis and Discussion

This section presents the performance metrics of our proposed IoT-enabled shopping system in terms of cost and power consumption. For this purpose, we have compared our system to a similar framework proposed in [9]. Table 1 lists the power consumption values of the components in the device connected to the cart in our proposed system against the one proposed in [9]. Based on the values, our proposed system incurs more cost compared to the one proposed in [9]. However, it is important to note that the system proposed in [9] uses a bar code scanner, where as our proposed system uses RFID technology, which outperforms the bar codes as discussed in [7]. Most importantly, our proposed system does not rely on a customer mobile phone and is a stand-alone system that can be easily used by anyone. Considering the simplicity, usability, and reliability associated with our proposed system, the system is cost effective compared to existing frameworks.

Table 2 lists the power consumption values of the components used in our proposed system and the system proposed in [9]. Based on these values, our system consumes more power than the one in [9]. However, it is important to note

Table 1. Cost comparison [20]

	Our Proposed System		System Proposed in [9]	
	Component	Cost (INR)	Component	Cost(INR)
Device in the cart	Arduino Nano 33 IoT	2469	ESP8266	270
	LCD	199	HX711	453
	RFID	549		
Total		3217		723

Table 2. Power consumption comparison

	Our Proposed System		System Proposed in [9]	
	Component	Power (W)	Component	Power (W)
Device in the cart	Arduino Nano 33 IoT	0.02	ESP8266	0.5
	LCD	1	HX711	0.2
	RFID	2		

that these values are comparatively low and as discussed before, the system does not rely on customer mobile phone which otherwise can add another dependency and power consuming element to the system.

The proposed system also outperforms similar existing frameworks due to the technology stack used in the design such as Firebase, MySQL, and Simple Mail Transfer Protocol (SMTP). The Firebase real-time database offers a distinct advantage by providing immediate access to data, thereby reducing delays in the billing process. The use of MySQL, combined with password-authenticated wallet payments, sets our system apart as many frameworks lack this dual-layered security approach, relying solely on basic authentication methods. The additional security measures in our framework significantly enhance the overall safety of payment transactions. Further, the use of SMTP-enabled email notifications directly integrated into the system offers a more streamlined and real-time approach to keeping customers informed. This direct integration surpasses frameworks that require external communication platforms, providing a more cohesive and user-friendly experience. Most importantly, existing frameworks often lack the integration of physical security components. The use of Arduino nano to communicate with the security system, triggering alarms for unscanned RFID tags, is a unique feature. Many frameworks may not incorporate such physical security measures, leaving them susceptible to unauthorized removal or theft. The advanced security measures of the proposed system ther4fore go beyond software-centric approaches, ensuring a more comprehensive security solution.

5 Conclusion

The design framework proposed in this work uses a web-based IoT architecture and low-cost embedded devices. The system provides reliable and cost-effective end-to-end solution to cater for wide range of consumers and business owners. The framework has the capability to easily upgrade the existing trolleys or baskets available at stores with smart IoT nodes and get them connected with a real-time automated billing counter. Further, in the future, dependency on Wi-Fi networks can be overcome by integrating alternative technologies such as Bluetooth low energy (BLE) for communication between the smart IoT nodes and the billing counter. By implementing AI algorithms in this framework, a recommendation system can be added to provide product recommendations for the users based on their previous shopping behaviors. This will further lead to research on data storage, latency-sensitive communication, and communication architectures in smart shopping environments.

References

1. Mogale, S.U., Dalvi, D.C.S.: A study on retailing in India: growth and challenges. Int. J. Trend Sci. Res. Dev. Special Issue(Special-FIIIIPM2019), 95–97 (2019). https://doi.org/10.31142/ijtsrd23073
2. Bascur, C., Rusu, C.: Customer experience in retail: a systematic literature review. Appl. Sci. (Basel, Switzerland) **10**(21), 7644 (2020). https://doi.org/10.3390/app10217644
3. Perera, S., Ekanayake, L., Bandara, A., Shakya, D., Oruthota, U.S.: IoT based smart shopping system. In: 10th International Conference on Information and Automation for Sustainability (ICIAfS), Negambo, Sri Lanka, pp. 225–229 (2021). https://doi.org/10.1109/ICIAfS52090.2021.9606124
4. Rani, S., Ahmed, S. H., Rastogi, R.: Dynamic clustering approach based on wireless sensor networks genetic algorithm for IoT applications. Wirel. Netw. **26**(4), 2307–2316 (2020).https://doi.org/10.1007/s11276-019-02083-7
5. Kumar, A., Sharma, S., Goyal, N., Singh, A., Cheng, X., Singh, P.: Secure and energy-efficient smart building architecture with emerging technology IoT. Comput. Commun. **176**, 207–217 (2021). https://doi.org/10.1016/j.comcom.2021.06.003
6. Ali, Z., Sonkusare, R.: RFID based smart shopping: an overview. In: 2014 International Conference on Advances in Communication and Computing Technologies (ICACACT 2014), Mumbai, India (2014). https://doi.org/10.1109/EIC.2015.7230698
7. Das, T.K., Tripathy, A.K., Srinivasan, K.: A smart trolley for smart shopping. In: 2020 International Conference on System, Computation, Automation and Networking (ICSCAN), Pondicherry, India (2020). https://doi.org/10.1109/ICSCAN49426.2020.9262350
8. Jeyanthy, P.A., Mohamed, S.R., Alarfaj, M.M., Reddy, S.S., Jenin, K.S.: Design and development of a smart shopping cart system. In: 2022 Third International Conference on Intelligent Computing Instrumentation and Control Technologies (ICICICT), pp. 184–188, Kannur, India (2022). https://doi.org/10.1109/ICICICT54557.2022.9917860

9. Mekruksavanich, S.: Design and implementation of the smart shopping basket based on IoT technology. In: 14th International Joint Symposium on Artificial Intelligence and Natural Language Processing (iSAI-NLP), pp. 1–6, Chiang Mai, Thailand (2019). https://doi.org/10.1109/iSAI-NLP48611.2019.9045336
10. Siva Rao, I.S., Rao, M.A, Teja, M.S., Pavan, N.T.S., Krishna, B.S., Raviteja, D.: Shop GO: an IoT based solution for smart shopping. In: 2020 International Conference on Computer Science, Engineering and Applications (ICCSEA), pp. 1–6, Gunupur, India (2020). https://doi.org/10.1109/ICCSEA49143.2020.9132889
11. Nastiti, M. D., Abdurohman, M., Putrada, A. G.: Smart shopping prediction on smart shopping with linear regression method. In: 7th International Conference on Information and Communication Technology (ICoICT), pp. 1–6, Kuala Lumpur, Malaysia (2019). https://doi.org/10.1109/ICoICT.2019.8835271
12. Vidya, K.P., et al.: Virtual cart: novel approach for revamping smart shopping experience. In: 2018 IEEE Distributed Computing, VLSI, Electrical Circuits and Robotics (DISCOVER), pp. 135–140, Mangalore, India (2018). https://doi.org/10.1109/DISCOVER.2018.8674117
13. Singh, R., Verma, S., Kriti, M.: RFID and IR based smart shopping mart management system. In: 2018 International Conference on Advances in Computing, Communication Control and Networking (ICACCCN), pp. 536–540, Greater Noida, India (2018). https://doi.org/10.1109/ICACCCN.2018.8748820
14. Rouniyar, F.P.R., Saxena, S.P., Sahoo, T.A.K.: SSAS: RFID-BASED smart shopping automation system. In: 2020 International Conference on Communication and Signal Processing (ICCSP), pp. 1016–1020, Chennai, India (2020). https://doi.org/10.1109/ICCSP48568.2020.9182354
15. Pradeepkumar, G., Ramesh, V., Karthika, A., Karthikeyan, A., Sacithraa, R., Vanitha, K.: Smart shopping trolley based on IoT with mobile application. In: 2023 International Conference on Sustainable Computing and Data Communication Systems (ICSCDS), pp. 1221–1225, Erode, India (2023). https://doi.org/10.1109/ICSCDS56580.2023.10104651
16. Megalingam, R.K., Vishnu, S., Sekhar, S., Sasikumar, V., Sreekumar, S., Nair, T.R.: Design and implementation of an android application for smart shopping. In: 2019 International Conference on Communication and Signal Processing (ICCSP), pp. 0470–0474, Chennai, India (2019). https://doi.org/10.1109/ICCSP.2019.8698109
17. Li, R., Song, T., Capurso, N., Yu, J., Cheng, X.: IoT applications on secure smart shopping. In: 2016 International Conference on Identification, Information and Knowledge in the Internet of Things (IIKI), pp. 238–243, Beijing, China (2016). https://doi.org/10.1109/IIKI.2016.25
18. Chi, H.C., Sarwar, M.A., Daraghmi, Y.A., Lin, K.W., Ik, T.U., Li, Y.L.: Smart self-checkout carts based on deep learning for shopping activity recognition. In: 2020 21st Asia-Pacific Network Operations and Management Symposium (APNOMS), pp. 185–190, Daegu, Korea (South) (2020). https://doi.org/10.23919/APNOMS50412.2020.9237053
19. Kothavale, S., Pawar, S., Kankarej, S., Patil, S., Raut, R.: Smart indoor navigation, shopping recommendation and queue less billing based shopping assistant using AI. In: 2021 International Conference on Artificial Intelligence and Machine Vision (AIMV), pp. 1–4, Gandhinagar, India (2021). https://doi.org/10.1109/AIMV53313.2021.9670948
20. www.robu.in

A Multi-objective Virtual Machine Placement Optimization in Sustainable Cloud Environment

Smruti Rekha Swain[1(✉)], Anshu Parashar[1], Ashutosh Kumar Singh[1], and Chung Nan Lee[2]

[1] National Institute of Technology Kurukshetra, Kurukshetra, India
{smruti_62000063,anshuparashar,ashutosh}@nitkkr.ac.in
[2] National Sun Yat-sen University, Kaohsiung, Taiwan
cnlee@mail.cse.nsysu.edu.tw

Abstract. Virtual Machine Placement involves the selection of the optimal physical machine for deploying a requested virtual machine within extensive cloud data centers. VM placement must be performed strategically by considering different factors of the available resources for optimal exploitation of them. Numerous methods have been devised to address this issue. Nevertheless, existing solutions only account for a restricted set of resource types, leading to an uneven workload distribution that triggers the activation of unnecessary PMs within the data center. This work introduces a Non-dominated sorting genetic algorithm II that integrates various resource-constraint metrics to determine the optimal PMs for deploying VMs in a cloud environment. This algorithm maximizes resource utilization while minimizing the data center's energy consumption and carbon footprints. The algorithm's performance assessment is conducted using the Google Cluster Data set, and the outcomes are compared with established methodologies. The results showcase a substantial reduction in energy consumption, carbon footprints, and number of active PMs by 19.29%, 47.50%, and 79.81%, respectively. Furthermore, there is a noteworthy enhancement in resource utilization, reaching up to 67.28%.

Keywords: Physical Machines · Virtual Machines · Energy Consumption · Resource Utilization · Carbon Footprint

1 Introduction

Cloud computing has gained popularity for its reliability, cost-effectiveness, and scalability in hosting and delivering software solutions across diverse industries. Data centers, which are vital infrastructures, are increasingly adopting cloud platforms to deploy many applications and cater to the service needs of cloud users. The effectiveness of a data center is significantly contingent on how virtual

Supported by NIT Kurukshetra.

D. Pastor-Escuredo et al. (Eds.): ICDLAIR 2023, LNNS 1001, pp. 415–426, 2024.
https://doi.org/10.1007/978-3-031-60935-0_38

machines (VMs) are managed and strategically positioned. Ensuring the efficient operation of cloud data centers requires a substantial energy expenditure. The elevated energy consumption primarily stems from the limited average utilization of the available computing resources. It's crucial to efficiently handle these cloud resources to reduce energy consumption, costs, and carbon emissions. Enhancing resource utilization presents a promising avenue for curbing data centers' substantial energy consumption and environmental impact. One predominant method for achieving improved resource utilization is implementing virtualization technology. Employing a hardware virtualization mechanism, VM consolidation techniques are employed to operate VMs on a minimal number of physical machines (PMs). Furthermore, operational expenses are diminished by utilizing VM management for dynamically provisioning cloud resources in an optimized manner. Numerous research endeavors have addressed Virtual Machine Placement (VMP) challenges, focusing on resolving energy and resource utilization concerns. Numerous resource management methods have recently been suggested to ensure adherence to Service Level Agreements (SLAs) and enhance the overall Quality of Service (QoS). Current endeavors to address the VMP issue are predominantly centered on minimizing server power usage, decreasing the count of active servers, and enhancing data center resource utilization through service consolidation. However, these methodologies often focus solely on primary data center constituents, such as networking, storage, and cooling systems, contributing to a substantial portion (approximately 32%) of the total power consumption in Cloud Data Centers. Consequently, establishing an energy-efficient cloud infrastructure demands a comprehensive resource management strategy encompassing all computing resources, including network, storage, servers, memory, and cooling components.

For a precise solution to the multi-objective and multi-constrained VM allocation problem, this study introduces a server consolidation framework based on Non-Dominated Sorting Genetic Algorithm II (NSGA-II). This framework employs rank-based, non-dominated sorting to generate Pareto-optimal solutions for multiple objectives. Its primary goals are to maximize resource utilization and minimize power consumption and carbon emissions. The Random Fit (RF) and First Fit Decreasing (FFD) algorithms are employed for initial VM-to-PM mapping. At the same time, NSGA-II is harnessed to thoroughly explore the most viable and optimal VM allocations.

Key Contributions

- A multi-objective VM consolidation strategy is introduced to allocate the user-requested VMs onto the available PMs while prioritizing maximum resource utilization, minimum energy consumption, and carbon emission.
- A new multi-objective algorithm named M-NSGA II is introduced to identify feasible and optimal allocations that strive to achieve a balanced overall performance for the data center.
- We have encoded VM allocations within genetic chromosomes and then selected the most optimal VM allocation through a fitness evaluation process based on ranking.

- We conducted experimental evaluations in static and dynamic scenarios to validate the effectiveness of the proposed scheme.

2 Related Work

Saxena et al. [1] introduced a secure and multi-objective virtual machine placement approach, leveraging the NSGA-II algorithm and whale optimization mechanism. This framework aimed to optimize resource utilization concurrently, minimize energy consumption, and reduce communication costs within the VMP process. Sharma et al. [2] combined the principles of Genetic Algorithm and Particle Swarm Optimization to create the HGAPSO mechanism, which aimed to decrease resource inefficiencies and Service Level Agreement violations in VMP processes. They employed a binary representation using bit values to indicate the presence or absence of VMs. When a PM was active, the bit value was set to 1; otherwise, it was 0. Han et al. [3] presented a Security-aware Multi-Objective Optimization-based virtual machine Placement strategy, which relied on a modified genetic algorithm and a weighted sum approach to achieve optimal VM placement. This approach focused on enhancing resource utilization, minimizing network traffic, and mitigating security risks in the placement process. Singh et al. [4] introduced a load-balancing framework known as Secure and Energy-efficient Load-Balancing, primarily focusing on security and energy efficiency. This approach utilizes Genetic Algorithm principles to optimize resource usage, reduce power consumption, and minimize conflicting servers. Tseng et al. [5] presented a technique for forecasting resource demand within a cloud data center by leveraging historical data. This mechanism incorporates Genetic Algorithm techniques and employs the prediction results to facilitate the placement of VMs. Beloglazov et al. [6] put forward the MBFD (Modified Best Fit Decreasing) algorithm as a solution to the VMP challenge, considering energy consumption and adherence to Service Level Agreements. The algorithm operates by arranging virtual machines (VMs) in decreasing order of their processing capacity and physical machines (PMs) in ascending order. Subsequently, the FFD algorithm assigns VMs onto energy-efficient PMs. However, a limitation of this method is its scalability requirement and its reliance on a single objective for VMP. Singh et al. [7] utilize NSGA II and Flower Pollination Optimization algorithm to address VMP challenges in cloud data centers. This approach takes a multi-objective standpoint for VMP, seeking to minimize energy consumption and carbon emissions and maximize resource utilization of the data center. Singh et al. [8] introduced a meta-heuristic Virtual Machine Placement (VMP) technique to minimize the power consumption of the cloud data center. An initial mapping of VMs to PMs was carried out using the RF algorithm, followed by the Flower Pollination Optimization algorithm to achieve optimal allocations. Saxena et al. [9] devised an innovative framework named Fault Tolerant Elastic Resource Management designed to address server overloading and underloading challenges, thereby reducing resource and power inefficiencies in data centers. The techniques involving random-fit (RF), best-fit (BF), and first-fit (FF) [11–13] are employed to reduce energy consumption in data centers by utilizing a reduced number of active physical machines (APMs). These approaches are expected to provide globally optimal solutions for multi-objective challenges. However, the extensibility of the Best-fit heuristic is constrained due to its extended convergence period.

3 Proposed Model

The proposed model considers a data center infrastructure consisting of a set of P host machines, denoted as PM_1, PM_2, ..., PM_P belonging to the set PM. Additionally, there are Q VMs labeled as VM_1, VM_2, ..., VM_Q, belonging to the set VM, and M users represented as CU_1, CU_2, ..., CU_M, belonging to the set U, as illustrated in Fig. 1. Each physical machine (PM_i) has CPU (PM_i^{CPU}), Memory (PM_i^{Mem}), and RAM (PM_i^{RAM}) capacities. Similarly, each VM (VM_j) possesses CPU (VM_j^{CPU}), Memory (VM_j^{Mem}), and RAM (VM_j^{RAM}) capacities. Ever PM operates in two modes: active mode ($\gamma = 1$) or sleep mode ($\gamma = 0$). We define a 2D matrix w with dimensions $P \times Q$, where P and Q denote the total number of PMs and VMs respectively. If PM_i hosts VM_j, we set $w_{ji} = 1$; otherwise, $w_{ji} = 0$. The Cloud Service Provider consistently aims to find optimal VM placement for VMs, improving resource utilization while

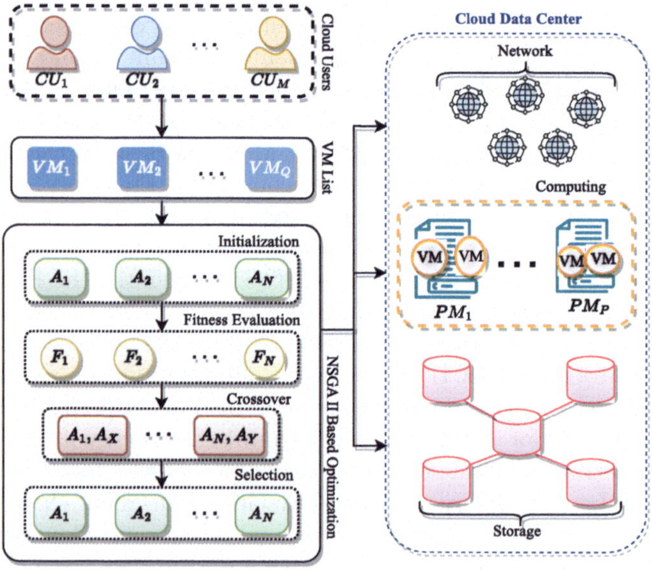

Fig. 1. Proposed Model

minimizing energy consumption and carbon footprint. Consequently, the multi-objective VMP problem is defined as presented in the following Eq. (1)

$$\sum_{i=j=1}^{P,Q} \omega_{ji} = Minf\left(\sum_{i=j=1}^{P,Q} \omega_{ji} \times (PW, CF, -RU)\right), \tag{1}$$

PW, CF, and RU represent Power Consumption, Carbon Footprint, and Resource Utilization for the jth PM. Every VM allocation is feasible only when it complies with the constraints specified in Eq. (2).

$$\sum_{j=1}^{Q} VM_j^r \times \omega_{ji} \leq PM_i^r \quad r \in RAM, CPU, Mem. \tag{2}$$

To accomplish this multi-objective Virtual Machine Placement (VMP), the following objective models have been formulated:

3.1 Resource Utilization

The resource utilization within the cloud data center can be calculated using Eqs. (3) and (4), where 'R' represents the number of resources.

$$RU_{dc} = \int_{t_1}^{t_2} \left(\frac{RU_{dc}^{CPU} + RU_{dc}^{Mem}}{|R| \times \sum_{i=1}^{P} \gamma_i}\right) dt \tag{3}$$

$$RU_{dc}^r = \sum_{i=1}^{P} \frac{\sum_{j=1}^{Q} \omega_{ji} \times VM_j^r}{PM_i^r} \qquad r \in RAM, CPU, Mem, etc. \tag{4}$$

3.2 Energy Consumption

The power consumption for the ith server can be expressed as PW_i, and the total power consumption, PW_{dc}, during the time interval [t1, t2] is represented in Eq. (5)

$$PW_{dc} = \int_{t_1}^{t_2} \left(\sum_{i=1}^{P} ([PW_i{}^{max} - PW_i{}^{min}] \times RU + PW_i{}^{idle})\right) dt, \tag{5}$$

where RU is resource utilization, PW_i^{max}, PW_i^{min} and PW_i^{idle} are maximum, minimum, and idle state power consumption for i^{th} server.

3.3 Carbon Footprint

This study assesses the power usage of typical CRAC (Computer Room Air Conditioning) units. To evaluate the cooling system's efficiency, we employ the CoP (Coefficient of Performance) metric, which quantifies how effectively the cooling system handles the cooling load. CoP is mentioned as the ratio of heat removal (d) to the energy expended (w) for removing that heat. A higher CoP value indicates superior efficiency, indicating the system can remove a significant amount of heat with minimal energy consumption. The CoP metric can be represented in Eq. (6)

$$CoP = \{0.0068T_{SUP}^2 + 0.0008T_{SUP} + 0.458\} \tag{6}$$

Carbon Footprint (CF) at time t is evaluated in Eq. (7).

$$CF(t) = PUE_d \times CFR \times PW \tag{7}$$

In Eq. (8), the dynamic power usage effectiveness (PUE) is used to calculate the data center's energy efficiency.

$$PUE_d = \left(\frac{Total\ facility\ power}{ITpower}\right) \tag{8}$$

In Eq. (9), the Total facility power is expressed.

$$Total\ Facility\ power_d = PW + OP \tag{9}$$

In Eq. (10), the variable OP represents the cumulative overhead power as determined.

$$OP = \frac{PW}{CoP} \tag{10}$$

Here, PW represents the energy consumption by all PMs within the data center.

4 NSGA II Based VM Allocations

The VMs are assigned using a Multi-objective NSGA II (M-NSGA II) approach, which combines the RF and FFD algorithm with the Pareto-optimal selection procedure from the NSGA-II. M-NSGA II involves four main steps: (i) initialization, (ii) Fitness Evaluation, (iii) Crossover followed by Mutation, and (iv) Selection as depicted in Fig. 1. At first, we generate N random VM allocations using the RF and FFD algorithms, represented as A_i for the placement of the ith VM (where $i \leq$ N), and encode them as chromosomes/solutions. These solutions are computed by using a fitness function η^g= [$f(A^g)_{PW}$, $f(A^g)_{CF}$, $f(A^g)_{RU}$] associated with resource utilization [Eq. (4)], energy consumption [Eq. (5)], and carbon-foot rate [Eq. (7)], respectively. Subsequently, We utilize one-point crossover and mutation operations to traverse the entire search space, enhancing solutions by relocating VMs from servers with lower energy efficiency to carefully chosen, more energy-efficient servers. The resulting solutions may initially violate VM allocation constraints, but we rectify this by rearranging the infeasible VM placements to make them feasible. Furthermore, we evaluate the fitness of the updated solutions and perform selection using the NSGA-II algorithm. This process involves replacing the least fit solutions with better ones.

4.1 Time Complexity Analysis

The time complexity analysis for Algorithm 1 can be summarized as follows: Lines 1 -26 generate N instances of VM-PM allocations by employing the RF and FFD algorithms and consume $O(N \times P \times Q)$ time complexity. Line 27 initializes N random solutions with O(1) time complexity. Lines 28-39 involve nested loops, particularly lines 31-37 represent the inner loop responsible for updating N solutions through crossover and mutation operations, which takes O(N) time per iteration. Line 30 applies non-dominated sorting, with its time complexity depending on the number of objectives (ob), the solution size (N), and their complexity, resulting in $O(ob \times N^2)$ time complexity. The outer loop (Line 28) iterates (G_{max}) times. Considering these factors, the overall time complexity can be articulated as: $O(ob \times N^2 \times P^2 \times Q^2 \times \text{Gmax})$.

5 Performance Evaluation

5.1 Experimental Set-Up

The experimental evaluations were performed on a server system with 2 Intel® Xeon® Silver 4114 CPUs, each comprising forty cores running at a clock speed of 2.20 GHz. This computation device has 128 GB of primary memory and runs on 64-bit Ubuntu 16.04 LTS. The data center environment comprises a cluster of servers organized in a linear topology. This setup includes three distinct types of physical machines (PMs), and the study considers four different categories of virtual machines (VMs). The configurations of PM-VM pairings are outlined in Tables 1 and 2. The resource attributes, such as power consumption (P_{max}, P_{min}), MIPS, RAM, and Memory, are sourced from actual server configurations: IBM and Dell. The specific models involved are 'ProLiantM110G5XEON3075', 'IBMX3250Xeonx3480', and 'IBM3550Xeonx5675', corresponding to PM_1, PM_2, and PM_3 respectively. The configurations of VM instances draw inspiration from the setups of VMs available on the Amazon website. To evaluate the proposed scheme's performance and compare it with other approaches, we utilize the Google Cluster Dataset [14].

5.2 Results

The overall performance of the proposed work is detailed in Table 3, encompassing various data center sizes with different combinations of PMs and VMs.

Algorithm 1. Proposed M-NSGA II based VM Allocations

Require: $List_{PM}$, $List_{VM}$, P,Q
Ensure: Optimized VM Allocations

```
for each k = (1, 2, ..., N) do
    for i = 1, 2, ..., P do
        for j = 1, 2, ..., Q do
            while List_PM ≠ Empty do
                for each VM in List_VM do
                    choose VM_j from List_VM
                    choose PM_i from List_PM at random
                    for each ℝ ∈ {Mem, CPU, RAM} do
                        if VM_j^R ≤ PM_i^R then
                            Assign VM_j to PM_i
                            set w_ji=1
                            Compute PM_i^R=PM_i^R - VM_j^R
                            Delete VM_j from List_VM
                        else
                            VM_j is not assigned
                            set w_ji =0
                        end if
                    end for
                end for
            end while
        end for
    end for
end for
for i = 1, 2, ..., P do
    Utilization of FFD algorithm for placement of remaining VMs.
end for
Initialize N random VM allocations (A_1, A_2, ..., A_N).
for i = 1, 2, ..., G_max do
    [PW, CF, RU]=ηA^i
    [A_Non-dominatedsort= NSGA II(A^i)]
    for i = 1, 2, ..., N do
        rn = rand(1, X), crossover point = rand(1, Z), where Z is randomly generated crossover-point
        Cr_1 = [A_i(1 : crossover point), A_rn(crossover point + 1 : Z)]
        Cr_2 = [A_rn(1 : crossover point), A_i(crossover point + 1 : Z)]
        Cr = [Cr, μ(Cr_1), μ(Cr_1)]
        VM^Optimal =Optimal VM Allocation
        [PW,CF,RU] = η(VM^Optimal)
    end for
end for
```

Table 1. PM Configuration

Server	PE	MIPS	RAM(GB)	storage(GB)	P_{max}	P_{min}/P_{idle}
PM_1	2	2660	4	160	135	93.7
PM_2	4	3067	8	250	113	42.3
PM_3	12	3067	16	500	222	58.4

Table 2. VM Configuration

VM Type	PE	MIPS	RAM(GB)	Secondary storage(GB)
VM_S	1	500	0.5	40
VM_M	2	1000	1	60
VM_L	3	1500	2	80
VM_{XL}	4	2000	3	100

Resource utilization (RU) was approximately 70% across all data center sizes. Energy consumption (PW) gradually increased with more VMs, illustrating a linear correlation between run time and the number of requested VMs. Concurrently, the number of active PMs (APMs) increased with the growing number of VMs and users, resulting in a higher rate of carbon emissions (CF).

Table 3. Performance Metrics for Proposed Model

VMs	PMs	User	RU (%)	PW (KW)	CF (tons)	APMs	Exec. time (ms)
50	30	20	69.68	3.93E+3	22.94	13	5.42
100	60	40	69.78	4.72E+3	48.16	20	10.46
200	120	80	69.80	1.57E+4	96.26	29	46.35
400	240	160	69.32	2.52E+4	143.91	70	149.32
600	360	240	69.20	3.42E+4	192.27	87	223.360
800	480	320	69.86	4.30E+4	240.294	101	322.95
1000	600	400	69.38	5.12E+4	287.815	125	394.54

5.3 Comparative Analysis in Dynamic VMP

For comparison, the existing approaches selected are WOGA [1], FF [13], BF [12], and RF [11]. The implementation of VMP has been carried out for all three performance metrics within two separate schemes: static and dynamic VMP. For dynamic VMP, VM configurations are determined during run time, enabling the dynamic allocation and deallocation of VMs. This procedure entails reallocating and removing VMs as necessary, transferring them from servers experiencing over-utilization or under-utilization to the most appropriate and efficient servers.

Various performance metrics are examined during the simulation, including RU, PW, CF, the number of APMs, and operational time. Based on Fig. 2, it is apparent that the proposed scheme attains an RU rate of as high as 69.86%., surpassing several other algorithms' performance. Specifically, it outperforms WOGA, FF, RF, and BF by margins of 11.69%, 59.12%, 62.81%, and 67.28%, respectively. Figure 3 demonstrates that as the data center size increases, the PW also increases. The proposed approach shows the lowest PW compared to the existing techniques while nearly on par with the WOGA method at a data center size of 200. Figure 4 illustrates the CF, which exhibits a linear growth pattern with data center size. Finally, in Fig. 5, a comparison is made regarding the number of APMs during VMP. It is evident that the proposed approach significantly reduces the number of APMs, demonstrating its efficiency compared to existing techniques.

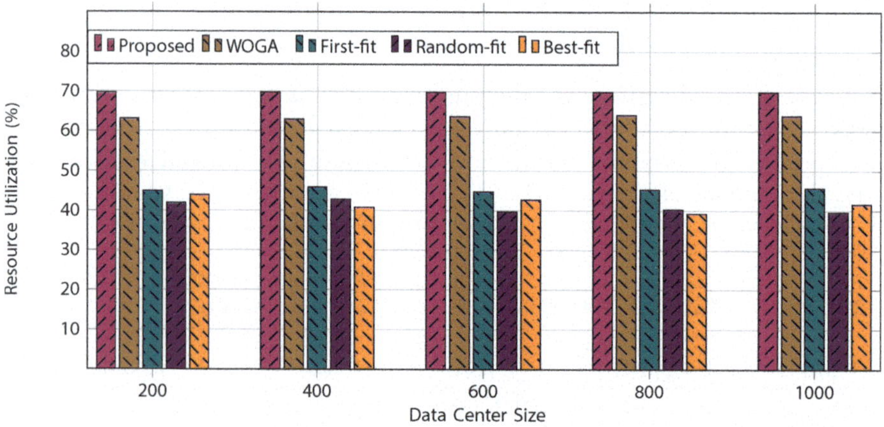

Fig. 2. Resource Utilization

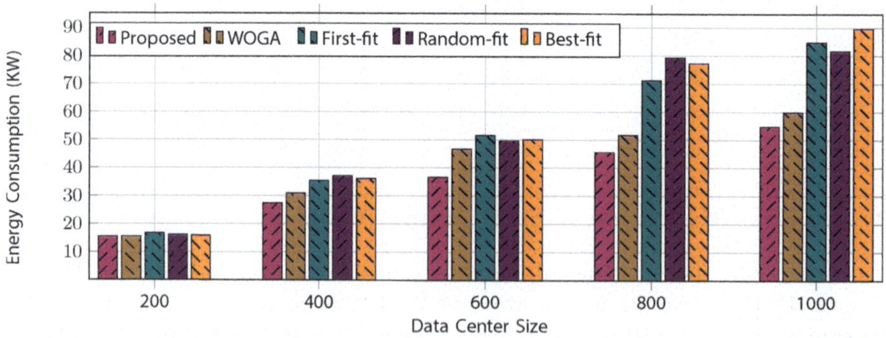

Fig. 3. Energy Consumption

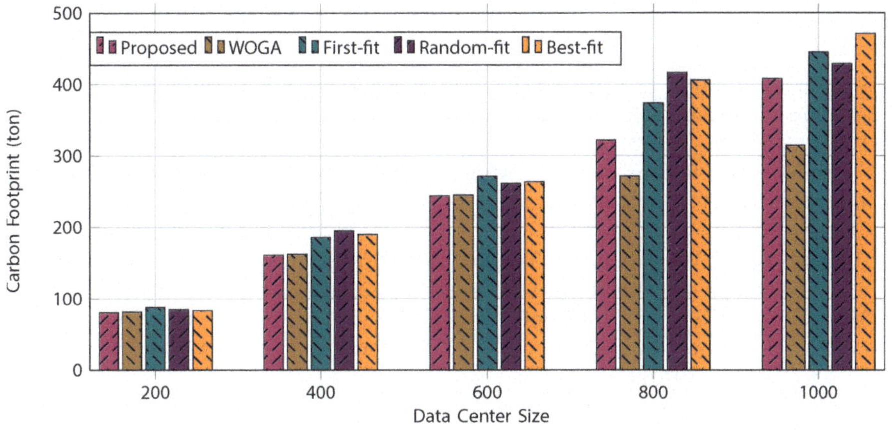

Fig. 4. Carbon Footprint

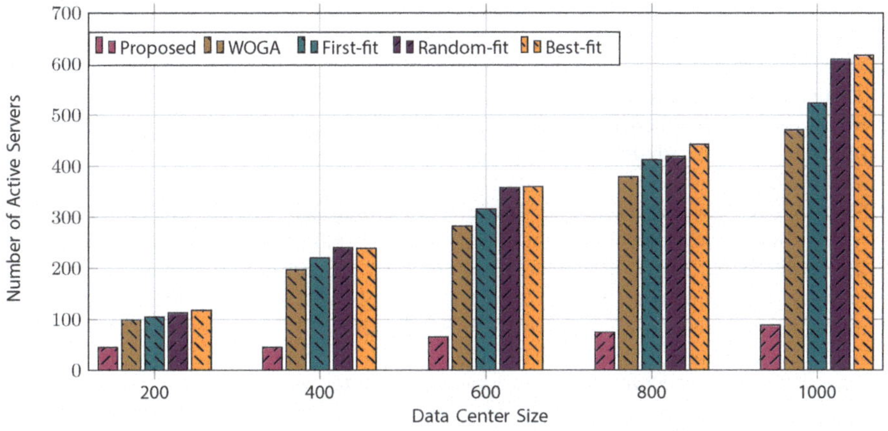

Fig. 5. Number of Active PM

6 Conclusion

This article presents a multi-objective Virtual Machine Placement (VMP) approach to address VM consolidation challenges. Based on NSGA II, the proposed optimization scheme seeks optimal VM-PM mapping in dynamic and static scenarios. The results indicate significant reductions in power consumption and carbon emissions through effective VM placement, coupled with improvements in resource utilization. A comparative analysis with established methods highlights the superiority of the proposed approach. Additionally, there is potential to extend the framework by integrating additional objectives such as reliability, security, and trust.

References

1. Saxena, D., Gupta, I., Kumar, J., Singh, A.K., Wen, X.: A secure and multiobjective virtual machine placement framework for cloud data centers. IEEE Syst. J. **16**(2), 3163–3174 (2021)
2. Sharma, N.K., Reddy, G.R.M.: Multi-objective energy-efficient virtual machine allocation at the cloud data center. IEEE Trans. Serv. Comput. **12**(1), 158–171 (2016)
3. Han, J., Zang, W., Chen, S., Yu, M.: Reducing security risks of clouds through virtual machine placement. In: Data and Applications Security and Privacy XXXI: 31st Annual IFIP WG 11.3 Conference, DBSec 2017, Philadelphia, PA, USA, July 19-21, 2017, Proceedings 31, pp. 275–292. Springer, Cham (2017). https://doi.org/10.1007/978-3-319-61176-1_15
4. Singh, A.K., Kumar, J.: Secure and energy-aware load balancing framework for cloud data center networks. Electron. Lett. **55**(9), 540–541 (2019)
5. Tseng, F.H., Wang, X., Chou, L.D., Chao, H.C., Leung, V.C.: Dynamic resource prediction and allocation for cloud data center using the multiobjective genetic algorithm. IEEE Syst. J. **12**(2), 1688–1699 (2017)
6. Beloglazov, A., Buyya, R.: Managing overloaded hosts for dynamic consolidation of virtual machines in cloud data centers under quality of service constraints. IEEE Trans. Parallel Distrib. Syst. **24**(7), 1366–1379 (2012)
7. Singh, A.K., Swain, S.R., Saxena, D., Lee, C.N.: A bio-inspired virtual machine placement toward sustainable cloud resource management. IEEE Syst. J. **17**, 3894–3905 (2023)
8. Singh, A.K., Swain, S.R., Lee, C.N.: A metaheuristic virtual machine placement framework toward power efficiency of sustainable cloud environment. Soft. Comput. **27**, 3817–3828 (2023). https://doi.org/10.1007/s00500-022-07578-8
9. Saxena, D., Gupta, I., Singh, A.K., Lee, C.N.: A fault-tolerant elastic resource management framework toward high availability of cloud services. IEEE Trans. Netw. Serv. Manage. **19**(3), 3048–3061 (2022)
10. Abdel-Basset, M., Abdle-Fatah, L., Sangaiah, A.K.: An improved Lévy based whale optimization algorithm for bandwidth-efficient virtual machine placement in cloud computing environment. Clust. Comput. **22**, 8319–8334 (2019)
11. Jangiti, S., Sri Ram, E., Shankar Sriram, V.S.: Aggregated rank in first-fit-decreasing for green cloud computing. In: Mallick, P., Balas, V., Bhoi, A., Zobaa, A. (eds.) Cognitive Informatics and Soft Computing: Proceeding of CISC 2017, pp. 545–555. Springer Singapore (2019). https://doi.org/10.1007/978-981-13-0617-4_53
12. Shirvastava, S., Dubey, R., Shrivastava, M.: Best fit-based VM allocation for cloud resource allocation. Int. J. Comput. Appl. **158**(9), 25–27 (2017)
13. Jung, G., Hiltunen, M. A., Joshi, K. R., Schlichting, R. D., Pu, C.: Mistral: dynamically managing power, performance, and adaptation cost in cloud infrastructures. In: 2010 IEEE 30th International Conference on Distributed Computing Systems, pp. 62–73. IEEE (2010)
14. Reiss, C., Wilkes, J., Hellerstein, J.L.: Google cluster-usage traces: format+schema. Google Inc. White Paper **1**, 1–14 (2011)

Fake News Detection Using Heterogeneous Information from Multimedia Content

Avantika Saklani, Shailendra Tiwari(✉), and H.S. Pannu

Thapat Institute of Engineering and Technology, Patiala, India
shailendra@thapar.edu

Abstract. In today's digital world, most people prefer to read the news online, and they only find it accessible if the news includes images, text, audio, or video. In recent years, the news on social media platforms has been becoming increasingly crowded with multimedia reports and knowledge. The widespread availability of multimedia news on social media has contributed to the dissemination of fake information. In this paper we propose a Multimodal Cross Attention CNN Network (MCACN) method for multimodal fake news detection that learns the correlation between the textual and the visual content of the news. We employ the cross attention mechanism to fuse the features from the visual and the textual modalities. The attention mask for the visual features is completely dependent on the textual feature of the news and vice-versa. The findings of this research indicate that the suggested technique works better than the methods that just rely on lingual or visual information alone. Additionally, the proposed method provides a mechanism that can be used for detecting comparable disinformation operations on multiple contexts within the same social domain.

Keywords: Fake News Detection · Multimodal Fusion · Image and text fusion · Cross Attention

1 Introduction

In today's world, citizens have access to an ideal platform for obtaining knowledge thanks to social media. Online social media are an integral part of our lives, with a worldwide audience of 4 billion people who collectively spend 2.5 h every day on these platforms. It makes a big difference in the real world as it provides people to access information which are of important and unimportant things. Due to the fact that its pace of information dissemination is significantly superior to that of traditional media, an increasing number of individuals are able to access a wide variety of information through it. Unfortunately it also provides a large space to fake news. The term 'fake news' refers to both misinformation - misleading information that is generated and disseminated unintentionally - and disinformation campaigns that attempt to confuse and manipulate people in an intentional and organised manner.

D. Pastor-Escuredo et al. (Eds.): ICDLAIR 2023, LNNS 1001, pp. 427–437, 2024.
https://doi.org/10.1007/978-3-031-60935-0_39

(a) (b)

Fig. 1. Few Example of Fake News from the social media

The proliferation of fake news has the potential to create massively detrimental impacts, such as creating negative impact on society or manipulating public events. There are numerous research investigations that have been done to discern the prevalence of fake news across various domains including politics [1], entertainment [2], science [3], health [4]. Prominent social media platforms such as Twitter, Facebook, and blogs also exert a significant influence in the dissemination of false information. During the Covid-19 pandemic, there were a lot of fake stories that got more attention than the real ones, and many people started to believe the fake ones. Because of this, there is a lot of psychological fear in the society [5]. In 2016, Facebook users believed and shared 37 million fake news stories to promote either candidate in the US presidential election [6]. Figure 1 shows some of the example of fake news from the Twitter dataset. Looking at the text or image individually of Fig. 1a, it is not clear to judge that the news is fake or not. As on examining the text in relation with image, we find that the news is fake. The second Fig. 1b tweet depicts the real image which was of a plane crash in Sicily but the text associated with it shows the fakeness of the news. Therefore it is of crucial importance to look in to the multimodal news and design some automated systems that can easily predict out the false information of the news. Identifying and preventing the spread of misinformation at an early stage is crucial to curb its impact on the wider audience.

There exist numerous conventional approaches, and more recently, with the emergence of deep learning models, more ways have become viable for the purpose of detecting fake news. But principally the attention is focused on text e.g. [7,8]. Several recent studies [6,9] have attempted to acquire knowledge on the common characteristics and interdependence of textual and visual elements.

However, these methodologies assign equal significance to all components of both text and image, which may not be the most appropriate approach. In lieu to these shortcomings and insufficiency to reliable feature extraction we propose an efficient deep learning based framework i.e. Multimodal Cross Attention Cnn Network (MCACN) model for fake news detection. The research contribution of the study can be summarized as below:

- We propose an cross attention based MCACN framework that captures the information from the modalities keeping in view the dependencies between the modalities.
- We integrates the features from the modalites and then apply a Multi Layer Perceptron model for fake news detection.
- The proposed work is experimented on benchmark dataset to validate the performance of the model. Experimental results shows that the proposed study outperforms several state-of-the-art methods.

2 Related Work

The proliferation of fake news is currently recognised as a significant peril to democratic systems, the field of journalism, and economic stability. The objective of the task designed to detect fake news is to determine whether or not a certain item of news is reliable [10]. Fake news detection has many related tasks such as spam detection [11,12] and rumor detection [13–15]. We review the present work in the succeeding two areas - fake news detection based on single modality data and fake news detection based on multimodal data.

Fake News Detection Based on Single Modality Data: Textual based early studies for fake news detection extracts linguistic features by the manual selection of the textual cues [16,17]. Chen et al. applied soft-attention and RNN to learn selectively temporal feature representation of post series [18]. This was done in order to capture the long-range dependency among variable length sequential information. In a similar manner, Yu et al. introduced a model that is based on convolutional neural networks (CNN). This model is used to extract low-level local-global features from the input sequences and then construct high-level interactions among essential features and then used to generate high-level features from the important one [19].

With the growth of social media news with images are grabbing one's attention. The study in [20] indicates that almost 86% of fake images comes from the retweets. Images can play a crucial role in detecting automatic news verification task in micro blogs. The researchers in [21] came to an understanding that real and fake news images were spread out in different ways. As a result, they came up with a number of visual and statistical ways to describe these different patterns. When compared to more conventional methods, single-modal deep learning-based false news detection has a higher accuracy rating but these method ignores news as a collection of multimedia data.

Fake News Detection Based on Multimodal Data: Recently there are many researches that deal with the problem of multimodal fake news detection. Jin et al. [22] in their study proposed a deep learning method in which they have employed attention mechanism to fuse the features the multimodal and social context features. Wang et al. [6] proposes an event adversarial neural network framework that has the ability to learn transferable features for events that have not yet occurred. In spite of this, both of these models only uncover details from the top most layer, ignoring any additional information that may exist across the modalities. To overcome these limitations we propose the MCACN model that learns the mutual correlation between visual and lingual modalities. To learn a joint representation of the text and the image, Khattar et al. [9] employ auto encoders in conjunction with a detector. However, they do not take into account the properties that fake images have at the physical level and the fused features that they have learned do not have correlations across many modalities. In the study [23], the authors have proposed multimodal Co-Attention network that fuses the image and text features for fake news detection. The model stacks multiple co-attention layer for the fusion. Despite the fact that the model is capable of learning the inter dependencies between different models, it has not yet been fully exploited in multimodal tasks.

3 Methodology

This sections explains the framework of the proposed model MCACN for classification of fake news detection. It takes visual-lingual pair such as multimodal news generated on social media as input and then after fusing the features from the individual modalities with cross attention mechanism inspired by [24], classifies the news. The framework is illustrated in Fig. 2, where Inception and Roberta are from [25,26].

3.1 Visual Model for Feature Map Extraction

Since nowadays people are more interested towards the visual content of the images as these are more easy to understand as compared to textual content. Therefore it is thus necessary to fully capture the visual features of the news to accurately determine the fake news from it. So, we apply a transfer learning technique to extract the visual features from the multimodal image. To achieve this objective we have used pretrained Inception-V4 model as the architectural design of Inception-v4 having numerous inception modules exhibits a higher degree of uniformity and simplification. In order to design it for our model we have excluded the top most layer from the default model. The study utilises Inception-v4 for the purpose of predicting fake news, as it has demonstrated remarkable performance in the ILSVRC 2012 competition, achieving an error rate of 3.08% across three residual networks [25]. Additionally, the model has demonstrated a top-1 accuracy rate of 80%, surpassing both Inception v3 and GoogLeNet. Image representation for each image is given as:

$$V_f^{(v)} = IV4(v_i) \tag{1}$$

In this equation $V_f^{(v)} \in \mathbb{R}^{(D_f)}$ represents the visual expression for v_i image and D_f is the dimension with $H \times W \times C$ feature maps.

3.2 Lingual Model For Embedding Extraction

The input textual sequence is tokenised into tokens as $S = [s_1, s_2, ..., s_n]$ where n is the number of tokens. For this we have employed RoBerta having dynamic masking technique which performs better than other techniques like BERT, XLNet [27] etc. During the training process, RoBERTa makes use of a dynamic masking strategy, which enables the model to learn more accurate and generalizable representations of words. For each lingual input the embeddings can be represented as:

$$L_f^l = RoBerta(t_i) \tag{2}$$

In this equation t_i is the sequence of input tokens with dimensions $L_f^l \in \mathbb{R}^{d_f \times 1}$ and $d_f \times 1$ is the dimensional vector of the textual representation.

3.3 Fusion of Visual and Lingual Features

It is stated that to better classify your mutltimodal systems there should be some correlations between the modalities. To this end we have applied an innovative cross attention mechanism [28] between the visual and the lingual modlaity to better learn the correlation among them. Recently there has been a breakthrough in the field of deep learning i.e. the attention mechanism [29]. Similar to the way humans tend to concentrate on the central aspect of an image or text, neural networks also employ attention mechanisms to prioritise the major component of an image or text. The proposed model uses the cross attention mechanism in which the visual attention mask is dependent on the textual embeddings and lingual attention mask is dependent on visual embeddings. This can be represented as:

$$\alpha_{v_i} = \sigma(W_v^N V_f^{(v)} + b_V) \tag{3}$$

$$\alpha_{lti} = \sigma(W_v^N L_f^{(l)} + b_V) \tag{4}$$

In this equation σ is the sigmoid function. After having attentions masks α_{v_i} and α_{lti} we integrated them by using concatenation operation. Lastly the concatenated features are finally fed to the 2-layer Multi layer Perceptron (MLP) for the final classification.

4 Dataset and Experiments

4.1 Dataset

For the experimental work we have used the benchmark Twitter dataset [30]. The dataset is designed by the researchers for high-quality multimodal fake news

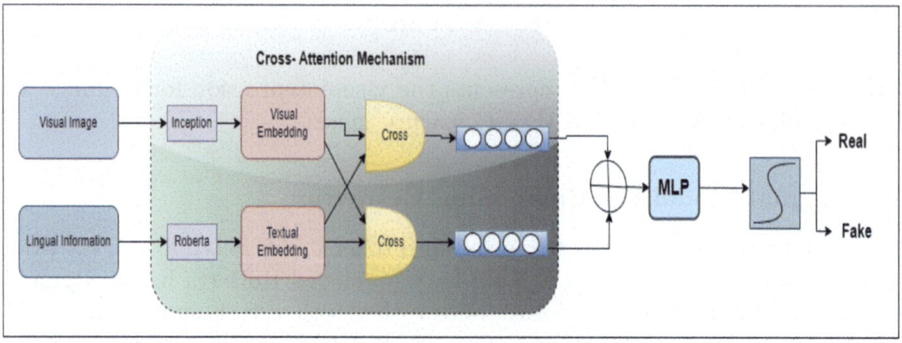

Fig. 2. Illustration of the Proposed Framework. The proposed work uses the cross attention mechanism for fusion from the visual and textual embeddings. The proposed model uses the two-layer MLP for final binary classification.

detection. For comparison with the previous work we have used this benchmark dataset for training our model. The dataset contains the text and its related images from the different events thus there is negligible chances of overlapping between the train and test sets. The dataset contains almost 18,264 tweets and among this training set includes 14,483 tweets and test set have 3781 tweets. The training set is further divided in to three classes - real, fake and humor. And out of these three classes we have not considered the humour class as the test set does not consist of this particular class. From the total training set 5001 tweets are real and 6848 tweets are fake. The data distribution for the real and fake class is further illustrated in Table 1.

Table 1. Dataset Statistics

Dataset	Train		Test	
	Fake	Real	Fake	Real
Twitter	6841	5009	2564	1217

4.2 Experimental Results and Analysis

The experimental setup utilised the NVIDIA RTX A400, a 16 GB Graphics Processing Unit (GPU), an Intel Xeon Gold Processor 6142 with a maximum turbo clock of 3.70 GHz, and 32 GB of RAM. The models were trained and tested using Python 3.10.5, Keras 2.9.0, and TensorFlow 2.9.1.

Evaluation Metrics. We evaluate the model in this paper using four metrics- accuracy, precision, recall and F1-score. These metrics are widely used for fake news detection. Accuracy is the measure of the ratio of correct predicted news to the total number of tested news.

$$Accuracy = \frac{\#ofcorrectpredictednews}{\#oftestednews} \tag{5}$$

Precision is defined as the ratio of true positive sample news to the all the predicted positive samples and is formalised as:

$$Precision = \frac{\#_{tp}}{\#_{tp} + \#_{fp}} \tag{6}$$

where $\#_{np}$ stands for the predicted number of positive samples and $\#_{fp}$ stands for the predicted number of false positive samples.
The recall ratio is the proportion of original positive samples that are confirmed as being correct.

$$Recall = \frac{\#_{tp}}{\#_{tp} + \#_{fn}} \tag{7}$$

where $\#_{fn}$ is the estimated number of false positive samples. The F1-score is a metric used in evaluating classification models, which is calculated as the harmonic mean of the precision and the recall. It may be mathematically expressed as:

$$F1 = \frac{2 * Precision * Recall}{Precision + Recall} \tag{8}$$

Table 2 shows the qualitative comparison results for the proposed methodology MCACN. The table demonstrates that the suggested strategy outperforms both the existing methods and the baseline models in terms of performance. Also it is evident to say that textual model performs better than the visual model. This phenomenon may be attributed to the fact that images can contain noisy data, but language often provides more reliable information. Based on our study's primary research objective, it is evident that the integration of images with text yields superior performance compared to employing either text or images. The learning curve for the Twitter dataset is depicted in Figs. 3a and 3b, respectively. The loss curve decreases gradually during the learning process until it reaches a stable state, indicating that the model has effectively learned information.

Table 2. Experimental classification results of the proposed model with the existing ones.

Dataset	Model	Accuracy	Fake News			Real News		
			Precision	Recall	F-Score	Precision	Recall	F-Score
Twitter	Textual	0.725	0.763	0.661	0.708	0.677	0.774	0.722
	Visual	0.643	0.523	0.580	0.551	0.652	0.601	0.624
	MVAE[9]	0.745	0.801	0.719	0.758	0.689	0.777	0.730
	SpotFake [31]	0.777	0.751	0.900	0.820	0.832	0.606	0.701
	AMFB[32]	0.883	0.890	**0.950**	0.920	0.870	0.760	0.810
	MCACN	**0.897**	**0.922**	0.936	**0.929**	**0.913**	**0.894**	**0.903**

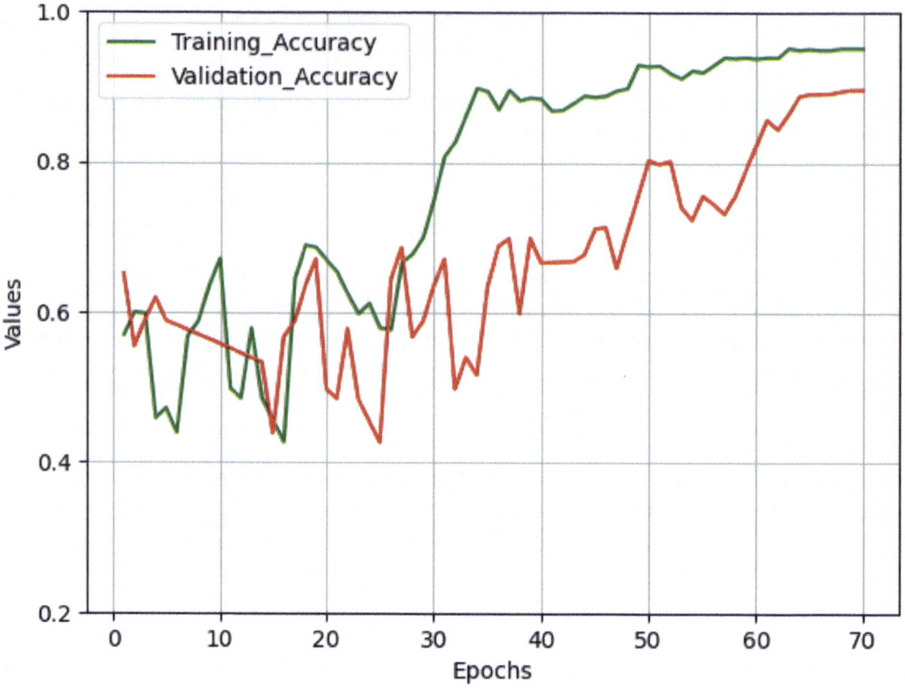

(a) Accuracy

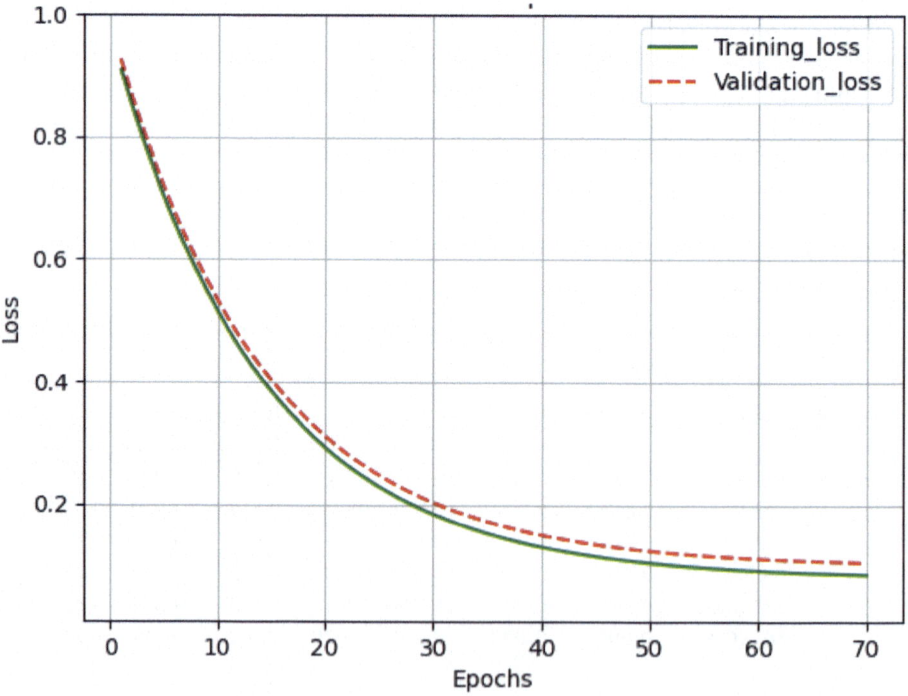

(b) Loss

Fig. 3. Learning Curves for Twitter Dataset

5 Conclusion and Future Work

In this research we have proposed the framework that efficiently captures the information from one modality keeping in consideration the other modality. The technique proposed takes the visual and lingual information of the multimedia post to detect whether the news is fake or real. For visual feature extraction we have employed the Inception, a CNN model and for lingual feature extraction we have employed the Roberta model. The joint representation from the individual representation is experimented using the cross attention mechanism and then fed to the Multi layer Perceptron (MLP) for the final classification of fake or real news. Experimental results shows that the proposed approach performs better than the several state-of-the-art methods.

Although the proposed work shows certain improvements over the existing but also have many possibilities to extend it further. The work can be further investigated to explore the semantic alignments between the visual and the lingual modalities for the better fusion mechanism. Many social media comprises of videos which can further provide an extension to the existing work by including other modalities such as audio, video etc.

References

1. Guess, A., Nagler, J., Tucker, J.: Less than you think: prevalence and predictors of fake news dissemination on facebook. Sci. Adv. **5**(1), eaau4586 (2019)
2. Shu, K., Wang, S., Liu, H.: Understanding user profiles on social media for fake news detection. In: 2018 IEEE Conference on Multimedia Information Processing and Retrieval (MIPR), pp. 430–435. IEEE (2018)
3. Del Vicario, M., et al.: The spreading of misinformation online. Proc. Natl. Acad. Sci. **113**(3), 554–559 (2016)
4. Ghenai, A., Mejova, Y.: Fake cures: user-centric modeling of health misinformation in social media. Proc. ACM on Hum. Comput. Interact. **2**(CSCW), 1–20 (2018)
5. Zeng, J., Zhang, Y., Ma, X.: Fake news detection for epidemic emergencies via deep correlations between text and images. Sustain. Urban Areas **66**, 102652 (2021)
6. Wang, Y., et al.: EANN: event adversarial neural networks for multi-modal fake news detection. In: Proceedings of the 24th ACM SIGKDD International Conference on Knowledge Discovery & Data Mining, pp. 849–857 (2018)
7. Jiang, G., Liu, S., Zhao, Y., Sun, Y., Zhang, M.: Fake news detection via knowledgeable prompt learning. Inf. Process. Manage. **59**(5), 103029 (2022)
8. Zubiaga, A., Liakata, M., Procter, R.: Exploiting context for rumour detection in social media. In: Ciampaglia, G., Mashhadi, A., Yasseri, T. (eds.) Social Informatics: 9th International Conference, SocInfo 2017, Oxford, UK, September 13-15, 2017, Proceedings, Part I 9, pp. 109–123. Springer, Cham (2017). https://doi.org/10.1007/978-3-319-67217-5_8
9. Khattar, D., Goud, J.S., Gupta, M., Varma, V.: MVAE: multimodal variational autoencoder for fake news detection. In: The World Wide Web Conference, pp. 2915–2921 (2019)
10. Zhou, X., Zafarani, R., Shu, K., Liu, H.: Fake news: fundamental theories, detection strategies and challenges. In: Proceedings of the Twelfth ACM International Conference on Web Search and Data Mining, pp. 836–837 (2019)

11. Kaghazgaran, P., Caverlee, J., Squicciarini, A.: Combating crowdsourced review manipulators: a neighborhood-based approach. In: Proceedings of the Eleventh ACM International Conference on Web Search and Data Mining, pp. 306–314 (2018)
12. Wang, X., Liu, K., Zhao, J.: Handling cold-start problem in review spam detection by jointly embedding texts and behaviors. In: Proceedings of the 55th Annual Meeting of the Association for Computational Linguistics (Volume 1: Long Papers), pp. 366–376 (2017)
13. Guo, C., Cao, J., Zhang, X., Shu, K., Yu, M.: Exploiting emotions for fake news detection on social media. arXiv preprint arXiv:1903.01728 (2019)
14. Imran, M., Ofli, F., Caragea, D., Torralba, A.: Using AI and social media multimodal content for disaster response and management: opportunities, challenges, and future directions. Inf. Process. Manage. **57**, 102261 (2020)
15. Arora, S., Rani, R., Saxena, N.: Modified valence aware dictionary for sentiment reasoning classifier for detection and classification of COVID-19 related rumors from social media data streams. Concurrency Comput. Pract. Experience **34**(21), e7124 (2022)
16. Sharma, K., Qian, F., Jiang, H., Ruchansky, N., Zhang, M., Liu, Y.: Combating fake news: a survey on identification and mitigation techniques. ACM Trans. Intell. Syst. Technol. (TIST) **10**(3), 1–42 (2019)
17. Ruchansky, N., Seo, S., Liu, Y.: CSI: a hybrid deep model for fake news detection. In: Proceedings of the 2017 ACM on Conference on Information and Knowledge Management, pp. 797–806 (2017)
18. Chen, T., Li, X., Yin, H., Zhang, J.: Call attention to rumors: Deep attention based recurrent neural networks for early rumor detection. In: Ganji, M., Rashidi, L., Fung, B., Wang, C. (eds.) Trends and Applications in Knowledge Discovery and Data Mining: PAKDD 2018 Workshops, BDASC, BDM, ML4Cyber, PAISI, DaMEMO, Melbourne, VIC, Australia, June 3, 2018, Revised Selected Papers 22, pp. 40–52. Springer,Cham (2018). https://doi.org/10.1007/978-3-030-04503-6_4
19. Yu, F., Liu, Q., Wu, S., Wang, L., Tan, T.: A convolutional approach for misinformation identification. In: IJCAI, pp. 3901–3907 (2017)
20. Gupta, A., Lamba, H., Kumaraguru, P., Joshi, A.: Faking sandy: characterizing and identifying fake images on twitter during hurricane sandy. In: Proceedings of the 22nd International Conference on World Wide Web, pp. 729–736 (2013)
21. Jin, Z., Cao, J., Zhang, Y., Zhou, J., Tian, Q.: Novel visual and statistical image features for microblogs news verification. IEEE Trans. Multimedia **19**(3), 598–608 (2016)
22. Jin, Z., Cao, J., Guo, H., Zhang, Y., Luo, J.: Multimodal fusion with recurrent neural networks for rumor detection on microblogs. In: Proceedings of the 25th ACM International Conference on Multimedia, pp. 795–816 (2017)
23. Wu, Y., Zhan, P., Zhang, Y., Wang, L., Xu, Z.: Multimodal fusion with co-attention networks for fake news detection. In: Findings of the Association for Computational Linguistics: ACL-IJCNLP, pp. 2560–2569 (2021)
24. Abavisani, M., Wu, L., Hu, S., Tetreault, J., Jaimes, A.: Multimodal categorization of crisis events in social media. In: Proceedings of the IEEE/CVF Conference on Computer Vision and Pattern Recognition, pp. 14679–14689 (2020)
25. Szegedy, C., Ioffe, S., Vanhoucke, V., Alemi, A.: Inception-v4, inception-ResNet and the impact of residual connections on learning. In: Proceedings of the AAAI Conference on Artificial Intelligence, vol. 31, no. 1 (2017)
26. Liu, Y., et al.: Roberta: a robustly optimized BERT pretraining approach. arXiv preprint arXiv:1907.11692 (2019)

27. Rajapaksha, P., Farahbakhsh, R., Crespi, N.: BERT, XLNet or RoBERTa: the best transfer learning model to detect clickbaits. IEEE Access **9**, 154704–154716 (2021)
28. Rajan, V., Brutti, A., Cavallaro, A.: Is cross-attention preferable to self-attention for multi-modal emotion recognition? In: ICASSP 2022-2022 IEEE International Conference on Acoustics, Speech and Signal Processing (ICASSP), pp. 4693–4697. IEEE (2022)
29. Vaswani, A., et al.: Attention is all you need. In: Advances in Neural Information Processing Systems, vol. 30 (2017)
30. Boididou, C., et al.: Verifying multimedia use at mediaeval 2015. In: MediaEval 2015, vol. 1436. CEUR-WS (2015)
31. Singhal, S., Shah, R.R., Chakraborty, T., Kumaraguru, P., Satoh, S.: SpotFake: a multi-modal framework for fake news detection. In: 2019 IEEE Fifth International Conference on Multimedia Big Data (BigMM), pp. 39–47. IEEE (2019)
32. Kumari, R., Ekbal, A.: AMFB: attention based multimodal factorized bilinear pooling for multimodal fake news detection. Expert Syst. Appl. **184**, 115412 (2021)

Comparative Analysis of CNN and Transformers on Malicious Intent Detection in HTTP

Kanishka Tiwari, Anmolpreet Singh Bhatia, Nibhrit Garg, Ishaan Arora, and Poonam Saini(✉)

Department of Computer Science and Engineering, Punjab Engineering College, Chandigarh, India
poonamsaini@pec.edu.in

Abstract. The utilization of web-based applications has greatly increased in recent years. This, in turn, has raised security concerns and data breaches leading to losses of millions of dollars. A large part of these attacks is carried out through modifications in HTTP requests and are often too complex to be recognised through simple rule-based checks. Thus, it is vital to utilize machine learning approaches in order to protect the applications against such attack vectors. Additionally, given the pace at which web servers need to operate, these machine learning methods need to be extremely fast at processing the requests. The paper presents a comparison of two recent deep learning models to recognize the web application attacks namely, a convolution neural network (CNN) and a Transformer-based model. We compare the prediction accuracies of both the models as well as their prediction time to identify the best model suited for being deployed in a real-life scenario. Based on implementation and results, it is observed that for our dataset, CNN tends to perform better in terms of both, accuracy and prediction time.

Keywords: Web application security · Web vulnerabilities · Machine learning · Deep learning · Word vectorisation · Transformers · Convolutional networks

1 Introduction

The usage of internet-based applications where the requests are being sent to servers for the exchange of critical information has increased a lot on a daily basis. This also poses a very high risk of security issues and data breaches which could result in losses of millions of dollars. A variety of attacks of increasing complexity are being used for malicious purposes. Some of the most common attack vectors are made using the Hypertext Transfer Protocol (HTTP) requests exchanged between the servers and the clients. Conventional methodologies to identify and prevent such attacks employ firewalls based on manual rules, usually formed using conditional statements. A lot of domain expertise and human intervention is needed to make these firewalls secure and effective. Due to constant exploration of new security bugs or zero-day vulnerabilities in the application, these firewalls also have to be constantly updated and monitored. A wide number of applications/scripts can construct executable commands, Structured Query Language (SQL)

D. Pastor-Escuredo et al. (Eds.): ICDLAIR 2023, LNNS 1001, pp. 438–453, 2024.
https://doi.org/10.1007/978-3-031-60935-0_40

injections, Cross Site Scripting (XSS), buffer overflow and other types of attacks that can be inserted in the Uniform Resource Locators (URL). Moreover, machine learning algorithms are also being widely used to produce more sophisticated and varied attacks to combat these rule-based defenses.

Our focus is to utilize natural language processing where the proposed model uses the URL and Post Body (if present) of an HTTP request and detect if there is any unusual behavior or malicious intent with the request. Since the request is in the form of text, therefore, pre-processing is done before applying the machine learning model to determine the attack type. For this, word2vec has been used to convert the URL sequence into a vector representation and thereafter data can be used to train the models. We propose a comparison of two techniques, a convolution neural network (CNN) model and a Transformer model.

To evaluate and train these models we used the CSIC 2010 dataset. This dataset contains 36,000 normal requests and more than 25,000 anomalous requests. Support Vector Machine (SVM) based models and CNN-based models have been used to tackle this problem in literature, however, Transformers have not been used for this problem before. Transformers are state-of-the art natural language processing (NLP) models and are widely used in NLP applications. The Transformer was a natural language processing model put forward as a replacement for the recurrent neural networks, presented as an upgrade on the serialized recurrent networks. Transformer is a solution for the lack of the parallelizability of the RNNs. The RNNs, for all their impressive results in NLP, posed a unique problem; From a computational efficiency point of view, the fact that the computations in recurrent neural networks (RNN) are sequential, and not parallel, was presenting itself as a bottleneck. Even if a model is accurate, its recall time has to be quick, otherwise it isn't of much use. A relatively recent discovery in NLP was the use of a technique called attention. It was first employed in the RNNs and it proved to be very useful in tasks like machine translation and question answering. The basis of this technique was that some parts of a sentence affect the meaning of a word more than others. The attention allows the network to focus on these parts of the sentence to come up with the right representation of the sentence, which benefits the performance on tasks like question answering and machine translation. While the mechanism of self-attention does not involve any learnable parameters (which are necessary for any machine learning algorithm to learn), such parameters are introduced to the self-attention mechanism, while keeping the complexity to a minimum to produce better results.

In our work, we use the Transformer model for the application of detecting malicious URLs. Additionally, we also create a CNN model to comparatively study the performance and accuracy of a larger model like the Transformer model to better understand which types of models would work better in the fast-paced and rapidly evolving environment of the internet where a model might potentially need to process data extremely quickly. Along with this, a special focus is also kept on the false positive rates specifically since we don't want too many non-malicious requests being flagged out by the system resulting in user distress. To achieve this, we create a complete pre-processing pipeline and put forth a novel method to analyze features recognised by the model which leads to the URL being tagged malicious.

2 Literature Review

A considerable amount of research has already been conducted on the use of deep learning for vulnerability and attack detection on web applications as also surveyed by Alaoui et al. [1]. There are a wide range of attacks a user with malicious intent can perform and a lot of research has been conducted in detecting individual attacks. One such attack is the SQL injection attack [2–5] which is one of the most common web attacks. The researchers have used different types of models such as convolutional neural networks (CNN) [2, 3], random forest, deep feature fusion network [4], and multi-layered perceptron [5].

Another type of attack is XSS injection attacks. Kadhim et al. [6] address this attack by using a CNN and long short-term memory (LSTM) network. The authors use a dataset in which malicious requests are obtained from www.xssed.com. An interesting feature of their dataset generation is that the URLs and path information are removed for privacy reasons, keeping only the payload for further processing. Through this, the work achieves considerably high accuracies. Another group [7] uses a multilayer perceptron to once again detect XSS attacks and the authors proposed a novel way of generating non-malicious requests in which the dataset represented different types of sites with corresponding requests that exist on the web. The authors scrape sites using random walks and jumps and use the page ranks of the sites to determine their probability of appearing in the dataset. For malicious requests, www.xssed.com and Open Bug Bounty were scraped and shown substantial results in terms of accuracy.

In determining the scope of our work, we found two main issues with the above-mentioned works. Firstly, generating robust datasets for the application was extremely difficult and without using the same dataset and without knowing the reliability of the dataset, it was difficult to compare the results of a particular work. Secondly, these works handle a singular instance of a web attack, however, a detection system may face many different types of attacks in general. Many different techniques have been used to detect malicious intent in datasets with multiple attack vectors. We take the example of Wang et al. [8], who compared CNN, LSTM, CNN+LSTM and decision tree (DT), K-Nearest Neighbor (KNN), neural network (NN), Linear SVM, Naive Bayes (NB). The authors use the technique of separating the URL into discrete tokens to make the URL look more like a sentence with words and then generating vector embeddings for these tokens using an n-gram learning model with one hidden layer. In their experiments, CNN, LSTM and its combinations gave much higher accuracies compared to other techniques. The similar results were observed in various works thereafter and found that CNN and RNN networks perform well on text classification. The same models can be used on URLs as well, given that correct preprocessing techniques are being applied.

In recent literature, few authors have proposed the use of CNN and Gated Recurrent Unit (GRU) [9] which also gives a high degree of accuracy. One of the findings of this paper was that while in regular NLP applications, the window size for a CNN network should not be more than five, the window size for this application should be as large as possible. This is because the larger the window size, the greater the amount of large-level information that can be extracted by the CNN model. Given that most URLs when broken down into tokens have semantic relations with far away tokens as well (like for example the query parameter and its value), this is very important to keep in mind. The majority of these researches which focus on multiple attack vectors have been conducted on the CSIC 2010 dataset, but a few have been conducted on private datasets like [10], or a mixture of private and CSIC dataset like [9, 11, 12]. Many papers like [12] also tried to do multi-class classification by manually extracting classes from the CSIC dataset or their private datasets. Given the prevalence of the CSIC dataset, we decided to go forward with the same in our work.

Additionally, in all the work we surveyed, the relatively newer Transformer model had not been used. Given its vast application in NLP-based problems, we felt that it was essential to apply it to this problem to understand the potential tradeoffs of such a model. The Transformer that we use in this paper and the associated concept of attention was first talked about by Vaswani et al. [13]. Transformers were applied by Maneriker et al. [14] in detecting URLs that contained phishing attacks. This showed that it is indeed possible to apply Transformers to URLs for security-based applications to success.

3 The Proposed Work

3.1 Overall Workflow and Process

The whole workflow is split into the following three steps: preprocessing, generation of embeddings and finally passing these embeddings into various models. Figure 1 shows the overall workflow of the proposed research work.

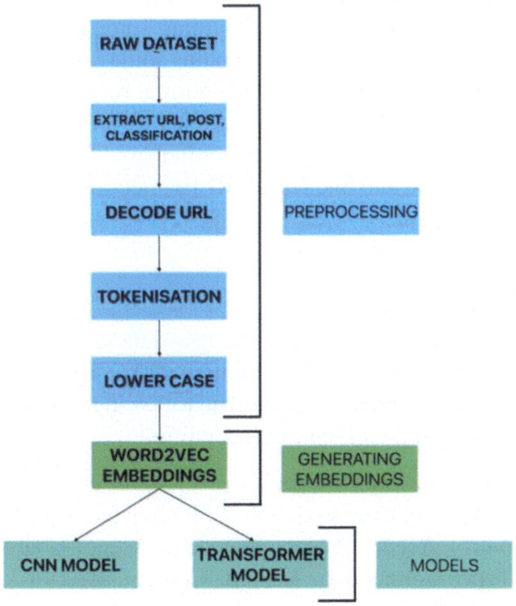

Fig. 1. Overall workflow of the proposed research work

A. Preprocessing: The preprocessing step involves the creation of URL tokens from the CSIC 2010 dataset. This is done by removing irrelevant features from the dataset, decoding the URLs, breaking them up into tokens, converting the tokens to lowercase and then adding the post bodies to the URLs in a similar way to query parameters.

B. Generate Embeddings: This step takes the token string formed from the raw data and uses it to generate embeddings for the tokens which can be used by the ML models. The tokens are generated using the word2vec model which makes embeddings such that the cosine product of embeddings of tokens occurring together often is close to one.

C. Models: We train two models on the dataset to predict malicious intent. The first one is a 1-D CNN model and the other is a state-of-the-art transformer model that has not yet been used for this type of problem. Both are compared for their accuracies, false positive rates and performance metrics. We also use a novel technique to analyze which parts of the URL lead to the malicious intent.

3.2 Dataset Details

The HTTP dataset CSIC 2010 contains thousands of web requests that are automatically generated. It can be used for the testing of web attack protection systems and was developed at the "Information Security Institute" of CSIC (Spanish Research National Council).

The HTTP dataset CSIC 2010 contains the generated traffic targeted to an e-commerce web application developed at CSIC. In this web application, users can buy items using a shopping cart and register by providing some personal information. As it is a web application in Spanish, the data set contains some Latin characters. The dataset is generated automatically and contains 36,000 normal requests and more than 25,000 anomalous requests. The HTTP requests are labeled as normal or anomalous and the dataset includes attacks such as SQL injection, buffer over-flow, information gathering, files disclosure, carriage return and line feed (CRLF) injection, XSS, server-side include, parameter tampering and so on.

3.3 Data Preprocessing

The overall aim of preprocessing is to take the raw dataset, extract the relevant fields, convert the URLs and POST bodies into a sentence-like string of tokens, and then to encode these tokens into an embedding space. The following steps helps to feed the URL data into the model for training and feature recognition:

i) The first step involves selecting the fields of the dataset that were relevant to our objective. The CSIC Dataset consisted of the following features: Method, User-Agent, Pragma, Cache-Control, Accept, Accept-Encoding, Accept-Charset, Language, Host, Cookie, Content-Type, Connection, Length, Content, URL, Classification. Out of these, almost all the features had the same values or values that are insignificant to our work. The only useful values were *URL*, *content* and *classification*, where:

   - The URL was one to which the request was made along with the query parameters that are present in the URL.

- The content was basically the POST Body of a post request. It was essentially a list of key-value pairs which are sent in the POST body.
- The classification was a binary classification which stated if the request was malicious or not.

A sample from the dataset has been shown in Fig. 2 along with all the attributes provided in the dataset. For this application, POST bodies and URL query parameters serve very similar logical functions since there is no binary data present in the POST bodies and no specific security implications which normally lead one to use POST to send data. Thus, we decided to include the POST data in the form of query parameters after the URL with the two separated by a '?' for uniformity in analysing the data.

```
Classification : 1
Method : GET
User-Agent : Mozilla/5.0 (compatible; Konqueror/3.5;
Linux) KHTML/3.5.8 (like Gecko)
Pragma : no-cache
Cache-Control : no-cache
Accept :
text/xml,application/xml,application/xhtml+xml,text/html;
q=0.9,text/plain;q=0.8,image/png,*/*;q=0.5
Accept-encoding : x-gzip, x-deflate, gzip, deflate
Accept-charset : utf-8, utf-8;q=0.5, *;q=0.5
language : en
host : localhost:8080
cookie :
JSESSIONID=B94733D7AC80E8CFC59A200EB8C8F156
content-type : NaN
connection : close
length : NaN
content : NaN
URL:http://localhost:8080/tienda1/publico/autenticar.jsp?
modo=entrar&login=benefiel&pwd=duc9i89&remember=b
ob%40%3CSCRipt%3Ealert%28Paros%29%3C%2FscrIPT%
3E.parosproxy.org&B1=Entrar HTTP/1.1
```

Fig. 2. Example input

ii) The next step was to decode the URL from 'cp1252' encoding. URL encoding, officially known as percent-encoding, is a method to encode arbitrary data in a Uniform Resource Identifier (URI) using only the limited US-ASCII characters legal within a URI. Special characters used in the URL are replaced by "%" followed by two hexadecimal digits. For example, '&' is represented by '%26'. These URLs may also be encoded multiple times because the '%' can also be encoded as '%25'. This leads to '%2526' being decoded first to '%26', which is further decoded to '&'. Thus, a single decoding step was not sufficient since some URLs in the dataset were encoded multiple times. Figure 3 shows a sample URL from the dataset decoded twice along with the original URL.

```
Original URL:
http://localhost:8080/tienda1/publico/vaciar.jsp?B2=any
%250D%250ASet-cookie%253A%2BTamper%253D57652
05567234876235
Decoded Once:
http://localhost:8080/tienda1/publico/vaciar.jsp?B2=any
%0D%0ASet-cookie%3A+Tamper%3D5765205567234876
235
Decoded Twice:
http://localhost:8080/tienda1/publico/vaciar.jsp?B2=any
\nSet-cookie:+Tamper=5765205567234876235
```

Fig. 3. Decoded URL

iii) After the decoding phase, the URL string had to be broken into tokens since the ML models used work on there being certain discrete units in the input. While character-level tokens may be used, it will generally be much harder for the algorithm to recognise not only the words the tokens formed because of their variable length but also the relationship between words which is important for such an application. URLs naturally contain certain alphanumeric characters which has a special meaning and is used to separate different units of a URL. The tokens were thus created by breaking the URL on the following set of alphanumeric characters:

? & = @ \ . % + < > ( ) / ˜ ; ' - * ! # , : \r \n

We also preserved these alphanumeric characters as tokens on their own as well.

iv) Next, we converted all the English alphabet characters to lowercase for uniformity and also because a lot of attackers try to prevent detection by simpler systems by changing the cases of characters in certain keywords. Additionally, each URL was padded so that it would be of equal length.

```
URL before preprocessing :
http://localhost:8080/tienda1/publico/autenticar.jsp?mod
o=entrar&login=benefiel&pwd=duc9i89&remember=bob%
40%3CSCRipt%3Ealert%28Paros%29%3C%2FscrIPT%3E.p
arosproxy.org&B1=Entrar

URL after preprocessing :
['http', ':', '/', '/', 'localhost', ':', '8080', '/', 'tienda1', '/',
'publico', '/', 'autenticar', '.', 'jsp', '?', 'modo', '=', 'entrar', '&',
'login', '=', 'benefiel', '&', 'pwd', '=', 'duc9i89', '&', 'remember',
'=', 'bob', '@', '<', 'script', '>', 'alert', '(', 'paros', ')', '<', '/',
'script', '>', '.', 'parosproxy', '.', 'org', '&', 'b1', '=', 'entrar']
```

Fig. 4. Tokenised URL

v) Finally, we have finally tokenised the URL after doing the above-mentioned pre-processing. An example is shown in Fig. 4 where the original URL as well as its tokenised form is shown. This tokenised URL can be used in the next step, which is to generate the embeddings.

3.4 Generate Embeddings

The issue with inputting text into a mathematical model is that there is no other way for the model to directly understand the text input. Hence, it has to be converted into a mathematical format and this process is called generating embeddings. There exist multiple ways to do this, like techniques that randomly map a word to a number. However, such techniques are not useful for our application because the context of a word has implications for the meaning of the word. We therefore use the word2vec technique for generating embeddings, specifically the continuous skip-gram variant with negative sampling. The model first establishes a network with a hidden layer to predict the context of a word and then extracts the parameters of the hidden layer as the word vectors. The number of neurons k in the hidden layer is exactly the word vector dimension. The model assumes the cosine distance of two-word vectors (embeddings) to be the determiner of word closeness. This cosine distance is then modified to a probability value using it as an input to a softmax function, and that probability is the probability of the context word being present in the context, given the actual word. This creates a loss function which the model can then try to optimise, and the trained weights form the embeddings. Figure 5 shows this process of embeddings.

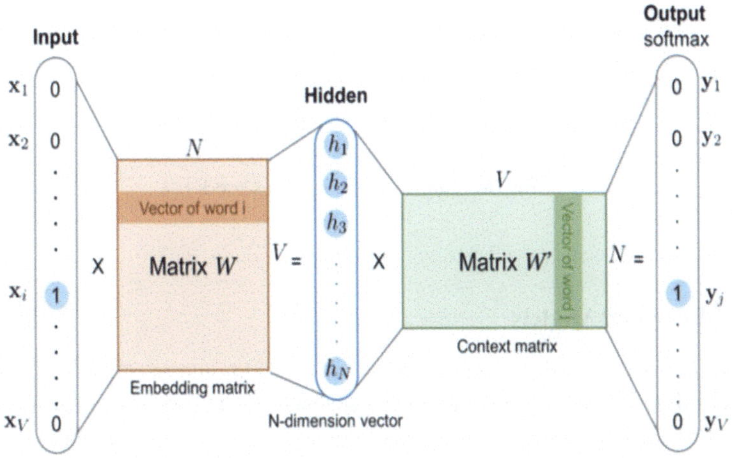

Fig. 5. Generate Embeddings Model

3.5 Visualisation

Having created these vectors, it was not entirely clear whether a model that was created for natural language text processing would work on URLs. Therefore, as a sanity check step, we created 'URL vectors' by averaging all the vectors (without the padding) of the tokens in a URL and then, we plotted these 'URL vectors' after applying principal component analysis (PCA) on them and retrieving results as shown in Fig. 6. It is observed that the malicious examples (Red) are distinctly separable from the non-malicious examples (Green). Therefore, we can conclude that the word2vec vectors do seem to capture some contextual information about the URLs in much the same way as regular text. Now, once the embeddings are finally generated, the machine (deep) learning models are explored further.

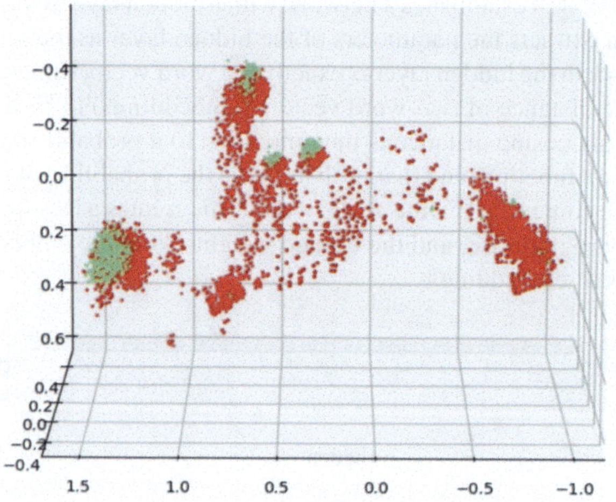

Fig. 6. Generate embeddings model

3.6 Deep Learning Models

1. CNN Model: Convolutional Neural Networks also known as CNNs or ConvNet, is a class of neural networks that specializes in processing data that has a grid-like topology, such as an image. The model works at different scopes of resolution on the input, where the local nuances of the input are handled by the initial layers of the CNN, which are then passed onto the later layers of the CNN for contextualizing and forming the big picture. Usually made for 2-D data, which has sequential patterns in both dimensions, we use a 1-D CNN since we need to process text-like data.

The first layer of our model is an embedding layer which allows us to retrain and fine-tune the word2vec embeddings for this model. This layer is followed by three blocks consisting of two convolutional layers and one max-pooling layer. The first two blocks use 64 filters for their convolutional layers and the last block uses 128 filters. The kernel size is six. The last layer of the model is densely connected. Allowing retraining of the base word2vec embeddings is done because even though with and without retraining the accuracy is basically the same, the number of epochs it takes to reach the same accuracy is lower with retraining allowed. Thus, the number of trainable parameters for this algorithm is 8,918,273.

2. *Transformer Model:* The model was designed to extract pertinent patterns local to different parts of the URL while being able to comprehend the big picture by contextualizing based on observed patterns. Several approaches used a recurrent neural network to achieve the aforementioned goal. The idea was that the RNN would be able to represent the sequential data by using a single vector which would then be passed to a dense network to predict the maliciousness of a URL.

CNNs are a viable option for solving this problem as well, and because of that, the first layer of the model was a convolutional block, which was employed as an alternative to the linear transformation of the input into a different vector space. The block consisted of a 1-D convolutional layer with a kernel size of five, which was run across the input size of (309, 256). The layer used the ReLU activation as the non-linearity, and used a total of 256 kernels, with the padding set to 'same'. This was followed by a 1-D max pooling layer with strides two and pool size two as well. Finally, the output was normalized using layer normalization.

This single CNN block transformed the input from the dimensions of (309, 256) to (154, 256). The output was then passed onto the transformer. Figure 7 shows the full architecture of the model.

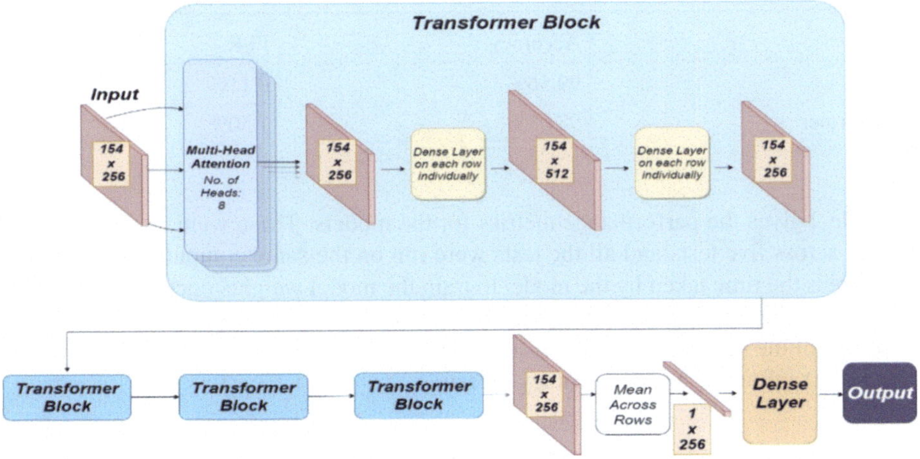

Fig. 7. Architecture of the transformer-based model

The transformer is a self-attention based neural network that uses the self-attention mechanism multiple times to arrive at a well-contextualized representation of the input. The multiple attention units are known as attention heads. These attention heads allow the network to attend to different parts of the input, hence, enabling the network to construct a well-rounded representation of the input. The Transformer model consists of four transformer blocks. Each block consists of first a multi-head attention layer, which consists of eight heads, followed by the normalization of the output of the multi-head attention layer using layer normalization. The output was then passed into a dense layer with an output size of 512, and a dropout of 0.1. After applying the rectified linear units (ReLU) non-linearity, the output was linearly transformed using another dense layer, and normalized in the end using layer normalization. The total number of trainable parameters is 9,838,081.

4 Results and Discussion

The most important metrics for a malicious intent detection system on the web are its accuracy, its false positive rate (*i.e.*, how many malicious requests it lets through) and its performance (its prediction time and training time). Therefore, our analysis also pertains to the same. For training and evaluation of the models, a train-test split of 80-20 was defined which meant 48852 training examples and 12213 testing examples. For both models, training was done for four epochs and the Adam optimizer was used. In the Transformer, training began with a learning rate of 1e-5, which then was reduced to 1e-6 after two epochs, which then was further reduced to 1e-7 after the third epoch. The accuracies obtained are given in Table 1.

Table 1. Results of the proposed model

Model	Accuracy	FPR
CNN	99.56%	0.11%
Transformer	98.03%	1.50%

Table 2 gives the performance metrics for the models. These were calculated as an average across five tests and all the tests were run on the same computer. The training time here is the time taken by the model to train the model weights once pre-processing has been completed. The prediction time here is the time taken by the model to evaluate the results of the validation set. The training and prediction times were calculated on the same 80-20 split used for the accuracy measures.

Table 2. Performance measures

Model	Training Time	Prediction Time	No. of Parameters
CNN	785.94 s	18.75 s	8,918,273
Transformer	62 min	6 min	9,838,081

From our data, we have the following observations:

- The accuracy for CNN is higher than the accuracy for the Transformer. We attribute this to the fact that for this particular application, the Transformer model may require an even greater amount of training data to learn the patterns present in the data, given that it is a more complex model.
- The time taken by CNN to both train and predict the testing data is far lower than by the Transformer model due to its complexity. Given the rapid nature of HTTP requests and the need to potentially re-train the model on new data, training and testing speeds matter greatly.

Given the outcome, we see that out of the two models, for the given application space, the simpler CNN model may be more performant than the Transformer model due to its complexity. This may also be noted from the fact that when compared with the accuracies provided in [15] which used a CNN, [16] which used a bi-directional LSTM and [17] which used a CNN + LSTM networks, our CNN model has higher accuracy metrics and is also trained in a fraction of the time. Thus, further work may be done to refine simpler models given their great performance benefit in this application space. Our Transformer model also provided better accuracy than some other complex models like [17] which used a CNN + LSTM model showing that attention-based models may perform better than combination models.

It is worth mentioning that the false positive rate (FPR) rate of our CNN model is better than both the Transformer model and the above-mentioned work. Given that FPR is a very important metric to minimize, especially where high security is required, this is another great advantage of the CNN model. Another important point was to handle the URL padding. The URLs in the dataset were of greatly varying length. The average length was much lower than the longest URL. Additionally, the added length did not contain any relevant information for most of the part but was instead random characters. However, due to the nature of the models, the URLs had to be padded to the length of the longest URL. One of the ways to address this was to average some number of the URL end-tokens so that information is not lost but at the same time, it is easier for the model to understand that the information is not as relevant. This method did yield marginally higher accuracies, but its applicability in a real-world context may be different since random tokens for confusing a machine learning algorithm don't need to be present at the end unlike the dataset here.

http://localhost:8080/tienda1/publico/anadir.jsp?id=
2&nombre=Jamón+Ibérico&precio=85&cantidad=';+
DROP+TABLE+usuarios;+SELECT+*+FROM+datos+
WHERE+nombre+LIKE+'%&B1=Añadir+al+carrito

(a)

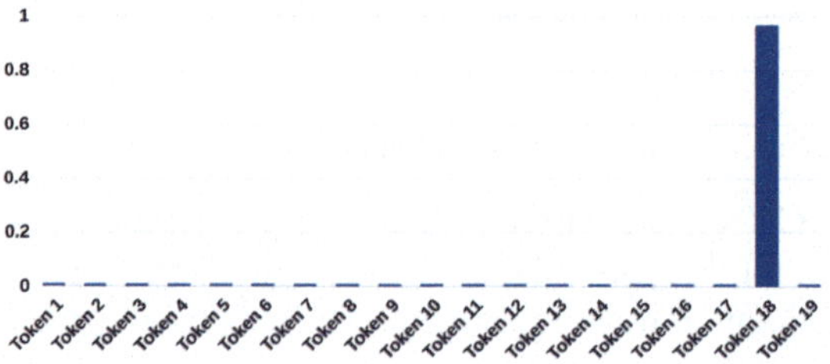

(b)

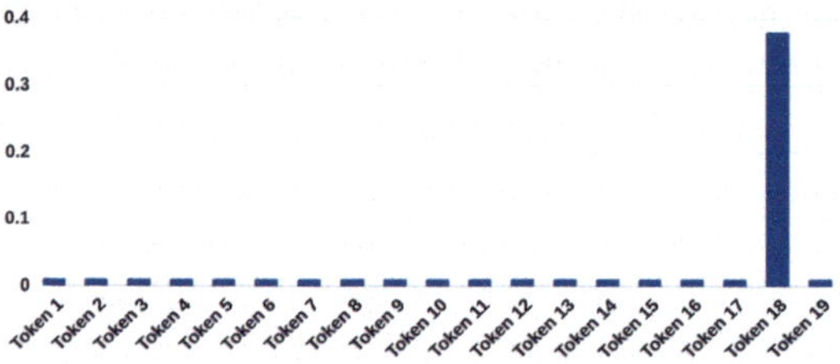

(c)

Token Number	Token Value
1	http
2	:
3	/
4	/
5	localhost
6	:
7	8080
8	/
9	tienda1
10	/
11	publico
12	anadir
13	.
14	jsp
15	id=2
16	nombre=jamón+ibérico
17	precio=85
18	cantidad=';+drop+table+usarios;+select+*+from+datos+where+nombre+like+'%
19	b1=añadir+al+carrito

(d)

Fig. 8. a) Decoded Test URL containing SQL injection attack; b) Token recognised by CNN; c) Token recognised by Transformer; d) Table mapping the token number on the chart with the respective token

http://localhost:8080/tienda1/publico/entrar.jsp?errorMsg=Credenciales+incorrectas<SCRIPT>alert("Paros");</SCRIPT>

(a)

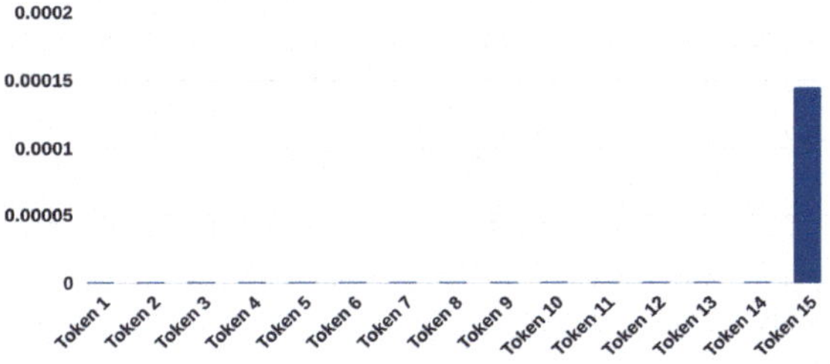

(b)

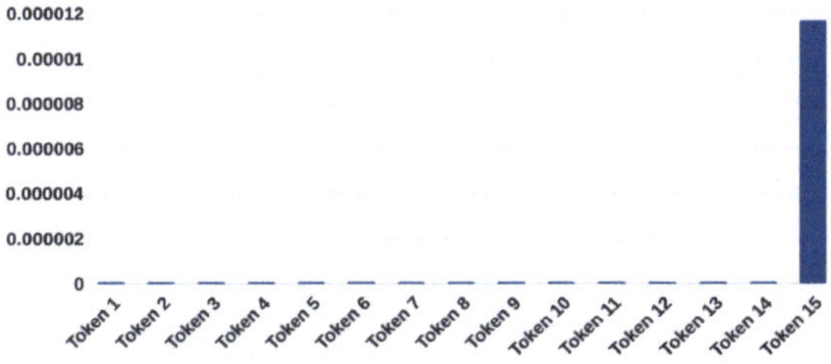

(c)

Token Number	Token Value
1	http
2	:
3	/
4	/
5	localhost
6	:
7	8080
8	/
9	tienda1
10	/
11	publico
12	entrar
13	.
14	jsp
15	errormsg=credenciales+incorrectas <script>alert("paros");</script>

(d)

Fig. 9. a) Decoded Test URL containing XSS attack b) Token recognised by CNN c) Token recognised by Transformer d) Table mapping the token number on the chart with the respective token

4.1 More Findings

Most of the attack vectors which utilise the HTTP medium are generally done through the URL query parameters or the POST body content. Both these are essentially a key-value pairing. Thus, we may hypothesise that in order to identify malicious intent in the URL, the model must somehow detect which of the key-value pairs within the URL has the malicious content. Hence, by removing the particular malicious key-value pair, the model's score prediction for the particular URL must decrease i.e., the model must predict it to be less malicious. With this hypothesis in mind, we devised a technique where a URL is represented by the initial URL tokens up to the query parameters, and then it is followed by the key-value pairs as singular tokens. We then take a baseline prediction score given by the model for the original URL and then remove each of these new tokens (essentially the key-value pair) one at a time, and compare the model's new prediction scores to the baseline score.

In general, we observed that the model accurately identifies the tokens of interest within the URLs. We provide a few examples here comparing the results of both the CNN and Transformer models on the same URLs. The first example (Fig. 8) is comparing an SQL injection request with the following token recognised:

cantidad=';+DROP+TABLE+usuarios;+SELECT++FROM +datos+WHERE+nombre+LIKE+'%*

The second example (Fig. 9) contains an XSS attack which is also recognised from the following token:

errorMsg=Credenciales+incorrectas<SCRIPT>alert("Paro s");</SCRIPT>

The graphs show the URL tokens on the horizontal axis and the difference in the model's predicted score on the vertical axis for each element. It is evident that there is no difference in the score whenever a regular element is removed from the URL. However, the difference is seen when the element with malicious intent is removed. The amount of change in score may vary from values with an order of magnitude as low as $1e-5$ to values very close to one.

5 Conclusion

Due to the high usage of web-based applications and related threat of security breaches and data loss essentially requires reliable protection mechanisms with less prediction time and more accuracy. The traditional rule-based checks may not handle many attacks due to their complexity, particularly those involving modifications in HTTP requests. The paper has presented the use of current machine learning approaches that can safeguard web applications against such sophisticated attack vectors. Our study compared Convolutional Neural Network (CNN) and a Transformer-based model for the security of web applications. The prediction accuracies and processing times of both the approaches have been critically analysed. Given the outcome, it has been found that for a given application space, the simpler CNN model may be more performant than the

Transformer model due to its complex architecture. In general, we observed that the CNN model accurately identifies the tokens of interest within the URLs and achieves higher accuracy rate with less training time, thereby, making it a faster technique to identify threats in web-based applications.

References

1. Alaoui, R.L., et al.: Deep learning for vulnerability and attack detection on web applications: a systematic literature review. Future Internet **14**(4), 118 (2022)
2. Luo, A., et al.: A CNN-based approach to the detection of SQL injection attacks. In: IEEE/ACIS 18th International Conference on Computer and Information Science (ICIS) (2019)
3. Zhang, K.: A machine learning-based approach to identify SQL injection vulnerabilities. In: 34th IEEE/ACM International Conference on Automated Software Engineering (ASE) (2019)
4. Tripathy, D., et al.: Detecting SQL injection attacks in cloud SaaS using machine learning. In: 2020 IEEE 6th International Conference on Big Data Security on Cloud (BigDataSecurity), IEEE International Conference on High-Performance and Smart Computing (HPSC) and IEEE International Conference on Intelligent Data and Security (IDS) (2020)
5. Sheykhkanloo, N.M.: Employing neural networks for the detection of SQL injection attack. In: Proceedings of the 7th International Conference on Security of Information and Networks, 2014 (2014)
6. Waheed Kadhim, R., et al.: A hybrid of CNN and LSTM methods for securing web application against cross-site scripting attack. Indonesian J. Electr. Eng. Comput. Sci. **21**(2), 1022 (2021)
7. Mokbal, F.M., et al.: MLPXSS: an integrated XSS-based attack detection scheme in web applications using multilayer perceptron technique. IEEE Access **7**, 100567–100580 (2019)
8. Wang, J., et al.: Evaluating CNN and LSTM for web attack detection. In: Proceedings of the 2018 10th International Conference on Machine Learning and Computing (2018)
9. Yang, W., et al.: Detecting malicious URLs via a keyword-based convolutional gated-recurrent-unit neural network. IEEE Access **7**, 29891–29900 (2019)
10. Zhao, J., et al.: Classifying malicious URLs using gated recurrent neural networks. In: Barolli, L., Xhafa, F., Javaid, N., Enokido, T. (eds.) Innovative Mobile and Internet Services in Ubiquitous Computing. IMIS 2018. Advances in Intelligent Systems and Computing, vol. 773, pp. 385–394. Springer, Cham (2018). https://doi.org/10.1007/978-3-319-93554-6_36
11. Dawadi, B., Adhikari, B., Srivastava, D.: Deep learning technique-enabled web application firewall for the detection of web attacks. Sensors **23**(4), 2073 (2023). https://doi.org/10.3390/s23042073
12. Luo, C., et al.: A novel web attack detection system for the Internet of Things via ensemble classification. IEEE Trans. Industr. Inf. **17**(8), 5810–5818 (2021)
13. Vaswani, A., et al: Attention is all you need. arXiv e-prints arXiv:1706.03762 (2017)
14. Maneriker, P., et al.: URLTran: improving phishing URL detection using transformers. arXiv e-prints arXiv:2106.05256 (2021)
15. Luo, C., Su, S., Sun, Y., Tan, Q., Han, M., Tian, Z.: A convolution-based system for malicious URLs detection. Comput. Mater. Continua **62**, 399–411 (2020)
16. Althubiti, S., et al.: Applying long short-term memory recurrent neural network for intrusion detection. In: SoutheastCon (2018)
17. Gong, X., et al.: CECOR-NET: a character-level neural network model for web attack detection. In: 2019 IEEE International Conference on Smart Cloud (SmartCloud) (2019)

Comparison of Sensor-MAC (S-MAC), Timeout-MAC (T-MAC), Berkeley-MAC (B-MAC), and Modified ML-MAC (M^2L-MAC) for Wireless Sensor Networks

Raghuveer Singh Dhaka(✉), Manju Khurana, and Sonu Lamba

CSED, Thapar Institute of Engineering and Technology, Patiala 147004, Punjab, India
rdhaka_phd21@thapar.edu

Abstract. Synchronous medium access control (MAC) protocols are crucial in wireless sensor networks since they provide energy-efficient data exchanges. This study introduces a modified multi-layer MAC protocol to optimize network lifetime, energy consumption, throughput, and latency of the network. In this article, the nodes inside each wireless subnet are partitioned into five levels by randomly distributing one-fifth of the nodes from each subnet to each layer. In this, initially, the nodes specified in all layers i.e. 1, 2, 3, 4, and 5 are in a dormant state. When the communication process begins, the nodes specified in layer 1 switch to an active state, while the nodes in other layers remain in a dormant state. In a similar fashion, when the nodes in layer 2 communicate, the nodes in the alternate layers i.e. 1, 3, 4, and 5 remain inactive. This active and inactive state is carried out throughout the entire process for all the successor layers assumed in the proposed method. The utilization of sensor nodes in the modified ML-MAC protocol is characterized by a significantly limited duration of listening/active periods, which effectively mitigates energy consumption during the communication process. The protocol is implemented and simulated using MATLAB. When compared to the different MAC protocols, our proposed modified ML-MAC protocol shows considerable improvements in network lifespan, latency, energy consumption, and throughput.

Keywords: Wireless Sensor Networks · Medium Access Control · Multi-Layer MAC · Sensor-MAC · Timeout-MAC · Berkeley-MAC · Modified Multi-Layer MAC

1 Introduction

Wireless sensor networks are cutting-edge systems deployed on a massive scale in challenging environments [1]. They are made up of several individual sensor nodes that are placed randomly for environmental sensing and gathering

D. Pastor-Escuredo et al. (Eds.): ICDLAIR 2023, LNNS 1001, pp. 454–471, 2024.
https://doi.org/10.1007/978-3-031-60935-0_41

the data. They exchange data through a radio link with a limited range to a sink node for analysis [2]. Industrial applications, smart buildings, healthcare, military surveillance, and monitoring of the environment are some of the most common application areas that uses wireless sensor networks (WSNs), which are comprised of an enormous amount of autonomous sensor nodes [3]. The majority of sensor nodes depend on batteries as their primary power source, and because of their deployment in hostile and isolated environments, they can't be recharged. Thus, energy conservation is a crucial factor for the longevity of networks. To reduce power consumption, this work focuses on contention-based protocols that employ an active/sleep state structure [4]. Achieving a healthy equilibrium between a great energy economy and good network performance is a significant challenge in the domain of wireless sensor network applications. The intrinsic decentralization of sensor networks adds an additional level of intricacy to the task of designing energy-efficient protocols. In the context of sensor networks, network lifetime is often considered to be the most important metric for evaluation. To reduce energy consumption, it is essential to optimize communication protocols and computational operations. The wireless interface, which encompasses both computation and radio, is the main energy consumer in sensing devices. The Media Access Control (MAC) layer functions as the immediate interface to the physical layer and is responsible for managing radio communication. Therefore, the optimization of the Medium Access Control (MAC) layer assumes a crucial significance in the advancement of energy-efficient protocols for Wireless Sensor Networks (WSNs) with the objective of prolonging the operational duration of the network [5]. In WSN, for the high throughput network, algorithms that are efficient with energy are of the utmost significance. It is essential to build a Medium Access Control (MAC) protocol for a WSN to be able to offer high throughput in an effective manner that conserves energy [6]. The MAC protocol is an important factor in the overall success of efforts to extend the lifespan of WSNs. The most common method for conserving energy in sensor nodes is known as radio's duty cycling, and the MAC layer protocol is the one in charge of determining when the radio should be turned on and off [7]. The conventional architecture of a wireless sensor network includes numerous sensor nodes in addition to a single sink node. Sensor nodes are of utmost importance in the process of gathering and relaying data to the central sink node. The basic sensor network architecture is shown in Fig. 1.

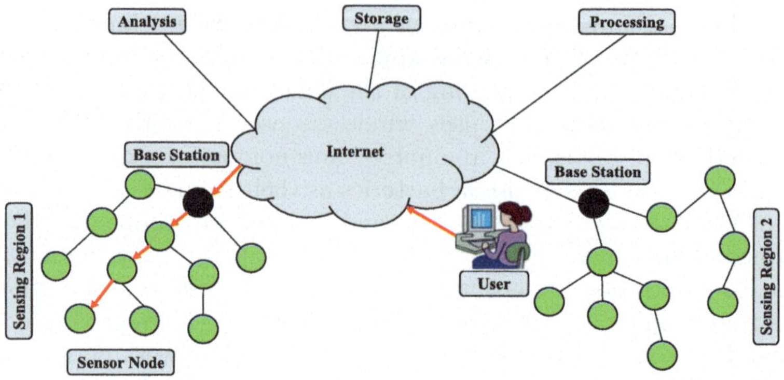

Fig. 1. A typical sensor network consists of sensor nodes, a controller, and a communication system

Sensor nodes acquire unprocessed data by sensing the environment [8]. Through local processing, sensors establish communication with one another, and if needed, they perform data aggregation. The data that has been gathered is aggregated and afterward sent to a central location. The retrieval of information collected by sensors may be accomplished by establishing an internet connection with the central destination node [9]. Sensor networks commonly employ multi-hop communication and prioritize data-centric approaches over node/address-centric ones. The prevalent many-to-one traffic pattern in sensor networks often gives rise to a challenging "hot spot" issue. Consequently, sensor networks must uphold the specific quality of service requirements [10,11]. Designing efficient protocols for Wireless Sensor Networks (WSNs) is challenging due to constraints such as high network density, low radio range, limited power, and the dynamic nature of the environment. WSNs often need to function autonomously for extended periods without human intervention. However, the limited energy resources of sensor nodes can lead to network performance degradation as nodes gradually become disconnected. Thus, ensuring a longer network lifetime is vital for sustaining optimal network performance. The energy consumption of sensors is greatly influenced by MAC protocols. The MAC layer is a constituent of the Data Link Layer in the Open Systems Interconnection (OSI) paradigm and is situated directly above the Physical layer. As the MAC layer governs the radio, which is the primary energy consumer, MAC protocols have a significant impact on the distribution of energy, ultimately determining the lifespan of the nodes [12].

The basic purpose of MAC protocols is to ascertain the mechanism by which nodes get exclusive access to a shared communication medium, guaranteeing that only one node may use the channel at any given point in time. Furthermore, MAC protocols are tasked with the management of channel sensing, and their meticulously crafted architectures have the capability to significantly mitigate occurrences of collisions. In addition, MAC protocols play a crucial role in minimizing idle listening and conserving energy resources by establishing the duty cycle for sensors. Therefore, the implementation of an energy-efficient Medium Access Control (MAC) protocol has the potential to greatly enhance the longevity of sensor networks in operation. In response to the increasing demand, a plethora of Medium Access Control (MAC) protocols have been devised to cater to the requirements of Wireless Sensor Networks (WSNs). Before commencing on the construction of new MAC protocols, it is crucial to thoroughly understand the properties of current ones [13].

1.1 Motivation

The use of battery power in nodes occurs throughout the processes of data sensing, data transfer, and data processing. However, it is worth noting that the act of communication requires a greater amount of power compared to the tasks of detecting and processing data. The longevity of sensor nodes or networks is heavily influenced by the lifespan of their batteries. The manual replacement of batteries in outdoor settings is deemed unfeasible. The paramount concern in the design of Wireless Sensor Networks (WSNs) is the optimization of energy efficiency. This research highlights the significance of contention-based procedures and their utilization of an active/sleep mode frame in order to effectively reduce energy usage. There are four main forms of energy inefficiency, including collision, overhearing, control packet overhead, and idle listening. In WSN, the power consumption during the idle listening process is nearly identical to the data reception process. Thus, it can be stated that idle listening is the primary source of energy waste as opposed to others. In the context of active/sleep cycle systems, sensor nodes intermittently deactivate their radio modules and transition into a sleep state, therefore effectively mitigating idle listening. Previously the work was done only for 3 layers in the 1500 s. This research introduces a design approach for energy-efficient wireless sensor networks by employing the multi-layering medium access control (MAC) concept. The strategy is specifically designed to extend the lifespan of a network by implementing a low-duty cycle and reducing the occurrence of collisions inside the network.

1.2 Contributions

The following are the article's major contributions:

1. In this article, the nodes inside each wireless subnet are partitioned into five levels by randomly distributing one-fifth of the nodes from each subnet to each layer. Communication between nodes in each tier is restricted to a time interval of one-fifth, equivalent to 500 s.
2. Initially, in the proposed strategy, the nodes specified in all layers i.e. 1, 2, 3, 4, and 5 are in a dormant state. When the communication process begins, the nodes specified in layer 1 switch to an active state, while the nodes in other layers remain in a dormant state. In a similar fashion, when the nodes in layer 2 communicate, the nodes in the alternate layers i.e. 1, 3, 4, and 5

remain inactive. This active and inactive state is carried out throughout the entire process for all the successor layers assumed in the proposed method.
3. The proposed modified multi-layer MAC protocol optimizes the network lifetime, energy consumption, throughput, and latency of the network. When compared to the different MAC protocols, the proposed modified ML-MAC protocol shows considerable improvements in all the above-stated aspects.

1.3 Organization of Paper

The present paper is organized in the following manner. The related literature is presented in Sect. 2, followed by the proposed methodology in Sect. 3. The simulation and results are in Sect. 4. The study is concluded in Sect. 5.

2 Literature Survey

The survey in [14] describes the strengths and weaknesses of MAC protocols concerning communication patterns. The description outlines the challenges involved in designing a MAC protocol for WSNs in [15]. In order to enhance the operational effectiveness of Wireless Sensor Networks (WSNs), the authors presented a multicast transmission protocol that leverages the TDMA technique [16]. This protocol aims to improve reliability, while also introducing a modified channel access mechanism to enhance network throughput [17].

The authors in [18] suggest an algorithm for adapting the size of Guaranteed Time Slots (GTS) in order to optimize their usage. The authors in [19,20] conducted a study on Wireless Highway Addressable Remote Transducer Protocol (HART)'s performance, an architecture for centralized network management in data communication aimed at optimizing real-time performance, reliability, and delay. However, their focus was solely on periodic data flows, and they specifically analyzed the TDMA-based approach. Sami *et al.* [21] proposed a novel cooperative medium access control (MAC) protocol with the objective of minimizing energy usage via the efficient use of leftover energy reserves in the relay node.

The authors [22] have suggested a relay selection method on the basis of success rate, link quality, remaining battery power, and neighboring nodes. However, this technique overlooks crucial factors such as buffer capacity, service rate, and arrival, which are vital for achieving low-latency data transmission. Consequently, the authors [23] initiated an investigation into the super-frame architecture of the slotted MAC, using mathematical frameworks in order to mitigate delay by considering packet creation rates. The authors [24] suggest an enhanced shortcut tree routing protocol for enhancing the network's lifespan. Nonetheless, they have failed to take into account the interference problems among the nodes.

To improve the network's reliability, the authors [25] developed a multichannel, priority-aware adaptive framework that is only applicable to tree topologies. Multichannel MAC facilitates parallel transmission while minimizing interference. To mitigate interference issues, the authors [26] propose a dynamic MAC protocol that identifies the optimal communication channel for facilitating coexistence among wireless nodes operating within the same ISM spectrum. Nevertheless, the synchronous duty-cycled MAC protocol outperforms the asynchronous counterpart in terms of energy efficiency. As a result, the authors [27] suggest an analytical model of Sensor Medium Access Control (SMAC). The study [28] also examines the effects of introducing a MAC-level finite buffer on the performance of WSNs.

Nevertheless, the existing research primarily concentrates on the star topology. The study [29] focuses on evaluating the performance of cross-layer IEEE 802.15.4/4e/4g Low Rate Wireless Personal Area Network (LR-WPAN). The authors [30] have devised a novel cross-layer framework that effectively combines the medium access control (MAC) and physical layers in unslotted IEEE 802.15.4 networks. The objective of this study [31] is to examine the efficiency of single-hop and multi-hop networks in using the unslotted IEEE 802.15.4 standard MAC protocol. However, the issue of enhancing the network's lifespan by strategically selecting the next node for transmitting data has not been discussed. To address the challenges posed by bursty data and ensure high reliability, the authors [32] propose a MAC protocol that prioritizes the receiver's perspective. This protocol aims to minimize collisions and achieve a significant improvement in throughput.

The IEEE 802.15.4e standard describes the DSME MAC, a deterministic and synchronous multichannel extension, which enables short-range communication, low power, and low data rate. The DSME MAC mode changes IEEE 802.15.4 from a single channel operation to a multiple operation for data transfers. This is done to mitigate the impacts of mutual interferences caused by nodes that are different from each other.

A recently developed Deterministic and Synchronous Multichannel Extension (DSME) MAC protocol has meant to [33] keep to prevent channel congestion resulting from the transmission of acknowledgment packets, eliminating the need for additional CCA. The evaluation of performance in Deterministic and Synchronous Multichannel Extension (DSME) enabled IEEE 802.15.4e standard WSNs, has been investigated [34] in terms of aggregate throughput and frame error rate. However, the mesh topology has not been studied. The authors [35] examine the energy usage and throughput of the Deterministic and Synchronous Multichannel Extension (DSME) MAC protocol in the context of the IEEE 802.15.4e standard. However, in dense networks, DSME MAC is unable to prevent beacon collision despite its capability for synchronous multichannel communication that improves network performance and decreases latency. The authors [36] suggest an improved multichannel MAC protocol that uses less energy and cuts down on delay. In order to accommodate varying traffic patterns, the authors [37] suggest a hybrid MAC approach that switches between Carrier Sense Multiple Access (CSMA) mode and Time Division Multiple Access (TDMA) mode depending on the traffic level.

Dong *et al.* did a comprehensive analysis of the issues that arise in relation to mobility at the link layer. The research also included presenting the results obtained from a comparative examination of many medium access control (MAC) protocols that have been specifically intended to address the issue of mobility in [38]. In [39], Doudou *et al.* looked at the latency problem that arises with certain asynchronous MAC protocols in wireless sensor networks (WSNs) and compared them to the latency of a static WSN. Zhao *et al.* [40] examined several WSNs MAC protocols, with the primary objective of assessing their benefits and drawbacks in a wireless environment. The conclusion by the authors in [40] suggests that a self-adaptive MAC layer [41] could be well-suited for managing challenging environments such as WSNs. Chen *et al.*, offered analysis of various different types of MAC protocols for use in an underwater acoustic environment and also brought attention to the difficulties in [42]. As previously stated [43], contemporary MAC protocols have been developed with the primary objective of optimizing energy consumption and improving the efficient transmission of data in densely populated networks. In order to have a deeper understanding of the subject matter, Ju *et al.* [44] performed a comprehensive survey on cooperative medium access control (MAC) protocols in wireless communication. The objective of the study was to assess the performance of MAC protocols when they interact with other levels of the network.

3 Proposed Methodology

The proposed approach, known as the Modified multi-layer MAC (M^2L-MAC) protocol, aims to further decrease node power consumption compared to other MAC protocols. The system under consideration is a distributed contention-based Medium Access Control (MAC) mechanism, in which nodes ascertain the presence of neighboring nodes by evaluating the strength of their radio signal.

To further expand on the idea of M^2L-MAC, the ideology can be better understood by simulating a topology interconnecting a base station with sensors that are deployed geographically apart. The topology in the Fig. 2 consists of a base station, an array of 28 sensors monitored by administrator systems, connected via switches, and a star topology of routers. In the topology, each sensor is represented by 'S-x' with x as its index number, and administrator systems are represented by 'AD-x' with x as their respective serial numbers. Each sensor has a MAC-Address associated with itself, and to facilitate the hardware interconnect, a fast-ethernet port is attached to each sensor (allowing manual static IP-Address assignment). All sensors connected to each independent router perform a specific purpose and are allotted IP addresses such that there is no overlap among the networks of adjacent routers for easy differentiation. Following general networking standards, the sensors send the monitored data to the base station, and when inactive, the base station pings all sensors. This ping is forwarded to every sensor through switches. When a specific set of sensors needs to be functioning, the pings are received by all sensors, which increases the overall power consumption within the network and reduces the life span.

Considering a specific scenario shown in Fig. 3 where we have to collect data from sensors with serial numbers 24-28 due to an M^{2}L-MAC implementation, the activation message can be sent directly to the respective subnetwork without activating other sensors.

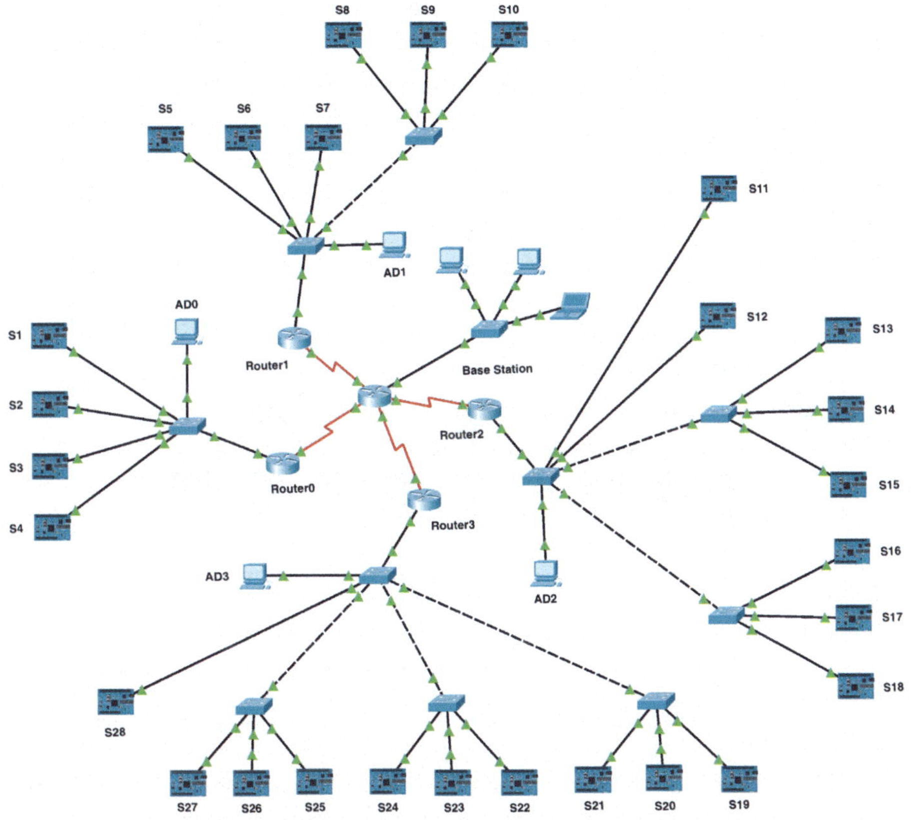

Fig. 2. Simulated Network Topology

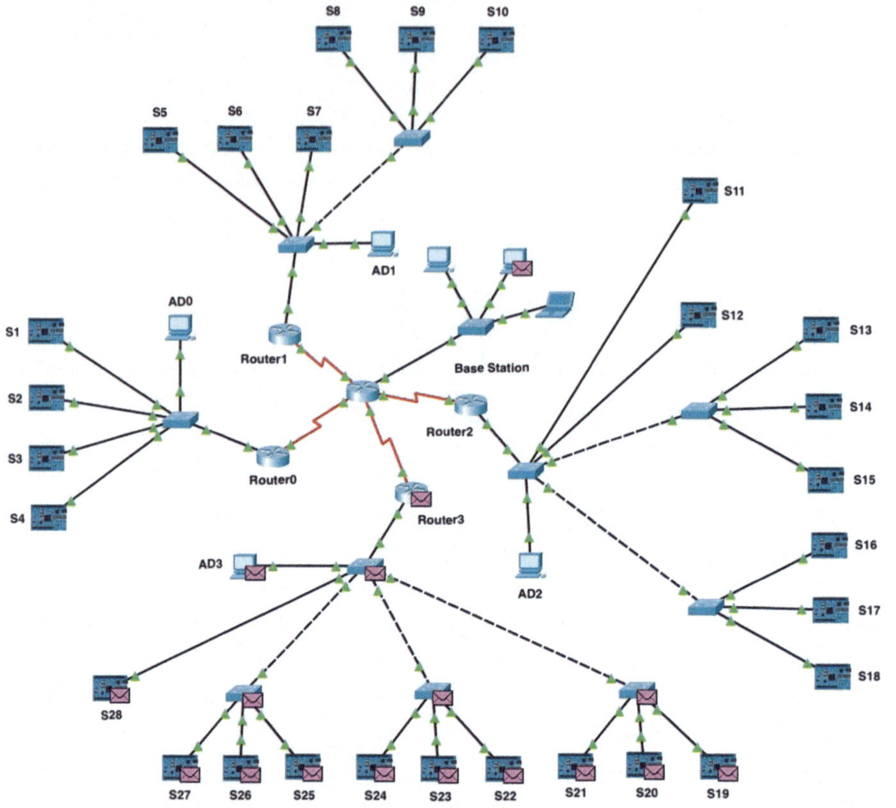

Fig. 3. Nodes receiving activation ping sent by base station; intended for sensors 24 to 28

If we consider the M^2L-MAC implementation within the network, due to sensors being allotted IPs of differentiated subnets - the frames in their active phase can contain the activation message/ping for the specific set of sensors that must be functioning without disturbing the other sensor networks. By this, the other sensors are not disturbed and stay in sleep mode, not consuming any power.

The only overhead in this case is five more active sensors, compared to 23 getting activated by a general networking protocol. The image above shows the complete through which the activation message passes and the sensors it reaches. This hardware implementation shows that M^2L-MAC proves to be beneficial for similar use cases over other MAC protocols in many aspects, like reducing power consumption and the overall latency within the network. The advantages M^2L-MAC has are also prominent in networks with a higher number of nodes; the performance for the same has been measured using MATLAB and is documented below.

Additionally, this MAC protocol is characterized by its self-organizing nature since it does not need the presence of a central node to oversee and regulate the activities of the individual nodes. Design of Modified ML-MAC (M^2L-MAC) is shown in Fig. 4. The nodes inside each wireless subnet are partitioned into five levels by randomly distributing one-fifth of the nodes from each subnet to each layer. Communication between nodes in each tier is restricted to a time interval of one-fifth, equivalent to 500 s. During the communication process between nodes in layer 1, nodes in layers 2 and 3 are in a dormant state. In a similar vein, during layer 2 communication, layers 1 and 3 remain inactive. The alteration of device start time and device stop time in the MAC layer is performed for every node inside the network.

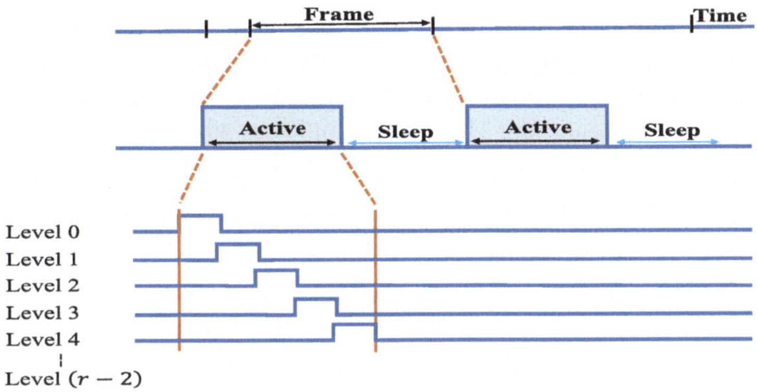

Fig. 4. Design of Modified ML-MAC (M^{2}L-MAC)

Modified ML-MAC (M^{2}L-MAC) in Fig. 4 consists of frames, whereby each frame is further divided into two distinct phases, namely the active period and the sleep period. The active period is partitioned into r distinct levels that do not overlap as shown in Fig. 5. The nodes are dispersed throughout this particular assortment of layers, with each layer comprised of nodes that adhere to a temporal schedule for periods of activity and rest. Notably, the schedules of these layers are temporally offset from one another.

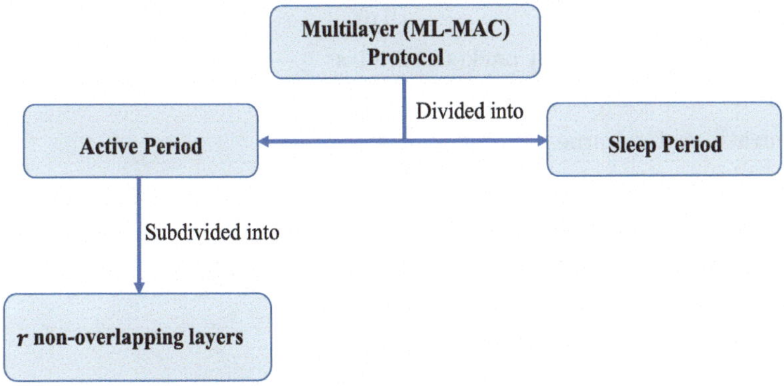

Fig. 5. Modified ML-MAC (M^{2}L-MAC)

Consequently, the listening durations of the nodes in distinct layers do not overlap. In the Modified ML-MAC (M^2L-MAC) protocol, a node is designed to awaken just its designated layer. Hence, Modified ML-MAC (M^2L-MAC) exhibits a lower energy consumption compared to Sensor-MAC (S-MAC)) due to the shorter listen length of a node in Modified ML-MAC (M^2L-MAC).

4 Simulation Results

For the numerical simulation in the proposed work, the values of the parameter are shown in Table 1. The primary aim of this simulation is to assess the performance of the Modified ML-MAC (M^2L-MAC) protocol in a network with the goal of minimizing energy usage and enhancing the network lifetime. The simulations were conducted using the MATLAB simulator, which enables the execution of scalable simulations for wireless sensor networks (WSNs). Within the context of the Modified ML-MAC (M^2L-MAC) architecture, the nodes are categorized into several tiers, and each tier functions according to its own distinct ON/OFF schedule.

In the unsynchronized Modified ML-MAC (M^2L-MAC) scheme, time frames are allocated to different groups and layers. Specifically, group 1 layer 1 is

Table 1. Simulation parameters

Parameter	Value
Terrain	200 m × 200 m
Simulation time	2500 s
Number of nodes	400
Layers drift timing	5–505, 506–1006, 1007–1507, 1508–2008, 2009–2509
Message rate	1 packet/sec
Message size	40 bytes
Transmission range	10 m
Average time between packets	2–11 s
Layers of access, in a count	1–10
Duration of a frame	1 s
Duration of layer	0.5/L s
Transmission power of node	24.75 mW
Listening power of node	13.5 mW
Sleeping power of node	15 μW
Transmission datarate of node	19.2 kbps

assigned the time frame of 5-505 sec, group 2 layer 1 is assigned 506-1006 sec, group 3 layer 1 is assigned 1007-1507 sec, group 4 layer 1 is assigned 1508-2008 sec, and group 5 layer 1 is assigned 2009-2509 sec. By using this approach, we are reiterating the time intervals for each layer. However, only one layer is actively functioning at any given moment, while the remaining layers are in a state of dormancy. In the given situation, a total of 400 nodes are distributed in a random manner. The dimensions of the terrain area are considered to be 200 m by 200 m. The simulation lasts for 2500 sec.

4.1 Evaluation of Network Longevity

In Fig. 6, a comparison of the network lifetime for Sensor-MAC (S-MAC), Timeout-MAC (T-MAC), Berkeley-MAC (B-MAC), and Modified ML-MAC (M^{2}L-MAC) is depicted.

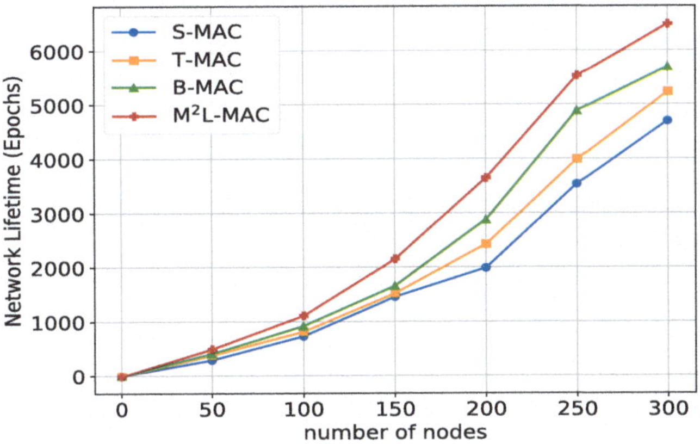

Fig. 6. Network Lifetime Comparison for Sensor-MAC (S-MAC), Timeout-MAC (T-MAC), Berkeley-MAC (B-MAC), and Modified ML-MAC (M^{2}L-MAC)

Figure 6 provides a succinct visual representation of the comparisons in network lifespan between the M^{2}L-MAC protocol presented in this study and other alternative protocols. The M^{2}L-MAC protocol demonstrates superior performance in comparison to the other protocols under consideration, specifically in relation to network longevity. On the other hand, S-MAC has a relatively reduced network lifespan. B-MAC exhibits a notable extension in network longevity when the number of nodes exceeds 150.

The estimation of network lifespan in a sensor network may be achieved by taking into account three parameters: complete capacity of the battery (C_B), duration of the simulation (D_S) and aggregate amount of energy used in L layers (AE_L). The total energy consumption may be determined using Eq. 1, which accounts for the combined energy usage during transmission, reception, idle, and sleep periods.

$$AE_L = E_{tr} + E_{rec} + E_{id} + E_{sl} \tag{1}$$

Thus, network lifetime $N_L(sec)$ will be given as:

$$N_L = \frac{(D_S * C_B)}{AE_L} \tag{2}$$

The energy consumption for Sensor-MAC (S-MAC), Timeout-MAC (T-MAC), Berkeley-MAC (B-MAC), and Modified ML-MAC (M^2L-MAC) is shown in Fig. 7. In the Modified ML-MAC (M^2L-MAC) protocol, the duration of listening periods is comparatively lower than that of other MAC protocols, leading to decreased energy usage.

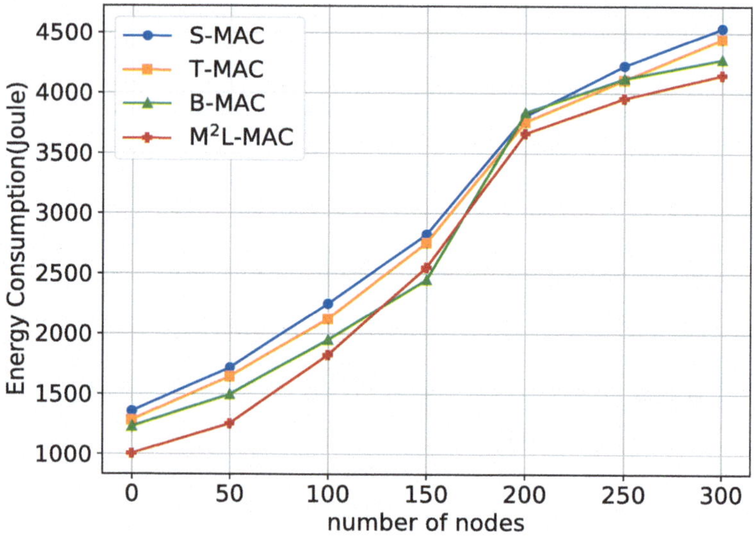

Fig. 7. Energy Comparison for Sensor-MAC (S-MAC), Timeout-MAC (T-MAC), Berkeley-MAC (B-MAC), and Modified ML-MAC (M^2L-MAC)

Figure 7 provides a clear depiction of the energy consumption comparison between the proposed M^2L-MAC protocol and the other protocols being evaluated. It is evident that the energy consumption of the M^2L-MAC protocol is lower than that of all the other protocols being considered. Moreover, when doing an analysis of the energy consumption outcomes for T-MAC and B-MAC, it is evident that B-MAC has superior performance compared to T-MAC.

As seen in Fig. 8, throughput is compared for Sensor-MAC (S-MAC), Timeout-MAC (T-MAC), Berkeley-MAC (B-MAC), and Modified ML-MAC (M^2L-MAC). In WSN, throughput mainly indicates the number of packets per second received at the destination. It is primarily calculated as:

$$Throughput = \frac{T_D}{T_T} \tag{3}$$

where,

$T_D \longrightarrow$ Total Data Transmitted
$T_T \longrightarrow$ Total Time Taken

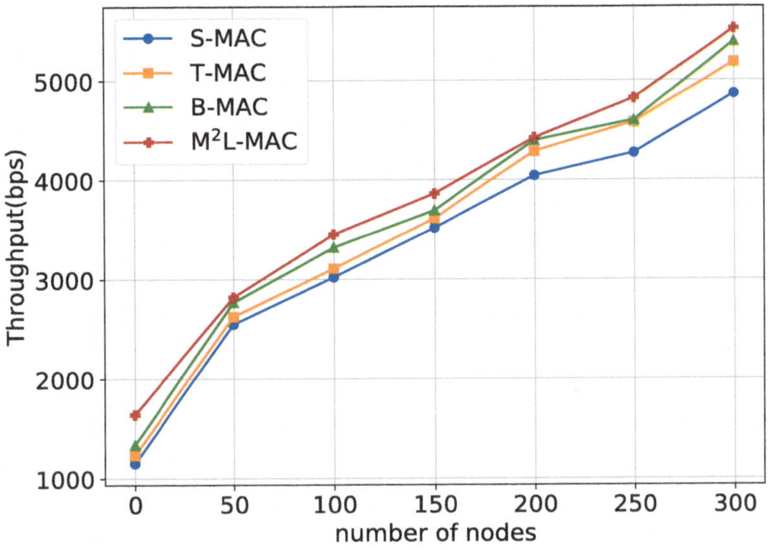

Fig. 8. Throughput Comparison for Sensor-MAC (S-MAC), Timeout-MAC (T-MAC), Berkeley-MAC (B-MAC), and Modified ML-MAC (M^{2}L-MAC)

It is clearly visible from Fig. 8 that the throughput of the proposed M^{2}L-MAC is better than that of all the considered protocols. Right from the beginning, the proposed protocol yields better throughput with respect to the number of nodes discussed in the article. In addition, if we examine the throughput results for T-MAC and B-MAC, it is evident that B-MAC yields lower throughput initially, but as the number of nodes increases, its performance appears to be better than that of the T-MAC protocol.

The Comparison of latency for Sensor-MAC (S-MAC), Timeout-MAC (T-MAC), Berkeley-MAC (B-MAC), and Modified ML-MAC (M^{2}L-MAC) is shown in Fig. 9. Latency in WSN refers to the amount of time required for a data packet to travel from one location to another. It is mainly calculated as:

$$Latency = \frac{T_{PA}}{T_{PD}} \tag{4}$$

where,

$T_{PA} \longrightarrow$ Time Taken for Packet Arrival
$T_{PD} \longrightarrow$ Time of Packet Departure

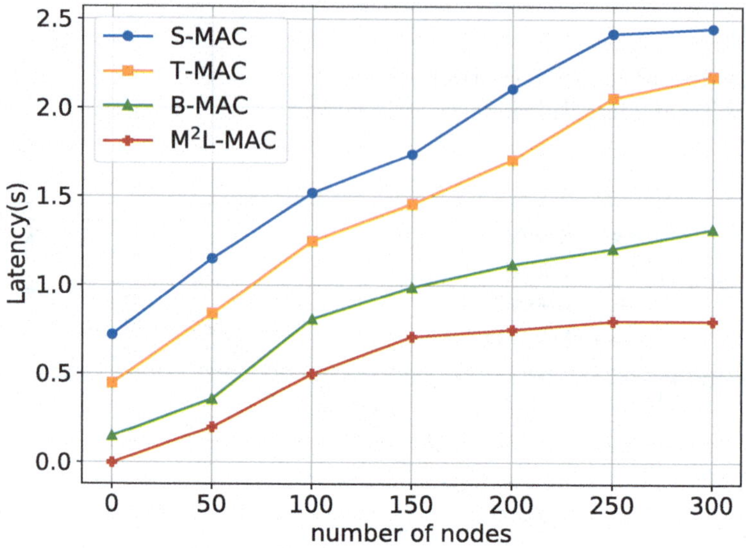

Fig. 9. Latency Comparison for Sensor-MAC (S-MAC), Timeout-MAC (T-MAC), Berkeley-MAC (B-MAC), and Modified ML-MAC (M^2L-MAC)

The B-MAC protocol demonstrated decreased latency in relation to the T-MAC protocol as the scale of the network expanded, as seen in Fig. 9. Significantly, the M^2L-MAC protocol had the lowest latency as the size of the network expanded, distinguishing it from the other protocols.

4.2 State of the Art Techniques

During the evaluation of the suggested methodology, a thorough study was undertaken to compare it with different approaches. The findings highlight the notable efficacy of the suggested methodology, constantly surpassing the performance of all evaluated approaches. Table 2 presents a comprehensive evaluation of the proposed system in comparison to standard protocols, considering several pertinent criteria.

The Modified ML-MAC (M^2L-MAC) we suggest places a high priority on energy efficiency, boasts a prolonged for the network, guarantees robust throughput, and minimizes latency.

Table 2. Performance Metrics

Protocol	Network Lifetime(Epochs)	Energy Consumption (joule)	Throughput (bps)	Latency(s)
S-MAC	4695	4545	4868	2.45
T-MAC	5241	4457	5189	2.18
B-MAC	5696	4286	5395	1.32
Proposed M^2L-MAC	**6495**	**4156**	**5524**	**0.80**

5 Conclusions

This research presents a novel approach for enhancing energy efficiency in wireless sensor networks (WSNs) through the use of the Modified ML-MAC (M^2L-MAC) protocol. The suggested protocol is employed to evaluate and analyze many key characteristics, including network lifespan, latency, energy consumption, and throughput, in relation to Modified ML-MAC (M^2L-MAC). Henceforth, it is found that the network lifetime of the proposed multi-layer protocol is 27.71% more as compared to S-MAC, 19.31% more than Timeout-MAC (T-MAC), and 12.30% more than the Berkeley-MAC (B-MAC) protocol. The proposed protocol has also shown its superiority in the case of energy consumption, throughput, and latency compared to alternative MAC protocols. In future work, we can increase the number of nodes in the network to check the efficiency of the suggested protocol and can use any optimization technique to enhance the performance of the Modified ML-MAC (M^2L-MAC) protocol.

References

1. Pundir, M., Sandhu, J.K.: A systematic review of quality of service in wireless sensor networks using machine learning: recent trend and future vision. J. Netw. Comput. Appl. **188**, 103084 (2021)
2. Nayak, P., Swetha, G., Gupta, S., Madhavi, K.: Routing in wireless sensor networks using machine learning techniques: challenges and opportunities. Measurement **178**, 108974 (2021)
3. Sahoo, P.K., Pattanaik, S.R., Wu, S.-L.: A novel synchronous MAC protocol for wireless sensor networks with performance analysis. Sensors **19**(24), 5394 (2019)
4. Jha, M.K., Pandey, A.K., Pal, D., Mohan, A.: An energy-efficient multi-layer MAC (ML-MAC) protocol for wireless sensor networks. AEU Int. J. Electron. Commun. **65**(3), 209–216 (2011)
5. Maitra, T., Roy, S.: A comparative study on popular mac protocols for mixed wireless sensor networks: from implementation viewpoint. Comput. Sci. Rev. **22**, 107–134 (2016)
6. Kumar, A., Zhao, M., Wong, K.-J., Guan, Y.L., Chong, P.H.J.: A comprehensive study of IoT and WSN mac protocols: research issues, challenges and opportunities. IEEE Access **6**, 76228–76262 (2018)

7. Khurana, M., Thalore, R., Raina, V., Jha, M.K.: Improved time synchronization in ML-MAC for WSN using relay nodes. AEU Int. J. Electron. Commun. **69**(11), 1622–1626 (2015)
8. Szewczyk, R., Mainwaring, A., Polastre, J., Anderson, J., Culler, D.: An analysis of a large scale habitat monitoring application. In: Proceedings of the 2nd International Conference on Embedded Networked Sensor Systems, pp. 214–226 (2004)
9. Zúniga, M., Krishnamachari, B.: Integrating future large-scale wireless sensor networks with the internet. Tech. Rep., USC Computer Science (2003)
10. Kumar, A., Shwe, H.Y., Wong, K.J., Chong, P.H.: Location-based routing protocols for wireless sensor networks: A survey. Wirel. Sens. Netw. **9**(1), 25–72 (2017)
11. Akyildiz, I.F., Su, W., Sankarasubramaniam, Y., Cayirci, E.: A survey on sensor networks. IEEE Commun. Mag. **40**(8), 102–114 (2002)
12. Langendoen, K., Halkes, G.: Energy-efficient medium access control. Embed. Syst. Handb. **6000**, 34 (2005)
13. Lin, L., Wong, K.-J., Kumar, A., Lu, Z., Tan, S.-L., Phee, S.J.: Evaluation of a TDMA-based energy efficient MAC protocol for multiple capsule networks. EURASIP J. Wirel. Commun. Netw. **2011**, 1–12 (2011)
14. Demirkol, I., Ersoy, C., Alagoz, F.: Mac protocols for wireless sensor networks: a survey. IEEE Commun. Mag. **44**(4), 115–121 (2006)
15. Yadav, R., Varma, S., Malaviya, N., et al.: A survey of mac protocols for wireless sensor networks. UbiCC J. **4**(3), 827–833 (2009)
16. Bhatia, A., Hansdah, R.: TRM-MAC: a TDMA-based reliable multicast mac protocol for WSNs with flexibility to trade-off between latency and reliability. Comput. Netw. **104**, 79–93 (2016)
17. Sahoo, P.K., Pattanaik, S.R., Wu, S.L.: A reliable data transmission model for IEEE 802.15. 4e enabled wireless sensor network under wifi interference. Sensors **17**(6), 1320 (2017)
18. Zhan, Y., Xia, Y., Anwar, M.: GTS size adaptation algorithm for IEEE 802.15. 4 wireless networks. Ad Hoc Netw. **37**, 486–498 (2016)
19. Saifullah, A., Xu, Y., Lu, C., Chen, Y.: Real-time scheduling for wirelesshart networks. In: 2010 31st IEEE Real-Time Systems Symposium, pp. 150–159. IEEE (2010)
20. Saifullah, A., Xu, Y., Lu, C., Chen, Y.: End-to-end delay analysis for fixed priority scheduling in wirelesshart networks. In: 2011 17th IEEE Real-Time and Embedded Technology and Applications Symposium, pp. 13–22. IEEE (2011)
21. Sami, M., Noordin, N.K., Hashim, F., Subramaniam, S., Akbari-Moghanjoughi, A.: An energy-aware cross-layer cooperative mac protocol for wireless ad hoc networks. J. Netw. Comput. Appl. **58**, 227–240 (2015)
22. Laurindo, S., Moraes, R., Nassiffe, R., Montez, C., Vasques, F.: An optimized relay selection technique to improve the communication reliability in wireless sensor networks. Sensors **18**(10), 3263 (2018)
23. Yan, H., Zhang, Y., Pang, Z., Da Xu, L.: Superframe planning and access latency of slotted mac for industrial WSN in IoT environment. IEEE Trans. Industr. Inf. **10**(2), 1242–1251 (2014)
24. Wadhwa, L., Deshpande, R.S., Priye, V.: Extended shortcut tree routing for ZigBee based wireless sensor network. Ad Hoc Netw. **37**, 295–300 (2016)
25. Patti, G., Bello, L.L.: A priority-aware multichannel adaptive framework for the IEEE 802.15. 4e-lldn. IEEE Trans. Ind. Electron. **63**(10), 6360–6370 (2016)
26. Correia, L.H., Tran, T.-D., Pereira, V.N., Giacomin, J.C., Silva, J.M.S.: DynMAC: a resistant MAC protocol to coexistence in wireless sensor networks. Comput. Netw. **76**, 1–16 (2015)

27. Martinez-Bauset, J., Guntupalli, L., Li, F.Y.: Performance analysis of synchronous duty-cycled MAC protocols. IEEE Wireless Commun. Lett. **4**(5), 469–472 (2015)
28. Al-Anbagi, I., Khanafer, M., Mouftah, H.T.: MAC finite buffer impact on the performance of cluster-tree based WSNs. In: 2013 IEEE International Conference on Communications (ICC), pp. 1485–1490. IEEE (2013)
29. Sum, C.-S., Lu, L., Zhou, M.-T., Kojima, F., Harada, H.: System evaluation of a practical IEEE 802.15. 4/4e/4g multi-physical and multi-hop smart utility network. IET Commun. **9**(5), 665–673 (2015)
30. Di Marco, P., Fischione, C., Santucci, F., Johansson, K.H.: Modeling IEEE 802.15. 4 networks over fading channels. IEEE Trans. Wireless Commun. **13**(10), 5366–5381 (2014)
31. Di Marco, P., Park, P., Fischione, C., Johansson, K.H.: Analytical modeling of multi-hop IEEE 802.15. 4 networks. IEEE Trans. Veh. Technol. **61**(7), 3191–3208 (2012)
32. Huang, P., Wang, C., Xiao, L.: RC-MAC: a receiver-centric mac protocol for event-driven wireless sensor networks. IEEE Trans. Comput. **64**(4), 1149–1161 (2014)
33. Sahoo, P.K., Pattanaik, S.R., Wu, S.-L.: A novel IEEE 802.15. 4e DSME MAC for wireless sensor networks. Sensors **17**(1), 168 (2017)
34. Lee, J., Jeong, W.C.: Performance analysis of IEEE 802.15. 4e DSME MAC protocol under WLAN interference. In: 2012 International Conference on ICT Convergence (ICTC), pp. 741–746. IEEE (2012)
35. Jeong, W.-C., Lee, J.: Performance evaluation of IEEE 802.15. 4e DSME MAC protocol for wireless sensor networks. In: 2012 The First IEEE Workshop on Enabling Technologies for Smartphone and Internet of Things (ETSIoT), pp. 7–12. IEEE (2012)
36. Liu, Y., Liu, H., Yang, Q., Wu, S.: RM-MAC: a routing-enhanced multi-channel MAC protocol in duty-cycle sensor networks. In: 2015 IEEE International Conference on Communications (ICC), pp. 3534–3539. IEEE (2015)
37. Zhuo, S., Wang, Z., Song, Y.-Q., Wang, Z., Almeida, L.: A traffic adaptive multi-channel mac protocol with dynamic slot allocation for WSNs. IEEE Trans. Mob. Comput. **15**(7), 1600–1613 (2015)
38. Dong, Q., Dargie, W.: A survey on mobility and mobility-aware mac protocols in wireless sensor networks. IEEE Commun. Surv. Tutorials **15**(1), 88–100 (2012)
39. Doudou, M., Djenouri, D., Badache, N.: Survey on latency issues of asynchronous mac protocols in delay-sensitive wireless sensor networks. IEEE Commun. Surv. Tutorials **15**(2), 528–550 (2012)
40. Zhao, Y.Z., Miao, C., Ma, M., Zhang, J.B., Leung, C.: A survey and projection on medium access control protocols for wireless sensor networks. ACM Comput. Surv. (csuR) **45**(1), 1–37 (2012)
41. Sha, M., Dor, R., Hackmann, G., Lu, C., Kim, T.-S., Park, T.: Self-adapting MAC layer for wireless sensor networks. In: 2013 IEEE 34th Real-Time Systems Symposium, pp. 192–201. IEEE (2013)
42. Chen, K., Ma, M., Cheng, E., Yuan, F., Su, W.: A survey on mac protocols for underwater wireless sensor networks. IEEE Commun. Surv. Tutorials **16**(3), 1433–1447 (2014)
43. Huang, P., Xiao, L., Soltani, S., Mutka, M.W., Xi, N.: The evolution of MAC protocols in wireless sensor networks: a survey. IEEE Commun. Surv. Tutorials **15**(1), 101–120 (2012)
44. Ju, P., Song, W., Zhou, D.: Survey on cooperative medium access control protocols. IET Commun. **7**(9), 893–902 (2013)

Automatic Code Scraping Based Test Case Generation

Arunkumar Chinnaswamy(✉) and B. A. Sabarish

Amrita School of Computing, Tamil Nadu, Coimbatore 641112, India
c_arunkumar@cb.amrita.edu

Abstract. Software testing is a critical phase in the software development life cycle, ensuring the reliability and functionality of applications. The major challenge in this process is the generation of effective test cases, which demands a comprehensive understanding of the underlying codebase. This paper presents a novel approach called Code Scraping Based Test case generation, leveraging automated techniques to extract relevant information from source code and subsequently generate comprehensive test cases. The CSBT methodology involves code parsing and algorithmic analysis to identify key components to infer the expected behavior of the program and hence ensuring the generation of relevant and meaningful test cases. Dynamic analysis techniques (Continuous Integration and Continuous Deployment) are integrated to validate the generated test cases against the actual code execution, enhancing the accuracy and coverage of the testing process. The results demonstrate notable improvements in both the efficiency and effectiveness of the testing process, showcasing CSBT's ability to adapt to various programming languages and project complexities.

Keywords: Software Testing · Code Scraping · Test case generation · Abstract Syntax Tree

1 Introduction

Development of software products includes a list of activities to satisfy the customer requirements. In the recent technological environment where Continuous Integration and Continuous Development gives flexibility to change the requirements during any stage of development. The quality of the project does not rely on any one stage of development, depends on reduced cost, contradicting requirements, deadlines on delivery. Quality can be revealing faults and readiness of product. Software testing is a crucial aspect of Software Development Life Cycle (SDLC) aimed at identifying and eliminating errors in the software product before deployment. With the advent of agile and DevOps methodologies, automated testing has become integral to software development. Continuous push and pull requests to the code repository (git/docker) and minor changes in the code generation of a new set of test cases every push to the source code repository. Generation, updating, and removing redundant test cases are useful in ensuring software quality as applications are becoming increasingly complex and dynamic due to changing

D. Pastor-Escuredo et al. (Eds.): ICDLAIR 2023, LNNS 1001, pp. 472–482, 2024.
https://doi.org/10.1007/978-3-031-60935-0_42

user requirements (user stories). However, manual testing methods have limitations on time duration, optimizing the number of test cases, and lack of skilled testers. Testing is carried out under time constraints, making it challenging to determine priority by understanding the requirements and identifying the relationship between the modules. Automated test case generation algorithms and methods can effectively detect faults and defects in artifacts during the early stages of software development. The available methods are Model-Based Testing, Symbolic Execution, Random Testing, Search-Based Testing, Fuzz Testing, Combinatorial Testing, Mutation Testing etc.

1.1 Motivation

Testing is the costly and time-consuming process in SDLC. Test case generation is the major process to identify and generate the number of test cases to cover all the requirements specified in the document, which leads to have some deviation in terms of implementation which misses out some part of the codes untested. To solve this challenge, the research work in this paper focuses on extracting information from codebases for test case generation which includes legacy codebases, maintainability, coverage improvement and knowledge transfer.

1.2 Organization of the Paper

Remaining of the paper is organized as, Sect. 2 describes state of the art techniques in test case generation challenges and research gaps in literature survey. Section 3 gives the description of the problem statement derived from the literature, Sect. 4 gives proposed approach for the problem statement whereas Sect. 5 evaluates the proposed model in terms of test coverage and Sect. 6 gives the conclusion, future enhancements can be carried out.

2 Related Work

This section describes the state-of-the-art techniques in the automatic test case generation process for improving the coverage and efficiency in identifying bugs to produce quality products. A systematic review of the test case generation using the UML model is carried out considering different methodologies, where a major part of the work is carried out using the behavioral diagram, including the state, activity, and sequence diagram of the project. Many of the methodologies used a traversal model for generating the test cases [1]. Metamorphic testing compares a program's output to its expected output, allowing for more accurate evaluation of faults. Metamorphic testing is becoming an effective technique for Oracle problem alleviation and automated data generation [2]. A Study on test case generation methods, minimization, selection, and prioritization techniques for improving software quality is carried out. It suggests that optimizing test cases can lead to faster execution, more coverage, and bug detection [3]. Formal test case generation techniques for conformance testing use labeled transition systems, finite state machines, and Petri-nets. Hybrid behavioral models can be robust, combining input from one model and enhancing details from another to create more complete test cases [4]. EvoSuite is a Java code tool that generates test cases with assertions using a hybrid approach. It optimizes test suites by adding concise assertions to detect deviations and protect against future defects [5]. Surveying application developers on test generation tools and test cases revealed requirements for successful technology transfer from academic research to industrial practice, including integration with popular build tools and improved test readability [6]. The study uses bytecode instrumentation to separate code from its environmental dependencies and extends the EVOSUITE Java test generation tool. Experiments on 100 Java projects showed improved code coverage by up to 80%/90% and reduced unstable tests by over 50% [7]. A multi-objective optimization approach implementing a framework with two representations for procedural and object-oriented code is proposed. Combinations of three objectives are experimentally evaluated [8]. Multi objective firefly algorithm with surrogate model (MOFA-SM) to solve complex optimization problems is proposed that initializes using chaotic mapping. The algorithm constructs an external archive using preference sorting and lightweight clustering pruning that shows improved convergence iteration and stability [9]. AlphaCode, an AI system for code generation, achieved an average ranking of 54.3% in simulated evaluations on the Codeforces platform. It generates millions of diverse programs using trained transformer-based networks, filtering and clustering them to a maximum of 10 submissions, marking the first time an AI system has performed competitively in programming competitions [10]. CodeXGLUE is a benchmark dataset designed to accelerate machine learning research in programming language tasks. The goal is to support model development and increase software developer productivity [11]. To ensure quality during significant updates, automated regression testing becomes essential. By employing a traceability matrix, both onward and backward tracking is facilitated that optimizes regression testing by eliminating repetitive test cases [12]. An algorithm designed for creating call graphs tailored to lambda structures is presented that retains method signatures to form a static call graph and uses an intermediary Abstract Syntax Tree (AST) like format, which is subsequently refined into the desired call graph structure [13]. Full-text search offers a superior and versatile solution, constructing indexes around words to bolster match accuracy. PostgreSQL's full-text search is amplified to increase efficiency by integrating semantic operators, thereby rendering search operations both robust and nuanced [14]. The main objective is to conduct an in-depth examination of source code written by developers to identify these clones. This examination considers numerous factors such as syntax, functions, methods, variables, data types, keywords, spacing, author information, imports, comments, line numbers, and last modified dates. The degree of code resemblance, which aids in efficient code oversight is visualized [15].

3 Problem Statement

Software quality is challenging in terms of changing nature of changing requirements between sprints in CI/CD approach in development. Identifying the optimal number of testcases to cover the complete scenarios of project and reducing the manual intervention and time in the testing process which is calculated using the cyclometric complexity measure. The Research Questions are: How can code scrapping based test case automation supports dynamic changes in the implementation which affects the behavior and test cases? How to identify the optimum number of test cases to adequately cover the entire project which may change with respect to changes in code? How AST based approach help in reducing the number of code smells and improving the quality of the development process by providing a complete traversal through the program? Using cyclometric complexity as a benchmark to compare and identify the optimal number of test cases for coverage analysis of the project?

4 Proposed Work

A code-based model is proposed to propose solution for the research questions evolved from the literature how automated test case generation from code rather than from UML model helps in improving the quality in dynamic programming environment. Unlike traditional approaches, code scraping model for test case generation doesn't need external inputs and intervention of human because in proposed methodology automatically generates test case for each push to the repository. The proposed model reads code and constructs the tree structure using AST. AST constructs a tree structure by programming constructs which include loops, boundary validation and input validation. Code scrapping-based test case generation is white-box testing that tries to uncover all possible traces.

4.1 Abstract Syntax Tree (AST) Generation

Parsing is used to extract information from the input file using the Abstract Syntax Tree (AST) for the program as shown in Fig. 1. The AST module allows parsing source code into a tree-like structure that represents the main function as the node, conditional statements, and looping statements as their child nodes (depends on the program) forming the recursive structure. It then uses ast.parse to parse the input string as an AST object. The AST object allows traversing and analyzing the structure of the code. By leveraging the AST, the code can dynamically extract information from the code itself, such as identifying variable names or function parameters, without relying on external inputs or data. This approach is useful for tasks such as identifying code smells, generating code metrics, or performing automated refactoring.

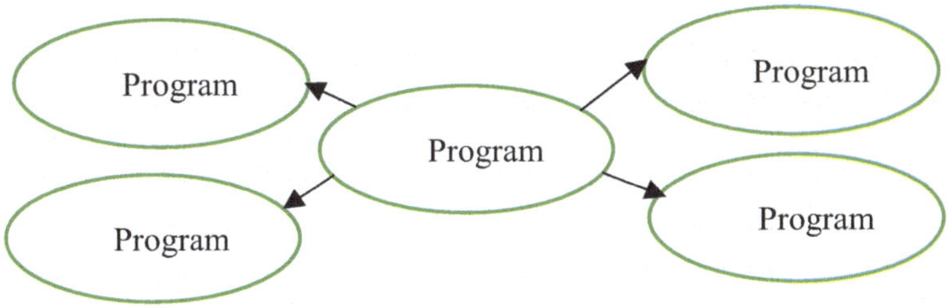

Fig. 1. Program is parsed into list of sub-functions

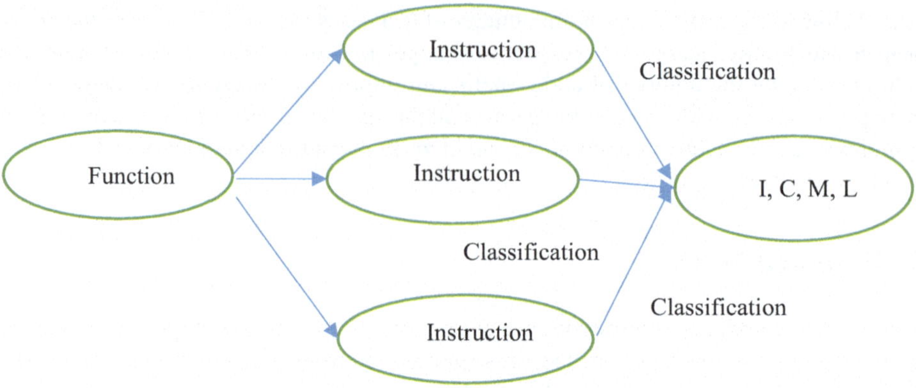

Fig. 2. Traversal and Classification

4.2 AST Traversal

AST traversal refers to the process of visiting each node in the AST tree structure. It allows us to analyse and extract information from the code represented by the AST. The AST traversal technique used here is Depth-First Search (DFS). Each function is a sequence of instructions. Each instruction is classified either as Initialization, Execution, Conditions or Looping as shown in Fig. 2. DFS explores the tree structure by visiting a node and then recursively traversing its child nodes before backtracking. The AST traversal follows a Depth-First Search approach, which explores the tree structure by visiting nodes in a depth-first manner.

4.3 Test Case Generation

During the AST traversal, specific nodes and their corresponding information can be used to generate test cases. For example, function nodes can be examined to identify function signatures and parameter types, which can be used to create test inputs. Similarly, condition nodes can be analysed to create test cases that cover different branches or conditions. During the AST traversal, various nodes in the abstract syntax tree can provide valuable information that can be used to generate test cases.

a. **Function Nodes**: When traversing function nodes, it is possible to extract important information such as function signatures and parameter types. This information can be used to create test inputs that cover different scenarios and edge cases. For instance, if a function accepts an integer parameter, it is possible to generate test cases with different integer values, including boundary values and special cases like zero or negative numbers. By analysing the function node, you can ensure that the test cases cover a wide range of input possibilities.
b. **Conditional Nodes:** During AST traversal, when encountering conditional nodes (e.g., if statements), it is possible to analyse the conditions and branches to create test cases that cover different conditions and ensure branch coverage. For example, if there is an if statement with multiple conditions, it is possible to generate test cases

that cover each condition individually, as well as combinations of conditions. This ensures that all possible branches and conditions within the conditional statements are covered by the generated test cases.

By leveraging the information extracted from function nodes and conditional nodes, it is possible to create test cases that provide comprehensive coverage of the code. This approach ensures that different paths and conditions are exercised and improves the overall quality of the software.

4.4 Test Case Prioritization

Generated test cases can be prioritized based on various criteria, such as code coverage, complexity, or potential risk areas. This ensures that the most critical parts of the code are thoroughly tested. The choice of prioritization criteria may vary depending on the specific requirements and goals of the testing process. When prioritizing test cases, the decision of whether to prioritize conditional statements, loops, or other code constructs depends on the context of the software being tested. The prioritization strategy should align with the goals and objectives of the testing effort. Regarding the type of loops, both conditional loops and iterative loops (sentimental looping) can be important for test case prioritization. Conditional loops involve executing a loop based on a specific condition, while iterative loops execute a block of code repeatedly for a specified number of iterations. In terms of prioritization criteria, some of the terms are defined below:

a. **Code Coverage**: Test cases that cover a larger portion of the code, such as branches or statements, can be prioritized to ensure maximum code coverage. This criterion ensures that critical parts of the code are thoroughly tested.
b. **Risk Analysis**: Test cases targeting high-risk areas of the code, such as complex algorithms, critical functionalities, or error-prone sections, can be prioritized. This approach focuses on mitigating potential risks associated with these areas.

Ultimately, the choice of prioritization criteria and the order in which different constructs are prioritized depend on the specific context and goals of the testing process.

4.5 Cyclomatic Complexity for Coverage Analysis

Cyclomatic complexity is calculated based on the number of decision points or control flow structures in the code, such as loops, conditionals, and function calls. In the context of code scraping-based test case generation, the cyclomatic complexity can be used to estimate the number of test cases required to cover all possible execution paths through the code. Cyclomatic complexity can be calculated using the formula $V = E - N + 2$, where V is the cyclomatic complexity, E is the number of edges or control flow paths in the code, and N is the number of nodes or decision points. The pseudo code for cyclomatic complexity is given in Table 1 below.

4.6 Decision Points Detection

Code coverage analysis is an important technique for evaluating the adequacy and effectiveness of software testing. Cyclomatic complexity is a widely used metric for measuring code coverage. It is calculated based on the number of decision points in the

Table 1. Cyclomatic Complexity Pseudo code

```
calculateCyclomaticComplexity(filename):edges = 0
nodes = 0
file = open(filename, "r")
for line in file:
line = trim(line)
if startsWith(line, "if") or startsWith(line, "else if") or startsWith(line, "while") or
startsWith(line, "for") or startsWith(line, "do"):
edges = edges + 2
nodes = nodes + 1
else if startsWith(line, "catch") or startsWith(line, "finally"):
edges = edges + 1
nodes = nodes + 1
else if startsWith(line, "case") or startsWith(line, "default"):
edges = edges + 1
file.close()
complexity = edges - nodes + 2
return complexity
```

code, which are represented by conditional statements (if, else if, switch, while, for) and loop statements. In addition to counting the decision points, it is also important to consider the entry and exit points of the code. To calculate the number of decision points, entry points, and exit points, we can perform a static analysis of the code. This involves analysing the code without actually executing it, and looking for specific patterns that indicate decision points, entry points, and exit points. The pseudo code for decision point detection is given in Table 2 below.

Table 2. Decision point detection Pseudo code

```
count_decision_points(code):
decision_points = 0
for line in code:
cleaned_line=strip_whitespace_and_comments (line)
ifstarts_with_decision_point(cleaned_line):
decision_points = decision_points + 1
return decision_points
strip_whitespace_and_comments(line):
# Implement code to remove whitespace and comments from the line
 return cleaned_line
starts_with_decision_point(line)
```

5 Result Analysis

When a program is given, the first step is typically to analyse the code to identify the different programming constructs that are used in the program. For example, if the program is written in an object-oriented language such as Java, the program may contain classes, objects, methods, and properties. Once the programming constructs have been identified, the next step is to determine how the constructs can be scraped to generate test cases. For example, if the program contains a class with several methods, each method can be scraped to identify the possible input values and expected output values. Similarly, if the program contains loops or conditional statements, each branch or iteration of the loop or conditional statement can be scraped to identify the possible input and output values. To generate test cases, the scraped information can be used to create input data and expected output data for the program. For example, if a method takes an integer input parameter and returns a Boolean value, test cases can be generated to cover different input values, such as zero, negative values, positive values, and null values, and to verify the expected output values. Consider a simple program for the addition of two numbers given in Table 3 as shown below,

Table 3. Program for largest of three numbers

```
// program to find the largest among three numbers
// take input from the user
const num1 = parseFloat(prompt("Enter first number: "));
const num2 = parseFloat(prompt("Enter second number: "));
const num3 = parseFloat(prompt("Enter third number: "));
let largest;
// check the condition
if(num1 >= num2 && num1 >= num3) {
   largest = num1;
}
else if (num2 >= num1 && num2 >= num3) {
   largest = num2;
}
else {
   largest = num3;
}
// display the result
console.log("The largest number is " + largest);
```

From the above program, which is parsed to generate the AST by identifying various sections of the program including environmental section, Variable declaration section and Execution section of the program. A Snippet of the JSON representation of AST is shown below in Fig. 3 with various sections of the program.

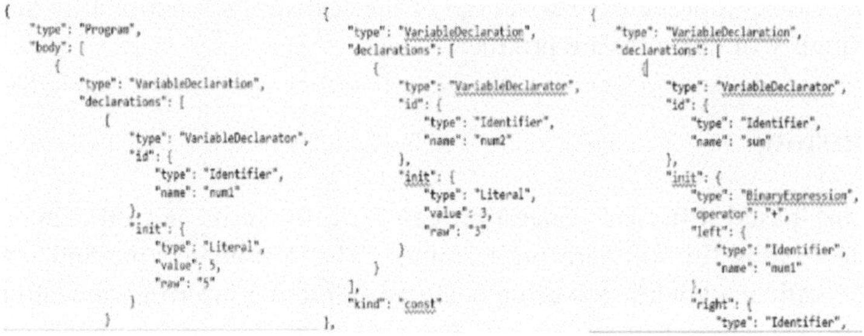

Fig. 3. AST in JSON format

From the JSON representation the program is converted to a tree structure to represent the execution part in form postfix expression, where the expression and variables are represented. From tree representation, random values are generated based on the data type and constraint specified. The tree is traversed to calculate the cyclometric complexity to uncover all the paths including cycles are uncover the bugs if available. The Number of test cases needed to be generated should be equal to the cyclometric complexity to ensure the coverage of the project. Proposed model makes sure the traversal of the program is completed for all nodes and generates the appropriate the values as shown in Fig. 4. For the sample program of largest of three number it generated three test cases to ensure all three variables are printed as largest in each iteration as shown in Fig. 5.

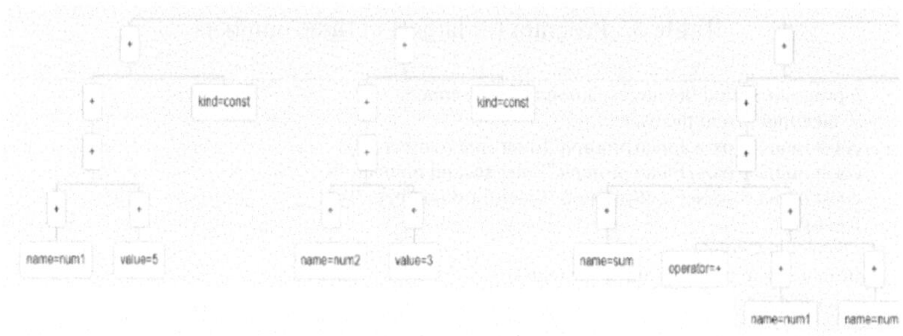

Fig. 4. AST in tree

Input: **5, 8, 3** Expected Output: **8**	Input: **23, 78, 12** Expected Output: **78**	Input: **37, 72, 89** Expected Output: **89**

Fig. 5. Generated Test cases

The analysis showed that code scraping can be an effective technique for automatic test case generation. White box testing is used in this process because it involves the analysis of the internal structure of the software system, such as the source code, which is necessary for code scraping. This technique requires knowledge of the software's internal workings and can help uncover issues that may not be detected through other testing methods. By utilizing white box testing with code scraping, the resulting test cases can provide comprehensive coverage of the software's functionality and improve the overall quality of the software product.

6 Conclusion

Code scraping-based test case generation can be a low-cost and efficient method of producing test cases for software programmes. The Automated method lowers error and improves efficiency while lowering costs by extracting information from the source code, such as the amount of decision points and loops, and using that knowledge to produce test cases. The code structure is dynamically analysed via code scraping-based test case automation. It uses techniques like as AST analysis, loop identification, the code coverage to adapt generated test cases to dynamic changes in implementation. The optimal number of test cases to be developed for code coverage is evaluated using the Cyclomatic complexity measure. The model can be extended along with CI/CD tools like Jenkins and identify impacted test case and optimize the number of test cases to be executed during each push/pull request to git or docker.

References

1. Jin, K., Lano, K.: Generation of test cases from UML diagrams-a systematic literature review. In: 14th Innovations in Software Engineering Conference (Formerly Known as India Software Engineering Conference), pp. 1–10 (2021)
2. Segura, S., Fraser, G., Sanchez, A.B., Ruiz-Cortés, A.: A survey on metamorphic testing. IEEE Trans. Software Eng. **42**(9), 805–824 (2016)
3. Hooda, I., Chhillar, R.: A review: study of test case generation techniques. Int. J. Comput. Appl. **107**(16), 33–37 (2014)
4. Shirole, M., Kumar, R.: UML behavioral model based test case generation: a survey. ACM SIGSOFT Software Eng. Notes **38**(4), 1–13 (2013)
5. Fraser, G., Arcuri, A.: Evosuite: automatic test suite generation for object-oriented software. In: Proceedings of the 19th ACM SIGSOFT Symposium and The 13th European Conference on Foundations of Software Engineering, pp. 416–419 (2011)
6. Almasi, M.M., Hemmati, H., Fraser, G., Arcuri, A., Benefelds, J.: An industrial evaluation of unit test generation: finding real faults in a financial application. In: 2017 IEEE/ACM 39th International Conference on Software Engineering: Software Engineering in Practice Track (ICSE-SEIP), pp. 263–272. IEEE (2017)
7. Arcuri, A., Fraser, G., Galeotti, J.P.: Automated unit test generation for classes with environment dependencies. In: Proceedings of the 29th ACM/IEEE International Conference on Automated Software Engineering, pp. 79–90 (2014)
8. Pinto, G.H., Vergilio, S.R.: A multi-objective genetic algorithm to test data generation. In: 2010 22nd IEEE International Conference on Tools with Artificial Intelligence, vol. 1, pp. 129–134. IEEE (2010)
9. Zhang, W., Zhou, Q., Jiao, C., Ting, Xu.: Multi-objective firefly algorithm for test data generation with surrogate model. In: Zeng, J., Qin, P., Jing, W., Song, X., Zeguang, Lu. (eds.) ICPCSEE 2021. CCIS, vol. 1451, pp. 283–299. Springer, Singapore (2021). https://doi.org/10.1007/978-981-16-5940-9_22
10. Li, Y., et al.: Competition-level code generation with alphacode. Science **378**(6624), 1092–1097 (2022)
11. Lu, S., et al.: Codexglue: a machine learning benchmark dataset for code understanding and generation. arXiv preprint arXiv:2102.04664 (2021)
12. Athira, B., Samuel, P.: Traceability matrix for regression testing in distributed software development. In: Abraham, A., Mauri, J.L., Buford, J.F., Suzuki, J., Thampi, S.M. (eds.) ACC 2011. CCIS, vol. 191, pp. 80–87. Springer, Heidelberg (2011). https://doi.org/10.1007/978-3-642-22714-1_9
13. Bedadala, P., Manasa, D., Nair, L.S.: Generation of call graph for java higher order functions. In: 2020 5th International Conference on Communication and Electronics Systems (ICCES), pp. 1151–1156. IEEE (2020)

14. Chaitanya, B.S.S.K., Reddy, D.A.K., Chandra, B.P.S.E., Krishna, A.B., Menon, R.R.: Full-text search using database index. In: 2019 5th International Conference On Computing, Communication, Control And Automation (ICCUBEA), pp. 1–5. IEEE (2019)
15. Tukaram, D.: Design and development of software tool for code clone search, detection, and analysis. In: 2019 3rd International conference on Electronics, Communication and Aerospace Technology (ICECA), pp. 1002–1006. IEEE (2019)

Prediction of AQI for Urban Metropolis Using MLSTM-GRU Model

Nairita Sarkar(✉), Pankaj Kumar Keserwani, and Mahesh Chandra Govil

Computer Science and Engineering, National Institute of Technology Sikkim, Ravangla, India
nairita.nairita@gmail.com

Abstract. Air pollution is caused by the existence of detrimental substances in atmosphere, which poses serious risks to both the environment and human health. Particulate matter, nitrogen oxide, sulphur dioxide, ozone, and volatile organic compounds are just a few of the contaminants in the air that have an adverse effect on air quality, climate change and various health problems. This study seeks to predict Air Quality Index (AQI) of Kolkata, a rising metropolis with substantial air pollution issues. In order to accomplish this, air quality data of various stations of Kolkata were collected and preprocessed accordingly. The preprocessed dataset was then used to train the prediction model. A Modified LSTM-GRU (MLSTM-GRU) model is proposed in this work and the experimental results demonstrated that the MLSTM-GRU model outperformed the other deep learning models by achieving superior result in terms of Mean Squared Error (MSE), Root Mean Squared Error (RMSE) and R-Squared (R^2) values. The findings of this study have significant consequences for methods used to regulate air quality.

Keywords: Air Quality · Air Pollutants · Deep Learning · Meteorological Parameter · MLSTM-GRU

1 Introduction

Air pollution in metropolitan cities is an acute trouble, exacerbated by urbanization, vehicular emissions, and industrial activities. As a heavily populated and industrialised metropolis, Kolkata confronts considerable issues in terms of air pollution management. The sources of pollution include traffic emissions, industrial activities, constructions, biomass burning etc. The geographical location of the city and weather patterns contribute a big responsibility to buildup the pollutants specially in the winter months when pollutant concentration level becomes high due to low temperature. The National Ambient Air Quality Standards categorizes the air pollution magnitude into six levels: fair (0–50), adequate (51–100), medium (101–200), unfair (201–300), very unfair (301–400), severe (401–500) [7]. Due to this different levels of AQI the individual faces various types of health related issues. In Kolkata, polluted air has a significant damaging influence on

D. Pastor-Escuredo et al. (Eds.): ICDLAIR 2023, LNNS 1001, pp. 483–492, 2024.
https://doi.org/10.1007/978-3-031-60935-0_43

daily life of human being. High concentrations of particulate matter as well as other pollutants may cause asthma, increase respiratory diseases, and also raise the risk of cardiovascular issues. Furthermore, air pollution is more frequently linked to lung cancer and other chronic diseases. People who have had diseases in the past are more unprotected to the harmful effects of air pollution on their health. Additionally, the cost of treating diseases brought on by air pollution, affects the healthcare system and reduces city productivity. To reduce the adverse effect of air pollution in Kolkata several comprehensive strategies such as: strict restrictions on emissions, encouraging the use of cleaner technology, wearing mask, and increasing public awareness are used. Effective actions might be taken to enhance the city's air quality, safeguard the public's health, and enhance standard of living.

AQI is measured by a mathematical formula based on the concentrations of many important air pollutants. The presence of the meteorological parameters also modify the density of air contaminants and thus also influence the AQI value. There are numerous artificial intelligence based methods such as: machine learning (ML), deep learning (DL) that have been developed to predict the quality of air of a location [2]. Each of the method has their own advantages and disadvantages over each other. This work has undertaken a thorough investigation including data gathering, preprocessing, and the use of cutting-edge DL approaches to forecast the AQI for the city Kolkata. Firstly, the AQI data and weather data of different air quality stations across Kolkata are accumulated from website of Central Pollution Control Board (CPCB). A thorough preprocessing has been conducted on data gathered in order to deal with missing values, outliers, and noises. Again the data is normalized to ensure the data consistency and to eliminate any biases that could have existed during model training. The simplicity and effectiveness in capturing temporal patterns make GRU based models a popular choice when working with time series data, such as the historical records of air quality measurements needed for predicting the AQI. Additionally, GRU model is a category of recurrent neural network (RNN) that can efficaciously handle sequential information without having the overfitting and vanishing gradient problem that can occur in traditional RNNs as well as other DL models. So, a combination of modified LSTM (Long-Short Term Memory) and GRU (Gated Recurrent Unit) i.e. MLSTM-GRU has been implemented in this work to predict AQI from the preprocessed dataset. Combining LSTM and GRU models allows for improved long-term dependency capture and computational efficiency by utilising the advantages of both architectures in air quality prediction. The integrated model improves accuracy and accelerates training convergence for robust AQI forecasts by utilising the memory retention capabilities of LSTM and the simpler structure of GRU. The effectiveness of the MLSTM-GRU model's implementation is then evaluated by contrasting its performance with that of other standalone and hybrid DL models like Recurrent Neural Network (RNN), LSTM, GRU, hybrid RNN-GRU. The experiment finds that the MLSTM-GRU model gives the finest outcome with regard to MAE, MSE, RMSE, and R^2 value.

1.1 Organization of the Paper

The following sections are organized as follows: Sect. 2 explores recent related research in the field, Sect. 3 provides details about the required materials and proposed model, Sect. 4 presents the results of experiments conducted, and Sect. 5 summarizes the conclusions of the work and outlines potential future directions.

2 Related Work

In today's world, the requirement of air quality forecasting is increasing because of understanding the impact of air quality on human health and environment. Present research direction recommends that the researchers employ several ML, DL and hybrid models for AQI prediction. A spatial and temporal deep neural network model is suggested by Soh et al. [11] to forecast air quality. Future $PM_{2.5}$ concentration is produced using historical data, including temporal information, pollutant concentrations and meteorological features. The authors collected data from 76 sites of Taiwan and Beijing and applied LSTM model to capture the long term prediction and CNN layer to extract the delay factor from the attribute. Again, study accomplished by Sigamani et al. [10] created a multiple linear regression (MLR) framework to forecast AQI based upon pollutant gases and meteorological factors. When compared to the current techniques, MLR predicted AQI with a better degree of accuracy. Xiang et al. [14] accumulated the Beijing air quality data over 8 years for training ensemble ML models and predict air quality. Authors utilized RF (random forest), SLR (simple linear regression), SVR (support vector regressor) to create the model and assessed the efficacy of the model based on MAE, RMSE, R^2 values. The MAE and RMSE values of the proposed model exist within the range (0.0128–0.0194) and (0.0230–0.0326) respectively. Zaini et al. [15] conducted a comprehensive review of the existing literature on AQI prediction using deep learning technologies along with the basic theoretical concept of DL, its advantages and disadvantages etc. The survey also focused on various hyperparameter tuning techniques used in air quality forecasting process. Gilik et al. [4] implemented a transfer learning based CNN+LSTM model to forecast the air quality of Barcelona, Kocaeli, Istanbul. The authors also performed the hyperparameter tuning on the proposed model to get least error rate. Zhang et al. [17] employed bidirectional LSTM (BiLSTM) network to predict the time series of air quality. The hourly $PM_{2.5}$ concentrations from 12 stations in Beijing, were gathered as part of air quality data between 2013 to 2017. The BiLSTM model performs better than the current models in terms of error functions. One year later, J Zhang et al. [16] analysed the air quality of Beijing by using CNN-LSTM model and got promising result. M Jamei et al. [6] evolved a novel blended DL framework to forecast the air quality of Australian city. They used LSTM network, extreme gradient boosting (XGBoost), classification and regression tree (CART), and best subset regression (BSR) to implement the model. C Wu et al. [13] used integrated residual neural network, graph convolutional network and BiLSTM to assess the hourly collected air quality data of Shanghai. Tariq et al. [12] invented a distance adaptive graph convolutional gated network to offer simultaneous predictions of main air contaminants at multiple locations. The predicted model reduced prediction errors for the $PM_{2.5}$ forecast for the next 12 h by 52.59%. Sarkar et al. [9] integrated LSTM along with GRU framework to analyse the air quality of Delhi with enhanced accuracy. J Hu et al. [5] explored an optimized hybrid CNN-BiLSTM-GRU model to assess the density of O_3 and $PM_{2.5}$ in the air and achieved very promising result.

The literature review reveals that prior researchers have employed different approaches, including ML, DL and hybrid models to predict AQI. However, they have encountered challenges in achieving high prediction accuracy. In response to these challenges, this research introduces a novel hybrid DL model designed to predict AQI of the city Kolkata.

3 Proposed Work

The presented methodological framework begins with the gathering of unprocessed data. Numerous data pre-processing techniques have been used to make the dataset suitable for modelling. Then the proposed MLSTM-GRU model has been trained using 80% of training samples and leftover 20% of samples are employed for validating the performance of suggested model. Flow diagram of the experiment is displayed in Fig. 1.

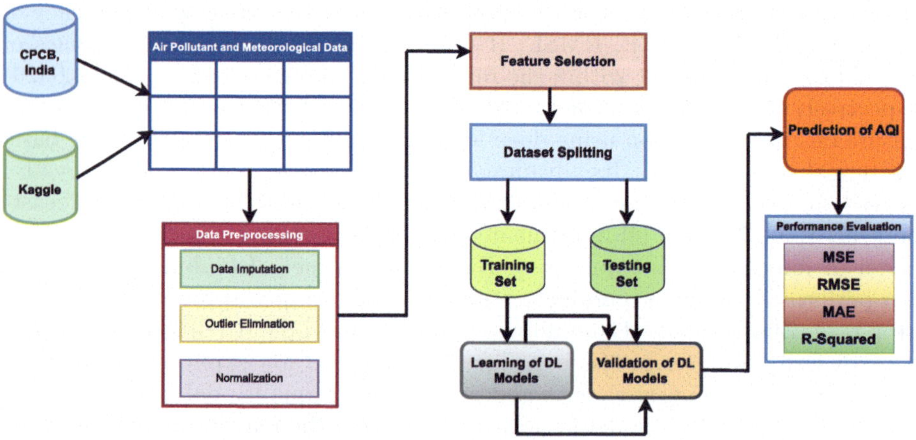

Fig. 1. Flowchart of the System Model

3.1 Collection and Description of Dataset

The dataset comprises of air pollution and weather data i.e. the concentration values of different pollutants, temperature and humidity for the city Kolkata in West Bengal. The information about the pollutants are gathered from CPCB, India website and information regarding temperature and humidity are collected from Kaggle that have been merged on the basis of date. The obtained dataset has gathered between January 1, 2018, and December 31, 2020 in hourly basis and comprised of 26,305 number of instances and thirteen attributes. The collected pollutants are: $PM_{2,5}$, PM_{10}, NO, NO_2, NO_x, NH_3, CO, SO_2, O_3. In the dataset the column AQI is considered as the target attribute. The first twelve attributes help to predict the behaviour of the target attribute.

3.2 Preprocessing and Visualization of Data

Preparing and cleaning the collected raw data is a vital step in data preprocessing which enhances the quality and utility of data. Data pre-processing practices used in the study are handling missing values, recognition and elimination of outlier values, normalization etc. The missing data are eliminated using the mean imputation approach whereas to remove the outlier values the Z-score method is employed in this work. To normalize the data min-max normalization technique has been used. The fluctuation of AQI values with respect to meteorological data is extremely important since it is well known that meteorological factors have a significant role in variations of air quality. Figure 2 shows the variation of AQI values according to the changes of temperature and it is clear from the figure that the air pollution level becomes high in winter time when temperature is low.

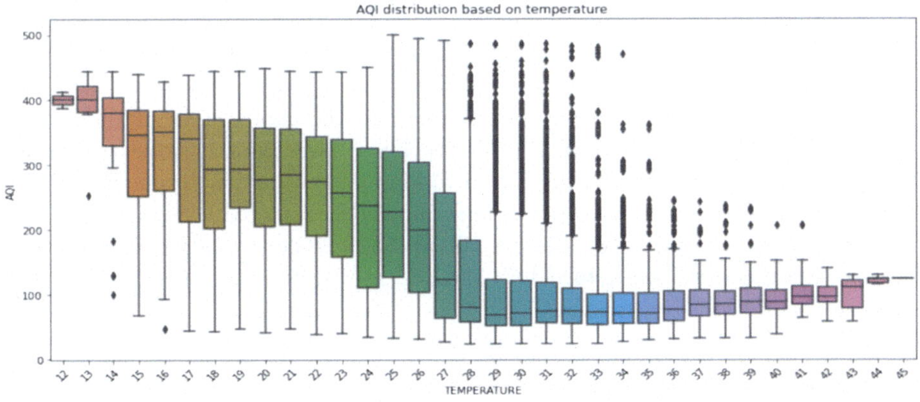

Fig. 2. Variation of AQI value with temperature

3.3 Feature Selection

Feature selection is the process of lowering the number of input variables. It plays an important role by decreasing the computational expense of modelling and improving the prediction accuracy. To find relevant features for evaluating a predictive model, the current study uses a correlation-based feature selection strategy which calculates the correlation value of every pair of features and target attribute. The features having the highest correlation to the target variable are then selected as relevant features and are used for future research. According to the correlation-based feature selection strategy CO and O_3 are lightly correlated with the AQI values, so these two columns are discarded.

3.4 Modified LSTM-GRU (MLSTM-GRU) Model

This study has developed a two-stage hybrid structure for predicting the air quality data based on pollutant gases, temperature and humidity of the city Kolkata. In this study, a modified hybrid LSTM-GRU model (MLSTM-GRU) which represents the cutting-edge approach to forecast air quality data is proposed. The model basically combines the strength of two powerful recurrent neural network architectures. This innovative fusion capitalizes on the LSTM network's ability to capture long-term dependencies and the GRU network's efficiency in handling shorter sequences. Due to these characteristics the MLSTM-GRU model can skillfully handle the dynamic and complex nature of air quality and meteorological data. Moreover, its adaptability also enables the incorporation of additional features (meteorological data, traffic patterns, and geographical information etc.), weakens the vanishing gradient problem, guarantees better training and enhances accuracy.

After the stages of data preprocessing and feature selection the whole set of data is bisected into training and testing part by 80:20 proportion. The suggested MLSTM-GRU model is trained using training data. In this model the LSTM is responsible for accumulating data inputs and transferring them to GRU framework. To avoid the overfitting problem dropout with value 0.4 is added to the layers. A set of 9 input features are supplied to the input layer of the LSTM structure and 150 neurons from the hidden layer are coupled with these nodes through the 'Relu' activation function. A hidden layer of fifty neurons, the "Relu" activation function, and 0.4 dropouts make up GRU framework, in contrast. The value of used epoch and batch size are 100 and 32 respectively. By hyper parameter tuning procedure the hybrid LSTM-GRU model [9] is getting modified and known as MLSTM-GRU model.

4 Result and Discussion

4.1 Performance Evaluation Metrics

1. **Mean Square Error (MSE):** The average squared difference between the measured and predicted values in a dataset is known as MSE and mathematically presented by Eq. 1 [9].

$$MSE = \frac{1}{n}\sum_{i=1}^{n}(M_{ACT(i)} - N_{PRED(i)})^2 \tag{1}$$

2. **Root Mean Square Error (RMSE):** RMSE value is obtained from the square root operation of MSE value and illustrated by Eq. 2 [9].

$$RMSE = \sqrt{\frac{1}{n}\sum_{i=1}^{n}(M_{ACT(i)} - N_{PRED(i)})^2} \tag{2}$$

3. **Mean Absolute Error (MAE):** Average magnitude of inaccuracy in anticipated values is measured by MAE and the mathematical representation is given by Eq. 3 [3].

$$MAE = \frac{1}{n}\sum_{i=1}^{n}|M_{ACT(i)} - N_{PRED(i)}| \tag{3}$$

4. **R-squared ($\mathbf{R^2}$):** A statistical metric known as R^2, sometimes referred to as the coefficient of determination used for assessing quantity of variance of dependent variable is taken into account by independent variables in a regression model [8]. R^2 is obtained from the mathematical representation denoted by Eq. 4.

$$R^2 = 1 - \frac{RSS}{TSS} \tag{4}$$

where, RSS represents the sum of square of residuals and TSS is total sum of square.

Note: $M_{ACT(i)}$, $N_{PRED(i)}$ and n are actual outcome, predicted outcome and total number of instances respectively.

4.2 Experimental Result and Comparative Analysis

To accomplish the implementation of the suggested MLSTM-GRU model, the Python programming environment is used for simulation on an Intel Core i7 CPU @ 2.40 GHz system with 16 GB of RAM running Microsoft Windows 10 Professional. The efficiency of MLSTM-GRU model is assessed on the basis of R^2 value, MSE, MAE and RMSE values. Scatter plot for the actual and forecast data points of the MLSTM-GRU model is exhibited in Fig. 3. The X-axis and Y-axis are representing the actual and forecast values of AQI. After that comparisons are made between the performance of MLSTM-GRU and other standalone and hybrid DL models. RNN, LSTM, GRU and hybrid RNN-GRU models are trained on the same dataset and the performance of all these models are also evaluated on account of R^2 value, MSE, MAE and RMSE values. Table 1 depicts comparison analysis of MLSTM-GRU model with other DL models and it is clear from the table that the suggested MLSTM-GRU model is giving the best performance in terms of all the taken evaluation metrics i.e. MSE, RMSE, MAE and R^2 values.

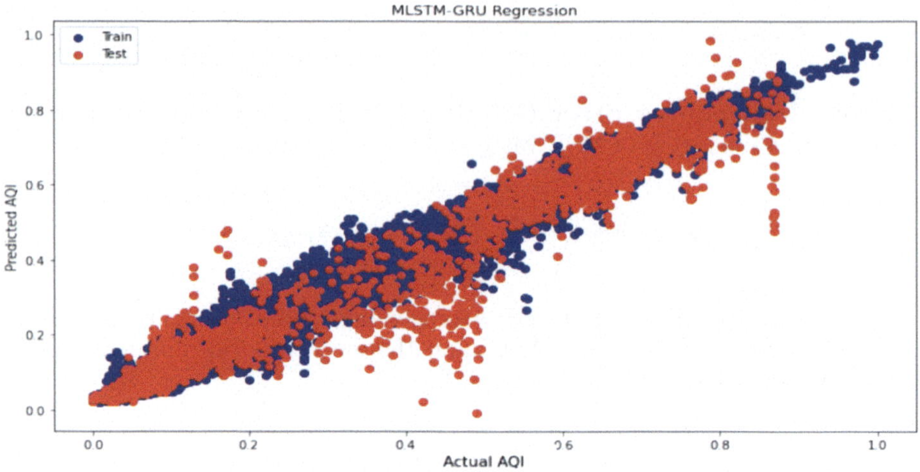

Fig. 3. Scatter Plot of Actual and Predicted Data Points Using MLSTM-GRU Model

Table 1. Comparison of MLSTM-GRU model with other models trained on the same dataset

Model	MSE	RMSE	MAE	R^2
RNN	0.0024	0.052	0.029	0.94
LSTM	0.0028	0.052	0.030	0.94
GRU	0.0026	0.053	0.029	0.94
RNN-GRU	0.0031	0.55	0.31	0.95
MLSTM-GRU	**0.0022**	**0.047**	**0.026**	**0.97**

From Table 1 it is clear that the suggested MLSTM-GRU model has least MSE, RMSE and MAE values which depict that the MLSTM-GRU model has the lowest error rate. R^2 measurement of the model is 0.97 which is highest among all the implemented DL models. The highest R^2 indicates minor difference between the actual and predicted data points and also suggests that the model is a better fit for the data. The hybrid RNN-GRU model has the second highest R^2 value but the values of other loss functions are greater than the other DL models. After analysing the performance of the standalone DL models it is observed that the RNN model has lowest MSE value, both the RNN and LSTM have the lowest RMSE value and the RNN and GRU models have achieved the same and MAE values. Figure 4 illustrates the scatter plots of the actual and predicted data points of AQI values on the grounds of the other utilized models. The execution outcome of proposed MLSTM-GRU model is contrasted with the execution outcome of recent existing models and presented in Table 2.

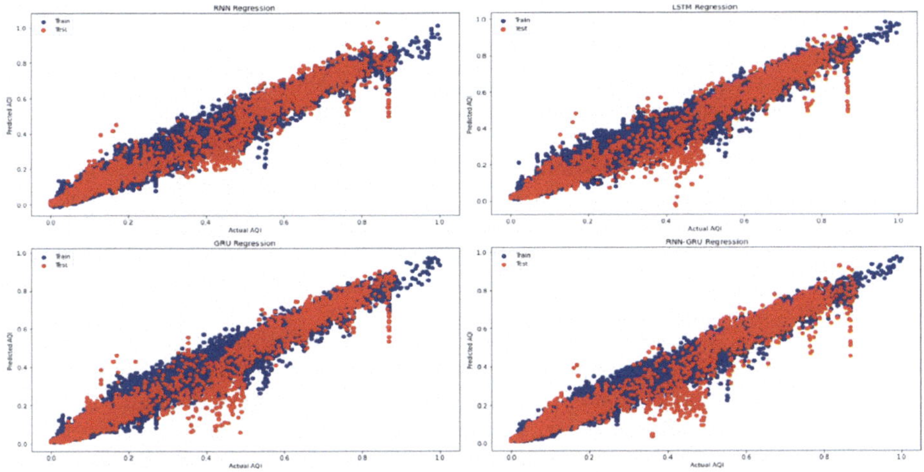

Fig. 4. Model-wise scatter plot of actual and predicted AQI values

Table 2. Comparison of MLSTM-GRU model with other existing models

Sl. No.	Model Name	R^2
1	RNN-LSTM [1]	0.902
2	CNN-LSTM [16]	0.71
4	**MLSTM-GRU(Proposed)**	**0.97**

5 Conclusion and Future Scope

More accurate prediction of AQI is not only crucial for safeguarding the public health for urban areas but also it is contributing a valuable impacts on environmental development to reduce the negative consequences of air pollution. This work introduces a blended and modified version of LSTM and GRU model known as MLSTM-GRU model designed for AQI prediction in Kolkata. The MLSTM-GRU model's performance is evaluated against that of other DL models namely: RNN, LSTM, GRU, RNN-GRU. The outcome designates that MLSTM-GRU performs better than the other models and highlighting its superiority. Moreover, this proposed approach can be enhanced to assess AQI in different locations. Hereafter, researchers can work with larger datasets containing diverse features gathered from various locations to enhance experimentation using different feature creation, selection, and classification techniques, including ensemble methods.

References

1. Balaraman, S., Partheeban, P., Elamparithi, P.N., Manimozhi, S.: Application of LSTM models in predicting particulate matter ($pm_{2.5}$) levels for urban area. J. Eng. Res. **10**(3B) (2022)
2. Begum, A., Fatima, F., Sabahath, A.: Implementation of deep learning algorithm with perceptron using TenzorFlow library. In: 2019 International Conference on Communication and Signal Processing (ICCSP), pp. 0172–0175. IEEE (2019)
3. Brassington, G.: Mean absolute error and root mean square error: which is the better metric for assessing model performance? In: EGU General Assembly Conference Abstracts, p. 3574 (2017)
4. Gilik, A., Ogrenci, A.S., Ozmen, A.: Air quality prediction using CNN+LSTM-based hybrid deep learning architecture. Environ. Sci. Pollut. Res. 1–19 (2022)
5. Hu, J., et al.: An optimized hybrid deep learning model for PM2.5 and O3 concentration prediction. Air Qual. Atmos. Health **16**(4), 857–871 (2023)
6. Jamei, M., Ali, M., Malik, A., Karbasi, M., Sharma, E., Yaseen, Z.M.: Air quality monitoring based on chemical and meteorological drivers: application of a novel data filtering-based hybridized deep learning model. J. Clean. Prod. **374**, 134011 (2022)
7. Kumar, K., Pande, B.: Air pollution prediction with machine learning: a case study of Indian cities. Int. J. Environ. Sci. Technol. **20**(5), 5333–5348 (2023)
8. Rights, J.D., Sterba, S.K.: R-squared measures for multilevel models with three or more levels. Multivariate Behav. Res. 1–28 (2021)
9. Sarkar, N., Gupta, R., Keserwani, P.K., Govil, M.C.: Air quality index prediction using an effective hybrid deep learning model. Environ. Pollut. **315**, 120404 (2022)
10. Sigamani, S., Venkatesan, R.: Air quality index prediction with influence of meteorological parameters using machine learning model for IoT application. Arab. J. Geosci. **15**(4), 340 (2022)
11. Soh, P.W., Chang, J.W., Huang, J.W.: Adaptive deep learning-based air quality prediction model using the most relevant spatial-temporal relations. IEEE Access **6**, 38186–38199 (2018)
12. Tariq, S., Tariq, S., Kim, S., Woo, S.S., Yoo, C.: Distance adaptive graph convolutional gated network-based smart air quality monitoring and health risk prediction in sensor-devoid urban areas. Sustain. Urban Areas **91**, 104445 (2023)
13. Wu, C., et al.: A hybrid deep learning model for regional o_3 and no_2 concentrations prediction based on spatiotemporal dependencies in air quality monitoring network. Environ. Pollut. 121075 (2023)
14. Xiang, X., Fahad, S., Han, M.S., Naeem, M.R., Room, S.: Air quality index prediction via multi-task machine learning technique: spatial analysis for human capital and intensive air quality monitoring stations. Air Qual. Atmos. Health **16**(1), 85–97 (2023)
15. Zaini, N., Ean, L.W., Ahmed, A.N., Malek, M.A.: A systematic literature review of deep learning neural network for time series air quality forecasting. Environ. Sci. Pollut. Res. 1–33 (2022)
16. Zhang, J., Li, S.: Air quality index forecast in Beijing based on CNN-LSTM multi-model. Chemosphere **308**, 136180 (2022)
17. Zhang, Z., Zeng, Y., Yan, K.: A hybrid deep learning technology for $pm_{2.5}$ air quality forecasting. Environ. Sci. Pollut. Res. **28**, 39409–39422 (2021)

Land Use Land Cover Classification Using Multi-spectral Satellite Imagery

Neelofar Jaha Mohammad, Radhesyam Vaddi, B. V. D. Soujitha Neelam(✉), and Hemanth Cheepulla

Department of IT, VR Siddhartha Engineering College, Vijayawada, India
neelofarjaha191202@gmail.com, syam.radhe@vrsiddhartha.ac.in, durgasoujitha234@gmail.com, hem4nthgee@gmail.com

Abstract. Land cover change detection using multispectral images is a critical application in remote sensing. It involves monitoring and assessing alterations in the Earth's surface's physical and biological characteristics over time. Multispectral images acquired by satellite sensors, which capture data across various electromagnetic spectrum bands, enable the identification and tracking of diverse land cover types, including forests, crops, urban areas, water bodies, and barren land. The fundamental concept behind change detection is the comparison of two temporally separated satellite images to identify discrepancies between the two timestamps. To achieve this, two distinct approaches have been implemented. The first approach utilizes K-Means clustering, which categorizes pixels in the images into clusters based on their spectral attributes, generating color maps that highlight areas of change. In the second approach, Gaussian Mixture Models (GMM) are employed to model pixel value distributions, resulting in similar color maps that identify regions of land cover change. These algorithms play a crucial role in producing visual representations of where alterations in land cover have occurred, facilitating applications in environmental management, urban planning, and agriculture. In summary, land cover change detection through multispectral images and remote sensing techniques is essential for monitoring Earth's surface changes over time. Two approaches, K-Means clustering and Gaussian Mixture Models, have been implemented to identify changes and generate colour maps that highlight areas of interest in land cover transformations.

Keywords: change detection · image classification · image segmentation · object-based image classification · land cover · k-means · google colab · multispectral Images

1 Introduction

Environmental monitoring depends heavily on the employ of land cover change detection. It involves an analysis of multispectral satellite imagery captured within the same geographical area at several points in time. Through the identification and quantification of mutate in land cover types, this research will offer important new information about environmental changes, urban growth, agricultural dynamics, and other topics.

D. Pastor-Escuredo et al. (Eds.): ICDLAIR 2023, LNNS 1001, pp. 493–502, 2024.
https://doi.org/10.1007/978-3-031-60935-0_44

The acquisition of the necessary satellite is often the foremost step in the change detection process, which is then followed by a few pre-processing pro cedures for ensuring accurate and consistent data. The practice of object-based or classification-based techniques, depending on the objectives and data at hand, is the core of change detection. These techniques make it simpler to identify areas of change by comparing land cover objects between the two time periods [8]. Major challenges are involved in mapping land cover, specifically in dynamic regions where changes happen regularly.

Multispectral image analysis is an essential element in remote sensing, aiming to derive significant insights from data collected by satellites. Multispectral imagery provides rich data which have leveraged for change identification tasks, forming it a beneficial tool for understanding alterations over time. Therefore, monitoring and understanding changes in agricultural lands, encompassing land utilization and land coverage, is of paramount importance.

2 Related Work

A Land Use/Land Coverage Change Identification and NDVI Estimation in Pakistan's Southern Punjab Province (2023) [1]:

Here, authors wore Landsat visual data for Change detection. They approached NDVI analysis and made us familiarized with it. The results include an increase in agriculture and a decrease in forests and barren land.

Land Utilization and Land Coverage Mapping using Multi-temporal Sentinel-1 Data and Machine Learning (2023) [2]:

The authors utilize a machine learning-based methodology to process multi-temporal data, incorporating feature extraction and classification algorithms for accurate land mapping. They assess their approach using metrics such as accuracy, precision, recall, and F1-score to confirm the caliber of their outcomes.

Land Coverage and Land Utilization Classification Approach to Maintain the Green Environment in Smart Village (2023) [3]:

The authors present an innovative approach for land classification, with a specific focus on maintaining a green environment in smart villages. This study applies a classification method utilizing a transfer learning model to categorize sentinel-2 images that result in creating a thematic map. Accuracy attained is 96%.

A Deep Transfer Learning Framework Using Teacher-Student Structure for Land Coverage Classification of Remote Sensing Imagery (2023) [4]:

In order to categorize land coverage using remote sensing, the paper introduces a learning framework of deep novel with a teacher-student structure, the teacher model has learned from studying large datasets to help the student model be improved upon using smaller datasets. This method boosts classification efficiency.

Assessing land coverage using satellite information in detail. (2023) [5]:

The authors analyze high-resolution satellite imagery. They use the architecture of convolutional neural networks (CNNs) to automatically extract features and categorize different types of land cover. To improve model performance, the authors also use data augmentation and preprocessing techniques. Used metrics like overall accuracy, precision, recall, F1-score, and confusion matrix.

Multiscale Context-Aware Feature Fusion Network for Land-Cover Classification of Urban Scene Imagery (2023) [6]:

The Multiscale Context-Aware Feature Fusion Network is a design of Deep learning, Where the authors proposed in this paper. This network is made to efficiently combine contextual data to enhance the precision of land-cover classification while analyzing urban scene imagery at various scales. It supports applications like infrastructure development, environmental monitoring, and urban planning.

3 Datasets

We obtained our input data using Google Earth Engine (GEE). We used Sentinel-2 data. Sentinel-2 is a satellite that takes pictures of Earth for the European Space Agency's Copernicus program. It includes environmental monitoring, forestry, agriculture, urban planning, and more.

To acquire ground truth data comprising latitude and longitude coordinates through Google Earth Engine, begin by accessing the code editor. From there, select the "Add a marker" option from the tool menu and tap on the map to pinpoint your desired location. If you require multiple samples, simply return this procedure for different areas of interest. After marking all the relevant locations, access the "Imports" menu and click on "Show generated code" to reveal the script. Copy this script and paste it into a text document, such as Notepad. To further process the data, open the text document in Microsoft Excel and utilize the "Data > Text to Columns" tool to convert it into an Excel spreadsheet format. Finally, save the resulting Excel file to your drive. This comprehensive procedure enables the extraction of precise latitude and longitude coordinates for research (Fig. 1 and Table 1).

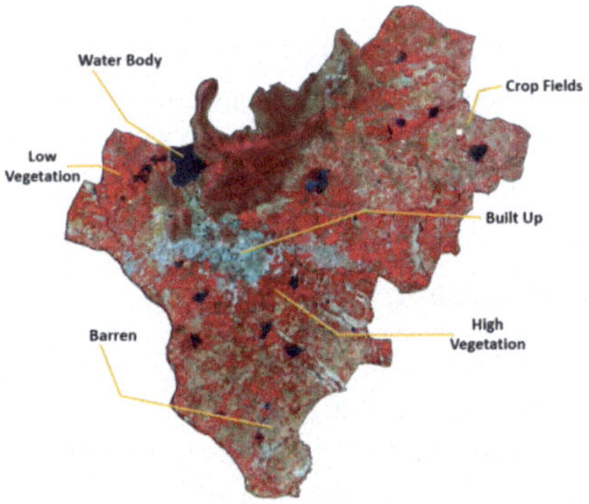

Fig. 1. Training Samples for Chimakurthy, Prakasam District, Andhra Pradesh

Table 1. Test Area details in Prakasam district

AOI	Latitude	Longitude	Name Of AOI	Area in Sq. Km
Test Area 1	15.585538	79.867181	Chimakurthy	50.67
Test Area 2	15.404765	79.506714	Kanigiri	50.67
Test Area 3	15.340253	80.036616	Tanguturu	50.67

We considered places called Kanigiri, chimakurthy, tangutur from Prakasam District. We took two timestamp images from each place of years 2019 and 2023 (Figs. 2 and 3).

Fig. 2. Prakasam District in 2019

Fig. 3. Prakasam District in 2023

4 Proposed Work

4.1 Design Methodology

The K-means algorithm is an unsupervised learning, plays a pivotal role in diverse applications, encompassing tasks like pattern recognition, and anomaly detection.

The Gaussian Mixture Model (GMM) algorithm assumes that data points arise from a mixture of Gaussian distributions, each representing a distinct cluster or mode.

4.2 System Architecture Diagram

The images were cropped to the correct size. The 2 multispectral images are sent into the K-means Algorithm to produce Color Maps. Time stamp T1's image generates the color map 1. The picture of timestamp T2 similarly generates the color map 2. These two color maps undergo Change Detection and results two Land Cover Pie charts [8]. Finally, output is a Rate of change of different Land covers between the Pie charts [7] (Fig. 4).

4.3 Description of Algorithms

K-means is a popular way to arrange data into groups basing on how similar they are. It's a step-by-step process used to cluster data and compress images.

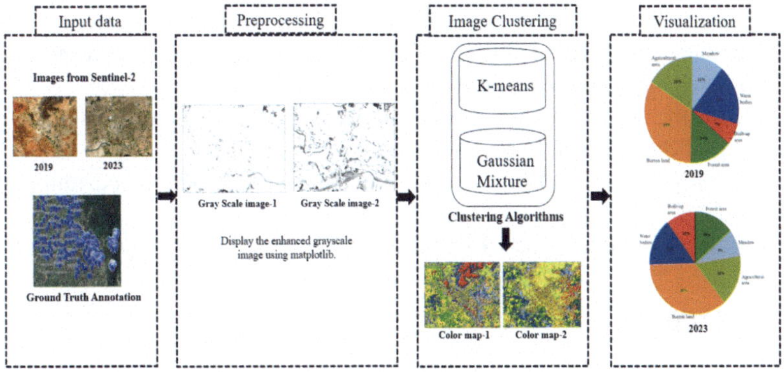

Fig. 4. System Architecture

Algorithm: K-Means Clustering
Input: k- how many groups to divide the data into
D- a collection of data points
Output: A segmentation of k points
Method:

1. Randomly pick k points from dataset to starting cluster centers.
2. Repeat until the clusters don't change anymore.
3. Put each point in the group whose center is closest to it.
4. Calculate the new center by averaging all the data points in cluster.

The algorithm aims to choose a centroid that minimizes inertia, or within-cluster sum-of-squares criterion.

$$\sum_{i=0}^{n} min(||x_i - u_j||)(||x_i - u_j||) \tag{1}$$

After assigning the new data point, the position of updated centroid is as follows (X, Y):

$$X = \frac{(x_1 + x_2 + x_3 + \ldots x_n)}{\text{n}} \tag{2}$$

$$Y = \frac{(y_1 + y_2 + y_3 + \ldots y_n)}{\text{n}} \tag{3}$$

Algorithm: Gaussian Mixture
Input: k: How many groups to divide the data into
D: A group of data points
Output: A gathering of k groups of data points that are similar and represented by their parameters (mean, covariance)
Method:

1. Initialize the parameters: Choose k initial clusters with their corresponding parameters: mean vectors (μ), covariance matrices and mixing weights (π)

2. Expectation-Maximization (EM) algorithm: Repeat until convergence:
   a. Calculate the responsibilites of each cluster using the current parameter estimates. Update the responsibilty matrix
   b. Update the mean vectors, covariance matrices, and mixing weights for each group of data points. Normalize the mixing weights.
   c. Compute the log-likelihood pertaining to the data given the current parameter estimates.
3. Output: Return the final set of k clusters, represented by their parameters.

5 Result Analysis

Image classification is a fundamental task of computer vision, involving the training of algorithms or models to automatically categorize images into predefined classes or categories [9]. This process enables the generation of precise classifications for data and the development of well-trained models (Figs. 5, 6, 7, 8, 9 and 10).

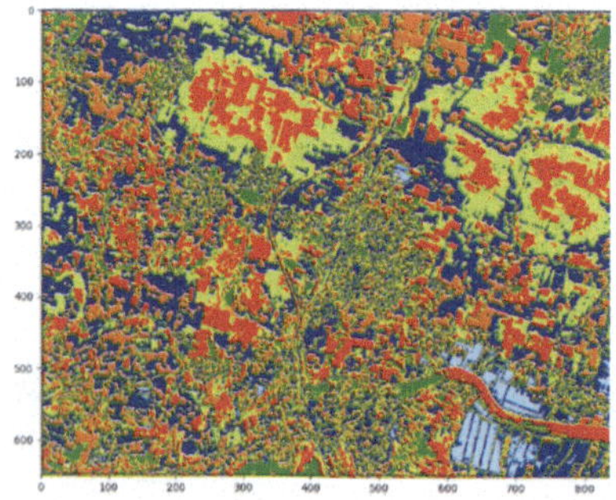

Fig. 5. Classified Image of Tangutur (2019)

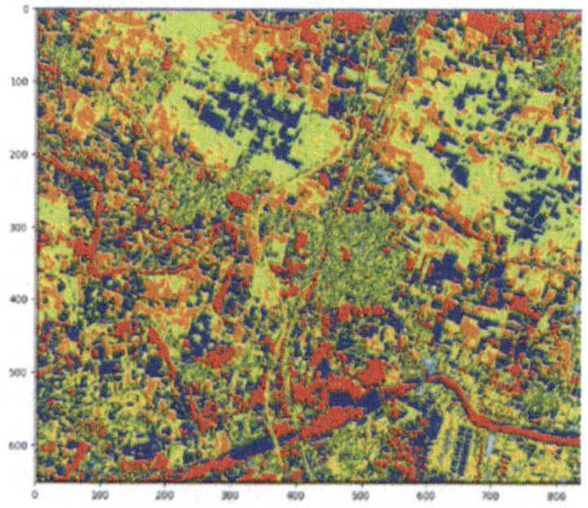

Fig. 6. Classified Image of Tangutur (2023)

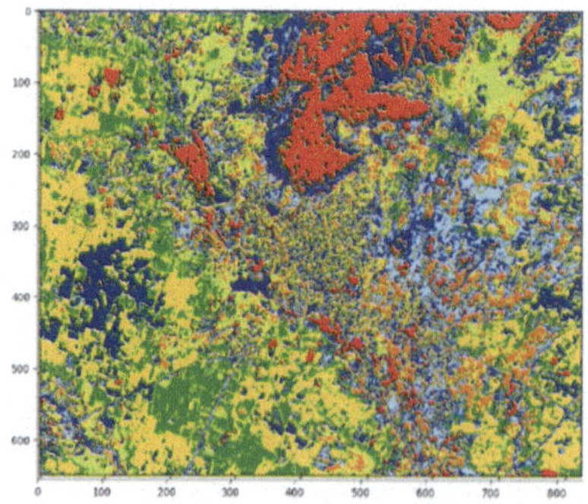

Fig. 7. Classified Image of Kanigiri (2019)

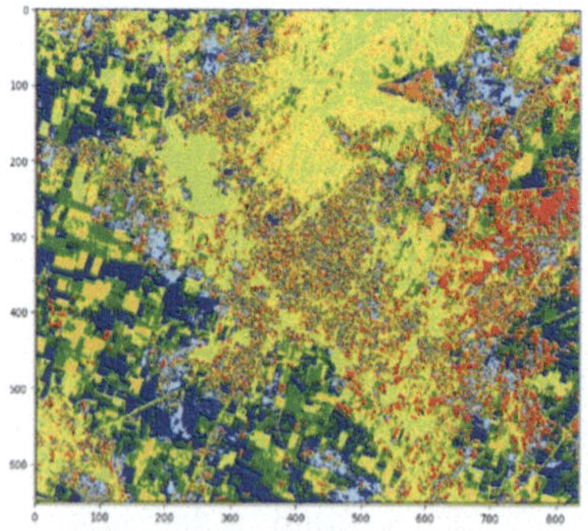

Fig. 8. Classified Image of Kanigiri (2023)

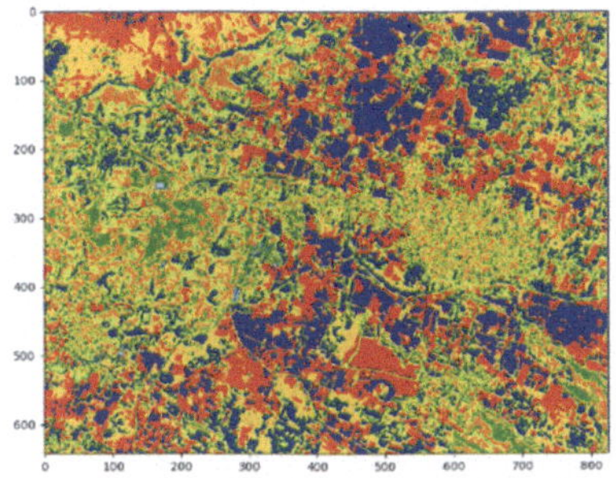

Fig. 9. Classified Image of Chimakurthy

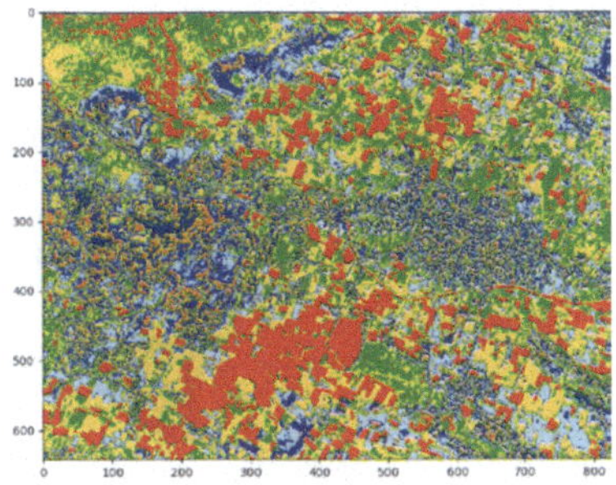

Fig. 10. Classified Images of Chimakurthy

To detect changes, photos from 2019 and 2023 are used. To ascertain the percentage change in each land-cover, all that is needed is a pie chart showing the fractional coverage of each land. K-means clustering was utilized to group colors into several clusters (Figs. 11, 12, 13, 14, 15 and 16).

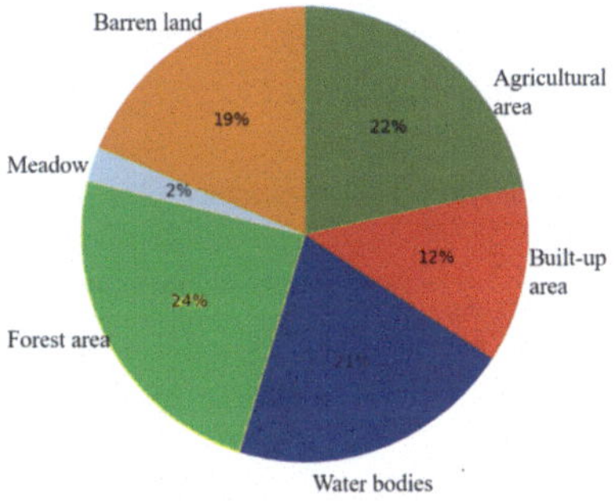

Fig. 11. Visualization of Tangutur (2019)

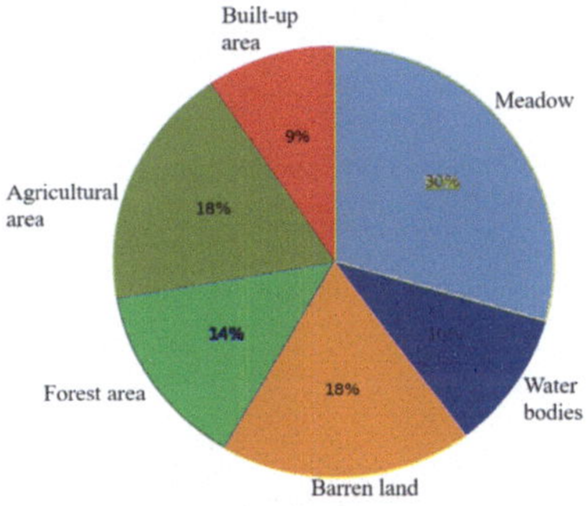

Fig. 12. Visualization of Tangutur (2023)

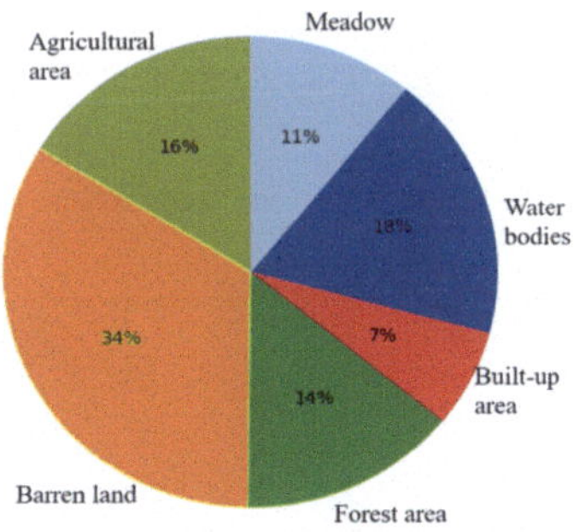

Fig. 13. Visualization of Kanigiri (2019)

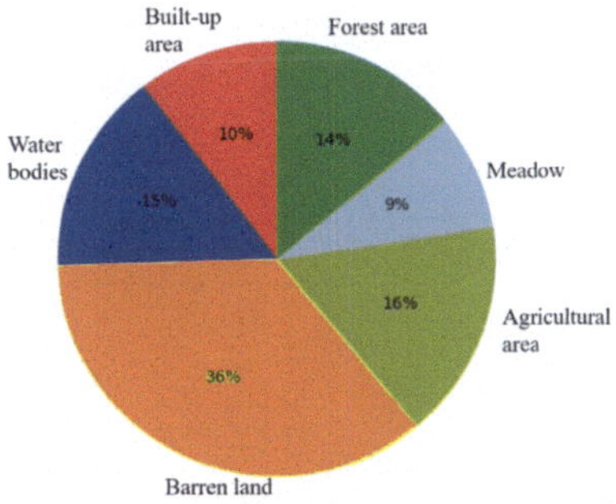

Fig. 14. Visualization of Kanigiri (2023)

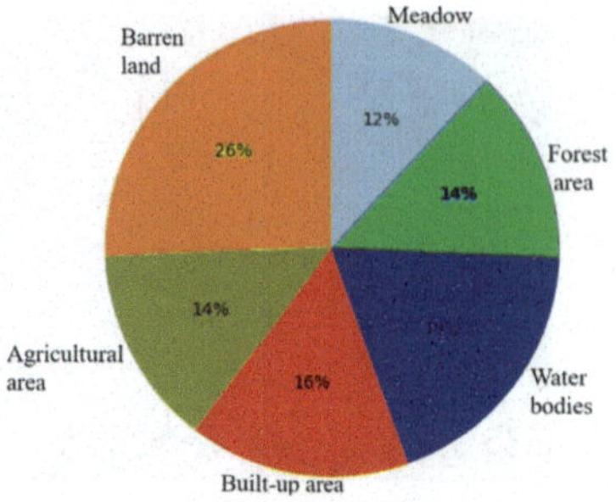

Fig. 15. Visualization of Chimakurthy (2019)

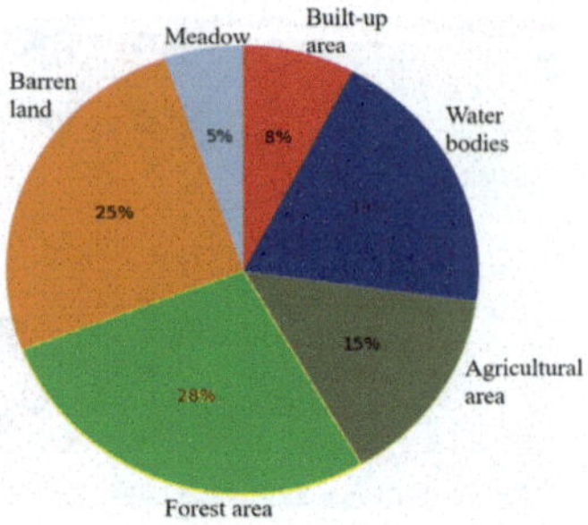

Fig. 16. Visualization of Chimakurthy (2023)

The proportional coverage of land cover is visualized in the time-stamped classified imagery. Red denotes a buildup area, blue denotes water bodies, green denotes a forest area, light blue denotes a meadow, Scrap green denotes Agriculture land and orange denotes barren land. The development pertaining to the alteration of land cover chart for the time stamps is as indicated below (Tables 2, 3, 4 and 5).

Table 2. Changes Detection for Tangutur (Test Area-1)

Land Cover	2019	2023	%Alteration in Surface Cover
Agricultural area	22	18	4
Barren land	19	18	1
Buit-up area	12	9	3
Forest area	24	14	10
Meadow	2	30	28
Water bodies	21	10	11

Table 3. Changes Detection for Kanigiri (Test Area-2)

Land Cover	2019	2023	%Alteration in Surface Cover
Agricultural area	16	16	0
Barren land	34	36	2
Buit-up area	7	10	3
Forest area	14	14	0
Meadow	11	9	2
Water bodies	18	15	3

Table 4. Changes Detection for Chimakurthy (Test Area-3)

Land Cover	2019	2023	%Alteration in Surface Cover
Agricultural area	14	15	1
Barren land	26	25	1
Buit-up area	16	8	8
Forest area	14	28	14
Meadow	12	5	7
Water bodies	19	19	0

Table 5. Comparison between Algorithms

Area	Gaussian Mixture	K-Means
Tangutur	92%	87%
Kanigiri	94%	89%
Chimakurthy	91%	88%

In summary, here we made comparisons between algorithms for three places Tangutur, Kanigiri and Chimakurthy [10]. In every comparison, we observed that Gaussian Mixture algorithm has outperformed the K-Means algorithm. The result is a collection of spatiotemporal representations of land cover change capable of distinguishing different change types and unchanged regions.

6 Conclusion

In the evaluation of clustering algorithms for the given dataset, it is evident that Gaussian Mixture outperforms K-Means in terms of accuracy. Specifically, Gaussian Mixture achieves an impressive accuracy rate of 92%, surpassing the 87% accuracy achieved by K-Means for Tangutur place. This outcome suggests that Gaussian Mixture is efficient this dataset. In addition to accuracy, it's worth that Gaussian Mixture models outperform K-Means clustering in many applications, due to their additional capabilities. Gaussian Mixture allows for the modeling of data as a combination of Gaussian distributions. This flexibility makes Gaussian Mixture a strong choice while conducting data analysis and overlapping clusters.

References

1. Sahu, M., Dash, R.: Land Coverage and Land Usage Classification Approach to Maintain Green Environment in Smart Village (2023)
2. Cahapin, E.L., Fajardo, A.C.: Land Usage and Land Coverage Mapping using Multi-temporal Sentinel-1 Data and Machine Learning (2023)

3. Srivastava, S., Sharma, Y., Prakash, A., Gupta, G.: Deep learning-based land cover evaluation utilizing high-resolution satellite imagery (2023)
4. Fan, F., Shi, Y., Zhu, X.X.: Categorizing Urban Land Cover from Sentinel-2 Images with Quantum-Classical Network (2023)
5. de Jong, K.L., Bosman, A.S.: Unsupervised Identification of Change in Unlabeled Optical Remote Sensing using CNN (2021)
6. Florez-Ospina, J.F., Benitez-Restrepo, H.D.: Exploiting variational inequalities for generalized change identification on graphs (2023)
7. Samudrala, S., Pranvithanand, M., Vaddi, R.: Change detection in land use-land cover using convolutional neural network (2019)
8. Lasya, M., Phanitha Sai Lakshmi, V., Anjum, T., Vaddi, R.: Change Detection Using Multispectral Images for Agricultural Application (2023)
9. Phaneendra Kumar, B.L.N., Vaddi, R.: A Framework for Hyperspectral Image Classification. https://doi.org/10.1109/ICAISS58487.2023.10250544
10. Aparna, G., Phaneedra Kumar, B.L.N.: Comparison of Feature Reduction Techniques for Change Detection in Remote Sensing (2022)

An Ensemble Learning Based Career Prediction Model

Aman Wakade[1(✉)], Amay Wakde[2], Roshni Maywade[1], Atharv Pandey[1], Jatinder Kumar[1], and Ashutosh Kumar Singh[1,3]

[1] National Institute of Technology Kurukshetra, Thanesar, Haryana, India
aman_52121204@nitkkr.ac.in
[2] National Institute of Technology Calicut, Kattangal, Kerala, India
[3] Indian Institute of Technology, Bhopal, India

Abstract. Selecting an appropriate career path is a pivotal decision, and in today's world of expanding career choices, it has become increasingly daunting for individuals seeking employment. Unfortunately, resources like career counsellors are often limited in availability and can be expensive, creating barriers for many students to access them. In response to this challenge, we suggest the development of a prediction model for a career that employs an ensemble learning technique i.e. voting classifier which includes a classifier such as K-Nearest Neighbour, Support Vector Machine, Stochastic Gradient Descent, Random Forest, and Decision Tree. The first set of voting classifier includes the K-nearest neighbour, Support Vector Machine, Stochastic Gradient Descent, and Decision Tree and the second set B contain the K-nearest neighbour, Support Vector Machine, Stochastic Gradient Descent, and Random Forest. The set with the highest accuracy and lowest fluctuation is set A with classifier K-Nearest Neighbour, Support Vector Machine, Stochastic Gradient Descent, and Decision Tree with an accuracy of 89.45%.

Keywords: Ensemble Learning · Career Prediction · voting Classification

1 Introduction

In Today's Society which is advancing itself in technology, the level of competition is increasing to its next level. To compete with such a level a student must be prepared from the beginning of their educational journey. There are many crucial tasks to be kept in mind such as assessing their progress, finding the areas of their interest, and checking again and again if they are on the right path or not in their career journey. The research analysis says that many new career options have emerged in many different fields. It is very easy to set goals but the difficult part is to find a way to achieve them. In the journey of career, the student has gone through much confusion in finding the right path and the reason for this confusion is that the students are not aware of available resources and because of this students fall into the well of various mental health issues and depression anxiety [1].

D. Pastor-Escuredo et al. (Eds.): ICDLAIR 2023, LNNS 1001, pp. 503–512, 2024.
https://doi.org/10.1007/978-3-031-60935-0_45

These issues about students will be demolished if the counsellor assists the student appropriately during the critical phase of their life and shows them the right path to achieve the desired goal. Students nowadays have a limited number of employment options, such as those for database administrators, business process analysts, developers, testing managers, networks managers, data scientists, and other positions. There is a method to find if the student has an interest in these types of roles or other roles or not, the candidate's counselling. The counsellor identifies the most appropriate career for them. The candidate will be placed in a job that complements their talents and interests thanks to this promise.

According to the report of The Council of Scientific & Industrial Research (CSIR) 40% of the students are confused about their career choices [2]. The Ratio says that there is 1 counselor only over 3000 students in INDIA [3]. A major question arises how to choose the right career and get the right path to reach it? As we know success is slow progress and it can never be achieved in a night. Students frequently ask their elder siblings or more seasoned peers for advice at various stages of their lives. There is still no clear way, though, and assurance is elusive. People on the other hand suggest ideas according to their expertise which may be not in your interest [4]. Even if there are many professional options, picking the right one is always a difficult task. Students take counseling from many places, which will always leave them in a dilemma and self-doubt. There are some online environments from which students can gather counseling. After investigating this platform it is found that students are still confused about their careers.

1.1 Contribution

After analyzing these platforms we have some drawbacks to discuss:

1. Some platforms demand costly student subscriptions, making it challenging to afford. Additionally, they impose a rule where students must purchase a course before accessing counseling, forcing them to subscribe even if they don't need it.
2. While these platforms offer one-on-one counseling sessions, the practicality is limited due to a high student-to-counselor ratio, making it challenging to secure appointments.
3. Furthermore, counselors, being human, are susceptible to making decisions that could potentially impact students' careers negatively.
4. These platforms offer a mix of courses, test modules, counselors, notes, and more, leading to potential confusion and distractions for students.

Rectifying the Flaws of the existing works by Proposed work?

1. The suggested study employs ensemble learning algorithms to foresee students' greatest career prospects and offers a thorough manual for picking the most appropriate courses.

2. Various classification algorithms are employed to determine the best-fit algorithms for predicting career choices.
3. Career prediction employs Ensemble learning algorithms i.e. Voting Classifier with a focus on empirical analysis.

1.2 Organisation

The rest of this essay is structured as follows: We describe the data collecting, pre-processing, algorithm flowchart, training, and testing aspects of the proposed model in Sect. 2. Section 3 provides an operational viewpoint. Our simulation findings, which include a comparison of all three techniques, are presented in Sect. 4. Finally, we wrap off this article with some closing thoughts in Sect. 5.

2 Proposed Model

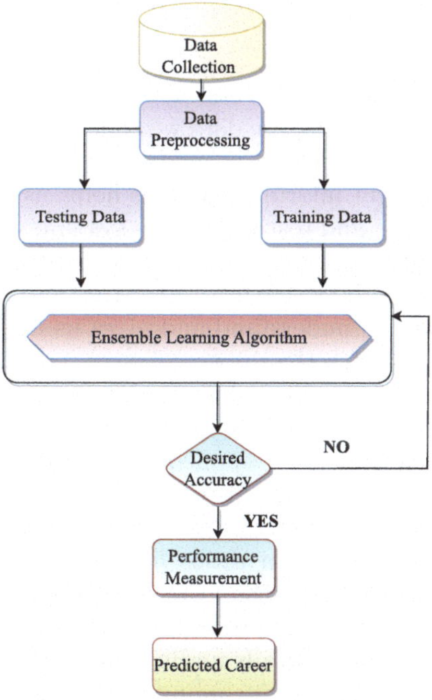

Fig. 1. Activity flow of proposed Ensembelled Model

The proposed model explained by these subsections includes a data collection in which the data is been collected and then collected data is preprocessed(all the null value is removed and not filled data is filled with random values) in data pre-processing which is followed by ensemble classification and classifiers which have all the classifier used in this career prediction model. In the sections that follow, the functions of these components are explained (Fig. 1).

2.1 Data Collection

Collecting data plays a pivotal and essential role in contemporary machine-learning endeavors. The data set, represented as $d = (d_1, d_2, d_3, ..., d_n) \in$ d, is initially gathered and then separated into two subsets: the training data (T_{train}) $= (d_1, d_2, d_3, ..., d_p)$ and the test data (T_{test}) $= (d_{P+1}, d_{P+2}, d_{P+3}, ..., d_n)$. The model is trained using the training data, and its performance is assessed using the test data. Usually, between 70% and 80% of the complete data set is made up of training data. The precision and caliber of the obtained data are closely related to the accuracy and efficacy of the algorithm.

The data set has around 9,000 entries and a total of 16 characteristics. These characteristics cover a variety of things, such as specializations, coding and processing prowess, technical knowledge, cybersecurity experience, AI/ML prowess, graphic design prowess, book interests, and more. These factors are pivotal in assessing a student's progression within various career areas.

2.2 Data Pre-processing

Collecting information is important, but making sure that it's organized and clean is just as crucial. Sometimes, the data we collect can be messy, with missing or wrong information. So, we clean it up by removing the messy parts and replacing them with the right information. This helps us have good-quality data that we can use for training and testing. In this system, we use something called "one-hot encoding" [12] to handle certain types of information. It's like turning yes-or-no questions into numbers. We use 1 for "Yes" and 0 for "No". This makes it easier for the computer to understand and gives us better results when we're trying to figure things out with the data.

The training procedure involves educating the computer and giving it the capacity to make predictions based on the information it has learned. To do this, algorithmic formulae must be applied to a particular training data set. The model that is developed during this training phase is then used for testing. To train the machine learning model, the original data set is condensed and used. It's normal to divide them in ratios like 80:20, 70:30, or 90:10 because the training data set is often larger than the testing data set. Since the training data forms the basis of its learning process, the model gets quite familiar with it. Essentially, training a model entails finding the best values for all models by being exposed to labeled instances, including weights and biases.

2.3 Ensemble Classification and Classifiers

Ensemble learning leverages a trio of algorithms designed to harness the power of synergy among multiple machine-learning models. The ensemble learning algorithm works by running a base learning algorithm multiple times and forming a vote out of the resulting hypothesis. This approach involves artfully amalgamating two or more classification techniques to create an innovative methodology tailored for addressing specific computational challenges within the realm of machine learning. This collaborative technique capitalizes on the strengths of individual models while mitigating their weaknesses, ultimately fostering more robust and accurate solutions for a diverse array of machine-learning tasks (Fig. 2).

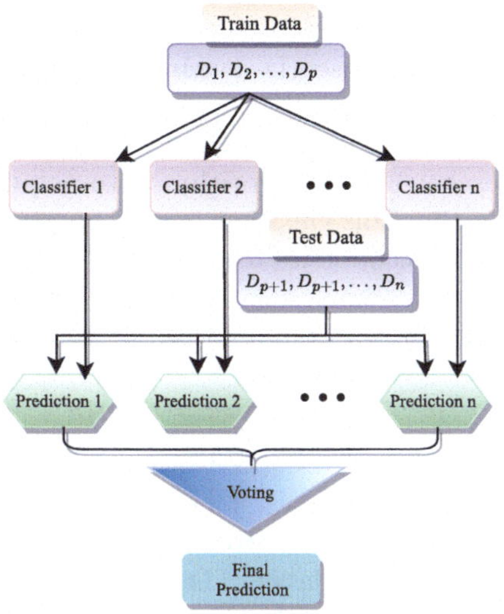

Fig. 2. Ensemble Voting Classifier Working

Voting Classifier. A Voting Classifier, referenced is a machine learning model that leverages an ensemble of models to make predictions by selecting the class with the highest probability of occurrence [10]. This classifier is commonly employed alongside other classifiers to enhance model performance and accuracy, characterizing it as a prominent ensemble method. Its primary function involves aggregating predictions from multiple models, with the voting process contingent on the individual performance of machine learning algorithms.

KNN [5]**.** The K-NN method maintains all of the available data and classifies a new data point by determining how similar it is to the data already there [8]. It is sometimes referred to as a "lazy learner" since, while being trained, it only stores input and makes no calculations [8].

$$d = \sqrt{(X_2 - X_1)^2 + (Y_2 - Y_1)^2} \tag{1}$$

SVM [6]**.** This is frequently applied to classification and regression problems [9]. Its principal uses are for a variety of categorization jobs. In a typical process, each data point is first mapped onto an n-dimensional space, where n is the number of features and each feature's value determines where it is located [13]. Following this, the classification is carried out by identifying the precise hyperplane that separates the two classes.

$$f(X) = b_0 + sum(A_I * (X, X_I)) \tag{2}$$

$$G(W, B) = Y_I(W, X + B - 1) \tag{3}$$

SGD [7]. With a training data-set D consisting of n labelled instances (X_1, Y_1), (X_2, Y_2), ..., (X_n, Y_N), where X_I represents the feature vector of the ith instance and y_i indicates its binary class label (+1 or −1), the objective of the Stochastic Gradient Descent (SGD) algorithm is to acquire knowledge about a linear classifier f(X; W) [13]. This classifier is used to predict the class of a new observation X, taking into account its feature vector and the weight values W that have been learned.

$f(X; W)$ is defined as a linear combination of the feature vector X and the weight vector W:

$$w = w - \alpha \times \nabla w(J(w, b))3 \tag{4}$$

DT. A classification decision tree is a hierarchical structure that recursively splits data based on feature conditions to assign class labels to instances, with each leaf node representing a class label. The decision tree is a simple machine-learning model for getting started with regression tasks [11].

$$\sum Entropy(s) = P_i log_2 P_i \tag{5}$$

RF. The Random Forest classifier, a versatile Supervised Machine Learning algorithm, is employed for Classification and Regression tasks [10]. It constructs decision trees from diverse samples, using majority voting for classification and averaging for regression. It excels in handling data sets with continuous variables in regression and categorical values in classification, making it suitable for challenging scenarios. Random Forest uses an ensemble approach, combining decisions from base models, typically simple decision trees, enhancing predictive performance. Its adaptability to various data types and problem domains makes it a versatile choice across applications.

$$F(x) = 1/N \sum_{i=1}^{N} h_i(x) \tag{6}$$

3 Operational Summary

Algorithm 1 provides an overview of the operational steps for the proposed work. In this model, the input data (D) is denoted as $d_1, d_2, ..., d_n$, and it undergoes evaluation based on accuracy, precision, recall, and F1 score through three algorithms: Voting and Random Forest.

The model's execution begins by initializing with the complete data set d (step 1). Subsequently, d is partitioned into separate training and testing data sets (step 2), and these data sets are further divided into training and testing subsets (step 3). The model then proceeds to the training phase employing the Voting and Random Forest algorithms with the training data set (d_{Train}) until the desired level of accuracy is attained (steps 4–5). Finally, the model undergoes testing with the testing data set (d_{Test}), and its performance is evaluated in terms of accuracy, precision, recall, and F1 score, individually for each of the Voting and Random Forest algorithms (steps 7–8).

Here are the formulas for these metrics:

Accuracy: By calculating the proportion of properly categorized examples to all cases, accuracy evaluates the overall accuracy of the model's predictions.

$$\text{Accuracy} = \frac{t_P + t_n}{t_p + f_p + t_n + f_n} \tag{7}$$

where f_n is False Negative and t_p is True Positive,f_p is False Positive and t_n is True Negative. Precision: By calculating the proportion of accurately anticipated positive instances to all expected positive cases, precision measures the positive predictive value.

$$\text{Precision} = \frac{t_p}{t_p + f_p} \tag{8}$$

where t_p is True Positive, and f_n is False positive Recall: Recall, sometimes referred to as the true positive rate, is the percentage of real positive cases that the model accurately recognized as positive. Equation Eq is utilized in the recall computation (9),

$$\text{Recall} = \frac{t_p}{t_p + f_n} \tag{9}$$

where f_n is False Negative and t_p is True Positive.

F1 score: The F1 score exhibits a harmonic balance between recall and accuracy, making it appropriate for circumstances where this balance is sought.

$$\text{F1 Score} = 2 * \frac{\text{Precision} * \text{Recall}}{\text{Precision} + \text{Recall}}. \tag{10}$$

The loop will continue running until the required result is obtained. After achieving the intended outcome, we will break out of the loop, arrange the data into a table, and then produce a related graph.

4 Results and Discussion

4.1 Experimental Setup and Data Set

The research was conducted using a computer operating on the Windows platform, specifically a machine powered by an AMD Ryzen 5 5500U processor with Radeon Graphics, clocked at 2.10 GHz. The computational tasks were carried out on a 64-bit Windows 11 system equipped with 8 GB of RAM. Python version

3.7.4 was employed as the programming environment to implement the proposed work. The study focused on the application of Voting, all of which were implemented using the sklearn library. The data set used for the study was sourced from a GitHub repository and consisted of 9000 instances, encompassing a total of 16 distinct profession categories, each serving as labels for the data set [14].

4.2 Performance Analysis

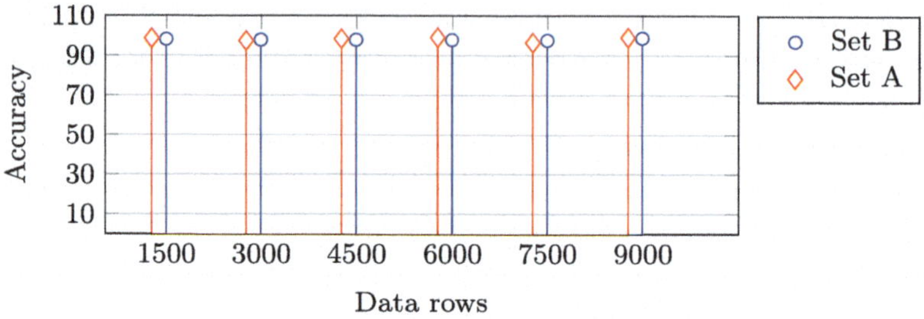

Fig. 3. Accuracy comparison graph

Figure 3 illustrates the accuracy comparisons for the data set across different intervals of 1500. Among the two sets, set A achieved the highest accuracy of 89.629 within the interval of 6000. On the other hand, Set B reached its peak accuracy of 79.11 within the interval of 9000. Despite both sets maintaining accuracy rates ranging from 75% to 90%, set A exhibited the lowest rate of fluctuation, establishing itself as the most stable classifier.

Figure 4(a) and Fig. 4(b) show the confusion matrix to visualize the performance of set A and set B providing a detailed summary of the model's predictions by breaking down the true positive, true negative, false positive, and false negative predictions, enabling better understanding of the model's accuracy and error rates for different classes.

```
Confusion Matrix:
[[106   0   0   0   0   0   0   0   0   0   0   0   0   0   0   0   0]
 [  3 102   0   0   0   0   0   0   0   1   0   4   0   0   0   0   0]
 [  0   0 113   0   0   0   0   0   0   0   0   2   0   0   0   0   0]
 [  1   1   0  90   0   1   0   0   0   1   1   2   1   0   0   1   0]
 [  4   0   0   1  53   0   0   0   0   0   0  56   0   1   0   0   0]
 [  1   0   0   0   0 101   0   0   0   0   0   1   0   0   0   0   0]
 [  3   1   0   0   0   0  61   0   0   0   0  40   0   0   0   0   0]
 [  0   0   0   0   0   0   0  64   0   0   0  55   0   0   0   0   0]
 [  0   0   0   0   0   0   0   0 113   0   0   0   0   0   0   0   0]
 [  0   0   0   0   1   0   0   0   1  93   0   7   0   0   0   0   1]
 [  2   3   1   1   2   0   0   1   0   0  96   2   1   1   0   0   0]
 [  0   0   0   0   0   0   0   0   0   0   0 114   0   0   0   0   0]
 [  3   0   0   0   0   0   0   0   0   0   0  38  62   0   0   0   0]
 [  6   0   0   0   0   0   0   0   0   0   0  52   0  52   0   0   0]
 [  7   0   0   0   0   0   0   2   0   0   0  49   0   0  50   3   0]
 [  1   0   0   0   0   0   0   0   0   0   0  33   0   0   0  74   0]
 [  3   0   0   0   0   0   1   0   0   0   0   0   0   0   0   0  88]]
```

(a) SET A

```
Confusion Matrix:
[[106   0   0   0   0   0   0   0   0   0   0   0   0   0   0   0   0]
 [  0 110   0   0   0   0   0   0   0   0   0   0   0   0   0   0   0]
 [  0   0   0   0   6   0   0 100   0   0   0   9   0   0   0   0   0]
 [  0   0   0  99   0   0   0   0   0   0   0   0   0   0   0   0   0]
 [  0   0   4   0   0   0   0 101   0   0   0  10   0   0   0   0   0]
 [  0   0   0   0   0 103   0   0   0   0   0   0   0   0   0   0   0]
 [  0   0   0   0   0   0 105   0   0   0   0   0   0   0   0   0   0]
 [  0   0   9   0  10   0   0  74   0   0   0  26   0   0   0   0   0]
 [  0   0   0   0   0   0   0   0 113   0   0   0   0   0   0   0   0]
 [  0   0   0   0   0   0   0   0   0 103   0   0   0   0   0   0   0]
 [  0   0   0   0   0   0   0   0   0   0 110   0   0   0   0   0   0]
 [  0   0   5   0   6   0   0 103   0   0   0   0   0   0   0   0   0]
 [  0   0   0   0   0   0   0   0   0   0   0   0 103   0   0   0   0]
 [  0   0   0   0   0   0   0   0   0   0   0   0   0 110   0   0   0]
 [  0   0   0   0   0   0   0   0   0   0   0   0   0   0 111   0   0]
 [  0   0   0   0   0   0   0   0   0   0   0   0   0   0   0 108   0]
 [  0   0   0   0   0   0   0   0   0   0   0   0   0   0   0   0  92]]
```

(b) SET B

Fig. 4. Confusion Matrix

Table 1. Performance in terms of Recall, Precision, and F1 score

Attributes	Voting Classifier Set A			Voting Classifier Set B		
	Precision	Recall	F1 Score	Precision	Recall	F1 Score
AI ML Specialist	0.51	1.00	0.68	1.00	1.00	1.00
API Integration Specialist	0.83	1.00	0.91	1.00	1.00	1.00
Application Support Engineer	0.90	0.54	0.67	0.00	0.00	0.00
Business Analyst	0.97	1.00	0.99	1.00	1.00	1.00
Customer Service Executive	0.99	0.92	0.95	0.00	0.00	0.00
Cyber Security Specialist	0.95	1.00	0.98	1.00	1.00	1.00
Database Administrator	0.95	0.87	0.91	0.20	0.65	0.31
Graphics Designer	0.96	0.92	0.94	1.00	1.00	1.00
Hardware Engineer	1.00	1.00	1.00	1.00	1.00	1.00
Helpdesk Engineer	0.96	0.93	0.94	1.00	1.00	1.00
Information Security Specialist	0.99	0.64	0.78	0.00	0.00	0.00
Networking Engineer	1.00	0.86	0.93	1.00	1.00	1.00
Project Manager	0.94	0.95	0.95	1.00	1.00	1.00
Software Developer	0.96	0.85	0.90	1.00	1.00	1.00
Software Tester	0.97	0.86	0.91	1.00	1.00	1.00
Technical Writer	0.98	0.96	0.97	1.00	1.00	1.00

Table 1 lists the precision, recall, and F1 Score performance data for three classifiers: Voting and Random Forest. The labels in the graph (1, 2, 3,..., 16) correspond to various skills, including those for Database Administrator, Graphics Designer, Hardware Engineer, Help-desk Engineer, Information Security Specialist, Networking Engineer, Project Manager, Software Developer, and others. The results for Recall, Precision, and F1 Score are calculated using a formula based on information supplied by the user. In the above table, Set A obtains the greatest Precision, F1 Score, and Recall values of 1.00 each, while Set B receives the lowest Precision, F1 Score, and Recall values of 0.51, 0.67, and 0.54, respectively.

The greatest Precision, Recall, and F1 Score values for Set B are all 1.00, and the lowest Precision, Recall, and F1 Score values for Set B are 0.00.

5 Conclusion

This paper introduces the AN Ensemble Learning Model for Career Prediction, which is an ensemble learning-based solution for career prediction. The system aims to assist job seekers in predicting suitable career paths by employing ensemble learning techniques i.e. Voting techniques and classifiers like KNN, SVM, SGD, DT, and RF. The performance of the most accurate classifier is used as the final result. Additionally, online job seekers can benefit from career-related guidance through this system. It involves presenting job seekers with a set of questions, and based on their answers, it offers personalised recommendations for their career choices.

References

1. Piper, A., Blanchflower, D.G., Bryson, A.: Is pain associated with subsequent job loss? A panel study for Germany. Kyklos **76**(1), 141–158 (2023)
2. Shankhdhar, A., Agrawal, A., Sharma, D., Chaturvedi, S., Pushkarna, M.: Intelligent decision support system using decision tree method for student career. In: 2020 International Conference on Power Electronics & IoT Applications in Renewable Energy and its Control (PARC), pp. 140–142. IEEE (2020)
3. Mietzel, G., Rüssmann-Stöhr, C., Mason, E.: School psychology in the Federal Republic of Germany. J. Sch. Psychol. **25**(3), 255–265 (1987)
4. Reddy, M.M.: Career prediction system. Int. J. Sci. Res. Sci. Technol. **8**(4), 54–58 (2021)
5. Pernkopf, F.: Bayesian network classifiers versus selective k-NN classifier. Pattern Recogn. **38**(1), 1–10 (2005)
6. Abdullah, D.M., Abdulazeez, A.M.: Machine learning applications based on SVM classification a review. Qubahan Acad. J. **1**(2), 81–90 (2021)
7. Prasetijo, A.B., Isnanto, R.R., Eridani, D., Soetrisno, Y.A.A., Arfan, M., Sofwan, A.: Hoax detection system on Indonesian news sites based on text classification using SVM and SGD. In: 2017 4th International Conference on Information Technology, Computer, and Electrical Engineering (ICITACEE), pp. 45–49. IEEE (2017)
8. Singh, A.K., Kumar, J.: A secure and privacy-preserving data aggregation and classification model for smart grid. Multimed. Tools Appl. 1–19 (2023)
9. Kumar, J., Gupta, R., Saxena, D., Singh, A.K.: Power consumption forecast model using ensemble learning for smart grid. J. Supercomput. 1–22 (2023)
10. Reddy, K., Akhila Reddy, M., Kaur, V., Kaur, G.: Career guidance system using ensemble learning. In: Akhila and Kaur, Veerpal and Kaur, Gagandeep (eds.) Career Guidance System Using Ensemble Learning, 14 July 2022 (2022)
11. Chaudhary, D., Prajapati, H., Rathod, R., Patel, P., Gurjwar, R.K.: Student future prediction using machine learning. Int. J. Sci. Res. Comput. Sci. Eng. Inf. Technol. **5**(2), 1104–1108 (2019)
12. Jie, L., Chen, J., Zhang, X., Zhou, Y., Lin, J.: One-hot encoding and convolutional neural network based anomaly detection. J. Tsinghua Univ. (Sci. Technol.) **59**(7), 523–529 (2019)
13. Wakade, A., Qureshi, U., Parmar, D., Kumar, J., Swain, S.R., Kumar Singh, A.: MARG-DARSHAK: a machine learning approach for career prediction. In: 2023 International Conference on Computer, Electronics & Electrical Engineering & Their Applications (IC2E3), Srinagar Garhwal, India, pp. 1–6 (2023)
14. Git hub. Intelligent Career Guidance System. https://github.com/sayaliwangane/INTELLIGENT-CAREER-GUIDANCE-SYSTEM/blob/5f6fd35677ca711caff7540903719417f365f781/dataset9000.csv

Feature Extraction Analysis in a Speaker Identification System

Deeksha Goel, Shreya Sharma(✉), Pooja Gambhir, Kiran Malik, and Poonam Bansal

Indira Gandhi Delhi Technical University, New Delhi, India
shreya2701sharma@gmail.com, pooja001phd19@igdtuw.ac.in

Abstract. Human interaction and communication both rely heavily on language. To emphasize the properties of communication signals, speech recognizers use a parametric signal structure. There are numerous methods for removing elements from audio streams. In this study, four characteristic extraction techniques - the Mel frequency cepstral coefficient (MFCC), the bark frequency cepstral coefficient (BFCC), the Gamma frequency cepstral coefficient (GFCC), and the Linear Predictive Coding (LPC)—are used to assess the performance of the Gaussian mixture model (GMM) in speech recognition. The research analyses the computation time, the number of speakers to be researched, and the recognition speed to determine the optimal feature extraction method. The findings show that GMM-based MFCC performs better in recognition than BFCC, LPC, and GFCC.

Keywords: Speaker identification · Feature Extraction · MFCC · BFCC · GFCC · LPC · GMM

1 Introduction

An important topic of research is speaker recognition, which aims to recognize speakers by their distinctive speech patterns. To improve the precision and effectiveness of speaker identification systems, a lot of work has been done in this area in recent years.

Extraction of features from the speech stream and modelling of the speaker's speech patterns using these features is a popular strategy for speaker recognition. The MFCC and the BFCC are two widely used feature extraction techniques. GFCC and LPC are also employed for this purpose.

The next step in the speaker recognition process is speaker modelling, which involves using the extracted features to create a model that can be used to identify the speaker. One popular modelling method is Gaussian Mixture Model (GMM), which has been used successfully in many speaker identification systems (Fig. 1).

D. Pastor-Escuredo et al. (Eds.): ICDLAIR 2023, LNNS 1001, pp. 513–523, 2024.
https://doi.org/10.1007/978-3-031-60935-0_46

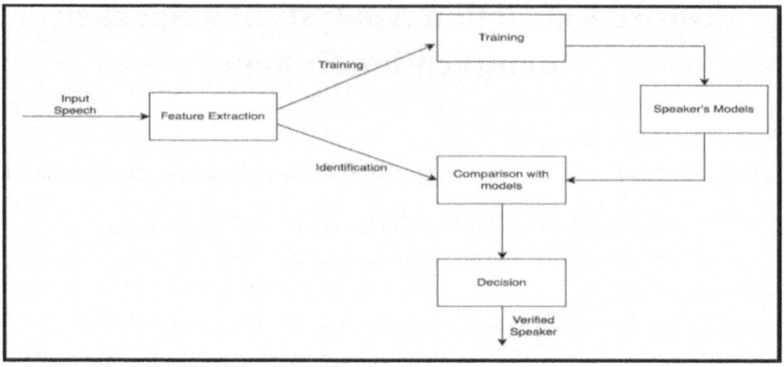

Fig. 1. Speaker Identification

1.1 About the Paper

This study aims to investigate the state of voice recognition technology today, considering the most recent developments in machine learning and signal processing. The basic drawbacks and shortcomings of speaker identification systems are also covered, along with their possible uses and social implications. This work seeks to contribute to the future development of more effective and moral speech recognition systems by giving a thorough review of the area.

1.2 Motivation

The motivation comes from the fundamental importance of language and speech recognition in human interaction and communication. Speaker recognition plays an important role in various real-world applications such as voice-controlled systems, security applications and forensic investigations. Choosing an optimal feature extraction technique is paramount to achieve accurate and reliable speaker detection. Therefore, this study aims to comprehensively evaluate four widely used techniques, namely MFCC, BFCC, GFCC, and LPC. By analyzing the performance of these techniques using the GMM, the research aims to identify the most effective and efficient approach for distinguishing between different speakers. The findings of this investigation hold the potential to advance speaker identification systems, enhance security measures, and pave the way for more robust and accurate speech recognition technologies across diverse applications, benefiting both industry and society.

1.3 Novelty and Contributions

1. This study aims to shed light on the advancements made and the potential for further exploration in Hindi language processing.
2. In our research paper, we conducted verification tasks that involved analyzing continuous speech, not merely isolated words, and without relying on specific predefined texts.

3. In many research studies, the Gaussian Mixture Model (GMM) is commonly used together with Mel Frequency Cepstral Coefficients (MFCC), followed by accuracy assessment. However, in our research paper, we have paired GMM not only with MFCC but also with other techniques such as Bark Frequency Cepstral Coefficients (BFCC), Gammatone Frequency Cepstral Coefficients (GFCC), and Linear Predictive Coding (LPC). This novel combination expands the scope beyond the typical approach, offering a broader spectrum of techniques to enhance the analysis and evaluation process.

2 Literature Survey

Chandar Kumar et al. [1] in their work "Analysis of MFCC and BFCC in a Speaker Identification System," conducted a comparative analysis of two prominent speaker identification systems, one utilizing Mel-frequency Cepstral Coefficients (MFCC) and the other bark frequency Cepstral coefficients (BFCC). Their study meticulously assessed these systems in terms of gender classification, identification rates, and computational efficiency. Mr. Yoghesh Dawande et al. [2], in their work "Analysis of different feature extraction techniques for speaker recognition system: a review," introduced a speaker identification-based security system. Their employment of MFCCs for feature extraction contributes to the reduction of data volume, and their paper extensively discusses the methodologies and methods employed for feature extraction in the realm of speaker recognition. Roma Bharti and Priyanka Bansal [3] presented a real-time speaker recognition system that effectively integrates MFCC and the Vector Quantization Technique. This research not only showcases the practical application of feature extraction in real-time scenarios but also enhances our understanding of speaker recognition. Yeni Astuti et al. [4] meticulously analyzed the performance of feature extraction-based speaker identification systems. They thoughtfully selected and tested three feature extraction techniques: Fast Fourier Transform (FFT), MFCC, and Discrete Wavelet Transform (DWT). Xinxing Jing et al. [5] introduced a novel approach for extracting mixed characteristic parameters using Principal Component Analysis (PCA). Their methodology draws inspiration from the widespread utilization of PCA and K-means clustering in image and voice signal processing, offering an innovative perspective. S K Das et al. [6], in their study titled "Performance Analysis of Speech Features for Speaker Verification in Emotional Conditions," conducted a thorough investigation into the influence of emotional states on speaker verification. Their research enhances our comprehension of the intricacies and challenges involved in speaker recognition under varying emotional conditions. Doreen Nongrum and Fidalizia Pyrtuh [7], in their "Comparative Study on the Effect of Temporal Phase for Speaker Verification," shed light on the significance of temporal phase in the context of speaker verification. Their study provides valuable insights into the temporal aspects of voice signal processing. Taabish Gulzar et al. [8] delved into the recognition of spoken words and the selection of feature extraction techniques, utilizing Artificial Neural Networks as classifiers. This research adds depth to our understanding of the practical application of these techniques in the recognition of spoken language (Table 1).

Table 1. Literature Review

Research Paper	Abstract	Findings
[1] Analysis of MFCC and BFCC in a Speaker Identification System	Speech recognition relies on signal parameters to identify crucial communication features. Various extraction methods capture these distinctions. The study evaluates GMM using MFCC and BFCC for speaker identification. Results indicate GMMbased MFCC as the optimal technique, considering identification rate, Speaker count, gender, and computational time	Comparing GMM-based speaker ID systems with MFCC and BFCC showed a 12% decrease in identification rates with increased speakers (4 to 20). Optimal performance was at 20 coefficients, where MFCC outperformed BFCC in both identification rates and efficiency
[2] Analysis of Different Feature Extraction Techniques for Speaker Identification - A Review	This paper provides an overview of speaker recognition's evolution, encompassing methods from spectral speech traits to behavioral features like prosody and conversation patterns. Starting with a brief on Speech Recognition, it compares MFCC, LPC, and PLP as feature extraction techniques	This paper demonstrates various speaker identification feature extraction methods through experiments. MFCC is a widely recognized technique in speaker recognition for describing signal characteristics
[11] Speaker Identification Using Mel Frequency Cepstral Coefficients	This paper introduces a speaker identification security system utilizing Mel frequency Cepstral Coefficients (MFCCs) for feature extraction. It employs vector quantization to reduce the volume of data processed	The study discovered that combining Mel frequency with the Hamming window yields the most optimal performance. Additionally, it recommends increasing the number of centroids as the quantity of speakers rises to achieve satisfactory results
[3] Real Time Speaker Recognition System using MFCC and Vector Quantization Technique	This paper introduces a robust mathematical algorithm for Automatic Speaker Recognition (ASR) in the digital realm, employing MFCC and vector quantization techniques. These methods are highly favored and promising in advancing voice recognition technology, reflecting significant advancements in the field	A real-time speaker recognition system employing MFCC and vector quantization, analyzing experimental outcomes via MATLAB. Conducted with 120 speakers using the TIDIGIT database, the study achieved 91% accuracy under normal environmental conditions at 20dB SNR
[4] Comparison of Feature Extraction for Speaker Identification System	This study evaluates speaker identification systems by comparing different feature extraction methods: Fast Fourier Transform (FFT), MFCC, and Discrete Wavelet Transform (DWT). These techniques, applied to recognize speakers based on spoken words, utilize Dynamic Time Warping (DTW) as the classifier	Tests in noisy environments favored DWT feature extraction with DTW classification over MFCC with DTW and FFT with DTW, showing better recognition performance. DWT combined with DTW outperformed others in this context

(*continued*)

Table 1. (*continued*)

Research Paper	Abstract	Findings
[6] Performance Analysis of Speech Features for Speaker Verification in Emotional Conditions	This study explores the impact of emotional conditions on speaker verification. The study investigates various speech features to assess their performance under emotional variations	The research reveals significant variations in the effectiveness of speech features for speaker verification in emotional conditions. Specific features demonstrate robust performance, while others exhibit susceptibility to emotional changes, highlighting the importance of tailored feature selection in enhancing speaker verification accuracy in emotionally diverse scenarios
[7] A Comparative Study on Effect of Temporal Phase for Speaker Verification	This study investigates the impact of temporal phase variations on speaker verification. The research aims to provide insights into the role of temporal phase in enhancing or degrading the performance of speaker verification systems	The comparative analysis conducted in this research elucidates the nuanced effect of temporal phase on speaker verification. The study identifies optimal temporal phase conditions that significantly contribute to improved verification accuracy, emphasizing the importance of considering temporal dynamics in designing robust speaker verification systems
[8] Comparative Analysis of LPCC, MFCC and BFCC for the Recognition of Hindi Words using Artificial Neural Networks, International Journal of Computer Applications	This study conducts a comprehensive comparison of LPCC, MFCC, and BFCC features for Hindi word recognition. The study employs Artificial Neural Networks (ANN) to assess the effectiveness of these features in the context of Hindi speech recognition	The investigation reveals distinctive performances of LPCC, MFCC, and BFCC features in the recognition of Hindi words using Artificial Neural Networks. The study contributes insights into the strengths and weaknesses of each feature set, aiding in the optimization of Hindi speech recognition systems
[9] Speaker Recognition Techniques: A Review	This study provides a comprehensive review of speaker recognition techniques. The authors present an overview of various methodologies employed in speaker recognition, offering valuable insights into the advancements and challenges in this domain	The review highlights the diverse array of speaker recognition techniques and provides a nuanced understanding of their strengths and limitations. This comprehensive analysis serves as a valuable resource for researchers and practitioners in the field of speaker recognition, aiding in the identification of effective approaches for different applications
[10] Text-independent speaker recognition using LSTM-RNN and speech enhancement Multimedia	This study focuses on text-independent speaker recognition. It employs Long Short-Term Memory Recurrent Neural Networks (LSTM-RNN) and speech enhancement techniques to enhance the accuracy and robustness of speaker recognition systems	The research demonstrates the efficacy of employing LSTM-RNN and speech enhancement for text-independent speaker recognition. The combination of these technologies exhibits promising results, paving the way for improved performance in challenging acoustic environments and contributing to advancements in speaker recognition technology

3 Proposed Methodology

A set of features that highlight both the speech and the speaker should be useful for computer speech recognition work. The most important step in the identification process is to obtain relevant information to ensure accurate identification. Through the process of feature extraction, the original speech signal can be efficiently transformed, resulting in a more stable and distinctive representation compared to the initial signal. Using the frame and window technique, the speech signal is processed before feature extraction.

3.1 MFCC

Mel Frequency Cepstral Coefficients (MFCCs) represent a method for extracting features from audio data. MFCCs partition the frequency spectrum into sub-bands based on the MEL scale, and subsequently employ the Discrete Cosine Transform (DCT) to derive the Cepstral Coefficients. The utilization of the MEL scale is advantageous in sound processing, as it closely mirrors the way humans perceive different frequencies (Fig. 2).

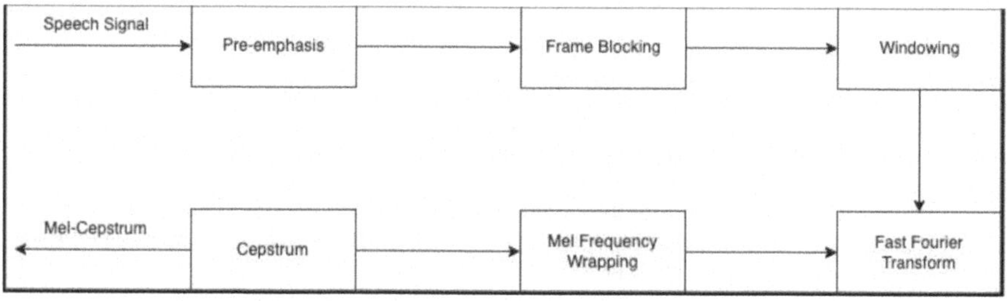

Fig. 2. Mel Frequency Cepstral Coefficients

3.2 BFCC

BFCC is another method for extracting features of speech sounds. BFCC technique is like the MFCC technique but uses a different frequency scale called the Bark scale. Based on psychoacoustic research, the Bark scale attempts to model how the human ear perceives sound. To compute the BFCC, the audio signal's frequency spectrum is initially partitioned into overlapping segments based on the Bark scale (Fig. 3).

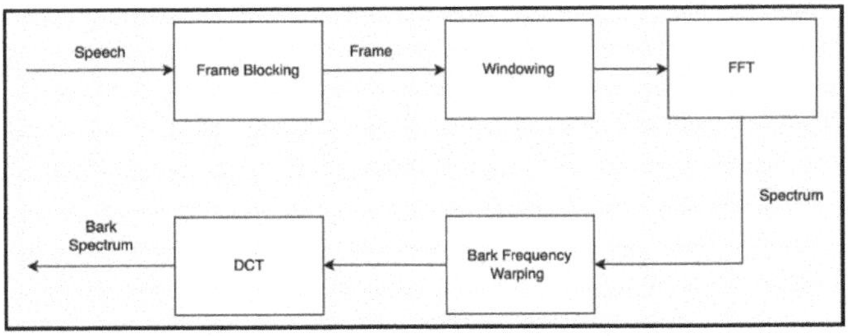

Fig. 3. Bark Frequency Cepstral Coefficients

3.3 GFCC

GFCC is a feature extraction technique like MFCC and BFCC, but it employs a different filter bank design based on the gammatone function, which simulates the frequency response of the human cochlea. GFCC is calculated by first dividing an audio signal's frequency spectrum into several overlapping bands using the gammatone filter bank (Fig. 4).

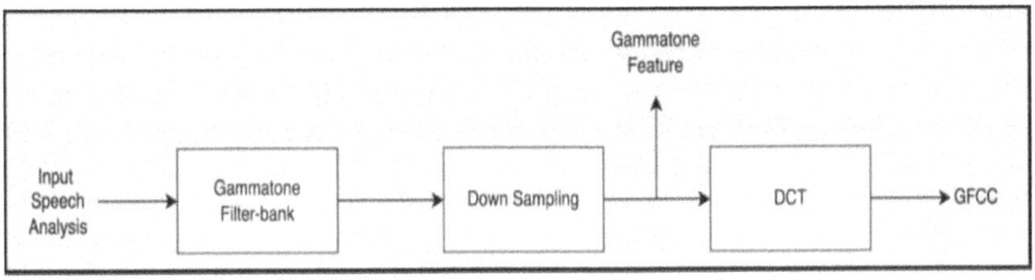

Fig. 4. Gammatone Frequency Cepstral Coefficients

3.4 LPC

Linear forecasting is a mathematical approach that involves the linear integration of large amounts of past data. LPC is the most used method to assess the salient features of speech. It is a powerful computational speech model that provides accurate estimation of speech parameters (Fig. 5).

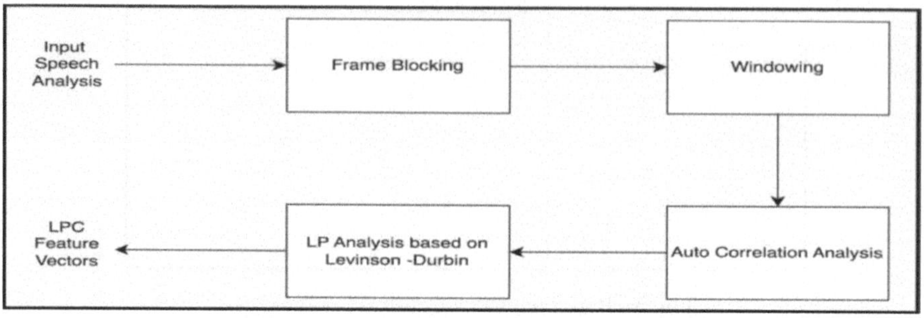

Fig. 5. LPC Feature Extraction Method

3.5 GMM

A Gaussian mixture model is a probabilistic model in which it's assumed that every data point was produced by combining a limited number of Gaussian distributions with unidentified characteristics. It is relatively independent for different probabilistic models. The extracted features of each speaker's audio signal are modeled using GMM. The results are represented by the sum of all linear components (i.e., Gaussian distribution) called the mixture. Python's sklearn.mixture library allows the use of Gaussian mixture models. It uses a combination of a finite number of Gaussian distributions with unknown parameters that uses probability to generate all data points. To combine information about covariance structure and information about hidden Gaussian centers, one could think of mixture models as generalized k-means clustering.

4 Experimental Setup

In real-world scenarios, data often exhibits noise, contains gaps, and can be in an unstructured format, rendering it unsuitable for direct utilization by machine learning and deep learning models. Data preprocessing becomes essential to cleanse and transform the data, ensuring it aligns with the requirements of machine learning and deep learning models. This process not only enhances model accuracy but also boosts its overall efficiency.

4.1 Dataset Used

Our dataset contains 600 audio samples from 100 speakers and is 600mb in size. Which is broken into two sections: training audio and testing audio. Training audio contains five speech samples from each speaker that can be trained using any model, while testing audio provides a single audio sample that can be used to verify the model. The dataset recorded in Hindi and yields good results.

4.2 Training Model

The GMM method was used to train the model. Python's sklearn. Mixture library enables Gaussian mixture models. The GMM model, rooted in probability theory, generates individual data points by blending them from a limited set of Gaussian distributions with unspecified parameters. Mixture models can be conceptualized as an extension of the k-means clustering approach, where they incorporate information about both the data's covariance structure and concealed Gaussian centroids. Scikit-learn uses several classes to estimate Gaussian mixture models corresponding to discrete estimation methods.

5 Results and Outcomes

The extraction methods used in the proposed method were MFCC, GFCC, BFCC and LPC based on the experimental findings shown in Table 2. Using MFCC properties, the proposed technique achieved 94% accuracy with an accuracy of 0.938. 0.935 and F1 points 0.936. The results show that the proposed technique performed best with MFCC.

Table 2. Comparative Result Analysis

Techniques	Accuracy	Precision	Recall	F1 Score
MFCC	94%	0.938	0.935	0.936
GFCC	91%	0.914	0.912	0.913
BFCC	88%	0.878	0.876	0.877
LPC	82%	0.824	0.825	0.824

6 Discussions and Future Aspects

These findings regarding the significance of MFCC in speaker recognition hold broad implications. It suggests advancements in security measures, more efficient voice controlled systems, and potential applications in forensic investigations. The focus on MFCC as the primary feature extraction method sets the groundwork for enhancing speech recognition technology across various fields, opening doors for future research and innovation. This emphasizes the need for refining MFCC-based models, exploring hybrid techniques, and integrating machine learning for more accurate speaker recognition. Additionally, applying these findings to real-world scenarios could lead to the development of robust systems in security, voice-controlled devices, and forensic voice analysis, contributing significantly to multiple industries reliant on voice technology.

7 Conclusion

The study's investigation into feature extraction methods for speaker recognition highlighted Mel Frequency Cepstral Coefficients (MFCC) as the most effective technique. Compared to Bark Frequency Cepstral Coefficients (BFCC), Gamma Frequency Cepstral Coefficients (GFCC), and Linear Predictive Coding (LPC), MFCC demonstrated superior performance in accuracy, precision, recall, and F1 score. This outcome emphasizes

MFCC's reliability and proficiency in distinguishing between speakers within recognition systems, indicating its potential for enhancing the precision and robustness of such systems. This finding suggests a promising direction for further refining speaker recognition technology by leveraging the strengths of MFCC or exploring its integration with other methods to advance accuracy and efficiency. Overall, the study underscores MFCC's pivotal role in driving advancements in speaker recognition technology.

Acknowledgement. We, the authors, would like to thank the Department of Artificial Intelligence and Data Science and Indira Gandhi Delhi Technical University for providing the necessary infrastructure to conduct the research.

References

1. Kumar, C., urRehman, F., Kumar, S., Mehmood, A., Shabir, G.: Analysis of MFCC and BFCC in a speaker identification system. In: iCoMET (2018)
2. Dawande, Y., Dhopeshwarkar, M.: Analysis of different feature extraction techniques for speaker recognition system: a review. IJATER (2015)
3. Bharti, R., Bansal, P.: Real time speaker recognition system using MFCC and vector quantization technique. IJCA (2015)
4. Astuti, Y., Hidayat, R., Bejo, A.: Comparison of feature extraction for speaker identification system. IEEE (2020)
5. Jing, X., Ma, J., Zhao, J., Yang, H.: Speaker recognition based on principal component analysis of LPCC and MFCC. IEEE (2014)
6. Das, S.K., Bhattacharjee, U.: Performance analysis of speech features for speaker verification in emotional conditions. YMER (2022)
7. Nongrum, D., Pyrtuh, F.: A comparative study on effect of temporal phase for speaker verification. COMSYS (2021)
8. Gulzar, T., Singh, A.: Comparative analysis of LPCC, MFCC and BFCC for the recognition of Hindi words using artificial neural networks. Int. J. Comput. Appl. (2014)
9. Todkar, S.P., Babar, S.S., Ambike, R.U., Suryakar, P.B., Prasad, J.R.: Speaker recognition techniques: a review. IEEE (2018)
10. El-Moneim, S.A., Nassar, M.A., Dessouky, M.I., Ismail, N.A., El-Fishawy, A.S., Abd El-Samie, F.E.: Text-independent speaker recognition using LSTMRNN and speech enhancement. Multimed. Tools Appl. **79**, 24013–24028 (2020)
11. Hasan, Md.R., Jamil, M., Rabbani, Md.G., Rahman, Md.S.: Speaker Identification Using Mel Frequency Cepstral Coefficients (2004)
12. Ibrahim, N.S., Ramli, D.A.: I-vector Extraction for Speaker Recognition Based on Dimensionality (2018)
13. Liu, J.C., Leu, F.Y., Lin, G.L., Susanto, H.: An MFCC-based text independent speaker identification system for access control (2018)
14. Panda, A.K., Sahoo, A.K.: Study of Speaker Recognition Systems, pp. 1–60 (2011)
15. Praveen Kumar, P.S., ThimmarajaYadava, G., Jayanna, H.S.: Text Independent Speaker Identification: A Review (2018)
16. Sagaya Mary, M.N.J., Katta, S.V., Umesh, S.: S-vectors: Speaker Embeddings based on Transformer's Encoder for Text-Independent Speaker Verification (2020)
17. Sumithra, M.G., Devika, A.K.: À study on feature extraction techniques for text independent speaker identification (2012)

18. Saon, G., Chien, I.T.: Large-vocabulary continuous speech recognition systems: a look at some recent advances (2012)
19. Sarkar, A.K., Sarma, H., Dwivedi, P., Tan, Z.H.: Data augmentation enhanced speaker enrolment for text-dependent speaker verification (2020)
20. Shi, Y., Chen, M., Huang, Q., Hain, T.: T-vector Weakly Supervised Speaker Identification Using Hierarchical Transformer Model (2020)
21. Zhao, X., Wang, D.: Analyzing noise robustness of MFCC and GFCC features in speaker identification (2013)
22. Garhwal, R., Wang, R.: Speaker identification features extraction methods: a systematic review (2017)

Breast Cancer Prognosis Based on Machine Learning Model

Isha[1], Md. Ashraful Babu[2(✉)], Premananda Sahu[1], Srikanta Kumar Mohapatra[3], Prakash Kumar Sarangi[4], and Jayashree Mohanty[5]

[1] School of Computer Science and Engineering, Lovely Professional University (LPU), Phagwara, Punjab, India

[2] Department of Physical Sciences, Independent University, Bangladesh, Dhaka, Bangladesh
ashraful388@gmail.com

[3] Chitkara University Institute of Engineering and Technology, Chitkara University, Rajpura, Punjab, India

[4] Department of Computer Science and Engineering (AI & ML), Vardhaman College of Engineering, Hyderabad, India

[5] Department of CSE-AIT, Chandigarh University, Mohali, Punjab, India

Abstract. Early detection is crucial for prompt and effective treatment of breast cancer, which is a serious health risk. Breast Cancer in particular, frequently held cancer among women fraternity. Breast cancer is a condition where the cells of mammary gland grow excessively and abnormally, forming a mass known as a tumor. Increasing numbers of cancer diagnosis and death due to cancer make it a most important medical problem. Though, in initial stage, the cancer detection process may avoid many deaths and critical cancer related issues. In order to create the prognostic classification techniques, this may be utilized for forecasting results for specific cancer patients. For this purpose, increasing number of Machine Learning techniques has played a vital role. In this work, we present the prognosis of breast cancer based upon machine learning and image processing techniques. The most common method for detecting breast cancer is mammography. Wisconsin Diagnostic Breast Cancer dataset (WDBC) is the widely used cancer dataset for early cancer predictions. Some frequently used techniques with cancer datasets are Naive Bayes, Logistic Regression, Decision Tree, Random Forest, K-Nearest Neighbour, Support Vector Machine along with Extreme Gradient Boosting has utilized for designing the model. Here we have focused machine learning technique, as SVM has secured the accuracy of 99.10%. In this article, we have also studied the outcome of the above techniques and compared their performance. Moreover, our experiment with WDBC shows that SVM outperforms other techniques.

Keywords: Machine Learning · Mammography · WDBC · SVM

1 Introduction

The world health organization estimated that there may be 943000 deaths of women around the world in 2022 [1]. If women are strong, then the world is going to be strong, but if women are getting weak, then there will be some issues have arisen. Women aged

D. Pastor-Escuredo et al. (Eds.): ICDLAIR 2023, LNNS 1001, pp. 524–534, 2024.
https://doi.org/10.1007/978-3-031-60935-0_47

40 to 55 are most likely to get breast cancer, which is the second leading cause of cancer death in women after lung cancer. Breast cancer cells can be malignant (cancerous) or benign (non-cancerous). Early diagnosis of the disease is the most efficient way for reducing numbers of death from cancer in breast. A precise or reliable diagnostic formula enables doctors to distinguish between benign and malignant breast tumors without immediately proceeding with a biopsy. A fine spine objective biopsy entails taking a sample of material or fluid from a nodule, followed by a microscopic examination of the tissues. Medical imaging methods like mammography or ultrasonography can be used to detect and identify the type of cell. These methodologies has considered as a potent techniques that can be used to treat the cancer in an effective manner. Mammogram screening programmers for women and early detection of cancer in breast in its initial stage can be both lessen the adverse effects of required therapies and improved patient's chances of survival. If the services are given with the highest level of quality, these outcomes are feasible [2]. Faster tumor detection and identification not only improves the accuracy of the diagnosis of breast cancer, but also offers a more effective technique to identify symptoms for a more precise diagnosis of illnesses.

In this work, the authors have mainly focused into the breast cancer which has classified as cancerous or non-cancerous at early stage, so that it may be treated then after. As Machine Learning have the potential to identify minor changes of tissues in breast that may indicate to the malignancy prior to the growth of tumor tissue which revealed on a mammography, as currently the situation. By integrating the precision and computational prowess of machine learning with ultrasound imaging, we aim to develop a system that can assist medical professionals in making more accurate and timely diagnoses. Initially the authors have taken Wisconsin Diagnostic Breast Cancer (WDBC) dataset for all the validation work. The Wisconsin Diagnostic Breast Cancer Dataset (WDBC) from Kaggle Repository is used to predict breast cancer. Here Python's programming environment and certain common Python modules are used to process the input and create the output. Given that the goal classes (B and M) for classifying detected tumors have already been established. For categorization, the SVM which is ML technique would be utilized. Therimary benefit of using SVM is its assistance in locating the hyper plane in N-dimensional space, which enables the extraction of N distinct characteristics. SVM aids in categorizing the provided data points into two unique classes, making it simple to distinguish between the affected and unaffected breast regions. It also uses less computer power to produce more precision. It manages the data with such an arrangement that the margins provide boundaries provide separate classes. A comparative study is undertaken to get a conclusion once the various algorithms have successfully finished the categorizing. It has been observed that SVM has given the best accuracy among all the above said techniques. The encouraging results not only advocate for a broader clinical integration of such methodologies but also inspire continued exploration and innovation in medical image analysis.

Rest of the work has organized in the following manner: Sect. 2 has expressed literature work, Sect. 3 has expressed materials and methods, Sect. 4 has expressed as result and discussion, Sect. 5 has expressed as conclusion followed by future work.

2 Literature Exploration

Earlier numerous researchers have worked on the breast cancer by means of machine learning techniques. Some of the approaches have given below:

OJ Egwom et.al [3] has developed a model related to breast cancer by the help of various machine learning techniques. In her work, they have taken the approaches like SVM, KNN, and DT and so on, but she has claimed that SVM has secured higher accuracy among all other methods. MA Naji et.al [4] has diagnosed breast cancer based upon various machine learning techniques. In their article, they have used WDBC dataset for all of the designing work. Here they have used mainly five ML techniques like RF, DT, LR, KNN along with SVM technique for classification of breast cancer as cancerous or non-cancerous. According to them, SVM has secured the accuracy level as 97.2%. Q. Huang et.al [5] has classified breast cancer based upon mining and adaboost techniques. They have also designed a CAD system for breast cancer, but according to them, CAD system has some issues for proper investigating the cancer identification process. Finally they have concluded that AdaBoost technique has achieved the accuracy as 94.2%.

T.A. Assegie [6] identified breast cancer based on KNN approach. In his work, he has applied grid search technique for determining k factor. Regarding the performance of the projected model, accuracy has measured from testing data. His model has secured 94.35% of classification accuracy. V. Nemade and V. Fegade [7] have predicted breast cancer by the help of various machine learning techniques. They have also used WDBC dataset for all validation work. They have also compared with six ML approaches like DT, RF, KNN, LR, XGBOOST and our proposed technique SVM, but according to them XGBOOST has achieved highest accuracy i.e. 97%.

H. Li et.al [8] has forecasted the model associated to breast cancer by Naïve Bayes algorithm. They have taken one factor i.e. Pathologic Complete Response by which they got good response and also they have compared their model with SVM, RF along with KNN. After calculating the performance measuring factors, they got 84.5% as their sensitivity which is not so far from accuracy. S. Chaudhury et.al [9] has forecasted breast cancer detection by exploring Principal Component Analysis as feature extraction process. Here a variety of classifiers can have their hyper parameter settings has adjusted to speed up their forecasting and training process with no sacrificing accuracy. They have also suggested that PCA can be used to quickly and accurately extract the optimal feature combinations to feed into classifiers for high accuracy scores. L. Liu [10] has examined breast cancer process by the help of logistic regression approach. In his work, he has mainly focused into the classification and this has done by the help of 2 properties like maximum perimeter and maximum texture. He has also used WDBC dataset where 32 properties have used for both training and testing purpose. After all the validation process, he has secured 96.5% of classification accuracy. Y. Ono and Y B. Mitani [11] have classified breast cancer based upon feature extraction technique PCA. Their study has worn WDBC dataset to give an assessment for obtaining of feature extraction methods with collaborative learning for breast cancer categorization. J.X. Tian et.al [12] has identified breast cancer based on feature extraction by PCA method and C5.0 decision tree algorithm. They have suggested that previously there have been some issues raised at the time of choosing the dataset which is imbalance and for this purpose the authors have used PCA for the feature extraction process. N. Thakur et.al [13] has explored breast cancer across medical images based upon both machine and deep learning techniques. In their work, they have explored many articles from 2010 to 2021. They have indicated that Support Vector Machine with some variation of Convolutional Neural Network has given immense accuracy. R. Rabiei et.al [14] has forecasted breast cancer based upon machine learning approaches. In their work, they have compared many machine learning approaches. Among all the approaches, Random Forest has secured 80% of accuracy.

From the above literature survey it has been observed that previously SVM has secured less accuracy due to applying of old ML techniques, but here we are going to apply more advance version of SVM along with the feature extraction technique by PCA.

Now all the literature work has depicted in Table 1.

Table 1. Literature Exploration

Reference No	Authors	Year	Method
3	OJ Egwom et.al	2022	SVM
4	M A Naji et.al	2021	SVM
5	Q. Huang et.al	2019	AdaBoost Technique
6	T.A. Assegie	2021	KNN
7	V. Nemade and V. Fegade	2023	XGBoost
8	H. Li et.al	2023	Naïve Bayes
9	S. Chaudhury et.al	2022	PCA as Feature Extraction
10	L. Liu	2018	Logistic Regression
11	Y. Ono and Y B. Mitani	2022	PCA as Feature Extraction
12	J.X. Tian et.al	2022	C5.0, Decision Tree
13	N. Thakur et.al	2023	SVM with CNN
14	R. Rabiei et.al	2022	All the Comparison based ML approaches

3 Materials and Methods

As the authors have clearly mentioned that breast cancer classification model only, so the various classification techniques have been depicted in Fig. 1 in our projected model.

Our proposed model has designed according to the below work flow diagram and the steps have described in the following manner:

- Initially the mammography images have taken from the above said dataset.
- The images are needed to be preprocessed for enhancing the properties of images.
- Then the dataset will be dividing into training and testing dataset as per the requirement.

- Forecasting the data onto the chosen component.
- Then after classification will be going on by comparing various ML methods like RB, DT, KNN, LR and our proposed approach SVM
- Finally the predicted result has found based on benign and malignant.

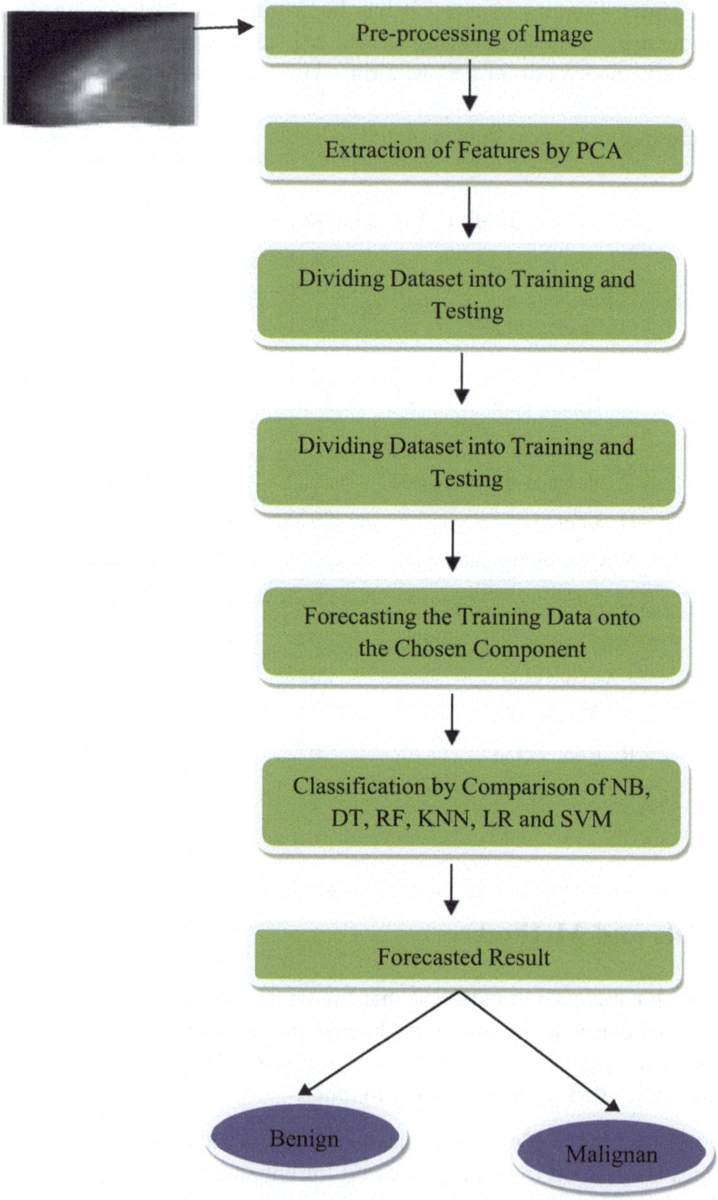

Fig. 1. Control Flow Diagram of Projected Model

3.1 Preprocessing of Image

Preprocessing plays a crucial role in breast cancer detection, as it involves transforming and preparing the data to improve the accuracy and effectiveness of the analysis and in our work, WDBC dataset has used for both the training and testing phase. Moreover, it eradicates all the omitted data which may be found out from the human being; as a result, the accuracy may be increased.

3.2 Feature Extraction

Basically, this is used for reduction of dimensions that means the number of attributes present in the dataset. If a greater number of attributes will present in the dataset, then the designing work will be quite difficult for predicting the final output. For this purpose, we have used feature extraction technique in our work and PCA approach has worn for the same.

The above said approach has utilized as dimension diminution technique where a new set of features has discovered from an existing attribute set [15]. Various machine learning algorithms on this approach are also used in different types of disease predictions [16–21]. The mechanism behind this technique is explained by following steps of execution

- Initially the dataset i.e. WDBC has organized.
- Estimate the covariance matrix for the attributes (m, n) and the formula is:

$$Cov_{m,n} = \frac{1}{z-1}\sum\nolimits_{i=0}^{z}(m_i - m_p)(n_i - n_p) \quad (1)$$

  where I is identity matrix which equals to 0, 1, 2.... z and mi, ni are the value of ith attribute, mp and np are mean value,
- Then after evaluate eigen value and vector for the covariance matrix.
- Finally, the first element is expressed by the most significant eigen vector corresponding to the maximum eigen value, while the next elements are determined in decreasing order of value.

3.3 Naïve Bayes

It is a type of classification technique based upon machine learning approach. It belongs to a family of algorithms that are all based on the idea that very pair of qualities that are being categorized is distinct from one another. Here basically multinomial model attributes have occurred which has determined as:

$$p(M/N) = \frac{P(N/M)P(M)}{P(N)} \quad (2)$$

where p is probability, M and N are the attributes.

3.4 Decision Tree (DT)

A potent machine learning technique called Decision Tree (DT) is employed for both classification and regression. DT may take the form of a tree-like structure, with the terminal nodes serving as the class or the prediction value to be forecasted and the inside nodes serving as test conditions for the vector's continued movement. DT works well when there are just a few class labels to classify, but it fails when there are numerous classes and few training observations. Additionally, the cost of computationally training DTs might be high.

3.5 Logistic Regression (LR)

It is also a type of machine learning approach that has used to predict whether a patient has suffered from a particular disease or not. Basically, it has determined by a given data of independent variable, so that the successful observation has occurred from the probability o and 1. This is a logistic regression model's distinguishing characteristic. However, the predictions must be encoded in order for each occurrence to be directly compared with the labels in the test data in order for us to determine whether the predictions are true.

3.6 Random Forest (RF)

It is also a type of machine learning approach which has used for classification of any type of diseases. It uses ensemble learning, a method that combines a number of classifiers to discover solutions to challenging issues. Numerous decision trees make up a random forest algorithm. Using bagging or bootstrap aggregation, the random forest technique's "forest" is trained. Machine learning systems' accuracy is improved by bagging, an ensemble meta-algorithm.

3.7 K-Nearest Neighbour (KNN)

According to some, the most basic and uncomplicated categorization technique is K-Nearest Neighbour (KNN). Like the majority of machine learning algorithms, K-NN simply uses the training data points to identify the K number of closest neighbours to each data point using Euclidean distance and classifies it into the category with the first K neighbours that are closest to it. It is a non-parametric algorithm that means it does not create any hypothesis on essential data.

3.8 Support Vector Machine (SVM)

It is a supervised based machine learning approach which has also used for the classification purpose. This approach has frequently used for the binary classification. In this approach, each attribute has represented as a precise coordinate and each amount of data which has used for both training and testing as n-dimension space. After that the identification of the hyper plane that mostly divides into two groups results for classification.

In order to get the label of the unknown picture, it compares the new data to the most effective image from the training data which are the properties of kernel function that used in SVM. In most of the cases the data may be partitioned into linearly or non-linearly which produces very accurate results. The SVMs produce a hyperplane that separates when the input data can be linearly separable, thereby maximises the degree of separation among classes as measured along a line perpendicular to the hyperplane.

4 Result and Discussion

In this section, the complete implementation has covered. Initially, the dataset we have used here is WDBC where 569 rows and 32 attributes have taken which has taken from UCI machine learning repository. The implementation has carried out by python programming language along with NumPy and pandas library function and google co-lab for coding description [22].

Here confusion matrix has also used to determine the comparison of actual and predicted values. As confusion matrix has determined in terms of heat map, so a heatmap has values for each value to be plotted that indicate varying tints of the same colour. In this case Correlation shows how closely connected changes in the two characteristics in the dataset are to one another. Positive correlation exists between two features that change in the same direction, whereas negative correlation exists between traits that change in the opposite direction. It displays the traits that are highly correlated with one another. This is important to understand because certain machine learning algorithms may perform poorly if the data contains input qualities that are strongly linked. Now the heatmap has depicted in Fig. 2.

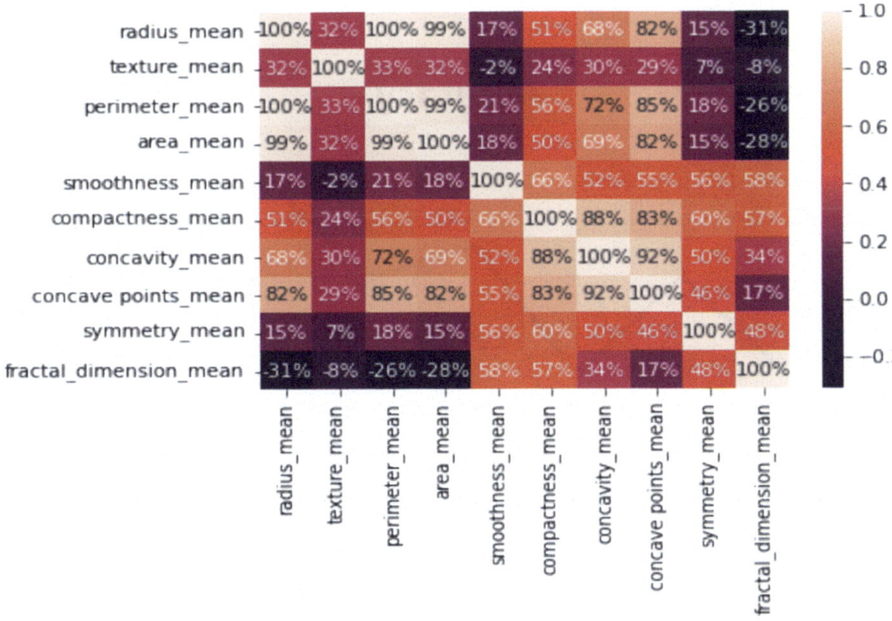

Fig. 2. Heatmap Depiction Graph

After determination of the confusion matrix successfully, now the comparison of all the classification models in terms of B and M class has expressed in Table 2.

Table 2. Comparison of All Classification Models

Algorithms	Class	Precision	Recall	F1-score	Support	Accuracy
NB	B	0.94	0.96	0.95	67	93.85
	M	0.93	0.91	0.92	47	
LR	B	0.96	0.99	0.97	67	96.49
	M	0.98	0.94	0.96	47	
DT	B	0.94	0.96	0.95	67	93.85
	M	0.93	0.91	0.92	47	
RF	B	0.96	1.00	0.98	67	97.36
	M	1.00	0.94	0.97	47	
KNN	B	0.94	1.00	0.97	67	96.49
	M	1.00	0.91	0.96	47	
SVM	B	0.99	1.00	0.99	67	99.10
	M	1.00	0.96	0.98	47	

The classification report graph for various classification techniques has been depicted in Fig. 3.

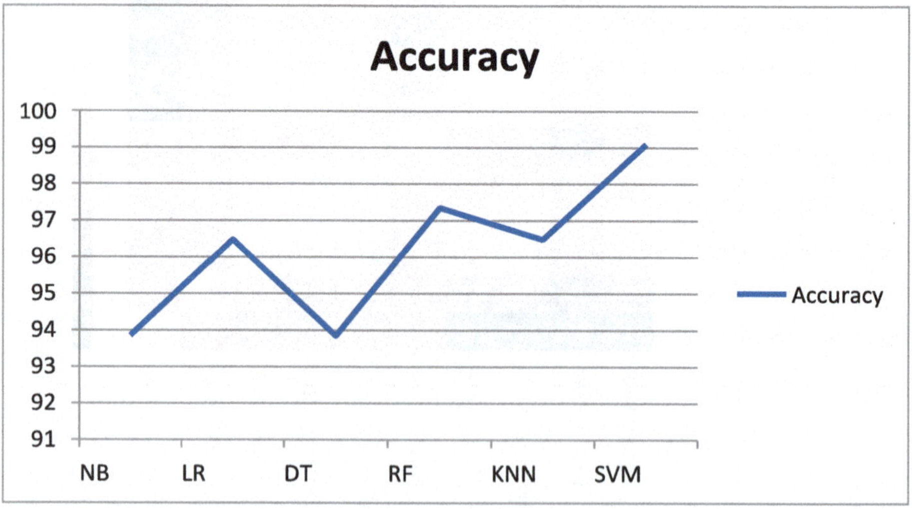

Fig. 3. Comparison Graph of All Classification Model

Finally, after all the investigation work, the comparison of all the techniques along with our proposed model has been depicted in Table 3.

Table 3. Comparison of All Techniques

S No	Reference	Methodology	Accuracy (%)
1	[5]	AdaBoost with CAD system	94.2
2	[6]	KNN	94.35
4	[7]	XGBoost	97
5	[8]	Naïve Bayes	84.5
6	[10]	LR	96.5
7	[14]	Random Forest	80
8	**Our Proposed Model**	**SVM**	**99.10**

5 Conclusion

In this research, we employed machine learning techniques to develop a model capable of distinguishing between benign and malignant with remarkable accuracy. The results, as evidenced by the Simplified Classification Report, are a testament to the model's robustness and precision. Initially, the dataset has taken by WDBC for the validation purpose. PCA has used for removal of old features with some new properties has added. All the classification technique's accuracy has been compared in terms of B and M class with each other where in the proposed model for B class, the precision, recall, F1 score and support are 0.99, 1.00, 0.99 and 67 respectively and for M class, the precision, recall, F1 score and support are 1.00, 0.96, 0.98 and 47 respectively. It has been noticed that SVM has secured highest accuracy among of all i.e. 99.10% where the model demonstrates its potential as a reliable tool in medical diagnostics.

In future, if the dataset size will increase that means if some more parameters related to breast cancer like histologic size, subtype and the status of lymph node can be added for the validation work and if some hybrid techniques that means if some ensemble methods of machine and deep learning techniques will be implemented to our proposed model, then the accuracy may be improved.

References

1. https://www.who.int/news-room/fact-sheets/detail/maternal-mortality
2. Mehmood, M., et al.: Machine learning enabled early detection of breast cancer by structural analysis of mammograms. Comput. Mater. Contin. **67**, 641–657 (2021)
3. Egwom, O.J., Hassan, M., Tanimu, J.J., Hamada, M., Ogar, O.M.: An LDA–SVM machine learning model for breast cancer classification. BioMedInformatics **2**(3), 345–358 (2022)
4. Naji, M.A., El Filali, S., Aarika, K., Benlahmar, E.H., Abdelouhahid, R.A., Debauche, O.: Machine learning algorithms for breast cancer prediction and diagnosis. Procedia Comput. Sci. **191**, 487–492 (2021)

5. Huang, Q., Chen, Y., Liu, L., Tao, D., Li, X.: On combining biclustering mining and AdaBoost for breast tumor classification. IEEE Trans. Knowl. Data Eng. **32**(4), 728–738 (2019)
6. Assegie, T.A.: An optimized K-Nearest Neighbor based breast cancer detection. J. Robot. Control (JRC) **2**(3), 115–118 (2021)
7. Nemade, V., Fegade, V.: Machine learning techniques for breast cancer prediction. Procedia Comput. Sci. **218**, 1314–1320 (2023)
8. Li, H., et al.: NetPSig: computationally guided neutrophil extracellular trap-associated signature significantly predicts the prognosis and therapeutic response in breast cancer (2023)
9. Chaudhury, S., Shelke, N., Rashid, Z.M., Sau, K.: Effect of grid search and hyper parameter tuned pipeline with various classifiers and PCA for breast cancer detection. Curr. Signal Transduct. Ther. **17**(3), 45–56 (2022)
10. Liu, L.: Research on logistic regression algorithm of breast cancer diagnoses data by machine learning. In: 2018 International Conference on Robots & Intelligent System (ICRIS), pp. 157–160. IEEE (2018)
11. Ono, Y., Mitani, Y.: Evaluation of feature extraction methods with ensemble learning for breast cancer classification. In: 2022 IEEE 4th Global Conference on Life Sciences and Technologies (LifeTech), pp. 194–195. IEEE (2022)
12. Tian, J., Zhang, J.: Breast cancer diagnosis using feature extraction and boosted C5.0 decision tree algorithm with penalty factor. Math. Biosci. Eng. **19**(3), 2193–2205 (2022). https://doi.org/10.3934/mbe.2022102
13. Thakur, N., Kumar, P., Kumar, A.: A systematic review of machine and deep learning techniques for the identification and classification of breast cancer through medical image modalities. Multimed. Tools Appl. 1–94 (2023)
14. Rabiei, R., Ayyoubzadeh, S.M., Sohrabei, S., Esmaeili, M., Atashi, A.: Prediction of breast cancer using machine learning approaches. J. Biomed. Phys. Eng. **12**(3), 297 (2022)
15. Behar, N., Shrivastava, M.: Pixel-level feature extraction model for breast cancer detection. Comput. Mater. Continua **74**(2) (2023)
16. Garg, R., Garg, H., Patel, H., Ananthakrishnan, G., Sharma, S.: Role of machine learning in detection and classification of leukemia: a comparative analysis. In: Solanki, A., Naved, M. (eds.) GANs for Data Augmentation in Healthcare, pp. 1–20. Springer, Cham (2023). https://doi.org/10.1007/978-3-031-43205-7_1
17. Sharma, O.: Prediction and analysis of heart attack using various machine learning algorithms. In: 2023 International Conference on Artificial Intelligence and Smart Communication (AISC), pp. 786–790. IEEE (2023)
18. Mohapatra, S.K., Jain, A.: Predictive analysis of stroke prediction by using machine learning implementations. In: 2023 IEEE International Conference on Contemporary Computing and Communications (InC4), vol. 1, pp. 1–6. IEEE (2023)
19. Jaroli, P., Singla, C., Bhardwaj, V., Mohapatra, S.K.: Deep learning model based novel semantic analysis. In: 2022 2nd International Conference on Advance Computing and Innovative Technologies in Engineering (ICACITE), pp. 1454–1458. IEEE (2022)
20. Mohapatra, S.K., Jain, A., Sahu, P.: Comparative approaches by using machine learning algorithms in breast cancer prediction. In: 2022 2nd International Conference on Advance Computing and Innovative Technologies in Engineering (ICACITE), pp. 1874–1878. IEEE (2022)
21. Mohapatra, S.K., et al.: Segmentation and classification of encephalon tumor by applying improved fast and robust FCM algorithm with PSO-based Elm Technique. Comput. Intell. Neurosci.. Intell. Neurosci. **2022**, 1–9 (2022). https://doi.org/10.1155/2022/2664901
22. Sahu, P., Mohapatra, S.K., Sarangi, P.K., Srivastava, S., Sharma, S.K.: Detection of diabetic retinopathy based on various machine learning algorithms and histogram equalization. In: 2022 International Conference on Machine Learning, Computer Systems and Security (MLCSS), pp. 6–10. IEEE (2022)

PRESENT Block Cipher Defined over Galois-Field (2^8)

Monika Jangra[1], Sudesh Rani[2(✉)], and Buddha Singh[3]

[1] Shri Vishwakarma Skill University, Palwal, Haryana, India
[2] Punjab Engineering College (Deemed to be University), Chandigarh, India
sudeshrani@pec.edu.in
[3] Jawaharlal Nehru University, New Delhi, India

Abstract. PRESENT block cipher, well known for its hardware efficiency among tiny computing environments, safeguards data from adversaries and various cyber attacks. To support lightweight implementation, it utilizes the 4-bit S-box, making it unsuitable for 8-bit word architecture based embedded devices. As per NSA reports, it is evident that such reductions in size reduces the security levels. In this paper, a new variant of PRESENT is proposed that optimizes the basic structure of PRESENT with 4-bit S-box to 8-bit S-box and is based on the Galois-Field (2^8). This new variant of PRESENT provides the highest security strength among existing PRESENT in Galois-Field (2^4) and other contemporary approaches in Galois-Field (2^8), evaluated through Avalanche of Change. The cipher text it generates demonstrates optimal randomness, validated by NIST testing suite. Further, the proposed variant minimizes memory requirements, standing out among Galois-Field (2^8) block ciphers.

Keywords: Cryptography · Encryption · PRESENT · S-box

1 Introduction

With the establishment of very strong ciphers like Advanced Encryption Standard, Elliptic Curve Cryptography, Hash function based ciphers etc. [1–3], the need of more ciphers is greatly diminished as these are applicable to many network-based applications. However, the resource requirement of such methods restricts them to be utilized in constrained environments such as Internet of Things (IoT) [4,5]. Understanding the need of lightweight cryptographic solutions PRESENT [6] was developed. PRESENT is an ultra lightweight block cipher in terms of hardware efficiency and competitive with today's lightweight ciphers. It has been adopted by International Organization for Standardization (ISO) for its ultra-lightweight implementation along with other two lightweight ciphers [7]. The architecture of PRESENT block cipher is efficient in terms of hardware implementation; since it occupies minimal Gate Equivalents (GEs) of around 1500. As stated by the study of National Institute of Standards and

D. Pastor-Escuredo et al. (Eds.): ICDLAIR 2023, LNNS 1001, pp. 535–543, 2024.
https://doi.org/10.1007/978-3-031-60935-0_48

Technology (NIST) [8], there exists a trade-off between size of substitution box (S-box), key-size or block-size etc. and security levels. Trade-off states that reduction in sizes limits the computational requirements but lowers the security levels. The PRESENT uses a 4-bit S-box which saves the space but at same time it compromises with security. For 4-bit S-box, more number of memory accesses is needed. For example, to encrypt a 64-bit block, the non-linear component S-box is accessed 16 times, whereas 8-bit S-box will take only 8 memory accesses. Moreover, there are multiple embedded devices that work efficiently with 8-bit cryptographic architecture.

In this paper, a PRESENT block cipher is redefined over the Galois Field, Galois-Field (2^8). The key-schedules and other operations in the round functions are redefined as per requirement. The objective is to propose a new variant of PRESENT over 8-bit field, with an improved security with least compromise in resource requirements. The 8-bit architecture of PRESENT provides an alternative and more secure choice for the stakeholder to opt the block ciphers as per organizational needs.

The paper organization is as follows: Sect. 2 is an insight to the related work; Sect. 3 is the redefinition of the PRESENT cipher over Galois-Field(2^8); Sect. 4 is the statistical testing of the proposed PRESENT cipher to ensure randomness in the $cipher - text$ generated; Sect. 5 is the comparative analysis among proposed PRESENT and the 4-bit PRESENT and some other contemporary 8-bit S-box based cryptographic mechanisms. Later Sect. 6 presents the conclusion of results and future scope of the research.

2 Related Work

The efforts of cryptographers have nurtured this area with many mechanisms that are well suited for low-resource environments. PRESENT [6], CLEFIA [9], HIGHT [10], PICCOLO [11], ZORRO [12] are some examples, however, they support a limited range of platforms. PRESENT and CLEFIA are the lightweight encryption schemes particularly used for resource-constraint environments such as IoT. PRESENT encrypts data in a block size of 64-bit and utilizes 1570 GE only. It has substitution-permutation based crypto-architecture. The different security levels are possible with two key sizes i.e. 80-bit and 128-bit. The substitution layer substitutes data via non-linear 4-bit S-box and permutation layer uses linear bit-wise permutation respectively in 31 rounds. The reduction in block and S-box sizes manage the memory limitation of low power devices. The computational details of S-box and permutations can be referred to in [6]. The S-box used in PRESENT is given in Table 1. The design of PRESENT in $GF(2^8)$ was presumed to work for smaller amount of data.

Table 1. S-box $F_2^4 \rightarrow F_2^4$

a:	0	1	2	3	4	5	6	7	8	9	A	B	C	D	E	F
S-box[a]:	C	5	6	B	9	0	A	D	3	E	F	8	4	7	1	2

CLEFIA [9,13] is a 128-block cipher that is certified by ISO as lightweight implementation along with PRESENT. It provides the security of different levels with key sizes varying from 128-bit to 256-bit. CLEFIA is a fiestal structure based cryptosystem. The generalized fiestal structure has 4 data lines and two 32-bit F-functions with Diffusion Switch Mechanism (DSM). CLEFIA uses two different S-boxes. Each S-box has different algebraic properties that embodies CLEFIA with extra algebraic immunity.

SIMON and SPECK [14] are the family of lightweight block ciphers that can be applied across an assortment of different hardware and software architecture. These were developed by the US National Security Agency to provide encryption facility for low-power applications. These were accepted by NIST Lightweight Cryptography Workshop, July 2015. SIMON and SPECK mechanisms are the collection of simple bitwise operations. The simple operations used within SIMON and SPECK are Modular Addition and Subtraction, Bitwise XOR, Bitwise AND, Left Circular Shift, and Right Circular Shift. These simple operations have an efficient bias towards both software and hardware implementations.

3 Proposed Method

PRESENT is an adopted lightweight block cipher by ISO. It has inherited the structure from Data Encryption Standard (DES) and features of Advanced Encryption Standard (AES) finalist candidate SERPENT. It is the combination of strong features itself; therefore, instead of developing a new encryption method for tiny computing devices, it has been decided to optimize the existing PERSENT structure for security efficiency with least compromise in resource requirement. To achieve the objectives, the 4-bit S-box is converted into 8-bit S-box and the round functions are re-implemented accordingly. Similar to PRESENT the proposed variant of PRESENT over Galois-Field (2^8) supports security levels of 80-bit and 128-bit with 31 rounds and 64-bit block size. The round functions of PRESENT are depicted in Fig. 1.

3.1 S-Box Layer

In the proposed method, 4-bit structure of PRESENT's S-box is converted to 8-bit S-box. To achieve this, the 8-bit word is broken into two nibbles lower word and upper word, then, respective 4-bit S-box translations is augmented to generate a final 8-bit S-box word. The 8-bit S-box ($S : F_2^8 \rightarrow F_2^8$) in hexadecimal notation is given in Table 2.

In the optimized S-box, the current state is taken as a block of 64 bits, notated as $b_{63}b_{62}b_{61}b_{60} \ldots\ldots\ldots b_1 b_0$. These 64 bits are the combination of eight 8 bits words, written as $w_8 w_7 \ldots\ldots\ldots w_0$, where,$w_i$ is defined in Eq. 1.

$$w_i = b_{8*i+7} \parallel b_{8*i+6} \parallel b_{8*i+5} \parallel b_{8*i+4} \parallel b_{8*i+3} \parallel b_{8*i+2} \parallel b_{8*i+1} \parallel b_{8*i+0} \quad (1)$$

The 8-bit word w_i is translated into $S(w_i)$ using Table 2.

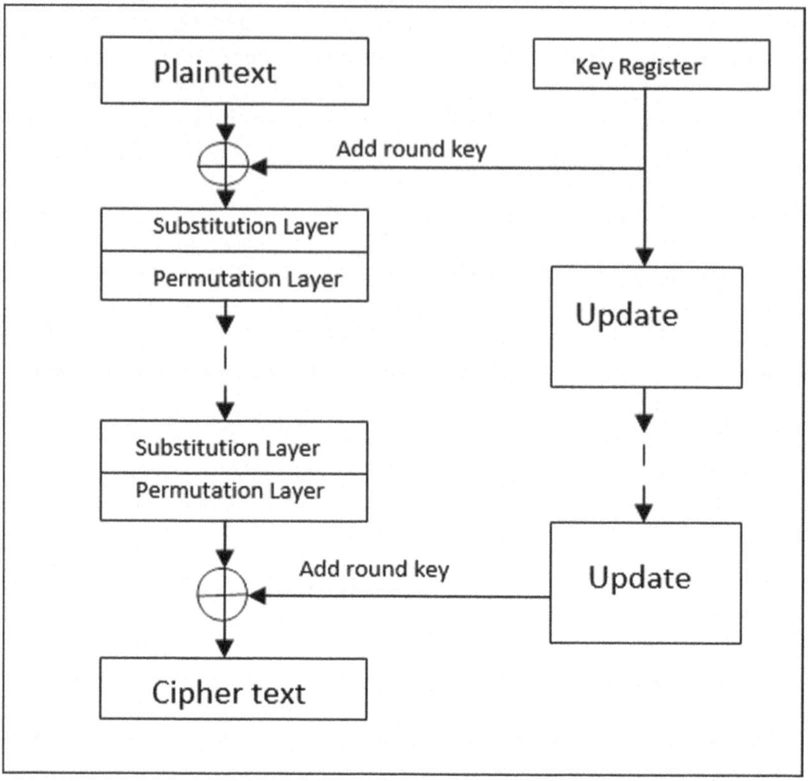

Fig. 1. PRESENT Block Cipher [6]

3.2 Permutation Layer

Bit permutations are considered faster for embedded devices. The p-layer of PRESENT works bitwise and the bit permutation used by the round function is independent of the S-box, therefore, p-layer is utilized in the optimized version is same as used in standard version. The Table 3 defines the p-layer.

3.3 Key Schedule

The key Schedule generates the round keys for each round. The key schedule is optimized according to 8-bit S-box. Say Key k $(k_{79}k_{78}\ldots\ldots\ldots k_0)$ is supplied by user and stored in register K. The round key generation is explained in following steps.

Step I. Key Extraction. At round i, The 64 most significant bits of 80-bit data in register K is taken as round key r_i represented in Eq. 2.

$$r_i = [k_{79}k_{78}\ldots\ldots\ldots k_{16}] \tag{2}$$

Table 2. The 8-bit S-box ($S : F_2^8 \rightarrow F_2^8$) for optimized PRESENT S-box Layer

	00	01	02	03	04	05	06	07	08	09	0a	0b	0c	0d	0e	0f
00	0xcc	0xc5	0xc6	0xcb	0xc9	0xc0	0xca	0xcd	0xc3	0xce	0xcf	0xc8	0xc4	0xc7	0xc1	0xc2
10	0x5c	0x55	0x56	0x5b	0x59	0x50	0x5a	0x5d	0x53	0x5e	0x5f	0x58	0x54	0x57	0x51	0x52
20	0x6c	0x65	0x66	0x6b	0x69	0x60	0x6a	0x6d	0x63	0x6e	0x6f	0x68	0x64	0x67	0x61	0x62
30	0xbc	0xb5	0xb6	0xbb	0xb9	0xb0	0xba	0xbd	0xb3	0xbe	0xbf	0xb8	0xb4	0xb7	0xb1	0xb2
40	0x9c	0x95	0x96	0x9b	0x99	0x90	0x9a	0x9d	0x93	0x9e	0x9f	0x98	0x94	0x97	0x91	0x92
50	0x0c	0x05	0x06	0x0b	0x09	0x00	0x0a	0x0d	0x03	0x0e	0x0f	0x08	0x04	0x07	0x01	0x02
60	0xac	0xa5	0xa6	0xab	0xa9	0xa0	0xaa	0xad	0xa3	0xae	0xaf	0xa8	0xa4	0xa7	0xa1	0xa2
70	0xdc	0xd5	0xd6	0xdb	0xd9	0xd0	0xda	0xdd	0xd3	0xde	0xdf	0xd8	0xd4	0xd7	0xd1	0xd2
80	0x3c	0x35	0x36	0x3b	0x39	0x30	0x3a	0x3d	0x33	0x3e	0x3f	0x38	0x34	0x37	0x31	0x32
90	0xec	0xe5	0xe6	0xeb	0xe9	0xe0	0xea	0xed	0xe3	0xee	0xef	0xe8	0xe4	0xe7	0xe1	0xe2
a0	0xfc	0xf5	0xf6	0xfb	0xf9	0xf0	0xfa	0xfd	0xf3	0xfe	0xff	0xf8	0xf4	0xf7	0xf1	0xf2
b0	0x8c	0x85	0x86	0x8b	0x89	0x80	0x8a	0x8d	0x83	0x8e	0x8f	0x88	0x84	0x87	0x81	0x82
c0	0x4c	0x45	0x46	0x4b	0x49	0x40	0x4a	0x4d	0x43	0x4e	0x4f	0x48	0x44	0x47	0x41	0x42
d0	0x7c	0x75	0x76	0x7b	0x79	0x70	0x7a	0x7d	0x73	0x7e	0x7f	0x78	0x74	0x77	0x71	0x72
e0	0x1c	0x15	0x16	0x1b	0x19	0x10	0x1a	0x1d	0x13	0x1e	0x1f	0x18	0x14	0x17	0x11	0x12
f0	0x2c	0x25	0x26	0x2b	0x29	0x20	0x2a	0x2d	0x23	0x2e	0x2f	0x28	0x24	0x27	0x21	0x22

The column is determined by the lower 4 bits, and the row by the upper 4 bits.
For example, the value $9a_{16}$ is converted into ef_{16}.

Step II. Shifting. This is step begins the updating of key register K. It rotates the contents of register K by 61 bit positions to the left as given in Eq. 3.

$$[k_{79}k_{78}.........k_0] = [k_{18}k_{17}.........k_{20}k_{19}] \tag{3}$$

Step III. S-Box. It replaces the 8 most significant bits of the Key Register to with the corresponding S-box data. This step adds the confusion to the round keys and provides a greater relative change in the subsequent round keys. Therefore, it makes related round key attacks harder to implement. The Eq. 4 utilizes the entries in Table 1.

$$[k_{79}k_{78}k_{77}k_{76}k_{75}k_{74}k_{73}k_{72}] = S[k_{79}k_{78}k_{77}k_{76}k_{75}k_{74}k_{73}k_{72}] \tag{4}$$

Table 3. Permutation Box: bit at index b in 64-bit data block is kept at index pbox[b]

b:	0	1	2	3	4	5	6	7	8	9	10	11	12	13	14	15
pbox[b]:	0	16	32	48	1	17	33	49	2	18	34	50	3	19	35	51
b:	16	17	18	19	20	21	22	23	24	25	26	27	28	29	30	31
pbox[b]:	4	20	36	52	5	21	37	53	6	22	38	54	7	23	39	55
b:	32	33	34	35	36	37	38	39	40	41	42	43	44	45	46	47
pbox[b]:	8	24	40	56	9	25	41	57	10	26	42	58	11	27	43	59
b:	48	49	50	51	52	53	54	55	56	57	58	59	60	61	62	63
pbox[b]:	12	28	44	60	13	29	45	61	14	30	46	62	15	31	47	63

Step IV. Salting. Salting step ex-ores the defined consecutive 5 bits in the key register K with round counter i as given in Eq. 5.

$$[k_{19}k_{18}k_{17}k_{16}k_{15}] = [k_{19}k_{18}k_{17}k_{16}k_{15}] \oplus i \tag{5}$$

4 Statistical Testing

The objective of generating cipher text is to protect the information from an attacker. The degree of randomness added to the $plain - text$ defines the higher confusion towards an attacker. There are many methods to testify the randomness of the generated cipher text. The NIST Testing Suite [15] is a collection of such methods. Table 4 represents the results of Statistical Testing passed by proposed variant of PRESENT.

Table 4. Randomness Testing

Test	p-value	Result
Frequency Test	0.14	Random
Frequency Test within a Block	0.78	Random
Run Test	0.46	Random
Longest run of ones in a block	0.44	Random
Discrete Fourier Transform	0.03	Random
Template Matching	0.37	Random
Linear Complexity Test	0.98	Random
Serial Test	0.39	Random
Approximate Entropy Test	0.99	Random
Cumulative Sums (Foreword) Test	0.15	Random
Cumulative Sums (Reverse) Test	0.20	Random
Random Excursions Test	0.78	Random
Random Excursions Variant Test	0.66	Random

5 Performance Analysis

The paper proposes a variant of PRESENT over Galois-Field (2^8). The main objective of the work is to provide a variant that can suit the architectures that work efficiently on 8-bit fields and to provide a variant that can provide higher level of security with least compromise in resource utilization. There are multiple ways to define crypt analyze the block ciphers [15–17]. The proposed method has been evaluated for its security and memory requirement.

5.1 Security

The security of new variant of PRESENT, existing PRESENT and other contemporary 8-bit mechanisms have been computed using Avalanche of Change metric. It gives the degree of change in the $cipher - text$ if the corresponding $plain - text$ is changed by single bit only. The metric is computed per block size i.e. 64 bit. The SIMON and SPECK has family of block ciphers; for comparative analysis, the versions with 64 bit block size are taken in account. For CLEFIA, the block size is 128 bit that encrypts 64 bit of $plain - text$ for the analysis. The security comparison is shown in Fig. 2. It is observed that proposed variant offers the highest security among all; since it preserves all the confusion properties from the 4-bit architecture and enhances the non-linearity by augmenting the outcomes from 4-bit S-box to form 8-bit S-box.

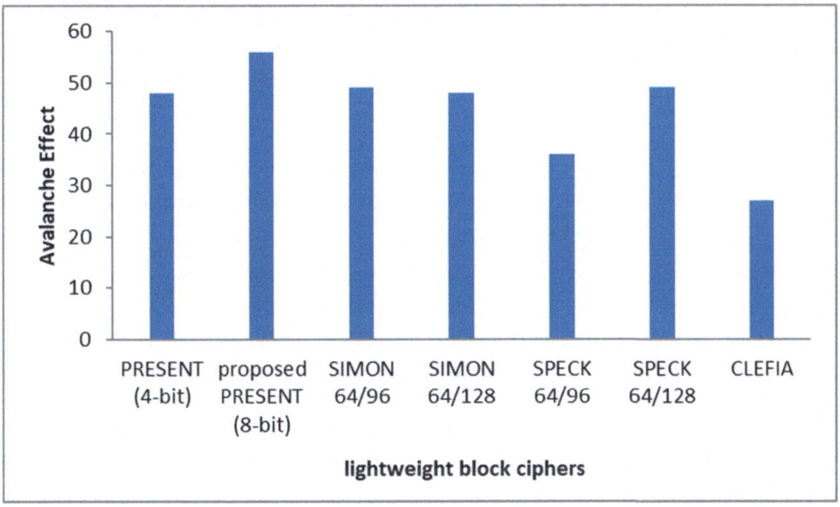

Fig. 2. Security Comparison of Proposed and contemporary Block Ciphers

5.2 Resource Requirement

The memory footprint of different cryptographic schemes is given in Table 5. It can be observed that resource requirement of PRESENT is smallest and the proposed variant of PRESENT costs a little extra memory requirement, however, it is still smallest among other contemporary block ciphers.

Table 5. Memory footprints of different block ciphers

PRESENT (4-bit)	Proposed PRESENT	SIMON 64/96	SIMON 64/128	SPECK 64/96	SPECK 64/128	CLEFIA
7 kb	8 kb	14 kb	14 kb	11 kb	11 kb	19 kb

6 Conclusion

Various cryptographic schemes have been developed so far that safeguard data in communication from various cyber attacks. One such popular scheme is PRESENT which is Substitution-Permutation-Network based block cipher that utilizes the 4-bit S-box. In the proposed approach, the 4-bit S-box is converted to 8-bit S-box and the complete architecture of 4-bit present is defined over Galois-Field (2^8). The augmentation of 4-bit S-box not only preserves the non-linearity and randomness but enhances it. The security offered by proposed PRESENT is highest among its contemporary cryptographic mechanisms. The memory requirement is slightly higher than 4-bit PRESENT but smallest among different 8-bit block ciphers. Therefore, proposed PRESENT is best choice among other 8-bit contemporary block ciphers. Moreover, it offers a flexible choice for the security experts to utilize it for embedded devices with 8-bit architecture.

The proposed PRESENT variant can be simulated with other dynamic and lightweight substitution-layer implementations. The simulation on different platforms can give better insight to resource requirements and other computational costs.

Acknowledgment. We are highly grateful to the *MeitY, Govt. of India*, for granting the fund for the research work under the Visveswarya Fellowship for Electronics and IT scheme.

References

1. Stallings, W.: Cryptography and Network Security, 4/E. Pearson Education India (2006)
2. Matsui, M.: Linear cryptanalysis method for DES cipher. In: Helleseth, T. (ed.) EUROCRYPT 1993. LNCS, vol. 765, pp. 386–397. Springer, Heidelberg (1994). https://doi.org/10.1007/3-540-48285-7_33
3. Daemen, J., Rijmen, V.: AES Proposal: Rijndael (1999)
4. Bertino, E., Choo, K.-K.R., Georgakopolous, D., Nepal, S.: Internet of Things (IoT) smart and secure service delivery (2016)
5. Sha, K., Wei, W., Yang, T.A., Wang, Z., Shi, W.: On security challenges and open issues in Internet of Things. Future Gener. Comput. Syst. **83**, 326–337 (2018)
6. Bogdanov, A., et al.: PRESENT: an ultra-lightweight block cipher. In: Paillier, P., Verbauwhede, I. (eds.) CHES 2007. LNCS, vol. 4727, pp. 450–466. Springer, Heidelberg (2007). https://doi.org/10.1007/978-3-540-74735-2_31
7. ISO/IEC 29192-2:2019 - Information security - Lightweight cryptography - Part 2: Block ciphers. https://www.iso.org/standard/78477.html. Accessed 2023
8. McKay, K., Bassham, L., Sönmez Turan, M., Mouha, N.: Report on lightweight cryptography (No. NIST Internal or Interagency Report (NISTIR) 8114 (Draft)). National Institute of Standards and Technology (2016)
9. Shirai, T., Shibutani, K., Akishita, T., Moriai, S., Iwata, T.: The 128-bit blockcipher clefia (extended abstract). In: Biryukov, A. (ed.) FSE 2007. LNCS, vol. 4593, pp. 181–195. Springer, Heidelberg (2007). https://doi.org/10.1007/978-3-540-74619-5_12

10. Hong, D., et al.: HIGHT: a new block cipher suitable for low-resource device. In: Goubin, L., Matsui, M. (eds.) CHES 2006. LNCS, vol. 4249, pp. 46–59. Springer, Heidelberg (2006). https://doi.org/10.1007/11894063_4
11. Shibutani, K., Isobe, T., Hiwatari, H., Mitsuda, A., Akishita, T., Shirai, T.: Piccolo: an ultra-lightweight blockcipher. In: Preneel, B., Takagi, T. (eds.) CHES 2011. LNCS, vol. 6917, pp. 342–357. Springer, Heidelberg (2011). https://doi.org/10.1007/978-3-642-23951-9_23
12. Gérard, B., Grosso, V., Naya-Plasencia, M., Standaert, F.X.: Block ciphers that are easier to mask: how far can we go? In: Bertoni, G., Coron, J.S. (eds.) CHES 2013. LNCS, vol. 8086, pp. 383–399. Springer, Heidelberg (2013). https://doi.org/10.1007/978-3-642-40349-1_22
13. Akishita, T., Hiwatari, H.: Very compact hardware implementations of the blockcipher CLEFIA. In: Miri, A., Vaudenay, S. (eds.) SAC 2011. LNCS, vol. 7118, pp. 278–292. Springer, Heidelberg (2012). https://doi.org/10.1007/978-3-642-28496-0_17
14. Beaulieu, R., Shors, D., Smith, J., Treatman-Clark, S., Weeks, B., Wingers, L.: The SIMON and SPECK lightweight block ciphers. In: Proceedings of the 52nd Annual Design Automation Conference, pp. 1–6 (2015)
15. Rukhin, A., et al.: A statistical test suite for random and pseudorandom number generators for cryptographic applications, vol. 22. US Department of Commerce, Technology Administration, National Institute of Standards and Technology Gaithersburg, MD (2001)
16. Ding, L., Jin, C., Guan, J., Wang, Q.: Cryptanalysis of lightweight WG-8 stream cipher. IEEE Trans. Inf. Forensics Secur. **9**(4), 645–652 (2014)
17. Jangra, M., Singh, B.: Performance analysis of CLEFIA and PRESENT lightweight block ciphers. J. Discret. Math. Sci. Cryptogr. **22**(8), 1489–1499 (2019)

Mammograms Image Quality Enhancement Using Center Adaptive Median Filter (CEAMF) for Noise and Artifact Removal

Neha Thakur(✉), Pardeep Kumar, and Amit Kumar

Department of Computer Science and Engineering, Jaypee University of Information Technology, Solan, HP, India
numt206@gmail.com

Abstract. Noise and artifact removal is a crucial phase of image preprocessing. Digital mammogram images contain many noises (i.e., salt and pepper, speckles) and artifacts (i.e., opacity, markers, chest wall, date, and background). Artifacts, illumination, and fewer pixel resolutions lead to poor-quality images, reducing segmentation and classification accuracy. This work proposes a center adaptive median filter (CEAMF) for noise and artifact removal to improve image quality. The proposed CEAMF filter performs pixel-wise traversal to remove the noise and artifacts. The proposed method is evaluated on collected and DDSM datasets using the peak signal-to-noise ratio (PSNR) and mean square error (MSE) as parameters and compared with median, mean and Gaussian filters. It is found that the MSE value for the mammogram images is reduced using the proposed method compared to other filters. This reduction in MSE leads to an increase in PSNR and improves the image quality of mammograms.

Keywords: Breast cancer · Image preprocessing · Digital mammograms · Noise and artifacts · Center-adaptive median filter

1 Introduction

In recent years, breast cancer has consistently ranked among the top contributors to elevated mortality rates, as the International Cancer Association (ICA) reported [1]. Early detection of breast cancer has been pivotal in reducing mortality and improving recovery rates. In recent times, digital mammogram images, known for their high-resolution and detailed depiction of internal breast anatomy, have played a crucial role in identifying early signs of breast cancer [2–4]. An inaccurate interpretation of a mammogram can lead to a higher rate of false diagnoses, as distinguishing cancerous masses from surrounding tissue is frequently challenging and prone to errors [5]. Various types of noise, such as Gaussian and salt-and-pepper noise, can impact mammograms, causing inaccurate classifications.

Additionally, mammograms frequently exhibit numerous artifacts that detrimentally impact the detection of breast cancer [6]. The presence of noise and artifacts in mammograms always affects the detection rate. Hence, the researchers performed preprocessing,

D. Pastor-Escuredo et al. (Eds.): ICDLAIR 2023, LNNS 1001, pp. 544–554, 2024.
https://doi.org/10.1007/978-3-031-60935-0_49

which includes noise and artifact removal, normalization and contrast enhancement etc., before the initialization of segmentation and classification. Previous studies utilized various filters during de-noising, such as the median, wiener, wavelet, and more [7, 8].

1.1 Motivation and Research Gaps

Breast cancer is a significant global health concern that contributes substantially to mortality rates. Various noises, including salt-and-pepper and speckles, alongside artifacts like opacity, markers, chest wall, date, and background, collectively contribute to poor-quality images. Despite the advancements in early detection, digital mammography faces persistent challenges in interpretability. Some of them are addressed as follows;

- Existing preprocessing techniques for digital mammogram images lack comprehensive strategies to handle the diverse range of noises and artifacts.
- There is a lack of comprehensive noise and artifact removal methods that address the unique characteristics of digital mammograms.
- Issues like variable illumination and limited pixel resolutions emphasize the need for innovative approaches to image preprocessing.

This research is motivated by the essential need to enhance the preprocessing methodologies, mainly focusing on the crucial noise and artifact removal phase. The proposed CEAMF offers a systematic means to eliminate noise and artifacts, through its pixel-wise traversal mechanism. The main aim of this research is to improve the dependability of digital mammography by precisely removing noise and artifacts. This work may align with the global initiative to advance earlier and more accurate diagnoses of breast cancer.

2 Related Work

Kaur et al. [9] introduced a novel method for breast cancer detection by analyzing mammogram images. The initial phase involved employing mean and median filters to effectively diminish image noise. Their simulation outcomes demonstrated that their proposed approach outperformed existing methods regarding accuracy, sensitivity, specificity, and the receiver operating characteristics curve (ROC). Toz et al. [10] introduced an automated segmentation technique to identify potentially problematic areas within breast images. The initial step involved the removal of noise and artifacts from the images by applying a median filter and wavelet transform. Subsequently, contrast enhancement was performed using the contrast-limited adaptive histogram equalization (CLAHE) algorithm. They proposed an adaptive multi-thresholding approach for detecting suspicious regions, which relied on the Otsu thresholding method. Almalki et al. [11] utilized a noise-removal method to improve image quality. Subsequently, they employed the region-growing technique to eliminate the pectoral muscle. The experimental results revealed that their proposed approach delivered outstanding performance in peak signal-to-noise ratio (PSNR), mean squared error (MSE), and accuracy. Mohiyuddin et al. [12] incorporated various preprocessing steps, which encompassed the removal of white borders, artifacts and correcting illumination and pectoral muscle presence. To eliminate illumination issues, they employed the CLAHE algorithm. Comparative analysis of the outcomes demonstrated that their proposed methodology surpassed other approaches, excelling in accuracy, sensitivity, specificity, true positive rate, true negative rate, and f-measure.

Almalki et al. [13] introduced an image enhancement method that followed a precise sequence of steps. They initially converted mammograms from the JPEG format to RGB images to extract channel-specific properties. Illumination issues were then addressed to improve contrast. Subsequently, they applied the principal component analysis (PCA) algorithm to convert the RGB image to grayscale, effectively removing the image background. They employed the region growth technique to refine the images further to eliminate the pectoral muscle region. Finally, the coherency of the cancerous region was considered to enhance the visualization of the segmented area.

There are several challenges related to image quality to improve breast cancer classification approaches. The noise and artifact removal of the original mammogram plays a vital role in obtaining accurate segmentation and feature extraction process for identifying the risk of breast cancer. Therefore, this study focuses on improving image quality with the proposed CEAMF filter. The proposed method performs pixel-wise traversal of breast images (mammograms) to remove the noise and artifacts.

3 Proposed Work

In this work, A Center Adaptive Median Filter (CEAMF) is proposed to enhance image quality by effectively removing noise and artifacts. The CEAMF employs pixel-wise traversal to eliminate unwanted elements systematically. The proposed method is evaluated on collected and DDSM datasets, with the peak signal-to-noise ratio (PSNR) and mean square error (MSE) serving as key performance parameters. Comparative analyses are conducted against traditional median, mean, and Gaussian filters. A comprehensive analysis of the proposed methodology is provided in the subsequent sections.

3.1 Image Dataset

In this study, images are acquired from two datasets, i.e., a collected dataset and a public dataset, i.e., a digital database for screening mammography (DDSM).

Collected Dataset: It is from Fortis Hospital [14] in Kangra district in Himachal Pradesh, India. It contains digital mammograms that are available in.jpeg format.

DDSM [15]: It is maintained by the University of South Florida. It is made available through the collaboration of Massachusetts General Hospital, Wake Forest University School of Medicine, and Sacred Heart Hospital. The DDSM dataset contains digitized mammography film images in the compressed.jpeg format.

The sample images of the collected dataset and DDSM are shown in Fig. 1.

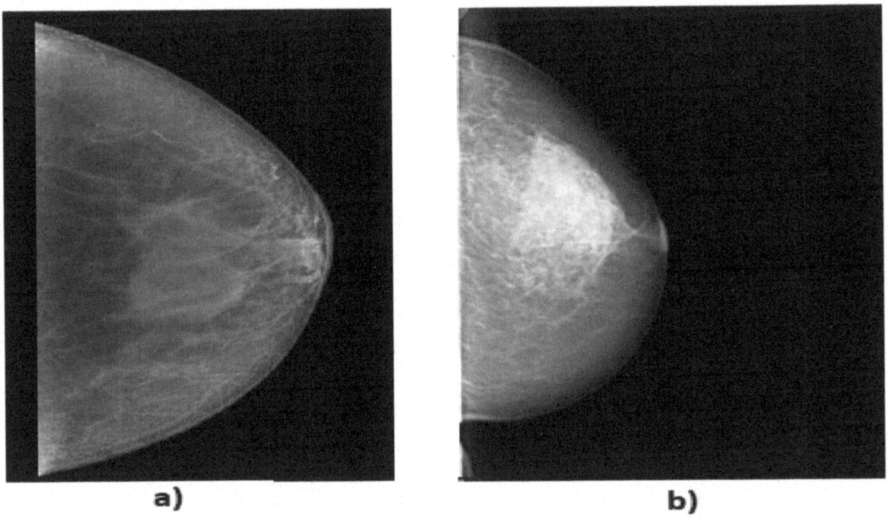

Fig. 1. Mammogram samples a) Collected dataset b) DDSM dataset

3.2 Methodology

The artifacts and noise in the acquired mammogram images must be eliminated to obtain a clear-cut view of the information in the image. The artifacts include background, opacity, objects laid on the breast, motion artifacts, data information, and artifacts due to machines. The noise includes Gaussian, speckle, salt and pepper, quantum, and impulse noise. The proposed CEAMF filter performs pixel-wise traversal to remove the noise and artifacts.

Let the input image be y with the size of $Z \times M$ in which y_{ji} denotes the pixel position of the image (j, i) where $(j, i) \in B \equiv 1, \ldots, Z \times 1, \ldots, M$. The filter size might increase the computation time, so we have set the size limit U, in which the size cannot exceed $U \times U$. The upper limit size can be formulated as,

$$U = \frac{\min\limits_{1 \le k \le x} \{\min(he_k, wi_k)\}}{4} \quad (1)$$

where, he_k and wi_k denotes the height and width of the light spot, respectively. On the other hand, we have initialized the size of the lower limit to the CEAMF as the smaller size of the CEAMF did not provide the desirable noise and artifact elimination effect. It is ensured that the lower limit size must not be less than 3×3. The size of U can be formulated as,

$$U = \text{mini}\left(2\left(e^{a_{ji}^{near}(y)/sdis_{ji}^{near}(y)} - 1\right) + 3, \text{Sdis}\right) \quad (2)$$

From the above equation, $sdis_{ji}^{near}$ denotes the distance between the center of the light spot (i.e., noisy/artifact) and the boundary area of the noise/artifact. From that, the

$sdis_{ji}^{near}(y)$ can be formulated as,

$$sdis_{ji}^{near}(y) = \frac{(j - cen_y^{near})^2 + wi_{near}^2}{4(i - cen_x^{near})^2} \tag{3}$$

where, wi_{near}^2 is the light spot width, cen_y^{near}, cen_x^{near} denotes the center co-ordinates of the light spot, and (j, i) denotes the pixel co-ordinates. In a case, if the present pixel is within the frame of the light spot, then the U can be formulated as,

$$U = mini\left(2\left\lfloor \log_{10}\left(\frac{sdis_{ji}^{near}(y)}{a_{ji}^{near}(y)}\right)\right\rfloor + 3, Sdis\right) \tag{4}$$

Let us assume that the size of window U × U is centered at the (j, i). The window size of the set is equal to the CEAMF $y_{j,i}(U)$ that can be expressed as,

$$y_{j,i}(U) = \left\{(q, w) : |q - j| \le \frac{U - 1}{2}, |w - i| \le \frac{U - 1}{2}, (q, w) \in B\right\} \tag{5}$$

From the above equation, the pixel median value can be presented as $y_{j,i}^{medi}(U)$. Henceforth, the CEAMF working can be formulated as,

$$y_{j,i} = y_{j,i}^{medi}(U) \tag{6}$$

Finally, the images can be reconstructed to remove noise and artifact removed image. The algorithm, for the CEAMF based noise and artifact removal can be provided below,

Algorithm 1: CEAMF based Noise/Artifact Removal

Input: Image y from the dataset
Output: Noise/Artifact removed image
Begin
For every pixel $(j, i) \in B$**do**
  Set $U = 3$
  Calculate $sdis_{ji}^{near}(y)$, $a_{ji}^{near}(y)$, and $y_{j,i}^{medi}(U)$
If$sdis_{ji}^{near}(y) < a_{ji}^{near}(y)$**then**
  Compute $U = mini\left(2\left(e^{a_{ji}^{near}(y)/sdis_{ji}^{near}(y)} - 1\right) + 3, Sdis\right)$
Else
  Compute $U = mini\left(2\left(e^{a_{ji}^{near}(y)/sdis_{ji}^{near}(y)} + 3, Sdis\right)\right)$
  Compute CEAMF working rule $y_{j,i} = y_{j,i}^{medi}(U)$
 End if
 End For
End

3.3 Performance Metrics

Mean square error and peak signal-to-noise ratio parameters could be used to measure the productiveness of computing enhancement strategies [7].

Mean Square Error (MSE): It represents the average of the squared differences between estimated and actual values. A lower MSE value indicates reduced errors, which leads to improved image quality. MSE is offered by:

$$MSE = \frac{1}{x \times y} \sum_{i=0}^{x-1} \sum_{j=0}^{y-1} [I(i,j) - K(i,j)] \quad (7)$$

where I(i,j) is the input image, m n is the size of the image and K(i,j) is the output image.

Peak Signal-to-Noise Ratio (PSNR): It is a term used to calculate the ratio of the maximum possible power of the signal and the power of noise that is corrupted in nature. The measurement is represented on a logarithmic decibel scale. The higher PSNR value indicates a lower error rate, leading to an enhanced product quality or the resulting image. PSNR is offered by:

$$PSNR = 10\log_{10}\left\{\frac{I_{MAX}^2}{MSE}\right\} \quad (8)$$

4 Results Analysis

This section details the proposed experimental results based on the evidence collected during implementation and comparative analysis. Matlab R2019b is utilized as a simulation tool to improve the simulation results' performance. The proposed algorithm is applied to 10 images from each dataset (CDI symbolizes the image from the collected dataset, and DDI symbolizes the image from the DDSM dataset). MSE and PSNR values are computed for mammograms before removing the noise and artifacts. Subsequently, the proposed method is applied to assess the enhancement in image quality.

The output of the simulation results is shown in Fig. 2.

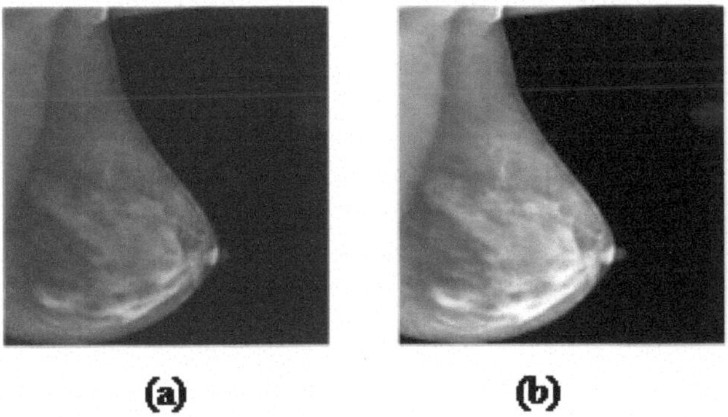

Fig. 2. Simulation results a) Input image b) Image after the noise and artifact removal

Gaussian and median filters are also employed to remove the noise and artifacts from mammograms, and their outcomes are compared with those of the proposed CEAMF. Tables 1 and 2 show the results of the MSE and PSNR values of the Gaussian filter, median filter, and proposed CEAMF for noise and artifact removal, respectively.

Table 1. MSE values for Gaussian filter, median filter and proposed CEAMF

Dataset	Image	Gaussian Filter	Median Filter	CEAMF
Collected Dataset	CDI_1	0.24	0.21	**0.13**
	CDI_2	0.35	0.26	**0.14**
	CDI_3	0.23	0.20	**0.09**
	CDI_4	0.23	0.21	**0.09**
	CDI_5	0.18	0.15	**0.07**
	CDI_6	0.21	0.18	**0.07**
	CDI_7	0.19	0.16	**0.07**
	CDI_8	0.19	0.15	**0.06**
	CDI_9	0.21	0.17	**0.07**
	CDI_{10}	0.19	0.17	**0.07**
DDSM	DDI_1	0.22	0.18	**0.08**
	DDI_2	0.22	0.18	**0.07**
	DDI_3	0.23	0.18	**0.08**
	DDI_4	0.18	0.13	**0.04**
	DDI_5	0.16	0.12	**0.04**
	DDI_6	0.17	0.12	**0.04**
	DDI_7	0.19	0.13	**0.04**
	DDI_8	0.18	0.13	**0.05**
	DDI_9	0.21	0.16	**0.06**
	DDI_{10}	0.19	0.14	**0.08**

Table 2. PSNR values of Gaussian filter, median filter and proposed CEAMF

Dataset	Image	Gaussian Filter	Median Filter	CEAMF
Collected Dataset	CDI_1	27.365	32.590	**55.485**
	CDI_2	22.870	30.870	**52.970**
	CDI_3	27.810	32.850	**58.931**
	CDI_4	27.781	32.978	**58.670**
	CDI_5	30.180	35.880	**59.0104**
	CDI_6	26.210	32.991	**59.231**
	CDI_7	29.101	34.456	**59.621**
	CDI_8	29.250	34.230	**59.987**
	CDI_9	27.690	32.981	**59.670**
	CDI_{10}	29.201	34.880	**59.197**
DDSM	DDI_1	27.365	34.590	**58.485**
	DDI_2	27.970	34.970	**59.978**
	DDI_3	26.810	33.050	**58.980**
	DDI_4	30.551	37.158	**61.637**
	DDI_5	32.980	39.880	**61.564**
	DDI_6	31.215	38.991	**61.310**
	DDI_7	29.250	37.311	**61.025**
	DDI_8	30.250	38.230	**60.920**
	DDI_9	28.690	35.890	**59.670**
	DDI_{10}	29.012	36.870	**58.141**

It is found that the MSE value reduces with the proposed method as the MSE value reduces and the PSNR value increases; it improves the image quality.

To provide a more subtle understanding, average values for PSNR and MSE are calculated for ten images from the collected dataset and ten from DDSM. Table 3 represents the comparative analysis of average values of PSNR and MSE of Gaussian, median filter and proposed CEAMF.

It is found that the proposed CEAMF gives better results than the Gaussian and median filters in terms of MSE and PSNR, as shown in Fig. 3 and 4, respectively.

Table 3. MSE values of Gaussian filter, median filter and proposed CEAMF

Dataset	Average MSE value		
	Gaussian Filter	Median filter	Proposed CEAMF
Collected dataset	0.22	0.19	0.09
DDSM	0.19	0.15	0.06
	Average PSNR value		
	Gaussian Filter	Median filter	Proposed CEAMF
Collected dataset	27.75	33.47	58.28
DDSM	29.41	36.69	60.17

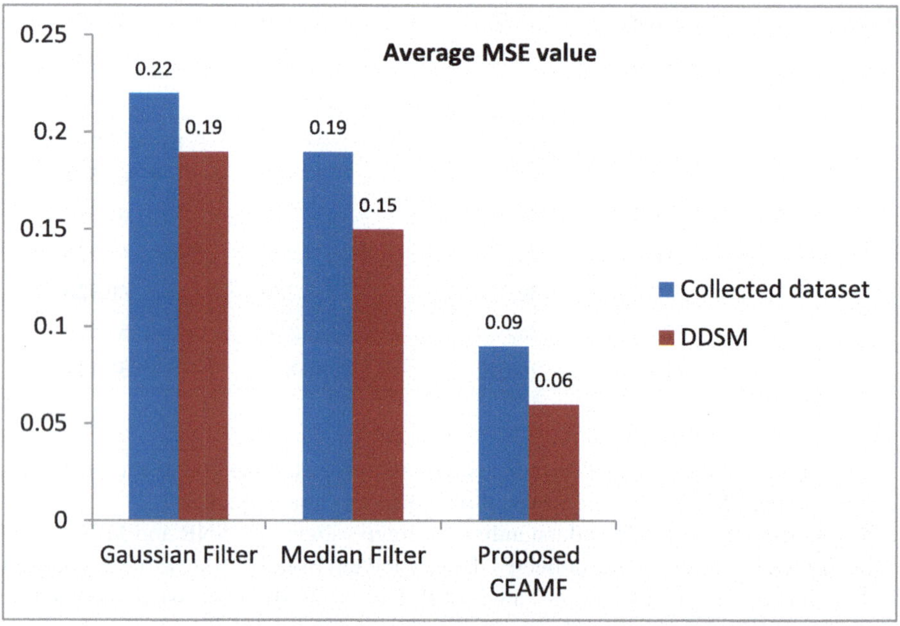

Fig. 3. Average MSE value for ten images from collected and DDSM dataset

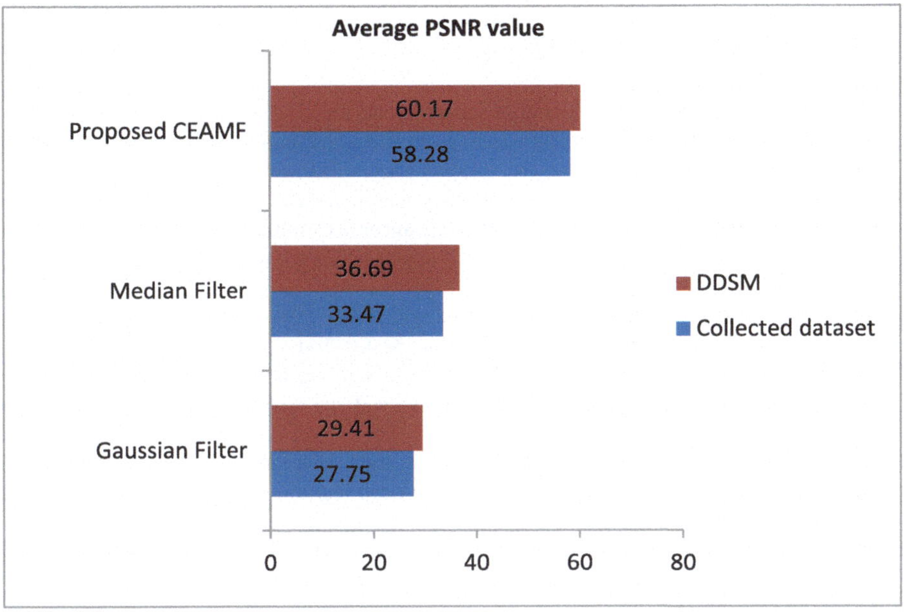

Fig. 4. Average PSNR value for ten images from collected and DDSM dataset

5 Conclusion

This work proposes a center adaptive median filter (CEAMF) to remove noise and artifacts and improves mammogram image quality for breast cancer detection. High-quality images help to improve the subsequent stages of breast cancer detection, like segmentation and classification. The proposed method removes noise as well as artifacts from mammograms. It is found that the MSE value for the mammogram images is reduced using the proposed method, and this reduction in MSE leads to an increase in PSNR and improves the image quality of mammograms. It is found that the proposed methods outperform the existing Gaussian and median filters for noise and artifact removal.

This work can be extended further for medical imaging modalities like ultrasound and MRI. Also, this work can be enhanced by improving the contrast of breast images.

References

1. Sechopoulos, I., Teuwen, J., Mann, R.: Artificial intelligence for breast cancer detection in mammography and digital breast tomosynthesis: state of the art. In: Seminars in Cancer Biology, vol. 72. Academic Press (2021)
2. Mahmood, T., Li, J., Pei, Y., Akhtar, F., Imran, A., Rehman, K.U.: A brief survey on breast cancer diagnostic with deep learning schemes using multi-image modalities. IEEE Access **8**, 165779–165809 (2020)
3. Meenalochini, G., Ramkumar, S.: Survey of machine learning algorithms for breast cancer detection using mammogram images. Mater. Today Proc. **37**, 2738–2743 (2021)

4. Nagpure, R., Chandak, S., Pathak, N.: Breast cancer detection using neural network mammogram. In: 2020 International Conference on Convergence to Digital World-Quo Vadis (ICCDW). IEEE (2020)
5. Montaha, S., et al.: BreastNet18: a high accuracy fine-tuned VGG16 model evaluated using ablation study for diagnosing breast cancer from enhanced mammography images. Biology **10**(12), 1347 (2021)
6. Dabass, J., Dabass, M.: Segmentation of noisy mammograms using hybrid techniques. In: Hura, G.S., Singh, A.K., Siong Hoe, L. (eds.) Advances in Communication and Computational Technology. LNEE, vol. 668, pp. 1371–1382. Springer, Singapore (2021). https://doi.org/10.1007/978-981-15-5341-7_104
7. Singla, C., Sarangi, P.K., Sahoo, A.K., Singh, P.K.: Deep learning enhancement on mammogram images for breast cancer detection. Mater. Today Proc. **49**, 3098–3104 (2022)
8. Sajjadnia, Z., Khayami, R., Moosavi, M.R.: Preprocessing breast cancer data to improve the data quality, diagnosis procedure, and medical care services. Cancer Inform. **19**, 1176935120917955 (2020)
9. Kaur, P., Singh, G., Kaur, P.: Intellectual detection and validation of automated mammogram breast cancer images by multi-class SVM using deep learning classification. Inform. Med. Unlocked **16**, 100151 (2019)
10. Toz, G., Erdoğmuş, P.: A novel hybrid image segmentation method for detection of suspicious regions in mammograms based on adaptive multi-thresholding (HCOW). IEEE Access **9**, 85377–85391 (2021)
11. Almalki, Y.E., Soomro, T.A., Irfan, M., Alduraibi, S.K., Ali, A.: Computerized analysis of mammogram images for early detection of breast cancer. In: Healthcare, vol. 10, no. 5, p. 801. MDPI (2022)
12. Mohiyuddin, A., et al.: Breast tumor detection and classification in mammogram images using modified YOLOv5 network. Comput. Math. Methods Med. **2022**, 1–16 (2022)
13. Almalki, Y.E., Soomro, T.A., Irfan, M., Alduraibi, S.K., Ali, A.: Impact of image enhancement module for analysis of mammogram images for diagnostics of breast cancer. Sensors **22**(5), 1868 (2022)
14. Fortis Hospital Kangra. https://www.fortishealthcare.com/india/fortis-hospital-in-kangra-himachal-pradesh
15. Heath, M., Bowyer, K., Kopans, D., Kegelmeyer, P., Moore, R., Chang, K., Munishkumaran, S.: Current status of the digital database for screening mammography. In: Karssemeijer, N., Thijssen, M., Hendriks, J., van Erning, L. (eds.) Digital Mammography: Nijmegen, 1998, pp. 457–460. Springer, Dordrecht (1998). https://doi.org/10.1007/978-94-011-5318-8_75

Fertilizer Recommendation with Stress Monitoring in Agriculture

G. Kalyani, P. Pravali, and J. Haswitha(✉)

Department of Information Technology, Velagapudi Ramakrishna Siddhartha Engineering College, Vijayawada, Andhra Pradesh, India
gkalyani@vrsiddhartha.ac.in, haswithajuluru2003@gmail.com

Abstract. Stress monitoring in agriculture is essential to increase farm production, encourage sustainability, and assist farmers while making choices that were ethical to food security and economic stability. The purpose of this study is to create a machine learning model to determine whether the plant is under stress or not. And if the plant experiences stress, what deficiency caused it, and recommend the appropriate fertilizer. We utilized the crop recommendation dataset and the fertilizer recommendation dataset as two datasets for this. Pair of classifiers to suggest fertilizer for the crop under stress, random forest and ANN are employed. As input parameters for classifiers, nitrogen, potassium, phosphorus, temperature, humidity, moisture, and crop kind are all employed. The Artificial Neural Networks (ANN) classifier has the highest accuracy (84%) when the overall performance of all the classifiers is considered. This study demonstrates algorithms can identify changes in the surroundings of the plant and forecast how the plant will react to these changes by analyzing the data. By comparing information produced using features and segments, our farmers can classify the stress and administer fertilizer. The automation of the stress detection process will boost agricultural productivity and benefit society.

Keywords: Classification · random forest · artificial neural networks (ANN) · machine learning · stress · agriculture

1 Introduction

Plants form the base of terrestrial ecosystems; they are vulnerable to a collection of abiotic and biotic stresses that might intervene in their regular biological functions. The accurate identification of plant stress, which is necessary for the early deployment of mitigation techniques and ensuring sustainable crop yields, is one of the biggest issues encountered by farmers. A primary pressure that adversely impacts a plant health is nutrient shortage, which can result in reduced development, poor crop quality, and increased susceptibility to disease [1].

D. Pastor-Escuredo et al. (Eds.): ICDLAIR 2023, LNNS 1001, pp. 555–565, 2024.
https://doi.org/10.1007/978-3-031-60935-0_50

It is impossible to overestimate the significance of proper nutrition levels for plants' healthy growth and development. Several micronutrients, including nitrogen, phosphorus, and potassium, serve as crucial building blocks for plant tissues. Traditional approaches for determining nutrient shortages in plants have frequently been arbitrary and open to mistakes. The ability of machine learning tools to analyze enormous volumes of data, spot trends, and generate precise predictions can be crucial in lowering plant stress.

Understanding and treating nutrient deficiency in plants can help us conserve resources, improve fertilization techniques, and lessen the negative effects of excessive nutrient runoff on the environment. In conclusion, the ability to identify plant stress is crucial in a variety of sectors, such as climatic protection, and plant research. It acts as a catalyst for creativity, scientific progress, and the creation of useful applications. Overall, to determine whether or not the plants are under stress is the primary goal of this endeavor. By comparing the data in the data set with our statistical computations, we may ascertain this. The plants will be healthy if the readings fall within our range; otherwise, they will be stressed. Additionally, using machine learning algorithms, to suggest fertilizer for stressed plants.

2 Literature Survey

Here are the available models the, that Researchers proposed and developed. They proposed different methodologies for crop recommendation and fertilizer recommendation.

Shilpa and B. R Aishwarya [2] they extrapolated findings for the areas of Karnataka and Maharashtra from a few chosen methods. Type of crop, period, season, type of soil, region, are the factors employed by algorithms. The exactness of the predicted crop production for each of the chosen algorithms is compared. With a 95% accuracy rate, the Random Forest method proved to be ideal for the data set provided. Selected ML techniques including ANN, Random Forest are used to forecast agricultural yield. The limits of current technologies and their usefulness for forecasting production were emphasized in this article. In the provided information obtained from states of Maharashtra and Karnataka, they built and evaluated a number of machine learning strategies, such as ANN, SVM, MLR, and KNN.

Mr. Santosh Mahagoankar and Devdatta [3] Python is utilised to design the entire system. Numerous statistics, including those for crops, crop yield, location, soil and crop nutrients, and fertilizer, are obtained from several sources, including agricultural publications and websites. Here, identifying the soil classification and predicting crop output are the key objectives. Random Forest's accuracy is 86.35%, whereas Support Vector Machine's accuracy is 73.75%, making it a good method for classifying soil. The following features may be added to their work to make it even better. By sharing images of farms, mobile applications may be created to assist farmers. The ability to anticipate agricultural production based on location and the effective use of algorithms has demonstrated that a greater crop yield is possible.

Dr. D John Aravindhar and JuhiReshma S R [4] In this study, the prediction of soil nutrients is made utilizing support vector machines, multiple linear regression, and artificial neural networks. The findings indicated that the svm models' average prediction accuracies were 83% and 77.87%, respectively. Additionally, the general regression neural network's average prediction accuracy is 92.86%, indicating that SVM and GRNN models are utilized to calculate the soil nutrient level using the proper dependent variables. To prevent overusing soil fertilizers, there is a growing need to assess soil fertility. Through the provision of an overview to suggest the quantity of fertilizers necessary for the crop of bananas, they have here offered a structure to address the misuse of fertilizers.

Dhruvi Gosai and Chintal raval [5] This literature survey focuses on agriculture's significance in India and its impact on crop production. The suggested framework utilizes IoT and ML, incorporating sensors for soil testing and observing various factors such as moisture, pH, and NPK nutrients. By analyzing the detector data with ML algorithms, such as random forest, the system suggests appropriate crop growth strategies to maintain crop health. Additionally, the survey explores the use of a CNN to identify potential plant diseases.

Numerous distinct technologies employed to accomplish this for fertilizer recommendation, according to a study of diverse prediction methodologies. The presentation and contrast of all the suggested approaches exposes certain benefits and drawbacks. The application of multiple ML algorithms for forecasting crop production and recommending fertilizer is covered in the article. With a 95% accuracy rate, Random Forest was proven being foremost reliable algorithm for fertilizer suggestion. To prevent overuse, the suggested model additionally takes fertilizer application time into account. The usage of GRNN and SVM models for calculating soil nutrient levels is also covered in the text.

3 Proposed Methodology

The conceptual technique is Fig. 1 illustrates, which uses machine learning to identify crops that are stressed and suggest the right fertilizer to treat them [6]. To ensure that undertake a comparative analysis, several ML algorithms have been used to contemporary datasets of fertilizer. We used Python and pandas, a tool that comes with Python, to produce ML models for the tests. Here, the ML algorithms Random Forest and Ann are used on the fertilizer data set.

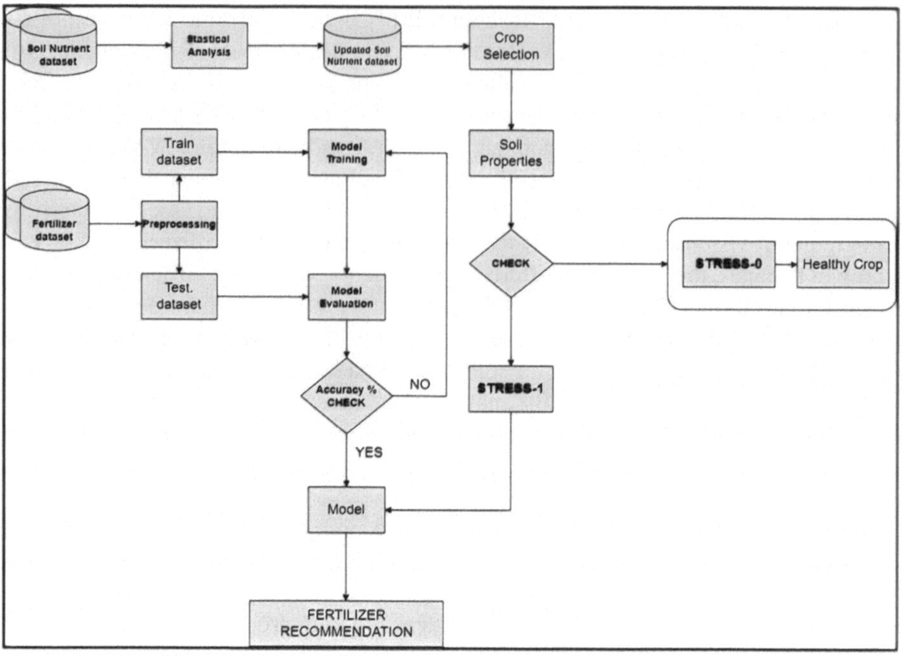

Fig. 1. Structure for classifying Stress and Fertilizer recommendation

Figure 1 depicts the conceptual methodology for Identifying whether the plants are under stress or not [7]. The dataset for crop recommendations was taken from Kaggle each attribute's average as well as its standard deviation in the dataset were then established for every label. The ranges for each attribute for the corresponding label are built by considering the range as [mean - standard deviation, mean + standard deviation]. A new dataset named ranges dataset was prepared after acquiring the ranges. The ranges dataset is then used to compare each record in the crop recommendation dataset. If any of the attributes are outside of the acceptable range, the crop is classified as under stress; otherwise, it is classified as a healthy crop. After that, the fertiliser suggestions dataset was considered and preprocessed. Testing and training sets of data came from the dataset. The dataset was instructed using a variety of techniques, including ANN and random forest. The method that provided the highest accuracy suits the model the best. The crops under stress will be given as input to the fertilizer recommendation dataset. So, the appropriate fertilizer will be determined when the match occurs.

3.1 Statistical Analysis

We have a variety of unique crops in the crop dataset, including rice, maize, bananas, etc. The normative range i.e., [Mean-Standard Deviation, Mean + Standard Deviation], determines whether a crop is healthy or stressed depending on whether its attribute values fall within these ranges. So, each attribute's average as well as its standard deviation in the dataset are determined for each label. The minimum and maximum values of an attribute are [Mean-Standard Deviation] and [Mean + Standard Deviation], respectively [8]. The Examination of Data in the different attributes for stress Monitoring in the crops is displayed in Table 1. For example, consider the crop as rice then for all the attributes we will calculate the min and max value.

1. let's consider label = rice and attribute = Nitrogen
Mean value of Nitrogen = sum of all Nitrogen values of rice label/Total number of rice labels
= 7989/100
= 79.89
Standard Deviation = (sum of squares of all nitrogen values/Total number of rice labels)$^{0.5}$
= (14184.81 / 100)$^{0.5}$
= 11.91

Table 1. Statistical Analysis for stress Monitoring

Label	N	P	K	T	H	ph	rf
Rice	79.89 ± 11.91	47.58 ± 7.90	39.87 ± 2.95	23.68 ± 2.03	82.27 ± 1.41	6.42 ± 0.76	236.18 ± 34.29
Maize	77.76 ± 11.94	48.44 ± 8.01	19.79 ± 2.94	22.38 ± 2.67	65.09 ± 5.44	6.24 ± 0.41	84.76 ± 15.54
Chickpea	40.09 ± 12.15	67.79 ± 7.49	79.92 ± 3.26	18.87 ± 1.16	16.86 ± 1.71	7.33 ± 0.80	80.05 ± 7.94
Kidneybeans	20.75 ± 10.83	67.54 ± 7.57	20.05 ± 3.10	20.11 ± 2.59	21.6 ± 0.20	5.74 ± 0.14	105.91 ± 26.11
Pigeonpeas	20.73 ± 11.84	67.73 ± 7.29	20.29 ± 2.81	27.74 ± 5.71	48.06 ± 10.96	5.79 ± 0.83	149.45 ± 32.98
Mothbeans	21.44 ± 11.34	48.01 ± 7.54	20.23 ± 3.04	28.19 ± 2.20	53.16 ± 6.99	6.83 ± 1.86	51.19 ± 13.75
Mungbean	20.99 ± 11.51	47.28 ± 7.87	19.87 ± 3.14	28.52 ± 0.84	85.49 ± 2.85	6.72 ± 0.28	48.4 ± 7.11
Blackgram	40.02 ± 12.66	67.47 ± 7.15	19.24 ± 3.18	29.97 ± 2.68	65.11 ± 2.79	7.13 ± 0.37	67.88 ± 4.18
Lentil	18.77 ± 12.19	68.36 ± 7.33	19.41 ± 2.96	24.5 ± 3.32	64.8 ± 2.93	6.92 ± 0.54	45.68 ± 5.63
Pomegranate	18.87 ± 12.61	18.75 ± 7.38	40.21 ± 3.03	21.83 ± 2.20	90.12 ± 2.82	6.42 ± 0.49	107.52 ± 2.89
Banana	100.23 ± 11.10	82.01 ± 7.69	50.05 ± 3.38	27.37 ± 1.42	80.35 ± 2.80	5.98 ± 0.27	104.62 ± 9.39
Mango	20.07 ± 12.32	27.18 ± 7.66	29.92 ± 3.09	31.2 ± 2.65	50.15 ± 2.75	5.76 ± 0.70	94.7 ± 3.33
Grapes	23.18 ± 12.46	132.53 ± 7.60	200.11 ± 3.2	23.84 ± 9.73	81.87 ± 1.17	6.02 ± 0.29	69.61 ± 2.95
Watermelon	99.42 ± 12.56	17 ± 7.53	50.22 ± 3.26	25.59 ± 0.85	85.16 ± 2.95	6.49 ± 0.28	50.78 ± 5.86
Muskmelon	100.32 ± 12.17	17.72 ± 7.18	50.08 ± 3.21	28.66 ± 0.86	92.34 ± 1.50	6.35 ± 0.23	24.68 ± 2.77
Apple	20.8 ± 11.86	134.22 ± 8.12	199.89 ± 3.3	22.63 ± 0.82	92.33 ± 1.45	5.92 ± 0.26	112.65 ± 7.10
Orange	19.58 ± 11.94	16.55 ± 7.69	10.01 ± 3.05	22.76 ± 7.34	92.17 ± 1.43	7.01 ± 0.57	110.47 ± 5.71
Papaya	49.88 ± 12.21	59.05 ± 7.05	50.04 ± 3.09	33.72 ± 6.26	92.4 ± 1.42	6.74 ± 0.14	142.62 ± 64.39
Coconut	21.98 ± 11.76	16.93 ± 8.35	30.59 ± 2.99	27.4 ± 1.38	94.84 ± 2.68	5.97 ± 0.28	175.68 ± 29.47
Cotton	117.77 ± 11.62	46.24 ± 7.34	19.56 ± 3.16	23.98 ± 1.13	79.84 ± 3.05	6.91 ± 0.62	80.39 ± 11.23
Jute	78.4 ± 10.96	46.86 ± 7.19	39.99 ± 3.31	24.95 ± 1.18	79.63 ± 5.50	6.73 ± 0.44	174.79 ± 15.07
Coffee	101.2 ± 12.34	28.74 ± 7.27	29.94 ± 3.24	25.54 ± 1.50	58.86 ± 5.83	6.79 ± 0.41	158.06 ± 25.69

Note: N-Nitrogen, P-Phosphorus, K-Potassium, T-Temperature, H-Humidity, Ph-Potential of Hydrogen, Rf-Rainfall

3.2 Fertilizer Recommendation

The series of features is passed to the ML algorithms RF and ANN [9]. The following is a description of how these algorithms really function in detail.

Random forest (RF)
Random Forest is the supervised classification technique that is most often employed. Building a decision tree from a dataset is chosen by random forests. The bagging concept serves to illustrate how Random Forest works. It selects a new dataset and creates a decision tree particularly for it for each iteration. To put it another way, Random Forest creates a decision tree for each individual smaller datasets it splits. In the prediction step, each decision tree uses its forecast. Finally, it selects the anticipated result because of the majority vote. Random Forest is viewed as an ensemble classifier that may solve the issue of overfitting and has a greater prediction rate than decision trees since the results are averaged [10]. Figure 2 displays the Diagrammatic working representation. This is the Random Forest (RF) Classification Algorithm:

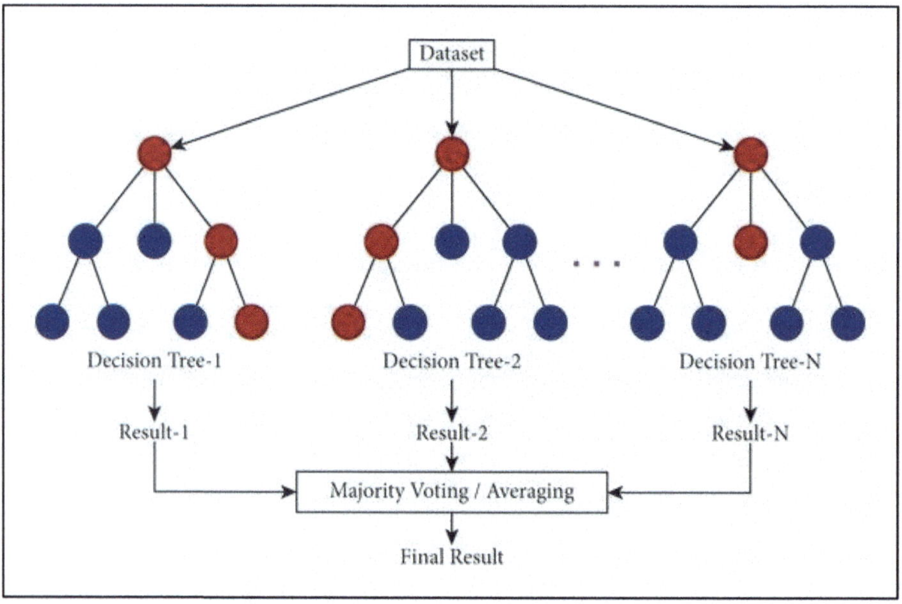

Fig. 2. Random Forest implementation process

Artificial Neural Networks (ANN)
Artificial neural networks are referred to as "ANN" algorithms in machine learning. A type of computational model known as artificial neural networks (ANN) is based on the structure and functionality of biological neural networks, including the human brain. They are often used in a variety of machine learning applications, including pattern recognition, classification, regression, and others. The core of an ANN is the artificial neuron, sometimes known as a perceptron on occasion. Perceptron's take in inputs, weigh them, and then process the weighted sum through an activation function to produce an output. The configuration of these neurons is then stacked to be able to create a network. The usual components of ANNs are an input layer, both or any hidden layers, and an output layer. The input is supplied by the top layer, processing is done by the hidden

levels, and the result is produced by the bottom layer. Every neuron in a layer connects to every other neuron in every layer above it to create a dense network.

4 Experimental Results

The following section examines the datasets that we used and the formulae necessary to calculate the metrics and result analysis using bar graph.

4.1 Dataset

Two datasets were utilized.

1. Crop Recommendation Dataset: This contains a variety of characteristics, including N, P, K, temperature, moisture, rainfall, and ph, with crop type serving as the final class designation. The dataset contains 20 distinct types of crops.[11]
2. Fertilizer Recommendation Dataset: This contains a set of characteristics, including N, K, and P, as well as temperature, moisture, rainfall, crop kind, and the final class label, fertilizer name.[12]

4.2 Evaluation Metrics

We employ the following criterion to evaluate how well machine approaches function in recommending fertilizers for the crops which are under stress.

1. Accuracy $= \frac{(TP+TN)}{(TP+FP+TN+FN)}$
   The percentage of correctly identified instances among the total instances.
2. Error rate = 1-Accuracy
   The proportion of misclassified instances relative to the total instances.
3. Precision $= \frac{TP}{(TP+FP)}$
   The ratio of accurately predicted positive cases to the total predicted positive cases.
4. Recall $= \frac{TP}{(TP+FN)}$
   The ratio of accurately predicted positive cases to the overall actual positive cases.
5. F1-score $= \frac{(2*Precision*Recall)}{(Precision+Recall)} = \frac{2TP}{(2TP+FP+FN)}$
   The harmonic mean of precision and recall, offering a balanced evaluation of a model's effectiveness.

Where, TP = Number of True Positives
TN = Number of True Negatives
FP = Number of False Positives
FN = Number of False Negatives

4.3 Discussion on Results

Figure 3 displays the outcomes from the statistical calculations. It provides categorization outcomes that show if the crop is healthy or under stress. Moreover, if the crop is experiencing stress, the attribute whose deficiency caused the stress to exist was also recognized.

Here in the Fig. 3, the 1st row the stress for the rice label is occurred due to potassium because the potassium value is 43 in the dataset. It is not in between minimum (36.92) and maximum values (42.81). So, in the next steps this data will be compared with the fertilizer recommendation data and appropriate fertilizer will be recommended. Similarly for the below rows based on the stress attribute comparisons with the fertilizer recommendation data will be done. If suppose there is no attribute causing stress then the crop is not under the stress so, we can say that it as a healthy crop.

```
       label   status stress_attr
0       rice   stress           K
1       rice   stress           P
2       rice   stress        N, K
3       rice   stress           P
4       rice  healthy
...      ...      ...         ...
2095  coffee  healthy
2096  coffee   stress           P
2097  coffee   stress           N
2098  coffee   stress        N, K
2099  coffee   stress           P

[2100 rows x 3 columns]
```

Fig. 3. Identification of stress and stress caused attribute

Table 2 displays the classification outcomes of the fertilizer recommendation. We only recommend fertilizer for the crop which has stress. The used two algorithms are Random Forest and ANN. Among the two classifiers better results are with ANN classifier that are Accuracy = 84%, Precision = 88%, Recall = 88% and f1-score = 87%. For a particular crop which is under stress those numbers are contrasted with the fertilizer recommendation dataset and then appropriate fertilizer is recommended.

Table 2. Classification results of fertilizer recommendation

Algorithms	Precision	Recall	F1-Score
Random Forest	0.78	0.74	0.74
ANN	0.88	0.88	0.87

The accuracy analysis between ANN and Random Forest is shown in Fig. 4. The algorithms are shown on the x-axis and accuracy is measured on the y-axis. As seen by the graph ANN has the maximum accuracy.

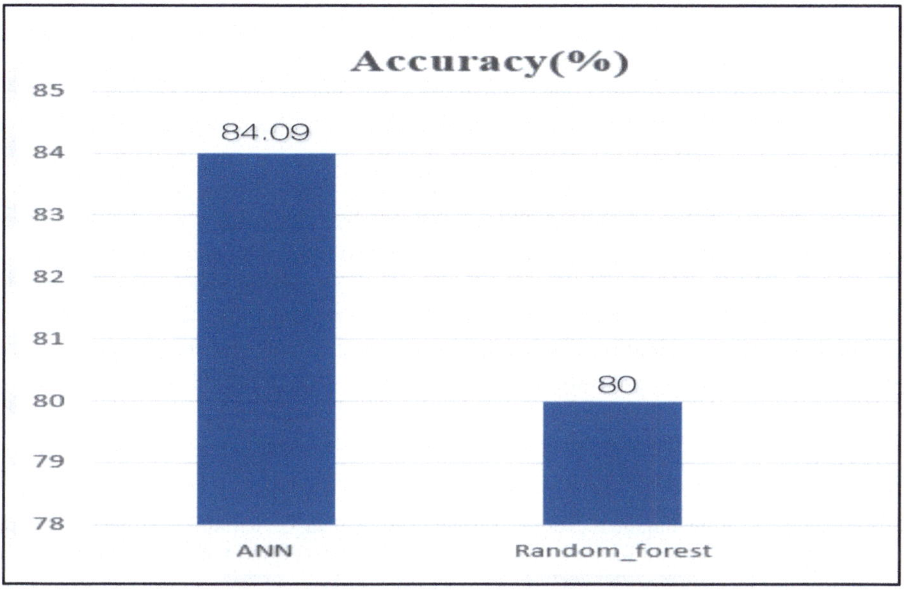

Fig. 4. Comparison of Accuracy for fertilizer recommendation

Following instruction, testing was done. On the first row of the dataset, which represents the rice crop that is under stress as there was no potassium. These datasets were contrasted with Fertilizer Recommendation dataset that recommends urea as a fertilizer (Fig. 5).

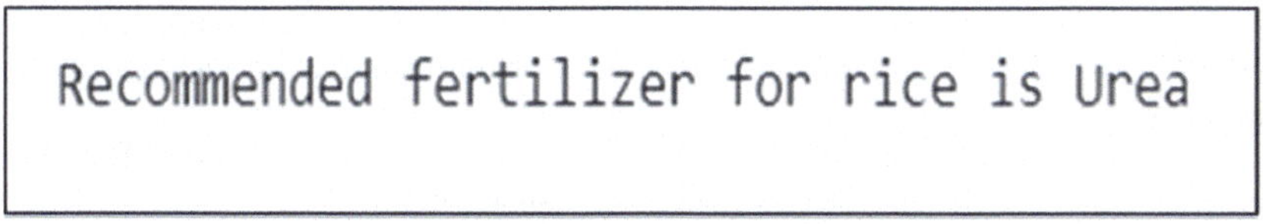

Fig. 5. Fertilizer Recommendation

5 Conclusion and Future Scope

The main objective of its goal is to make fertilizer recommendations for stressed crops. In our study, we considered two datasets: the crop recommendation dataset and the fertilizer recommendation dataset, to be able to construct an appropriate classification model regarding the forecast of fertilizer. The crop recommendation dataset determines if a crop is stressed or not, and the fertilizer recommendation dataset suggests the best fertilizer for determining crop stress. We employed random forest and ANN classifiers for this purpose, with the ANN classifier achieving the highest accuracy of 84%. This outcome can assist farmers in preventing or reducing crop loss, saving money on crop loss-related expenses, and boosting earnings. In order to acquire more precise and reliable results, the study of this project could be expanded by integrating sensor technologies to gather real-time data derived from crop. It's vital to remember that the accuracy and dependability of the monitoring technology utilized, as well as the farmers' desire to adopt and follow data-driven practices, are all elements that influence how well stress monitoring works in agriculture. To better illustrate and enhance the advantages of stress monitoring in agriculture, ongoing research and development in this area is crucial.

References

1. Pudumalar, S., Ramanujam, E., Harine Rajashree, R., Kavya, C., Kiruthika, T., Nisha, J.: Crop recommendation system for precision agriculture. In: 2016 Eighth International Conference on Advanced Computing (ICoAC), pp. 32–36. IEEE (2017)
2. Pande, S.M., Ramesh, P.K., Anmol, A., Aishwarya, B.R., Rohilla, K., Shaurya, K.: Crop recommender system using machine learning approach. In: 2021 5th International Conference on Computing Methodologies and Communication (ICCMC), pp. 1066–1071. IEEE (2021)
3. Bondre, D.A., Mahagaonkar, S.: Prediction of crop yield and fertilizer recommendation using machine learning algorithms. Int. J. Eng. Appl. Sci. Technol. **4**(5), 371–376 (2019)
4. JuhiReshma, S.R., John Aravindhar, D.: Fertilizer estimation using deep learning approach. NVEO-Nat. Volatiles Essential Oils J. | NVEO 5745–5752 (2021)
5. Gosai, D., Raval, C., Nayak, R., Jayswal, H., Patel, A.: Crop recommendation system using machine learning. Int. J. Sci. Res. Comput. Sci. Eng. Inf. Technol. **7**(3), 558–569 (2021)
6. Dhal, A., Lodhi, D., Gupta, R., Singh, V., Adak, I., Goel, P.K.: Integrating artificial intelligence and machine learning for crop yield prediction, fertilizer recommendation and disease detection
7. Azimi, S., Kaur, T., Gandhi, T.K.: A deep learning approach to measure stress level in plants due to Nitrogen deficiency. Measurement **173**, 108650 (2021)
8. Ransom, C.J., et al.: Statistical and machine learning methods evaluated for incorporating soil and weather into corn nitrogen recommendations. Comput. Electron. Agric. **164**, 104872 (2019)
9. Reddy, D.A., Dadore, B., Watekar, A.: Crop recommendation system to maximize crop yield in ramtek region using machine learning. Int. J. Sci. Res. Sci. Technol. **6**(1), 485–489 (2019). https://doi.org/10.32628/IJSRST196172
10. Ajoodha, R., Mufamadi, T.O.: Crop recommendation using machine learning algorithms and soil attributes data. In: Mathur, G., Bundele, M., Tripathi, A., Paprzycki, M. (eds.) Artificial Intelligence: Advances and Applications, pp. 31–41. Springer, Singapore (2023). https://doi.org/10.1007/978-981-19-7041-2_3

11. https://www.kaggle.com/datasets/atharvaingle/crop-recommendation-dataset
12. https://www.kaggle.com/datasets/gdabhishek/fertilizer-prediction

Surface Water Mapping and Analysis in SANGAMAM of River Godavari and Krishna Rivers Using Google Earth Engine

Parvataneni Chaitanya Swaroop(✉), T. Anuradha, and Pali Uday Manikanta

Department of Information Technology, Velagapudi Ramakrishna Siddhartha Engineering College, Vijayawada, India
swaroop6p@gmail.com, anuradha_it@vrsiddhartha.ac.in, 208w1a1241@vrsec.ac.in

Abstract. An Innovation with Analysis in the Krishna River Region Using Google Earth Engine seeks to provide comprehensive insights into surface water dynamics within the Krishna River basin, with a particular focus on the convergence region with the Godavari River, known as "SANGAMAM." The primary aspect of this SANAGAMAM involves studying the PATTISEEMA lift irrigation project, primarily aimed at enhancing irrigation facilities near the Krishna River delta. The study delves into variations in surface water levels, changes, and water quality, analyzing trends over the period from 2015 to 2021 through a meticulous examination of satellite images and remote sensing data. Researchers employ a diverse array of methodologies and algorithms to accurately delineate water bodies from remote sensing data, allowing for a precise assessment of surface water dynamics. This approach facilitates detailed surface water mapping, comprehensive time-series analysis, and aids in effective water resource management and environmental preservation. The sophisticated capabilities of Google Earth Engine serve as a crucial tool, enabling the visualization and analysis of complex data sets to better understand the dynamics of surface water and its interaction with the environment. By utilizing these advanced technologies, the project aims to contribute to sustainable water resource management and safeguarding the ecosystem in the Krishna River region.

Keywords: NDWI · MNDWI · Water Management · SVM · Naive Bayes · NDCI · NDTI

1 Introduction

The project conducted in the Krishna River basin from 2015 to 2020 [1] signifies an exceptional fusion of advanced technology and astute water resource management. At its heart lies the Pattiseema Lift Irrigation Project (PLIS), which harnesses a rich dataset consisting of satellite imagery and remote sensing data spanning several years to comprehensively analyze surface water dynamics [3]. This initiative is underpinned by a crucial mission: to transfer water sourced from the Godavari River in Pattiseema to the

D. Pastor-Escuredo et al. (Eds.): ICDLAIR 2023, LNNS 1001, pp. 566–579, 2024.
https://doi.org/10.1007/978-3-031-60935-0_51

Polavaram Project Right Main Canal [2], thereby benefiting the farmers residing in the Krishna River delta. The inception of the PLIS project is rooted in intricate inter-state agreements and the Bachawat tribunal, which paved the way for the diversion of 100 tmc of water from the Godavari River to the Krishna River, an agricultural lifeline [4]. Despite initial opposition and concerns regarding its storage capacity, the project persevered under the visionary leadership of Chief Minister Nara Chandrababu Naidu. It emerged as a vital lifeline for farmers cultivating an extensive 1.3 million acres in the arid Navin Delta [5], alleviating critical water shortages that typically afflict the region from June to August.

The operation of the PLIS is a remarkable engineering feat, involving the pumping of water from the Godavari River into the canal [6], a journey spanning approximately 160 kilometers and taking 7 to 8 days by gravity alone before reaching the Prakasam Barrage [7]. During the dry season from December to June, when the Polavaram right canal cannot receive water by gravity due to lowered reservoir levels [8], the Pattiseema Lift efficiently transfers dead storage water from the Polavaram reservoir to the Godavari River, ultimately supplying additional Krishna water to the Rayalaseema region [9]. Complementing this vital infrastructure is the project's reliance on Google Earth Engine's (GEE) advanced capabilities to analyze surface water dynamics within the Krishna River basin [10].

Moreover, the project stands out for its commitment to in-depth time-series analysis, enabling continuous monitoring of surface water changes over time [11]. This reveals seasonal patterns and enduring trends, offering invaluable insights into how surface water responds to climatic variations and human activities [14]. Google Earth Engine's capabilities facilitate the incorporation of various parameters and advanced algorithms to detect changes in water levels during specific time periods, contributing to a comprehensive understanding of surface water dynamics [16]. In essence, this innovative project epitomizes the potential of advanced technology to address complex water resource challenges [19], ultimately contributing to sustainable development and environmental protection.

Table 1. Literature Review

Ref. No	Authors	Year	Methodologies	Outcome
[1]	P. D. Sreekanth, P. Krishnan, N. H. Rao, S. K. Soam and Ch. Srinivasarao	2021	Normalized Difference Vegetation Index (NDVI), Normalized Difference Water Index (NDWI), modified Normalized Difference Water Index (mNDWI) *Google Earth Engine*	Analysis on water change

(*continued*)

Table 1. (*continued*)

Ref. No	Authors	Year	Methodologies	Outcome
[2]	Ruimeng Wang, Haoming Xia, Yaochen Qin, Wenhui Niu, Li Pan, Rumeng Li, Xiaoyang Zhao, Xiqing Bian and Pinde Fu	2020	mNDWI, Enhanced Vegetation Index (EVI) and NDVI *Google Earth Engine*	Analysis on water change during climate change
[3]	Haoming Xia, Jinyu Zhao Yaochen Qin, Jia Yang, Yaoping Cui, Hongquan Song, Liqun Ma, Ning Jin and Qingmin Meng	2019	mNDWI, EVI and NDVI *Google Earth Engine*	Different patters analysis of Landsat images
[4]	Chao Wang, Mingming Jia, Nengcheng Chen and Wei Wang	2018	NDVI, NDWI, mNDWI *Google Earth Enhine*	Water change detection at a river region
[5]	Jiacong Huang, Yinjun Zhang, Haijian Bing, Jian Peng, Feifei Dong, Junfeng Gao, George Arhonditsis	2019	Automated Water Extraction Index (AWEI), Normalized difference turbidity Index (NDTI), and Normalized difference chlorophyll index (NDCI) *Google Earth Engine*	Estimating the water quality during different conditions
[6]	Delpla, A.-V. Jung, E. Baures, M. Clement, O. Thomas	201 7	Water Quality Index, CCME Method, Weighted Method *Google Earth Engine*	Analyzing climate change as well as water quality
[7]	Chang Huang, Yun Chen, Shiqiang Zhang, Jianping Wu	201 8	NDWI, mNDWI, AWEI, Water Extraction Index (WI) *Google Earth Engine*	Pixel un-mixing and reconstruction, and spatial- temporal fusion approaches to enhance surface water monitoring
[8]	Sukanya Ghosh, Deepak Kumar, Rina Kumari	2020	Histogram Sampling, Otsu Thresholding, Image Classification *Google Earth Engine*	The impact of flood inundation over the vegetation cover with a cloud-based computing

(*continued*)

Table 1. (*continued*)

Ref. No	Authors	Year	Methodologies	Outcome
[9]	Sansar Raj Meena, Akshansha Chauhan, Kushanav Bhuyan & Ramesh P. Singh	2021	NDWI, NDTI and NDCI *Google Earth Engine*	Pronounced changes in water quality and flood plains using Sentinel data
[10]	Hongxing Liu, Min Xu, Richard A. Beck	201 8	NDTI and NDCI *Google Earth Engine*	An Ensemble Approach to Retrieving Water Quality Parameters from Multi spectral Satellite Imagery

2 Literature Review

The authors in Table 1 have used different Technologies for analyzing surface water. The analysis outcomes that are mentioned in [1–10] are Analysis on water change, Analysis on water change during climate change, Using different patters analysis of Landsat images, Water change detection at a river region, Estimating the water quality during different conditions, Analyzing water during climate change as well as water quality, Pixel immixing and reconstruction, and Spatiotemporal fusion approaches to enhance surface water monitoring, The impact of flood inundation over the vegetation cover with a cloud-based computing, Pronounced changes in water quality and flood plains using Sentinel data, An Ensemble Approach to Retrieving Water Quality Parameters from Multispectral Satellite Imagery. Some of the authors created an App to display the status of the parking slots but some of them just stored the data in the cloud.

[5, 6] and [9] had used for detecting the water quality using the different types of parameters such as NDCl i.e. chlorophyll index and [1–4, 7, 8, 10] used for analyzing the water changes and also analyzing the water during the climate change, and some of them are used to know the drawbacks of different parameters or methodologies to strengthen our project and also some authors used the optical sensors for analyzing the water flow.

By considering all the above research papers i.e.; [1–10], we just analyzed the parameters and methodology that we need to use in our project and we do both quality check and also the surface water analysis and we used NDCI for the Quality check and NDWI, mNDWI for the analysis of the surface water at a certain time lap.

3 Methodology

3.1 Description

1. **Remote Sensing Image Collection**: Start the procedure by gaining access to remote sensing images from satellites like Landsat-8 and Sentinel-2 using the vast dataset of Google Earth Engine. These satellites collect multispectral data, which is essential for mapping water.
2. **Data Preprocessing:** Prior to analysis, clean up cloud cover and other atmospheric disturbances from the remote sensing data. Tools are available in Google Earth Engine to create cloud masks and apply them to images.
3. **Training Data Generation:** Create training datasets by picking sample data points inside the ROI that are either categorized as "water" or "non-water." The training of categorization models requires this data.
4. **Model Training:** Using the supplied training data, train models using machine learning classifiers such as Support Vector Machines (SVM), Classification and Regression Trees, and Naive Bayes. These models acquire the capability to classify pixels as either "water" or "non-water".
5. **Pixel Classification:** Use the trained classifiers to categories every pixel inside the ROI as either "water" or "non-water," producing a binary classification map.
6. **Calculation of Vegetation and Water Indices:** Calculate spectral indices for each pixel contained in the categorized imagery, such as the Normalized Difference Water Index and modified NDWI. These indicators are useful for separating vegetation from water bodies.
7. **Visualization using Google Earth Engine**: Map and visualize the surface water and vegetation indices by utilizing the visualization features of Google Earth Engine. Data understanding can be aided by interactive maps, time series animations, and other visualization techniques.
8. **Thorough Analysis:** Examine the categorized and indexed material in-depth. Within the ROI, derive statistical measures, track trends, and obtain knowledge about the dynamics of the vegetation and the surface water. Additional statistical and geographical methods may be used at this step.

3.2 Architecture Diagram

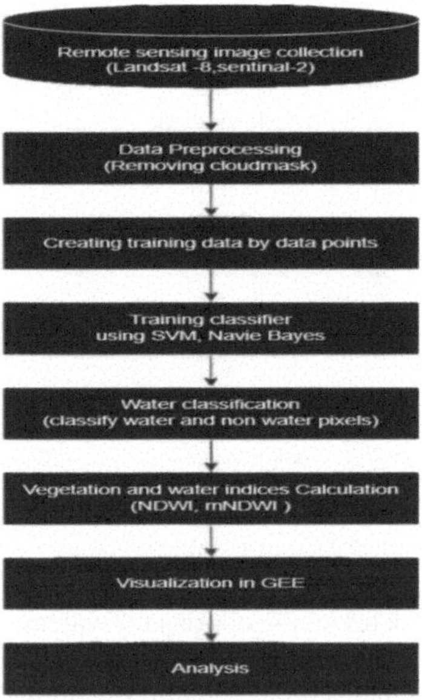

Fig. 1. Architecture Diagram-1

3.3 Description

1. **The Remote Sensing Image Collection:** In the first phase, Google Earth Engine is used to retrieve remote sensing data from sources like Landsat-8 and Sentinel-2. These satellites give multispectral data, which is essential for analyzing the water quality.
2. **Calculation of water indicators (NDWI, mNDWI):** We next calculate water-related spectral indices for each pixel inside our Region of Interest (ROI), including the Normalized Difference Water Index (NDWI) and modified NDWI (mNDWI). These indices are essential for locating water features in the gathered images.
3. **Pixel Masking:** We use pixel masks to remove unwanted or noisy data points in order to improve the accuracy of our analysis. Cloud masks or other pertinent masks can be used for this to take non-water characteristics out of our computations.
4. **NDCI and NDTI (Chlorophyll and Turbidity Indices**): Through the Normalized Difference Chlorophyll Index (NDCI) and Normalized Difference Turbidity Index (NDTI) methodologies, our analysis includes the estimate of water quality indicators, notably chlorophyll content and turbidity.
5. **Observing Turbidity Changes:** Examining turbidity variations over time or between various ROI sections is an important part of our investigation. This research indicates water quality patterns, which might vary depending on the amount of sediment present and the clarity of the water.

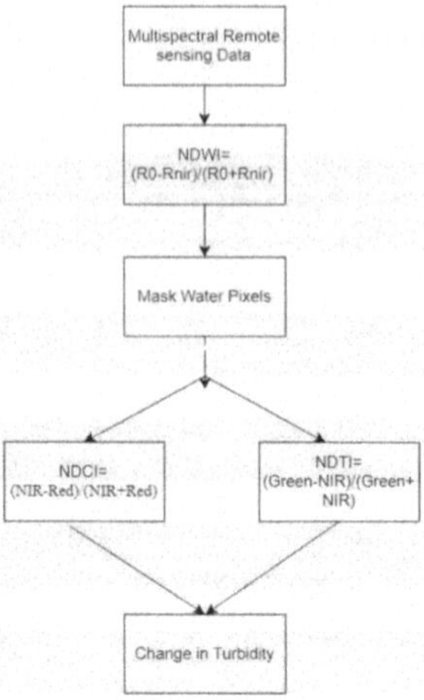

Fig. 2. Architecture Diagram-2

The different lessons that may be drawn from Fig. 1 include: Launch Google Earth and construct a polygon. Think about the options for water and no-water bodies. In order to improve the image quality, we must remove the cloud mask. Following that, we shall train both the water and non-water bodies. Then, we divide the land into water bodies and non- water bodies using the SVM algorithm. Then figure out the different indices, such as NDWI, MNDWI, NDTI, and NDCI. Visualize water categorization and NDWI, MNDWI, NDCI, and NDTI, and you're done (Fig. 2).

3.4 Description of Algorithms

SVM Algorithm

- Classification: Searches for optimal hyperplane/decision boundary separating the tuples of one class from another.
- The decision boundary of a Support Vector Machine (SVM) is defined as a linear combination of the support vectors.

$$y = w_0 + w_1x_1 + w_2x_2 + w_3x_3 + \cdots$$
$$y = w0 + \sum_{i=1}^{m} w_ix_i$$
$$y = w_0 + w^TX$$
$$y = b + w^TX$$

- If Feature vector is 2D then the linear equation represents straight line.
- If Feature vector is 3D then the linear equation represents hyper plane.
- W represents orientation & b represents position.
- g(x1) = wtx1 + b > 0 implies x1 belongs to C1
- g(x1) = wtx1 + b < 0 implies x1 belongs to C2

Naive Bayes Algorithm
Along with decision trees, neural networks, nearest nbr, one of the most practical learning methods is Naïve Bayes Algorithm.

$$v_{MAP} = \arg\max_{y_j \in V} P(a_1, a_2, ..a_n|vj)$$

$$v_{MAP} = \arg\max_{y_j \in V} \frac{P(a_1, a_2, ..a_n|vj)P(v_j)}{P(a_1, a_2, ..a_n)}$$

$$v_{MAP} = \arg\max_{y_j \in V} P(a_1, a_2, ..a_n|vj)P(v_j)$$

$$P(a_1, a_2, ..a_n|vj) = \prod_i p(a_n|vj)$$

$$v_{N\,B} = \arg\max_{y_j \in V} P(v_j) \prod_i p(a_n|vj)$$

Naïve_Bayes_Learn(examples)

For each target value v_j

$P(v_j) \leftarrow$ estimate $P(v_j)$

For each attribute value a_i of each attribute a

$P(a_j|v_j) \leftarrow$ estimate $P(a_j|v_j)$

Classify_New_Instance(x)

$$v_{N\,B} = \arg\max_{v_j \in V} p(v_j) \prod_i (a_i|v_j)$$

When to use

- Moderate or large training set available
- Attributes that describe instances are conditionally independent given classification

Successful applications:

- Diagnosis
- Classifying text documents.

4 Experimental Results

Data Collection
The Google Earth Engine and Java script were used for all computational and Programming operations. The LandSat-8 satellite is utilized to gather the data for this research.

Water-Bodies from 'SANGAMAM' make up the area under consideration.

Experimental Metrics

Normalized Difference Water Index (NDWI)

$$NDWI = \frac{(Green - NIR)}{(Green + NIR)}$$

Modified Normalized Difference Water Index (MNDWI)

$$MNDWI = \frac{(Green - SWIR)}{(Green + SWIR)}$$

Normalized Difference Chlorophyll (NDCI)

$$NDCI = \frac{(NIR - RED)}{(NIR + RED)}$$

Normalized Difference Turbidity Index (NDTI)

$$NDTI = \frac{(GREEN - NIR)}{(GREEN + NIR)}$$

Discussion on Results

The Fig. 3 indicates the surface water bodies which have been gradually changed through the years since 2015 till 2023 and this represents the history of water.

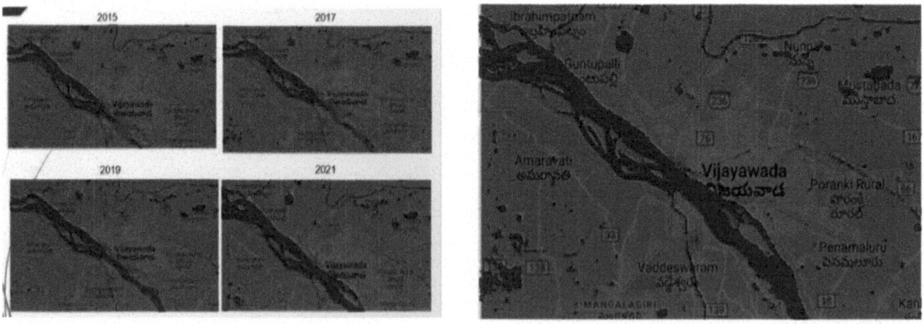

Fig. 3. Yearly Water History

The Fig. 4 indicates the Images that represents the lost water bodies in the region (2023) by comparing with the previous year's images.

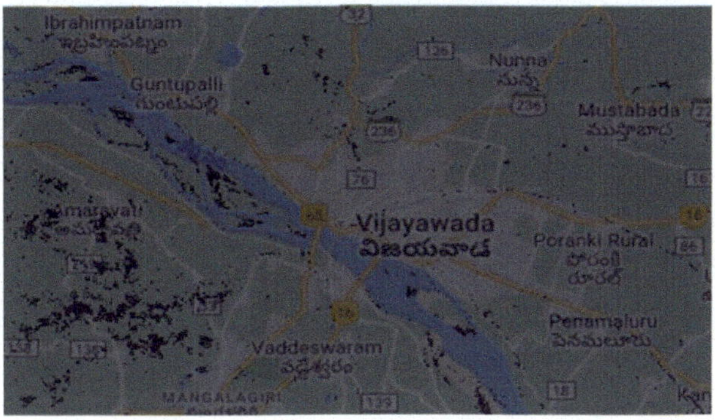

Fig. 4. Lost Water Bodies

The Fig. 5 indicates the image that is represented through NDWI in the region of SANGAMAM in 2023 (Figs. 6, 7 and 8).

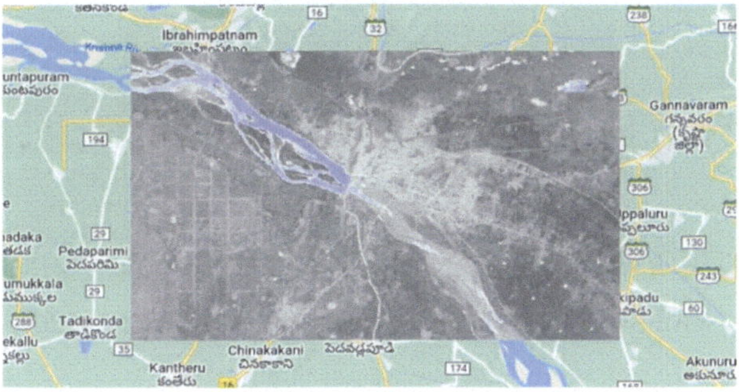

Fig. 5. NDWI in the region

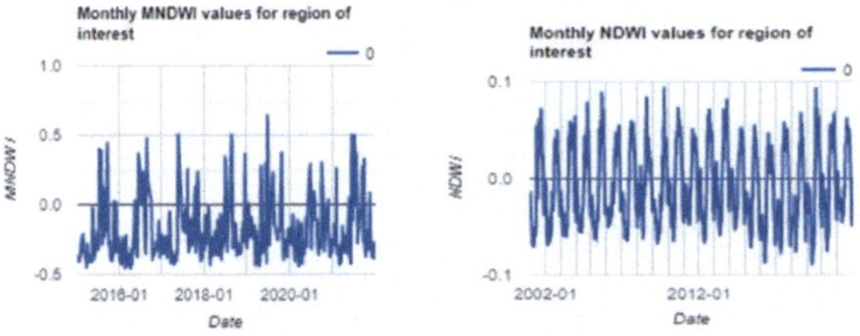

Fig. 6. NDWI and MNDWI Values yearly plot

The Fig. 9 Indicates the percentage of NDCI values, Blue color indicates less NDCI values, Orange color indicates average NDCI Values and Red color indicates the High NDCI values.

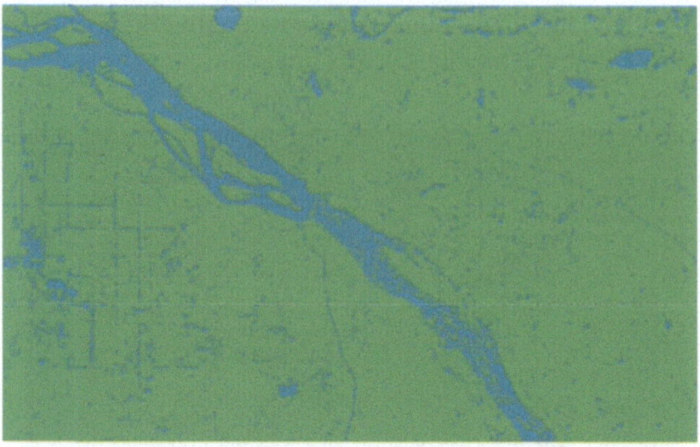

Fig. 7. SVM classifier plotted image over the period of 2015–2023

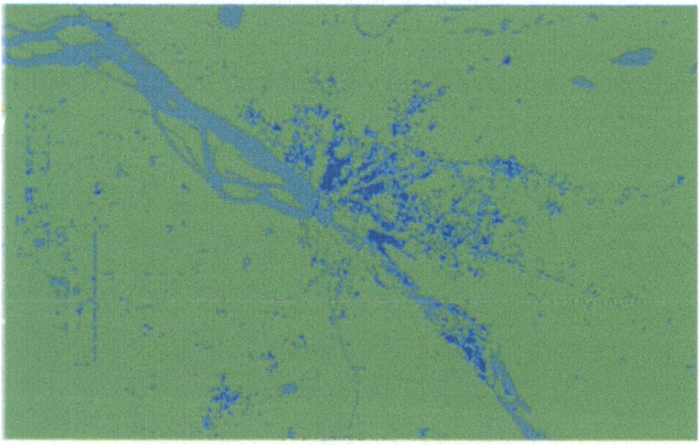

Fig. 8. Naive Bayes classifier plotted image over the period of 2015–2023

Fig. 9. NDCI in the period of 2015–2020

The Fig. 10 indicates the percentage of NDTI values, Blue color indicates low NDTI values, Red color indicates high NDTI values and the orange color indicates average NDTI values.

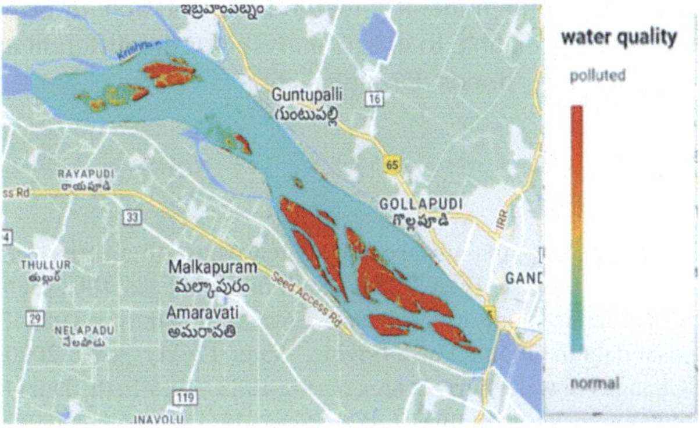

Fig. 10. NDTI in the period of 2015–2020

The images are those which are obtained from the final analysis of our project. By calculating the co-efficient matrix and accuracy by comparing the values we can find the best method (Fig. 11).

```
Confusion matrix NaiveBayes:
[[0,0,0],[0,7,2],[0,1,12]]

Overall Accuracy NaiveBayes:
0.8636363636363636

Confusion matrix SVM:
[[0,0,0],[0,8,1],[0,0,13]]

Overall Accuracy SVM:
0.9545454545454546
```

Fig. 11. Classifiers evaluation

5 Conclusion

Google Earth Engine's visualization tools can help in creating maps, charts, and graphs to visualize the results of the analysis. This can aid in identifying the changes in surface water area over time and understanding the spatial distribution of surface water bodies. Overall, the project demonstrates the changes occurred in the Krishna river basin over the time period of 2015–2020. It also demonstrates the capability of the gee to analyze multispectral images and detect changes in the water cover over time. This project can also demonstrate the quality of the water in the Krishna river basin over the time period of 2015–2020.

In this paper, after identifying the values of various indexes, we have to specifically go to that particular area and then identify the problem and able to give the analysis of the water recourses and we can able to use to analysis for the different applications like power generation, agriculture, floods, Droughts etc.

References

1. Sreekanth, P.D., et al.: Mapping surface-water area using time series landsat imagery on Google Earth Engine: a case study of Telangana, India. Curr. Sci. (00113891) **120**(9) (2021)
2. Wang, R., et al.: Dynamic monitoring of surface water area during 1989–2019 in the hetao plain using landsat data in Google Earth Engine. Water **12**(11), 3010 (2020)
3. Xia, H., et al.: Changes in water surface area during 1989–2017 in the Huai River Basin using Landsat data and Google earth engine. Remote Sens. **11**(15), 1824 (2019)
4. Wang, C., Jia, M., Chen, N., Wang, W.: Long-term surface water dynamics analysis based on Landsat imagery and the Google Earth Engine platform: a case study in the middle Yangtze River Basin. Remote Sens. **10**(10), 1635 (2018)
5. Huang, J., et al.: Characterizing the river water quality in China: recent progress and on-going challenges. Water Res. **201**, 117309 (2021)
6. Delpla, I., Jung, A.V., Baures, E., Clement, M., Thomas, O.: Impacts of climate change on surface water quality in relation to drinking water production. Environ. Int. **35**(8), 1225–1233 (2009)
7. Huang, C., Chen, Y., Zhang, S., Wu, J.: Detecting, extracting, and monitoring surface water from space using optical sensors: a review. Rev. Geophys. **56**(2), 333–360 (2018)
8. Kandekar, V.U., et al.: Surface water dynamics analysis based on sentinel imagery and Google Earth Engine Platform: a case study of Jayakwadi dam. Sustain. Water Res. Manag. **7**(3), 44 (2021)
9. Meena, S.R., Chauhan, A., Bhuyan, K., Singh, R.P.: Chamoli disaster: pronounced changes in water quality and flood plains using Sentinel data. Environ. Earth Sci. **80**(17), 601 (2021)
10. Liu, H., Xu, M., Beck, R.: An ensemble approach to retrieving water quality parameters from multispectral satellite imagery. In: IGARSS 2018–2018 IEEE International Geoscience and Remote Sensing Symposium, pp. 9284–9287. IEEE (2018)
11. Ashraf, M., Nawaz, R.: A comparison of change detection analyses using different band algebras for Baraila wetland with NASA's multi-temporal landsat dataset. J. Geogr. Inf. Syst. **7**, 1–19 (2015)
12. Mizuochi, H., Hiyama, T., Ohta, T., Nasahara, K.N.: Evaluation of the surface water distribution in north-central Namibia based on MODIS and AMSR series. Remote Sens. **6**, 7660–7682 (2018)
13. McFeeters, S.K.: The use of the normalized difference water index (NDWI) in the delineation of open water features. Int. J. Remote Sens. **17**, 1425–1432 (1996)
14. DeVries, B., Huang, C., Armston, J., Huang, W., Jones, J.W., Lang, M.W.: Rapid and robust monitoring of flood events using Sentinel-1 and Landsat data on the Google Earth Engine (2020)
15. Gillespie, T.W., Foody, G.M., Rocchini, D., Giorgi, A.P., Saatchi, S.: Measuring and modelling biodiversity from space. Prog. Phys. Geogr. **32**, 203–221 (2008)

16. Shelestov, A., Lavreniuk, M., Kussul, N., Novikov, A., Skakun, S.: Exploring Google Earth Engine platform for big data processing: classification of multi-temporal satellite imagery for crop mapping (2017)
17. Han-Qiu, X.A.: Study on information extraction of water body with the modified normalized difference water index (mNDWI) (2005)
18. Karsli, F., Guneroglu, A., Dihkan, M.: Spatio-temporal shoreline changes along the southern Black Sea coastal zone. J. Appl. Remote. Sens. **5**, 1–14 (2011)
19. Acharya, T.D., Subedi, A., Lee, D.H.: Evaluation of water indices for surface water extraction in a Landsat 8 scene of Nepal. Sensors **18**, 2580 (2018)
20. Nandi, D., Chowdhury, R., Mohapatra, J., Mohanta, K., Ray, D.: Automatic delineation of water bodies using multiple spectral indices. Int. J. Sci. Res. Sci., Eng. Technol. 4, 498–512 (2018)

Spyware Alert System and Image Steganography for Military Application

Nandigama Sushma(✉), Gandikota Prashanth Satya Sai Kiran, Esa Vinay, and Subba Reddy Chavva

Department of IT, VR Siddhartha Engineering College, Vijayawada, India
nsushma27@gmail.com

Abstract. Cybersecurity is increasingly threatened by the growing prevalence of spyware and other malicious software intended to enter computer systems and modify data. This research paper proposes a ground- breaking amalgamation of steganography techniques with spyware detection procedures to counter these risks and secure against unwanted access and covert data exfiltration. This study proposes a novel strategy to strengthen military cybersecurity by utilizing advanced image steganography techniques and creating a specialized Spyware Alert System. In the context of military operations, the Spyware Alert System uses cutting-edge threat detection algorithms and anomaly analysis to give real-time alerts and effective solutions against spyware threats. The paper also incorporates image steganography, a potent method for encrypting sensitive information and hiding it within digital photographs to protect the security and integrity of mission-critical data. By ensuring the secrecy and accuracy of critical information during military operations, this comprehensive solution strives to strengthen military data protection. The article provides a thorough approach for fending against the constantly changing panorama of cyber threats by outlining the aims, methodology, and anticipated contributions to military cybersecurity. This project aims to improve the security and resilience of military systems in the face of persistent digital attackers by pioneering this innovative amalgamation of technologies.

Keywords: Steganography · Spyware · Threat · Privacy · Cyber Defense · Keylogger · Trojan Horse

1 Introduction

The protection of digital assets and personal information is becoming more and more important in a time when everything is digital and connected. Our initiative focuses on two crucial facets of cybersecurity: the covert embedding of data into photos using steganography and the diligent monitoring of spyware threats like keyloggers and the avoidance of malicious file downloads. Image steganography uses technology to covertly send a cover image while concealing secret information within it. This allows for covert communication without raising any red flags [1]. Our study focuses on using steganography's strength for both defensive and offensive military purposes. Our first goal is to create a cutting-edge steganography technique that is specifically suited to military requirements. With the help of this technique, military personnel can securely and covertly

D. Pastor-Escuredo et al. (Eds.): ICDLAIR 2023, LNNS 1001, pp. 580–600, 2024.
https://doi.org/10.1007/978-3-031-60935-0_52

insert one image into another. The focus is on using cutting-edge data embedding techniques and military-grade encryption techniques to strengthen steganography's covert capabilities and make it resistant to cutting-edge detection technologies. According to various concealment strategies, the currently used image steganography can be divided into classic, coverless, and deep-learning based systems [2]. Our project incorporates a multifaceted defense strategy for spyware detection and prevention that is especially suited to military settings. Keyloggers and other espionage threats pose serious risks to delicate military operations and intelligence missions. Spyware is a sort of software that collects and distributes personal data without the owners' knowledge or agreement to advertising, data businesses, or outside users [2]. Keylogging software, commonly known as keyloggers, is a category of malware that secretly records user keyboard input in an effort to steal private information. Capturing typed characters or numbers is known as keystroke logging, commonly referred to as key logging. Passwords, user IDs, document content, and other potentially sensitive information are some examples of the data that can be gathered [3]. Our strategy combines watchful behavioral analysis with ongoing military system monitoring to recognize and effectively counter these threats. The confidentiality and integrity of vital military information are guaranteed by this instantaneous intervention. We seek to strengthen military information security, safeguard secret data, and uphold national interests by utilizing picture steganography's capabilities for secure data hiding and implementing proactive spyware detection techniques. By providing workable solutions to guarantee the secrecy and security of vital military information, our research and conclusions are positioned to significantly advance the developing field of military cybersecurity. The necessity to combat cyber espionage threats and stop data breaches in critical military operations is what spurred the development of a military spyware alert system. By quickly identifying and reducing spyware risks, this system seeks to improve operational security while protecting the privacy of classified data. Threat detection in real time is critical in the high-speed military setting. Moreover, the integration of image steganography detection is imperative in mitigating the use of clandestine communication techniques by adversaries, thereby strengthening the system's resilience against a wide range of cyber threats. Security is crucial because it keeps spies out. The Spyware Alert System acts as a guard dog, spotting and neutralizing any nefarious spyware that might attempt to access sensitive military data. It is a powerful defense against constantly evolving cyberthreats because it operates in real-time and responds quickly to any new threats. By limiting access to sensitive military data to only those who are authorized, the system lowers the possibility of insider issues. DCT image steganography conceals crucial information within seemingly unremarkable photos, acting as a kind of secret keeper to give crucial military details an additional degree of security. This integrated project's primary areas of focus are threat prevention and data privacy. In this paper, we developed a Spyware Alert System that leverages the YARA Framework and rules to analyze document and link patterns to determine the maliciousness of the content. We have established rules for analysis, such as hash-based and string- based, which are loaded into files that finish in the YARA Extension. The input file is then made available in order to ascertain its spyware content. Using the rules loaded in the YARA file as a guide, it examines the input file and determines whether spyware is present. This phase is all about averting dangers. Next, we will discuss DCT image steganography and its applications in the military, where highly sensitive data such as secret geographic locations, national security plans, and images are handled. This crucial information needs to be protected and kept confidential. By enclosing the original data within undesirable data and maintaining all pertinent information, DCT image steganography plays a critical role in safeguarding sensitive data in this situation.

2 Literature Review

Steganography of Images with Generalizability A method based on convolutional neural networks (CNNs) combines the power of CNNs with steganography, a technique for concealing information in digital images. In this method, CNNs are used to embed sensitive information into photographs while preserving the host image's visual quality. It stands out for its capacity to generalize, which enables it to operate on a range of images without becoming overfit to a particular dataset. This makes it a reliable and flexible method for hiding information in photographs, which can be beneficial for applications involving secure communication and data security [1]. The Histogram of Oriented Gradient (HOG) feature extraction method and the PVD-LSB (Pixel Value Differencing - Least Significant Bit) technique are combined in the steganographic technique known as an adaptive image steganography method in order to hide information in digital images. This technique can be used to cover up information in a variety of images because it adjusts to the host image's properties by taking gradients and texture aspects into account. It offers a reliable and safe steganography solution for data-hiding applications by fusing HOG and PVD-LSB [2]. With the help of multimedia steganography, presents an original technique that improves counting-based secret sharing. It enhances the efficiency and robustness of the counting-based confidential division method, enabling the concealment of shares within multimedia images. This research contributes to the field of multimedia steganography by offering a fresh approach to the secure and effective secret sharing of multimedia images [3]. The methodology that extends picture steganography by utilizing quantum substitution boxes is a revolutionary image steganography technology. The security and durability of data hidden within digital images are improved by using quantum substitution boxes, the quantum counterparts of conventional cryptography S-boxes. With the extra feature of quantum-level security, this technique makes it possible to conceal data transmission inside of images. This technology represents a promising advancement for secure data communication and protection using image steganography [4]. Authentic Steganography for 2D images Recaman's Sequence is used to strengthen the security of 2D picture steganography in a novel steganographic method called using Recaman's Sequence. Recaman's Sequence is a mathematical sequence that produces non-repeating numbers and is used to choose the pixel places for secret data embedding in a method that is both safe and impervious to statistical analysis. This technique guarantees the covert transmission of data within 2D images while upholding a high level of security, making it an exciting advancement in the realm of steganography for safe and private data communication and protection [2]. The efficiency of several mathematical sequences is compared to be assessed in the context of 3D image steganography. The goal of the study is to evaluate how various mathematical sequences might be used to conceal data in 3D images while upholding security and data integrity.

The research sheds light on the usefulness of these sequences for hiding data in 3D images by contrasting how well they function, ultimately advancing secure and reliable steganographic approaches for three-dimensional data [5]. Detailed analysis of steganographic techniques that conceal data in digital photos via Least Significant Bit (LSB) modification. The analysis in this paper takes considering a lot of important considerations include these approaches' ability to embed data, the security of the hidden data, the effect on the host image's visual quality, and the effectiveness of the embedding and extraction processes. Additionally, the study investigates real-world applications and contrasts LSB-based steganography with alternative methods to highlight the benefits and drawbacks of each approach. Overall, this study offers scholars and practitioners in the field of data security and concealment within digital imaging insights for understanding and evaluating the strengths and shortcomings of LSB-based image steganography [6]. Techniques for steganography that target the spatial aspect of digital images. This survey provides an overview of several spatial domain steganographic techniques, illuminating their core ideas, methodology, and distinguishing features. With an emphasis on methods that do not rely on frequency domain alterations, it examines a variety of techniques used to conceal information within images. The paper also explores spatial domain image steganography applications and prospective use cases. This survey is beneficial for researchers, practitioners, and those who are interested in data security and concealment inside images since it provides a thorough overview of the present state of spatial domain steganography [7]. The solution described in this [8] research uses deep learning AI to find and stop malware in user apps. It focuses on locating spyware and putting security controls in place for applications. The study emphasizes the usage of deep learning models for efficient spyware detection and stresses how crucial it is to protect user applications from harmful threats. A cutting- edge method for spyware detection in software programs. The main technique is setting up a controlled virtual environment so that software may be thoroughly examined for spyware activity without endangering the host system's security. Researchers [9] can carefully observe the software's operations in this controlled environment, spotting any odd behavior or coding patterns linked to spyware. The benefits of using virtual environments are emphasized in the paper, including increased security, reproducibility of studies, and protection of the host machine from potential spyware compromise. This research provides a workable and forward-thinking technique for spyware detection within software, ultimately leading to improved cybersecurity measures in the digital age where software security is of utmost importance.[10] proposes an innovative technique that uses irreversible feature modifications to detect Android malware. It aims to improve the detection and prevention of malicious software on Android devices, addressing a critical issue in the field of mobile security. The paper explores the fundamental idea of feature transformation, placing special emphasis on how it makes features in the Android environment hard to reverse engineering. The strategy successfully thwarts harmful attempts to examine and compromise the system by making sure these modifications are both reliable and irreversible. The study emphasizes the need of discovering patterns of malicious behavior while assessing the modified features to detect malware. In a time when Android smartphones are widely used and frequently attacked by hackers.[11] The paper delves into the specifics of Android spyware, offering insights into its design, functionality, and the methods it employs for unauthorized data collection and monitoring of user activities. Researchers rely on a data-driven approach to analyze a wide range of spyware samples, discerning common features, behavior patterns, and mechanisms used for obfuscation. This data-driven characterization provides a valuable foundation for understanding the evolving landscape of Android spyware, enabling better defenses against these invasive threats and safeguarding user privacy and security on Android devices.

3 Proposed Method

3.1 Architecture Diagram

The process starts with the input image. This could be any digital image in a common format, like PNG, JPG, or BMP. Apply any necessary pre-processing to the input image. This may involve resizing or formatting the image, depending on what the steganography method requires. Apply the Discrete Cosine Transform (DCT) to the image. DCT converts geographical data into frequency data by applying mathematics. It is widely used in picture compression algorithms such as JPEG. The DCT's coefficients are quantized. Quantization is the process of mapping a large range of input values to a smaller range. The amount of data that needs to be retained on file is reduced as a result of this procedure. The steganography algorithm embeds the secret data into the quantized DCT coefficients. Specific DCT coefficients can be changed to represent the hidden data bits. The changes must be made in a way that keeps the image from noticeably altering. To obtain the modified DCT coefficients, reverse the quantization process. The Inverse Discrete Cosine Transform (IDCT) can be used to convert the modified DCT coefficients back into the spatial domain. In this step, the image is rebuilt using the embedded data. Apply any necessary post- processing to the stego-image. This can mean altering the image format or quality, depending on the desired end result. The hidden data is contained in the final stego-image that is produced. Finding the right balance between the amount of information you want to hide and the aesthetic appeal of the image is necessary for effective DCT image steganography. Substantial alteration or over-embedding of data can result in detectable artifacts that could be signs of hidden information. However, if you embed too little data, steganography may not function as intended (Fig. 1).

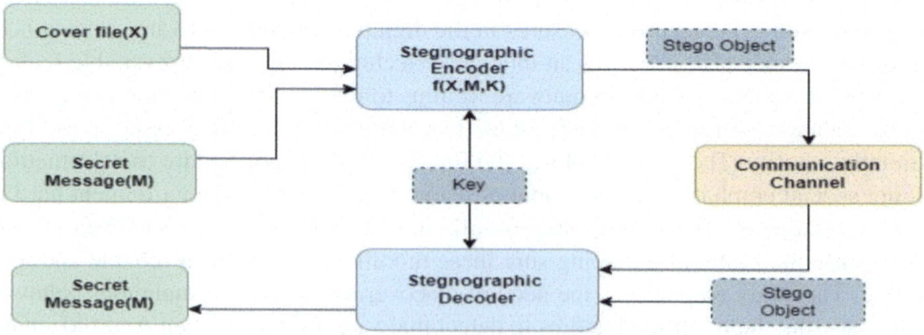

Fig. 1. Architecture Diagram

3.2 Description of Algorithms

3.2.1 Discrete Cosine Transform (DCT)

The DCT coefficients of particular blocks of frequencies are altered during DCT steganography to conceal information. To do this, the DCT coefficients are often modified by adding or removing a minor value.

This is how the procedure is summed up:

1. Divide the image into N x N chunks that don't overlap.
2. Transform each block using DCT algorithm.
3. To embed the hidden data, choose particular DCT coefficients within the blocks.
4. Make the necessary adjustments to these coefficients to encrypt the concealed data.
5. Use the hidden data to reconstruct the image using the inverse DCT(IDCT)

3.2.2 Steganography in Frequency Domain

Using the Fourier transform, the method most usually used to separate an image into its sine and cosine components. We employ DCT, DWT, and DFT among other frequency domain approaches that are readily available. The signal or image is converted from the spatial to the frequency domain using the discrete cosine transform (DCT), DCT separates the picture of low, middle, and high-frequency coefficients (Fig. 2).

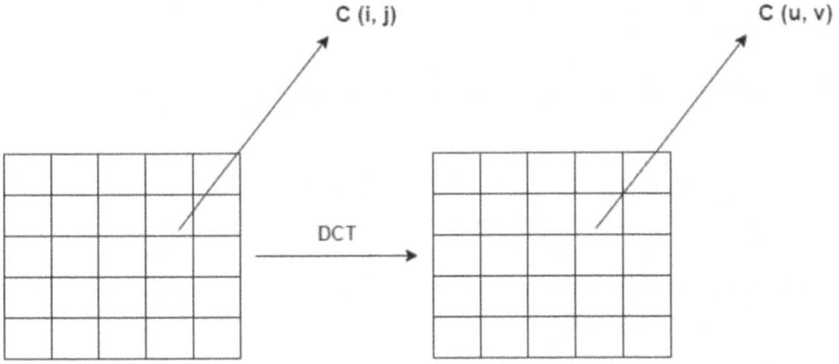

Fig. 2. DCT Transform

Let I(x,y) represent a cover image that is 8 bits and has the values 1,2, 3,...m and 1,2, 3,...n. Each of the eight 8 x 8 blocks in this m x n cover image is subjected to a 2-D DCT.

$$C(\mathbf{u},\mathbf{v}) = \mathbf{a}(\mathbf{u})\mathbf{a}(\mathbf{v})\sum\nolimits_{x=0}^{N-1}\sum\nolimits_{x=0}^{N-2} f(x,y)\mathbf{cos}\left[\frac{(2x+1)u\pi}{\cdot 2N}\right]\cos\left[\frac{(2y+1)u\pi}{2N}\right] \quad (1)$$

For u= 0, 2….7 and v = 0,1 …..7 (3) 1
Where a(k) = { $\sqrt{}2$ for K = 0}
Otherwise, 1
IDCT's mathematical formula is

$$F(\mathbf{x}, \mathbf{y}) = \sum_{x=0}^{N-2} \sum_{x=0}^{N-2} C(u)C(v)\cos\left[\frac{(2x+1)u\pi}{2N}\right]\cos\left[\frac{(2y+1)u\pi}{2N}\right] \quad (2)$$

The secret data is first encrypted for added security, and the ASCII values that have been encrypted are then converted into binary form [3]. Due to its advantages, which include its high ratio of compression, low bit error rate, strong ability to integrate information, and strong. Due to its ability to simulate objects and impact calculations, DCT is well-regarded in a variety of applications.

3.2.3 Algorithm to Embed the Text Message

Algorithm:

1. Examine the cover image.
2. Read a coded message and convert it to binary.
3. The cover image is divided into 8 × 8-pixel blocks.
4. To get rid of 128 in each pixel block, subtract 128 while moving from top to bottom and left to right.
5. Each block is given a DCT.
6. Each block is compressed using a quantization table.
7. Assign the secret message to each bit in the LSB of the DC coefficient.
8. Compose a stego image.

3.2.4 To Retrieve Text Messages

Algorithm:

1. Examine the Stego image.
2. An 8 by 8 block of pixels makes up the Stego image.
3. Subtract 128 from top to bottom and left to right in each pixel block.
4. Every block receives a DCT.
5. The quantization table compresses each block.
6. Figure out each DC coefficient's LSB.
7. Obtain each 8-bit character by retrieving them.

3.2.5 Algorithm for DCT Image Steganography

Input:

- Cover image, original image.
- Secret image: the hidden message in the cover image.

Output:

- Stego image: the image with hidden message

Procedure:

```
EmbedMessageInImage(cover_image,secret_image):
{
block_size = DefineBlockSize()
quantization_matrix = DefineQuantizationMatrix()
message_index = 0
While message_index < length(secret_message) Do
{
   Divide the cover_image into non-overlapping blocks of size block_size
   For each block in cover_image Do
   {
      Apply DCT to the block
      Flatten the DCT coefficients into a 1D array
      For each coefficient in the array Do
         {
           binary_charecter=ConvertNextCharecterToBinary(secret_message)
           ModifyCoefficientWithBit(coefficient,binary_charecter)
         }
      }
           Inverse DCT to get the block
           Replace the block in the cover_image with the modified block
   }
         stego_image=cover_image
         Return stego_image
   }
```

3.2.6 Procedure for extracting Message from Image

```
ExtractMessageFromImage(stego_image):
{
  block_size = DefineBlockSize()
  message= " "
  message_length=0

 \\ Separate the stego image into non-overlapping chunks of the size block_size.
  For each block in stego_image Do
  {
     Apply DCT to the block
     Flatten the DCT coefficients into a 1D array
     For each coeffiecient in the array Do
     {
       Extracted_bit=ExtractLeastSignificantBit(coefficient)
       Append extracted_bit to message
       Increment message_length until it reaches the required length
       }
   }
   extracted_message=ConvertBinaryToText(message)
    Return extracted_message
  }
  Function DefineBlockSize():
  {
        Return a chosen block size
  }
  Function DefineQuantizationMatrix():
  {
        Return a predefined quantization matrix
  }
  Function ConvertNextCharacterToBinary(secret_message):
{
        Convert the next character in secret_message to binary representation
        Return the binary representation

}
  Procedure ModifyCoefficientWithBit(coefficient, bit):
{

        Replace the specified bit in the coefficient's least significant bit.

  }
  Function ExtractLeastSignificantBit(coefficient):
  {
        Extract and return the least significant bit of the coefficient
  }
  Function ConvertBinaryToText(binary_message):
  {
        Convert binary message to its original text representation
        Return the text message

   }
```

3.2.7 Quantization

Quantization, a lossy compression method used in image processing, compresses reducing a group of values to a unique quantum or comparatively tiny discrete value. With fewer discrete symbols, a given stream becomes more compressible. For instance, by using fewer colors to represent an image, the digital image's file size can be lowered. To achieve quantization, a matrix similar to the one below is used:

$$Q = \begin{bmatrix} 16 & 11 & 11 & 16 & 24 & 40 & 51 & 61 \\ 12 & 12 & 14 & 19 & 26 & 58 & 60 & 55 \\ 14 & 13 & 16 & 24 & 40 & 57 & 69 & 56 \\ 14 & 17 & 22 & 29 & 51 & 87 & 80 & 62 \\ 18 & 22 & 37 & 56 & 68 & 109 & 103 & 77 \\ 24 & 35 & 55 & 64 & 81 & 104 & 113 & 92 \\ 49 & 64 & 78 & 87 & 103 & 121 & 120 & 101 \\ 72 & 92 & 95 & 98 & 112 & 100 & 103 & 99 \end{bmatrix}$$

3.2.8 Comparative Analysis

In Spyware Detection current method a brief explanation of common techniques used in current spyware detection systems, including machine learning algorithms, heuristic analysis, and signature-based detection. The Proposed System emphasized the special features of your Spyware Alert System and how it either deviates from or enhances current practices. Discussed how your system compares to conventional methods in terms of efficiency, accuracy, and real-time capabilities.

Steganography Techniques: Current Strategies Provide an overview of the most widely used steganography methods in military applications, such as frequency domain techniques or LSB (Least Significant Bit) embedding. The DCT image steganography methods that are specifically used in your system are described in Proposed Techniques. Emphasize any innovations, improvements, or special characteristics that set your method apart from other steganography techniques. We achieved the highest accuracy in hiding the secret image in the cover image in our paper. The DCT algorithm is the most resilient to compression; however, in contrast to LSB, DCT exhibits greater susceptibility resilient.

3.3 Spyware Detection through Alert Message using Regular Expression

Threat hunting is a proactive strategy to discovering and managing cyber threats that have already infiltrated a company's network. It entails actively looking for signs of compromise (IOC) and evidence of malicious behavior that may have gone undetected by more standard security measures such as firewalls or antivirus software. To uncover and pinpoint potential hazards, threat hunters employ a variety of techniques, including the examination of log files, network traffic, and system configurations. They may also employ tools such as malware analysis programs, security incident and event management (SIEM) systems, and threat intelligence feeds to aid in the discovery of potential threats. Yara is a tool that is used to detect and classify malware and other undesirable applications. It achieves this by utilizing a set of rules known as Yara rules, which are expressed in a unique syntax. These rules define the characteristics of the malware or other potentially harmful software that Yara is looking for. A Yara rule, for example, can specify that a given spyware family can be identified by a specific file signature or that it has a specified collection of text strings in its code. When Yara is executed, it compares the properties specified in the Yara rules to the characteristics of the software being analyzed. Yara flags software that meets the requirements mentioned in the rule as potentially dangerous. Using Yara rules, we can detect a wide range of malicious software, including Trojan horses, viruses, and other spyware.

Fig. 3. Flow diagram of Spyware Detection

Using the YARA framework, spyware detection entails building custom rules that match particular patterns or traits connected to recognized spyware signatures. YARA is an effective tool for matching patterns and for classifying and detecting malware based on rules. Defining the traits or patterns connected to spyware by creating a YARA rule file (.yar). This entails giving specific strings, hex patterns, or other spyware file-specific indicators.

```
rule SpywareRule {
strings:
$string1 = "unique_string1"
$string2 = "unique_string2"
$hex_pattern = { 90 90 ?? ?? 90 }
condition:
$string1 or $string2 or $hex_pattern
}
```

"unique_string1" and "unique_string2" should be changed to actual strings or patterns connected to spyware. Make sure your YARA rule correctly detects spyware by testing it on sample files. The yara command-line tool is available for testing.

```
yara - r your_ rule_ file.yar path_ to_ sample_ files/
```

Spyware changes all the time, so to keep your YARA rules working, make sure you update and improve them frequently. This entails examining fresh spyware samples, identifying distinctive patterns, and incorporating them into your rule set. Incorporating the YARA rules into applications or security systems that handle file scanning. This can involve specialized security solutions, intrusion detection systems, or antivirus software. Using programs or YARA generators to automate the process of creating rules based on sample analysis. These tools guarantee a more thorough coverage of spyware variants and can expedite the rule creation process. YARA rules are transferable among security community members. Information on intelligence and security. Effectively addressing these issues is the suggested system, which combines a DCT image steganography with a Spyware Alert System. Engage in communities and platforms that share threat intelligence to gain access to common guidelines and enhance spyware detection capabilities overall. Putting in place and keeping an eye on systems that periodically search for matches in accordance with your YARA rules. Set up an incident response system to do the necessary actions, like quarantining or notifying security personnel, when a match is discovered. Continually instruct and train security staff on how to read YARA rule matches and handle possible spyware incidents. This guarantees a well-coordinated and efficient reaction to security risks. Information security and privacy are critical in military applications because the data frequently contains critical plans, intelligence, and security details. The suggested system is made to address these issues by combining a DCT image steganography with a Spyware Alert System.

3.3.1 Description of the Spyware Detecting Process

- Load YARA Rules: To start the spyware detection process, load the YARA rules that you've developed or acquired. These guidelines specify the patterns or traits you wish to look for while searching through files.
- Compile Rules: To make the loaded YARA rules more effective in finding patterns in files, they are converted into a binary format.
- Input Files: The files or folders you wish to check for spyware are represented in this stage. Any files that you believe may contain malware can be included here.
- Scan Files: The built YARA rules are used to scan the input files. The YARA engine examines the file contents to check whether any of the stated patterns are present.
- Matching the Rules? The system checks to see whether any YARA rules matched after scanning each file. When one or more conditions are met, it indicates that spyware traits have been found in the file.
- Report Matches: If rules are in agreement, the system creates a report or log detailing which rules were compatible with which files. You may use this information to spot potentially harmful files.
- Display or Log: The user has the option of seeing the matched results or logging them for further analysis and action.

Here's an example of a simple Yara rule that might be used to identify a certain type of malware:

```
rule example_rule {
strings:
$a = "malware string 1"
$b = "malware string 2"
condition:
any
}
```

3.3.2 Types of Yara Rule

- To recognize and categorize malware and other harmful software, you can utilize a variety of Yara rules. Some instances are as follows:
- Malware can be identified using rules that use text strings as literal values or regular expressions. A string-based criterion could be used to detect malware that has a specified string of characters in its source code, for example.
- File metadata-based rules: These rules examine the data present in the files being scanned in order to detect malware. A file metadata-based rule, for example, could be used to find files of a specific file type or files that were recently generated or updated.
- Rules based on hashes These laws use cryptographic hashes to detect malware. A file's contents are uniquely represented by a cryptographic hash, which will alter if any aspect of the file does. The use of hash-based criteria allows for the detection of malware that has been altered with or camouflaged in order to avoid detection.

Network-based rules: These policies detect malware by analyzing network traffic data such as IP addresses or ports. For example, it may use a network-based rule to detect malware that communicates with a specific IP address or port.

3.3.3 String Based Yara Rules

The text strings connected to the virus that you are attempting to detect must be specified when writing a string-based Yara rule. A straightforward string-based Yara rule is demonstrated here:

```
rule example_string_rule
{
  strings:
    $a = "malware string 1"
    $b = "malware string 2"
  condition:
  any of them
}
```

The two strings $a and $b in this Yara rule are linked to the virus you're looking to find.

3.3.4 File Metadate-Based Yara

To create a file data-based Yara rule, you must define the metadata properties of the files associated with the virus you're looking for. The following is an example of a simple Yara rule based on file information:

```
rule example_metadata_rule
{
  condition:
    file.extension == "exe" and
    file.size > 100KB
}
```

3.3.5 Hash Based Yara

To create a hash-based Yara rule, you must specify the cryptographic hashes of the files associated with the virus you're looking for. Here's an example of a basic Yara rule that makes use of a hash rule:

```
example_hash_rule :
        {
          strings:
            $a = {0A0B0C0D0E0F0A0B0C0D0E0F0A0B0C0D}
          condition:
            $a
        }
```

4 Results and Observations

4.1 Result and Analysis of Image Steganography

In an experiment involving the usage of cover images with steganography, the objective is to obfuscate as much information as you can while undergoing the least perceptible alteration to the stego-image. The difference in Mean Squared Error (MSE) and Peak Signal to Noise Ratio (PSNR) between the cover picture and the stego-image. Serve as an objective and practical operational measurement for image distortion. MSE, or Mean Square Error It is a measurement of how much the initial image and the deformed or noisy image have changed. This formula is used to determine mean square error:

$$\mathbf{MSE} = \frac{1}{mxn}\sum_{i=0}^{n-1}\sum_{j=0}^{n-1}(o(i,j) - s(i,j))^2 + (\mathbf{2n} + \mathbf{m}) \tag{3}$$

$$\mathbf{MSE} = \frac{1}{mxn}\sum_{i=0}^{n-1}\sum_{j=0}^{n-1}(o(i,j) - s(i,j))^2 \tag{4}$$

where "S" and "O" signify the stego-image and the original picture, respectively, and their image sizes are (m x n). Where m and n are the number of rows and columns. Peak signal to Noise Ratio (PSNR), a measure of image quality, can be found by comparing the original image and the stego- image. The quality of the stego image is assessed using PSNR once the secret message has been encoded in the cover image.

$$\mathbf{PSNR} = \mathbf{10}\log_{10}\frac{\mathbf{256^\wedge 2}}{MSE} \tag{5}$$

where the wide range of bits that make up each picture pixel is represented by the number 256. As an example, the depth region of the images is where the PSNR parameter is calculated (Tables 1, 2 and Fig. 4).

Table 1. PSNR and MSE

Cover Image	DCT technique	
	PSNR (db)	MSE (db)
Img1.jpg	58.454	0.577
Img2.jpg	53.56	0.695
Img3.jpg	49.22	0.375
Img4.jpg	50.42	0.378

Table 2. Parameter Analysis of DCT Method

Features	DCT
Invisibility	High
Cover image capacity	Medium
Robustness against image manipulation	Medium
PSNR	High
MSE	Low

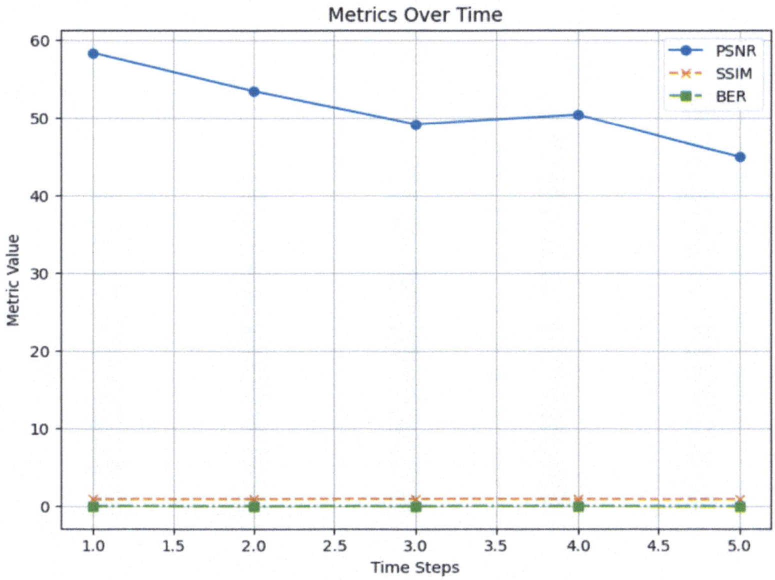

Fig. 4. Result of Image Steganography

4.1.1 Performance Metrics for Image Steganography

Three crucial performance metrics—Peak Signal-to-Noise Ratio (PSNR), Structural Similarity Index (SSIM), and Bit Error Rate (BER)—are visually represented in Figure 3 as they have changed over time. The effectiveness and precision of various signal processing or communication systems are frequently evaluated using these measures over a number of time steps. The graph shows the following salient characteristics: The time steps from 1 to 5, which indicate various occurrences or intervals across which the metrics were measured, are shown on the X-axis. The metrics values for each time step, which serve as critical gauges of the system's performance, are shown on the y-axis. Each of the three measures is shown by a separate bar, and is highlighted by a different colour Peak Signal-to-Noise Ratio (PSNR) PSNR readings are shown in blue, declining from an initial high of a decrease in signal fidelity with time can be seen at time steps 1 to 45 at time step 5, for example. The orange-coloured SSIM (Structural Similarity Index) value ranges from 0.90 at time step 1 to 0.94 at time step 3, indicating substantial structural similarity in the processed data. These values consistently remain high and are comparatively stable. Bit Error Rate, or BER The BER values, shown in green, show a small variation, starting at 0.02 at time step 1, temporarily peaking at 0.022 in time step 4, and then reverting to 0.014 in time step 5.The picture offers a clear comparison of how these measures vary over time, providing useful insights into the system's performance dynamics. Signal processing and communication system researchers and practitioners can use this visualization to assess the strengths and weaknesses of the system at different time intervals (Figs. 5 and 6).

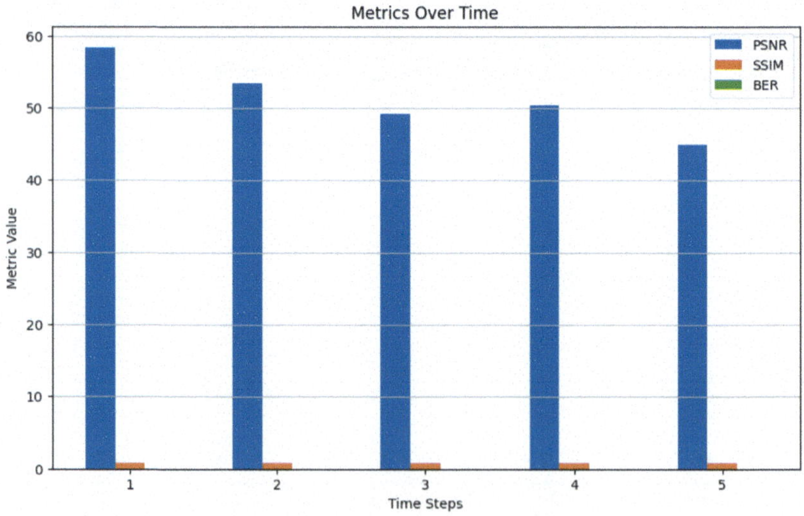

Fig. 5. Temporal Evolution of Performance Metrics

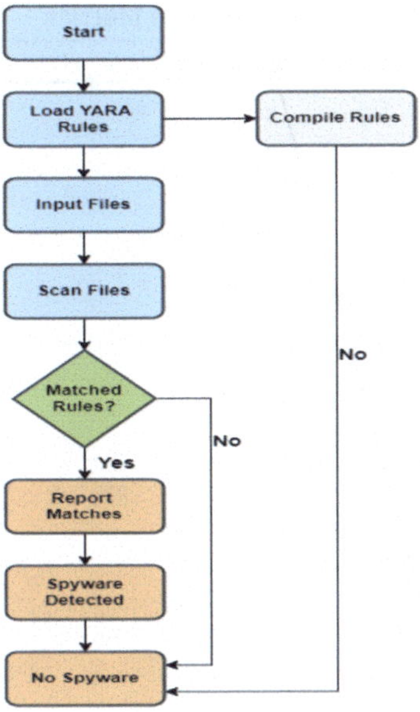

Fig. 6. Bar graph that shows the temporal evolution of PSNR,SSIM,BER

4.1.2 Performance Metrics for Spyware Detection

This statistic, which has a high value of 0.94, shows how well the system can recognize instances of spyware. This statistic, which has a low value of 0.04, assesses how frequently the system misclassifies benign software as spyware. The proportion of genuine positive detections to all positive detections is known as precision. It is 0.959 in this instance, demonstrating the system's accuracy in detecting spyware. The F1 Score achieves a score of 0.949, indicating a strong balance between these two key factors, by balancing the trade-off between precision and recall. This score, which measures the classifications made by the system overall accuracy, is exceptionally high (0.95). Specificity, which assesses the capacity to correctly identify instances of non-spyware, registers at 0.96 (Figs. 7 and 8).

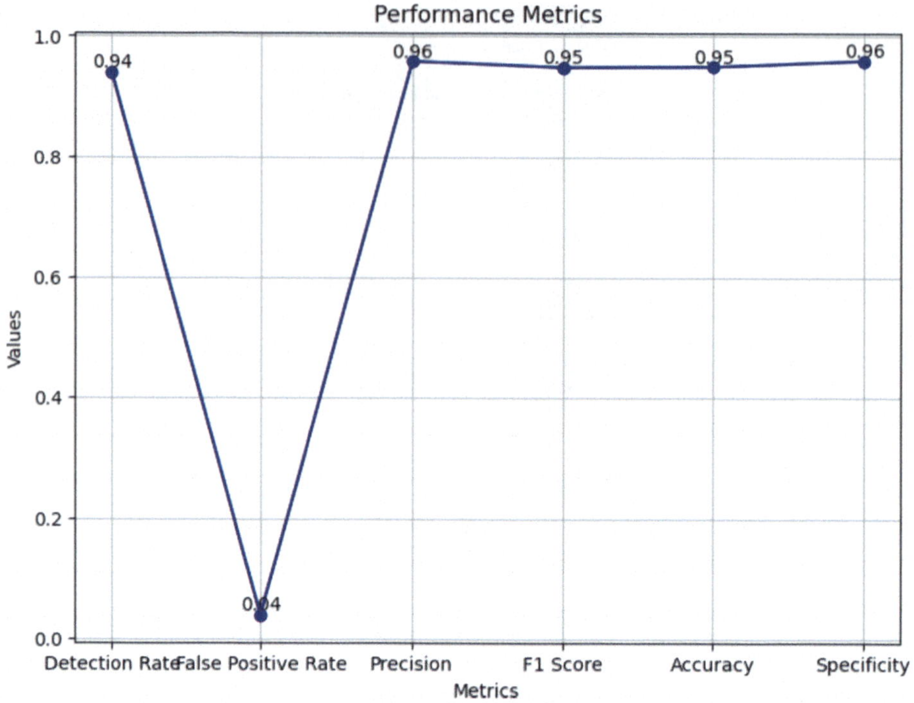

Fig. 7. A spyware detection system's major performance parameters are represented graphically in the figure.

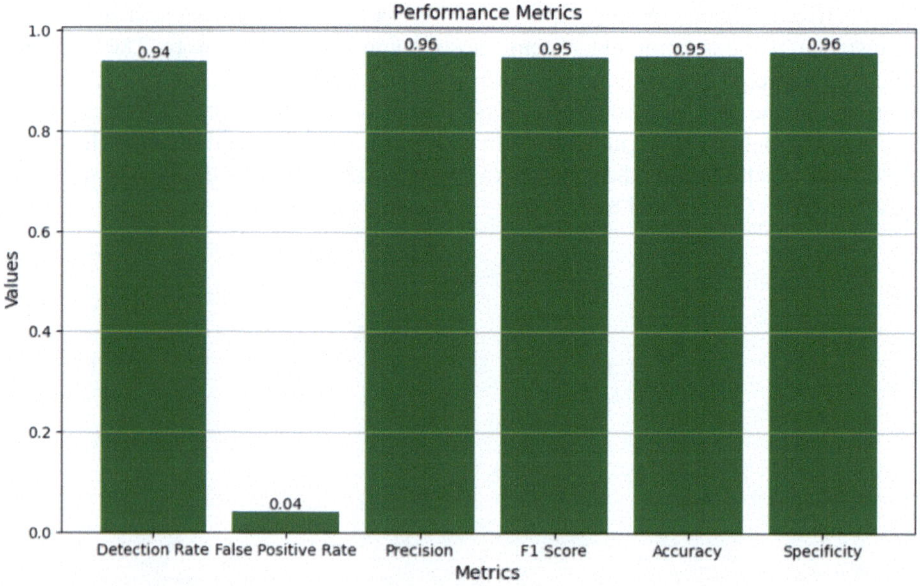

Fig. 8. Performance metrics for spyware detection systems are displayed in a bar graph

5 Conclusion

Using a method known as image steganography, data can be masked in photographs. Numerous uses exist for this, including data storage and covert communication. Spyware alert systems can hide information regarding spyware infections, such as the type of malware, the location of infected files, and the identity of the attacker, using picture steganography. After then, a central server may receive this data for analysis and response. The addition of photo steganography to malware alert systems has various benefits. First of all, it might help make spyware invasions more covert. Behavioral analysis or signatures are widely used by conventional spyware alarm systems to detect infestations. However, attackers can easily circumvent these detection methods by altering their malware. When spyware detection information is cloaked in images utilizing picture steganography, it is much more difficult for attackers to find. The necessity to combat cyber espionage threats and stop data breaches in critical military operations is what spurred the development of a military spyware alert system. By quickly identifying and reducing spyware risks, this system seeks to improve operational security while protecting the privacy of classified data. Threat detection in real time is critical in the high-speed military setting. Moreover, the integration of image steganography detection is imperative in mitigating the use of clandestine communication techniques by adversaries, thereby strengthening the system's resilience against a wide range of cyber threats.

References:

1. Duan, X., Liu, N., Gou, M., Wang, W., Qin, C.: SteganoCNN: image steganography with generalization ability based on convolutional neural network. Entropy **22**(10), 1140 (2020)
2. Hameed, M.A., Hassaballah, M., Aly, S., Awad, A.I.: An adaptive image steganography method based on histogram of oriented gradient and PVD-LSB techniques. IEEE Access **7**, 185189–185204 (2019). https://doi.org/10.1109/ACCESS.2019.2960254
3. Farrag, S., Alexan, W.: Secure 2D image steganography using recamán's sequence. In: 2019 International Conference on Advanced Communication Technologies and Networking (2019)
4. Gutub, A., Al-Ghamdi, M.: Hiding shares by multimedia image steganography for optimized counting-based secret sharing. Multimedia Tools Appl. **79**(11–12), 7951–7985 (2020). https://doi.org/10.1007/s11042-019-08427-x
5. Abd EL-Latif, A.A., Abd-El-Atty, B., Venegas-Andraca, S.E.: A novel image steganography technique based on quantum substitution boxes. Optics Laser Technol. **116**, 92–102 (2019)
6. Alexan, W., El Beheiry, M., Gamal–Eldin, O.: A comparative study among different mathematical sequences in 3d image steganography. Int. J. Comput. Digital Syst. **9**(4), 545–552 (2020). https://doi.org/10.12785/ijcds/090403
7. Thangadurai, K., Sudha Devi, G.: An analysis of LSB based image steganography techniques. In: 2014 International Conference on Computer Communication and Informatics (ICCCI-2014) , Coimbatore, India, 03–05 January 2014 (2014)
8. Mahesh, V., Sumithra devi, K.A.: Spyware detection and prevention using deep learning AI for user applications. Int. J. Recent Technol. Eng. (IJRTE) **7**(5) (2019). ISSN: 2277–3878
9. Mallikarajunan, K.M.E.N., Preethi, S.R., Selvalakshmi, S., Nithish, N.: Detection of spyware in software using virtual environment. In: Third International Conference on Trends in Electronics and Informatics (ICOEI 2019) IEEE Xplore Part Number: CFP19J32-ART (2019). ISBN: 978-1-5386-9439-8

10. Han, Q., Subrahmanian, V.S., Xiong, Y.: Android malware detection via (somewhat) robust irreversible feature transformations. IEEE Trans. Inf. Forensics Secur. **15**, 3511–3525 (2020). https://doi.org/10.1109/TIFS.2020.2975932
11. Pierazzi, F., Mezzour, G., Han, Q., Colajanni, M., Subrahmanian, V.S.: A data-driven characterization of modern android spyware. ACM Trans. Manag. Inf. Syst. (TMIS) **11**(1), 1–38 (2020). https://doi.org/10.1145/3382158

Abstract. A Review on YOLO Algorithms for Social Distancing

Vijay Kumar(✉) and Mahendra Kumar Murmu

Computer Engineering Department, National Institute of Technology, Kurukshetra 136119, India
golla_32113117@nitkkr.ac.in

Abstract. Object detection is one of the appealing branch of the study in the fields of computer science. Object detection defines some popular object detection models in deep learning such as YOLO (You Only Look Once), Single Sort Detector, R-CNN etc. It can be applied to solve many important challenges in computer vision, human-computer interaction such as health-care monitoring, autonomous driving, anomaly detection etc. Social distancing terminology has been coined repeatedly in-light of COVID19. Social distancing is an inherent part of health-care systems. In general, nowadays, social distancing applications have become more important due to the transmission of various diseases like smallpox, Scabies, COVID19, etc. It happens due to not maintaining of specified distance between infected peoples. The paper proposes a review of real-time applications of YOLO based object detection in social distancing. The present work is useful to understand the overview of YOLO models that can be applied for social distancing. Additionally, comparisons between the various YOLO versions with respect to different performance metrics concerning speed, accuracy, and computational complexity. These results are vital for the selection of an optimal YOLO version for real-time social distancing application.

Keywords: Object Detection · YOLO · Surveillance system · Social Distancing · CNN · DeepSORT

1 Introduction

Visual image system is one of the mostly heard topics in computer science discipline. The visual image refers to the complex networks structure and processes. It allows an entity to perceive and interpret visual information from the environment. It is experienced that, the visual system is fast and accurate in various applications. The object detection is considered as a part of the visual image systems. For example, eyes may be the human visual system, which detect light and convert it into neural signals. Then after, the human's brain processes these signals and create our consciousness experience of seeing. In traditional computer system, computer vision techniques automatically detect and identify objects through digital images or videos.

D. Pastor-Escuredo et al. (Eds.): ICDLAIR 2023, LNNS 1001, pp. 601–617, 2024.
https://doi.org/10.1007/978-3-031-60935-0_53

In visual systems, nowadays, convolutional neural networks (CNN) is frequently used to process (e.g. analyze) the input objects such as images including videos. The CNN extracts features from the image, and used to identify and classify the objects. The visual system in object detection is used in various applications, such as surveillance, autonomous vehicles, robotics etc. The work is to mimic the human brain to a computer so that it can process the visual system accurately and fast like human brain. There are numerous algorithm exist in this direction such as YOLO (You Only Look Once), SSD (Single Shot Detector), R-CNN (Regions with Convolutional Neural Networks), Faster R-CNN, Fast R-CNN, Mask R-CNN, R-FCN (Region-based Fully Convolutional Networks) etc. Most of the object detection models uses a two-stage process. In the first stage, region proposals are generated, and in the second stage, the classification of the objects within those proposals are performed. In result, the systems perform slow because they requires multiple forward and backward passes of the networks, and they are limited in their ability to handle large and complex objects. As per as the YOLO object detection algorithm is concerned, it works faster and more accurately which is well-suited for real-time applications. YOLO performs object detection and classification in a single step.

The present works is mainly focuses on the study and analysis of YOLO series algorithms in view of social distancing. YOLO is an object detection algorithm, introduced in 2015 [1]. The architecture of YOLO is simple, efficient, and scalable. Due to this, it has been observed that it becomes a popular choice for object detection among researchers. There are several improvements and extensions of the systems exist such as YOLOv2, YOLOv3, YOLOv4,YOLOv5, YOLOv6, and YOLOv7. YOLO has significantly improved its accuracy and speed. YOLO is a state-of-the art object detection system model. It can process images and videos in real-time to detect objects that may include person, cars, bicycles, animals etc. YOLO is unique in a way that it can process the entire image in one forward pass of deep neural networks instead of dividing the image into smaller regions and processes each one individually. Due to this, YOLO can make fast predictions of object locations in real-time. The working of basic YOLO structure is given in Fig 1.

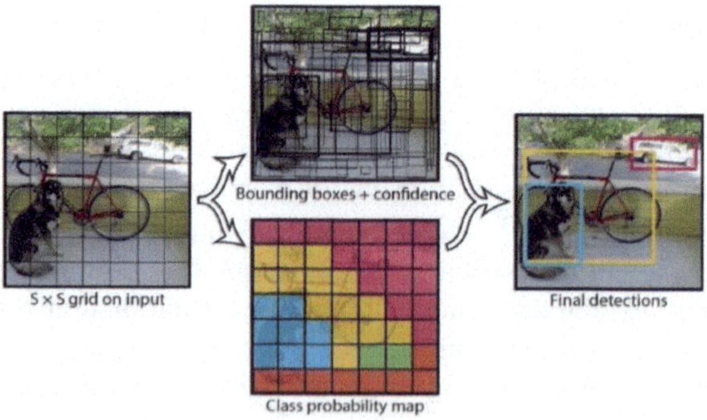

Fig. 1. Basic Working of YOLO [14]

YOLO uses a single neural network to directly predict object bounding boxes and class probabilities in a single forward pass. Nowadays, YOLO is popularly used in various applications, including surveillance systems, self-driving cars, security systems, etc. YOLO is considered one of the most important object detection systems available. One of the important applications where YOLO can be used is social distancing [2]. The literature shows that YOLO is useful in avoiding the transmissions of most respiratory infections.

Social distancing is a public health measure aimed at reducing the spread of infectious diseases, particularly those that are highly contagious, like COVID-19, Samll pox, scabies etc. The important care that must be taken in social distancing is to maintain a physical separation of at least 6 feet (about 2 m) between persons. The typical diagram shown in Fig. 2 describes a person maintaining a social distancing.

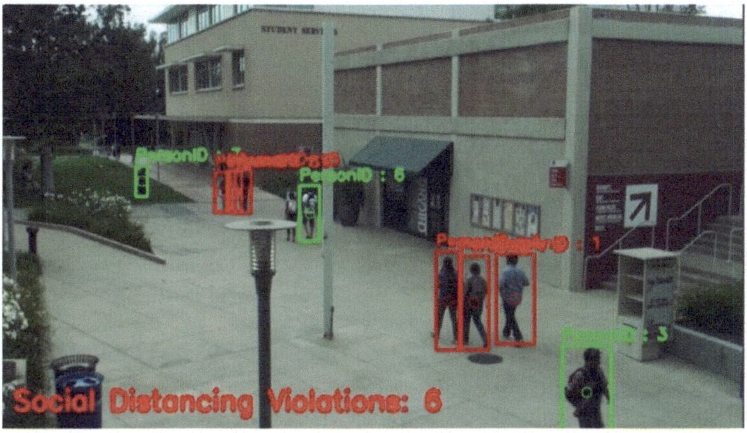

Fig. 2. Indicates a person maintaining social distancing with a green bounding box and a person violating social distance

The aim of social distancing is to reduce contact among the persons. If so, one can slow down the spread of the disease. In short, reduce the normal persons from infectious people who are seriously ill or die. The social distancing can also help to reduce the pressure on healthcare systems and protect vulnerable populations, such as older adults and people with underlying health conditions. Examples of social distancing measures include staying home as much as possible, avoiding crowded places, and avoiding close contact with people who are sick. When going out in public, wearing a mask is also recommended to prevent the spread of the virus.

1.1 Motivation

In recent years, the world has faced a disease called COVID-19 pandemic. COVID-19 is experienced as highly contagious disease caused by the novel corona virus [3]. The virus primarily spreads through respiratory droplets generated when an infected person talks, coughs, or sneezes, and it can cause a range of symptoms from mild to severe. In severe cases, COVID-19 can lead to pneumonia, acute respiratory distress syndrome (ARDS), and death. To take precautionary measures monitoring system for social distancing can help. The use of real-time object detection model may perform timely distancing measures and slow down the spread of infectious diseases, such as COVID-19 (novel corona virus) pandemic. Thus, we intend to study and analyze YOLO algorithms for object detection in application i.e., social distancing. Providing insights and comparisons among different versions of YOLO models in terms of various factors. The Yolo algorithms may help in reducing close contact between peoples, the virus has fewer opportunities to jump from one person to another, which in turn to decrease the overall rate of infections. This helps to "flatten the curve" of the cases which caused by infectious diseases. Our system may be helpful and reduces the burden on healthcare systems and saves lives. Additionally, social distancing measures can also be help to reduce the transmission of other infectious diseases, such as the flu or measles.

The highlights of the current research works are as follows:

- The workflow of the general social distancing model explains the overall system.
- Our aim is to apply object detection for social distancing purposes.
- Then after, the present study conducts insights of tracking methods for applying in social distancing.
- The tabulation of YOLO series, i.e., YOLO V3, YOLO V4, YOLO V5, YOLO V6, and YOLO V7 algorithms shows the developments and their usefulness.
- Finally, comparison results describe YOLO performances with respect to different factors.

The rest of the works have been organized as follows. Section 2 describes the related works. In Sect. 3, research model for social distancing has been explained. In subsequent Sect. 4, describes the problem of social distancing followed by object detection model and tracking model included in Sect. 5. Section 6 describes the details of deep learning based object detection and in next Sect. 7 illustrates results and comparisons. Finally, the conclusion is given in Sect. 8.

2 Related Works

There are remarkable literatures present for object detections, and few of them are available for social distancing applications. The works authored by Ren et al. [4] developed a faster R-CNN (convolutional neural networks) technique that has primarily addressed two issues: Region Proposal Networks (RPN) and fast production of candidate regions. Here, fast production of candidate regions refers to the model to quickly generate a large number of candidate regions. Afterwards, it is filtered and refined in the second stage to produce the final object detections. Faster R-CNN is a two stage detector which uses an

RPN to generate region proposals. Then after, it is passed to a fully convolutional networks(FCN) to classify and refine the object bounding boxes. The RPN is a lightweight CNN that runs in parallel with the FCN, and is trained to predict object bounds. This allows Faster R-CNN to generate region proposals in a single pass, which usually takes multiple passes in original R-CNN algorithm. Faster R-CNN has been used for object detection in images and videos.

The works of Liu et al. [5] proposed a single shot detector object detection algorithm. It is designed to detect objects in images and videos efficiently. It detects objects in an image or frame in a single pass. SSDs use a combination of CNNs and anchor boxes to detect objects, and are known for their real-time performance with high accuracy. They are commonly used in applications such as autonomous vehicles, surveillance systems, and video analytics. The author of [10], Redmon et al. [1], suggested a YOLO convolutional neural networks for real-time object detection. The algorithm only looks once the image, then after making prediction requires only forward propagation across the neural networks. The YOLO algorithm keeps balanced the real-time fast performance with high accuracy for object detection. In Dai et al. [6], region-based fully convolution networks is present. The algorithm outperforms RCNN and Faster R-CNN in terms of processing speed. Usually, R-CNN or Faster R-CNN is experienced a costly per region sub-networks several times. Ross presented a region-based convolutional networks approach [7] based on classifying object suggestions with a deep ConvNet. In result, accuracy of the object detection has been improved. However, it has significant limitations such as training is a multistage pipeline and expensive in terms of space and time, and object detection is slow [8]. As per as the social distancing problem solving algorithm is concerned, Ansari et al. have proposed in [9]. The algorithm is based on CNN object detector model to detect human presence. The social distancing has been solved under two main tasks: first, human detection and tracking; second, monitoring of social distancing among humans. Human detection is further classified into two stages i.e., localization of an object in the first stage and in the second stage the classification of the localized objects [10–12]. Here, the model uses video sequences as input. Afterwards, human detection and localization have been accomplished. By doing so, humans will be tracked and then it will calculate the social distance between persons. If it is less than the threshold value then it displays the red bounding boxes for violating the social distance, otherwise, it displays green bounding box for safe distance maintain between persons. The object detector's output is used to calculate favourable distance between each pair of humans. The works in Ahamad et al. [3] describes the detection of people in segmented regions in order to save energy. This boils down to a new goal known as a region of interest (ROI), which is intended in identifying the social distance of people in the zone of interest. They accomplished it by masking the region of interest, as well as performing person detection localization and social distancing. The Caffe deep learning model framework is utilized to run the object detection model. The model uses MobileNet SSD due to which gives shorter execution time. Also, threading has been used in order to reduce the execution time, required to process the object detection, per frame. In case, to runs the frames and perform the object detection at the same time, multithreading strategy is used. The accuracy observed in Mask R-CNN Inception Resnet version 2 Atrous and Faster R-CNN Inception version 2 COCO is 99 percent which higher over the 27 test images [13]. As per as the accuracy and execution time is concerned, SSD MobileNet V1 COCO is better for the social distancing. They achieved this by first identifying the people in the video frames and then apply an algorithm to calculate the distance between the peoples those are recognized. Due to the different viewpoints of the camera and objects plainly not caught by the camera this works with more accuracy indoors than outdoors [3]. The work authored by Punn et al. [2] is a YOLOv3 algorithm is based on real-time deep learning technique. It is powerful framework which automates the measurement of social distancing using object identification and tracking techniques. Participants are identified in real-time using bounding boxes. In this, the proximity attribute is measured which refers to the closeness of two or more objects. It is assessed using pairwise vectorized approach. The pairwise vectorized approach is a method which deals with enforcing social distancing guidelines. And, bounding boxes is generated. It is achieved by calculating the distances between all pairs of individual (or person) in a given population and identifying those who lies within the given threshold. The proposed model is able to distinguish humans from background objects. In this, bounding boxes have been used to track the people those are recognized.

3 Research Model for Social Distancing

In past recent years, especially during COVID time, the research community conducts speedy works in social distancing. Nowadays, as the transmission of disease increases due to the violation of social distancing, it has been observed that the said topic is one the most relevant field of study. To measure the physical distance between individuals in public areas, various social distance-detecting tools are available. In order to achieve social distancing, a person must have been detected within a frame using object detection models. For object detection, YOLO models can be used. In case if more than one person is detected in a frame, the measurement of distance between every other is accomplished. If measurement score is lower than the min threshold score then it display a red bounding box otherwise shows green bounding box. We can understand the present illustrations using a typical diagram, shown in Fig. 3 and Fig. 4.

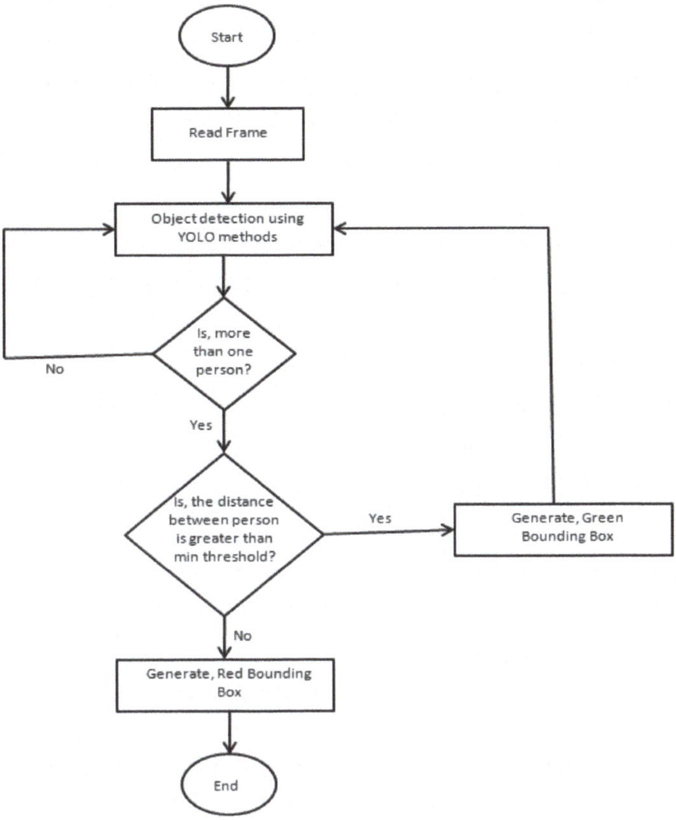

Fig. 3. Monitoring system for social distancing using YOLO series

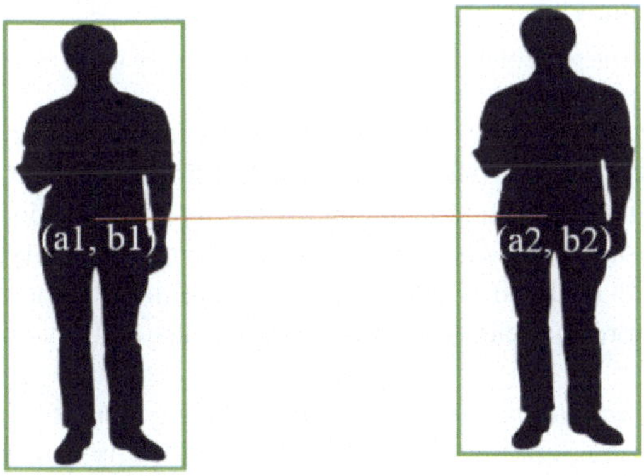

Fig. 4. Represents acentroid of two persons i.e., (a1, b1) and (a2, b2)

In order to find the centroid point of a person in a bounding box, we can use eq. (1)

$$centroid(a, b) = \left(\frac{a_{min} + a_{max}}{2}, \frac{b_{min} + b_{max}}{2}\right) \tag{1}$$

where,

amin = minimum width, *amax* =maximum width
bmin =minimum height, *bmax*=maximum height

We can calculate distance between two bounding boxes of detected persons using Eq. (1) as follows;

$$Distance = \sqrt{(a2 - a1)^2 + (b2 - b1)^2} \tag{2}$$

4 Problems, If Not Maintaining Social Distance

- Increased Transmission: It deals with the contagious diseases spread easily from infected person to another one. This happens, when a person does not maintain the social distancing. In result, higher infection rates and potential outbreaks.
- Overload On Healthcare Systems: It deals with challenge to provide adequate care. It includes, an increase in cases can overload healthcare facilities, leading to shortages of medical supplies, beds etc.
- *Higher Mortality Rates*: Increased transmission can lead to more severe cases and a higher death rate, particularly among vulnerable populations like those who have already been affected.
- *Strain on Public Health Resources:* Contact tracing and testing efforts become less effective when social distancing is not maintained, making it difficult to identify and isolate the cases quickly.
- *Risk to High-Risk Individuals:* In such diseases, elderly individuals are at greater risk for infection and severe illness.
- *Economic Impact:* The disease can impact all around our daily life. Frequent outbreaks and lockdowns resulting from insufficient social distancing can interrupt economic stability, causing job losses and financial instability for individuals and businesses.
- *Mutations and Variants:* This is an additional challenge of such type of diseases. Continued transmission of thevirus can increase the likelihood of new variants emerging. Thus, making it more challenging to control the disease and develop effective vaccines and treatments.
- *Global Health Implications:* In case, if the authority fails to control a contagious disease within one region, it can have a global health implication. The international travel and trades can be affected and facilitate the spread of the disease to other parts of the world.

5 Object Detection and Tracking Model

5.1 Object Detection

Object detection is a process to accomplish the following information such as object type, location of an object and class of an object. Object detection is achieved through the series of steps. In this, image classification, segmentation, image localization etc. are necessary to evaluate. The below given Fig. 5 represents the above explanation.

Fig. 5. Various objects detected under the bounding box with the label

The key component of object detection is the bounding box, which detects an object's borders and forms a rectangle (or square) around it with an object name. Object name is used to label the object that can be a person, car, or dog etc. There are various object detection models available, that are shown Fig. 6. The major steps involve in object detection are following.

Image Classification. It is part of object detection. Image classification predicts the class of an object detected. In other words, the detected object belongs to which class. Moreover, it indicates whether or not a specific object is present in a given image. The image classification does not offer an object's position.

Segmentation. The segmentation technique is also a part of object detection. It is a process to detect object by grouping pixels with their similar characteristics are found. It is important to note that, the segmentation does not utilize bounding boxes for object detection.

Object Localization. The object localization is also an important part of object detection. It shows where an object is located inside a picture. In other words, the location of an object inside the image is shown.

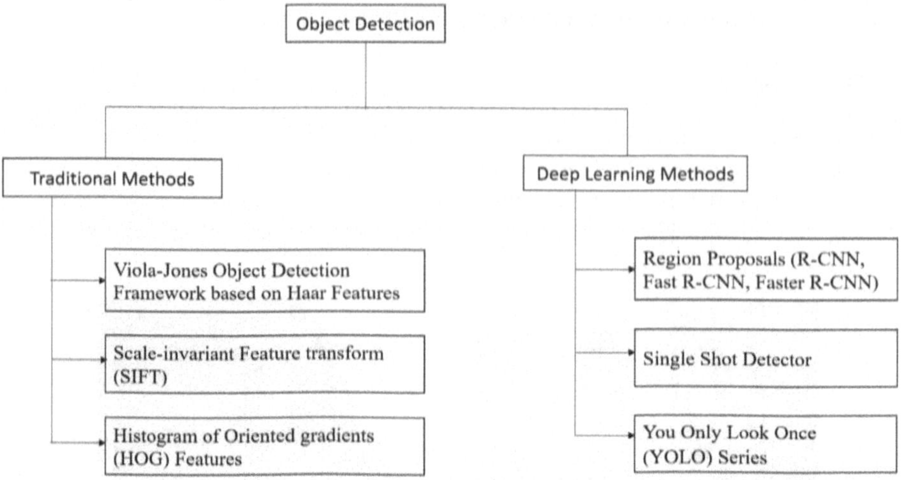

Fig. 6. Flow Chart of different techniques for object detection

5.2 Object Tracking

Once successful object detection is achieved, object tracking takes place. Object tracking is the process of identifying and locating an object of interest in a video stream. The objective of object tracking is to locate and follow a specific object of interest as it moves within a scene, and maintain its identity and trajectory over time. Object tracking typically involves several steps such as object detection, feature extraction, motion estimation, and data association. In the first step, an object detector is used to identify potential objects of interest in each frame of the video. In the second i.e., motion estimation step, the position and velocity of the object are estimated based on the movement of object features over time. Finally, data association algorithms are used to link the object's estimated position in each frame and to maintain the object's identity over time. Object tracking can be used in a variety of applications, such as surveillance, robotics, and autonomous vehicles. There are two broad classifications of object trackers such as single object trackers and multi object trackers.

Single Object Trackers. Single object trackers algorithms designed to track a single object in a video stream. They, first initializes the location of the item during the initial frame and afterward tracking it over the sequence of frames. These techniques for tracking are quite rapid. They are typically faster and efficient as they do not need to keep track of multiple objects at the same time. There are many examples of traditional computer vision algorithms which can be considered as single object trackers such as CSRT, KCF etc. As per as deep learning method single object tracker is concerned, deep learning-based trackers, on the other hand, have been observed as significantly more accurate than traditional trackers. Some popular example of deep learning - based single object trackers are SiamRPN and GOTURN.

Multi Object Tracker. Multi-object trackers algorithms are designed to track multiple objects in a video stream. Multiple object trackers are trained on a large amount of data. As a result, they are more accurate since they can track several entities, including various classes at once and maintaining reasonably good speed. Most popular multi object trackers algorithms are DeepSORT, JDE, and CenterTrack. The said algorithm solves the majority issues that the trackers may encounter. Following are the details of multi-object trackers.

DeepSORT. is an extension of SORT (Simple Online Real-time Tracking) algorithm. DeepSORT is a computer vision tracking approach that monitors entities or objects. Each object is uniquely identified by allocating an identifier value (ID). DeepSORT incorporates deep learning into the SORT algorithm by using a description of one's look to minimize identification swaps and hence improve tracking efficiency. The interested one can first learn the SORT algorithm followed by DeepSORT.

The SORT is comprised of four main components; detection, estimation, data association, and the creation and deletion of track IDs. SORT is a technique used to track the entity. On tracking, one can claims themselves better than many other internet trackers by using Kalman filters and Hungarian algorithms. It is observed that, SORT is effective interms of tracking accuracy and precision. However, when obstruction happens, SORT produces tracks with significant number of identifier changes and it fails. This occurs due to the use of association matrix during the process.

The extended form of SORT called DeepSORT performs tracking of the objects based on their velocity and motion. Also, it considers the appearance of object in the video frame. DeepSORT include large-scale human re-identification dataset in the context of tracking. In order to train the deep association metric model, DeepSORT employs a cosine metric learning technique. The cosine distance is a measure that assists the model in recovering identities when motion estimation fails. The cosine distance analyses appearance information of object is particularly important to recover identities after long-term occlusions. This is important, particularly when motion is less discriminative. Using these fundamental cosine metric learning technique, one can boost the tracker's power and accuracy [14].

6 Deep Learning Based Object Detection

YOLO (You Only Look Once) is, nowadays, one of the popular deep learning techniques for object detection. For tracking, YOLO is well suited with DeepSORT. It has been observed that, the YOLO with deep sort gives better results. There are different variants of basic YOLO algorithm discussed as follows.

6.1 You Only Look Once (YOLO V6)

YOLOv6 is an extension of YOLO series algorithms. YOLOv6 is a single-stage object detection framework that is specifically designed for industrial applications, for example security and surveillance, quality control in manufacturing and autonomous vehicles. It is observed that the YOLOv6 works faster and more accurate as compared to traditional two-stage object detection systems. Moreover, YOLOv6 introduces a new method called dense anchor boxes to generate the anchor boxes. Dense anchor boxes refer to a strategy that generate tight boundary box. This is important because the anchor boxes are placed densely over the image, with overlapping boxes at different scales and aspect ratios. This allows the model to more accurately localize objects of different sizes and shapes, particularly small objects that might be missed by larger anchor boxes. YOLOv6 uses the EfficientRep backbone consisting of RepBlock, RepConv, and CSPStackRep blocks. The implementation results of YOLO V6 indicate, RepBlock is more favorable to use for YOLOv6-Tiny, small, and nano models. However, for medium and large models, CSPStackRep backbone is suitable. YOLO V6 can achieve state-of-the-art results on various object detection benchmarks. YOLOv6 object detection model uses two loss functions that are VFL (Varifocal Loss) for classification loss and DFL (Distribution Focal Loss) along with SIoU (Scale-Invariant Intersection over Union) or GIoU (Generalized Intersection over Union) for box regression loss. In order to label assignment strategy, YOLOv6 models use the TAL (Task Aligned Assignment). Where, TAL is the best label assignment strategy for anchor-free object detection models.

YOLOv6 uses knowledge distillation in order to further improve the accuracy of the models for object detection. This can be achieved without requirement of a huge computation cost and save resources. Wherein knowledge distillation, a teacher model is used to train a student model which helps in improving the accuracy while performing object detection. The predictions of the teacher model act as soft labels along with the ground truth (or bounding box) to train the student model. Overall, YOLOv6 is a powerful and efficient object detection framework that is well suited for various industrial applications. It operate quickly and precisely, which makes it well-suited for use in realtime systems [15].

6.2 You Only Look Once (YOLO V7)

YOLOv7 is also an advanced version of YOLO series algorithms. YOLOv7 is a single stage, anchor-based detector that forecasts bounding boxes and class probabilities for objects in a picture using a convolutional neural network (CNN). YOLOv7 is a realtime algorithm to detect items (Entity or Object) from a given picture. The algorithm can detect objects from images as well as video frames. The backbone architecture of YOLOv7 remains same as of YOLOv4 and both are real-time algorithms in order to detect items [16]. However, as per the YOLO architecture is concerned, it contains backbone, neck and a head.

In YOLOv7, it contains multiple heads. In YOLOv7, the head is responsible for final output which is known as lead head. The head is used to assist the training in the middle layers called as Auxiliary Head. With the help of an assistant loss, the weights of the auxiliary heads are updated. The YOLOv7 uses E-ELAN (Extended Efficient Layer Aggregation Network) which is the computational block. E-ELAN architecture enables the framework to better learn the object. YOLOv7 introduced bag of freebies (bof). These are the method that enhances the performance of a model without increasing the training cost. Bag of freebies includes planned r-parameterized convolution technique that is used, after training, to improve the model. It also contains coarse for auxiliary and fine for lead loss (Table 1).

Table 1. Different YOLO Versions Comparisons and their Improvements

Author	Algorithm	Advantage	Disadvantage	Major Improvements
Redmon *et.al* [17]	YOLOv3	Multi-scale detection achieved by using FPN(Feature Pyramid Networks)	Comparatively low recall and more localization error	Introduces the concept of the feature pyramid networks. It uses new CNN Architecture called Darknet-53
Liao *et.al* [18]	YOLOv4	Performances of YOLOv4 twice than the EfficientDet	It is a challenge to detect small objects	It uses spatial pyramid pooling (SPP) which is used to divide the input image into multiple subregions, or pyramid. MISH Activation function and Hardswish Activation function are used
Jocher*et.al* [19]	YOLOv5	It is faster than previous versions of YOLO. Also, it closely matched the size and shape of the observed objects	There is article published for YOLOv5	Uses new architecture called EfficientDet. It generates anchor boxes using a novel technique known as dynamic anchor boxes
Chuyi*et.al* [15]	YOLOv6	YOLOv6 gives faster inference time and higher mean average precision (mAP) than any previous YOLO series	It could have been performed better, if matches stricter requirements and conditions. However, this is rarely occurs	It introduces a new method for generating the anchor boxes, called dense anchor boxes. It uses EfficientRep architecture as backbone
Chien *et.al* [16]	YOLOv7	Ideal for real-time applications where faster processing speed is common, including in self-driving cars and surveillance systems	Detectors with two stages, such faster R-CNN and mask R-CNN are more accurate than YOLOv7	In YOLOv7, 4introduction of bag of freebies (bof) increases the performance. YOLOv7usesE-ELAN as backbone

The YOLOv7 can process the frames at high speed in real-time. Thus, it is suitable for video surveillance and otherreal-time applications. Overall, YOLOv7 is a highly efficient and accurate object detection algorithm. YOLOv7 has been improvised in many ways such as multi-scale training, improved anchor design, and other techniques of previous YOLO series and has improved accuracy and performance [16].

7 Results and Comparisons

The given below table (Table 2) provides information regarding trends of YOLOv3, YOLOv4, YOLOv5, YOLOv6, and YOLOv7 during the years since 2018 to 2022 for object detection. Furthermore, the object detection results can be used for social distancing purposes. It includes web search, news search, image search and YouTube search data. This breakdown by years illustrates that the newer versions such as YOLOv5, YOLOv6, and YOLOv7 counts were increasing with respect to years (Fig. 7).

Table 2. YOLO Versions search interest breakdown by years for social distancing

Year/Technology	YOLO V3	YOLO V4	YOLO V5	YOLO V6	YOLO V7
2018	3882	1055	1300	1004	644
2019	5155	1318	1086	968	1661
2020	3423	2587	1413	1466	1193
2021	2556	2022	1761	1126	1008
2022	2919	2559	4612	1289	3210

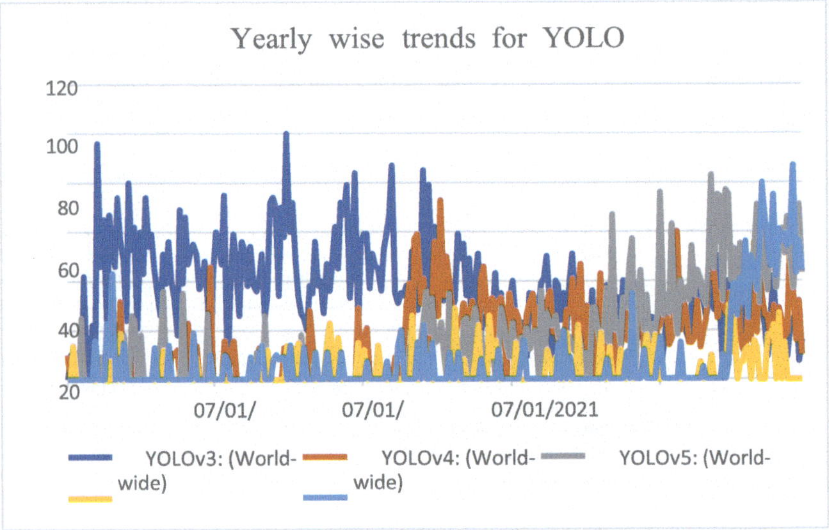

Fig. 7. Year wise trends for YOLO Versions with respective Web Search

The below Table 3, 4, 5 and 6 provides a comparison of the YOLOv4, YOLOv5, YOLOv6, and YOLOv7 object detection models on the COCO 2017 Val dataset, showcasing their Mean Average Precision (MAP), computational complexity measured in FLOPs, and frames per second (FPS).

The units of the given parameters below are size in pixels, MAP in percentage, Flops (Floating point operations per Second), and G stands for billion.

Table 3. Comparison of the MS COCO 2017 Validation (Val), dataset's batch size(bs) = 1 object detectors' speed andaccuracy

Model	Size	MAP	Flops (G)	FPS
YOLOv4 [18]	512	43.0	52	83

Here, batch size (bs) refers to the number of samples processed at once during a single iteration. In our assumption, bs =1 represents that it will consume less memory usage, fast convergence and more control on training process. As if we increase the batch size, the accuracy may be degraded and hence selected bs = 1, only.

Table 4. Using dataset COCO 2017 Val, comparisons with other YOLO-series detectors. On a Tesla T4 (Graphical Processing Unit) in a similar environment with TensorRT. FPS, and latency are measured at FP16 (Floating Point) precision

Model	Size	MAP	Flops (G)	FPS(bs = 1)
YOLOv5-N [19]	640	28.0	4.5	602
YOLOv5-S [19]	640	37.4	16.5	376
YOLOv5-M [19]	640	45.4	49.0	182
YOLOv5-L [19]	640	49.0	109.1	113

Table 5. Results of the YOLOv6 series MAP and speed are assessed using the TensorRT 7.2 (runtime) on the Tesla T4 and the COCO 2017 Val dataset with an input resolution of 640 × 640.

Model	Size	MAP	Flops (G)	FPS(bs = 1)
YOLOv6-N [20]	640	37.5	11.4	779
YOLOv6-S [20]	640	45.0	45.3	339
YOLOv6-M [20]	640	50.0	85.8	175
YOLOv6-L [20]	640	52.8	150.7	98

Table 6. Performance Metrics of the YOLOv7 series on COCO 2017 Val.

Model	Size	MAP	Flops (G)	FPS V 100
YOLOv7 [16]	640	51.2	104.7	161
YOLOv7-X [16]	640	52.9	189.9	114
YOLOv7-W6 [16]	1280	54.6	360.0	84
YOLOv7-E6 [16]	1280	55.9	515.2	56
YOLOv7-D6 [16]	1280	56.3	806.8	44
YOLOv7-E6E [16]	1280	56.8	843.2	36

8 Conclusion

In this paper, we have conducted a review of YOLO versions in social distancing for health care system. The YOLO algorithms have been implemented using deep learning technique. Deep learning has gained popularity in object detection. Here, we summarized and included comparison of YOLOv3, YOLOv4, YOLOv5, YOLOv6, and YOLOv7 for object detection. It has been observed that, in every version, inference time and MAP increases. Inference time shows the execution time of the prediction of the model. As fast as the algorithm, the social distancing can be timely measure and avoid from infections. The MAP (mean average precision) indicates the accuracy of the model. It can help the person to not violate the social distancing. In all the versions, it is observed that the algorithm remains with some common features such as single stage detector, grid based detector, convolution neural networks etc. For tracking purposes, DeepSORT have been used to achieve social distancing. In this direction, YOLOv3 for social distancing using deep sort has been implemented. Therefore, the higher versions of YOLO for social distancing using deep sort opens the research scope largely.

References

1. Redmon, J., Divvala, S., Girshick, R., Farhadi, A.: You only look once: unified, real-time object detection. In: Proceedings of the IEEE Conference on Computer Vision and Pattern Recognition, pp. 779–788 (2016)
2. Punn, N.S., Sonbhadra, S.K., Agarwal, S., Rai, G.: Monitoring COVID-19 social distancing with person detection and tracking via fine-tuned YOLO v3 and Deepsort techniques. arXiv preprint arXiv:2005.01385 (2020)
3. Ahamad, A.H., Zaini, N., Latip, M.F.A.: Person detection for social distancing and safety violation alert based on segmented ROI. In: 2020 10th IEEE International Conference on Control System, Computing and Engineering (ICCSCE), pp. 113–118. IEEE (2020)
4. Ren, S., He, K., Girshick, R., Sun, J.: Faster r-cnn: towards real-time object detection with region proposal networks. Adv. Neural Inf. Process. Syst. **28** (2015)
5. Liu, W., et al.: SSD: single shot multibox detector. In Computer Vision–ECCV 2016: 14th European Conference, Amsterdam, The Netherlands, 11–14 October 2016, Proceedings, Part I 14, pp. 2137. Springer, Heidelberg (2016)
6. Dai, J., Li, Y., He, K., Sun, J.: R-fcn: Object detection via region-based fully convolutional networks. Adv. Neural Inf. Process. Syst. **29** (2016)
7. Chen, C., Liu, M.Y., Tuzel, O., Xiao, J.: R-CNN for small object detection. In: Computer Vision– ACCV 2016: 13th Asian Conference on Computer Vision, Taipei, Taiwan, 20–24 November 2016, Revised Selected Papers, Part V 13, pp. 214–230. Springer, Heidelberg (2017)
8. Girshick, R.: Fast r-cnn. In: Proceedings of the IEEE International Conference on Computer Vision, pp. 1440–1448 (2015)
9. Ansari, M.A., Singh, D.K.: Monitoring social distancing through human detection for preventing/reducing COVID spread. Int. J. Inf. Technol. **13**(3), 1255–1264 (2021)
10. Abughalieh, K., Alawneh, S.: Pedestrian orientation estimation using cnn and depth camera (No.2020-01-0700). SAE Technical Paper (2020)
11. Karpagavalli, P., Ramprasad, A.V.: An adaptive hybrid GMM for multiple human detection incrowd scenario. Multimedia Tools Appl. **76**, 14129–14149 (2017)
12. Bahri, H., Chouchene, M., Sayadi, F.E., Atri, M.: Real-time moving human detection using HOG and Fourier descriptor based on CUDA implementation. J. Real Time Image Process. **17**, 1841–1856 (2020)
13. Yadav, N., Binay, U.: Comparative study of object detection algorithms. Int. Res. J. Eng. Technol. (IRJET) **4**(11), 586–591 (2017)

14. Wojke, N., Bewley, A., Paulus, D.: Simple online and real time tracking with a deep association metric. In: 2017 IEEE International Conference on Image Processing (ICIP), pp. 3645–3649. IEEE (2017)
15. Li, C., et al.: YOLOv6: a single- stage object detection framework for industrial applications. arXiv preprint arXiv:2209.02976 (2022)
16. Wang, C.Y., Bochkovskiy, A., Liao, H.Y.M.: YOLOv7: Trainable bag-offreebies sets new state of-the-art for real-time object detectors. arXiv preprint arXiv:2207.02696 (2022)
17. Redmon, J., Farhadi, A.: Yolov3: an incremental improvement. arXiv preprint arXiv:1804.02767 (2018)
18. Bochkovskiy, A., Wang, C.Y., Liao, H.Y.M.: Yolov4: optimal speed and accuracy of objectdetection. arXiv preprint arXiv:2004.10934 (2020)
19. Jocher, G.: (2022). https://ultralytics.com/yolov5. Accessed Feb 2023
20. Chuyi, L., et.al.: (2022). https://github.com/meituan/YOLOv6

A Hybrid Approach for Detecting Software Refactoring Sequencing

Ritika Maini[1(✉)] and Amandeep Kaur[2]

[1] Department of Computer Science, Sri Guru Granth Sahib World University, Fatehgarh Sahib, India
maini_ritika@rediffmail.com

[2] Department of Computer Engineering, NIT Kurukshetra, Kurukshetra, India
amandeep1426@nitkkr.ac.in

Abstract. The process of refactoring improves the software that is created by making changes to the composition of a design, while maintaining the framework of the software. However, addressing these smells without setting suitable priorities may not yield the desired outcomes. Code smells significantly increase maintenance costs and hinder the evolution of systems. To address these issues, refactoring sequencing techniques are utilized to improve a system's design by modifying its internal structure while preserving its outer behavior. However, in huge-scale structure, the variety of code smells that need to be fixed can be substantial, and not all of them can be resolved automatically. Therefore, it becomes crucial to prioritize the order of the code smells according to various criteria such as the risk as well as importance of the affected category. In this paper work, a novel hybrid approach is presented, leveraging the Hybrid Spotted Hyena and Emperor Penguin (HSHEP) optimizer-based algorithm. The primary objective of this approach is to maximize the number of fixed threatful code smells based on the maintainer's opinions and requirements. For evaluating the impact of the proposed approach, it is analyzed on small as well as large-sized open-source systems, targeting five different kinds of code smells.

Keywords: Software Refactoring · Code Smell · Sequencing · Optimization · Spotted Hyena Optimizer · Emperor Penguin Optimizer

1 Introduction

Code smells are structural defects within the software, indicative of poor software design, making it difficult to maintain [1]. They point out design issues in software applications. The presence of code smells hampers code comprehension and potentially leads to increased changes and susceptibility to faults. Detecting and removing code smells from the code origin enhances the programmer's understanding of the computer code. In large-sized systems, identifying refactoring operations for the detected code smells can be complex, and not all code smells can be automatically fixed [2]. Maintenance and evolution of large-scale systems account for approximately 87% of the total software

D. Pastor-Escuredo et al. (Eds.): ICDLAIR 2023, LNNS 1001, pp. 618–625, 2024.
https://doi.org/10.1007/978-3-031-60935-0_54

costs [3]. These costs are attributed to tasks like adding new functionalities, bug corrections, and code modifications to enhance quality. However, these continuous changes can lead the software far away from its distinctive design and architecture, introducing code-smells, which are bad design effects [4–6].Code-smells have an adverse effect on quality characteristics like adaptability and durability and are often unintentionally brought in by software developers during very first development as the outcome of poor design choices or maintenance decisions. Therefore, it becomes necessary to prioritize the sequence of refactoring operations based on the maintainer's preferences for removing code smells in such systems.

1.1 Refactoring Sequencing

Refactoring Sequencing, a widely-used technique, is employed to address code-smells and improve its structure while retaining overall functionality and actions [7]. This process involves two primary steps: (1) Detecting portions of code that require improvement, such as code-smells, and (2) Identifying suitable refactoring sequencing techniques to achieve the desired improvements. The initial approach involves applying a novel optimization algorithm namely Hybrid Spotted Hyena and Emperor Penguin optimizer (HSHEP). This hybridized algorithm commonly aims to combine the strengths of both approaches. Once the results from the initial SHO algorithm are obtained, there the hybridized algorithm will be run to demonstrate the advantages of combining with EPO algorithm. This optimization-based code smell refactoring sequencing technique could help to ensure that code refactoring is completed efficiently and effectively, reducing the risk of introducing new bugs or affecting the code's functionality.

1.2 Hybridization of Algorithms

Hybridizing Spotted Hyena and Emperor Penguin (HSHEP) optimizer-based algorithm in software refactoring can improve the effectiveness of the optimization process, combining the advantages of both algorithms. The goal is to address code smells by iteratively applying refactoring to the code, seeking a better solution while avoiding getting trapped in local optima.

1.2.1 Spotted Hyena Optimizer (SHO)

Spotted hyenas [8] can become familiar with the site of prey and encircle them. To mathematically depict the behavior of spotted hyenas, the best search agent, representing the optimum, as the one aware of the prey's location. Other search firms build a network of reliable companions aligned with the direction of the best search agent. They maintain the best ways obtained thus far to update their current position in pursuit of their objective.

There are four steps considering the working of SHO:

1. The Encircling Prey Process
2. Hunting Process
3. Attacking the Prey Process
4. Searching the Prey Process

1.2.2 Emperor Penguin Optimizer (EPO)

To begin with, Emperor penguins randomly establish the perimeter of their huddle. Once established, Simultaneously, the distance between emperor penguin is calculated, which aids in further exploration and exploitation [9]. Subsequently, the optimal solution, known as the effective mover, is determined. This solution guides the search agents to better positions. Using the updated emperor penguin positions or search agents, the huddle boundary is then recomputed to enhance the overall effectiveness of the process.

There are four steps considering the working of EPO:

1. Achieve and gather complete chaos border
2. Temperature around gathering
3. The separation between emperor penguins
4. Relocating the mover

1.2.3 Algorithm for Refactoring Sequencing the Hybrid Spotted Hyena and Emperor Penguin

Step 1: Initialization Phase
Initialize the population with random solutions. Set the parameters such as size of the population, maximum iterations, crossover rate, and mutation and local search rate. Find the initial best solution in the population based on the fitness function.

Step 2: Main Optimization Loop
Repeat for a predefined number of iterations

Hunt and Capture Phase (Spotted Hyena Behavior):
Create offspring individuals by performing crossover and mutation operations on the population. Add the offspring to the population.

Information Sharing Phase (Spotted Hyena Behavior):
Share information between individuals to enhance global search capabilities.

Huddling Phase (Emperor Penguin Behavior):
Sort the population based on fitness and perform local search around the top individuals. Add the local search results to the population.

Individual Movement Phase (Emperor Penguin Behavior):
Generate new solutions by exploring new regions in the search space. Add the new solutions to the population.

Selection:
Select the best individuals from the population based on fitness, keeping the population size constant.

Update the Best Solution:
Update the best solution if a better solution is found during the iteration.

Step 3: Termination
End the main loop when the maximum number of iterations is reached or another stopping criterion is met.

Step 4: Bring back the best solution
Bring back the best solution found during the optimization process as the optimal solution for the problem.

In this theoretical description, each step outlines the main actions and phases performed in the Hybrid Spotted Hyena and Emperor Penguin Optimization (HSHEPO) algorithm. The specific details and implementation of functions may vary based on the problem domain and the algorithm's requirements. The algorithm combines emperor penguin social dynamics and spotted hyena hunting behavior to improve optimization outcomes.

2 Research Methodology

Prior to elaborating on the suggested framework for refactoring prioritization, we will now discuss the proposed research methodology, outlined in figure given below [10]. The primary objective of this paper is to explore the potential analogies between the behaviors and characteristics of emperor penguins with spotted hyenas and the process of refactoring code sequences. The methodology comprises six phases, each of which is introduced below (Fig. 1):

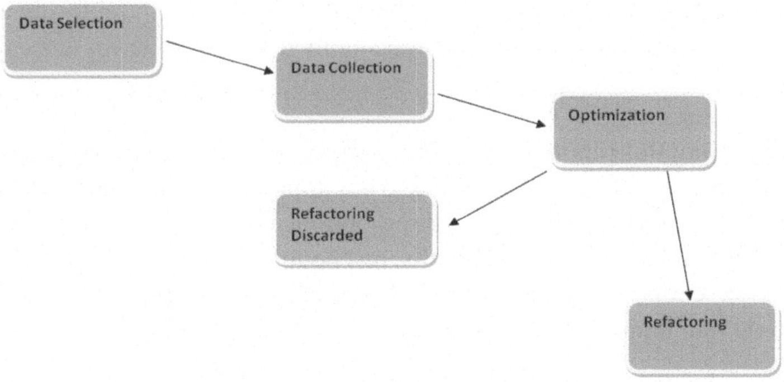

Fig. 1. Proposed Methodology of Code Smell Refactoring Sequencing.

Phase 1:

Data Selection: The data selection for performance evaluation and analysis involves choosing different open-source and publicly available Java datasets. These different datasets will be used to assess and analyze the proposed approach.

Phase 2:

Data Collection: The data and code smell is being collected. For this a software tool is built to extract metrics like that count number of classes coupled with main class(CBO),the relation between various methods and parameters of class is measured(LCOM), the complexity of class is being measured (RFC),the maintainability value and methods complexity is being measured. Five code smells like long parameter

Table 1. Data Set Metrics.

Metrics	Description
CBO	It maintains track of the number of classes connected to the primary class
LCOM	LCOM calculates the correlation between the class's numerous methods and its instance variable
RFC	The number of method calls is used to gauge the class's level of complexity
WMC	The number of method calls is used to gauge the class's level of complexity

list, long method, blob, feature envy and large class have been identified that is to be refactored. The following table gives detail of metrics and their description (Table 1).

Phase 3:

Optimization: The hybrid spotted hyena and emperor penguin algorithm (HSHEP) serves as the foundation for identifying the most effective refactoring sequences to completely eliminate the identified design smells. By drawing inspiration from the resourcefulness of spotted hyenas and the collective intelligence of emperor penguins, the algorithm proposes creative strategies to tackle code smells and optimize the refactoring process. These metaphorical mappings guide developers in selecting the optimal refactoring actions that result in the successful eradication of design smells from the code.

Phase 4:

Refactoring: During this phase, refactoring tasks are performed based on the sequences generated in the previous phase. According to our approach, to facilitate user interaction, software engineers are presented with three options to handle side effects. They can choose to address all newly emerged smells, reschedule the sequencing of smells, or opt to ignore the new smells altogether. These choices empower engineers to make informed decisions while maintaining control over the refactoring process.

The table given below shows description of various datasets used in this study. The datasets contain five open source java sets (Table 2).

Table 2. Open Source Program Parameters

Dataset name	Dataset Type	Programming Language	Total Number of Classes	Total Number of Smells
JHotDraw	Open Source	Java	398	1215
GanttProject	Open Source	Java	776	63
JEdit	Open Source	Java	5421	1365
JFreeChart	Open Source	Java	3257	793
Xerces	Open Source	Java	2551	5989

Phase 5:

Evaluation: This study evaluates the efficacy of the Hybrid Spotted Hyena and Emperor Penguin Algorithm (HSHEP) in rectifying model smells in diverse object-oriented software systems. Specifically, the paper focuses on one optimization attribute: software maintainability. The objective is to produce refactoring solutions that maximize programme quality while minimizing maintenance efforts (software modifications). The study seeks to decrease maintainability efforts while increasing the number of applied refactoring. All suggested solutions are automated and don't need any assistance from people, allowing for a smooth integration of refactoring algorithms and automation tools. Five well-known open-source software systems are chosen for study in order to benchmark the method, which have been extensively investigated in software engineering literature. The hybrid spotted hyena and emperor penguin algorithm (HSHEP) is assessed for its effectiveness in refactoring these systems to address any identified design smells [11, 12]. The projects that were chosen, which are listed in Table, include a variety of descriptive data, with a range of classes of 937 to 2600 and a range of recognized odours of 224 to 1440. The study takes into account well-known odours that represent several kinds of odours to examine their influence on software quality. For reproducibility, other smell types and corresponding refactoring tasks can be considered. The chosen projects span a variety of applications and discipline ensuring a comprehensive evaluation of the HSHEP algorithm's efficiency (Table 3).

Table 3. Programme Statistics.

System	#Classes	#All Smells	Fixed Smells
JHotDraw	472	294	32
GanttProject	1021	792	44
JEdit	1089	1024	72
JFreeChart	1059	480	80
Xerces	937	864	70

Prioritization for Code Smell Refactoring Sequencing Results

The initial step involved a comparison of the approach with two alternative algorithms that do not incorporate prioritization – specifically, Particle Swarm Optimization (PSO) [13] without prioritization and the method proposed by Genetic Algorithm (GA) [14]. In this comparison, a fitness function is employed that quantifies the number of corrected code-smells. This was done to ascertain the effectiveness of employing prioritization in the schema (Table 4).

Table 4. Refactoring Results: Code Smell

System	Approaches	Long Method (%)	Feature Envy (%)	Blob (%)	Long Parameter List (%)	Large Class (%)
JHotDraw	PSO GA Proposed Algorithm	91(5/9) 91(6/7) 92(7/7)	89(7/9) 91(6/7) 91(8/9)	93(28/30) 90(27/31) 93(29/30)	89(17/19) 91(6/7) 92(8/7)	93(28/30) 90(27/31) 93(29/30)
GanttProject	PSO GA Proposed Algorithm	91(7/9) 92(6/7) 92(7/7)	94(7/9) 91(6/7) 94(8/9)	93(28/30) 90(27/31) 93(30/30)	87(7/9) 91(6/7) 92(6/7)	94(28/30) 91(27/31) 94(29/30)
JEdit	PSO GA Proposed Algorithm	90(7/9) 92(6/7) 92(7/7)	90(7/9) 91(6/7) 92(7/8)	93(28/30) 90(27/31) 93(29/30)	92(7/9) 91(6/7) 92(8/9)	93(28/30) 90(27/31) 93(30/30)
JFreeChart	PSO GA Proposed Algorithm	92(7/9) 91(6/7) 92(7/7)	87(7/9) 91(6/7) 92(8/9)	90(7/9) 93(16/17) 93(9/9)	87(7/9) 90(6/7) 91(5/7)	87(7/9) 91(6/7) 91(7/7)
Xerces	PSO GA Proposed Algorithm	91(7/9) 93(6/7) 93(7/7)	91(7/9) 92(6/7) 92(6/7)	89(7/9) 91(6/7) 92(7/7)	87(7/9) 91(6/7) 91(8/9)	87(7/9) 90(5/7) 92(6/7)
Average		92	92	92	91	92

Furthermore, in order to guarantee the effectiveness and practicality of our approach, there conducted a manual verification of the feasibility of the various proposed refactoring sequences for each system. There encountered certain semantic errors in the program's behavior. When identifying such errors manually, the classified operations associated with these changes as suboptimal recommendations.

3 Conclusions

This article proposes a unique hybrid strategy based on the Hybrid Spotted Hyena Emperor Penguin algorithm that uses single strategies for objective optimization. The refactoring of design smells is handled by the solutions, in contrast to existing refactoring methodologies. The optimum refactoring order, which results in the highest quality with the least amount of maintainability, may be made using a variety of model smells. A sizable bespoke dataset was used to evaluate the effectiveness of the chosen refactoring strategies are created along dataset using five well-known open-source software datasets. The proposed technique is able to correct all observed code smells while effectively identifying the refactoring sequences that result in the best gains in software quality and maintainability.

References

1. Almogahed, A., et al.: A refactoring classification framework for efficient software maintenance. IEEE Access **11**, 78904–78917 (2023)
2. Kaur, M., Singh, D.: An intelligent code smell detection technique using optimized rule-based architecture for object-oriented programmings. In: International Conference on Artificial Intelligence and Sustainable Engineering: Select Proceedings of AISE 2020, vol. 1, pp. 349–363. Springer, Singapore (2022)
3. Erlikh, L.: Leveraging legacy system dollars for e-business. IT Prof. **2**(3), 17–23 (2000)
4. Brown, W.H., Malveau, R.C., McCormick, H.W.S., Mowbray, T.J.: AntiPatterns: Refactoring Software, Architectures, and Projects in Crisis. John Wiley & Sons Inc., Hoboken (1998)
5. Fenton, N., Pfleeger, S.L.: Software Metrics: A Rigorous and Practical Approach, 2nd edn. International Thomson Computer Press, London (1997)
6. Fowler, M., Beck, K., Brant, J., Opdyke, W., Roberts, D.: Refactoring: Improving the design of existing code, 1st edn. Addison-Wesley, Reading (1999)
7. Fowler, M.: Refactoring. Addison-Wesley Professional, Boston (2018)
8. Dhiman, G., Kumar, V.: Multi-objective spotted hyena optimizer: a multi-objective optimization algorithm for engineering problems. Knowl.-Based Syst. **150**, 175–197 (2018)
9. Dhiman, G., et al.: BEPO: a novel binary emperor penguin optimizer for automatic feature selection. Knowl.-Based Syst. **211**, 106560 (2021)
10. AbuHassan, A., Alshayeb, M., Ghouti, L.: Prioritization of model smell refactoring using a covariance matrix-based adaptive evolution algorithm. Inf. Softw. Technol. **146**, 106875 (2022)
11. Crosby, P.B.: Quality is free: the art of making quality certain (1979)][Juran, J. M. (1951). Quality-control handbook. *1974*
12. Juran, J.M.: Quality-control handbook 1974 (1951)
13. Poli, R., Kennedy, J., Blackwell, T.: Particle swarm optimization: an overview. Swarm Intell.Intell. **1**, 33–57 (2007)
14. Holland, J.H.: Genetic algorithms. Sci. Am. **267**(1), 66–73 (1992)

ADNet: An Attention Embedded DenseNet121 Model for Weed Classification

Akshay Dheeraj(✉), Sudeep Marwaha, Sapna Nigam, Md. Ashraful Haque, and Madhu

ICAR- Indian Agricultural Statistics Research Institute, New Delhi, India
akshaydheeraj.jmi@gmail.com

Abstract. Weed management is a critical factor that significantly impacts crop yield and overall crop productivity.Identification of the appropriate weed has been the big problem for the farmers over the few decades since weeds have the similar appearances with crops. This paper presents the vision-based weed classification system using convolutional neural networks and attention mechanism for maize crop. A model named ADNet has been presented in this study by employing DenseNet121 model with convolutional block attention mechanism (CBAM). ADNet yields the good performance with 99.58% accuracy for weed identification in maize crop. The proposed work was implemented with image dataset of maize crop with 6000 images and four weed species of corn crop. CBAM has been used on the feature map of DenseNet121 for extracting the distinguish features and focussing on the local regions. In addition, some well-known CNN models like EfficientNetB0, DenseNet121, MobileNetV2, InceptionResNetV2 and ResNet50V2 were also implemented in this work and their performance were compared with proposed work. Amongst all experimented CNN models, our proposed model ADNet (DenseNet121 + CBAM) outperformed all five CNN models implemented. The presented model shows the effectiveness of the attention mechanism embedded DenseNet121 model to significantly distinguish and classify the weeds in the maize crop.

Keywords: Attention module · Weed Classification · Deep learning · CBAM · Smart agriculture

1 Introduction

Corn is among the most extensively cultivated and consumed commodities globally, including in the USA. Its widespread popularity stems from its exceptional productivity and versatile applications in both industrial and agricultural domains, encompassing uses like animal feed, biofuel production, and the production of food sweeteners. One of the primary obstacles to achieving greater corn productivity is the effective control and management of weed growth. Weeds not only compete with corn for essential nutrients and agricultural resources but also introduce detrimental bacteria, viruses, and other microorganisms, leading to substantial yield losses. Weeds are undesirable plants that compete with the crop for water, nutrients, and sunlight. The current most-worldwide

D. Pastor-Escuredo et al. (Eds.): ICDLAIR 2023, LNNS 1001, pp. 626–638, 2024.
https://doi.org/10.1007/978-3-031-60935-0_55

used method to eradicate weeds from crop fields is the chemical method [1], due to its effectiveness and practicality. This method is causing severe environmental pollution of the soil and groundwater since it is usually sprayed uniformly within crop fields. Automated weed control for targeted herbicidal application using modern technologies such as machine learning and computer vision is gaining interest among the research community [2].

2 Related Work

Conventional machine learning rely on extraction of handcrafted features like color, shape and texture of weeds from images for identification of weed species [3–5]. However this conventional machine learning is not only time consuming but also bring biasness as experts may give weightage to various features. One of the fields of machine learning known as Deep learning, has an advantage of an automatic feature extraction using convolutional neural networks (CNNs). In smart agriculture, Deep learning with advancements in digital cameras and computer technology is being successfully used in plant disease and weed classification [6–12]. A study by Razfar et al. [13] employed 5 layers CNN model for detection of weeds in soybean crop and achieved overall accuracy of 97.7%. The work focussed on one category of weed. Another work by Subeesh et al. [14] used InceptionV3 model for identification of weeds in bell pepper and obtained an accuracy score of 97.7%. Dheeraj and Chand [11] developed the model for identification of weeds associated with wheat, maize and sugar beet at seedling stage. The authors used MobileNetV2 and DenseNet169 and achieved 94.92% and 97.14% accuracy respectively. In the study of Garibaldi-Márquez et al. [15], VGG16 model reported the superior performance than VGG19 and Xception model and machine learning model. The developed model achieved 97.83% accuracy for weed identification in initial growth stage of corn crop in natural environment. Tao and Wei [16] employed a hybrid CNN and SVM model for identification of rape seedlings and their associated four weed species in natural field. The author used VGG19 as feature extractor and SVM as classifier and yielded 92.1% accuracy. Their work outperformed other five models implemented. Jiang et al. [17] documented graph convolutional network based technique with CNN model for identification of weeds in lettuce, corn and radish crop. The authors developed a model with ResNet101 and GCN technique. Their work reported 99.37%, 97.8%, 98.93% accuracy on lettuce, corn and radish dataset respectively. The GCN combines both vertex features and graph structures and send them combined through various layers. Mu et al. [18] combined efficient channel attention (ECA) mechanism with DenseNet model for weeds associated with corn, wheat and sugar beet crop. The authors modified existing DenseNet model by inserting ECA mechanism after every dense connection to enhance feature weight and used local variance algorithm to reduce redundant background features from weed images. Their work achieved 97.98% accuracy rate.

The proposed research work has the following contributions:

1. We have developed a model named ADNet by employing DenseNet121 model with Convolutional Block Attention mechanism (CBAM) module for weed classification in maize crop.
2. The proposed ADNet has the overall accuracy score of 99.58% for weed classification.

3. This research study also employed pre-trained CNN models like EfficientNetB0, DenseNet121, MobileNetV2, InceptionResNetV2 and ResNet50V2 and their performance have been compared with that of ADNet.

Following the Related work, Sect. 3 elaborates the proposed methodology. Experiment results are discussed in Sect. 4. Finally, paper is concluded with future scope in Sect. 5.

3 Proposed Methodology

This section covers the description of dataset used and methodology employed in this research work.

3.1 Dataset

This research work used the corn weed dataset [17] which is publicly available. The dataset consists of five classes of weed and corn and have a total of 6000 images. Dataset cosnsists of four weed species named bluegrass, chenopodium album, cirsium setosum and sedge associated with corn crop. The dataset has been distributed in different classes as shown in Table 1.

3.2 Data Preprocessing

All the images were resized to 224 × 224, from the original image size of 800 × 600 before they are fed as input to CNN model. Image size of dimension 224 × 224 has been used for all the models experimented here. Distribution of the dataset among five classes has been shown in Table 1. Some samples of dataset from each category are shown in Fig. 1.

Table 1. Description of dataset used in research study.

S.No	Weed / Crop name	#images	#training images	#testing images	#validation images
1	bluegrass	1200	840	240	120
2	chenopodium album	1200	840	240	120
3	cirsium setosum	1200	840	240	120
4	corn	1200	840	240	120
5	sedge	1200	840	240	120

3.3 Architecture of the Proposed ADNet Model

Deep neural network is being widely used for weed detection in natural field in real time [19]. In this research work, we have employed DenseNet121, EfficientNetB0, MobileNetV2, ResNet50V2 and InceptionResNetV2 model to discriminate corn crop and its associated weed species. We integrated the Convolutional Block Attention Mechanism (CBAM) with DenseNet121 model [20] to develop a model named Attention embedded DenseNet121 (ADNet) to further enhance the overall performance of the DenseNet121 model. CBAM [21] module has been used on the feature map of DenseNet121 model which capture the important feature by focussing on the local region.

Fig. 1. Sample images from the dataset. The first row shows the samples of bluegrass, second row is for chenopodium album, third row is for cirsium setosum, fourth row is for sedge and last row is for corn crop.

DenseNet121 Architecture

In DenseNet architecture, each layer is connected to every other layer in feed forward manner. In a network of n layer, there will be $\frac{n(n+1)}{2}$ connections, whereas the conventional network has n connections. DenseNet121 architecture starts with basic convolution and pooling layer. Initially 7×7 convolution is performed on the input followed by 3×3 max pooling with a stride of 2. There are four dense blocks employed in DenseNet121 model. First dense block follows this in which two convolution operation with 1 × 1 and 3 × 3 dimension are performed. This process is repeated six times. First dense block is followed by first transition block which consists of 1 × 1 convolution followed by 2 × 2 average pooling layer. First transition layer is followed by second dense block where the same convolution and pooling operation are performed twelve times. Dense block 2 is followed by second transition block followed by third dense block where conv and pooling are performed 24 times. This dense block is followed by third transition block followed by fourth dense block where conv and pooling are performed 16 times. In the classification layer, global average pooling of 7 × 7 is used. Architecture of DenseNet121 model is shown in Fig. 2.

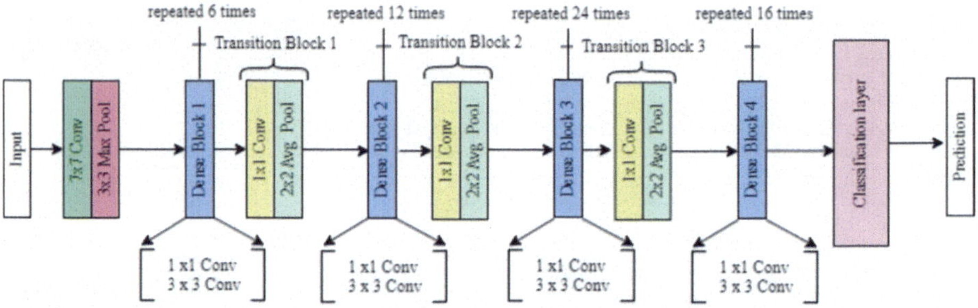

Fig. 2. Architecture of the DenseNet121 model

CBAM Module

The CBAM module gives attention to relevant features by focusing on the key part of the image and gives less important to useless information like background of the images. By doing so with the existing CNNs, efficiency of the pre-trained CNNs improves significantly. CBAM module consists of channel and spatial attention module. Channel attention module performs the global average pooling and global max pooling on the input to squeeze the spatial dimension of the input and then sent to multilayer perceptron. The output is merged, and then sigmoidal function is applied with the output and the input feature to give feature map of channel attention module. Channel attention output is given as the input to the spatial attention module followed by the global average pooling and global max pooling operations. Convolution operation with 7 × 7 dimension is performed on the merged output and then sigmoid activation is applied. Inner product of the resultant output with the channel attention map is performed to give the spatial attention map.

3.3.1 The Proposed ADNet Model

The proposed Attention embedded DenseNet121 (ADNet) model consists of DenseNet121 model integrated with CBAM module and a set of appended layers. The corn/ weed image is given as the input to the model. The DenseNet121 model yields the feature map which is then sent to CBAM module. CBAM nodule then applied to the feature map of DenseNet121 model after which few layers are employed to reduce the complexity and the overfitting of the model. After CBAM module, global average pooling is performed followed by 512 units dense layer. Dropout with 0.5 value has been used to reduce overfitting. SeLU as activation function has been used in the model which generally yields good performance for image classification tasks. A dense layer with 5 unit and softmax as classifier has been used in the last layer to classify the maize and the weed species. This set of additional layers reduces the overfitting and enhances the overall model performance. The architecture of the ADNet model is shown in Fig. 3.

Hyperparameters Finetuning

In this research study, transfer learning approach has been employed. On top of all CNN architectures used here, an additional set of layers has been added. Different hyperparameters with their values are defined in Table 2. The maximum number of epochs were defined to be 30 and Batch size was 16. Adam and categorical cross entropy were employed as optimizer and loss function. The learning rate is set to be 0.0001 and dropout rate is set to be 0.5 to reduce the overfitting. All the models were trained with the same set of hyperparameters.

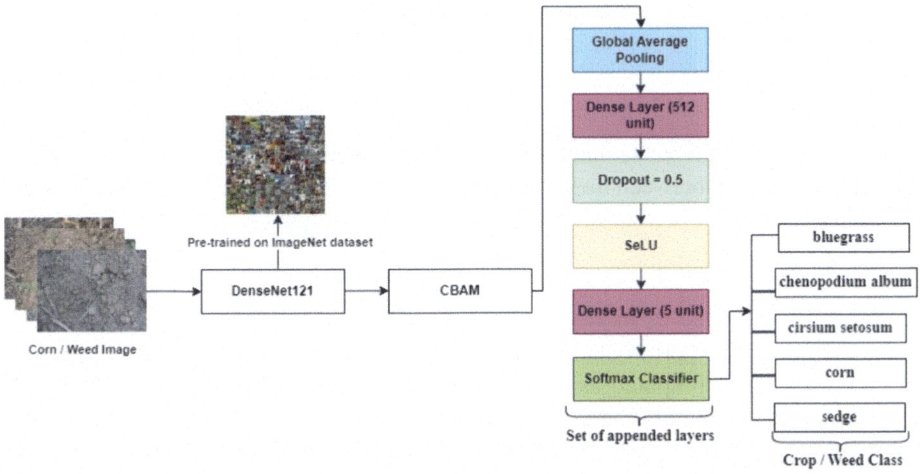

Fig. 3. Architecture of the proposed ADNet

Performance Metrics

All the models were tested with 1200 unseen images and different performance metrics, derived from confusion matrix, were employed to assess the performance of the models. Precision, recall, F1 score, and accuracy were used.

Accuracy refers to the overall classification performance of the crop and weed classification system and is calculated as:

$$Accuracy = \frac{TP + TN}{TP + TN + FP + FN} \tag{1}$$

Table 2. Settings of Hyperparameters

Hyperparameter	Value
Epoch	30
Optimize	Adam
Loss function	categorical cross entropy
Dropout	0.5
Learning rate	0.0001
Batch size	16

Precision is defined as a metric that measures the accuracy of positive predictions made by a classification system. It is calculated as:

$$Precision = \frac{TP}{TP + FP} \tag{2}$$

Recall, also known as sensitivity or true positive rate, measures the ability of the classification system to identify all instances of the positive class.

$$Recall = \frac{TP}{TP + FN} \tag{3}$$

F1 Score is a single value that takes both precision and recall into consideration and it provides a balance between these two metrics. It is particularly useful when you want to evaluate the overall performance of a classification system.

$$F1Score = \frac{TP}{TP + FN} \tag{4}$$

In addition to the performance metrics mentioned earlier, floating-point operations (FLOPs) were also computed using the "keras-flops" package in Python [22]. FLOPs is a measure that quantifies the total number of mathematical operations (additions and multiplications) required by a model to perform a classification task or any other computational task.

4 Experiment Results

This section delves into the experimental outcomes of all the CNN models utilized in the research study, while also conducting a comparative analysis with previous research findings.

4.1 Experiment Setup

Five well-known CNN models were tested on an image dataset containing corn and its associated weed species sourced from the corn weed dataset. The experiment employed a transfer learning approach, where all five models were trained using identical hyper-parameters. The implemented work utilized NVIDIA DGX GPU servers, each of which was equipped with 512 GB of RAM and Tesla V100 GPUs. Each GPU had a capacity of 32 GB. The work was implemented using the Python library Keras with TensorFlow as the backend.

4.2 Dataset Preparation

Dataset preparation consists of many steps. In the first step, dataset was partitioned into three distinct sets such as training, validation, and testing dataset in the ratio 70%, 10% and 20% respectively. In second step, all the images were resized to dimension of 224 × 224. In third step, all the images were normalized by dividing the intensity value of each pixel by 255 to reduce the computation complexity.

4.3 Performance Analysis of Proposed CNN Models

The performance of the proposed ADNet has been compared with five pre-trained CNN models named DenseNet121, EfficientNetB0, InceptionResNetV2, ResNet50V2, and MobileNetV2 and is demonstrated in Table 3. Among all the model experimented, the proposed ADNet has the highest overall performance with 99.58% followed by DenseNet121 model. ADNet has almost same number of parameters and FLOPs as that of DenseNet121. This means that the CBAM slightly increases the number of parameters but significantly enhances the overall performance of the ADNet. MobileNetV2 takes the least number of parameters and FLOPs but have the second worst performance among all the tested model here. ResNet50V2 is computationally expensive with 24M parameters and 6.99G FLOPs and is the worst performing model among all. Precision, recall and F1 score of ADNet are 99.59%, 99.58% and 99.58% respectively which show the satisfactory performance of ADNet.

Table 3. Comparison of all CNN models experimented.

Model Name	Accuracy (%)	Precision (%)	Recall (%)	F1 score (%)	#Parameter	FLOPS (G)
DenseNet121	99.50	99.51	99.50	99.50	7,564,869	5.7
EfficientNetB0	99.42	99.42	99.42	99.42	4,708,008	0.796
InceptionResNetV2	99.08	99.09	99.08	99.08	54,528,741	13
ResNet50V2	98.75	98.76	98.75	98.75	24,066,565	6.99
MobileNetV2	99.17	99.17	99.17	99.17	2,571,589	0.613
Proposed ADNet (DenseNet121 + CBAM)	99.58	99.59	99.58	99.58	7,631,655	5.7

Figure 4 demonstrates the accuracy and loss curve of the proposed ADNet model. As evidenced by the fig, both training and validation accuracy show similar patterns in their values as both values increase with decrease in the respective loss values. This shows that the model is properly trained and is not overfitted.

Figure 5 shows the confusion matrix for the proposed ADNet model. The model misclassifies five samples in the test dataset. One image in bluegrass has been classified as corn. One image in chenopodium album has been incorrectly classified as sedge. One image of corn crop has been misclassified as cirsium setosum and two images in sedge have been incorrectly classified as corn. ROC and Precision- Recall curve for comprehensive evaluation have been shown in Fig. 6.

AUROC value of less than 0.5 indicates that the model cannot distinguish a particular case. Here, the proposed model has a consistent AUC value of 1.00 for all the classes. AUPR curve particularly focuses on unbalanced data and incorrectly diagnosis cases. AUPR has a value of 1.00 for all the classes.

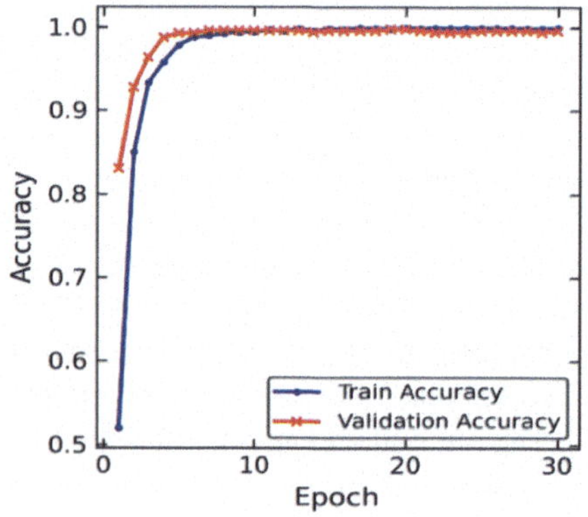

(a)

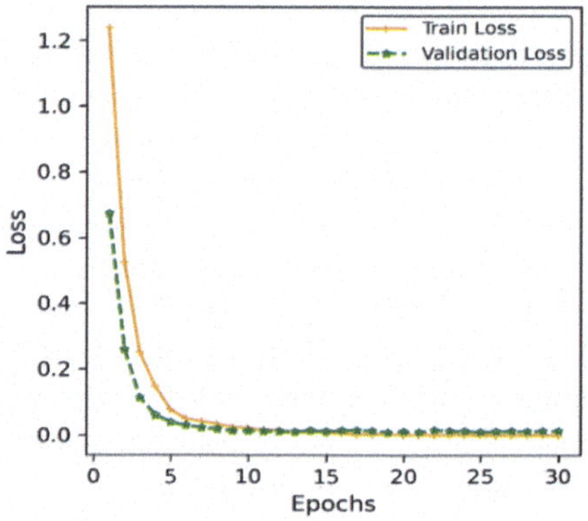

(b)

Fig. 4. (a) Accuracy curve (b) Loss curve of the proposed ADNet model

4.4 Performance Comparison with Existing Research Work

Table 4 illustrates the performance comparison of the proposed study with existing CNN model developed for crop and weed classification. Our proposed ADNet model outperforms existing research with remarkable performance with less number of parameters thus making ADNet the computationally inexpensive model for corn and weed classification. CBAM module integrated with the DenseNet121 model enhances the overall performance of the pre-trained DenseNet121 model. The ADNet takes 68% less parameters and yield better performance than work done by Subeesh et al. [14] because the proposed work utilized small pre-trained DenseNet121 model.

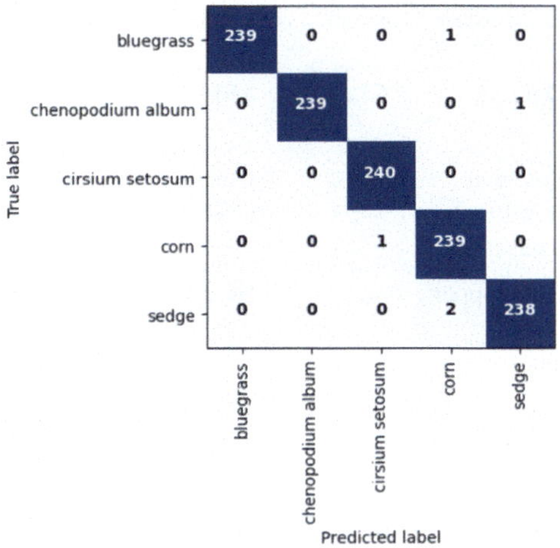

Fig. 5. Confusion matrix of the proposed ADNet model

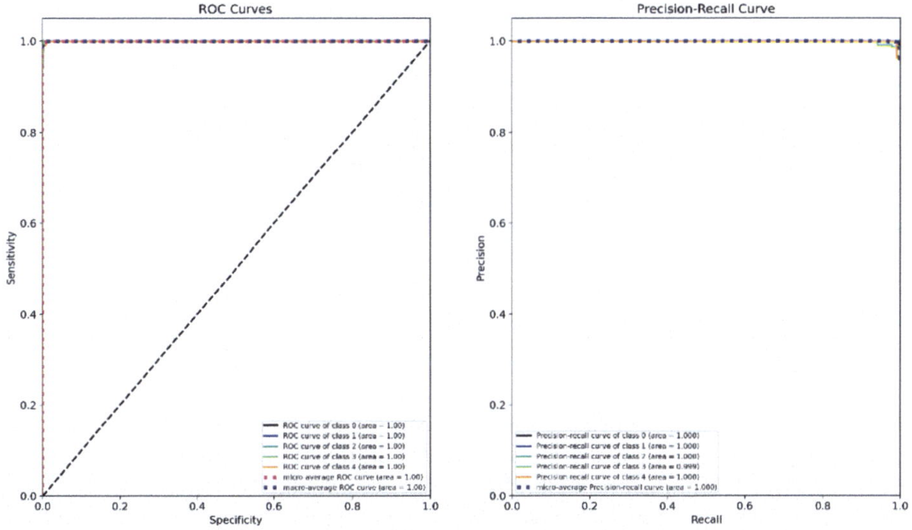

(a) (b)

Fig. 6. (a) ROC Curve (b) Precision- Recall curve of the proposed ADNet model

Table 4. Model performance comparison with existing work.

Study	Dataset	# Classes	Model	Accuracy (%)	#Parameter (million)
Razfar et al. [13]	Soybean weed dataset	Four	Custom 5-layer CNN	97.70	6.3
Subeesh et al. [14]	Bell pepper weed dataset	Two	InceptionV3	97.70	23.9
Garibaldi-Márquez et al. [15]	Own corn weed dataset	Three	VGG16	97.83	–
Tao and Wei [16]	Rapeseed weed dataset	five	Hybrid VGG-SVM	92.10	–
Jiang et al. [17]	Corn weed dataset	five	Graph convolutional network with ResNet101	97.80	–
Mu et al. [18]	V2 plant seedling dataset	twelve	Efficient channel attention with DenseNet	97.78	–
ADNet (Proposed)	Corn weed dataset	five	DenseNet121 + CBAM	99.58	7.6

5 Conclusion

To address the challenges associated with low accuracy and limited generalization ability in identifying weed species and crops in complex agricultural environments, this study introduced a weed identification method that leverages an enhanced version of DenseNet. With the DenseNet121 model, CBAM module and a set of few layers were integrated to enhance the overall performance. This CBAM module can extract the important weed features from the images. The set of additional layer appended can reduce the parameter count and mitigate the overfitting. The ADNet outperforms the existing pre-trained CNN models on the corn weed dataset with 99.58% overall accuracy score and 7.6M parameters amount. The proposed work also yields better performance than existing work reported in the literature with less computation cost and complexity. In the future, the proposed model can be employed with orther weed / crop datset to evaluate the generability of the model. There is potential to deploy this model on mobile devices or any other intelligent weeding equipment, enabling precise farmland detection for targeted weed control and enhancing overall corn crop production efficiency.

References

1. Hamuda, E., Glavin, M., Jones, E.: A survey of image processing techniques for plant extraction and segmentation in the field. Comput. Electron. Agric. **125**, 184–199 (2016)
2. R Shamshiri, R., et al.: Research and development in agricultural robotics: a perspective of digital farming (2018)
3. Venkataraju, A., Arumugam, D., Stepan, C., Kiran, R., Peters, T.: A review of machine learning techniques for identifying weeds in corn. Smart Agric. Technol. **3**, 100102 (2023)
4. Al-Badri, A.H., et al.: Classification of weed using machine learning techniques: a review—challenges, current and future potential techniques. J. Plant Dis. Prot. **129**(4), 745–768 (2022)
5. Zhao, X., Wang, X., Li, C., Fu, H., Yang, S., Zhai, C.: Cabbage and weed identification based on machine learning and target spraying system design. Front. Plant Sci. **13**, 924973 (2022)
6. Kamilaris, A., Prenafeta-Boldú, F.X.: Deep learning in agriculture: a survey. Comput. Electron. Agric. **147**, 70–90 (2018)
7. Hasan, A.M., Sohel, F., Diepeveen, D., Laga, H., Jones, M.G.: A survey of deep learning techniques for weed detection from images. Comput. Electron. Agric. **184**, 106067 (2021)
8. Hu, K., Coleman, G., Zeng, S., Wang, Z., Walsh, M.: Graph weeds net: a graph-based deep learning method for weed recognition. Comput. Electron. Agric. **174**, 105520 (2020)
9. Borhani, Y., Khoramdel, J., Najafi, E.: A deep learning based approach for automated plant disease classification using vision transformer. Sci. Rep. **12**(1), 11554 (2022)
10. Chug, A., Bhatia, A., Singh, A.P., Singh, D.: A novel framework for image-based plant disease detection using hybrid deep learning approach. Soft. Comput. **27**(18), 13613–13638 (2023)
11. Dheeraj, A., Chand, S.: Using deep learning models for crop and weed classification at early stage. In: Shakya, S., Du, KL., Ntalianis, K. (eds.) Sentiment Analysis and Deep Learning. Advances in Intelligent Systems and Computing, vol. 1432, pp. 931–942. Springer, Singapore (2023). https://doi.org/10.1007/978-981-19-5443-6_69
12. Dheeraj, A., Chand, S.: Deep learning model for automated image based plant disease classification. In: Sharma, H., Saha, A.K., Prasad, M. (eds.) Proceedings of International Conference on Intelligent Vision and Computing (ICIVC 2022). ICIVC 2022. Proceedings in Adaptation, Learning and Optimization, vol. 17, pp. 21–32. Springer, Cham (2023). https://doi.org/10.1007/978-3-031-31164-2_3
13. Razfar, N., True, J., Bassiouny, R., Venkatesh, V., Kashef, R.: Weed detection in soybean crops using custom lightweight deep learning models. J. Agric. Food Res. **8**, 100308 (2022)
14. Subeesh, A., et al.: Deep convolutional neural network models for weed detection in polyhouse grown bell peppers. Artif. Intell. Agric. **6**, 47–54 (2022)
15. Garibaldi-Márquez, F., Flores, G., Mercado-Ravell, D.A., Ramírez-Pedraza, A., Valentín-Coronado, L.M.: Weed classification from natural corn field-multi-plant images based on shallow and deep learning. Sensors **22**(8), 3021 (2022)
16. Tao, T., Wei, X.: A hybrid CNN–SVM classifier for weed recognition in winter rape field. Plant Methods **18**(1), 29 (2022)
17. Jiang, H., Zhang, C., Qiao, Y., Zhang, Z., Zhang, W., Song, C.: CNN feature based graph convolutional network for weed and crop recognition in smart farming. Comput. Electron. Agric. **174**, 105450 (2020)
18. Mu, Y., et al.: DenseNet weed recognition model combining local variance preprocessing and attention mechanism. Front. Plant Sci. **13**, 1041510 (2023)

19. Rakhmatulin, I., Kamilaris, A., Andreasen, C.: Deep neural networks to detect weeds from crops in agricultural environments in real-time: a review. Remote Sens. **13**(21), 4486 (2021)
20. Huang, G., Liu, Z., Van Der Maaten, L., Weinberger, K.Q.: Densely connected convolutional networks. In: Proceedings of the IEEE Conference on Computer Vision and Pattern Recognition, pp. 4700–47082017
21. Woo, S., Park, J., Lee, J.Y., Kweon, I.S.: CBAM: convolutional block attention module. In: Proceedings of the European Conference on Computer Vision (ECCV), pp. 3–192018
22. Tokusumi. Keras-flops Calculator (2020)

NRASV: Noise Robust ASV System for Audio Replay Attack Detection

Nidhi Chakravarty(✉) and Mohit Dua

Department of Computer Engineering, National Institute of Technology, Kurukshetra, India
nidhichakravarty03@gmail.com, er.mohitdua@nitkkr.ac.in

Abstract. Recent advancements in researching countermeasure systems against audio replay attacks have notably bolstered the resilience of Automatic Speaker Verification (ASV) Systems. However, current countermeasures face challenges in adapting to unknown attack variations. This limitation stems from the failure to capture context-specific details from the speech data, a crucial factor leading to the suboptimal performance of these systems against novel threats. To address this, our proposed anti-replay system integrates Gammatone features with t-distributed Stochastic Neighbor Embedding (t-SNE) for dimensionality reduction. In the classification phase, we employ diverse machine learning algorithms including Support Vector Machine (SVM), Random Forest (RF), K Nearest Neighbor (KNN), and Gradient Boost (Gboost). Our evaluation, conducted on the VSDC dataset in both controlled and noisy environments, introduces babble noise at varying levels (15dB, 10dB, 5dB, and 0dB). Remarkably, our model, GTCC-t-SNE with SVM, attains exceptional performance with Equal Error Rates (EER) of 2.2%, 3.3%, 3.8%, and 4.4% for the respective noise levels, demonstrating its robustness against audio replay attacks.

Keywords: ASV · GTCC · t-SNE · SVM · RF · KNN · Gboost

1 Introduction

Automatic Speaker Verification Systems (ASVs) have become crucial in various applications, notably in areas like phone-based banking. They offer a swift and reliable means of authenticating individuals through speech recognition. However, despite their efficiency, ASVs are not without vulnerabilities. A major concern is their susceptibility to spoofing attacks, where unauthorized users exploit system weaknesses to gain access to sensitive data. This security flaw emphasizes the urgent need to bolster ASVs against potential attacks, safeguarding the confidentiality and integrity of crucial information—especially vital in times like the present pandemic where disease control and secure remote authentication are paramount.

From the viewpoint of a verification problem, speech verification research spans mainly in two directions. First is feature extraction technique that is used at the front end, and the second is classifier which is applied at the back-end to classify the audio as fake or real. Some researchers have used traditional feature extraction techniques such

D. Pastor-Escuredo et al. (Eds.): ICDLAIR 2023, LNNS 1001, pp. 639–647, 2024.
https://doi.org/10.1007/978-3-031-60935-0_56

as Mel Frequency Cepstral Coefficient (MFCC), Perceptual Linear Prediction (PLP) and Linear Frequency Cepstrum Coefficients (LFCC), while others have used some advanced methods like Constant Q Cepstrum Coefficients (CQCC), Gammatone Cepstral Coefficients (GTCC) [1, 2], and Basilar-membrane Frequency-band Cepstral Coefficients (BFCC) [3] to extract meaningful information. However, every technique has its own merits and demerits.

Malik et al. [4] used K- Nearest Neighbour (KNN), Naïve Bayes (NB), and Support Vector Machines (SVM) to train a binary classifier. Also, deep learning techniques such as Convolution Neural Network (CNN), Long Short-Term Memory (LSTM), and their improved versions are also performing a very crucial role in developments of efficient classification models for ASV systems [5]. The proposed work in this paper implements front end of an ASV replay detection system by using GTCC [6] features along with t-SNE dimensionality reduction at front end and Machine learning algorithm at the back end.

The remaining part of the paper is laid out as follows. The related work is presented in Subsect. 1.1. Section 2 describes the proposed system architecture in detail. The experimental setup and results are given in Sect. 3 of the paper. Section 4 of the paper contains the discussion and comparative analysis, whereas Sect. 5 concludes the paper.

1.1 Related Work

This section provides the overcome of existing countermeasures developed by authors in recent years. Spoofing an authentication system is done to get a positive verification status for the claimed identity. The issue of replay spoofing detection is addressed through the ASVspoof 2017 Challenge.

Authors in [7] have created a replay attacks detection system that used an attention-based filtering mechanism to enhance feature representations in both time and frequency domains. AF Network is composed of Attentive Filtering and a Dilated Residual Network. The proposed model has achieved an EER of 8.54%. Haung et al. [8] was focused on implementing a replay detection device. The proposed model was created using three features MFCC, CQCC, and a fusion of these two features. GMM, CNN, and LSTM did not achieve the desired results, so the authors adopted the DenseNet structure to solve problems. The hybrid feature and DenseNet model achieved 6.69% EER for the development set of the ASVspoof2017 dataset, 15.66 for the evaluation set, and 13.36% for the combined development.

In [9] authors have proposed a new approach using a segment-based hybrid FE method MFCC, CQCC and their hybrid features[10]. The hybrid features were trained using various DL networks, including DenseNet, LSTM, and DenseNet-LSTM hybrid architectures. DenseNet-LSTM achieved 3.62%, 9.56%, and 8.84% EER for the development, evaluation set, and combination of development and evaluation set.

Joshi et al. [11] have proposed an ASV system that uses GTCC for FE and LSTM for classification. The model is trained and evaluated using the VSDC dataset. The proposed system has achieved an EER of 2.38% and 2.72% in 1PR and 2PR attacks, respectively, and an accuracy of 98% and 97.26% in the same attacks.

Dua et al. [12] studied three backend classification models: LSTM, CNN, and a hybrid. They tested the proposed model for detecting 0PR, 1PR, and 2PR audios. The

CNN + LSTM model outperformed standalone models in both experiments, achieving 97.6% accuracy, 0.98 recall, 0.98 precision, 0.98 F1-score, and 0.036 EER. The 0PR and 2PR hybrid models achieved 97% accuracy, 98% precision, 98% recall, 98% F1-score, and 0.0296% EER. Joshi et al. [13] aimed to create a solution to make ASV systems more resilient against multi-order replay assaults. The features are extracted from audio samples by combining MFCC with FDLP. A GRU classification model is used in the backend. The suggested system obtained an ERR of 2.99% and 1.6%, under 1PR and 2PR respectively.

Our proposed anti-replay system integrates Gammatone features with t-distributed Stochastic Neighbor Embedding (t-SNE) for dimensionality reduction. In the classification phase, we employ diverse machine learning algorithms including Support Vector Machine (SVM), Random Forest (RF), K Nearest Neighbor (KNN), and Gradient Boost (Gboost). The primary contributions of the suggested framework are outlined as follows.

1. We have created a model to detect audio replay attack under noisy environment. For this we have added babble noise [14] at different SNR value.
2. The Frontend of the proposed model uses GTCC features along with t-distributed Stochastic Neighbor Embedding (t-SNE).
3. The extracted features are individually used to train four classifiers: SVM, RF, KNN, and Gradient Boost.
4. The proposed model trained and tested on VSDC dataset.
5. The work tests all the models for deepfake partition of ASVspoof2021, and uses evaluation metrics such as Equal Error Rate (EER)[15], Accuracy, Precision, F1-score, and Recall[16] for the performance analysis.

2 Proposed Approach

The proposed model's architecture is depicted in Fig. 1. It utilizes both clean and replayed audio samples from the VSDC dataset. Our model is constructed in three key phases: Feature Extraction, Feature Reduction, and Classification. In the initial phase, GTCC features are extracted from the audio data. These GTCC features possess dimensions equal to the number of audio samples multiplied by 42. To evaluate the model's performance in noisy conditions, babble noise is introduced at Signal to Noise Ratios (SNR) of 0 dB, 5 dB, 10 dB, and 15 dB. Moving on to the second phase, we implement dimensionality reduction through t-distributed Stochastic Neighbor Embedding (t-SNE) to obtain the most salient features. This results in two optimal features. In the final phase, we employ four distinct machine learning models; SVM, RF, KNN, and Gradient Boost. These models are trained using the two best GTCC features, allowing us to classify audio samples into real and spoofed categories.

2.1 Feature Extraction

This section contains the brief about the techniques used at the front end. Frontend plays a vital role in ASV system. Feature extraction done at the frontend to extract information from the audios. In our proposed method we have used GTCC features.

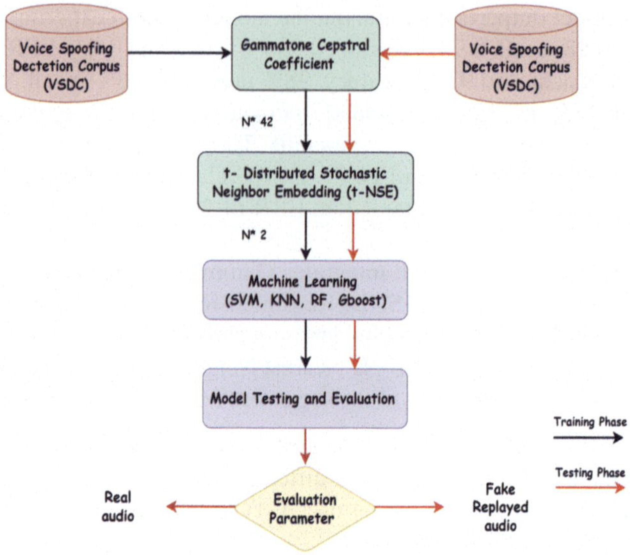

Fig. 1. Proposed Architecture

Dimensionality Reduction

t-distributed Stochastic Neighbour Embedding (t-SNE) is a potent dimensionality reduction technique used in machine learning and data visualization. It transforms high-dimensional data into a lower-dimensional space while preserving local relationships between data points. By modelling probabilities of point pairs in both high and low dimensions, t-SNE minimizes the Kullback-Leibler divergence to find an embedding that mirrors the original data's structure. This approach is especially adept at revealing non-linear patterns and complex clusters, making it invaluable for visualizing intricate datasets. However, its stochastic nature and computational demands should be considered when applying t-SNE to large-scale datasets.

2.2 Classification

In our proposed work, four separate classification models are created employing four different machine learning algorithms, including SVM, RF, KNN, and Gboost. VSDC training partition has been used to train and validate the model. The model has been evaluated using the VSDC testing partition.

- **SVM:** Because data is not linearly separable, a different kernel function was used to construct an SVM model. The recommended technique makes use of four kernel functions: "linear", "poly", "rbf", and "sigmoid". However, the "rbf" kernel outperformed the others.
- **RF:** The random forest technique was utilised to develop a classification model in our suggested study. After fitting several decision tree classifiers to diverse dataset subsamples, a random forest is a meta estimator that uses averaging to boost predicted accuracy and decrease overfitting. RF accepts the number of decision trees

as an input, and in our instance, the estimator value was set to a random integer between 10 and 100.

- **KNN:** The KNN model has been used to classify audio. KNN calculates the closest neighbour by calculating the distance between the data points. In our proposed technique, three variables were used to fine-tune KNN experiments: the number of neighbours (KN), the distance metric for finding the nearest neighbours, and the distance weights. Different neighbour values were chosen for the trials by creating a random number between 1 and 10, with the best results achieved when the neighbour value is 5, and the Euclidean distance metric and equal distance weight were also employed.
- **Gradient Boost:** Gradient Boosting, as implemented in Scikit-learn via the GradientBoostingClassifier, is a strong ensemble learning approach. It trains a sequence of decision trees consecutively, with each successive tree correcting the flaws of its predecessor. The approach emphasises misclassified data during training, effectively minimising the loss function. Users can alter settings like as learning rate, maximum tree depth, and tree count. Gradient Boosting is a computationally efficient algorithm that is commonly used for classification and regression problems. It excels at high predicted accuracy and strong performance over a wide range of datasets and domains.

3 Experiments and Result Analysis

In this section, we have provided a comprehensive overview of the experiments that have been carried out to assess the effectiveness of our voice replay detection methodology. Our evaluation has been conducted on the VSDC dataset. To implement our approach, we have used MATLAB 2022, while the back-end model has been written in Python 3.7 and implemented using Anaconda. The computational resources that have been employed for this task are a computer equipped with an 11th generation core i5 processor and 16GB of RAM. The detailed explanation of the dataset utilized for our proposed work and also describes the thorough experimental approach used to evaluate the effectiveness of our model.

3.1 Performance of Proposed Model Under Clean Environment

Table 1 shows the result of first experiment that is performed under clean environment. In this experiment we have used 2 best GTCC featured selected by t-SNE dimensional reduction technique. At backend we have used four ML model; SVM, RF, KNN, and Gboost. From Table 1, it can be concluded that SVM outperformed other ML model to detect replay attack in clean environment by achieving an EER of 1.2%.

3.2 Performance of Proposed Model Under Noisy Environment

This section describes the performance of our proposed model under noisy environment. We have used babble noise at different SNR rate such as 0 dB, 5 dB, 10 dB and 15 dB.

Table 1. Performance of proposed model under clean environment

Feature	Classification model	Accuracy (%)	Recall	Precision	F1-Score	EER
GTCC + t-SNE	SVM	99	99	99	99	1.2
	RF	97	97	97	97	3
	KNN	95	95	95	95	4.5
	Gboost	94	95	95	95	5

Performance of Proposed Model Using SVM Classifier. The second experiment, aimed at implementing the proposed model, is summarized in Table 2. This model's frontend integrates t-SNE and GTCC features and was rigorously tested under noisy conditions. The backend employed SVM for classification. The results indicate that our model achieved an impressive EER of 2.2% under a Signal to Noise Ratio (SNR) of 15dB. This signifies the robustness and effectiveness of our approach, particularly in challenging, noisy environments.

Table 2. Performance of proposed model using SVM Classifier

Feature	Noise	Accuracy (%)	Recall	Precision	F1-Score	EER
GTCC + t-SNE	BN[1] (15dB)	98	98	98	98	2.2
	BN (10dB)	95	96	97	97	3.3
	BN (5dB)	96	96	97	97	3.8
	BN (0dB)	93	95	95	95	4.4

Performance of Proposed Model Using RF Classifier. In the third experiment, which aimed to implement our proposed model, the findings are presented in Table 3. This model incorporates t-SNE and GTCC features at its frontend and underwent thorough testing in noisy conditions. For classification, the backend utilized RF. The results demonstrate that our model achieved an outstanding Equal EER of 4% at SNR of 15 dB.

Performance of Proposed Model Using KNN Classifier. Table 4 shows the results of the fourth experiment, which intended to put our suggested model into action. This model contains t-SNE and GTCC features at the frontend and was thoroughly tested in noisy environments. The backend used RF for categorization. The findings show that our model produced an excellent Equal EER of 8% at SNR of 15 dB.

Performance of Proposed Model Using Gradient Boost Classifier. Table 5 shows the results of the fifth experiment, which intended to put our suggested model into action.

[1] Babble Noise.

Table 3. Performance of proposed model using RF Classifier

Feature	Noise	Accuracy	Recall	Precision	F1-Score	EER
		(%)				
GTCC + t-SNE	BN[2] (15dB)	96	96	96	96	4
	BN (10dB)	82	90	88	89	9
	BN (5dB)	77	86	86	86	13
	BN (0dB)	80	89	86	88	10

Table 4. Performance of proposed model using KNN Classifier

Model	Noise	Accuracy	Recall	Precision	F1-Score	EER
		(%)				
GTCC + t-SNE	BN[3] (15dB)	92	92.4	92	92	8
	BN (10dB)	80	87	89	88	12
	BN (5dB)	79	87	87	87	12.6
	BN (0dB)	75	82	87	84	17

This model contains t-SNE and GTCC features at the frontend and was thoroughly tested in noisy environments. The backend used Gboost for categorization. The findings show that our model produced an excellent Equal EER of 3.7% at SNR of 15 dB.

Table 5. Performance of Proposed Model Using Gradient Boost Classifier

Model	Noise	Accuracy	Recall	Precision	F1-Score	EER
		(%)				
GTCC + t-SNE	BN[4] (15dB)	97	96	97	97	3.7
	BN (10dB)	84	94	88	91	5.6
	BN (5dB)	82	94	85	89	5.7
	BN (0dB)	81	92	86	89	7.9

[2] Babble Noise.
[3] Babble Noise.
[4] Babble Noise.

4 Discussion and Comparative Analysis

The main of this paper is to detect audio replay attack under clean and noisy environment. To build this model we have used GTCC features at frontend along with dimension reduction algorithm to reduce the complexity of our proposed ASV system. Four machine learning models SVM, RF, KNN and Gboost have been used and trained by reduced GTCC features. From Table 1 it can be concluded that in clean environment SVM outperformed another ML model. Table 2, 3, 4, 5 shows the result of experiments performed in noisy environment. From this result we can conclude that our proposed model achieved reasonable EER in noisy environment on high SNR rate. However, our model is not able to performed good for 0dB SNR. Table 6 shows the comparison of proposed model with existing techniques. We have chosen those studies we have used VSDC dataset. From Table 6 it can be concluded that our model outperformed existing techniques.

Table 6. Comparative Analysis with Existing techniques

Author	Frontend	Backend	Noise	EER (%)	[a]Ac*	[b]F1*	[c]Re*	[d]Pre*
Dua et al. [12]	CQCC, Mel spectrogram	LSTM, CNN	NO	2.9	97.7	98.5	98.3	98.8
Joshi et al. [13]	FDLP-MFCC	GRU	NO	1.6	97.9	×	×	×
Joshi et al. [11]	GTCC	LSTM	NO	2.38	98	×	×	×
Proposed Work	GTCC-t-SNE	SVM, RF, KNN, Gboost	Clean	1.2	99	99	99	99
			Noisy	2.2	98	98	98	98

[a] Accuracy*
[b] F1-score*
[c] Recall*
[d] Precision*

5 Conclusion

In this paper, we have proposed an innovative anti-replay system that amalgamates Gammatone features with t-distributed Stochastic Neighbor Embedding (t-SNE) for robust dimensionality reduction. In the subsequent classification phase, a diverse suite of machine learning algorithms including Support Vector Machine (SVM), Random Forest (RF), K Nearest Neighbor (KNN), and Gradient Boost (Gboost) are employed. Rigorous evaluation on the VSDC dataset under controlled and noisy conditions, featuring varying babble noise levels (15 dB, 10 dB, 5 dB, and 0 dB), showcases the extraordinary performance of our model. Particularly, the GTCC-t-SNE with SVM combination achieves striking Equal Error Rates (EER) of 2.2%, 3.3%, 3.8%, and 4.4% for the respective noise levels. This achievement underscores the robustness of our approach in combatting audio replay attacks. To further enhance the proposed model, it is recommended to incorporate more attacks and advanced acoustic modelling approaches in the backend.

References

1. Valero, X., Alías, F.: Gammatone cepstral coefficients: biologically inspired features for non-speech audio classification. IEEE Trans. Multimedia **14**, 1684–1689 (2012). https://doi.org/10.1109/TMM.2012.2199972
2. Liu, G.K.: Evaluating gammatone frequency cepstral coefficients with neural networks for emotion recognition from speech. arXiv Prepr. arXiv1806.09010 (2018)
3. Dua, M., Aggarwal, R.K., Biswas, M.: Optimizing integrated features for hindi automatic speech recognition system. J. Intell. Syst. **29**(1), 959–976 (2018). https://doi.org/10.1515/jisys-2018-0057
4. Malik, K.M., Javed, A., Malik, H., Irtaza, A.: A light-weight replay detection framework for voice controlled IoT devices. IEEE J. Selected Topics Signal Process. **14**(5), 982–996 (2020). https://doi.org/10.1109/JSTSP.2020.2999828
5. Mittal, A., Dua, M.: Static–dynamic features and hybrid deep learning models based spoof detection system for ASV. Complex Intell. Syst. 1–14 (2021)
6. Chakravarty, N., Mohit, D.: Data augmentation and hybrid feature amalgamation to detect audio deep fake attacks. Phys. Scr. (2023). https://doi.org/10.1088/1402-4896/acea05
7. Lai, J., Abad, A., Richmond, K., Yamagishi, J., Dehak, N.: Attentive filtering networks for audio replay attack detection (2018)
8. Huang, L., Gan, Y., Ye, H.: Audio-replay attacks spoofing detection for automatic speaker verification system. In: 2019 IEEE International Conference on Artificial Intelligence and Computer Applications (ICAICA), pp. 392–396 (2019)
9. Huang, L., Pun, C.-M.: Audio replay spoof attack detection using segment-based hybrid feature and densenet-LSTM network. In: ICASSP 2019–2019 IEEE International Conference on Acoustics, Speech and Signal Processing (ICASSP), pp. 2567–2571 (2019)
10. Chakravarty, N., Dua, M.: Spoof detection using sequentially integrated image and audio features. Int. J. Comput. Digit. Syst. **13**(1), 1 (2023)
11. Joshi, S., Dua, M.: LSTM-GTCC based approach for audio spoof detection. In: 2022 International Conference on Machine Learning, Big Data, Cloud and Parallel Computing (COM-IT-CON), vol. 1, pp. 656–661 (2022)
12. Dua, M., Sadhu, A., Jindal, A., Mehta, R.: A hybrid noise robust model for multireplay attack detection in Automatic speaker verification systems. Biomed. Signal Process. Control **74**, 103517 (2022). https://doi.org/10.1016/j.bspc.2022.103517
13. Joshi, S., Dua, M.: Multi-order replay attack detection using enhanced feature extraction and deep learning classification. In: Proceedings of International Conference on Recent Trends in Computing: ICRTC 2022, pp. 739–745 (2023)
14. Chakravarty, N., Dua, M.: Noise robust ASV spoof detection using integrated features and time delay neural network. SN Comput. Sci. **4**(2), 127 (2022)
15. Cheng, J.-M., Wang, H.-C.: A method of estimating the equal error rate for automatic speaker verification. In: 2004 International Symposium on Chinese Spoken Language Processing, pp. 285–288 (2004)
16. Novaković, J.D., Veljović, A., Ilić, S.S., Papić, Ž, Milica, T.: Evaluation of classification models in machine learning. Theory Appl. Math. Comput. Sci. **7**(1), 39–46 (2017)

Machine Learning for the Proactive Identification of Polycystic Ovary Syndrome (PCOS): Empowering Women's Health

Ardra Nair(✉) and Geetika Chatley

Lovely Professional University, Phagwara, Punjab, India
ardra333nair@gmail.com, Geetika.25107@lpu.co.in

Abstract. Polycystic Ovary Syndrome (PCOS) is a prevalent endocrine disorder affecting women of reproductive age, often remaining undiagnosed until complications like infertility or metabolic disturbances arise. Early detection is crucial for effective management. This research introduces an innovative PCOS diagnosis model employing machine learning algorithms. After a comparative study of machine learning algorithms KNN was showed highest accuracy of 91%. Utilizing user-provided responses to simple yes/no questions, the model offers accessible and user-friendly early warning signs of PCOS, enabling timely medical consultation and intervention. Doctors commonly face late-stage PCOS diagnoses, leading to complex treatments. This research addresses the need for early identification, reducing associated health risks and enhancing the quality of life for affected individuals. This paper proposes a method to track early indicators such as BMI, period health, skin health, and weight patterns, identifying negative patterns and notifying users early, before they manifest as PCOS. This proactive approach can prevent complications, including obesity, insulin resistance, diabetes, infertility, and cardiovascular issues, offering a comprehensive solution to a pressing healthcare concern. By identifying negative patterns and notifying users early, we aim to prevent complications such as obesity, insulin resistance, diabetes, infertility, and cardiovascular issues. This cost-effective method relies solely on non-invasive factors, making it a convenient and accessible tool for early PCOS detection, eliminating the need for laboratory tests or ultrasound results.

Keywords: Polycystic ovary syndrome · K-Nearest Neighbors · Machine Learning · Logistic Regression

1 Introduction

Women of reproductive age frequently suffer from Polycystic Ovary Syndrome (PCOS), an endocrine condition that sometimes goes untreated until problems like infertility or metabolic disorders occur. For management to be effective, early detection is essential. This study presents a novel machine learning-based PCOS detection model, with K-Nearest Neighbors (KNN) exhibiting the greatest accuracy of 91%. The concept provides accessible and user-friendly early warning signals of PCOS through the use of

D. Pastor-Escuredo et al. (Eds.): ICDLAIR 2023, LNNS 1001, pp. 648–657, 2024.
https://doi.org/10.1007/978-3-031-60935-0_57

user-provided answers to straightforward yes/no questions, facilitating prompt medical evaluation and intervention. One enduring problem is the delayed diagnosis of PCOS by medical practitioners, which frequently results in complicated therapies. The present study endeavors to tackle the pressing requirement of prompt detection, mitigating concomitant health hazards, and augmenting the standard of living for impacted persons. The study also emphasizes the underdiagnosis of PCOS in teenagers, highlighting the significance of early identification in this population. It highlights the stigma around talking about irregular menstrual periods, which can result in cases going untreated and have a negative impact on mental health. This study suggests a way to monitor early markers like body mass index, weight patterns, skin health, menstrual health, and BMI. The goal is to spot problematic trends early on and alert users before they become symptoms of PCOS. This proactive strategy offers a holistic answer to a critical healthcare issue by preventing complications such as obesity, diabetes, insulin resistance, infertility, and cardiovascular problems. In a time of lightning-fast technological advancement, machine learning is essential to the healthcare industry. The creation of models and algorithms for machine learning enables computers to examine massive datasets and forecast outcomes based on patterns found in them. Because this approach can handle large amounts of patient data rapidly and reliably, it has become more and more significant in medical diagnosis. The promise of machine learning to transform healthcare by improving diagnosis speed and accuracy makes it crucial for medical diagnosis. It makes it possible to spot patterns and trends that human practitioners could miss, which could result in the early discovery of conditions like PCOS. Additionally, machine learning-driven diagnostic tools frequently offer greater affordability, accessibility, and user-friendliness, enabling people to take charge of their health. In this work, a highly effective model was constructed to predict the existence or absence of PCOS using a labelled historical dataset. In order to help with early identification and prompt consultations, users can forecast medical issues by answering questions. By enhancing early detection and healthcare delivery, this research will help individuals and healthcare clinics through data selection, sophisticated analysis, and practical implementation. In the end, the health and wellbeing of women everywhere are improved by this research's contribution to the ongoing efforts to enhance the early detection and management of PCOS.

2 Literature Review

In the study by Pedersen et al., a validated questionnaire for the diagnosis of Polycystic Ovary Syndrome (PCOS) was introduced. This questionnaire provides a valuable tool for healthcare professionals to aid in the diagnosis of PCOS, potentially improving early detection and management of the condition. The study indicated that a history of irregular menstrual cycles, hirsutism, obesity, and acne were robust predictors of a PCOS diagnosis. Conversely, a history of unsuccessful pregnancy attempts did not prove to be a valuable indicator for identifying PCOS in this context [1]. The paper concludes that hirsutism is significantly correlated with anthropometric, metabolic, and hyperandrogenic disorders in PCOS-infertile women [2]. The study examines the application of machine learning algorithms to predict Polycystic Ovary Syndrome (PCOS). The study gathered a diverse dataset from various Indian regions, considering factors like age, lifestyle, and urban or rural settings to enhance representativeness. Five algorithms were tested. The Decision Tree Classifier outperformed others, achieving an 81% accuracy rate in PCOS prediction [3]. The paper presents a machine learning approach to diagnose Polycystic Ovary Syndrome (PCOS). They use Decision Tree, K Neighbors, and Random Forest classifiers with visible features like age, weight, and symptoms to assess PCOS likelihood. Random Forest achieved the highest accuracy at 90.44%. The user-friendly system asks for 15 easily observable inputs for diagnosis, providing clear results. The study balances accuracy and accessibility, making it a promising tool for PCOS diagnosis [4]. The study [5] addresses the challenge of PCOS detection. It proposes a system that employs machine learning, specifically the Random Forest Classifier (RFC), achieving an impressive 89.02% accuracy. In this study [6] a novel approach was employed for PCOS diagnosis, involving the utilization of an ensemble comprising six models, including SVM, logistic regression, and ridge classifiers, as well as bagging, gradient boosting, and random forest classifiers. These models' predictions, along with the initial dataset, were consolidated, and a neural network was trained on the merged data for the ultimate classification. This innovative model demonstrated an impressive accuracy rate of 90.74%. The study outlines a pioneering PCOS detection system harnessing the power of machine learning techniques. The paper [7] evaluates five machine learning algorithms to predict Polycystic Ovary Syndrome (PCOS). The Random Forest algorithm achieved the highest accuracy at 89%, followed by SVM (86.5%), KNN (84.3%), Naive Bayes (83.1%), and Logistic Regression (79%). This system has the potential to assist both doctors and women in early PCOS assessment. It also offers a user-friendly web application to facilitate easy access to medical consultations, improving patient care.

3 Methodology

This research study involves utilizing a dataset of features related to PCOS for the development of a machine learning model. It employs exploratory data analysis to uncover data patterns and factors that might be the primary contributors to PCOS. After preprocessing the dataset, it is divided into two subsets: a training set and a test set, with a split ratio of 75% for training and 25% for assessing the model's performance on unseen data. The model, which is fine-tuned with hyperparameters, is thoroughly trained, and tested. Similarly, the model is subjected to testing using various machine learning algorithms, ultimately selecting the model that exhibits the highest accuracy and recall. This chosen model is then prepared for deployment to predict PCOS disease. The research methodology and workflow are visually depicted in Fig. 1.

3.1 Data Set

The primary objective of this study is to predict the occurrence of PCOS in women based on their reported symptom. According to research findings, the primary confirmation of a PCOS diagnosis typically involves clinical examinations. Subsequently, to assess the extent of the condition, patients are recommended to undergo pelvic ultrasound procedures. Considering this, an electronic form has been pre-structured for recording

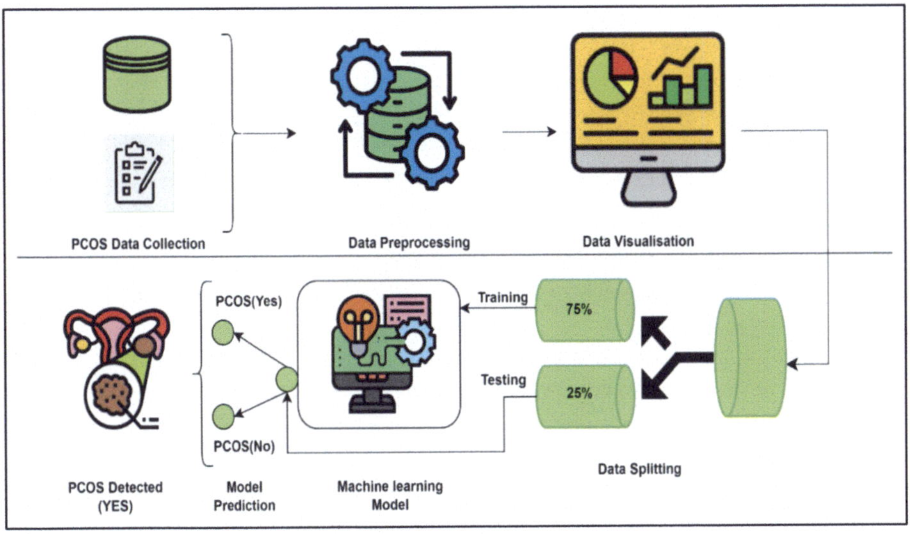

Fig. 1. Research Methodology

information and distributed to women aged 15 and above. These questionnaires included inquiries about age, PCOS status, age, as well as symptoms such as unusual hair growth, pimples, and others. Approximately 199 women responded to the survey.

Subsequently, this survey data is integrated with a dataset sourced from Kaggle, a renowned platform for scientific datasets often utilized by Machine Learning professionals. The Kaggle dataset, titled "Polycystic Ovary Syndrome," contains clinical and physical parameters for a total of 541 women, of which 364 were deemed healthy, while 177 had been diagnosed with PCOS. For reference, Fig. 2 illustrates the structure of the survey questionnaire.

3.2 Data Analysis and Visualization

The selection of an appropriate algorithm heavily depends on the type and nature of the collected data. Therefore, having a thorough understanding of the data is of paramount significance. We gained insights into the data by employing data visualization techniques in Python and Excel. Visualization proves invaluable in identifying meaningful associations between variables or categories.

Count plots are employed to visualize the instance distribution of both classes within the PCOS dataset. Figure 3 represents a count plot displaying the instances for both "no" and "yes" categories. The "no" category comprises 362 instances, while the "yes" category consists of 282 instances in this binary class dataset. Notably, "0" signifies the absence of PCOS (No PCOS), whereas "1" signifies the presence of PCOS (Yes PCOS). Figure 3 complements this by offering a pie chart illustrating the percentage composition of each class in the dataset. Specifically, 56.2% of the data belongs to the PCOS No class, and 43.8% belongs to the Yes class. This research approach identifies fourteen key features for PCOS prediction, enabling patients to input them conveniently

Structure of the survey questionnaire	
1	What is your current age? **(Age)**
2	What is your current weight in kilograms? (**Weight in kg**)
3	What is your current height in centimetres? (**Height in cm**)
4	Have you been medically tested for Polycystic ovary Syndrome(**PCOS** or **PCOD**)?
5	If yes, were you diagnosed with Polycystic ovary syndrome(**PCOS** or**PCOD**)
6	How long do your periods typically last? (In days, how many days does the bleeding occur?)
7	Do you find any difficulty in maintaining your **body weight?**
8	Are you facing hair thinning? (**hair loss**)
9	Are you experiencing **pimples/Acne**?
10	How often do you **exercise**?
11	How often do you **exercise**?
12	Have you observed any **excess facial or body hair** growth ? (in upper lip/ chin/facial hair/ chest/belly)
13	Have you observed any **excess facial or body hair** growth ? (in upper lip/ chin/facial hair/ chest/belly)
14	What is the measurement of your **waist in inches**?
15	What is the measurement of your **hip in inches**?

Fig. 2. Structure of the survey questionnaire

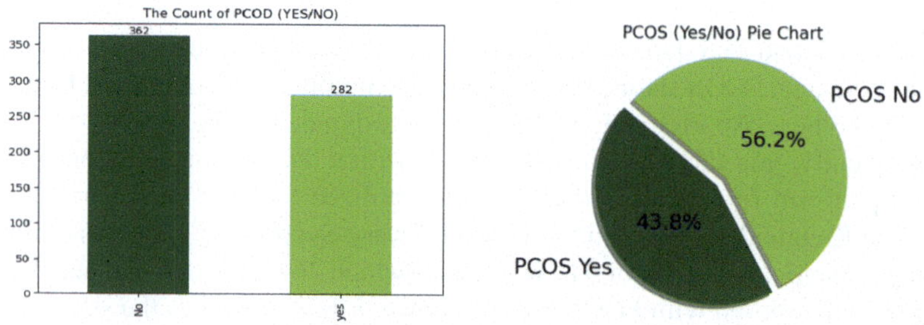

Fig. 3. A count plot and pie chart illustrate the frequency of occurrences for both classes.

without the need for a doctor or extensive laboratory tests. In Fig. 4, we present the results of our feature selection process, highlighting both the chosen features and their correlations (Fig. 5).

Insights from the bar plot suggest that increased weight, excessive hair growth, the presence of pimples, hair loss, frequent consumption of fast food, and skin darkening are indicative of a higher risk of PCOS. These characteristics can also be described as potential symptoms of PCOS.

3.3 Data Preprocessing

The received data contained Null values, which were addressed. Unnecessary attributes, such as patient record numbers and names, were removed. Subsequently, the data underwent normalization using a min-max scaler.

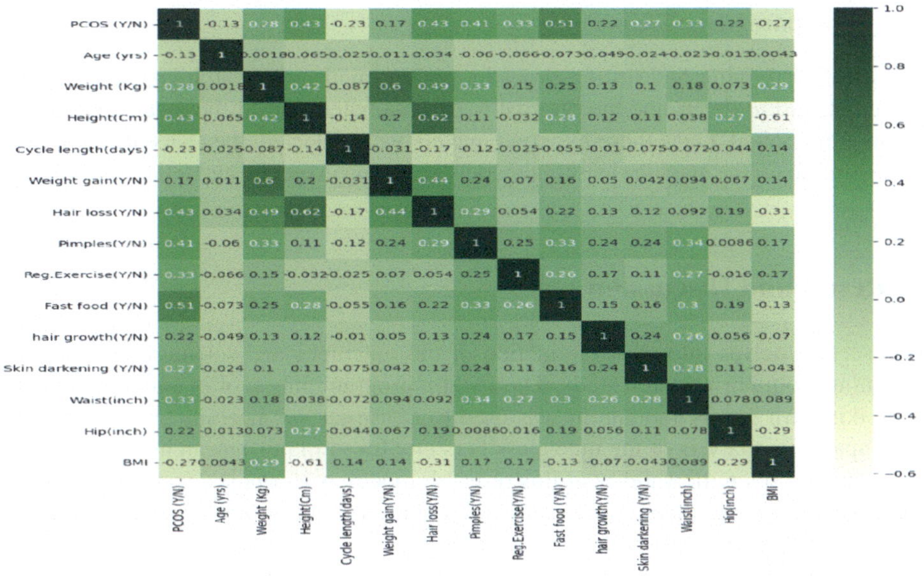

Fig. 4. The analysis of feature correlations within the selected datas

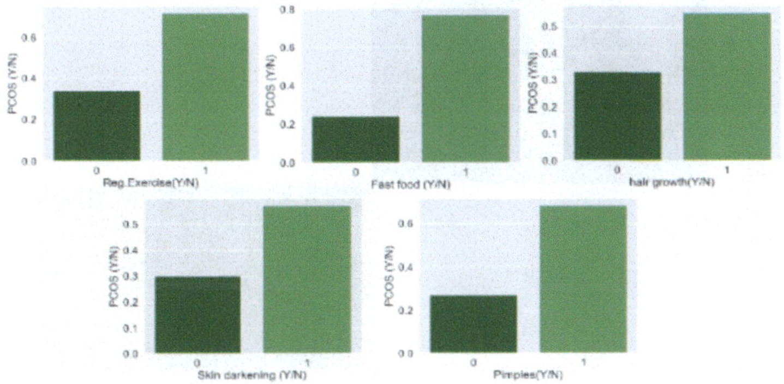

Fig. 5. Bar plot analysis examines how the distribution of PCOS "Yes" or "No" relates to certain chosen features.

3.4 Selection and Training the Model

This phase serves as the foundation for the study. Given that the prediction output is a class, it presents a binary classification problem. The research utilized the following classifier algorithms: KNN, Adaboost, Linear Discriminant Analysis, SVM, Logistic Regression, Random Forest, Catboost, Naïve Bayes and Decision Tree. In machine learning, the dataset is typically divided into two, specifically the training set and the testing set. The classifiers were trained using the training dataset, and their performance was assessed on a previously unseen test dataset. The split for the dataset is in a 7.5:2.5 ratio, allocating 75% to the training dataset and 25% to the test dataset.

3.5 Hyperparameter Tuning

The iterative training and testing procedure, along with grid search cross-validation, was employed to determine the optimal hyperparameters for all the machine learning techniques applied. Once these hyperparameters were selected, they served as the final configuration for the machine learning models, resulting in precise and reliable prediction outcomes.

4 Result and Discussion

This section delves into the examination of the outcomes and scientific assessments stemming from our research study. The Python programming tool in conjunction with the scikit-learn library module to was used to construct the machine learning model. A comparative analysis of the nine classifiers employed in this research paper is done. To ensure the scientific validity of this research model, performance metrics have been used, namely the accuracy score, sensitivity, training time in the evaluation process. Subsequently, the confusion matrices that were generated for each of these classifiers and the confusion matrix of the KNN model is shown in Fig. 6.

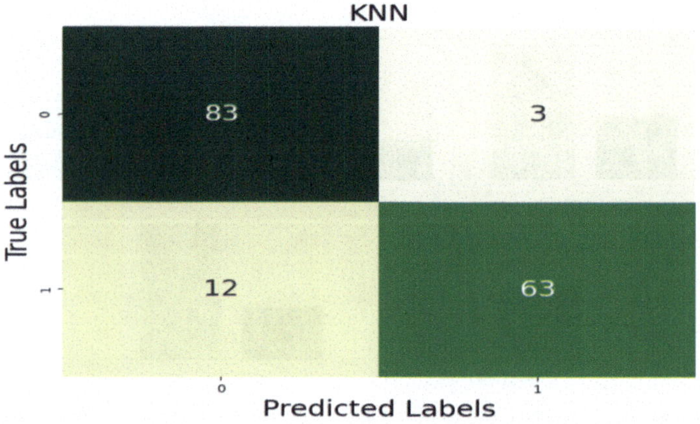

Fig. 6. Confusion matrix for KNN model

By comparing the accuracy, recall and training time of the different models it can be determined which model performs best.The accuracy,recall and training time comparison is shown in Table 1 and the graphical representation of the same is illustrated in Fig. 6, 9. The Fig. 9 shows that random forest takes maximum training time and Naïve bayes takes the least time compared to other algorithm and followed by KNN. The KNN outperforms the other models, with an accuracy of 91%, recall of of 90%, and training time of 0.005982 s (Fig. 7).

Table 1. Validation Parameters

SI. No	Method	Training time(seconds)	Accuracy (%)	Recall (%)
1	KNN	0.005982	91	90
2	Adaboost	0.107	91	90
3	Linear Discriminant Analysis	0.12169	91	90
4	SVM	0.0139	90	89
5	Logistic Regression	0.02	90	90
6	Random Forest	2.65	90	89
7	Catboost	1.939	89	89
8	Naïve Bayes	0.00583	85	84
9	Decision Tree	0.0075	81	81

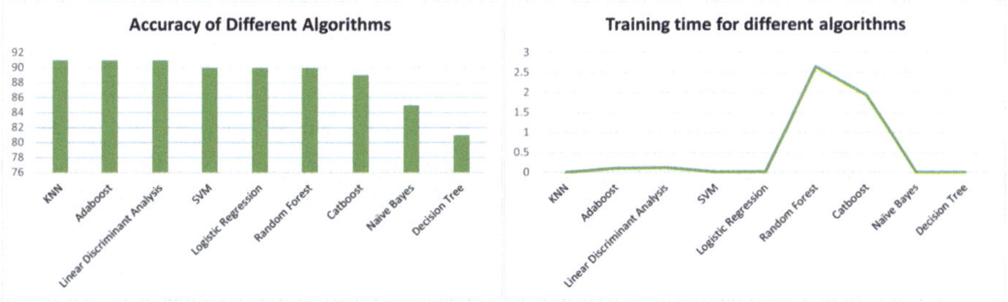

Fig. 7. Plot of Accuracy and training time in seconds.

5 Conclusion

This study offers a strategy for the early diagnosis and treatment of polycystic ovarian syndrome (PCOS). Initially, we leveraged a labelled historical dataset that we acquired from Kaggle, supplemented with expert medical insights. Our main goal was to create a highly accurate predictive model that could identify whether PCOS was present or not, with an emphasis on improving early detection and prompt medical consultations. Nine different machine learning classifiers were used on the dataset over the course of our investigation. The performance of the algorithms was contrasted in the end, we were able to determine that KNN had the best accuracy (91%), recall (90%), and training time (0.005982 s). Our study has a very clear practical application in mind for the future. This research will be used further to develop an intuitive mobile application designed specifically for medical clinics. With the help of this software, users will be able to anticipate their medical situations and provide relevant answers to queries. Our goal is to improve the overall quality of healthcare delivery by accelerating medical consultations and interventions through the facilitation of early identification.

References

1. Pedersen, S.D., Brar, S., Faris, P., Corenblum, B.: Polycystic ovary syndrome: validated questionnaire for use in diagnosis. Can. Fam. Phys. **53**(6), 1041–1047 (2007). PMID: 17872783; PMCID: PMC1949220
2. Khan, A., Karim, N., Ainuddin, J.A., Fahim, M.F.: Polycystic ovarian syndrome: correlation between clinical hyperandrogenism, anthropometric, metabolic and endocrine parameters. Pak. J. Med. Sci. **35**(5), 1227–1232 (2019). https://doi.org/10.12669/pjms.35.5.742. PMID: 31488983; PMCID: PMC6717481
3. Chauhan, P., Patil, P., Rane, N., Raundale, P., Kanakia, H.: Comparative analysis of machine learning algorithms for prediction of PCOS. In: 2021 International Conference on Communication information and Computing Technology (ICCICT), pp. 1–7. Mumbai, India (2021). https://doi.org/10.1109/ICCICT50803.2021.9510128
4. Karia, A., Poojary, A., Tiwari, A., Sequeira, L., Sokhi, M.K.: BeRedy (Period Tracker & PCOS diagnosis). In: 2023 International Conference on Communication System, Computing and IT Applications (CSCITA), pp. 142–147. Mumbai, India (2023).https://doi.org/10.1109/CSCITA55725.2023.10104755
5. Denny, A., Raj, A., Ashok, A., Ram, C.M., George, R.: i-HOPE: detection and prediction system for polycystic ovary syndrome (PCOS) using machine learning techniques. In: TENCON 2019 - 2019 IEEE Region 10 Conference (TENCON), pp. 673–678. Kochi, India (2019).https://doi.org/10.1109/TENCON.2019.8929674
6. Katarya, R., Srivastava, G., Chauhan, N.: A novel polycystic ovarian syndrome diagnostic system using machine learning. In: Abraham, A., Castillo, O., Virmani, D. (eds.) Proceedings of 3rd International Conference on Computing Informatics and Networks: ICCIN 2020. LNNS, vol.167, pp. 333–343. Springer, Singapore (2021). https://doi.org/10.1007/978-981-15-9712-1_47
7. Mittapalli, J.S., Khanna, K., Mutha, J.A., Nair, S.: A cloud-based prediction and self-diagnosis system for PCOS using machine learning models. In: Subhashini, N., Ezra, M.A.G., Liaw, SK. (eds.) Futuristic Communication and Network Technologies. LNNE, vol. 966, pp. 477–484. Springer, Singapore (2023). https://doi.org/10.1007/978-981-19-8338-2_39
8. Bedrick, B.S., Eskew, A.M., Chavarro, J.E., Jungheim, E.S.: Self-administered questionnaire to screen for polycystic ovarian syndrome. Women's Health Rep. **1**(1), 566–573 (2020)
9. Subha, R., et al.: Computerized diagnosis of polycystic ovary syndrome using machine learning and swarm intelligence techniques (2022)
10. Thomas, N., Kavitha, A.: Prediction of polycystic ovarian syndrome with clinical dataset using a novel hybrid data mining classification technique. Int. J. Adv. Res. Eng. Technol. **11**(11), 1872–1881 (2020)
11. Barber, T.M., McCarthy, M.I., Wass, J.A.H., Franks, S.: Obesity and polycystic ovary syndrome. Clin. Endocrinol. **65**(2), 137–145 (2006)
12. Pedersen, S.D., Brar, S., Faris, P., Corenblum, B.: Polycystic ovary syndrome: validated questionnaire for use in diagnosis. Can. Fam. Physician **53**(6), 1041–1047 (2007)
13. Soni, P., Vashisht, S.: Exploration on polycystic ovarian syndrome and data mining techniques. In: 2018 3rd International Conference on Communication and Electronics Systems (ICCES), pp. 816– 820. IEEE (2018)
14. Cheng, J.J., Mahalingaiah, S.: Data mining polycystic ovary morphology in electronic medical record ultrasound reports. Fertil. Res. Pract. **5**(1), 1–7 (2019)
15. Sinthia, G., Poovizhi, T., Khilar, R.: Analysis on polycystic ovarian syndrome and comparative study of different machine learning algorithms. In: Mohanty, M.N., Das, S. (eds.) Advances in Intelligent Computing and Communication. LNNS, vol. 430, pp. 191–196. Springer, Singapore (2022). https://doi.org/10.1007/978-981-19-0825-5_20

16. Vijayalakshmi, N., UmaMaheshwari, M.: Data mining to elicit predominant factors causing infertility in women. Int. J. Comput. Sci. Mob. Comput. **5**(8), 5–9 (2016)
17. . Chandra, S., Kaur, M.: Creation of an adaptive classifier to enhance the classification accuracy of existing classification algorithms in the field of medical data mining. In: 2015 2nd International Conference on Computing for Sustainable Global Development (INDIACom). IEEE (2015)
18. Neto, C., Silva, M., Fernandes, M., Ferreira, D., Machado, J.: Prediction models for Polycystic Ovary Syndrome using data mining. In: Antipova, T. (ed.) Advances in Digital Science: ICADS 2021. AISC, vol. 1352, pp. 210–221. Springer, Cham (2021). https://doi.org/10.1007/978-3-030-71782-7_19
19. Chatley, G., Kaur, S., Sohal, B.: Software clone detection: a review. Int. J. Control Theory Appl. **9**(41), 555–563 (2016)
20. Bulsara, J., Patel, P., Soni, A., Acharya, S.: A review: brief insight into polycystic ovarian syndrome. Endocr. Metab. Sci. **3**, 100085 (2021)

RADU-Net: A Fully Convolutional Neural Network for Efficient Skin Lesion Segmentation

Rajdeep Kaur(✉) and Sukhjeet Kaur Ranade

Department of Computer Science, Punjabi University, Patiala 147002, India
{rajdeep_rs23,sukhjeet}@pbi.ac.in

Abstract. Automatic skin lesion segmentation in dermoscopic images is a challenging task due to the presence of artifacts, fuzzy boundaries, variability in skin lesion appearance, low contrast between lesion skin and normal skin, and limited training data. In this paper, we introduced a robust, fully convolutional neural network for efficient and effective segmentation of skin lesions. Our proposed model, the Residual Attention Double U-Net (RADU-Net), is an encoder-decoder architecture that connects two U-Nets in a cascade manner. It comprises repeated convolutional and deconvolutional layers, connected with skip connections through the attention gates, and incorporates residual learning within convolution layers. Double U-Net captures fine details while persevering contextual information. The attention gates enable the model to preserve essential spatial information while ignoring irrelevant feature representations, enabling the model to focus on target regions. Additionally, the inclusion of residual connections enhances the training of a deeper network by ensuring a smoother gradient glow. We assessed the performance of the proposed model on the publicly available ISIC 2018 dataset, achieving a dice of 0.9074 and Jaccard of 0.8589. The proposed model consistently outperforms the baseline models for skin lesion segmentation.

Keywords: Skin lesion segmentation · U-Net · Attention mechanism · Residual learning

1 Introduction

Skin cancer is one of the deadliest forms of cancer, caused by the uncontrolled growth of irregular cells within the human body. It often has ability to reproduce, spread through the lymphatic and blood systems, and demolish healthy cells. According to the World Health Organization (WHO), in 2020, more than 1.5 million cases of skin cancer were diagnosed, leading to over 120,000 deaths worldwide [1]. Timely diagnosis significantly increases the chance of survival. The five-year survival rate of cancer patients with localized stage can be up to 99%, however, it dramatically decreases to 23% with distant stage. Such incoherence emphasizes the early diagnosis of skin cancer. Dermatologists and other medical professionals utilize the handheld device dermoscope to diagnose skin cancer. However, the manual examination of dermoscopic images is a time-consuming process and the diagnosis accuracy depends on the experience and perception of the dermatologist.

D. Pastor-Escuredo et al. (Eds.): ICDLAIR 2023, LNNS 1001, pp. 658–673, 2024.
https://doi.org/10.1007/978-3-031-60935-0_58

Due to technological advancement in past decades, many efficient Computer-aided diagnosis (CAD) systems have been developed to assist dermatologists in early diagnosis and accurate decision-making. Cancer diagnostic CAD systems involve a series of steps, including image capturing, pre-processing, segmentation, feature extraction, detection, and classification. The automatic segmentation of skin lesions plays a crucial role in the accurate extraction of relevant features for the detection and classification of skin cancer. However, segmentation of skin lesions from dermoscopic images presents a difficult challenge due to the following distinctive characteristics: a) the presence of artifacts such as hair, ink and ruler marks, water bubbles, and blood vessels, b) fuzzy boundaries of skin lesions, c) inter-class similarity between the lesion and normal skin, d) variability in skin lesion appearance in shapes, colors, sizes, patterns, and textures, e) patient-specific properties such as skin color, and f) insufficient training data.

Over the past few decades, a plethora of skin lesion segmentation methods have been introduced by numerous researchers. These methods can be categorized into five distinct groups to compare them: threshold-based [2], edge-based [3], region-based [4], artificial intelligence (AI) based [5], and other methods [6]. With the development of computer vision, AI-based methods, especially deep learning, had significant success in the field of image processing. These methods are more accurate, efficient, and consistent as compared to other methods. Convolutional neural network (CNN) layered architecture is one of the earliest deep-learning architecture consisting of several layers including the convolution layer, pooling layer, and fully connected layer. It efficiently integrates local and global contextual information to generate a label for each pixel, resulting in a segmentation mask that highlights the lesion region.

Fully convolutional network (FCN) [7] is one of the widely adopted CNN architectures in the field of medical image segmentation. It is a convolutional-deconvolutional layered architecture that is trained end-to-end at the pixel level to segment the images. Further, Ronneberger et al. [8] introduced U-Net architecture, the extended version of FCN architecture to improve the better prediction. The U-Net architecture consists symmetric encoder and decoder path connected with the skip connections which capture both high-level context and detailed localization information. In addition, the performance of U-Net has been significantly improved with the incorporation of residual learning [9], attention mechanism [10], and other non-adaptive functions in convolution layers.

Inspired by the success of U-Net, we proposed a robust U-Net-based architecture called Residual Attention Double U-Net (RADU-Net) that uses the properties of residual U-Net and attention gates. The major contributions of our proposed architecture are as follows:

1. The proposed architecture is an extension of the standard U-Net architecture, consisting of two U-Net connected in a cascade manner with the additional features of residual learning and attention mechanism.
2. U-Net-I generates the initial segmentation maps which, are further refined by the U-Net-II for more accurate semantic segmentation.
3. Residual blocks address the problem of vanishing gradients, leading to more effective training of the proposed model.
4. Attention gates allow the model to focus on the activation of relevant features while ignoring irrelevant information.

5. Experimental analysis of the proposed model is performed on ISIC 2018 to demonstrate that the proposed model outperforms its baseline models for skin lesion segmentation.

The rest of the paper structured as follows: In Sect. 2, we discuss existing techniques for skin lesion segmentation. Section 3 is dedicated to the exploration of the proposed technique. Moving forward, Sect. 4 details the experimental analysis, while Sect. 5 unfolds the results and discussion, encompassing both quantitative and qualitative performance analyses, Finally, last section encapsulated the conclusion, summarizing the findings and implications of the study.

2 Related Work

In the past decades, several skin lesion segmentation methods have been proposed based on deep learning. In this section, we discuss existing deep learning-based methods for skin lesion segmentation.

Al-masni et al. [11] introduced a semantic segmentation model called Full Resolution Convolutional Network (FrCN) for skin lesion segmentation. It comprises a stack of convolutional layers, each followed by the ReLU activation and dropout layer. The performance of FrCN is evaluated using publicly available datasets and achieved a dice of 0.917 and Jaccard of 0.847 on the PH2 dataset and a dice of 0.870 and Jaccard of 0.771 on the ISIC 2017 dataset. However, failed to segment the artifacts-affected images. Further, Li et al. [12] introduced encoder-decoder architecture, namely, a dense deconvolutional network (DDN) for efficient skin lesion segmentation. It consists of dense deconvolutional layers with additional features of residual learning and chained residual pooling to capture local and global contextual information. The exhaustive experiments are carried out on the ISIC 2016 and ISIC 2017 datasets without any pre- and post-processing. They achieved a Jaccard of 0.870 on ISIC 2016 and 0.765 on ISIC 2017. Dash et al. [13] modified standard U-Net architecture with the addition of batch-normalization to address the problem of covariate shift. They used combined binary cross-entropy and dice loss functions for better optimization of the model. The exhaustive experiments are performed on 5179 images collected from Psoriasis Clinic and Research Centre, Pune, and achieved a dice of 0.93. And Jaccard of 0.860. Their proposed model failed to segment the low-contrast images and fuzzy boundaries. Alom et al. [14] also proposed a similar architecture based on U-Net with the addition of residual learning and recurrent neural networks. They proposed two models, namely Recurrent Convolutional Neural Network (RCNN) and Recurrent Residual Convolutional Neural Network (RRCNN) for medical image segmentation. The performance of the proposed model is evaluated using three diverse datasets as ISIC 2017 dataset for skin lesions segmentation, which achieved an accuracy of 0.937 using RCNN and 0.942 using RRCNN, the CHASH_DBI dataset for blood vessel segmentation in retina images, which achieved accuracy of 0.962 using RCNN and 0.963 using RRCNN, and LUNA dataset for lung lesion segmentation with reported accuracy of 0.987 using RCNN and 0.988 using RRCNN. However, these models are computationally expensive as compared to standard U-Net. The combination of U-Net and attention mechanism is introduced by Abraham and Khan [15] for skin lesion segmentation. They used attention gates in skip connection Tversky loss function during the training process to preserve the spatial information for efficient semantic segmentation. Attention gates highlight the important features while ignoring the irrelevant features and the focal tversky loss function is used for better optimisation of the model. They evaluated the performance of the model using the ISIC 2018 dataset and reported dice of 0.856. Arora et. al. [16] proposed a similar architecture with the addition of group normalization instead of batch normalization to further improve the performance of skin lesion segmentation. The batch normalization operation normalizes the features in the groups before being passed to the convolution operation. It reduces the batch size dependency and increases the segmentation performance. They performed exhaustive experiments on ISIC 2018 and achieved a dice of 0.920 and a Jaccard of 0.860. However, their proposed model failed to accurately segment the skin lesions in low-contrast images. Furthermore, Tong et al. [17] introduced an extended version of attention U-Net with the addition of spatial and channel attention-based dual decoders for capturing the correlation between the activated features for accurate skin lesion segmentation. Their proposed model is evaluated using three datasets such as PH2, ISIC 2016, and ISIC 2017, and achieved an accuracy of 0.943, 0.954, and 0.926, respectively. However, their proposed model performs inaccurate skin lesion segmentation in the presence of artifacts. Wibowo et al. [18] introduced a lightweight U-Net-based architecture named MobileNetV3-UNet for skin lesion segmentation. They optimize the learning parameters using stochastic weight averaging (SWA) schema to enhance the performance of network. Additionally, they employed post-processing operations to improve segmentation results. The experimental results are evaluated using two datasets, ISIC 2017 and ISIC 2018, and reported dice of 0.877 and 0.802, respectively. Kaur and Ranade [19] proposed a deep learning architecture that incorporates group normalization and combined loss function to enhance the accuracy of skin lesion segmentation. The experimental analysis conducted using ISIC 2016 and ISIC 2017, reporting an accuracy of 0.947 and 0.9133, respectively. Similarly, Le et al. [20] presented a U-Net based architecture with the incorporation of anti-aliasing layers and attention gates in skip connections. The anti-aliasing layers are designed to prevent shift equivariance, while attention gates highlight the important features. They achieved dice of 0.881 and Jaccard of 0.809 on the ISIC 2018 dataset. However, their proposed architecture failed to accurately segment the lesion with fuzzy boundaries.

From the review of the aforementioned state-of-the-art methods, we observed that the existing methods failed to perform accurate segmentation in the presence of artifacts, due to low contrast between lesion and normal skin and fuzzy boundaries of lesion. Moreover, some methods perform pre-processing and post-processing operations which increase the computational complexity. To overcome these limitations, we proposed a robust convolutional neural network that utilizes the properties of standard U-Net, residual learning, and attention mechanism. The proposed method efficiently segments the challenged dermoscopic images and performs accurate skin lesion segmentation without any pre- and post-processing.

3 Proposed Method

Motivated by the success of standard U-Net in medical image segmentation, we introduced a novel method for accurate skin lesion segmentation called Residual Attention Double U-Net (RADU-Net). The general pipeline of the proposed method for skin lesion segmentation is illustrated in Fig. 1. Training images, along with their corresponding segmented masks, are input into the proposed model for training process. Subsequently, the trained model is used to predict the segmented images for test images.

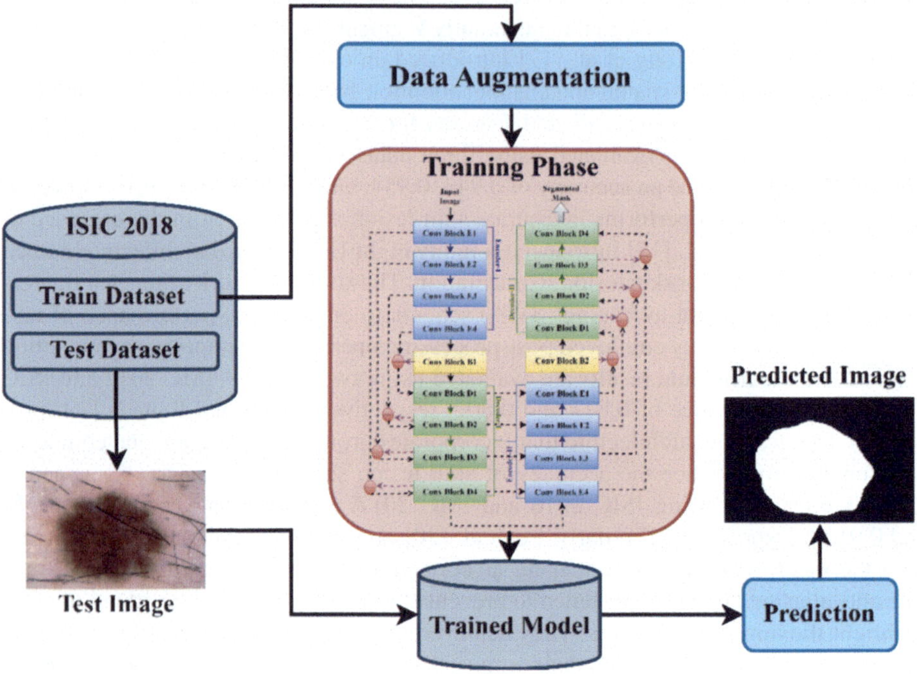

Fig. 1. General pipeline for the proposed method for skin lesion segmentation.

Figure 2 illustrates the architecture of the proposed model, comprising two U-Nets, namely, U-Net-I and U-Net-II, connected in parallel. Each U-Nets model consists of three parts main components such as encoder path, decoder path, and connecting bridge. Both the encoder and decoder paths are composed of four convolutional layers, while the bridge contains a convolutional layer to facilitate the transition between them. Skip connections link the layers of the encoder to their corresponding decoder layers. Additionally, the layers of the decoder of U-Net-I are connected to the corresponding encoder layers of U-Net-II through additional skip connections. A detailed description of various components is presented in the following subsections:

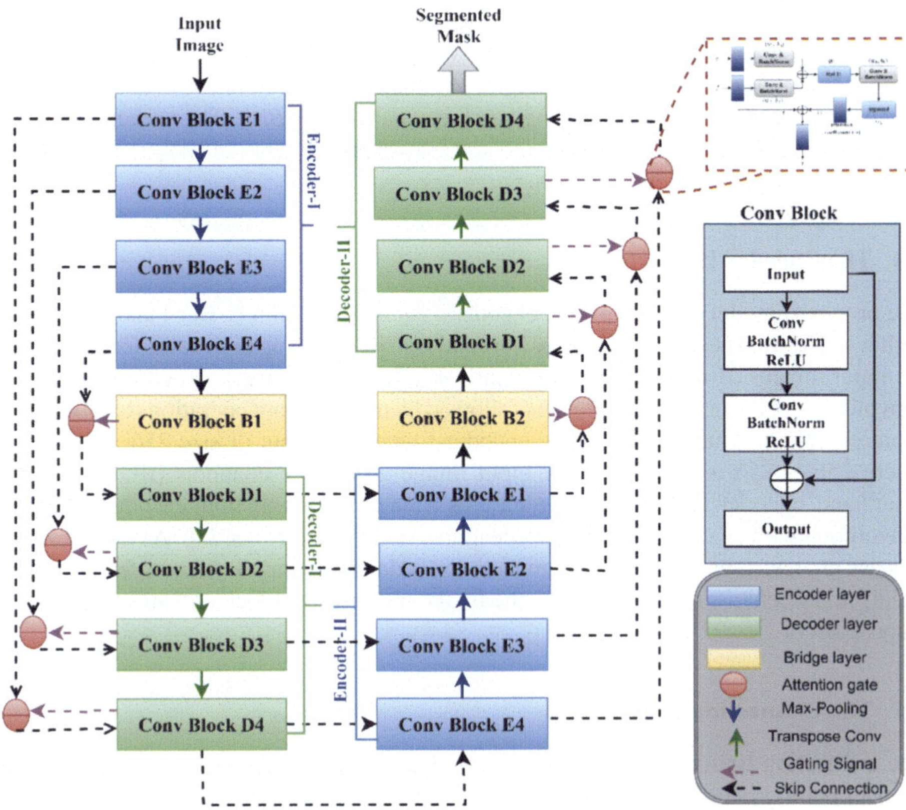

Fig. 2. Architecture of the proposed model.

3.1 Convolutional Block

The proposed model used the convolutional blocks in the encoders and decoders of both U-Nets for efficient feature extraction from the dermoscopic images for skin lesion segmentation. The proposed convolutional block consists of repeated convolution operations, batch normalizations, and activation function with the residual connection. The normalized features using batch normalization [21] are passed to the convolution layer and further activated by the ReLU [22] activation function. Every convolutional block performs these identical operations to generate the features and pass them to the next layers. The mathematical representation of convolutional blocks is as follows:

$$C_{m,n} = f\left(\sum_{i=0}^{k-1}\sum_{j=0}^{k-1}\left(\left(X_{m+i,n+j}\right)^{BN} * W_{i,j} + b_{m,n}\right)\right) \quad (1)$$

In Eq. (1), $C_{m,n}$ represents the output features obtained from a convolutional block, where $f(.)$ denotes the activation function, X represents the input features with dimensions $(H \times W \times C)$, H stands for height, W stands for width, and C stands for input channels, which are normalized using batch normalization (BN). The variable k represents the size of the kernel, $*$ represents the convolution operation, $W_{i,j}$ represents the weights, and $b_{m,n}$ represents the bias.

3.2 The Encoder Path

The proposed model consists of two encoders: encoder-I within U-Net-I and encoder-II within U-Net-II. These encoders play a crucial role in capturing high-level features from the input images. Both encoders are identical in layer structure, each comprising four convolutional blocks and each block is followed by max-pooling operations to efficiently capture contextual information. The encoder-I takes the input of size $256 \times 256 \times 3$, where 256×256 is the size of an input image with three channels (RGB). On the other hand, encoder-II receives input features from decoder-I. In both encoders, the first convolutional block consists of 64 feature maps, each of size 256×256. As the encoding process proceeds, the number of feature maps is doubled, and the size of feature maps is halved at each layer. Consequently, the final convolutional block contains 512 feature maps, each of size 32×32.

3.3 The Connecting Bridge

The connecting bridge serves as the central component that connects the encoder and decoder paths, enabling efficient communication and effective features transmission of the features between these paths to ensure accurate skin lesion segmentation. In the proposed model, the connecting bridge of both U-Nets consists of a convolutional block with 1024 feature maps, each of size 15×15. Generated feature maps passed to the first decoder for reconstructing the desired output to capture the more precise information.

3.4 The Decoder Path

In the proposed model, the decoders of both U-Nets perform reverse operations of the encoders to restore the spatial information and generate a pixel-wise segmentation map from these features. These decoders comprise four convolutional blocks, each followed by a convolution transpose [23] layer of size 2×2 for up-sampling operation. Each decoder layer is connected to the corresponding encoder through skip connections with attention gates. The output of each decoder layer is concatenated with the output of the attention gate and passed to the next decoder layer. The first decoder from the bottom contains 512 feature maps, each of size 32×32. During the decoding process, the number of feature maps is halved, and the size of feature maps is doubled. Consequently, the size of feature maps is up-sampled to the actual input size by the last decoder layer.

The feature maps from each layer of decoder-I passed to the corresponding layer of encoder-II, creating a connection between U-Net-I and U-Net-II. Further, refinement of features is carried out by U-Net-II to enhance the accuracy of skin lesion segmentation. Finally, the features from the final layer of decoder-II passed to the convolution layer and sigmoid activation function to generate the final segmented map.

3.5 Residual Connection

In the proposed model, residual connections are a critical component within the convolutional block to address the problem of vanishing gradient that often arises during the training process. These residual connections enable the proposed model to preserve information from the preceding layer, thereby facilitating improved gradient flow during training and enhancing the overall performance of skin lesion segmentation. The mathematical representation of the residual connection is as follows:

$$y = F(x) + x \tag{2}$$

Here, x represents the input to a convolution layer and $F(x)$ be the transformation applied within the block. The output y of the block with a residual connection. x is directly added to the output of the transformation $F(x)$. This additive operation allows the network to effectively learn the information and gradients, ensuring that important features are retained throughout the network.

3.6 Attention Gates

In the proposed model, the encoder path shares the information with the decoder path through the utilization of skip connections. These connections facilitate the transmission of valuable information from lower layers to higher layers. Nevertheless, an excessive influx of irrelevant information can degrade the performance of the proposed model. To address this challenge, attention gates are incorporated into the skip connections. These attention gates are designed to focus on skin lesion pixels while selectively disregarding other non-lesion pixels to enhance the accuracy of skin lesion segmentation.

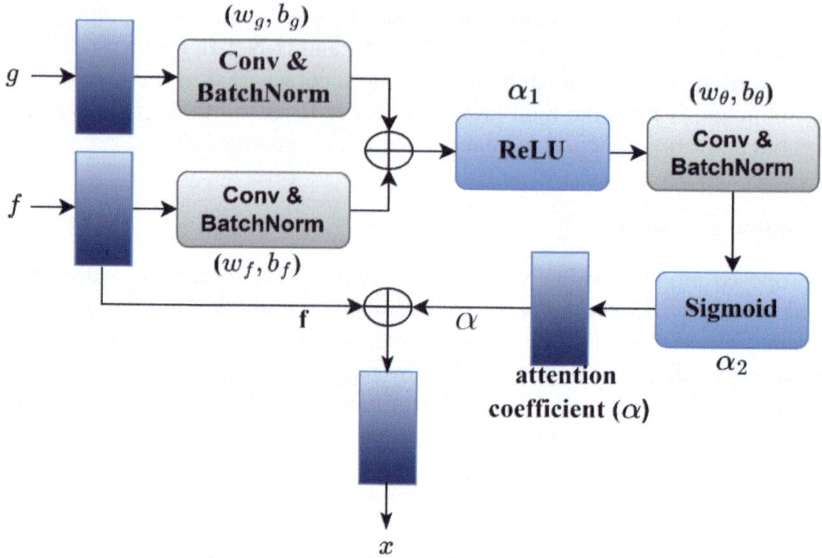

Fig. 3. Attention gate of the proposed model.

Figure 3 represents the architecture of attention gates that take the two inputs. The input feature maps denoted as (f), originate from the respective encoder layer, containing spatial and contextual information. Additionally, feature maps denoted as (g), come from the previous decoder layer, providing, a more refined representation of crucial features. These two input features are linearly transformed through successive convolutional operations (w_g, w_f) followed by batch normalization (b_g, b_f), and then merged pixel-by-pixel. The merged features are further activated using the Rectified Linear Unit $(ReLU, \sigma_1(x) = \max(0, x))$. Subsequently, the activated features are passed through convolution operation (w_θ) and batch normalization (b_θ) followed by the sigmoid activation function $\sigma_2 = \frac{1}{1+e^{(-x)}}$, and get the attention coefficient (α). Finally, the attention coefficients are computed using channel-wise multiplication with the features from the encoder to generate the final output (x) of the attention gate. To control the distribution of attention across different spatial locations, a learnable gating coefficient (α), is determined using an additive attention mechanism, as formulated below:

$$fm = \sigma_1\left[\left(w_f^T \times f + b_f\right) + \left(w_g^T \times g \times b_g\right)\right] \tag{3}$$

$$\alpha = \sigma_2\left(w_\theta^T \times fm + b_\theta\right) \tag{4}$$

$$x = f \times \alpha \tag{5}$$

The attention gates enable the proposed model to discern the relevant information from dermoscopic images and identify the essential features that need to be prioritized to achieve accurate skin lesion segmentation.

4 Experimental Analysis

In this section, we provide an overview of the dataset, detail the data augmentation process, and evaluation metrics used to assess the performance of the proposed model.

4.1 Dataset

International Skin Imaging Collaboration (ISIC) provides a large repository of dermoscopic images, encompassing various skin lesion challenges. They annually organize the "Skin Lesion Analysis toward Melanoma Detection" challenge for different tasks such as skin lesion segmentation, detection, and classification. In our study, we trained and tested the proposed model using the ISIC 2018 [24] dataset for skin lesion segmentation. This dataset contains 2594 training images, 100 validation images, and 1000 testing images. These images vary in size, ranging from 767 × 576 to 6682 × 4401 pixels. The ISIC 2018 dataset is available publicly at https://challenge.isic-archive.com/data/#2018.

4.2 Data Preparation and Augmentation

The ISIC 2018 dataset comprises dermoscopic images ranging in size from 767 × 576 to 6682 × 4401. In order to manage computation demands, we resized all dermoscopic images to 256 × 256. Furthermore, to improve the performance and robustness of our proposed model, we applied data augmentation techniques to dermoscopic images, effectively expanding the ISIC 2018 training dataset. These augmentation operations include flipping, rotation, height shifting, width shifting, and shearing, thus enhancing the diversity of our dermoscopic training images.

4.3 Experimental Setup

All the experiments are carried out on a system equipped with an Intel(R) Xeon(R) W-2195 CPU boasting 64 GB RAM, and an NVIDIA Quadro P4000 GPU with 8 GB of dedicated memory. The proposed model is implemented using Keras 2.11.0, trained up to 100 epochs, and optimized using an adaptive moment estimation (Adam) [25] optimizer with a learning rate of 0.001.

4.4 Evaluation Metrics

The performance of the proposed method is assessed through both qualitative and quantitative measures. Qualitatively, we conduct an evaluation by visually comparing the predicted images to the verified ground truth images. In terms of quantitative analysis, we utilize the standard evaluation metrics, including the dice coefficient, Jaccard index, accuracy, sensitivity, and specificity [26], which are computed as follows:

$$Dice\ Coefficient\ (DI) = \frac{2 * TP}{2 * TP + FP + FN} \tag{6}$$

$$Jaccard\ Index\ (JA) = \frac{TP}{TP + FP + FN} \tag{7}$$

$$Accuracy\,(AC) = \frac{TP + TN}{TP + TN + FP + FN} \tag{8}$$

$$Sensitivity\,(SP) = \frac{TP}{TP + FN} \tag{9}$$

$$Specificity\,(SE) = \frac{TN}{TN + FP} \tag{10}$$

Here, the terms TP, TN, FP, and FN denote the count of true positive, true negative, false positive, and false negative, respectively computed using a confusion matrix as illustrated in Table 1.

Table 1. Confusion matrix.

	Ground Truth		
		Lesion Pixels	Non-lesion Pixels
Prediction	Lesion Pixels	True Positive (TP)	False Positive (FP)
	Non-Lesion Pixels	False Negative (FN)	True Negative (TN)

TP represents the number of pixels correctly predicted as lesion pixels, while TN corresponds to the number of pixels correctly predicted as background pixels. FP includes the number of pixels that are predicted as lesion pixels but actually non-lesion pixels and FN includes the number of lesion pixels predicted as non-lesion pixels.

5 Results and Discussion

In this section, we present the ablation study of the proposed network, quantitative performance analysis followed by the qualitative performance analysis of the proposed method trained, validated, and tested on dermoscopic images for skin lesion segmentation task.

5.1 Ablation Study

Skin lesion segmentation is a prerequisite step in computerized skin cancer recognition from dermoscopic images. Despite the extensive research in the field, it remains a challenging task, primarily due to the considerable variations in dermoscopic images, the presence of artifacts, low contrast between lesions and normal skin, and the influence of patient-specific factors. In this paper, we proposed a deep convolutional network architecture that outperforms its baseline models for skin lesion segmentation. Training a deep network efficiently for skin lesion segmentation is a challenging task. As the depth of the network increases, issues such as vanishing gradients become more prominent, rendering it challenging to fine-tune the parameters of the initial layers. Additionally, achieving efficient training for a deeper network necessitates a substantial dataset to prevent the problem of overfitting. Unfortunately, the dataset for skin lesion segmentation task currently available is limited. This dilemma creates difficulties in training the network. To tackle these challenges, our proposed skin lesion segmentation method employs the following strategies: a) Instead of a patch-wise learning approach, we have adopted a fully convolutional model for semantic segmentation, enabling pixel-level predictions. b) The integration of batch normalization reduced the covariate shift, while Adam optimization permits the use of a relatively large learning rate, thereby expediting the training process. c) We effectively utilized dropout and image augmentation techniques to address the problem of overfitting. d) During network training, we utilize a combination of Jaccard and binary cross entropy as the loss function, which is more relevant to the task of semantic image segmentation.

5.2 Quantitative Performance Analysis

The quantitative performance analysis of the proposed model is conducted using standard evaluation metrics with the benchmark dataset ISIC 2018 for skin lesion segmentation task. Table 2 and Table 3 demonstrate the comparison of the proposed methods with its baseline methods and state-of-the-arts methods, respectively.

Comparison with Baseline Methods

As depicted in Table 2, the best segmentation results are achieved with the proposed method RADU-Net for several reasons. Firstly, the interconnection of two U-Nets effectively captures fine details while persevering contextual information. Secondly, the inclusion of residual connections directly propagates the gradients between layers, mitigating the issue of vanishing gradients. Thirdly, the incorporation of attention gates in the skip connection that highlights the important features, resulting in more accurate segmentation. Lastly, the combination of dice and binary cross-entropy allows for the simultaneous execution of localization and pixel-level classification tasks. Furthermore, data augmentation operations are employed to address the issue of overfitting. These features enabled the proposed model to undergo a more comprehensive and robust training process, resulting in highly accurate skin lesion segmentation.

Table 2. Comparison of our proposed method and their baseline methods on the test dataset.

Method	DI	JA	AC	SP	SE
U-Net	0.7451	0.6510	0.8891	0.7329	0.8542
Residual U-Net	0.7675	0.6612	0.8925	0.7802	0.8423
Attention U-Net	0.8339	0.7991	0.9107	0.9066	0.9158
RADU-Net	0.9074	0.8589	0.9597	0.9432	0.9714

Comparison With State-of-the-Art Methods

We also conducted a comparative analysis of the proposed model, RADU-Net, with state-of-the skin lesion segmentation methods, namely MAAU [20], and MobileNetV3-UNet [18]. These models are trained and tested in the ISIC 2018 datasets for thorough evaluation. The results in Table 3 demonstrate that our proposed method outperforms these state-of-the-art methos in the all parameters. Specifically, our proposed model achieved significantly higher performance in the term of dice and Jaccard as compared to MAAU. Similarly, when we MobileNetV3-UNet, our proposed model outperformed in terms of Jaccard, accuracy, specificity and sensitivity, except for dice, where MobileNetV3-UNet marginally performed better.

Table 3. Comparison of proposed method and state-of-the-art methods on the test dataset.

Method	DI	JA	AC	SP	SE
MAAU [20]	0.8810	0.8090	—	—	—
MobileNetV3-UNet [18]	0.9098	0.8344	0.9479	0.9089	0.9638
RADU-Net (Proposed)	0.9074	0.8589	0.9597	0.9432	0.9714

5.3 Qualitative Performance Analysis

We also assessed the segmentation performance of the baseline models and the proposed model qualitatively by analyzing the quality of the prediction images. Figure 4 illustrates the visual analysis of the predicted segmented images randomly selected from the ISIC 2018 test dataset. From the qualitative analysis, we observed that the predicted images by the proposed model closely resemble the ground truth images in comparison to those generated by predicted by other baseline models. The proposed model demonstrates its ability to efficiently segment the skin lesion in challenging conditions. Evidently, in rows a and d, the proposed model accurately segments the skin lesion in low-contrast images. Furthermore, it consistently and accurately segments the skin lesion in rows c and g, even in the presence of hair.

The visual comparison illustrates the effectiveness of the proposed model in skin lesion segmentation. It reduces the false positive rate by highlighting crucial features and more effectively segments the skin lesions in the presence of artifacts and the irregularly shaped lesions by learning the detailed feature representation.

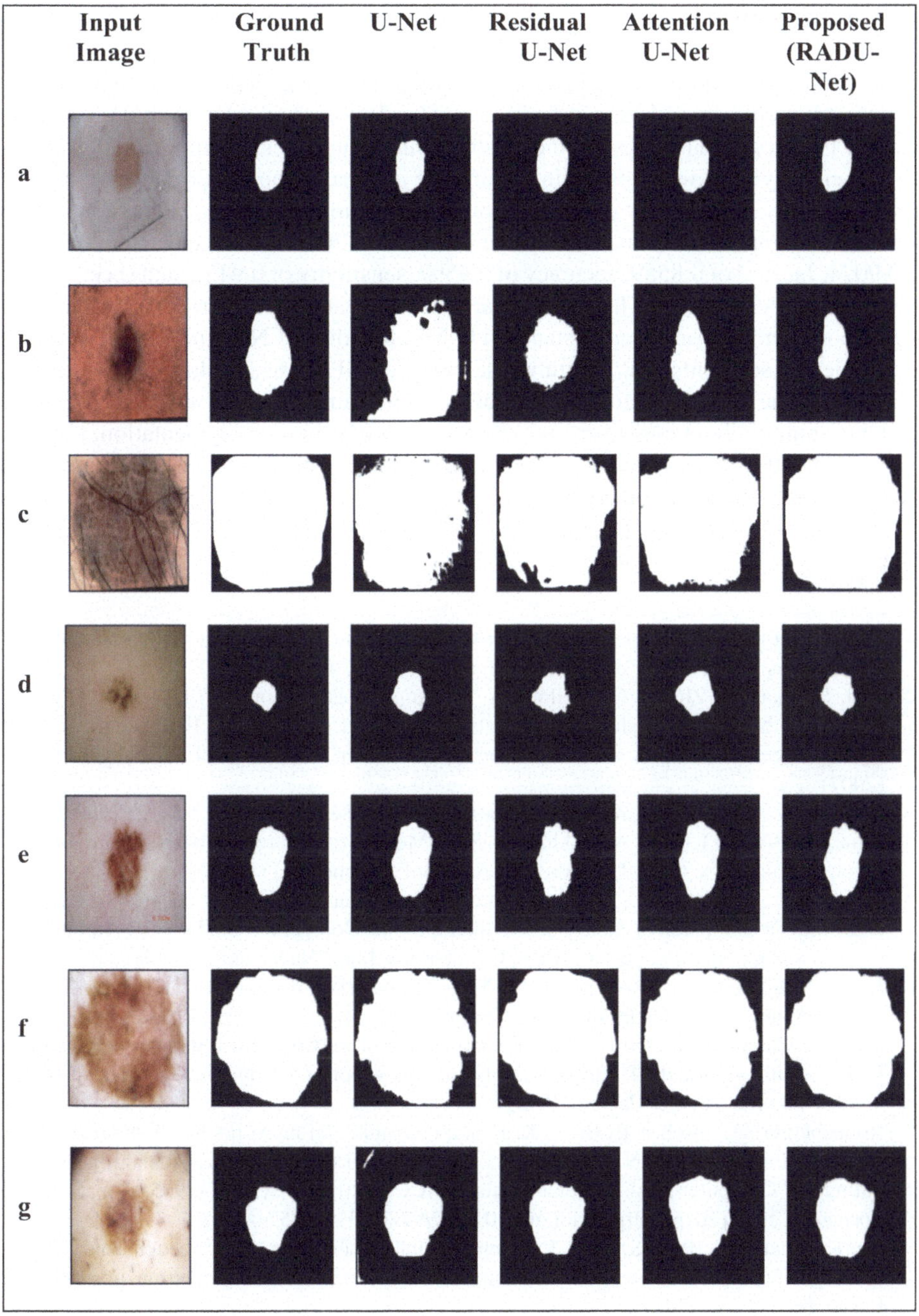

Fig. 4. Qualitative comparison between the ground truth images and predicted images by U-Net, Residual U-Net, Attention U-Net, and proposed RADU-Net.

6 Conclusion

In this paper, we introduced the Residual Attention Double U-Net, a convolutional neural network architecture designed for precise skin lesion segmentation in dermoscopic images. The proposed RADU-Net model streamlines the process of automatically detecting and classifying skin lesions with remarkable accuracy. The performance of the proposed model is significantly elevated through the incorporation of residual connection and attention gates. It is trained and tested on the ISIC 2018 dataset and achieved a dice of 0.9074, Jaccard of 0.8589, accuracy of 0.9597, sensitivity of 0.9432, and specificity of 0.9714 on the test dataset. This study clearly demonstrates the superior performance of the proposed model compared to standard U-Net, Residual U-Net, and Attention U-Net for skin lesion segmentation. In future research, we will aim to simplify the architecture by reducing the count of trainable parameters. This simplification will result in faster model training without compromising the accuracy of semantic segmentation. Furthermore, we will develop a versatile unified model capable of efficiently performing various medical image segmentation tasks.

References

1. Dyba, T., et al.: The European cancer burden in 2020: Incidence and mortality estimates for 40 countries and 25 major cancers. Eur. J. Cancer **157**, 308–347 (2021)
2. Zhu, S., Xia, X., Zhang, Q., Belloulata, K.: An image segmentation algorithm in image processing based on threshold segmentation. In: Third international IEEE conference on signal-image technologies and internet-based system, pp. 673–678, IEEE, Shanghai, China (2007)
3. Farag, A.A.: Edge-based image segmentation. Remote Sens. Rev. **6**(1), 95–121 (1992)
4. Raja, S.K., Abdul Khadir, A.S., Ahamed, S.R.: Moving toward region-based image segmentation techniques: a study. J. Theoretical Appl. Inf. Technol. 5(1) (2009)
5. Indira, S.U., Ramesh, A.C.: Image segmentation using artificial neural network and genetic algorithm: a comparative analysis. In: International Conference on Process Automation, Control and Computing, pp.1-6, IEEE, Coimbatore, Tamil Nadu, India (2011)
6. Hasan, M.K., Ahamad, M.A., Yap, C.H., Yang, G.: A survey, review, and future trends of skin lesion segmentation and classification. Comput. Biol. Med. **155**, 106624 (2023)
7. Long, J., Shelhamer, E., Darrell, T.: Fully convolutional networks for semantic segmentation. In: Proceedings of the IEEE conference on computer vision and pattern recognition, pp. 3431–3440, Waikoloa Village, Hawaii (2015)
8. Ronneberger, O., Fischer, P., Brox, T.: U-net: convolutional networks for biomedical image segmentation. In: Navab, N., Hornegger, J., Wells, W., Frangi, A. (eds.) Medical Image Computing and Computer-Assisted Intervention–MICCAI 2015. LNCS, vol. 9351, pp. 234–241. Springer, Cham (2015). https://doi.org/10.1007/978-3-319-24574-4_28
9. He, K., Zhang, X., Ren, S., Sun, J.: Deep residual learning for image recognition. In: Proceedings of the IEEE Conference on Computer Vision and Pattern Recognition, pp. 770–778, Las Vegas, Nevada (2016)
10. Vaswani, A., et al.: Attention is all you need. In: Advances in Neural Information Processing Systems, vol. 30 (2017)
11. Al-Masni, M.A., Al-Antari, M.A., Choi, M.T., Han, S.M., Kim, T.S.: Skin lesion segmentation in dermoscopy images via deep full resolution convolutional networks. Comput. Methods Programs Biomed. **162**, 221–231 (2018)

12. Li, H., et al.: Dense deconvolutional network for skin lesion segmentation. IEEE J. Biomed. Health Inform. **23**(2), 527–537 (2018)
13. Dash, M., Londhe, N.D., Ghosh, S., Semwal, A., Sonawane, R.S.: PsLSNet: automated psoriasis skin lesion segmentation using modified U-Net-based fully convolutional network. Biomed. Signal Process. Control **52**, 226–237 (2019)
14. Alom, M.Z., Hasan, M., Yakopcic, C., Taha, T.M., Asari, V.K.: Recurrent residual convolutional neural network based on u-net (r2u-net) for medical image segmentation. arXiv preprint arXiv:1802.06955 (2018)
15. Alahmadi, M.D.: Multiscale attention U-Net for skin lesion segmentation. IEEE Access **10**, 59145–59154 (2022)
16. Arora, R., Raman, B., Nayyar, K., Awasthi, R.: Automated skin lesion segmentation using attention-based deep convolutional neural network. Biomed. Signal Process. Control **65**, 102358 (2021)
17. Tong, X., Wei, J., Sun, B., Su, S., Zuo, Z., Wu, P.: ASCU-Net: attention gate, spatial and channel attention u-net for skin lesion segmentation. Diagnostics **11**(3), 501 (2021)
18. Wibowo, A., Purnama, S.R., Wirawan, P.W., Rasyidi, H.: Lightweight encoder-decoder model for automatic skin lesion segmentation. Inf. Med. Unlocked **25**, 100640 (2021)
19. Kaur, R., Ranade, S.K.: Improving accuracy of convolutional neural network-based skin lesion segmentation using group normalization and combined loss function. Int. J. Inf. Technol. **15**, 1–9 (2023)
20. Le, P.T., et al.: Anti-aliasing attention U-net model for skin lesion segmentation. Diagnostics **13**(8), 1460 (2023)
21. Bjorck, N., Gomes, C.P., Selman, B., Weinberger, K.Q.: Understanding batch normalization. In: Advances in Neural Information Processing Systems, vol. 31 (2018)
22. Agarap, A.F.: Deep learning using rectified linear units (relu). arXiv preprint arXiv:1803.08375 (2018)
23. Dumoulin, V., Visin, F.: A guide to convolution arithmetic for deep learning. arXiv preprint arXiv:1603.07285 (2016)
24. Codella, N., et al.: Skin lesion analysis toward melanoma detection 2018: A challenge hosted by the international skin imaging collaboration (isic). arXiv preprint arXiv:1902.03368 (2019)
25. Okewu, E., Misra, S., Lius, F.S.: Parameter tuning using adaptive moment estimation in deep learning neural networks. In: Gervasi, O., et al (eds.). ICCSA 2020. LNCS Part VI 20, vol. 12254, pp. 261–272, Springer, Cham (2020). https://doi.org/10.1007/978-3-030-58817-5_20
26. Müller, D., Soto-Rey, I., Kramer, F.: Towards a guideline for evaluation metrics in medical image segmentation. BMC. Res. Notes **15**(1), 1–8 (2022)

Smart Healthcare AI-Infused Saline Monitoring System with IoT- Empirical Study

Bhupesh B. Lonkar(✉), Shivali Bujade, Pratik Sonune, Chandni Khilwani, Varun Paunikar, Sushant Borkute, and Aditya J. Kadu

Department of Data Science, G. H. Raisoni Institute of Engineering and Technology, Nagpur, India
bhupesh.lonkar@gmail.com

Abstract. Integrating Internet of Things (IoT) technology into the healthcare system has revolutionized how medical data is collected, monitored, and managed. This system can recognize abnormal circumstances, alert the patient, and notify the doctor by SMS or email. Through a cloud-based and Machine Learning (AI) approach, the technology allows the doctor to monitor and share on the Android application simultaneously. At any given moment, more than one patient can be monitored using an Android app. In this paper, we study the Smart Healthcare AI–Infused Saline Monitoring System, consisting of level sensors that recognize the liquid status in the bottle, whether it is normal or a warning. In this paper, we conduct a comparative study based on different methodologies, their limitations and accuracy. Smart Healthcare AI – Infused Saline Monitoring System aims to enhance patient safety (Security), reduce nurse workload, and improve the overall efficiency of healthcare.

Keywords: IoT · Saline level · Alert · Machine Learning · Monitoring · Accuracy

1 Introduction

In today's fast-paced world, the healthcare industry constantly seeks innovative solutions to improve patient care and optimize medical resources. As our population is increasing daily, the scarcity of care is also growing. One such solution is the implementation of an Automated Saline Tracker using Internet of Things (IoT) technology. Today, saline infusion nurses have to continue monitoring the flow rate of saline physically, which may also interrupt the patients. By this tracker, the caretaker who will be relaxed will be appointed for Night Shifts, and our health care system will also be efficient for Shifts. It will help reduce the stress of continual monitoring by Doctors and Nurses at an affordable cost [1, 2]. Saline is provided when the human body is dehydrated or for heavy injections to rehydrate the body. Sometimes, if a bottle is empty and the needle is not withdrawn from the patient's vein, the blood flows outward into the bottle owing to pressure, which can result in an injury. Fortunately, we have an Infusion Pump capable of controlling the flow rate of infusion fluid, but here, the infusion pump doesn't

D. Pastor-Escuredo et al. (Eds.): ICDLAIR 2023, LNNS 1001, pp. 674–690, 2024.
https://doi.org/10.1007/978-3-031-60935-0_59

display on the screen and cannot be accessed by a phone application. Due to the high cost of infusion pumps, regular patients cannot use them; they can only be used in the ICU. For heart body temperature, saline Arduino-based development is necessary and important. By implementing the Automated Saline Tracker, we can monitor and control saline and drop rates. An empirical study on a saline tracker using IoT technology is a valuable exploration of healthcare innovation that combines the Internet of Things (IoT) and saline monitoring to enhance patient care and streamline medical processes. This research investigates the effectiveness, reliability, and real-world applications of IoT-based saline trackers in healthcare settings. This study seeks to provide insights into the practicality and potential benefits of implementing such systems by collecting and analyzing data from various case studies and scenarios [3, 4].

Now, let's talk about the Internet of Things or IoT. It is like connecting everyday things, even medical devices, to the internet. It's a big deal and is as the next big thing in technology. With over 30 billion devices expected to hit the market soon, it's changing how we do things. In healthcare, it's a game-changer. We can connect special medical tools to a big network, making it possible to check on patients from far away. It is important, especially after the recent pandemic, where staying away from sick people is crucial. IoT can help with that [5].

In simple terms, keeping an eye on IV fluids, especially saline, is important in healthcare. When you're sick, you might need these fluids through a needle in your vein. But it's tricky, and getting the right amount is important. Traditionally, nurses or even family members check on this every hour. It could be a better system and takes a lot of time. That's why we've developed new devices like pumps and sensors to do this job, but they can be expensive [6].

But here's the cool part: smart people are working on projects to improve healthcare, especially in places with fewer resources. They're creating an "Electrolyte Monitoring System" using local sensors to make it affordable. This worked shows how technology can help improve healthcare in areas without doctors, making it accessible to more people. It's all about using smart solutions to help patients get better care [7]. In Sect. 2, we do the review of literature & in Sect. 3, we do a comparative study based on the literature review.

2 Literature Review

Rajeshwari et al. [8] proposed saline Monitoring with Alert and Control. This cutting-edge system amalgamates various elements to improve the monitoring of saline levels in medical contexts. The system's signaling aspect comprises a buzzer and a GSM/GPRS module. At the same time, the stopping component incorporates a solenoid valve, as illustrated in Fig. 1. It streamlines the regulated saline administration into a patient's bloodstream via an IV tube connected to a saline bottle securely positioned on a stand. A notable feature of this design is the inclusion of a piezoresistive force sensor ingeniously affixed to the support structure of the saline bottle. In conjunction with a resistor, this sensor seamlessly integrates with an Arduino, forming a sophisticated monitoring system. Incorporating a buzzer within the circuit is an efficient alert mechanism. Should the saline level in the bottle fall below a pre-established threshold, the buzzer activates, enabling healthcare professionals to address the patient's needs promptly. The piezoresistive force sensor plays a pivotal role in this system, accurately gauging the force applied to the saline bottle. As the salty level diminishes, the pressure on the bottle fluctuates, and the sensor continually tracks these changes. Ensuring the sensor is appropriately positioned within the stand loop, with the bottle's top near the sensor's detection element, guarantees precise and dependable monitoring. Incorporating intravenous tubing connected to the suspended saline bottle enhances the practicality of the system. To safeguard the sensor and the stand's loop, they are shielded by a robust casing. Furthermore, the system benefits from an Arduino board fitted with a buzzer, conveniently connected to a computer via a portable link.

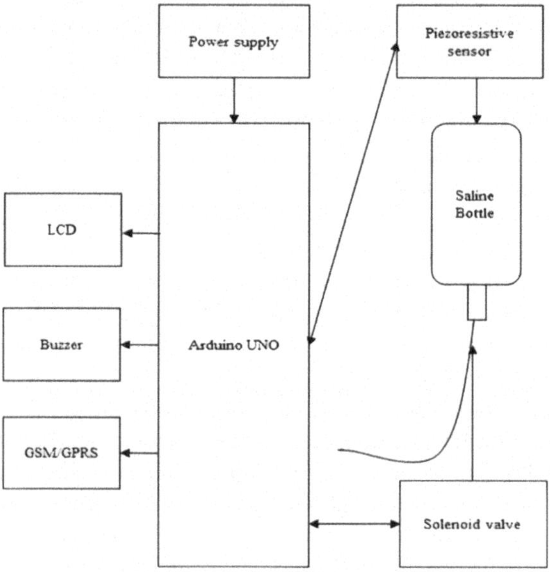

Fig. 1. Block diagram for novel methodology of alert and stopping system

Ranjani et al. [9] suggested that their study introduces a comprehensive model for monitoring essential health indicators like heart rate and body temperature in this Saline Infusion Monitoring and Control System. It utilizes an infrared (IR) sensor to assess the saline solution level in a bottle accurately. The sensor incorporates an LED light that activates when the container is full and deactivates when it's empty. An Arduino microcontroller processes the data from this sensor, displaying it on a serial monitor. Should the fluid level fall below a predetermined threshold, an immediate alert is sent to healthcare professionals via their mobile devices. Furthermore, the system employs a pulse sensor (MAX30100) to monitor the patient's heart and blood oxygen levels. A temperature sensor (LM35) is positioned on the patient's wrist to detect any unusual fluctuations in body temperature. Whenever abnormal readings are seen, healthcare professionals receive an alert. This monitoring system is connected to a Firebase database, allowing data access through a web application. A GSM modem, known as a Global System for Mobile Communications (GSM) modem, is a specialized device that accepts a SIM card and functions similarly to a mobile phone. A SIM card is necessary for the GSM module to operate. Once all the components and sensors are linked to the GSM modem, the system can alert nurses and doctors if any irregularities in the patient's health are detected. This system is categorized as an Internet of Things (IoT) based Saline Monitoring System. Moreover, patient historical data is stored in a Firebase database, allowing doctors to access it remotely, eliminating the need for physical presence. The saline infusion monitoring and control system comprises sensors, a control unit, and a monitoring device. These sensors, including those for heart rate, pulse oximetry, and temperature, are placed on the patient's body to collect crucial health data. They monitor the heart's electrical activity, blood oxygen levels, and body temperature. The control unit, which might be a Node MCU board, receives data from these sensors and adjusts the rate of saline infusion based on the patient's vital signs, ensuring that the patient gets the right amount of fluids and oxygen. The monitoring device, such as a computer or tablet, provides real-time displays of the patient's vital signs, allowing healthcare professionals to monitor the patient's condition and respond to any changes or irregularities that might indicate a health issue. This monitoring device can also be connected to a central monitoring system or cloud server for remote monitoring and analysis by authorized personnel. The system also incorporates a pump controlled by the control unit to infuse the saline solution into the patient's body. This system continually gathers vital sign data and makes adjustments as needed to ensure optimal fluid and oxygen delivery and enables a rapid response to any issues that might arise.

Kabir et al. [10] purported Intravenous Fluid Monitoring, Warning, and Reverse Flow Blocking System. A block diagram in Fig. 2 illustrates the intravenous fluid monitoring and reverse blood flow blocking system. This system is designed to provide automatic GSM-based alerts and prevent the reverse blood flow in intravenous (IV) therapy. It employs a load cell sensor to measure the weight of the saline bag (IV fluid bag). As the saline level decreases, the weight of the bag decreases accordingly. The load sensor continually provides weight data to the microcontroller (Arduino), and the display shows the percentage of liquid used from the bottle. The microcontroller assesses whether the fluid level is below the predetermined threshold (15% of the total bag weight) based on input data from the load sensor. If it falls below this threshold, the system responds locally by activating a buzzer and blinking an LED for notification. When the fluid level reaches around 10% of the total bag weight, a message is sent via phone to alert the nurse that the patient is running out of saline. In cases where the nurse doesn't respond promptly, and the saline level drops to 7%, an automatic arrangement using a solenoid valve is activated. This valve blocks the saline tube, preventing the reverse flow of saline and blood from the patient's veins into the bottle. The circuit design outlines a sophisticated system integrating multiple elements into a unified setup to oversee and manage saline infusion. The structure includes a load cell for measuring the saline bag's weight, and it incorporates a solenoid valve within the supply line for controlling the flow of saline. The load cell connects to an Arduino through an HX711 module, with predefined weight thresholds specified in the code. Whenever the weight surpasses these predetermined thresholds, the system transmits an alert to the hospital administration via a GSM module (GA6-B). In cases where there's a delayed response from the nurse, the system engages a solenoid valve to obstruct the saline tube, thus preventing any reverse flow. The solenoid valve operation is governed by a relay connecting to a 12 V battery power source. Real-time monitoring of the saline bag's weight is facilitated by an LED indicator and an LCD display, where the LED functions as an alarm when the weight falls below a specified threshold. The system is fully sealed to safeguard against fluid loss. This system unites diverse sensors and components for supervising and controlling saline infusion. It provides alerts to healthcare professionals and serves as a safeguard against reverse blood flow. It delivers ongoing monitoring and can be seamlessly integrated into healthcare settings to elevate the quality of patient care.

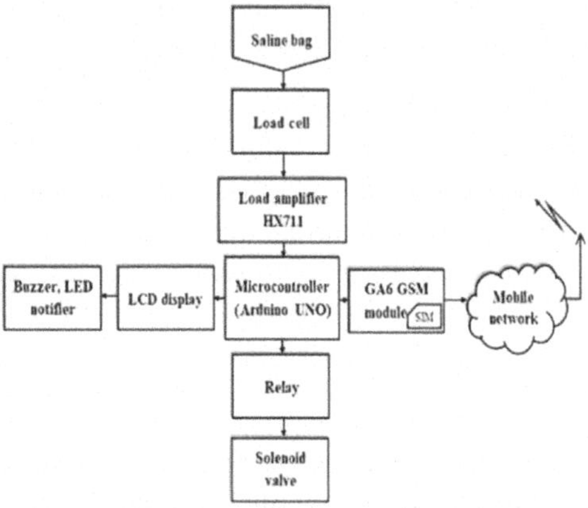

Fig. 2. Block Diagram Intravenous Fluid Monitoring System

Rafi et al. [11] designed to adjust the saline solution's flow rate automatically. This research aims to create a cost-effective and easily accessible autonomous controller for regulating the flow rate of saline. Instead of opting for more complex solutions like advanced devices or 3D printing, which don't align with the primary objectives of this study, we chose to employ straightforward, economical devices with specific advantages. These devices were assembled to meet the two primary criteria and effectively serve their purpose in the medical field. In line with this approach, we developed a customized device by combining a DC gear motor and a screw gauge, connecting both components with an aluminum shaft. In the final device, the spindle of the screw gauge was designed to move linearly by 0.5 mm in a single circular rotation, adhering to the screw gauge's specifications. By rotating the motor head 20 times, we achieved a precision of 1 cm for the spindle. This process facilitated the calibration of the device. A formula used by the health workers to determine the flow rate is given below Eq. 1:

$$\text{Flowrate (gtt/min)} = \text{Volume/Min} * \text{Drop Factor} \tag{1}$$

In this, we employed data from three different subjects to validate the authenticity of our experiment. These subjects varied in age. We measured their heart rates and then compared the obtained flow rate in drops per minute to a reference value. Additionally, we calculated the error if any discrepancies were identified in Eq. 2.

$$\text{Error} = \text{reference value} - \text{measured value}/\text{reference time} * 100 \tag{2}$$

Umchid et al. [12] proposed a system that measures the current volume of saline in a particular bottle, which is effectively displayed and communicated by this cutting-edge method. The device is prepared to send a prompt alarm when the saline level approaches a specified critical point. The impending low saline volume is promptly informed to medical personnel and caregivers using an auditory alarm that emanates from a speaker. This ground-breaking solution can also warn medical staff via a specialized mobile phone application that uses Bluetooth for seamless communication. This system is also reusable, as it may be used with subsequent saline bottles to avoid the need for ongoing replacement. It simplifies the medical professionals' monitoring procedure and helps reduce costs. Ultimately, this system enables medical practitioners to easily monitor and control saline levels remotely, improving the overall effectiveness and standard of patient care.

Rangsee et al. [13] proposed one of the most cost-effective and easy-to-use systems for measuring saline droplet levels in patient rooms of rural public hospitals. The Saline Droplet Measurement System (SDMS) is intended for use in a typical patient room within a rural general hospital. This system is designed to monitor the status of saline droplets at each patient's bedside and is divided into two main components: the patient and the nurse parts. The patient part includes two sensors responsible for detecting critical saline levels and the presence of saline droplets. A microcontroller processes the data from these sensors, and a wireless module transmits this information from the patient to the nurse. The SDMS consists of 18 patient parts, each corresponding to a patient's bed. In the nurse part, a wireless module receives sensor data from each patient part. The primary processor is responsible for calculating and processing information from all patient beds, allowing for real-time display and alerts. This system will enable nurses to monitor saline droplets' status, rate and remaining time on a computer screen. The system displays four states: Normal (GREEN light), Warning (yellow light), Error (RED light), and Change New Bag (blue light). This system allows nurses to remotely monitor the status of saline droplets, eliminating the need for repeated bedside checks. This makes it an ideal system for low-cost implementation in rural public.

Manam et al. [14] proposed that their prototype can monitor the salinity of a bottle continually and upload the information to the Thing Speak cloud. We have incorporated several key components in this proposed system, specifically the Node Mcu, Ultrasonic sensor, and a Breadboard. The Node MCU, a microcontroller with Wi-Fi capabilities, was chosen for its versatility. We've employed the Ultrasonic sensor to function as a level sensor. This sensor operates by emitting sound waves, which bounce back upon encountering an obstacle and are detected by a receiver. The system then calculates the time taken for the ultrasonic sound to travel and subsequently determines the distance using a specific in Eq. 3:

$$distance = time * \frac{0.5}{29} \tag{3}$$

Using the obtained distance value (d) in conjunction with the overall quantity of saline bottles (qs) and the height of each saline bottle (h), we can accurately measure the volume of saline mentioned in Eq. 4.

$$\text{quantity of saline}(q) = qs - \left(d * \frac{qs}{h}\right) \quad (4)$$

Joseph et al. [15] proposed that due to the surge in in-home care for older people with chronic illnesses, our solution addresses the real-time monitoring of patients using IV drips. In all the scenarios we've examined, it's been evident that continuous monitoring of the mass flow rate, the fluid level in the IV bag, and the patient's overall health condition is crucial. However, the decision regarding the rate at which the fluid should be administered is typically determined by the doctor or nurse based on the patient's condition. The proposed model focuses on two primary aspects:

- Manual control of the infusion rate through a mobile app
- Real-time monitoring of the IV fluid level in the bag and providing alerts to medical staff for timely bottle changes

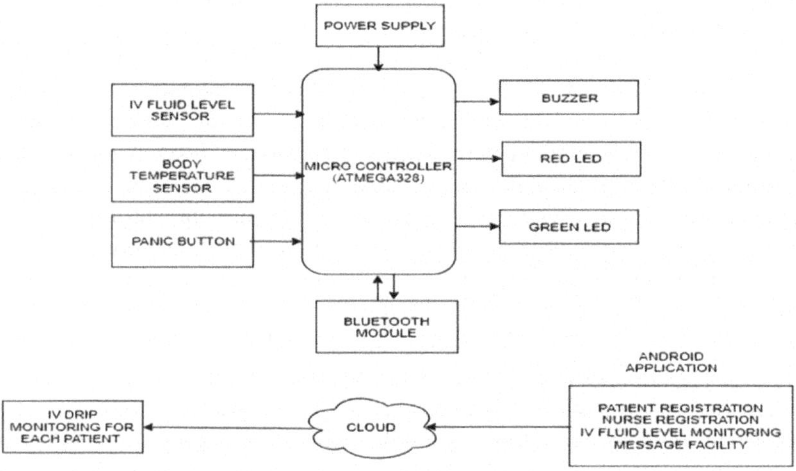

Fig. 3. Block Diagram Atmega328 Micro Controller

As per Fig. 3, the Arduino (ATMEGA328 Micro Controller) is programmed to capture analogue signals from sensors and package this data into a digital format. The Arduino, equipped with a Bluetooth Module, can connect to a smartphone application. Sensors collect data from the patient's body and the IV bag, transmitting this information to the application. Nurses and staff can then analyze this data and make informed decisions in real time, enabling effective monitoring and control.

Nagaraj et al. [16] planned an IoT-based Smart Hospital Saline Monitoring System. Our system features Auto/Manual Modes. In Auto mode, blood and saline flow are adjusted based on a heart sensor and monitored blood pressure. Manually, caregivers can remotely control saline flow via an IOT module. Two solenoid valves regulate saline flow according to heart rate; both open normally, but one closes if the heart rate or blood pressure deviates. Photocell sensors detect blood flow changes by monitoring visible light spectrum shifts. An algorithm on the MCU detects red color to alert clot leakage during hemodialysis, sending notifications to caregivers. The sensor includes an LED and a photodiode, with the LED emitting IR radiation. An IR sensor detects the altered light, and the LDR sensor's output is converted to digital signals. The PIC 16F877A MCU manages the system specialized for clot leakage and saline level detection. It provides a stable +5 V supply with a 7805 V regulator. It tracks salty levels using load cells and non-invasive detectors mounted on the patient's arm. The device modifies the saline flow rate based on the patient's temperature and heart rate. A warning is provided to caregivers using an IoT wireless transmission technique if blood leakage or low saline levels are discovered. This helps medical professionals to act quickly and reduce patient dangers.

Gnanadas et al. [17] suggested that the approach is intended to keep track of and regulate the saline in the infusion bag. The load cell sensor, which made it possible to determine the weight, allowed the monitoring to be accomplished within the bottle. Based on its level, the bottle's level is calculated. Weight and the ongoing saline level monitoring have been completed. It is possible to manipulate the catheter tube using a solenoid valve. This regulation is finished when a medical professional alters the flow of the webpage. Based on this, the shaft compresses the solenoid input provided by the doctor via the website. When the saline level reaches the end valve, the solenoid valve compresses. The system delivers saline to patients for recovery. The system automates this process with sensors to measure saline levels linked to a microcontroller. A solenoid valve controls the flow rate; the system is web-based for monitoring and adjustments. As the saline nears its endpoint, the system stops flow to prevent backflow. Data is recorded and sent to the cloud via Wi-Fi. This system blends healthcare and technology by accurately measuring saline bottle weight using a load cell sensor. The microcontroller displays the level on a real-time web page. It automatically stops flow when the endpoint is reached, and the solenoid valve can be adjusted via the web page. Real-time monitoring ensures constant saline flow and prevents backflow. Data is accessible via a user-friendly GUI. The system accurately monitors and controls saline levels. Future enhancements may include machine learning for prediction. The system provides accurate weight measurements and stops flow when needed, reducing costs by about 74%.

Digarse and Patil [18] suggested that the system measures various parameters and sends SMS notifications to doctors, nurses, or family members through an Android device. The estimated data is conveyed via SMS. This Arduino UNO and GSM-based wireless health monitoring system represents an advancement over existing technology due to its swift SMS transmission, user-friendliness, and long-distance coverage at a low cost. It relays data on heart rate, body temperature, and saline levels to doctors, enabling prompt action in critical situations. The microcontroller is the system's core, with Arduino Uno chosen for its specifications and ease of use. It features 14 digital pins (6 supporting PWM output), six analogue inputs, a DC power jack, a reset button, an ICSP header, a 16 MHz ceramic resonator, and a USB interface. An LM35 IC sensor was selected, emulating temperature data collection. This sensor continually measures ambient temperature and converts it into voltage levels received by the Arduino Uno. The ultrasonic transducer, working with ultrasound of a specific frequency, was chosen in its piezoelectric form for its cost-effectiveness and ease of use. The ultrasound velocity is influenced by temperature and the medium, with a rate of 343 m/s at 25 °C. The system requires a 1cm resolution, with ultrasound taking approximately 29.15 μs to travel 1cm. For a basic Centigrade temperature sensor, the rating is (+2°C to + 150°C), and the temperature vs. voltage relation is represented through Eq. 5:

$$\text{Vout} = 0\,\text{mV} + 10.0\,\text{mV}/^{\circ}\text{C} \tag{5}$$

Thus, the velocity in the air is designed using Eq. 6 below.

$$\text{V} = 340 + 0.6(\text{t} - 15)\text{m/s} \tag{6}$$

In Arduino programming, the critical saline level is set at 350 ml, given that a standard saline bottle contains 500 ml. When the saline level surpasses 350 ml, a red light blinks, and an alert SMS is sent to doctors or nurses. The system uses GSM (Global System for Mobile Communication) to transmit SMS alerts to doctors' mobile phones. The GSM module has an RS232 interface for serial communication and functions through AT commands. The microcontroller receives results and then retransmitted via HyperTerminal using a baud rate 9600.

Selvi et al. [19] present an automated system for monitoring and controlling saline fluid levels using a solenoid valve. Accessible via computer or smartphone, healthcare providers can remotely monitor saline levels, temperature, blood oxygen, and heart rate. The fully automated system minimizes human intervention, benefiting nurses, especially in busy hospitals. It's user-friendly and can be reused for subsequent saline bottles.

Sunil et al. [20] research that an automated saline monitoring system reduces nurses' need for continuous manual monitoring, making it very efficient and valuable, especially at night. An IR sensor detects infrared light emitted by nearby objects. Our system puts it in the saline bottle's drip chamber to monitor saline flow. An Arduino microcontroller is the system's brain, controlling a servo motor and buzzer. The servo motor adjusts position, speed, and acceleration using Pulse Width Modulation (PWM) signals, while the buzzer alerts nurses when the saline is empty. The Arduino is powered via USB, and the IR sensor watches for saline drops. If it doesn't detect drops for 30 s, it activates the servo motor to prevent backflow and sounds the buzzer.

Two scenarios exist:

1. Buzzer Alert: The buzzer notifies nurses if saline drops aren't detected.
2. Tube Rotation: The servo motor turns the tube and sounds the buzzer to prevent backflow.

Saha et al. [21] present an automated system for monitoring and controlling saline fluid levels using a solenoid valve. Accessible via computer or smartphone, healthcare providers can remotely monitor saline levels, temperature, blood oxygen, and heart rate. They have created a health monitoring system with Zigbee for wireless data transfer. It employs a DS18B20 temperature sensor with features like a 1-wire interface and operates on a power supply of 3.0 V–5.5 V. An Arduino Nano processes the data with an ATmega328 microcontroller. This system monitors patients' body temperature and other vital signs. The data is transmitted to a Central Control Unit (CCU) or computer using Zigbee for wireless connections and USB-shielded cables for close-range connections. Zigbee is chosen for its low power consumption, wide range (300 f to 40 miles), modest data rate (250 Kbit/s), and cost-effectiveness in healthcare applications.

Amarasekara et al. [22] introduced an E18-D80NK IR sensor to detect the saline drops going through the drip chamber. The drip chamber will be placed in front of the IR sensor. When a drop is felt through the section, the sensor detects that, and the count will be increased by one. The sample rate of this method is one minute, and the collected data are sent to the database.

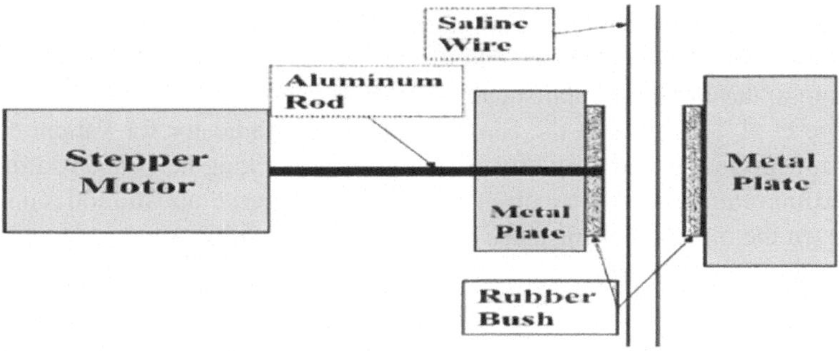

Fig. 4. Drop Rate Controller

As illustrated in Fig. 4, a stepper motor is fixed with a small aluminum rod with threads over the edge. When the motor starts to rotate, the aluminum cylinder moves horizontally. Therefore, the aluminum cylinder compresses the saline tube, similar to the manual procedure. That will cause an increase in the air pressure in the saline tube and reduce the drop rate falling through the drip chamber.

Kodali and Yenduru [23] suggested an IoT-based saline level monitoring system eliminates the need for manual involvement from doctors, nurses, and other hospital staff. This advancement ensures zero risk to patients by automating the entire process, eliminating the need for nurse intervention. This system offers a significant advantage, particularly during night-time, as it relieves hospital staff from the stress of continuously visiting patient rooms to check saline levels. Instead, the system sends alert notifications to the mobile devices of hospital personnel, including nurses and doctors, when the saline weight falls below critical levels. In contrast to traditional audible alerts, our prototype avoids using disruptive buzzers, as they go against hospital policies due to the potential disturbance they can cause, particularly to patients with heart conditions. We have implemented a Silent Notification Service (SNS) to send push notifications, which can be broadcast to many devices simultaneously. This automatic saline-level monitoring system introduces flexibility for nurses and doctors, ultimately enhancing patient care. It alleviates the constant monitoring burden on medical professionals, reducing stress and offering cost-effective solutions. Consequently, it not only saves patients from risks but also saves significant time for medical staff. The system is distinguished by its high-quality performance, security, and cost-effectiveness, making it a convenient solution for everyone involved. Moreover, the same setup can be easily reused for subsequent saline bottles without further modifications. A particularly noteworthy feature is the system's accessibility, enabling nurses, doctors, and attendants to monitor and check saline flow levels anywhere on the planet, even without an internet connection. It allows staff to access and modify saline data as needed, enhancing patient care and overall convenience.

Ghosh et al. [24] present a low-cost, sophisticated, affordable, intelligent saline-level monitoring system. This system allows for remote monitoring of the saline administered to a patient by individuals such as nurses, caregivers, hospital staff, and doctors. We have chosen to implement the MQTT-S protocol due to its efficiency with low-cost and low-power devices. Additionally, MQTT-S ensures the reliable delivery of messages, supporting asynchronous communication through message buffering. We are confident that this monitoring system will enable distant tracking of saline levels in the bottle, contributing to developing a sophisticated healthcare system.

Kishore et al. [25] present the Smart Saline Level Indicator for Patient Safety as a low-cost, low-energy, and very effective device for tracking how much saline is in a bottle. The importance of constant observation during saline administration is underlined because even the most minor mistakes could have fatal results. A valuable and cutting-edge technique for monitoring saline levels is introduced in this study. It enables medical staff members, such as nurses, doctors, and caretakers, to remotely monitor the quantity of saline given to the patient. The effectiveness of the MQTT-S protocol and its capacity to make communication between sender and receiver easier justifies its selection. The Buffering mechanism is used in circumstances of offline operation to guarantee message recovery. As an extra benefit to the intelligent healthcare service, it is clear that this established system enables remote monitoring of a patient's saline levels, offering vital support to healthcare personnel, including doctors, nurses, and caretakers.

Singh et al. [26] introduce an uncomplicated, cost-effective approach to monitoring blood pressure. Integrating the Internet of Things principles with an Arduino microcontroller and a pressure sensor creates a Blood Pressure Monitoring System utilizing wireless technologies. This work aims to establish a network that enables concerned individuals to access a patient's blood pressure readings remotely. Bluetooth and Wi-Fi technologies are employed to access these results through portable devices such as mobile phones, tablets, and laptops. The work also integrates a prediction algorithm using MATLAB software. Readings can be manually recorded over time and input into the program to create a data log (Table 1).

3 Comparative Study Based on Literature Review

Table 1. Comparative Study based on Literature Review

Year	Author	Prototype /framework description	Used sensor hardware components	Used Software Components	Communication module	Used prototype and framework
2023	Rajeshwari et al. [8]	A piezoresistive sensor is used in healthcare to monitor saline levels, with its high sensitivity, wide dynamic range, and durability. It's connected to a microcontroller, allowing healthcare workers to monitor and alert if saline levels exceed thresholds	Piezoresistive pressure sensor, Arduino UNO, Solenoid Valve, Buzzer, GSM /GPRS module	Not specify	GSM /GPRS module	The proposed solution offers a reliable, automated method for managing saline drip rates, thereby enhancing patient safety and preventing complications in hospitals and private homes
2023	Ranjani et al. [9]	The system utilizes sensors to monitor blood pressure and heart rate, sends remote alerts, and prevents vein damage by halting saline flow	IR senor, Heartbeat/pulse Sensor (MAX30100), Temperature Sensor (LM65), Node MCU	Web based application	GSM Module	The proposed IoT-based saline infusion monitoring system improves patient safety by continuously monitoring saline levels using a sensor and alerting healthcare providers if they fall below 10%
2023	Kabir et al. [10]	The system, which uses a microcontroller, load cell, switching valve, and GSM module, is designed for patient safety during saline feeding hours	Arduino Uno, Liquid Crystal Display (LCD), GA6-B, Weight Sensor, Solenoid Valve	Not specify	GSM Module	The proposed system utilizes a weight measuring sensor to monitor saline bottle weight, process it via a microcontroller, and shut off flow when empty for accurate intravenous treatment
2019	Rafi et al. [11]	This research aims to regulate intravenous saline solution flow-rate, utilizing heart rate as a parameter, to ensure patient safety in modern clinical practice	Pulse sensor SEN-11574, motor driver LN-298, Arduino UNO and saline set	Not specify	Wire connection	Arduino controls a customized device, motor driver, and Arduino to manipulate flow-rate using pulse sensor data, considering three patient data and calculating errors for authenticity
2018	Umchid et al. [12]	The study aims to design and build a saline monitoring system for medical staff, alerting them when the saline level is nearly empty or below the set value	Load cell (model YZC-133, Weight sensor amplifier module (model HX711), Input switches, Arduino UNO R3, Bluetooth module (model HC-05), Liquid crystal display (LCD), Speaker	Not specify	Bluetooth module	The developed system underwent four tests to validate its performance, with a 0.43% error rate in the first test, a 2.13% error rate in the second test, and a 0.43% error rate in the third test

(*continued*)

Table 1. *(continued)*

Year	Author	Prototype /framework description	Used sensor hardware components	Used Software Components	Communication module	Used prototype and framework
2014	Rangsee et al. [13]	The Saline Droplet Measurement System (SDMS) monitors saline droplet status in a rural public hospital, consisting of patient and nurse parts, sensors, and a microcontroller	IR Sensors, Microcontroller, Zigbee	Microsoft Visual Basic 2008	Zigbee wireless Technology	The proposed low-cost SDMS monitors patient saline droplet levels in rural public hospitals, displaying rate and remaining time in each patient's bed
2023	Manam et al. [14]	The research employs advanced techniques to monitor saline levels in healthcare, addressing the insufficient proximity of nurses to patients	ULTRASONIC SENSOR, Node MCU, Wi-Fi module	Android Application	Wi-Fi module	The IoT-based Automatic Saline Monitoring System monitors electrolyte levels, storing data in the cloud, and triggers an IFTTT applet when fluid completion is near, promoting healthcare industry development
2023	Joseph et al. [15]	The system controls saline flow rate, alerts nurses, and automatically changes the bottle after filling, reducing manual work and automating IV fluid injections	LM35 sensor, IV fluid level sensor, Arduino Uno, Bluetooth Module	Android Application	Bluetooth module	The model is designed to monitor patient's mass flow rate, IV fluid level, and body condition via a mobile app
2023	Nagaraj et al. [16]	Saline and blood leakage are critical blood vessel medical issues, particularly during dialysis therapy. Carelessness can lead to complications and necessitate remote monitoring and telehealth services	PIC Microcontroller, Driver relay circuit, solenoid valve, IOT Module Board, load sensor	Not specify	MQQTT's Protocol	The system controls saline levels and blood flow using a load cell detector, ESP32 wireless network, and SOC micro-controller, sending alert signals to caregivers via MQTTP protocol
2023	Gnanadas et al. [17]	Advancements in technology and population growth have digitized healthcare, combining principles with engineering for automated versions. Manual monitoring of saline in blood transfusions is revolutionized by infusion pumps.	ESP8266, Solenoid Valve, Hook Connector, Load Cell, Wifi	Not specify	Wi-Fi Module	The proposed project utilizes wireless transfer to automate saline bottle monitoring and control, reducing manual labor and physician visits during pandemic outreach
2017	Digarse and Patil [18]	About 20% of India's population dies due to interrupted health monitoring systems, which may cause critical health issues or death due to doctors visiting patients in both morning and evening shifts	Arduino UNO, Heart Beat Sensor, Temperature sensor, Saline level sensor, GSM Modem	Not specify	GSM Modem	The smart embedded system continuously monitors patients' health, detecting abnormalities in heart rate, body temperature, and saline liquid level, alerting doctors or care providers for corrective actions

(continued)

Table 1. *(continued)*

Year	Author	Prototype /framework description	Used sensor hardware components	Used Software Components	Communication module	Used prototype and framework
2022	Selvi et al. [19]	Advancements in clinical care, including sensors and micro-controllers, are enhancing COVID-19 treatment and nutrition, with saline solution being crucial, but persistent administration is necessary to prevent backflow issues	XKC-Y25, Pulse sensor MAX30100, Temperature sensor DHT11, Arduino ATMEGA328, Arduino mega, GSM module SIM900A and servo motor	Not specify	GSM Modem	The proposed saline level monitoring and automatic alert system is designed to safeguard patients during Covid-19 and ensure their safety during saline feeding hours
2020	Sunil et al. [20]	An IoT-based saline level monitoring system has been developed to ensure continuous monitoring of saline levels in medical treatment, preventing staff inattentiveness and blood backflow	IR Sensor, Arduino Microcontroller, Servo Motor, Buzzer	Not specify	Not specify	The model features a sensor that detects saline drops, alerting hospital staff when they fail to do so, thereby ensuring patient safety
2021	Saha et al. [21]	The internet of things has revolutionized real-time health monitoring through sensor-based systems, necessitating the creation of cost-effective, usable health monitoring systems for patient monitoring	DS18B20 temperature Sensor, Arduino Nano, ZigBee	Microsoft Visual Basic 6	ZigBee wireless Technology	The proposed prototype system uses a DS18B20 temperature sensor, Arduino Nano, and ATmega328 for real-time health monitoring, enabling timely preventive measures and saving user data for future reference
2018	Amarasekara et al. [22]	Hospitalized patients require proper nutrition and saline therapy, with assistants or nurses monitoring patient saline supply. However, busy schedules can cause mistakes, such as blood reflux. Traditional methods of supplying saline are inaccurate, as drop rate is adjusted based on drop speed in the drip chamber	E18-D80NK IR Sensor, Stepper motor, UPS, Wi-Fi	Apache web Server,	MQTT Protocol	The proposed system uses IoT to remotely monitor the saline system, enabling a sophisticated method for controlling saline drop rate
2023	Kodali and Yenduru [23]	An IoT-based prototype alerts hospital staff to low saline levels, preventing blood reversal and putting patients at risk, using ESP8266 NODEMCU, load sensor, and HX711 amplifier	ESP8266 nodemcu, HX711 Amplifier, The Load Cell	AWS IOT	MQQTT's Protocol	The system enables hospital staff to monitor saline levels and change bottles on the brink of emptying without visiting patients' rooms

(continued)

Table 1. (*continued*)

Year	Author	Prototype /framework description	Used sensor hardware components	Used Software Components	Communication module	Used prototype and framework
2018	Ghosh et al. [24]	Saline, a vital intravenous therapy, requires constant monitoring by nurses to prevent needle removal errors, necessitating remote monitoring for telehealth services	Load Sensor, ESP32 WiFi Chip	Not Specify	MQQTT's Protocol	The smart saline level monitoring device utilizes a load sensor and ESP32 WiFi SoC microcontroller to convert bottle weight to voltage, ensuring reliable healthcare delivery
2023	Kishore et al. [25]	The healthcare system is rapidly advancing, with saline, a vital intravenous therapy, being monitored closely by nurses and caretakers to prevent significant loss and death	The Liquid level switch contactless sensor, NodeMCU ESP8266	Not Specify	MQTT-S	The smart saline level monitor utilizes a low-power, ultra-low-cost liquid level switch contactless sensor and a NodeMCU ESP8266 Wi-Fi module for reliable communication in medical treatment
2019	Singh et al. [26]	This work aims to establish a network that allows remote access to patient blood pressure readings using Bluetooth and Wi-Fi technology on handheld devices	Pressure Sensor: BMP180, Node-MCU With Wi-Fi Technology, Bluetooth Technology With HC-05 Module, Internet Of Things, KPM27H Mini Air Pump Motor, L293d Dual Motor Driver H Bridge Module Board	MATLAB software	Wi-Fi Module	The proposed work introduces a user-friendly, cost-effective method for monitoring blood pressure using an Arduino microcontroller and pressure sensor, incorporating Internet of Things concepts

4 Conclusion

The Saline Monitoring System is a sophisticated medical technology that proficiently tracks and displays the current saline volume within a bottle. When the saline level nears a predetermined critical threshold, it activates an alarm through a built-in speaker. It simultaneously sends alert notifications to medical personnel via a dedicated mobile application. Notably, it empowers healthcare professionals to control the saline bottle valve remotely, eliminating the need for physical presence at the patient's bedside. This cutting-edge system allows remote supervision and effective management of a patient's saline levels, optimizing patient care. This innovation offers several compelling advantages. Firstly, it enhances cost-effectiveness by preventing saline waste and minimizing the need for constant manual monitoring. Its adaptability makes it suitable for various medical settings, from hospitals to home care. The system's easy deployment and high reliability ensure seamless integration into existing healthcare infrastructures, reducing disruption and ensuring consistent operation. The system's design also emphasizes sustainability by promoting reusability with subsequent saline bottles, aligning with environmental goals and enhancing its practicality as a long-term investment. The Saline Monitoring System revolutionizes saline administration in healthcare, providing accurate real-time information, early warning capabilities, remote control, and other benefits that collectively enhance patient care and streamline medical processes.

References

1. Reddy, D.R., Prakash, S., Dinakar, A., Kumar, S., Kodali, P.: Smart saline monitoring system for automatic control flow detection and alertness using IoT application. In: Sharma, H., Saraswat, M., Yadav, A., Kim, J.H., Bansal, J.C. (eds.) CIS 2020. AISC, vol. 1335, pp. 745–757. Springer, Singapore (2021). https://doi.org/10.1007/978-981-33-6984-9_59
2. Baluprithviraj, K.N., Harini, R., Janarthanan, M., Jasodhasree, C.: Design and development of smart saline level indicator for healthcare using IoT. In: 2022 Second International Conference on Artificial Intelligence and Smart Energy (ICAIS) (2022). https://doi.org/10.1109/icais53314.2022.9742794
3. Pantawane, R., Jyoti, A., Sheoran, S., Uikey, Y., Sonarkar, S., Thakare, R.: Implementation of IoT-Based health care and saline monitoring system using Arduino UNO. In: Shaw, R.N., Das, S., Piuri, V., Bianchini, M. (eds.) Advanced Computing and Intelligent Technologies. Lecture Notes in Electrical Engineering, vol. 914, pp. 513–526. Springer, Cham (2022). https://doi.org/10.1007/978-981-19-2980-9_41
4. Otanasap, N., Kanchanawongkul, V., Chalermsuk, C.: Design and development of Saline infusion administration system using IoT. ASEAN J. Sci. Technol. Rep. **25**(1), 41–49 (2022). https://doi.org/10.55164/ajstr.v25i1.244460
5. Lonkar, B.B., Kuthe, A., Shrivastava, R., Charde, P.: Design and implement smart home appliances controller using IOT. In: Garg, L., et al. (eds.) ISMS 2020. LNNS, vol. 303, pp. 105–117. Springer, Cham (2022). https://doi.org/10.1007/978-3-030-86223-7_11
6. Desai, M.N., Davande, R., Mahaadevan, K.R.: IoT-based saline controlling system. In: Choudrie, J., Mahalle, P., Perumal, T., Joshi, A. (eds.) IOT with Smart Systems. Smart Innovation, Systems and Technologies, vol. 312, pp. 677–687. Springer, Singapore (2023). https://doi.org/10.1007/978-981-19-3575-6_65
7. Babu, M., Shankaranarayanan, B., Sanjay, S., GA, S.K.: IoT based smart bottle for healthcare with an efficient SMS4-Bsk encryption transmission system. In: 2021 Second International Conference on Electronics and Sustainable Communication Systems (ICESC) (2021).https://doi.org/10.1109/icesc51422.2021.9532601
8. Rajeshwari, R., Ragu Prasath, M., Nikhil Manikandan, K., Tamilan, C.: Solitary saline monitoring with alert and control system using IoT. In: 9th International Conference on Advanced Computing and Communication Systems (ICACCS) (2023). https://doi.org/10.1109/icaccs57279.2023.10112864
9. Ranjani, J., Hariharan, S., Jeyashree, A., Aishwarya, E.: Saline infusion monitoring and control system using IoT. In: International Conference on Distributed Computing and Electrical Circuits and Electronics (ICDCECE) (2023).https://doi.org/10.1109/icdcece57866.2023.10150737
10. Kabir, M.F.I., Ahmed, S.S., Paul, L.C., Rani, T., Karaaslan, M.: Development of an intravenous fluid monitoring, warning, and reverse flow blocking system. In: IEEE International Women in Engineering (WIE) Conference on Electrical and Computer Engineering (WIECON-ECE) (2022).https://doi.org/10.1109/wiecon-ece57977.2022.10150612
11. Rafi, R.H., Ferdous, T.R., Islam, M.M., Ratan, Z.A.: Developing cost effective and automated saline flow rate controlling system for medical system. In: 2019 4th International Conference on Electrical Information and Communication Technology (EICT) (2019).https://doi.org/10.1109/eict48899.2019.9068824
12. Umchid, S., Kongsomboom, P., Buttongdee, M.: Design and Development of a Monitoring System for Saline Administration. In: Proceedings of the World Congress on Engineering 2018, vol. 1 (2018)

13. Rangsee, P., Suebsombut, P., Boonyanant, P.: Low-cost saline droplet measurement system using for common patient room in rural public hospital. In: 4th Joint International Conference on Information and Communication Technology, Electronic and Electrical Engineering (JICTEE) (2014). https://doi.org/10.1109/jictee.2014.6804111
14. Manam, P.S., Chalasani, R., Katikala, L.S., Kolli, S., Kondapalli, V.H., Nalluri, S.: Iot based automatic saline monitoring system using node MCU. In: 2021 2nd International Conference on Smart Electronics and Communication (ICOSEC) (2021).https://doi.org/10.1109/icosec51865.2021.9591873
15. Joseph, S., Francis, N., Adriani, J., Farha, B., Baby, A.: Intravenous drip monitoring system for smart hospital using IoT. In: 2nd International Conference on Intelligent Computing, Instrumentation and Control Technologies (ICICICT) (2019).https://doi.org/10.1109/icicict46008.2019.8993241
16. Nagaraj, P., Muneeswaran, V., Sudar, K.M., Ali, R.S., Someshwara, A.L., Kumar, T.S.: Internet of things based smart hospital saline monitoring system. In: 5th International Conference on Computer, Communication and Signal Processing (ICCCSP) (2021).https://doi.org/10.1109/icccsp52374.2021.9465512
17. Gnanadas, A.A., Indhumathi, K., Boopal, M., Saranya, R.S.: Internet of things based development of continuous saline monitoring and control system. In: 2022 6th International Conference on Electronics, Communication and Aerospace Technology (2022).https://doi.org/10.1109/iceca55336.2022.10009410
18. Digarse, P.W., Patil, S.L.: Arduino UNO and GSM based wireless health monitoring system for patients. In: International Conference on Intelligent Computing and Control Systems (ICICCS) (2017).https://doi.org/10.1109/iccons.2017.8250529
19. Selvi, M. C., Bhuvaneswari, T., Priyadarshin, R.N., Chitralekha, R., Ramya, R.: IoT based Saline monitoring system. J. Pharm. Negative Results, **13**(3) (2022). https://doi.org/10.47750/pnr.2022.13.S03.158
20. Sunil, G.K., Aluvala, S., Reddy, G.R., Sreeharika, V., Sindhu, P., Keerthana, S.: IoT based saline level monitoring system. In: IOP Conference Series, vol. 981, no. 3, p. 032095 (2020). https://doi.org/10.1088/1757-899x/981/3/032095
21. Saha, R., Biswas, S., Sarmah, S., Karmakar, S., Das, P.: A working prototype using DS18B20 temperature sensor and arduino for health monitoring. SN Comput. Sci. **2**(1), 33 (2021). https://doi.org/10.1007/s42979-020-00434-2
22. Amarasekara, C., Manage, R., Abeywickrama, C., Perera, R.L., Achchige, R., Wijekoon, J.: Poster abstract: smart saline management system. In: 17th ACM/IEEE International Conference on Information Processing in Sensor Networks (ISPN) (2018).https://doi.org/10.1109/ipsn.2018.00017
23. Kodali, R.K., Yenduru, P.M.: Saline water level monitoring using AWS. In: 2022 International Conference on Computer Communication and Informatics (ICCCI) (2022).https://doi.org/10.1109/iccci54379.2022.9740914
24. Ghosh, D., Agrawal, A., Prakash, N.K., Goyal, P.: Smart saline level monitoring system using ESP32 and MQTT-S. In: 20th International Conference on e-Health Networking, Applications and Services (Healthcom) (2018).https://doi.org/10.1109/healthcom.2018.8531172
25. Kishore, S., Abarnaa, K.P., Priyanka, S., Atchaya, S.S., Priyanka, P.L., Amala, J.J.N.: Smart saline level monitoring system using liquid level switch contactless sensor, NodeMCU and MQTT-S. In: 2022 International Conference on Applied Artificial Intelligence and Computing (ICAAIC) (2022). https://doi.org/10.1109/icaaic53929.2022.9792849
26. Singh, B., Urooj, S., Mishra, S., Haldar, S.: Blood pressure monitoring system using wireless technologies. Procedia Comput. Sci. **152**, 267–273 (2019). https://doi.org/10.1016/j.procs.2019.05.017

Music Recommendation System Using Psychological Scale

Nawal Ayesha Khan[1](✉), Sarah Jasim[1], Intisar Tahmid Naheen[1], and Ferdous Bin Ali[2]

[1] Department of Electrical and Computer Engineering, North South University, Dhaka, Bangladesh
{nawal.khan,sarah.jasim,intisar.naheen}@northsouth.edu
[2] Visie, Dhaka, Bangladesh

Abstract. Individual music preferences can generally depend on several factors, namely demographic data such as gender, age, etc., and more specific psychological factors like mood and personality. However, while mainstream music applications tend to consider the former, they usually ignore psychological factors that would otherwise allow for more accurate recommendations. In this paper, we attempt to develop a smart music recommendation system by comparing various machine-learning algorithms to recommend a genre based on the user's current mood and personality traits, determined using a psychological scale, along with their basic demographic data. A custom dataset was built, where the TIPI scale was used to identify users' personality traits, which we then fed into the models to generate recommendations. Based on the results, ensemble algorithms performed the best for the moods of "Happy", "Gloomy", and "Energetic", while semi-supervised algorithms achieved the best scores for the moods of "Stressed" and "Relaxed".

Keywords: Machine learning · Psychological scale · Recommendation systems · TIPI

1 Introduction

When we search for and consume music, we prefer to listen to music frequently connected with our unique emotional states. For example, when we are down, we may choose a piece of classical music by Beethoven. Similarly, when we are happy, we may prefer pop music by the likes of Taylor Swift [4]. With the growing number of choices regarding music on various platforms, users can often feel overwhelmed. Recommendation systems are thus effective in helping users navigate through various types of content, discover new music, and provide a more personalized experience. These systems extract information from users and use it to provide personalized recommendations. Most other music recommendation systems use more computationally intensive algorithms like Collaborative Filtering [11], Support Vector Machines, K-means clustering [12], Neural Networks [1],

D. Pastor-Escuredo et al. (Eds.): ICDLAIR 2023, LNNS 1001, pp. 691–700, 2024.
https://doi.org/10.1007/978-3-031-60935-0_60

etc. However, many of these are unsuitable for new users with no pre-existing preferential data or those who interact less frequently with the system, also known as the cold start problem [10]. In such cases, using more general information about the users, such as their age, gender, etc., may help such systems to infer which sort of music they are likely to prefer. Additionally, while the type of music preferred by a user can help ascertain their personality, we can similarly use the personality traits of a user to determine their taste in music. For this purpose, we can use psychological scales to determine a user's temperament, and then use that data for recommending music more generally to them. This process can be tuned as more of the user's musical preferences are revealed as they continue interacting with the system.

Typically, personality is defined as a combination of features, recurring behavioral patterns and personal processes that form a person's unique character, and frequently affects their interests and choices [3,9]. Mood is a transient state of consciousness or emotion. Even though emotions are undoubtedly involved, mood is typically more constant than a single emotion and should not be equated with them. The combination and interaction of a person's ideas, feelings, and behaviors make up their personality, and once created, it often lasts a lifetime and is more stable or consistent, unlike mood. Personality traits have been used to improve the performances of recommendation systems in deducing users' preferences, as has mood, but they are not commonly combined to recommend music to the user base of a system.

1.1 Our Contributions

While research has been conducted extensively regarding the connections between our musical tastes and our emotions, a computational procedure identifying both the user's general mood and personality traits would be beneficial in studying this link. Our approach to evaluating systems using such a method was to compare numerous models making personalized recommendations dependent on the demographic data, current mood, and combined with their Big-Five personality traits, as determined by a psychological scale known as the Ten-Item Personality Inventory (TIPI) scale.

In this paper, we describe a music recommendation system that utilizes a combination of the results of a psychological scale and the general emotions of a user. A dataset has been built from scratch for this study, and the TIPI scale has been incorporated into the dataset to determine their Big-Five personality traits. Classification models have been trained and tested on the dataset to generate results for each mood, which are then compared using classification metrics to determine the best model. Additionally, we aim to determine if such personalized recommendations can effectively recommend music instead of more complex algorithms, while also solving issues like the cold start problem.

The paper has been organized to first look into any related work, describe the dataset collected and the methodology used, followed by an analysis of the results, and finally, the conclusions that can be drawn from our findings.

2 Related Work

In this section, we briefly review several recent papers that have used machine-learning or deep-learning approaches to design recommendation systems based on mood recognition or other methods.

Some have chosen to use sentiment analysis to detect a user's mood. In [1], the system recommended music and movies based on the mood of the user by using SentiWordNet to perform sentiment analysis on user posts. An NLP was also used to perform a separate sentiment analysis on the lyrics of the songs, and the songs were categorized to suit different moods, after which the Random Forest algorithm then classified the lyrics into different groups (happy, sad, joyful). Finally, the system recommended the user movies and music of these specific categories based on their posts.

Multiple papers chose to employ physiological signals to determine a user's emotions. [2] designed an emotion-based music recommendation framework that uses physiological signals to learn the emotion of a user. The user's emotion was classified using a wearable computing device combined with physiological sensors, and this data was then fed to a collaborative or content-based recommendation engine as arousal and valence scores. Their results suggested a high correlation between emotion and the GSR PPG signals. They obtained higher accuracy rates for only GSR at 71.53% in arousal prediction and 71.04% in valence prediction.

Similarly, [3] explored music recommendation systems with multi-modal information sources, including physiological signals of listeners and acoustic features in music. These signals included users' personality traits as determined by the TIPI scale, and physiological signals registered through a wearable wristband. Based on the collected dataset, a machine-learning study with 4 regression algorithms (Decision Tree, LightGBM, Neural Network, and Support Vector Regression) was conducted to compare model performances across various combinations of feature sets. According to their results, the DT regression model attained the best performance among the models on feature group 3 (Combining user personalities, music acoustic features, and physiological signals), with an RMSE score of 0.718.

In a similar vein, [4] proposed a music recommendation system based on the personality traits, emotions, and moods of a single user, which were recognized through the behaviour analysis of users in certain social environments. The myPersonality5 and PsychoFlickr datasets were used to evaluate the effectiveness of user personality recognition, while the Deezer dataset was used to classify mood detection and recommendation. The system classified user personalities based on the Big Five personality traits- Openness, Conscientiousness, Extraversion, Agreeableness, and Neuroticism (OCEAN)- by comparing the performances of the Naive Bayes, Logistic Regression, SVM, etc. and networks such as AlexNet, ResNet50, etc. Among the models, Logistic Regression and Random Forest both performed the best in classifying the personality traits of the users.

Face recognition systems utilizing CNNs were also frequently used to extract emotions, as seen in [5]. The authors suggested a technique that would create a playlist for the user after extracting their facial expressions using a CNN. The user would give input into the text or the speech format, which was then converted for analysis to a server. The server then calculated the emotions and searched for keywords to return songs under the cooperation filter. The CNN facial micro-expression model obtained a 62.1% recognition rate in the dataset, following multiple parameter adjustment tests.

Likewise, [6] proposed a mood-based music player which could perform real-time mood detection via the system's camera and suggest a playlist of songs as per the detected mood. The FaceDetector class was used to detect faces from the images of their collected dataset, after which MobileNet was used with Keras to train and test their model in detecting the facial emotions among seven classes - happy, sad, neutral, surprise, angry, fear and disgust. Based on their results, their model achieved about 75% overall accuracy. However, it faced some limitations due to its limited dataset, namely overfitting, which led to fluctuating accuracy levels.

[7] captured a user's facial expressions and classified them into certain moods using a CNN. A music recommendation module then matched the detected moods to a corresponding playlist in the database. An image dataset from Kaggle was used to train the CNN, while an assortment of Bollywood songs was sorted for each mood and used to build the dataset. The validation and testing accuracy scores achieved by the CNN were 95% and 71% respectively. [8] proposed an emotion-aware hybrid approach for recommending music through deep learning (emoMR) that combined the approaches of content-based recommendation with that of collaborative filtering. It exploited the low-level features of the audio signals and metadata in music to predict the representations of emotion as a continuous form. The emoMR models' accuracy for the emotion representation model was 97.46%, while the accuracy of emotion state representation model was 88.81%.

3 Proposed Work

3.1 Data Collection

The dataset has been built using a survey that collected demographic data and personality attributes from survey respondents. The survey included the 10-item inventory developed for TIPI to evaluate the Big Five personality traits among respondents, where they were expected to rate how they viewed themselves from a scale of 1 to 7 for each item. It also included questions regarding their gender, age, region of living (city, suburbs, village, etc.), current profession (student, unemployed, job-holder), and preference levels for specific genres(Pop, Rock/Metal, Hip-hop, Dance/Electronica, Classical/Instrumental, Indie/Acoustic, Blues/Jazz, Country) when in certain moods (Happy, Gloomy, Stressed, Energetic, Relaxed).

The survey was posted on various online forums such as different Facebook groups, particularly those popular among university students, to ensure that the sample was chosen randomly. 150 responses were ultimately collected for the dataset, after which the responses were split into 5 rows for each mood with Genre being the target variable. Five more variables (Extraversion, Agreeableness, Conscientiousness, Emotional Stability, Openness to Experiences) were then added to the dataset using the 10-item inventory, as explained in the TIPI section. The dataset used for the project is cross-sectional, and has been displayed in Fig. 1.

Age	Gender	Region	Professior	Extraversi	Agreeable	Conscienti	Stability	Openness	Happy	Gloomy	Stressed	Energetic	Relaxed
23	Female	City	Student	2.5	3	4	5	5	Pop	Indie/Aco	Pop	Hip-hop	Classical/Instrumental
23	Female	City	Student	4.5	2	2	4.5	2	Dance/Ele	Indie/Aco	Blues/Jazz	Rock/Met	Rock/Metal
24	Female	City	Job-holde	5	7	4.5	6.5	1.5	Dance/Ele	Indie/Aco	Pop	Dance/Ele	Indie/Acoustic
23	Female	City	Student	5	7	6.5	5	7	Dance/Ele	Blues/Jazz	Indie/Aco	Dance/Ele	Indie/Acoustic
22	Male	City	Student	2.5	3	5	3	3	Pop	Rock/Met	Hip-hop	Dance/Ele	Blues/Jazz

Fig. 1. Dataset used for the study. The columns Extraversion, Agreeableness, Conscientiousness, Emotional Stability, and Openness to Experiences have been derived from the 10-item inventory.

3.2 TIPI Scale

TIPI (Ten-Item Personality Inventory) is a brief measure of the Big-Five personality attributes with somewhat reduced psychometric properties, as often observed in such relatively shorter measures. While many rating instruments have been developed to measure the Big-Five dimensions, most are too lengthy for general research purposes [13]. For this reason, we have chosen TIPI to assess and evaluate the personality traits of the survey respondents.

According to the Big-Five framework, most individual differences in human personality can be categorized into five major domains: Extraverted (sociable and enthusiastic, NOT reserved or quiet), Agreeable (generous and sympathetic, NOT aggressive or cold), Dependable (responsible, NOT careless or impulsive), Emotionally stable (relaxed and self-confident, NOT anxious or easily upset), and Open to experience (curious and open-minded, NOT conventional). To determine the TIPI score, 10 questions (Q1-Q10, as publicly available on [15]) are asked to the respondent, and each Big-Five personality domain is represented by an item and its corresponding reverse-scored item, i.e., Q1 with Q6 (Extraverted), Q7 with Q2 (Agreeable), Q3 with Q8 (Dependable), Q9 with Q4 (Emotionally stable), and Q5 with Q10 (Open to experience).

To calculate one of the Big-5 scores, we must first recode the reverse-scored items (Q2, Q4, Q6, Q8, and Q10) with their corresponding values (i.e., 7 with a 1, 6 with a 2, 5 with a 3, etc., while 4 remains as is). We then average the standard items with the recoded reverse-coded items to calculate the TIPI score. For example, using the Extraversion scale, suppose a participant scores 5 on Q1 (Extraverted, enthusiastic) and 2 on Q6 (Reserved, quiet). We first recode reverse-scored item Q6 by replacing the 2 with a 6. We then average the Q1 score of 5 and the recoded Q6 value of 6. Hence, the Extraversion score on the TIPI scale would be $(5 + 6)/2 = 5.5$.

3.3 Dataset Description

To understand the demographics, trends, relationships, and correlations in the dataset, Exploratory Data Analysis (EDA) was applied to the dataset before modeling it. For each of the five different moods, a correlation could be seen with the preferred genres of the participants, as shown in Fig: 2. For example, for the mood Happy, participants predominantly preferred upbeat music from the Pop or Dance/Electronic genres. Conversely, for the mood Gloomy, participants opted for soft or soothing music, such as Classic or Indie/Acoustic.

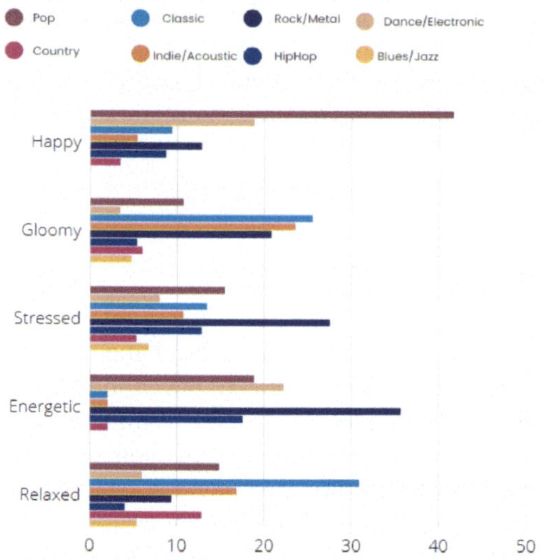

Fig. 2. Proportion of participants preferring certain genres (e.g., Pop, Jazz, etc.) for each mood (Happy, Gloomy, etc.)

Among the 150 survey participants, 54% were female, 44.67% were male, and 1.33% were of other genders or preferred not to say. The average age of the participants was around 23.7 years. Most participants were city dwellers, with 89.9% of the participants living in cities, 5.4% in suburbs, and only 4.7% in villages. Regarding the profession of the participants, 70.67% were Students, followed by Job-holders at 24.0%. Only a small percentage of the respondents, about 5.33%, were Unemployed.

For the five personality dimensions, the average score for Extraversion was 3.69, 4.81 for Agreeableness, 4.45 for Conscientiousness, 4.25 for Emotional Stability, and 4.8 for Openness. The mean scores of each dimension for the three genders have been displayed in Fig. 3.

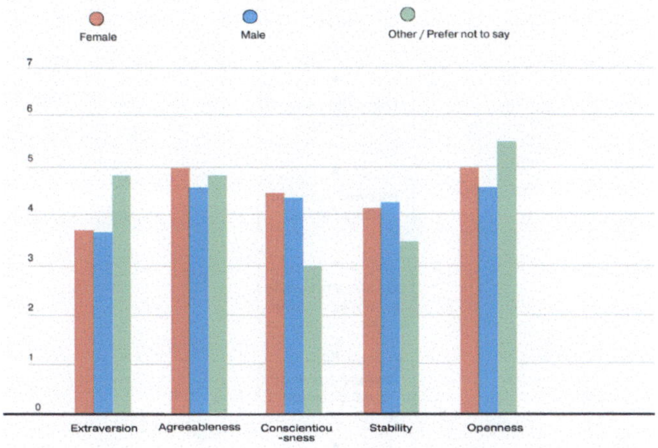

Fig. 3. Average score of each gender corresponding to each personality dimension.

3.4 Data Preprocessing

The data was cleaned and the categorical variables were encoded for the models to work with them easily. During the data analysis, we observed that the collected data was imbalanced, which could result in biased or inaccurate recommendations. The imbalance was mostly seen in the genre categories, which were subsequently balanced using an over-sampling method known as Synthetic Minority Over-sampling Technique (SMOTE) [14]. The SMOTE over-sampling was used to add synthetic data points to the minority class of the preferred genre for each mood. The newly generated data points increase data diversity and balance the class distribution, which helps improve the performance of the classifiers. The SMOTE algorithm was thus implemented on all 5 moods before the data was fed to the model for generating recommendations.

3.5 Recommendation Generation

We have employed a supervised classification algorithm called Lazy Classifier to assess the performance of several machine learning algorithms on the dataset without the extensive training needed for each of them. This approach was used in the genre prediction for each mood, providing a baseline to compare the performance of several machine learning algorithms on the testing dataset. A flowchart for the proposed final working sequence of the entire system has been displayed in Fig. 4. Performance metrics such as accuracy, F1-score, etc. were compared for each algorithm to determine which best predicted music genres to users.

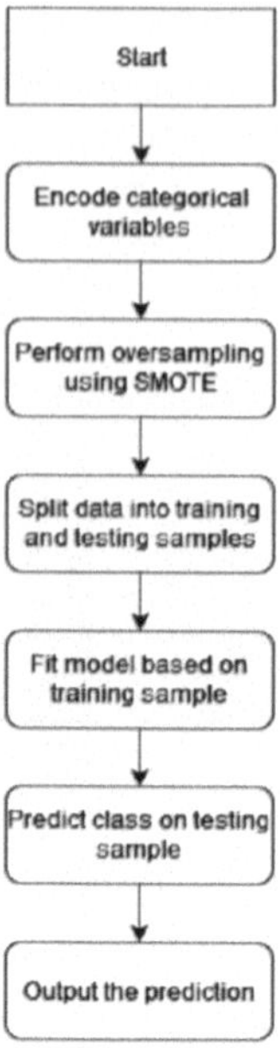

Fig. 4. Working sequence of the proposed recommendation system.

4 Result Analysis

The performances of the top three classifiers for each mood have been summarised in Fig. 5.

Based on the results, ensemble algorithms achieved the highest overall accuracy for the models predicting genres when the user is Happy', with ExtraTreesClassifier at 91.0%, RandomForestClassifier at 89.0%, and BaggingClassifier at 87.0%. Semi-supervised algorithms like XGBClassifier, LabelPropagation, and LabelSpreading all followed at 85.0%.

For the models predicting genres when the user is Gloomy', ensemble algorithms once again achieved the highest overall accuracy, with BaggingClassifier and RandomForestClassifier at 69.0%, and ExtraTreesClassifier at 67.0%. LGBMClassifier, XGBClassifier, and NuSVC all followed at 66.0%.

As for the models predicting genres when the user is Energetic', LGBMClassifier acheived the highest accuracy at 79.0%, followed by ExtraTreesClassifier at 77.0%, NuSVC at 76.0%, and XGBClassifier and DecisionTreeClassifier at 75.0%. On the other hand, among the models predicting genres when the user is Stressed', LabelPropagation, LabelSpreading, and ExtraTreesClassifier all achieved the highest accuracy at 71.0%, followed by LGBMClassifier and XGBClassifier, both at 67.0%.

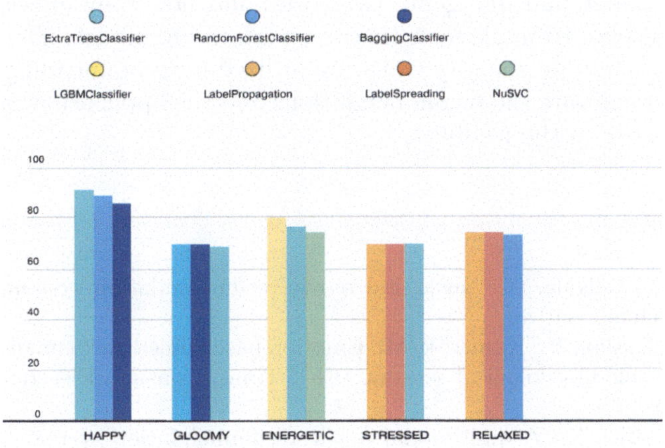

Fig. 5. Bar charts comparing accuracy scores of the top 3 best-performing models for each mood.

Finally, for models predicting genres when the user is Relaxed', LabelPropagation and LabelSpreading achieved the highest overall accuracy at 74.0%, followed closely by RandomForestClassifier at 73.0%, BaggingClassifier at 70.0%, and ExtraTreesClassifier at 69.0%.

5 Conclusion

With the abundance of options on numerous music platforms, users can often feel overwhelmed; recommender systems are thus effective in helping them navigate through various kinds of content, discover new music, and offer a more tailored experience. Although much research has been conducted on the relationships between the music we like and our emotions, a computational technique that can identify a user's overall mood at a specific time of day may be helpful in understanding this connection.

Based on our results, ensemble algorithms such as RandomForestClassifier, BaggingClassifier, ExtraTreesClassifier, and LGBMClassifier performed best most frequently, achieving the highest accuracy scores for the moods of Happy, Gloomy, and Energetic. On the other hand, semi-supervised algorithms such as LabelPropagation and LabelSpreading achieved the best scores for the moods of Stressed and Relaxed. Given that the models perform relatively well for most moods, we can conclude that recommendation systems are able to generate better recommendations for their users based on a combination of their demographic and psychological information. Furthermore, if such systems are to utilise these factors in their recommendations, issues such as the cold start problem can eventually be minimised.

In the future, we hope to grow the size of the dataset and incorporate more genres and other moods into the study, while also considering the nuances of music falling under multiple genre categories and how they affect user preferences. We also plan to implement the best performing models into applications where user data and personality traits are utilized in recommending music, after which the user can rate the recommendations and the application can alter their predictions based on the feedback.

References

1. Deshmukh, P., Kale, G.: music and movie recommendation system. Int. J. Eng. Trends Technol. (2018)
2. Deger, A., Yaslan, Y., Kamasak, M.: Emotion based music recommendation system using wearable physiological sensors. IEEE Trans. Consum. Electron. **64**(2), 196–203 (2018)
3. Ruilun, L., Hu, X.: A multimodal music recommendation system with listeners' personality and physiological signals. In: Proceedings of the ACM/IEEE Joint Conference on Digital Libraries in 2020 (2020). https://doi.org/10.1145/3383583.3398623
4. Moscato, V., Picariello, A., Sperli, G.: An emotional recommender system for music. IEEE Intell. Syst. **36**(5), 57–68 (2020)
5. Vidyasagar, S.B., Karwande, V.: Emotion based music recommendation system by using different ML approach. Int. J 6(6) (2021)
6. Mahadik, A., et al.: Mood based music recommendation system. Int. J. Eng. Res. Technol. (IJERT) **10** (2021)
7. Athavle, M., et al.: Music recommendation based on face emotion recognition. J. Inf. Electr. Electron. Eng. (JIEEE) **2**(2), 1–11 (2021)
8. Wang, S., et al.: A novel emotion-aware hybrid music recommendation method using deep neural network. Electronics **10**(15), 1769 (2021)
9. Schedl, M., Zamani, H., Chen, C.W., et al.: Current challenges and visions in music recommender systems research. Int. J. Multimedia Inf. Retrieval **7**(2), 95–116 (2018)
10. Briand, L., Salha-Galvan, G., Bendada, W., et al.: A semi-personalized system for user cold start recommendation on music streaming apps. In: Proceeding of 7th ACM SIGKDD Conference on Knowledge Discovery & Data Mining, pp 2601–2609 (2021). https://doi.org/10.1145/3447548.3467110
11. Shakirova, E.: Collaborative filtering for music recommender system. In: 2017 IEEE Conference of Russian Young Researchers in Electrical and Electronic Engineering (EIConRus)
12. Kim, D., Kim, K., et al.: A music recommendation system with a dynamic k-means clustering algorithm. In: 6th International Conference. https://doi.org/10.1109/ICMLA.2007.97
13. Gosling, S., Rentfrow, P., Swann, W., Jr.: A very brief measure of the Big-Five personality domains. J. Res. Pers. **37**, 504–528 (2003)
14. How to Deal with Imbalanced Datasets with Smote Algorithm. https://www.turing.com/kb/smote-for-an-imbalanced-dataset
15. Ten Item Personality Measure (TIPI). https://gosling.psy.utexas.edu/scales-weve-developed/ten-item-personality-measure-tipi/

Bidirectional Long-Short Term Memory Based Approach for Sampling in Dual Arm Locomotion

Deepankar Pande[1(✉)] and Mayur Sawant[2]

[1] Department of Manufacturing Engineering and Industrial Management, COEP Technological University, Pune 411005, Maharashtra, India
pandedc21.mfg@coep.ac.in

[2] Department of Mechanical Engineering, MIT ADT University, Pune 412201, Maharashtra, India

Abstract. Bimanual movement involves coordinated movement of both limbs, a multifaceted challenge in robotics and biomechanics. The efficiency of movement planning in bimanual movement is of utmost importance in many applications, from developing humanoid robots to designing rehabilitation devices. This paper presents an innovative method that uses Bi-LSTM (bidirectional long-short-term memory) networks to improve motion sampling strategies for bimanual movements. The proposed approach uses the temporal dependencies and bidirectional contextual information provided by Bi-LSTM networks to generate adaptive and naturalistic movement trajectories, effectively capturing the complex nuances of interlimb coordination. Results highlight the superior performance of our method over existing techniques. This study provides a promising solution to the complexities of movement planning for bimanual movement, which has wide applications in robotics, assistive devices, and healthcare. In addition, it is strategically aligned with an emerging trend in the manufacturing industry - the integration of collaborative robots (cobots) with 5 and 7 degrees of freedom - which improves production flexibility and agility. At the same time, it reiterates the core principles of Industry 4.0 and emphasizes the development of intelligent, interconnected production systems that use advanced robotics and automation to increase work efficiency and adaptability in the manufacturing environment.

Keywords: Bimanual Locomotion · Bidirectional Long-Short Term Memory · Deep Learning · Robotic Coordination · Industry 4.0 · Collaborative Robots

1 Introduction

The technique of dual-arm locomotion or manipulation is utilized in robotics to execute tasks by employing two robotic arms in a synchronized manner. This technique coupled with collaborative robots finds its application in various sectors,

D. Pastor-Escuredo et al. (Eds.): ICDLAIR 2023, LNNS 1001, pp. 701–713, 2024.
https://doi.org/10.1007/978-3-031-60935-0_61

such as manufacturing, healthcare, and home automation [1]. Two main categories within dual-arm manipulation exist: uncoordinated and coordinated. Further subcategorizing coordinated manipulation reveals two distinct approaches: bimanual and goal oriented. Bimanual manipulation refers to using both arms together in a synchronized manner to complete tasks, often requiring careful coordination. On the other hand, goal-oriented manipulation involves utilizing both arms to achieve specific objectives [2]. Dual arm manipulation allows robots to execute complex and delicate tasks that demand a high degree of accuracy and precision. For robots to engage in dual arm manipulation, they must possess the capability to strategize and carry out synchronized movements of both arms, all while considering the precise placement and alignment of the objects being manipulated. This process is known as robot motion planning. It involves finding a safe and obstacle-free route for a robot to travel from its initial position to a desired endpoint within specified surroundings. The aim is to discover a route that fulfils all limitations, like evading obstacles and achieving specific goals, such as reducing the path length, or the time consumed. This is known as path planning.

When it comes to robotics, path planning can be classified into two main categories of heuristics-based algorithms: search-based and sampling-based methods. Search-based algorithms systematically explore the configuration space to find a feasible path from a start point to a goal, often using techniques like A* search. On the other hand, sampling-based algorithms, such as Rapidly exploring Random Trees (RRT), generate random configurations probabilistically, connecting them to form a path. These algorithms provide flexibility when dealing with spaces that are high-dimensional and complex. The choice between these two approaches depends on the specific requirements and characteristics of the robotic task. A* is suitable for problems with well-defined heuristics and known environments, making it ideal for grid-based maps and known terrain. However, it can be computationally expensive for large, high-dimensional state spaces and may encounter challenges in dynamic or changing environments. RRT is particularly effective in handling unknown and dynamic environments as it quickly explores uncharted areas, adapting to new obstacles and changes in the environment [3]. It does not guarantee optimality, but it converges towards an optimal solution with increasing samples. RRT is well-suited for robots and vehicles with non-holonomic constraints, allowing them to explore feasible paths in challenging environments. Which more concisely mimics real life problems [4]. Meta heuristic approaches to path planning such as Genetic Algorithm, Differential Evolution, Particle Swarm Optimization, and Cuckoo Search Algorithms are also explored by researchers [5].

Two main types of sampling are commonly used in sampling-based algorithms: random sampling and goal-biased sampling. Configuration points are generated in the search space without any particular bias when conducting random sampling. This approach encourages exploration of the entire space, as shown in Fig. 1. In sharp contrast, goal-biased sampling takes a strategic approach by generating points that are more likely to be in proximity to the desired goal. This technique aims to speed up the process of reaching the solution by focusing on points that can lead to faster convergence. Focusing on specific regions that are more likely to contain the desired path enhances efficiency in finding a feasible or optimal path. The selection of sampling methods often relies on the particular problem and the desired trade-off between exploration and exploitation.

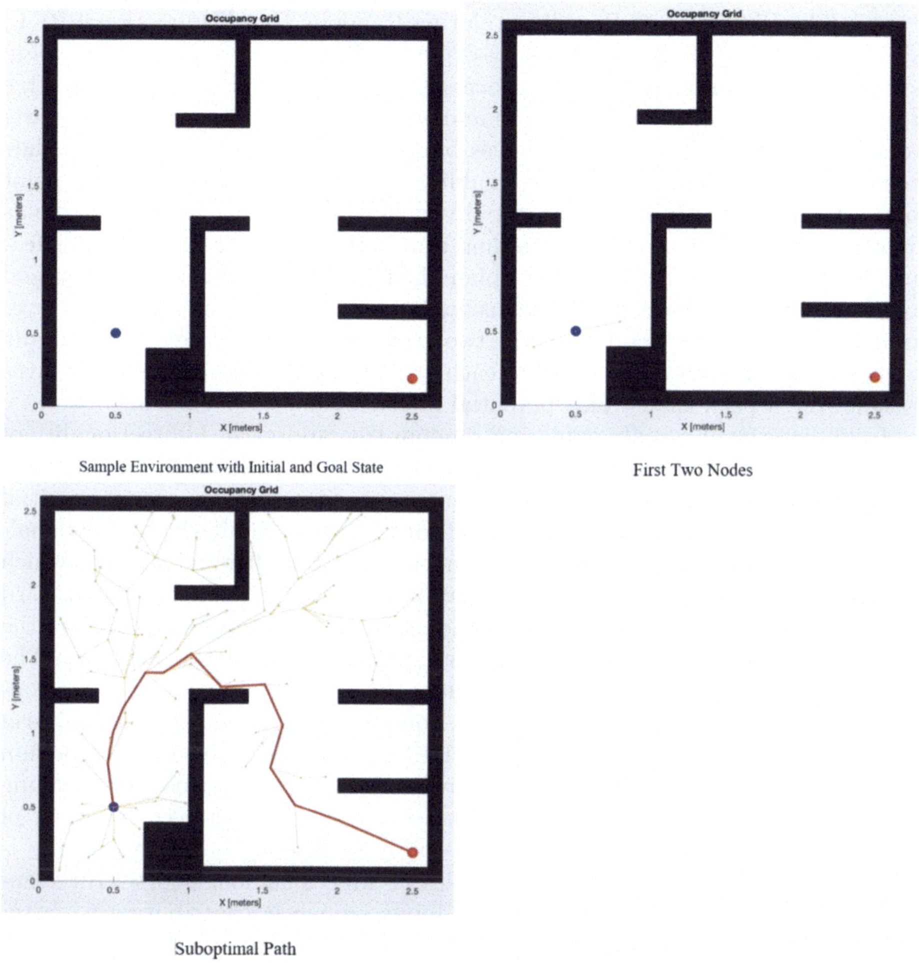

Fig. 1. Suboptimal path Generation with RRT using Random Sampling

Sampling efficiency in path planning algorithms can be improved by incorporating deep learning models like Long Short-Term Memory (LSTM) to enhance goal-biased sampling. LSTM can learn the underlying structure of the configuration space and efficiently bias the sampling process toward the goal. LSTM's predictive power can be enhanced by training it on historical data, enabling it to anticipate spatial areas more inclined to have a viable path [6]. By leveraging learned patterns, this goal-biased sampling technique expedites the convergence of path-planning algorithms, allowing robots to discover optimal paths towards their intended targets quickly. LSTM is a valuable tool for optimizing goal-biased sampling strategies in motion planning because it can capture complex dependencies and temporal information.

For goal coordinated dual arm manipulation, bidirectional RRT is more effective [7]. The Bidirectional Rapidly-exploring Random Tree (Bidirectional RRT) is a variant of the widely used Rapidly-exploring Random Tree (RRT) algorithm designed for motion planning. The process grows two trees at once, with one tree starting from the initial configuration and advancing, while the other tree starts from the goal configuration and moves backwards. When the trees meet, they connect in the configuration space, enabling a pathway from the starting point to the desired goal. Bidirectional Rapidly-exploring Random Tree (RRT) holds great potential as a valuable tool in dual-arm path planning scenarios. This feature lets both arms simultaneously plan their paths, increasing efficiency and coordination in intricate, multidimensional environments. This method proves to be particularly beneficial when the two arms must collaborate or prevent any possible collisions. It guarantees that each arm achieves its designated objectives by following a path that avoids potential interferences.

In bi-directional rapidly exploring random tree approach, bidirectionality of the algorithm is a one of the aspects that makes it more suitable for dual arm manipulation as described previously, that one arm starts to move from initial configuration while other starts from goal configuration. Similarly, bidirectionality of bi-directional long-short term memory algorithm offers advantages which includes incorporating contextual information by processing data in both forward and backward directions. This enables the model to simultaneously consider the entire path from start to end configurations, enhancing prediction accuracy. The model's proficiency in handling long sequences and maintaining long-range connections is particularly advantageous in complex path scenarios. In dual arm manipulation path planning, bi-directional long short-term memory algorithm is employed due to its ability to comprehend sequence relationships, leading to more effective sampling. The significance lies in the improved handling of long paths or intricate routes. Additionally, the efficiency boost in learning is notable when all time steps are available, as bi-directional long short-term memory algorithms train two LSTMs on the input sequence, contributing to faster learning. This approach is vital in dual-arm manipulation path planning, expanding workspace possibilities for complex manipulations beyond the capabilities of single-arm systems.

Going forward the paper progress with Sect. 2 for literature review, Sect. 3 for proposed methodology which consist for dataset specifications and Bi-LSTM projector architecture, Sect. 4 discussing the results and scope for future work.

2 Related Work

Elbanhawi, et al. has highlighted importance of sampling-based methods. It recognizes the challenges in motion planning and highlights the efficiency of sampling-based techniques in addressing these complexities. It concludes sampling is more efficient than search-based algorithms [8]. Ferguson, et al. has introduced "Anytime RRTs,". It utilizes rapidly exploring random trees (RRTs) to navigate such spaces. Unlike traditional algorithms, Anytime RRTs incrementally refine solutions by building upon information from previous trees, enhancing path quality. Biasing techniques expedite initial solutions, with ongoing improvements, ensuring subsequent solutions are better than previous ones within a user-defined improvement bound. Empirical evaluations in single-robot and multi-robot planning domains demonstrate the algorithm's versatility and real-world applicability [9]. Karaman, et al. has introduced PRM* and RRT*, which are provably asymptotically optimal, guaranteeing convergence to optimal solutions. These algorithms balance optimality and efficiency, and the analysis is grounded in connections between stochastic sampling-based algorithms and the theory of random geometric graphs [10]. Noreen, et al. has introduced RRT*-adjustable bounds (RRT*-AB), an extension of the RRT* algorithm, to enhance path planning in cluttered environments. RRT* offers probabilistic completeness and asymptotic optimality but has limitations like slow convergence and dense sampling. RRT*-AB addresses these issues with innovative strategies, including connectivity regions, goal-biased bounded sampling, and path optimization [11].

Devaurs, et al. has introduced the Multi-T-RRT, an extension of the Transition-based Rapidly-exploring Random Tree (T-RRT) algorithm, designed to address path planning challenges in continuous cost spaces. It leverages the properties of trees and parallel exploration to navigate continuous cost spaces more efficiently. Through rigorous evaluations and comparisons with other algorithms, the paper demonstrates the Multi-T-RRT's ability to outperform existing methods regarding efficiency and solution quality [12]. Kala. Et al, has presented a novel approach to robot path planning by generalizing Rapidly Exploring Random Trees (RRT) and Probabilistic Roadmaps (PRM) into Rapidly Exploring Random Graphs for multiple mobile robots. The essential contribution is simultaneous exploration by multiple robots, leading to a well-connected graph structure that enhances exploration efficiency-underscoring its practical relevance and efficiency in real-world scenarios [13]. Qureshi, et al. has introduced Intelligent Bidirectional Rapidly Exploring Random Trees (IB-RRT*), a novel variant of the RRT* algorithm designed for optimal motion planning in complex and cluttered environments. IB-RRT* leverages the bidirectional tree approach and an intelligent sample insertion heuristic informed by uniform sampling heuristics. The paper's theoretical analysis and empirical results demonstrate the superior efficiency of IB-RRT* compared to RRT* and B-RRT* in navigating complex environments. This empirical validation highlights the real-world applicability and advantages of IB-RRT* over existing techniques for motion planning [14].

Qureshi, et al. has introduced Motion Planning Networks (MPNet), a novel neural planner that bridges the gap between learning-based neural networks and classical motion planning techniques. MPNet employs neural networks to generate efficient paths between initial and goal configurations, using a bidirectional approach for both direct connections and near-optimal solutions [15]. Watt, et al. introduced PathNet, a deep neural network-based solution for 2D motion planning. PathNet efficiently generates obstacle-avoiding trajectories within complex environments by compressing environment maps into a latent vector and seamlessly connecting start and goal positions. A significant strength of PathNet is its adaptability to previously unseen environments, enhancing its practical utility in real-world applications. Notably, the network's computational efficiency, requiring only a single pass to generate trajectories, is advantageous for real-time decision-making in robotics and autonomous systems. Integrating deep learning with motion planning opens doors for applications in autonomous vehicles, robotics, and other domains requiring efficient and adaptable path planning [16]. Qureshi, et al. This research paper introduces Deep Sampling-based Motion Planner (DeepSMP), a novel neural network-based adaptive sampler designed to accelerate Sampling-based Motion Planners (SMPs). DeepSMP significantly enhances speed and scalability while maintaining efficiency in higher-dimensional scenarios [17]. Sherstinsky, et al. has provided a comprehensive and in-depth understanding of Recurrent Neural Networks (RNNs) and Long Short-Term Memory (LSTM) networks. It goes beyond superficial explanations by formally deriving the canonical RNN formulation from differential equations, enhancing comprehension and establishing a solid foundation [18]. Cui, et al. This paper introduces a deep stacked bidirectional and unidirectional LSTM (SBU-LSTM) neural network architecture for network-wide traffic speed prediction. It leverages forward and backward dependencies in time series data, incorporating bidirectional LSTM (BDLSTM) layers to capture spatial features and bidirectional temporal dependencies from historical data. Comparative analyses with classical and state-of-the-art models demonstrate that the SBU-LSTM neural network achieves superior prediction accuracy and robustness for the entire traffic network [19].

Chitta, et al., has provided a tutorial for capabilities of MoveIt plugin in ROS for validation of generated path on physical hardware [20].

3 Proposed Work

The dataset is generated and then solved using RRT* algorithm and used as training and testing of Bi-LSTM projector. 70% of the dataset is reserved for training and rest 30% is used for testing of the algorithms performance.

3.1 Dataset Generation

Here, for the training and testing of the LSTM projector, the dataset of 40 units by 40 units workspace is populated with 7 squares of 5 units' side. The reasoning behind choosing the dimension, if far too small obstacles are chosen, it increases the probability of existence of a straight line joining initial and goal stage. If the size and number of obstacles is taken too large the dataset might contain an increasing number of instances where no path between initial and goal state exists, as initial state and goal state are completely separated by obstacles Examples of which are given in Figure [2]. The dataset mainly consists of two types of work envelope, varying sets of obstacles with varying start and goal point and another one varying sets of start point and goal point in fixed set of obstacle locations. They are shown in Fig. 3 and Fig. 4. Five fixed obstacles dataset were generated, and 100 different start and end points were part of each fixed obstacle dataset.

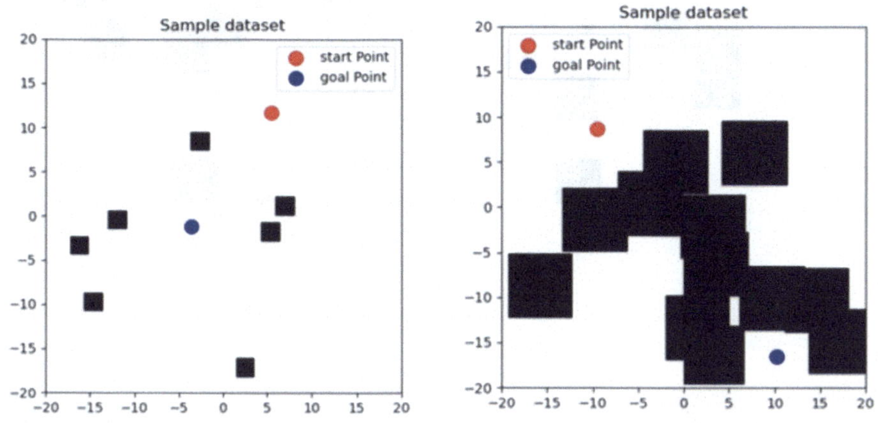

Small Obstacle Dataset Example

Large and higher no of obstacle

Fig. 2. Direct line of sight and No path possible condition in dataset

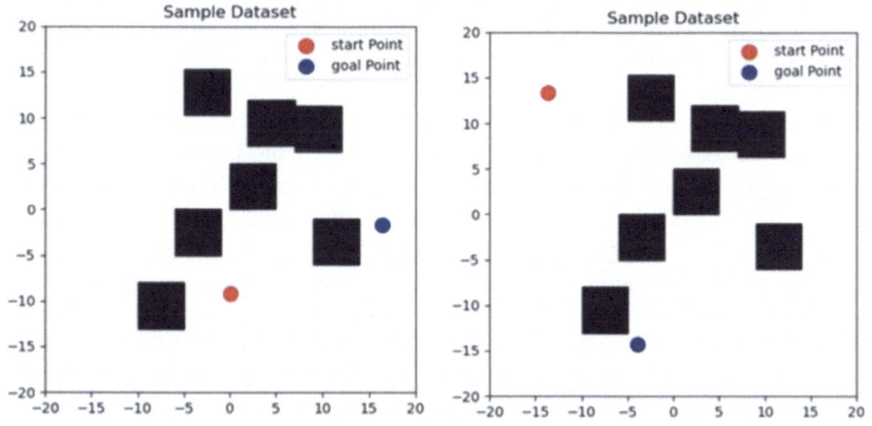

Fig. 3. Fixed obstacles with varying Start and Goal points

For the dataset generation, the scope of the workspace, dimensions and number of obstacles are defined. The dataset generation process can be explained as follows.

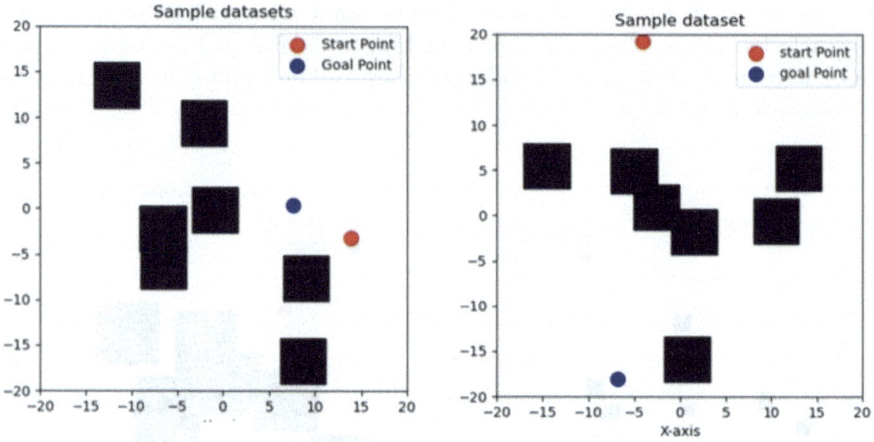

Fig. 4. Varying obstacles with varying start and goal points

At first, initialization of the Search Space Class is performed. In the SearchSpace class, an instance of the class is created to represent the search space. It involves specifying the dimensions of the space and optionally providing a list of obstacles. In the second step, R-tree Initialization and Obstacle Addition are done. Here, the SearchSpace class initializes an R-tree with properties such as dimensionality. If obstacles are provided, the class ensures they are valid and conform to the dimensionality of the search space. The obstacles are added to the R-tree using a dedicated function, which yields obstacle information for efficient insertion into the R-tree. Here 'rtree' library from Python is used.

The third step is obstacle generation. This step generates random obstacles within the search space with the obstacle-generating function. This function takes parameters such as edge length and n for the number of obstacles to be generated. It uses a while loop to create obstacles until the desired number is achieved. For each obstacle, it randomly generates a centre and allocates edge lengths for each dimension. In this case, 2D. Each obstacle is generated as a hyperrectangle by randomly determining the centre and specified edge lengths for each dimension. The generated obstacle centres are checked to ensure they are distinct from existing obstacle centres. The fourth step is start and end point generation. Here, start and end points are generated randomly with a function. The points are checked to ensure they do not lie within the obstacle hyperrectangles generated in the previous step. The last step involves adding obstacles to the R-tree. Valid obstacles plus start and goal point pairs are added to the R-tree using a generator function that yields tuples containing a unique identifier (UUID), the obstacle itself, and the obstacle's bounding box. This generator is designed to add obstacles to the R-tree one by one efficiently. Each start and goal point pair is represented by a unique identifier, its x and y coordinates.

The R-tree efficiently organizes these obstacles and starts goal points for spatial querying.

3.2 Finding Solution to the Dataset Using RRT*

The RRT* algorithm explores the configuration space and finds feasible paths from the start point to a goal point. As the RRT* algorithm explores, it constructs a tree structure representing possible trajectories consisting of vertices and edges. Each edge in the tree represents a path segment between two consecutive path points. This tree structure encapsulates successful paths through the environment, navigating around obstacles to connect the initial and goal states. These paths are then retained as a vital component of the dataset. The paths in the dataset are stored as sequences of path points. The second step involves encoding the obstacle configurations and corresponding trajectories into a format suitable for training a deep learning model, specifically a bidirectional long short-term memory model. Each obstacle configuration is transformed into a suitable representation (e.g., hyperrectangle vertices), and this information is encoded. Additionally, the paths obtained from the RRT* exploration are used to create trajectory data, capturing the sequence of path points that form a successful path. The encoded obstacle configurations and trajectories are combined into a comprehensive dataset. This dataset is subsequently employed to train a deep learning model, allowing the model to learn the relationships between obstacle configurations and the corresponding successful paths.

3.3 Training of the Bi-directional Long Short Term Memory Projector

During the training, the Bi-directional long short term memory model undergoes an iterative optimization process over the dataset for a predefined number of epochs. In each epoch, the model is updated using stochastic gradient descent (SGD) to minimize the disparity between the predicted next path point and the following path point from the dataset. The training objective involves employing a regression loss, such as Mean Squared Error (MSE), to quantify the dissimilarity between the predicted and actual path points. The model's parameters are adjusted through backpropagation, and this training loop repeats for multiple epochs until the Bi-directional long short term memory model converges, effectively capturing the underlying relationships within the obstacle configurations and their corresponding paths.

3.4 Goal Biased Sampling

The trained Bi-directional long short-term memory model is utilized for goal-biased sampling while generating new configurations for exploration. Specifically, the Bi-directional long short-term memory model is employed to generate goal-biased samples that guide the exploration towards regions of the configuration space more likely to lead to the goal point. The Bi-directional long short term memory model is conditioned on the historical information from the dataset, which includes obstacle configurations and corresponding successful paths. This training enables the Bi-directional long short term memory to learn patterns and relationships within the dataset, generating obstacle configurations biased towards successful paths.

During the execution of the Bi-RRT algorithm, when the algorithm needs to generate a new path point for exploration, it queries the trained Bi-directional long short-term memory model. The Bi-directional long short-term memory model produces a goal-biased sample, which is then used as a candidate path point for further exploration.

3.5 Bi-LSTM Projector Architecture

The dataset is encoded to be fed into Bidirectional LSTM Projector to get the next sampling point from start point. The Bi-RRT starts its sampling from both start and goal position. The robot hand following path from initial state is considered dominant hand. The path is to be planned in this one direction until it is unable to move any further due to being stuck or any other reason. In this case the tree starts to grow from its goal state to initial state. The Architecture of Bi-LSTM is shown in Fig. 5.

Fig. 5. Bi-LSTM architecture

3.6 Path Length and Time Required

The average path length in a particular obstacle configuration is obtained by averaging the path length of each initial and goal point combination in that particular obstacle configuration. The path length is calculated by calculating the Euclidean distance between two adjacent points in the final path in the RRT* and Bi-directional long-short term memory Bi-RRT algorithm. The time required is calculated using a simple 'time' module in Python, which calculates the time difference between when the code starts executing and when it has done its execution. Here, the dataset for each obstacle configuration space is separate. Therefore, the average time required to compute the path for all start and goad points in the particular obstacle configuration is obtained.

3.7 Setup

All the experimentation was performed on an MSI GE 62 2QE Apache system consisting of 4th Gen Intel i7 Quad core processor with hyperthreading enabled. 16 GB dual channel DDR3 Memory clocked at 1600 MHz and an NVDIA GTX 965 GPU with 2 GB video memory.

4 Results Analysis

Table 1 shows the average time required for Bi-LSTM BiRRT, RRT* and BiRRT to plan the path for all the start and goal point pairs in an individual obstacle configuration. Similarly, it also shows the average path length obtained in each obstacle configuration.

As RRT* goes on to iterate even after the suboptimal path is obtained to find a close-to-optimal path, it is evident from the results that the path length obtained in the case of the RRT* algorithm is the shortest in every case when compared to the other two algorithms as they converge after a suboptimal path is obtained. The average path length of all obstacle configurations is shortest in the case of RRT* as we use the RRT* algorithm's path as optimal.

As bidirectional RRT consists of two RRTs exploring simultaneously, it is evident from the results that the algorithm can obtain a feasible path quickly as compared to RRT*. The average time required for BiRRT to plan the path for all the start and goal point pairs in an individual obstacle configuration is consistently lower than the corresponding RRT*'s time, which is expected and evident from the result table.

As Bi-LSTM BiRRT uses goal-biased sampling, it is evident from the results that it takes less time to obtain a feasible path between the start and goal point than the other two algorithms. It is also evident from the results that the path obtained in the case of Bi-LSTM BiRRT is closer to optimal than BiRRT as it is less costly than BiRRT but more than RRT*. The path length in the case of Bi-LSTM BiRRT is consistently more than the path length obtained with RRT*. Interestingly, in average path length in the case of the 5th dataset, Bi-LSTM and BiRRT obtains a shorter path than RRT*, which could be an anomaly. It could result from a false, impossibly small path length in one of the start goal point configurations in the case of the 5th dataset. Nevertheless, it requires further investigation.

Table 1. Average Path Planning Time and Path Length

Set No.	Bi-LSTM BiRRT		RRT*		BiRRT	
	Time	Path Length	Time	Path Length	Time	Path Length
1	0.15 s	7.37 cm	0.26 s	7.21 cm	0.19 s	7.42 cm
2	0.20 s	8.52 cm	0.31 s	8.05 cm	0.25 s	8.67 cm
3	0.35 s	11.6 cm	0.42 s	10.2 cm	0.37 s	11.8 cm
4	0.18 s	8.06 cm	0.22 s	8.0 cm	0.19 s	8.44 cm
5	0.23 s	9.57 cm	0.29 s	10.3 cm	0.27 s	10.12 cm
Avg	0.222 s	9.024 cm	0.3 s	8.752 cm	0.254 s	9.29 cm

5 Conclusion

In summary, the novel Bi-directional Long Short-Term Memory (BiLSTM) approach incorporated into Bidirectional Rapidly-exploring Random Trees (RRT) for dual-arm manipulation has exhibited promising outcomes, surpassing the performance of conventional RRT* and BiRRT methods in terms of path time. It is noteworthy, however, that the BiLSTM-BiRRT approach falls behind RRT* in relation to path length while demonstrating improved performance compared to BiRRT in the same aspect.

The research, situated at the convergence of advanced technologies, explores the synergy of deep learning and robotics coordination in the context of Industry 4.0 and collaborative robots. Looking forward, a critical avenue involves refining error functions to evaluate algorithmic performance more precisely. Furthermore, expanding the algorithm's scope to handle three-dimensional datasets becomes imperative, ensuring its relevance in addressing path planning challenges within a 3D environment and broadening its applicability in real-world robotics and automation applications.

Notably, the experiment focused on a static environment, laying the foundation for prospective investigations into dynamic environmental applications. Adapting the algorithm to dynamic scenarios is crucial for addressing challenges associated with adaptive path planning in response to changing surroundings. These forward-looking considerations underscore the commitment to advancing path planning and robotics, aligning with the dynamic landscape of modern industrial practices. Ultimately, this research contributes valuable insights into robotic coordination and sets the stage for ongoing innovation and improvements in the field.

References

1. Aaltonen, I., Salmi, T.: Experiences and expectations of collaborative robots in industry and academia: barriers and development needs. Procedia Manuf. **38**, 1151–1158 (2019)

2. Smith, C., et al.: Dual arm manipulation-a survey. Robot. Auton. Syst. **60**(10), 1340–1353 (2012)
3. LaValle, S.: Rapidly-exploring random trees: a new tool for path planning. Res. Rep. 9811 (1998)
4. Mac, T.T., et al.: Heuristic approaches in robot path planning: a survey. Robot. Auton. Syst. **86**, 13–28 (2016)
5. Wahab, A., Nadhir, M., Nefti-Meziani, S., Atyabi, A.: A comparative review on mobile robot path planning: classical or meta-heuristic methods? Annu. Rev. Control. **50**, 233–252 (2020)
6. Hochreiter, S., Schmidhuber, J.: Long short-term memory. Neural Comput. **9**(8), 1735–1780 (1997)
7. Tsai, YC., Huang, H.P.: Motion planning of a dual-arm mobile robot in the configuration-time space. In: 2009 IEEE/RSJ International Conference on Intelligent Robots and Systems. IEEE (2009)
8. Elbanhawi, M., Simic, M.: Sampling-based robot motion planning: a review. IEEE Access **2**, 56–77 (2014)
9. Ferguson, D., Stentz, A.: Anytime rrts. In: 2006 IEEE/RSJ International Conference on Intelligent Robots and Systems. IEEE (2006)
10. Karaman, S., Frazzoli, E.: Sampling-based algorithms for optimal motion planning. Int. J. Robot. Res. **30**(7), 846–894 (2011)
11. Noreen, I., et al.: Optimal path planning in cluttered environment using RRT*-AB. Intell. Serv. Robot. **11**, 41–52 (2018)
12. Devaurs, D., Siméon, T., Cortés, J.: A multi-tree extension of the transition-based RRT: application to ordering-and-pathfinding problems in continuous cost spaces. In: 2014 IEEE/RSJ International Conference on Intelligent Robots and Systems. IEEE (2014)
13. Kala, R.: Rapidly exploring random graphs: motion planning of multiple mobile robots. Adv. Robot. **27**(14), 1113–1122 (2013)
14. Qureshi, A.H., Ayaz, Y.: Intelligent bidirectional rapidly-exploring random trees for optimal motion planning in complex cluttered environments. Robot. Auton. Syst. **68**, 1–11 (2015)
15. Qureshi, A.H., et al.: Motion planning networks: bridging the gap between learning-based and classical motion planners. IEEE Trans. Robot. **37**(1), 48–66 (2020)
16. Watt, A.M., Yoshiyasu, Y.: Pathnet: learning to generate trajectories avoiding obstacles. In: 2020 IEEE International Conference on Image Processing (ICIP). IEEE (2020)
17. Qureshi, A.H., and Michael, C.Y.: Deeply informed neural sampling for robot motion planning. In: 2018 IEEE/RSJ International Conference on Intelligent Robots and Systems (IROS). IEEE (2018)
18. Sherstinsky, A.: Fundamentals of recurrent neural network (RNN) and long short-term memory (LSTM) network. Physica D **404**, 132306 (2020)
19. Cui, Z., et al.: Deep bidirectional and unidirectional LSTM recurrent neural network for network-wide traffic speed prediction. arXiv preprint arXiv:1801.02143 (2018)
20. Chitta, S.: MoveIt!: an introduction. Robot Oper. Syst. (ROS) Complete Ref. **1**, 3–27 (2016)

Decoding Emotions: Unveiling Sentiments and Sarcasm Through Text Analysis

Ajay Indian(✉), Priyanka Manethia, Gaurav Meena, and Krishna Kumar Mohbey

Central University of Rajasthan, Ajmer 305817, India
{ajay.indian,gaurav.meena,kmohbey}@curaj.ac.in,
priyanka9460269619@gmail.com

Abstract. Sentiment analysis is a natural language processing technique that is used to determine the emotional tone or attitude communicated in a piece of text, often classifying it as positive, negative, or neutral. Whereas, sarcasm detection is employed to identify and determine whether a statement or piece of text is intended to convey a meaning conflicting to its literal interpretation, often with a mocking or ironic tone. In this study, two methods are proposed for sentiment analysis and one for Sarcasm Detection. One dataset i.e. Amazon Food Reviews is used for analyzing the sentiment and another dataset i.e. Sarcastic News Headlines is used for analyzing sarcasm. To overcome the shortcomings of some previously implemented methods, RoBERTa (Robustly Optimized BERT Approach) and VADER (Valence Aware Dictionary for Sentiment Reasoning) are proposed. In the study, RoBERTa performed better than VADER. RoBERTa has proven better due to its Transfer Learning approach, deep contextual understanding, domain adaptation, and extensive pretraining. Naive Bayes is proposed as it is sometimes considered better than other methods for sarcasm detection due to its simplicity, efficiency, and effectiveness in capturing certain linguistic patterns associated with sarcasm. Sentiments analysis model based on RoBERTa attained an accuracy of 88.95% and VADER attained an accuracy of 83.13%. Whereas, sarcasm detection model based on Naive Bayes attained an accuracy of 90.05%.

Keywords: Sentiment Analysis · Sentiment Polarity · RoBERTa · VADER · Sarcasm Detection · Naive Bayes

1 Introduction

1.1 Background and Motivation

The exponential growth of digital communication platforms has led to an unprecedented volume of textual content being generated and exchanged daily. Understanding communication's sentiments, emotions, and nuances has become paramount within the vast sea of text. Sentiment analysis, a branch of natural language processing (NLP), is crucial in uncovering insights from textual data, enabling organizations to comprehend customer opinions, track brand perception, and make informed decisions based on public sentiment [19].

D. Pastor-Escuredo et al. (Eds.): ICDLAIR 2023, LNNS 1001, pp. 714–731, 2024.
https://doi.org/10.1007/978-3-031-60935-0_62

The motivation for this research stems from the realization that while sentiment analysis has made remarkable strides, there are still challenges to overcome, especially when dealing with informal language, idiomatic expressions, and nuanced sentiments. In this context, advanced models such as RoBERTa, a transformer-based architecture renowned for capturing intricate contextual information, offer a promising avenue for achieving more accurate sentiment analysis outcomes. RoBERTa's fine-tuned language representation capabilities can uncover subtle linguistic cues, contributing to heightened sentiment prediction accuracy [9].

Furthermore, while deep learning models like RoBERTa exhibit exceptional performance, they often require substantial computational resources for training and inference. This consideration led us to explore an alternative approach—VADER, a lexicon and rule-based sentiment analysis tool tailor-made for analyzing social media content. VADER's capacity to handle informal language, emoticons, and slang expressions make it an attractive solution for sentiment analysis, particularly when computational resources are limited [20].

In addition to sentiment analysis, another intriguing dimension of textual communication is sarcasm, a form of expression characterized by irony and contrast between literal and intended meaning. Sarcasm detection is challenging due to the intricate interplay of linguistic cues and contextual dependencies. As such, our research is motivated by the desire to explore the application of the Naive Bayes classification algorithm—a simple yet effective approach—to detect sarcasm instances within textual content accurately. The ability of Naive Bayes to capture co-occurrence patterns and the 'naive' assumption of feature independence presents an interesting basis for investigating sarcasm identification.

By combining advanced deep learning models like RoBERTa and rule-based tools like VADER for sentiment analysis alongside applying Naive Bayes for sarcasm detection, this research aims to enhance sentiment understanding and identify nuanced communication patterns. The outcomes of this study hold implications for a wide range of fields, including marketing, customer service, brand management, and social media analysis. Ultimately, our work seeks to advance the capabilities of sentiment analysis and sarcasm detection, fostering more accurate and insightful analyses of human communication in the digital age.

1.2 Problem Definition

Sentiment analysis, commonly called opinion mining, is a discipline within natural language processing (NLP) that aims to ascertain a document's emotional tone or attitude. The primary goal of sentiment analysis is to categorize the sentiment conveyed in the text as Sentiment Polarity. Sentiment polarity refers to the direction or orientation of sentiment expressed in a text. It indicates whether the sentiment is positive, negative, neutral, or even a combination. Sentiment polarity classification is a key task in sentiment analysis, where the goal is to assign a sentiment label to text based on the emotional tone it conveys. This analysis can extract valuable insights into people's opinions, emotions, and attitudes toward specific topics, products, services, individuals, or events.

1.3 Problem Objectives

The first objective is to design a sentiment analysis model that can accurately classify customer reviews of food products into positive, negative, and neutral categories with compound scores using two different methods. In this study, sentiment analysis is proposed using two distinct approaches, RoBERTa and VADER, to automatically classify sentiment from a collection of textual data. RoBERTa is specifically designed for pre-training on large text corpora. As, it is known for its exceptional performance on different NLP tasks, comprising sentiment analysis, text classification, and more. The architecture's ability of RoBERTa to capture intricate contextual information in the data contributed to its high performance.

In addition to the advanced RoBERTa model, the study employed the VADER (Valence Aware Dictionary and sEntiment Reasoner) lexicon-based approach for sentiment analysis. VADER leverages a sentiment lexicon containing words assigned to sentiment scores, which are then used to compute the overall sentiment of a text. This approach is particularly effective for informal and social media text because it handles emoticons, slang, and informal language. Although VADER is more complex than RoBERTa, VADER's performance showcases its proficiency in analyzing sentiment in various text sources, including those with colloquial expressions and informal language [9, 20].

Sarcasm detection is a complex task in natural language processing (NLP) that focuses on identifying instances of sarcasm in text or speech. Sarcasm is a type of communication in which the intended meaning conflicts with or differs from the literal interpretation of the words used. It often involves irony, humor, or mockery, making it challenging to detect using automated methods due to its nuanced and context-dependent nature.

This leads to the second objective, which is to develop an effective sarcasm detection system capable of identifying instances of sarcasm in headlines. In this study, sarcasm detection was undertaken using a Naive Bayes classification algorithm, with the primary objective of automatically identifying instances of sarcasm within textual content. The Naive Bayes algorithm, a probabilistic classifier, was chosen for its simplicity and effectiveness in text classification tasks. The model leveraged the bag-of-words representation of text, treating each word as an independent feature. Despite the 'naive' assumption of feature independence, Naive Bayes has demonstrated proficiency in capturing sarcasm patterns due to its ability to identify the co-occurrence of certain words and phrases in sarcastic expressions.

Further, this study is organized into six sections to ensure a strong understanding of the study's progression. Section 2 reviews relevant literature and comparative analysis of some previously proposed methods. Section 3 presents the proposed methodology, dataset overview, algorithms overview, and overall workflow of the proposed study. Section 4 presents the findings of the proposed study, a comparative analysis with other states of art, and discusses the implications. Section 5 gives the concluding remarks for the study, limitations, and future directions. Section 6 lists the references used for the study.

2 Literature Survey

2.1 Sentiment Analysis

Sentiment analysis was studied by Prabhu et al. [1] using Long Short-Term Memory (LSTM) networks on a variety of datasets, including the SemEval' 13 dataset and Code Mixed Text of Twitter user comments on posts. The two LSTM models the author proposed, Char-LSTM and Subword-LSTM, have accuracy values of 59.8% and 69.7% on their dataset and 46.6% and 60.57% on SemEval's 13. A convolutional neural network (CNN), which includes user behavior data within a specific document (tweet), was proposed by Alharbi et al. [2] employed in their study, achieved an accuracy of 82.63% on two datasets from the SemEval-2016 workshop. The sentiment analysis is carried out by a study by Wongkar et al. [3] for the 2019 Republic of Indonesia presidential candidates using Twitter analysis. The study was conducted in several steps, including data collection using Python libraries, text processing, splinting the dataset into test and training samples, and text categorization. The accuracy values for Naive Bayes, Suppport Vector Machne (SVM), and K-Nearest Neighbours (K-NN) approaches were computed using RapidMiner and attained the accuracies of 75.58%, 63.99%, and 73.34% respectively.

The sentiment analysis (SA) is carried out on the dataset of Amazon's Fine Food Reviews in a big data setting in a paper by Ahmed et al. [4]. Several demos were performed utilizing the Spark ML package to analyze the sentiments in huge datasets using a variety of classification approaches, including NB, SVM, and Logistic Regression. SVM provided 88.38% accuracy, Naive Bayes 83.43% accuracy, and Logistic Regression 87.38% accuracy. This suggests that linear SVC performs better than Naive Bayes and logistic regression in performance.

Table 1. A Comparative Study of Prominent Other State-Of-Art for Sentiment Analysis

Proposed By	Year	Method	Dataset	Accuracy
Prabhu et al. [1]	2016	LSTM	Own	69.7%
Alharbi et al. [2]	2018	CNN	SemEval-2016	82.63%
Wongkar et al. [3]	2019	Naive Bayes	Twitter Analysis-2019	75.58%
Ahmed et al. [4]	2021	SVM	Amazon Fine Food Reviews	88.38%
Prabhu et al. [1]	2016	LSTM	Own	69.7%

The research conducted by Meena, G. et al. [18] concentrated on classifying the different reviews expressed by individuals of various indigenous groups. The experiments were done using real-time data acquired from social media networks. To categorize sentiments, sentiment polarities like positive, negative, and neutral were utilized. Moreover, the CNNs method was used to perform trials, and the outcomes were compared to other machine learning algorithms. The outcomes revealed that the CNN model can attain accuracy levels of up to 94.47%, 95.4%, and 94%, respectively, when assessed using

Phone, Laptop, and TV review datasets. A comparative study of other state-of-art is presented in Table 1.

As we can conclude from the above table, SVM has performed comparatively better than the other methods. This could be due to the technique of feature extraction as Ahmed et al. [4] performed feature extraction using Machine Learning classifiers through Spark MLib.

2.2 Sarcasm Detection

Misra et al. [5] used a hybrid NN for sarcasm detection on the News Headlines dataset. They reported the quantitative outcomes of the Baseline and the Hybrid Neural Network (HNN) using classification accuracy as their dataset was almost balanced. The baseline model gave 84.88% accuracy, and the HNN model gave 89.7% accuracy. Gupta et al. [10] use a Voting Classifier to classify the text as sarcastic or neutral after using NLTK to get the sentiment score. They have used the SemEval' 11 Dataset for their approach. In their first approach, the best accuracy of 74.59% is attained by the Support Vector Machine algorithm. In the second approach, the voting classifier attained the maximum accuracy of 83.53%.

A well-framed data set comprising sarcastic and non-sarcastic tweets was used to prepare the proposed model by Salim et al.[11]. They have employed Recurrent Neural Networks (RNNs), LSTM, and Word embedding for efficient sarcasm detection and to easily classify the statements from Twitter. Sarcasm and Non-Sarcasm accuracies were obtained as 85.23% and 86.47% respectively. In a study, Eke et al. [12] investigated the word vector model (GloVe). They studied the predictive performance by investigating sarcasm identification in tweets using a machine-learning method. The experimentation outcomes obtained better accuracy, illustrating the importance of GloVe embedding in sarcasm classification. The GloVe model requires a large volume of corpus in order to generate better predictive results. It bore an accuracy of 74.68% obtained with 46,796 features. In a paper, Hazarika et al. [14] proposed CASCADE (a ContextuAl SarCasm DEtector) that employed a hybrid method of content and context-driven modeling for sarcasm recognition in online social media dialogues using a large-scale self-annotated corpus (SARC) for sarcasm and obtained an accuracy of 0.77%.

Table 2. A Comparative Analysis of Prominent Other State-of-Art for Sarcasm Detection

Proposed By	Year	Method	Dataset	Accuracy
Hazarika et al. [14]	2018	CASCADE + CNN	SARC	77%
Gupta et al. [10]	2020	Voting Classifier	SemEval' 11	83.53%
Salim et al. [11]	2020	RNN-LSTM	Own	85.23%
Eke et al. [12]	2020	GloVe + RandomForest	Own	74.68%
Pan et al. [13]	2020	BERT	MMSDD	86.05%

Pan et al. [13] suggested a BERT architecture-based model for multi-modal sarcasm detection. They assessed the model using a publicly available multi-modal sarcasm detection dataset and attained an accuracy of 86.05%. A comparative analysis of prominent other state-of-art is presented in Table 2.

The table shows that the Hybrid Neural Network performed better than the other methods. It can be the result of using pre-trained embedding from the Word2Vec model.

3 Material and Proposed Methods

The following sections present the overview of the datasets used in the proposed study, the working of the proposed methods, and the implementation of the proposed methodology.

3.1 Dataset Overview

Amazon Food Reviews Dataset

Amazon Food Reviews dataset was collected from Kaggle [6]. This dataset comprises reviews of fine foods from Amazon. Further, the dataset consists of 568454 reviews with 10 columns for over 10 years. Reviews have information related to the respective product and the consumer, star ratings, and a plain text review provided by the consumer. Reviews from all other Amazon categories are also included. The first five rows of the Amazon Food Reviews Dataset are presented in Fig. 1 for reference.

Only useful features, such as ID, Score, and Text, are extracted to develop the proposed model. As ID is essential for Indexing, Score compares calculated results and text for detection. After the feature extraction, the shape of the dataset became 568454 rows and 3 columns. After which, the dataset looks as shown in Fig. 2.

	Id	ProductId	UserId	ProfileName	HelpfulnessNumerator	HelpfulnessDenominator	Score	Time	Summary	Text
0	1	B001E4KFG0	A3SGXH7AUHU8GW	delmartian	1	1	5	1303862400	Good Quality Dog Food	I have bought several of the Vitality canned d...
1	2	B00813GRG4	A1D87F6ZCVE5NK	dll pa	0	0	1	1346976000	Not as Advertised	Product arrived labeled as Jumbo Salted Peanut...
2	3	B000LQOCH0	ABXLMWJIXXAIN	Natalia Corres "Natalia Corres"	1	1	4	1219017600	"Delight" says it all	This is a confection that has been around a fe...
3	4	B000UA0QIQ	A395BORC6FGVXV	Karl	3	3	2	1307923200	Cough Medicine	If you are looking for the secret ingredient i...
4	5	B006K2ZZ7K	A1UQRSCLF8GW1T	Michael D. Bigham "M. Wassir"	0	0	5	1350777600	Great taffy	Great taffy at a great price. There was a wid...

Fig. 1. First Five Rows of the Amazon Food Reviews Dataset

	Id	Text	Score
0	1	I have bought several of the Vitality canned d...	5
1	2	Product arrived labeled as Jumbo Salted Peanut...	1
2	3	This is a confection that has been around a fe...	4
3	4	If you are looking for the secret ingredient i...	2
4	5	Great taffy at a great price. There was a wid...	5

Fig. 2. First Five Rows of the Amazon Food Reviews Dataset after Useful Feature Extraction

Fig. 3. First WordCloud of Dataset Amazon Food Reviews Dataset

Figure 3 shows the wordcloud plotted for the text column in the dataset, which consists of consumer reviews for a certain product. It displays the most frequently used words that appeared in the dataset. Figure 4 depicts a plot of stars given by consumers for particular products. As seen in the above plot, the frequency of 5-star ratings given by consumers is comparatively higher.

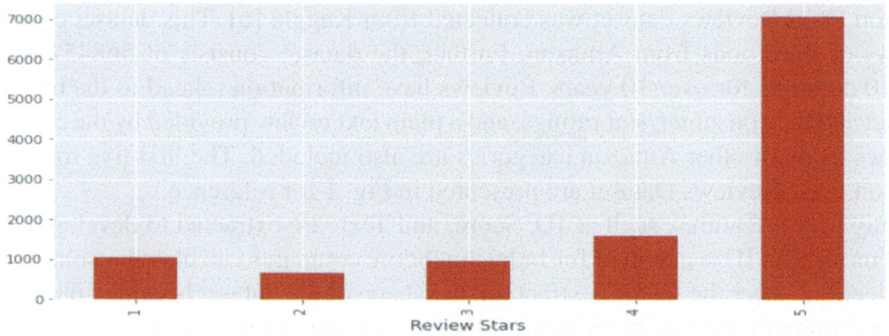

Fig. 4. Depicting a Plot of Review Stars Given By Consumers for Particular Products

News Sarcastic Headline Dataset

News Sarcastic Headline is also collected from Kaggle [7], formed by collecting headlines from two news websites. The first website, TheOnion, provides sarcastic versions of recent events. The other one is HuffPost, from which real and non-sarcastic news headlines have been gathered. The dataset consists of 55328 rows with 3 columns. The first 5 samples of the dataset are shown in Fig. 5.

	article_link	headline	is_sarcastic
0	https://www.huffingtonpost.com/entry/versace-b...	former versace store clerk sues over secret 'b...	0
1	https://www.huffingtonpost.com/entry/roseanne-...	the 'roseanne' revival catches up to our thorn...	0
2	https://local.theonion.com/mom-starting-to-fea...	mom starting to fear son's web series closest ...	1
3	https://politics.theonion.com/boehner-just-wan...	boehner just wants wife to listen, not come up...	1
4	https://www.huffingtonpost.com/entry/jk-rowlin...	j.k. rowling wishes snape happy birthday in th...	0

Fig. 5. First Five Rows of the News Sarcastic Headline Dataset

3.2 Proposed Methods Used for Sentiment Analysis and Sarcasm Detection

This section presents the basic introduction and working of proposed methods used for sentiment analysis- RoBERTa and VADER and for sarcasm detection Naïve Bayes classifier.

RoBERTa

RoBERTa (Robustly Optimized BERT Approach) is a natural language processing model that builds upon the architecture of BERT (Bidirectional Encoder Representations from Transformers). Researchers at Facebook AI introduced it in a 2019 paper. RoBERTa aims to improve BERT's pretraining methodology by optimizing various hyper-parameters and training strategies. It uses a longer batch size, more data to train, and longer training periods.

RoBERTa is trained on a much larger amount of text compared to BERT. This larger dataset helps the model learn more diverse language patterns and representations. RoBERTa uses dynamic masking during pretraining, where the positions of masked tokens are changed for each batch. This technique helps prevent the model from relying too much on the positions of masked tokens. RoBERTa is trained for significantly longer periods than BERT. This extended training allows the model to capture more nuanced linguistic features. RoBERTa uses a larger batch size during training, which can lead to improved generalization. To achieve better performance, RoBERTa optimizes hyper-parameters, for example, learning rate, batch size, and more [15].

Softmax activation is used in RoBERTa for classification. For the vector ($z = [z_1 z_2, \ldots, z_n]$), the probability ($p_i$) for each element ($z_i$) is calculated by Eq. (1). Softmax function can be given as:

$$p_i = \frac{e^{z_i}}{\sum_{j=1}^{n} e^{z_j}} \quad (1)$$

where e is the base of the natural logarithm, n is the number of classes.

VADER

VADER (Valence Aware Dictionary and sEentiment Reasoner) is a lexicon and rule-based sentiment analysis technique designed to determine a text's sentiment (positive, negative, or neutral) and sentiment intensity, typically at the sentence level. Researchers at the Georgia Institute of Technology developed it. VADER uses a predefined lexicon (list of words) annotated with sentiment scores. Each word in the lexicon is assigned a sentiment score based on its sentiment polarity (positive or negative) and intensity. The lexicon also considers words that indicate neutrality [16]. VADER takes into account capitalization and punctuation in text to determine sentiment intensity. For example, "GOOOOOD!!!" would be interpreted as conveying more intense positive sentiment than just "good."

VADER considers degree modifiers (like "very" or "extremely") that can modify the intensity of sentiment. These modifiers are used to adjust the sentiment score of words accordingly. VADER can identify negations in text and reverse the sentiment polarity of words that appear after negations. For instance, "not good" would be recognized as a negative sentiment. VADER is designed to handle informal language and often-used emoticons or slang terms that carry sentiment information. VADER computes a compound sentiment score for a text by combining the individual sentiment scores of words. This score represents the overall sentiment intensity of the text. VADER is particularly useful in sentiment analysis of short, informal text, such as social media posts, reviews, and comments. It also computes compound scores with polarity scores. The compound score is a single numeric value that summarizes the complete sentiment of the entire text. It is computed by combining the sentiment intensity scores of all words in the text while considering each word's valence. The compound score provides a holistic measure of the sentiment conveyed in the text. The range of compound scores and their interpretations are given below-

- The compound score range goes from -1 to 1.
- A compound score closer to -1 designates a highly negative sentiment.
- A compound score closer to 1 designates a highly positive sentiment.
- A compound score of around 0 suggests a neutral sentiment.

It can provide a quick way to gauge sentiment without requiring extensive training data and complex machine-learning models. However, it might not capture complex nuances in sentiment as effectively as more sophisticated models like deep learning-based approaches like RoBERTa. In VADER, the compound score (C) can be calculated using Eq. (2)-

$$\mathrm{C} = \frac{\sum_{i=1}^{n} S_i W_i}{\sum_{i=1}^{n} W_i} \tag{2}$$

where n is the number of words in the text, Si is the sentiment score of the ith word, and Wi is a weighting factor applied to the ith word to account for its intensity [16].

Naive Bayes

Naive Bayes classifiers are based on probability theory using Bayes' theorem with strong (naive) independence assumptions among the features. A Naive Bayes model is easy to construct and does not require time-consuming iterative parameter estimation, making it very beneficial in medicine for identifying heart patients. The Naive Bayesian classifier is popular because, despite its simplicity, it frequently outperforms more complex classification techniques [8]. Bayes' theorem computes the probability of a hypothesis (class label) given the evidence (features). It is represented using Eq. 3-

$$P(class|features) = (P(features|class) \cdot P(class))/P(features) \tag{3}$$

The "naive" assumption in Naive Bayes is that the features are conditionally independent, meaning that one feature's presence does not affect another's presence. This simplifies the calculations. Naive Bayes requires a labeled dataset to estimate the probabilities. It computes the prior probability of every class and the likelihood of features given to each class. For a given new instance of features (data point), Naive Bayes computes the probability of each class given those features and assigns the class with the highest probability for the given instance. There are different variants of Naive Bayes algorithms, such as Gaussian Naive Bayes (assuming features follow a Gaussian distribution), Multinomial Naive Bayes (used for text classification), and Bernoulli Naive Bayes (used for binary data). The selection of variation depends on the nature of the data being classified. For the mathematical form of Naiye Bayes, let:

- X represents a vector of features (e.g., words in a document).
- C Represents the class label (e.g., positive or negative sentiment).
- $P(C)$ is the prior probability of class C.
- $P(xi|C)$ is the conditional probability of the feature xi given class C.

The Naive Bayes algorithm computes the posterior probability $P\ (C|X)$, which is the probability of class C given the observed features X. It uses Bayes' theorem, incorporating the Naive assumption of feature independence; posterior probability is computed by using Eq. 4-

$$P(C \mid X) = (P(X|C) \cdot P(C))/P(X) \tag{4}$$

Since sarcasm detection is the classification problem, the posterior probabilities can be compared for different classes, with the class having the maximum probability. The Naive Bayes formula can be expanded using the Naive assumption using the following Eq. 5.

$$P(C|x1, x2, x3, \ldots xn) = (P(x1|C) \cdot P(x2|C) \ldots P(xn|C))/(P(x1) \cdot P(x2) \ldots P(xn)) \tag{5}$$

Given this, the Naive Bayes classifier assigns a new observation (X) to the class (C) for which the above probability is the highest. For text classification problems, the features are often words, and the probabilities (P (x_i | C)) are estimated from the training data using techniques like counting word occurrences or applying smoothing methods (e.g., Laplace smoothing) to handle unseen words [17].

3.3 Overall Workflow of Proposed Methodology and Implementation

This section presents the workflow of proposed methodologies adopted for sentiment analysis and Sarcasm Detection.

3.3.1 Sentiment Analysis

The complete workflow of the proposed sentiment analysis method using RoBERTA and VADER is presented in Fig. 6. All the necessary libraries are imported using Python, and the dataset is loaded using the Pandas library. For preprocessing of data EDA, Tokenizing, and pos_tagging have been performed. Then, using visualization, multiple graphs were plotted using Seaborn and matplotlib libraries, and the WordCloud figure was plotted using the wordcloud library. For tokenization, the nltk library was used. The proposed RoBERTa and VADER model has been implemented using the transformers library. At last, results were computed in terms of accuracy for measuring the performances of the proposed models.

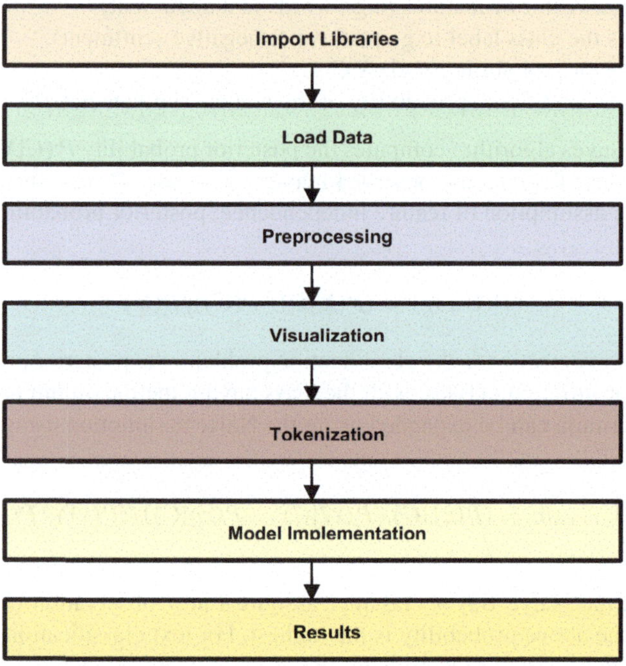

Fig. 6. Overall Workflow for Proposed Sentiment Analysis Models using RoBERTa and VADER

Sarcasm Detection

The complete workflow of the proposed model for sarcasm detection using the Naïve Bayes classifier is shown in Fig. 7, which includes the steps such as importing libraries, loading the dataset, applying preprocessing, applying CountVectorizer, partitioning the dataset into training and test samples followed by model implementation using Python programming environment and at last, finding and analyzing the outcomes of proposed model and comparing with other state-of-art.

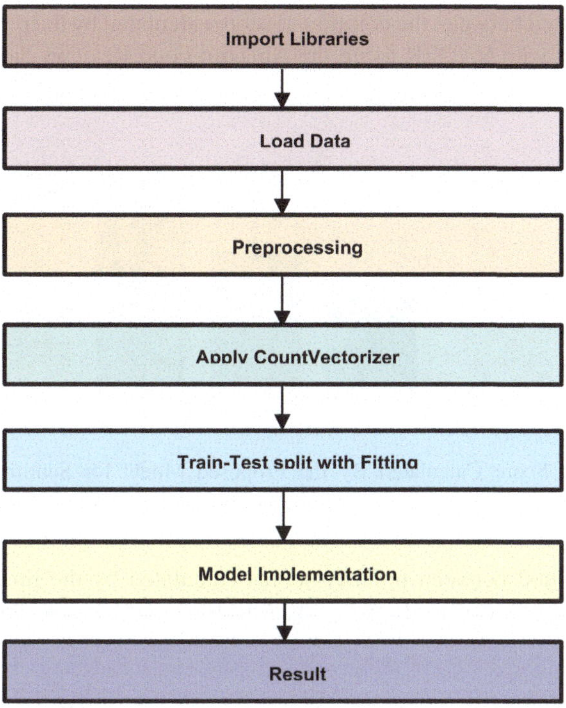

Fig. 7. Workflow for Proposed Model for Sarcasm Detection using Naïve Bayes Classifier

4 Results and Discussion

This section presents the results observed from the proposed models for sentiment analysis with RoBERTa and VADER and sarcasm detection with Naïve Bayes.

4.1 Sentiment Analysis

The graph plotted between polarity scores calculated by the suggested model for sentiment analysis with VADER using the Amazon Food Reviews dataset is presented in Fig. 8.

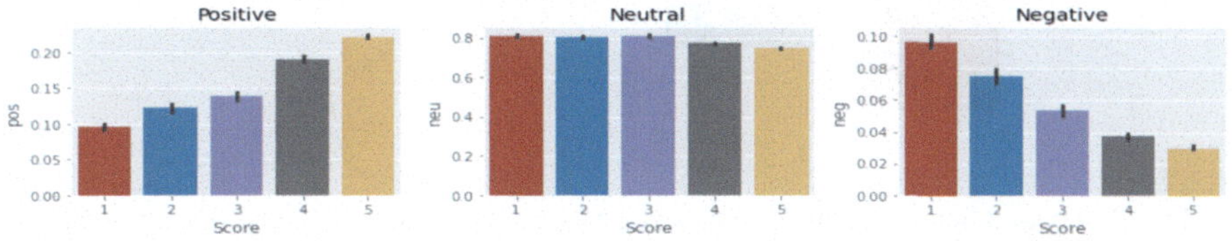

Fig. 8. Polarity Scores Calculated by the Proposed Sentiment Analysis Model with VADER

The graph plotted between the compound score calculated by the proposed model for sentiment analysis with VADER using the Amazon Food Reviews dataset is presented in Fig. 9.

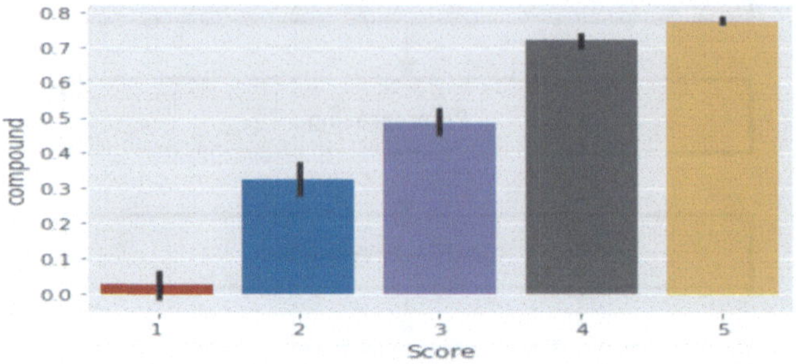

Fig. 9. Compounds Score Calculated By the Proposed Model for Sentiment Analysis with VADER

The graph plotted between polarity scores calculated by the proposed model for sentiment analysis with RoBERTa using the Amazon Food Reviews dataset is shown in Fig. 10.

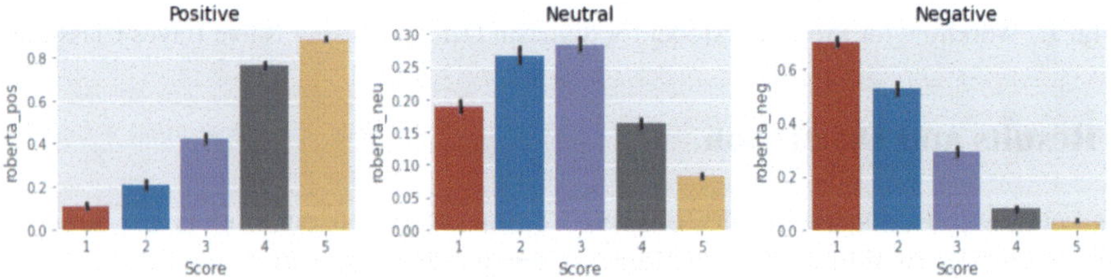

Fig. 10. Polarity Scores Calculated By the Proposed Model for Sentiment Analysis with RoBERTa

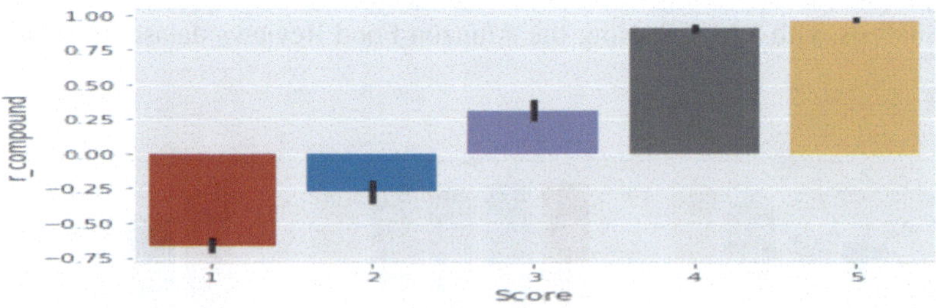

Fig. 11. Compound Score Calculated Proposed Model for Sentiment Analysis with RoBERTa

The graph plotted between the compound score calculated by RoBERTa and the Scores given in the Amazon Food Reviews dataset is presented in Fig. 11. In order to analyze the performance of the proposed Sentiment Analysis model with RoBERTa, two variables, weight_positive and weight_neutral, were adopted to compute the compound score. Different combinations of values for these two variables were considered to obtain the highest accuracy, as presented in Table 3.

Table 3. Variation in Accuracy Due to Changes in Neutral and Positive Values with RoBERTa

Weight Positive	Weight Neutral	Accuracy
0.5	0.5	88.75%
1	0.5	88.15%
0.5	1	88.55%
1	1	88.75%
2	**1.5**	**88.95%**
1.5	2	88.75%
2	2	88.75%
0.5	0.5	88.75%
1	0.5	88.15%

From the above table, the highest accuracy is obtained when the weight_positive value is 2, which is much higher than weight_neutral. As in the dataset, there are more data with higher star scores given by consumers, which can be seen in Fig. 5. Table 4 displays the comparative analysis of the outcomes of the proposed study with other state-of-the-art sentiment analyses.

The results shown in Table 4 show that RoBERTa performed better than VADER. RoBERTa is a transformer-based language model that benefits from pretraining on massive amounts of data. It enables it to capture nuanced language patterns effectively, allowing for better sentiment analysis on a dataset compared to VADER, a lexicon and rule-based approach. As portrayed in the table, one of the proposed methods, RoBERTa, performed comparatively better than others, giving 88.95% accuracy due to extensive pretraining, Pretrained Language Representation, Capturing Contextual Information, Hierarchical Representations, Transfer Learning, Out-of-Vocabulary Handling, and Fine-tuning.

Table 4. Comparative Analysis of Proposed Sentiment Analysis Models

Proposed By	Year	Method	Dataset	Accuracy
Prabhu et al. [1]	2016	LSTM	Own	69.7%
Alharbi et al. [2]	2018	CNN	SemEval-2016	82.63%
Wongkar et al. [3]	2019	Naive Bayes	Twitter Analysis-2019	75.58%
Ahmed et al. [4]	2021	SVM	Amazon Fine Food Reviews	88.38%
Proposed	2023	VADER	Amazon Fine Food Reviews	83.13%
Proposed	2023	RoBERTa	Amazon Fine Food Reviews	88.95%

As mentioned, one previously proposed method by Ahmed et al. [4] using SVM has also used the dataset as this study, but RoBERTa performed better. RoBERTa works better than SVM for sentiment analysis because it can capture complex language patterns, contextual information, and nuances in sentiment through pretraining on extensive text data, leading to more accurate sentiment predictions.

4.2 Sarcasm Detection

Experimentation showed that when the split ratio for training and testing was 80:20, and the random state was 200, the proposed model for sarcasm detection using the Naïve Bayes classifier obtained the highest accuracy of 90.05%. Although various combinations of the split ratio of the dataset and random state were adopted to analyze the proposed model's accuracy, the best-obtained accuracy was 90.05%, as presented in Table 5. A comparative study of the proposed approach with other state-of-arts adopted for sarcasm detection is presented in Table 6.

Table 5. Accuracy Table with Parameters for Sarcasm Detection

Parameters	
Method	Naive Bayes
Train and Test Split	80:20
Random State	200
Accuracy	90.05%

As presented in Table 6, the suggested model based on Naive Bayes performed comparatively better than other state-of-arts. The features that distinguish sarcasm are simple and directly related to word frequencies and probabilities, which aligns well with Naive Bayes' feature independence assumption.

Table 6. Comparative Analysis of Proposed Sarcasm Detection Model with Other State-of-Arts

Proposed By	Year	Method	Dataset	Accuracy
Hazarika et al. [14]	2018	CASCADE + CNN	SARC	77%
Gupta et al. [10]	2020	Voting Classifier	SemEval' 11	83.53%
Salim et al. [11]	2020	RNN-LSTM	Own	85.23%
Eke et al. [12]	2020	GloVe + Random Forest	Own	74.68%
Pan et al. [13]	2020	BERT	MMSDD	86.05%
Misra et al. [5]	2023	Hybrid NN	News Headlines	89.7%
Proposed Method	2023	Naive Bayes	News Headlines	90.05%

5 Conclusion and Future Scope

For Sentiment Analysis, VADER attained an accuracy of 83.13%. While effective for sentiment analysis, VADER has limitations in accurately handling complex language nuances and context. It struggles with identifying sarcasm, irony, and sentiment expressed through intricate language constructions. Additionally, its lexicon-based approach might misinterpret words in diverse contexts, leading to false positives or negatives. VADER's performance can be hindered by domain-specific language and emerging slang, affecting its precision on evolving and specialized texts.

For Sentiment Analysis, as observed from the outcomes, RoBERTa attained a higher accuracy of 88.95%. While RoBERTa excels in sentiment analysis due to its deep contextual understanding, it has limitations. One key area for improvement is its dependence on large amounts of labeled data for fine-tuning, making it less effective for domains with limited labeled sentiment data. Additionally, Roberta might need help with rare or domain-specific sentiment expressions well-represented in its pretraining data. Lastly, RoBERTa's complexity demands substantial computational resources, which can constrain smaller research or industry settings. Both accuracies conclude RoBERTa to be a better method than VADER. While RoBERTa's contextual understanding revolutionized sentiment analysis, exploration could focus on optimizing fine-tuning strategies for domain-specific sentiment nuances and minimizing data dependence. Moreover, enhancing VADER's rule-based framework might involve domain-specific lexicon expansion and refining sentiment lexicons for evolving online language. Integrating domain knowledge, user-specific context, and interpretability could elevate sentiment analysis accuracy and usability.

For Sarcasm Detection, Naive Bayes gave 90.05% accuracy. Whereas it still has limitations due to its assumption of feature independence. Sarcasm often relies on complex linguistic and contextual cues, which may lead to dependencies between features that Naive Bayes cannot capture, especially when they involve subtleties that challenge the feature independence assumption. After employing Naive Bayes for sarcasm detection in our research work, the next step in advancing sarcasm detection could involve exploring more sophisticated machine-learning models that capture complex linguistic nuances and contextual dependencies inherent in sarcasm.

References

1. Joshi, A., Prabhu, A., Shrivastava, M., Varma, V.: Towards sub-word level compositions for sentiment analysis of Hindi-English code mixed text. In Proceedings of COLING 2016, the 26th International Conference on Computational Linguistics: Technical Papers, pp. 2482–2491, December 2016
2. Alharbi, A.S.M., de Doncker, E.: Twitter sentiment analysis with a deep neural network: an enhanced approach using user behavioral information. Cogn. Syst. Res. **54**, 50–61 (2019)
3. Wongkar, M., Angdresey, A.: Sentiment analysis using the naive Bayes algorithm of the data crawler Twitter. In 2019 Fourth International Conference on Informatics and Computing (ICIC), pp. 1–5. IEEE, October 2019
4. Ahmed, H. M., Javed Awan, M., Khan, N. S., Yasin, A., & Faisal Shehzad, H. M. (2021). Sentiment analysis of online food reviews using big data analytics. Hafiz Muhammad Ahmed, Mazhar Javed Awan, Nabeel Sabir Khan, Awais Yasin, Hafiz Muhammad Faisal Shehzad (2021) Sentiment Analysis of Online Food Reviews using Big Data Analytics. Elementary Education Online, 20(2), 827–836
5. Misra, R., Arora, P.: Sarcasm detection using news headlines dataset. AI Open **4**, 13–18 (2023)
6. https://www.kaggle.com/datasets/snap/amazon-fine-food-reviews
7. Misra, R.: News headlines dataset for sarcasm detection. arXiv preprint arXiv:2212.06035 (2022)
8. Vembandasamy, K., Sasipriya, R., Deepa, E.: Heart disease detection using Naive Bayes algorithm. Int. J. Innov. Sci. Eng. Technol. **2**(9), 441–444 (2015)
9. Liu, Y., Ott, M., Goyal, N., Du, J., Joshi, M., Chen, D., et al.: RoBERTa: a robustly optimized BERT pretraining approach, arXiv:1907.11692 (2019)
10. Gupta, R., Kumar, J., Agrawal, H.: A statistical approach for sarcasm detection using Twitter data. In: 2020, 4th International Conference on Intelligent Computing and Control Systems (ICICCS), pp. 633–638. IEEE, May 2020
11. Salim, S.S., Ghanshyam, A.N., Ashok, D.M., Mazahir, D.B., Thakare, B.S.: Deep LSTM-RNN with word embedding for sarcasm detection on Twitter. In: 2020 International Conference for Emerging Technology (INCET), pp. 1–4. IEEE, June 2020
12. Eke, C.I., Norman, A., Shuib, L., Fatokun, F.B., Omame, I.: The significance of global vectors representation in sarcasm analysis. In: 2020 International Conference in Mathematics, Computer Engineering and Computer Science (ICMCECS), pp. 1–7. IEEE, March 2020
13. Pan, H., Lin, Z., Fu, P., Qi, Y., Wang, W.: Modeling intra and inter-modality incongruity for multi-modal sarcasm detection. In: Findings of the Association for Computational Linguistics: EMNLP 2020, pp. 1383–1392, November 2020
14. Hazarika, D., Poria, S., Gorantla, S., Cambria, E., Zimmermann, R., Mihalcea, R.: Cascade: Contextual sarcasm detection in online discussion forums. arXiv preprint arXiv:1805.06413 (2018)
15. Liu, Y., et al.: Roberta: a robustly optimized Bert Pretraining approach. In: International Conference on Learning Representations (ICLR 2020) (2019)
16. Elbagir, S., Yang, J.: Twitter sentiment analysis using natural language toolkit and VADER sentiment. In: Proceedings of the International Multiconference of Engineers and Computer Scientists, vol. 122, pp. 1–6, March 2019
17. Webb, G.I., Keogh, E., Miikkulainen, R.: Naïve Bayes. Encyclopedia Mach. Learn. **15**(1), 713–714 (2010)
18. Meena, G., Mohbey, K.K., Indian, A.: Categorizing sentiment polarities in social networks data using convolutional neural network. SN Comput. Sci. **3**, 116 (2022). https://doi.org/10.1007/s42979-021-00993-y

19. Pandey, K.K., Thorat, M., Joshi, A., Hussein, S.D.A., Alazzam, M.B.: Natural language processing for sentiment analysis in social media marketing. In: 2023 3rd International Conference on Advance Computing and Innovative Technologies in Engineering (ICACITE), Greater Noida, India, pp. 326–330 (2023). https://doi.org/10.1109/ICACITE57410.2023.10182590
20. Hutto, C., Gilbert, E.: VADER: a parsimonious rule-based model for sentiment analysis of social media text. Proc. Int. AAAI Conf. Web Soc. Media **8**(1), 216–225 (2014)

Rice Disease Identification Using Vision Transformer (ViT) Based Network

Md. Ashraful Haque[1(✉)], Chandan Kumar Deb[1], Sudeep Marwaha[1], Subrata Dutta[2], Mehraj Ul Din Shah[3], Ananta Saikia[4], and Abhishek Shukla[5]

[1] ICAR-Indian Agricultural Statistics Research Institute, New Delhi, India
ashraful.haque@icar.gov.in
[2] Bidhan Chandra Krishi Viswavidyalaya, Nadia 741252, India
[3] Sher-e-Kashmir University of Agricultural Sciences and Technology, Srinagar, India
[4] Assam Agricultural University, Jorhat 785013, India
[5] Navsari Agricultural University, Navsari 396450, India

Abstract. Rice crop plays a significant role in upholding India's and global food security. However, the potential yield of rice crops is continuously threatened by various types of diseases throughout its growth stages. Researchers are devising advanced & automated disease identification and management techniques instead of manual inspection to combat this issue. In this context, deep learning-based techniques appear as promising avenues, demonstrating impressive performance in disease identification using digital images. In this work, we implemented a vision transformer-based network, which was implemented to identify and classify the images of rice crops into 11 distinct predefined categories. To ensure the network's robustness, the rice crop images were collected from experimental fields of diverse locations across the country. We also applied augmentation techniques to expand the rice image dataset significantly. The proposed network was developed using four transformer encoders, each comprising a multi-head self-attention function and dense layer. The outcomes of the experimentations of the proposed network on the rice image dataset were remarkable, showcasing classification accuracy of 98.83% and the f1-score of 98.87%. Therefore, employing advanced technologies like vision transformer-based networks offers a lot of hope to the farm community for effectively identifying diseases in rice crops.

Keywords: Disease identification · Deep learning · Vision transformers · Multi-head self-attention · Rice diseases

1 Introduction

Rice is considered as one of the most valuable cereal crops in India and worldwide, playing a pivotal role in ensuring global food security [22]. The majority of India's population is dependent on rice crop as their staple food source [18]. Therefore, ensuring an ample supply of high-quality rice is crucial to meet

D. Pastor-Escuredo et al. (Eds.): ICDLAIR 2023, LNNS 1001, pp. 732–741, 2024.
https://doi.org/10.1007/978-3-031-60935-0_63

this substantial demand. Nevertheless, the growth of rice crops is vulnerable to various diseases, which leads to reduced yield and quality of the crop. In the most traditional way, the diseases can be controlled by manual inspection or laboratory-based experiments, which proved to be laborious, time-consuming and cost-intensive. Consequently, there is a growing interest in developing automated and data-driven techniques for plant disease identification [16].

In recent years, the advancement of Artificial Intelligence and computer vision techniques has revolutionized several domains, including agriculture [23]. Deep learning techniques, especially Convolutional Neural Networks (CNNs), have emerged as powerful techniques for image-based analysis tasks. CNNs have shown remarkable results in image-based disease identification in crops [9]. Researchers across the world have developed a series of CNN-based algorithms for plant disease detection. Recent studies showed that transformer-based algorithms are also being applied in the computer vision domain, popularly known as Vision-Transformers (ViTs) [5]. Considering this, an effective deep learning-based technique would be a viable solution to identify diseases of rice crop in an automated way.

In this work, we proposed a vision transformer-based (ViT) network for the identification and classification of diseased images of rice crops. The network architecture comprises four transformer encoders integrated with the multi-head attention function. The proposed network has been trained on a large image dataset of diseases of rice crop collected from experimental fields in broad daylight conditions and shows significant performance on the test dataset.

2 Related Works

During the past few years, the interest of researchers has shifted toward the use of deep learning techniques in the area of image-based disease identification in crops [20]. A plethora of research works are actively introducing a variety of techniques for disease detection by leveraging the fusion of image processing and deep learning. In 2016, Mohanty et al. [24] pioneered in introducing image-based plant disease detection using deep learning techniques. They applied pre-trained networks on a public dataset, PlantVillage [13], containing 54K images in 38 classes representing plant diseases. Ferentinos [6] applied transfer learning concept on the extended version of PlantVillage dataset containing 87,848 images with 58 categories. Too et al. [32] provided a comparative analysis of four SOTA networks viz. VGG16, Inception, ResNet, and DenseNet, applied to PlantVillage dataset. Lee et al. [19] studied three publicly available datasets: PlantVillage, IPM, and Bing and implemented VGG16 and InceptionNet using transfer learning approach for disease classification. Atila et al., [1] introduce the use of the EfficientNet architectures for disease detection in crops using the PlantVillage dataset. Meanwhile, Haque et al. [9–11] concentrated on detection of maize diseases based on deep learning techniques by utilizing PlantVillage dataset and custom maize datasets. Nigam et al. [25] worked on the identification of wheat rust diseases using EfficientNets networks. Gole et al. [7] proposed a lightweight deep learning model combining CNNs and vision transformers for disease detection in crops.

Lu et al., [21] proposed a custom CNN model for the detection of diseased images of rice crop captured from experimental fields. Patidar et al. [27] tackled the disease detection challenges using residual neural networks and applied them to rice disease identification. Jiang et al. [14] combined the CNN networks with SVM techniques to identify the diseases of rice crop. Sethy et al., [28] applied ResNet50 with SVM for the classification of Rice diseased images. Deng et al., [4] used an ensemble approach to develop a deep learning-based model for the identification of diseases in rice crop. Shrivastava et al. [29] conducted a comparative analysis of various pre-trained deep learning architectures for image-based classification of rice diseases. Chen et al. [2] incorporated the attention mechanism in the MobileNet-v2 network to detect the small lesions of diseases of rice crop. Pandi et al. [26] employed a dilated CNN network with GAP for the detection of disease-infected images of rice crop. Daniya & Vigneshwari [3] extracted deep features from the diseased images of rice crop using a segmented network and employed these features in a deep neural network. Sudar et al. [31] harnessed the power of SOTA ResNet50 and VGG16 networks for classifying paddy diseases. Simhadri & Kondaveeti [30] provided a comparative analysis of 15 pre-trained deep learning networks applied on rice-diseased images. Jiang et al., [15] proposed an improved version of DenseNet network by incorporating attention function in the base network and implementing it on a publicly available rice dataset. Haridasan et al. [12] used a combination of CNN and SVM methods for detecting paddy diseases. Zhang et al. [33] introduced the concept of vision transformer in the CNN networks for accurately classifying the diseases of rice crops.

3 Rice Disease Dataset

Around 12,322 images of diseased rice crops were collected from agricultural fields in different parts of the country: West Bengal, Jammu & Kashmir, Assam, and Gujarat. The disease-infected images were captured under normal background conditions using NIBPP mobile application. The rice disease dataset consists of images from early stage of infestation to the moderate stage of infestation. There are seven diseases in the image dataset, including the healthy ones. Sample images of the disease-infected rice crop are provided in Fig. 1. Then, the raw images were curated by domain experts to discard noisy, out-of-focus, blurred, and unwanted images. Following the curation, images were augmented with synthetically generated images by applying rotation, flipping, skewing, zooming, shearing, brightness adjustment, etc. augmentation techniques. These augmentation strategies increase the number of images, reducing the chance of overfitting and significantly improving the model's generalization capability [8]. The overall summary of the dataset is presented in Table 1.

Fig. 1. Sample images of all the classes of rice disease dataset

4 Proposed Network Architecture

In this work, we proposed a transformer-based network inspired by Vision Transformer (ViT) [5] and implemented it for accurately identifying the images of Rice crop diseases. Our proposed network comprises two key components: a patch extractor and a transformer encoder block. The proposed network is designed to effectively process the symptomatic characteristics of the images of rice crop diseases. The overall flow of the proposed approach is shown in Fig. 2.

The proposed network accepts the input images of size 384×384 with 3 channels for red (R), green (G), and blue (B) bands. First, the input images are divided into smaller patches of size 16×16, creating 576×3 image-patches from each image. Next, the corresponding positional information is embedded in each image-patch for capturing spatial information of the input image-patches. Then, the position-embedded image-patches *aka* image-tokens are passed to the transformer encoder block as input. A series of transformer encoders process the input image-tokens and generate the output probabilities.

In our proposed network, the transformer encoder block consists of four sequentially aligned transformer encoders. Each transformer encoder consists of three distinct functions: layer normalization, multi-head attention (MHA), and Dense layer, as shown in Fig. 3. The layer normalization function was implemented to normalize all the inputs of a layer at one go. The MHA included a two-headed self-attention function, which processed each image patch individually. The self-attention function maps the input image-tokens to Query (Q), Key

Table 1. Summary of rice disease dataset.

Disease classes	Original images	Augmented images	Total images
Bacterial leaf blight	839	5,034	5,873
Narrow brown spot	1,053	5,265	6,318
Sheath rot	876	5,255	6,131
Leaf blast	1,199	4,796	5,995
Neck blast	841	5,046	5,887
Sheath blight (leaf)	1,002	5,010	6,012
Sheath blight (sheath)	1,348	5,392	6,740
False smut	853	5,118	5,971
Brown spot	2,067	4,134	6,201
Tungro disease	1,308	5,232	6,540
Healthy	936	5,616	6,552
Total	12,322		68,220

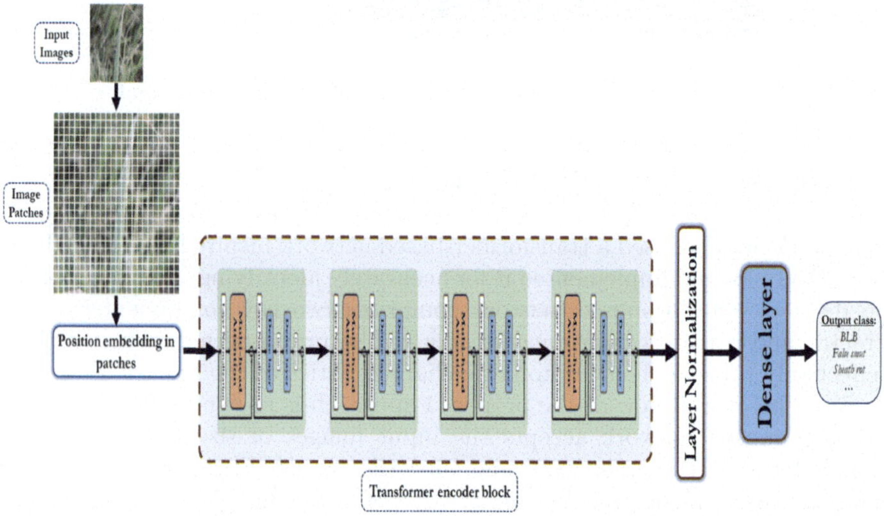

Fig. 2. Overall architecture of the proposed Rice disease identification network

(K), and Value (V) by multiplying them with three learnable weight-matrices such as W^Q,$W^K$and W^V [17]. Then, the Query (Q) and Key (K) components of the image-tokens were multiplied to generate a matching score between them. After that, the softmax function was applied to the matching score to generate attention scores of the image-tokens in the form of probability values [5]. Finally, the attention scores-weighted sum of Value (K) components were considered the self-attention output of the input image-tokens as shown in Eq. (1):

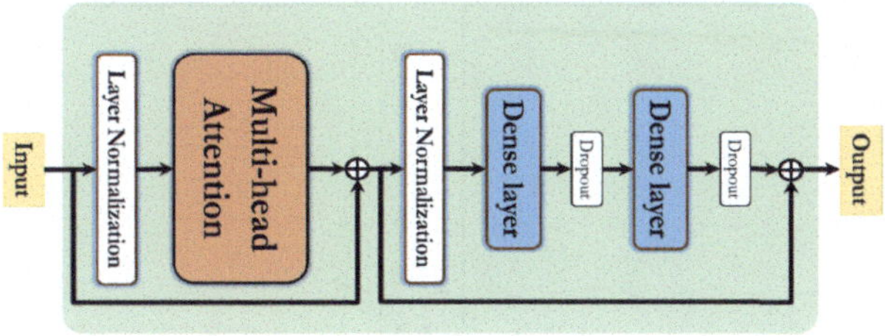

Fig. 3. Architecture of the transformer encoder

$$self - attention = \sum_j (softmax((Q \cdot K_j^T)/\sqrt{d_q})) \cdot V_j \qquad (1)$$

The whole experimentation was conducted on NIVIDIA-DGX machines configured at Krishimegh cloud infrastructure at ICAR-IASRI. We implemented the models using tensorflow frameworks using Python. The hyperparameter configurations of the proposed network are as follows: Image size: $384 \times 384 \times 3$; Patch size: 16×16; transformer encoder: 4; No of heads in MHA: 4; Batch size: 32; Epochs: 100. We compiled the network using Adam optimization function, Categorical cross-entropy loss function with a learning rate of 0.0001

5 Results and Discussion

In this research work, a vision-transformer-based network was developed and implemented for the identification of diseases in rice crop. Our proposed network consists of 8.6M trainable parameters, which is much less than the popular state-of-the-art network architectures. The rice disease dataset contains around 68K RGB images comprising 11 classes. The entire image dataset was divided into two subsets, viz., training set and testing set, in an 80:20 ratio. We measured the performance of the proposed network in terms of accuracy, precision, Recall and f1-scores, calculated on the testing dataset. Figure 4 illustrates the performance of the proposed network on the rice images during both the training and testing phases. The convergence of the network on the rice dataset is clearly depicted in the figure, showcasing the efficiency of the network. The proposed network performed magnificently on the rice disease dataset, obtained a training accuracy of 97.89% with a loss of 0.066 and a testing accuracy of 98.83% with a loss of 0.063. The decreasing trend of the loss values of the model over the iterations also implies no overfitting in the network for the rice image dataset.

Our proposed network obtained 98.32%, 98.51% and 98.87% as precision, recall and f1-score metrics, respectively. We further presented the class-wise experimental results of the proposed network in Table 2. The findings in the table reveal that precision, recall and f1-score metrics consistently vary within the 96–100% range across all the classes. Among all the classes, the proposed model performed best for the Tugro disease-infected images and least for the Bacterial leaf blight disease-infected ones. Despite these variations, the overall performance of the network was quite impressive.

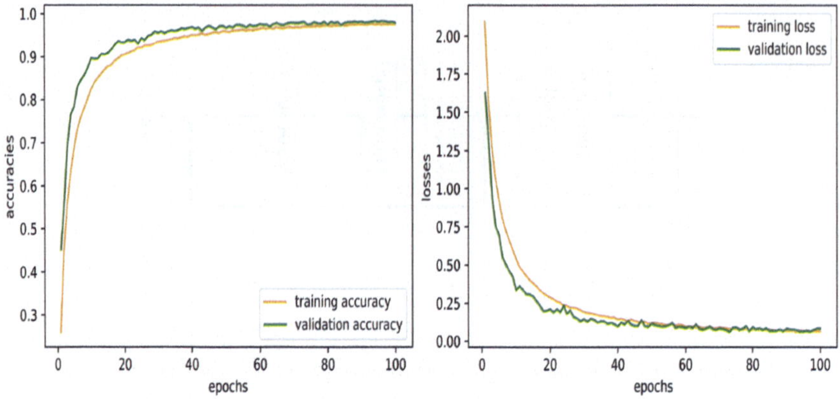

Fig. 4. Accuracy and loss curves of the proposed network

Table 2. Class-wise performance measure of the proposed network.

Disease classes	Precision	Recall	F1-Scores
Bacterial leaf blight	0.9634	0.9764	0.9748
Narrow brown spot	0.9839	0.9683	0.9761
Sheath rot	0.9885	0.9788	0.9836
Leaf blast	0.9824	0.9842	0.9883
Neck blast	0.9992	1.0000	0.9996
Sheath blight (leaf)	0.9792	0.9858	0.9975
Sheath blight (sheath)	0.9863	0.9978	0.9970
False smut	0.9727	0.9832	0.9779
Brown spot	0.9736	0.9936	0.9936
Tungro disease	1.0000	1.0000	1.0000
Healthy	0.9862	0.9677	0.9870

Next, we presented a comparative performance analysis between our proposed work and a few popular works related to rice disease detection in Table 3. Here, we primarily highlighted the key aspects like the deep learning network used, source of image dataset, nature of dataset collection, no of classes considered and respective performances in these datasets. In Table 4, it is clearly demonstrated that our proposed network achieved superior performance in rice disease identification across 11 classes, even for the images collected under natural conditions.

Furthermore, we also compared our proposed work with the one proposed by [33] to demonstrate the superiority of our proposed network, as presented in Table 4. In [33], authors proposed a ResNet-based ViT network to address rice disease classification within three distinct classes, achieving an impressive

Table 3. Performance comparative analysis

Works	Methodology	Dataset	Cls	Env	Acc
[27]	Custom CNN network	Public dataset	3	Controlled	95.83%
[28]	ResNet50 with SVM	Author collected	4	In-field	98.38%
[14]	CNN with SVM	Author collected	6	In-field	96.8%
[4]	Ensemble networks	Author collected	6	In-field	91%
[21]	Custom CNN network	Author collected	11	In-field	95.48%
[2]	Inception with MobileNet	Internet sources	11	Mixed	97.89%
Our work	**ViT-based network**	**Author collected**	**11**	**In-field**	**98.83%**

Table 4. Comparision of our work with work proposed by [33]

Works	Network	Dataset	Cls	Acc	F1	Paras
[33]	ResNet with ViT	Public dataset	3	99.87%	96%	11.3M
Our work	**Pure ViT-based**	**Author collected**	**11**	**98.83%**	**98.87%**	**8.6M**

accuracy of 99.87%. In contrast, our study focused on leveraging a purely ViT-based network for classifying the diseased images of rice into a wider range of 11 classes, obtaining an accuracy of 98.83%. Upon closer examination, it can be seen that the images used by [33] were captured under controlled conditions having uniform backgrounds, whereas we collected the images in natural conditions with complex backgrounds. As a result, our proposed network achieved a higher f1-score (98.87%) than that of [33] (96%). Additionally, the network proposed by [33] has 11.3 M trainable parameters, whereas our network is much lighter than that, having only 8.6M trainable parameters. In summary, while our network may have a slightly lower classification accuracy compared to [33], it is much more efficient, robust, and effective in detecting rice diseases in diverse real-life conditions.

6 Conclusion

In this study, disease-infected images of rice crop were captured from agricultural research fields in different locations across the country. Using these images, a vision transformer-based network has been developed for the automated identification and classification of rice diseases into 11 predefined classes. The proposed network comprises four transformer encoders, each having a multi-head attention mechanism and a fully connected layer. At first, the input images were converted into patches of 16×16 size and these patches were processed by the transformer encoders to extract the relevant patterns. The proposed network was able to generalize well on the rice image dataset, achieving a classification accuracy of 98.83% on the testing dataset. The experimentation findings demonstrated that the lightweight ViT networks are sufficiently effective in extracting the relevant features/patterns from the diseased images. Additionally, the comparative performance analysis of our proposed network with previously proposed networks demonstrated that our network is superior in terms of effectiveness, robustness and efficiency. Further, this study can be extended for image-based detection of the severity stages of the diseases.

References

1. Atila, Ü., Uçar, M., Akyol, K., Uçar, E.: Plant leaf disease classification using efficientnet deep learning model. Eco. Inform. **61**, 101182 (2021)
2. Chen, J., Chen, W., Zeb, A., Yang, S., Zhang, D.: Lightweight inception networks for the recognition and detection of rice plant diseases. IEEE Sens. J. **22**(14), 14628–14638 (2022)
3. Daniya, T., Vigneshwari, S.: Deep neural network for disease detection in rice plant using the texture and deep features. Comput. J. **65**(7), 1812–1825 (2022)
4. Deng, R., Tao, M., Xing, H., Yang, X., Liu, C., Liao, K., Qi, L.: Automatic diagnosis of rice diseases using deep learning. Front. Plant Sci. **12**, 701038 (2021)
5. Dosovitskiy, A., et al.: An image is worth 16×16 words: transformers for image recognition at scale. arXiv preprint arXiv:2010.11929 (2020)
6. Ferentinos, K.P.: Deep learning models for plant disease detection and diagnosis. Comput. Electron. Agric. **145**, 311–318 (2018)
7. Gole, P., Bedi, P., Marwaha, S., Haque, M.A., Deb, C.K.: Trincnet: a lightweight vision transformer network for identification of plant diseases. Front. Plant Sci. **14**, 1221557 (2023)
8. Haque, M.A., et al.: A lightweight convolutional neural network for recognition of severity stages of maydis leaf blight disease of maize. Front. Plant Sci. **13**, 1077568 (2022)
9. Haque, M.A., et al.: Image-based identification of maydis leaf blight disease of maize (zea mays) using deep learning. Indian J. Agric. Sci. **91**(9), 1362–7 (2021)
10. Haque, M.A., Marwaha, S., Deb, C.K., Nigam, S., Arora, A.: Recognition of diseases of maize crop using deep learning models. Neural Comput. Appl. **35**(10), 7407–7421 (2023)
11. Haque, M.A., et al.: Deep learning-based approach for identification of diseases of maize crop. Sci. Rep. **12**(1), 6334 (2022)
12. Haridasan, A., Thomas, J., Raj, E.D.: Deep learning system for paddy plant disease detection and classification. Environ. Monit. Assess. **195**(1), 120 (2023)
13. Hughes, D., Salathé, M., et al.: An open access repository of images on plant health to enable the development of mobile disease diagnostics. arXiv preprint arXiv:1511.08060 (2015)
14. Jiang, F., Lu, Y., Chen, Y., Cai, D., Li, G.: Image recognition of four rice leaf diseases based on deep learning and support vector machine. Comput. Electron. Agric. **179**, 105824 (2020)
15. Jiang, M., Feng, C., Fang, X., Huang, Q., Zhang, C., Shi, X.: Rice disease identification method based on attention mechanism and deep dense network. Electronics **12**(3), 508 (2023)
16. Kamilaris, A., Prenafeta-Boldú, F.X.: Deep learning in agriculture: a survey. Comput. Electron. Agric. **147**, 70–90 (2018)
17. Khan, S., Naseer, M., Hayat, M., Zamir, S.W., Khan, F.S., Shah, M.: Transformers in vision: a survey. ACM Comput. Surv. (CSUR) **54**(10s), 1–41 (2022)

18. Khush, G.S.: What it will take to feed 5.0 billion rice consumers in 2030. Plant Mol. Biol. **59**, 1–6 (2005)
19. Lee, S.H., Goëau, H., Bonnet, P., Joly, A.: New perspectives on plant disease characterization based on deep learning. Comput. Electron. Agric. **170**, 105220 (2020)
20. Liu, W., Yu, L., Luo, J.: A hybrid attention-enhanced densenet neural network model based on improved u-net for rice leaf disease identification. Front. Plant Sci. **13**, 922809 (2022)
21. Lu, Y., Yi, S., Zeng, N., Liu, Y., Zhang, Y.: Identification of rice diseases using deep convolutional neural networks. Neurocomputing **267**, 378–384 (2017)
22. Mahajan, G., Kumar, V., Chauhan, B.S.: Rice production in India. Rice Production Worldwide, pp. 53–91(2017)
23. Marwaha, S., Deb, C.K., Haque, M.A., Naha, S., Maji, A.K.: Application of artificial intelligence and machine learning in agriculture. In: Harohalli Masthigowda, M., Gopalareddy, K., Khobra, R., Singh, G., Pratap Singh, G. (eds.) Translating Physiological Tools to Augment Crop Breeding, pp. 441–457. Springer, Singapore (2023). https://doi.org/10.1007/978-981-19-7498-4_21
24. Mohanty, S.P., Hughes, D.P., Salathé, M.: Using deep learning for image-based plant disease detection. Front. Plant Sci. **7**, 1419 (2016)
25. Nigam, S., Jain, R., Marwaha, S., Arora, A., Haque, M.A., Dheeraj, A., Singh, V.K.: Deep transfer learning model for disease identification in wheat crop. Eco. Inform. **75**, 102068 (2023)
26. Pandi, S.S., Senthilselvi, A., Gitanjali, J., ArivuSelvan, K., Gopal, J., Vellingiri, J.: Rice plant disease classification using dilated convolutional neural network with global average pooling. Ecol. Model. **474**, 110166 (2022)
27. Patidar, S., Pandey, A., Shirish, B.A., Sriram, A.: Rice plant disease detection and classification using deep residual learning. In: Bhattacharjee, A., Borgohain, S., Soni, B., Verma, G., Gao, XZ. (eds.) Machine Learning, Image Processing, Network Security and Data Sciences, MIND 2020, Communications in Computer and Information Science, vol. 1240, pp. 278–293. Springer, Singapore (2020). https://doi.org/10.1007/978-981-15-6315-7_23
28. Sethy, P.K., Barpanda, N.K., Rath, A.K., Behera, S.K.: Deep feature based rice leaf disease identification using support vector machine. Comput. Electron. Agric. **175**, 105527 (2020)
29. Shrivastava, V.K., Pradhan, M.K., Thakur, M.P.: Application of pre-trained deep convolutional neural networks for rice plant disease classification. In: 2021 International Conference on Artificial Intelligence and Smart Systems (ICAIS), pp. 1023–1030. IEEE (2021)
30. Simhadri, C.G., Kondaveeti, H.K.: Automatic recognition of rice leaf diseases using transfer learning. Agronomy **13**(4), 961 (2023)
31. Sudar, K.M., Nagaraj, P., Yeshwanth, K.V., Kumar, Y.D., Kumar, V.S.J., Reddy, V.N.S.: Recognitionof diseases in paddy using deep learning. In: 2022 6th International Conference on Intelligent Computing and Control Systems (ICICCS), pp. 1458–1463. IEEE (2022)
32. Too, E.C., Yujian, L., Njuki, S., Yingchun, L.: A comparative study of fine-tuning deep learning models for plant disease identification. Comput. Electron. Agric. **161**, 272–279 (2019)
33. Zhang, Y., Zhong, L., Ding, Y., Yu, H., Zhai, Z.: Resvit-rice: a deep learning model combining residual module and transformer encoder for accurate detection of rice diseases. Agriculture **13**(6), 1264 (2023)

Image Based Rice Weed Identification Using Deep Learning and Attention Mechanisms

Sapna Nigam[1(✉)], Ashish Kumar Singh[2], Vaibhav Kumar Singh[3], Bishnu Maya Bashyal[3], Sudeep Marwaha[1], and Rajender Parsad[1]

[1] ICAR-Indian Agricultural Statistics Research Institute, New Delhi, India
sapna.nigam@icar.gov.in
[2] Indian Institute of Technology, New Delhi, India
[3] ICAR-Indian Agricultural Research Institute, New Delhi, India

Abstract. Weed management is a critical aspect of modern agriculture, directly impacting crop yield and quality. In rice cultivation, weeds can significantly reduce productivity, leading to economic losses and increased environmental impact due to the excessive use of herbicides. In this study, authors have proposed a model ResNet50 based on convolutional neural networks (CNNs) architecture and attention mechanism module to develop a robust and efficient system for weed detection and classification. A comprehensive dataset of images is also developed, containing two major weed species, *Cyperus difformis, and Echinochloa colona,* commonly found in rice fields and healthy rice plants, to train and validate our deep learning models. The proposed system is designed to detect the presence of weeds in real time by analyzing images captured within the rice fields. It can distinguish between weeds and healthy rice plants with a testing accuracy of 97.40% and an average F1 score of 96.5%. Additionally, our model can be deployed into mobile applications for weed detection for implementing targeted weed management strategies for farmers.

Keywords: Weed detection · rice weed classification · deep learning · attention mechanism

1 Introduction

Rice is a staple crop consumed by more than half of the world's population [1] faces a significant threat to its productivity in the form of weeds. Weeds not only pose substantial economic challenges but also compromise the quality of rice crops. Throughout the growth cycle of rice, weeds present a constant competition for vital resources, including water, sunlight, and nutrients [2, 3]. Early intervention in weed control offers multiple advantages, preventing potential crop yield losses of up to 34% and reducing the prevalence of pests and diseases. Consequently, chemical and non-chemical weed control methods have been widely deployed in rice fields. However, non-chemical techniques such as manual weeding are labor-intensive, expensive, and time-consuming. An alternative non-chemical approach involves mechanical weed control, where machine vision

D. Pastor-Escuredo et al. (Eds.): ICDLAIR 2023, LNNS 1001, pp. 742–751, 2024.
https://doi.org/10.1007/978-3-031-60935-0_64

technology plays a pivotal role in enabling real-time weed tracking by robotic mechanical weeding systems [4]. However, using herbicides to control weeds in fields has benefits but also comes with problems. The traditional method of spraying herbicides can harm the environment and safety. Therefore, we need to find more sustainable and automated ways to identify weeds and reduce herbicide usage.

Above challenge of automated weed detection can be resolved through computer vision and artificial intelligence-based smart systems. As per the literature, artificial intelligence techniques have shown prominent results in image-based automated disease classification [5–7]. Similarly, several studies have focused on camera-based methods for automatically identifying and classifying weeds. For instance, [8] employed a 101-dimensional feature extraction approach, encompassing color, shape, and texture, to enhance recognition accuracy in rice fields. They utilized deep belief networks with fusion features, achieving a recognition rate of 91.13%. On the other hand,[9] introduced two classification methods for assessing weed density in images. The first method employed a Support Vector Machine (SVM) based on the gray-level co-occurrence matrix (GLCM) and attained an accuracy of 73%. The second method utilized features like invariant scale and rotation moments in combination with a Random Forest classifier, achieving an accuracy of 86%. However, these techniques were limited as they did not address other types of weeds, such as broadleaf and sedges [10].

In [12], a practical and efficient method for maize detection achieved consistently high accuracy over three years, with overall accuracy rates of 90.19%, 92.36%, and 93.87%. [13] introduced a novel vision-based classification system that demonstrated remarkable performance, achieving an impressive F1 score of 99.29% for weed identification. The combination of fine-tuned DenseNet and Support Vector Machine yielded a micro F1 score of 99.29%, with other approaches consistently achieving F1 scores exceeding 95%. Furthermore, [14] detailed the development of a site-specific weed control system, effectively reducing herbicide use through a map-based approach. [15] presented a targeted pesticide spraying control system designed to maintain consistent lighting conditions and achieved real-time cabbage identification with 95.7% accuracy. Automation in precision farming practices for weed control, emphasizing machine learning and deep learning algorithms, was discussed in [16], particularly for high-accuracy weed mapping in rice fields using remote sensing. In [17], a deep learning model named WeedDet demonstrated exceptional accuracy with a mean average precision of 94.1% and an impressive 24.3 frames per second inference time, outperforming other models in weed detection within paddy fields. Finally, [18, 19] conducted a comprehensive review of various CNN architectures and segmentation models, contributing to the evolving landscape of weed detection research.

Most prior research efforts have predominantly focused on weed detection utilizing pre-trained convolutional neural networks (CNNs), resulting in improved performance compared to conventional machine learning methods. However, there remains unexplored potential in exploring other deep learning techniques like attention mechanisms to achieve even higher accuracy when dealing with the visual similarity in color, shape, and texture between weeds and crops. Furthermore, it is worth noting that the aforementioned studies primarily utilized publicly available datasets, many of which were not acquired under real-life field conditions. Given these limitations, our research aims to address this gap by using a pre-trained deep learning architecture integrated with attention mechanisms on a dataset of real field images, specifically focusing on two common weeds in rice crops. In summary, our main contributions in this paper are:

1. A natural field condition image dataset of weed images with three classes, namely *Echinochloa colona* and *Cyperus difformis* and healthy leaves from paddy fields.
2. A novel model for precise weed detection in paddy fields, employing transfer learning, the ResNet model, and a convolutional block attention module for weed classification.

2 Methodology

2.1 Dataset

The weed images were collected from the rice fields of ICAR-Indian Agricultural Research Institute, New Delhi. The rice crop's most prominent weeds during June-October 2023 are *Echinochloa colona* and *Cyperus difformis.* The *Echinochloa colona,* commonly called Barnyard Grass, is an annual weed found in agricultural regions, with a notable presence in rice fields. This weed, belonging to the Poaceae family, exhibits varying heights, elongated and slender leaves with a distinct midrib, and an inflorescence consisting of dense clusters of spikelets bearing multiple seeds. *Cyperus difformis*, commonly known as Small flower Umbrella Sedge, is a perennial weed species that belongs to the Cyperaceae family. They are relatively small in stature, typically growing to a height of 15 to 60 cm (approximately 6 to 24 inches). It has linear, grass-like leaves, often growing in clusters or tufts and producing characteristic umbrella-shaped inflorescences. These inflorescences consist of numerous small, densely packed spikelets with tiny flowers. Both weeds are known for their competitive nature and adverse effects on crops, especially rice, making it a significant concern for farmers seeking effective weed management strategies. Also, a few healthy rice plant leaves images were also collected (Fig. 1).

A total of 600 images of weeds, 200 images in each class of weed, were used for the experimentation after augmentation. These images were collected through a mobile device, iPhone 11, from 10:00 a.m. to 12:00 p.m. The device was focused at 25–30 cm from the weed with different angles.

2.2 Data Pre-processing

As mentioned earlier, the dataset created by the author consists of RGB images with arbitrary sizes. Firstly, all images are resized to a uniform size of 256*256 pixels. Subsequently, both model optimization and prediction are carried out on these resized images. Secondly, all pixel values are normalized by dividing them by 255 to align with the network's initial values. In the third step, training images are subjected to various random augmentations, including rotation, shearing, and flipping. These augmentations enhance the model's ability to generalize and prevent overfitting.

(a) Healthy rice leaves

(b) Echinochloa colona

(c) Cyperus difformis

Fig. 1. Rice weed images

2.3 Model Architecture

In this research, the authors introduced a model architecture for classifying prominent rice weed images, employing transfer learning and attention mechanisms. The fundamental structure of the model is illustrated in Fig. 3. The model is composed of a pre-trained ResNet-CBAM backbone, followed by global average pooling, batch normalization, and a fully connected layer. The model is constructed by initially utilizing the pre-trained ResNet model, which has been trained on the ImageNet dataset, as a starting point. Subsequently, a straightforward attention mechanism called CBAM is introduced and incorporated into the final ResNet residual block. Detailed explanations of the base ResNet50 architecture and CBAM are provided in the following sections.

2.3.1 Residual Network (ResNet 50) Architecture

ResNet50 belongs to the ResNet model family, specifically designed to address the vanishing gradients problem in deep neural networks. ResNet-50 is composed of a total of 50 layers, encompassing various types of layers such as convolutional, pooling, and fully connected layers [20]. The architecture of ResNet-50 is based on residual blocks, which allow the network to learn residual mappings instead of direct mappings. This helps to alleviate the vanishing gradient problem and enables the network to be trained more effectively. ResNet-50 has been shown to achieve state-of-the-art performance on various computer vision tasks, including image classification, object detection, and semantic segmentation. In this study, ResNet-50 is used as the backbone architecture for integrating CBAM module, which helps to improve the feature extraction of the model and reduce the impact of irrelevant or noisy features.

2.3.2 Convolutional Block Attention Module (CBAM)

CBAM is an attention mechanism that can be integrated into convolutional neural networks (CNNs) to improve their performance in various computer vision tasks [21]. CBAM consists of two separate sub-modules: the Channel Attention Module (CAM) and the Spatial Attention Module (SAM), as described in Fig. 2. The CAM performs channel attention by learning to weight the importance of each feature map channel. In contrast, the SAM performs spatial attention by learning to weigh the importance of each spatial location within each feature map. The CAM and SAM are combined to produce a joint attention map, which is used to modulate the feature maps produced by the CNN. The architecture is illustrated below:

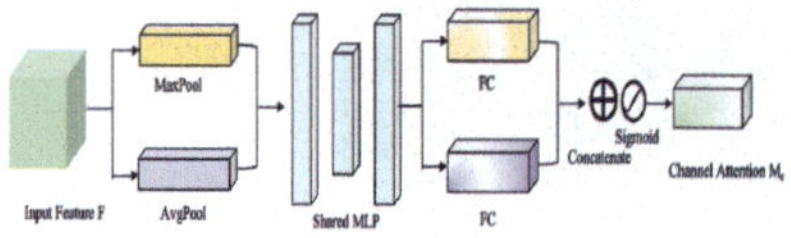

(a) Channel attention module (CAM)

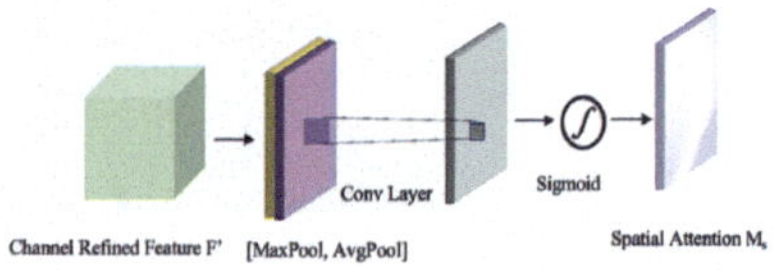

(b) Spatial attention module (SAM)

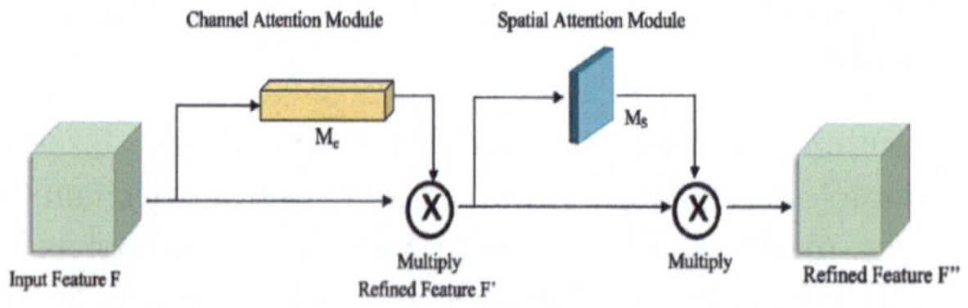

Fig. 2. Overall CBAM architecture

2.3.3 Proposed Model Framework

Our proposed ResNet-CBAM model is initialized with pre-trained weights from the ImageNet dataset and fine-tuned on the dataset used in this study. The global average pooling layer is used to reduce the spatial dimensions of the feature maps produced by the base model while preserving the channel-wise information. The fully connected layer maps the pooled features to the output classes, and the SoftMax activation function makes a probability distribution over the classes. All the parameters of the model, such as the ResNet, the CBAM, and the new FC layers, are retrained. The CBAM module's attention mechanism helps the model focus on the most informative features of the input image, which can improve the discriminative power of the model and reduce the impact of irrelevant or noisy features in major weed image classification in rice. The authors found that the transferred ResNet50-CBAM model achieved the best performance for weed classification, which is higher than the accuracy achieved by the ResNet50 model without CBAM. The improved sensitivity, precision, F1-score, and overall accuracy metrics Table 3 also demonstrated the effectiveness of CBAM. The CBAM module helps the model to focus on the most informative features of the input image, which can improve the discriminative power of the model and reduce the impact of irrelevant or noisy features.

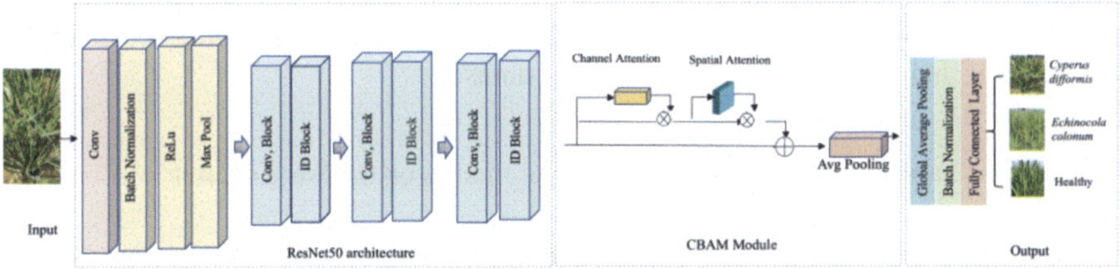

Fig. 3. Proposed model architecture

CBAM introduces a channel-wise attention mechanism, which allows the model to adaptively focus on different channels of the feature maps. Spatial attention enables the model to focus on specific spatial locations within the feature maps, allowing it to capture fine-grained details and improve localization accuracy. Thus, by incorporating both channel and spatial dependencies in the image data, it generates the better feature map. CBAM can be integrated anywhere in the CNN model. Here we have applied the CBAM over the feature map of ResNet50 which generates the better discriminative feature map.

2.4 Hyperparameters

Hyperparameters refer to parameters not learned by the model but manually adjusted by the programmer to ensure effective training. For the training of the ResNet50-CBAM model, the selected hyperparameters consisted of a batch size of 32, 50 epochs, a learning rate set at 0.001, utilization of the Adamax optimizer, and the application of categorical cross-entropy as the loss function. Additionally, a dropout rate of 0.15 was introduced to prevent overfitting, while a momentum of 0.9 was employed during experimentation.

2.5 Experimental Implementation

The experiments were carried out on a GPU server, utilizing the Keras and TensorFlow frameworks. The operating system employed was Ubuntu, operating on an Intel® Xeon® CPU. The GPU responsible for the computations was the NVIDIA Tesla A100-SXM, featuring a substantial 80 GB of memory (Table 1).

Table 1. Experimental setup specifications

Operating System	GPU	Memory	Frameworks	Programming Language	System
Ubuntu	NVIDIA Tesla A100 SXM	80 GB	Keras and TensorFlow	Python	Intel Xeon®

3 Results and Discussion

In this study, we conducted experiments utilizing various pre-trained CNN models, including VGGNet, InceptionNet, and ResNet50 architectures. Subsequently, we incorporated the attention mechanism CBAM into these models, using them as base models, and evaluated their performance on our weed dataset.

Table 2. Performance evaluation of models on the Weed dataset

Model	Training Accuracy	Testing accuracy
VGGNet	85.24	86.12
VGGNet + CBAM	87.98	87.10
InceptionV3	90.45	90.13
InceptionV3+CBAM	92.93	91.67
ResNet50	93.62	93.81
ResNet50-CBAM (Proposed model)	96.44	97.40

From Table 2, we observe that including the CBAM consistently led to improved testing accuracy across various model architectures, including VGGNet, InceptionV3, and ResNet. This increased accuracy is crucial for weed identification in rice fields, where precise classification is essential to differentiate between weeds and healthy crops. ResNet-CBAM, in particular, achieved the highest testing accuracy of 97.40%. This exceptional performance indicates that adding CBAM greatly enhances the model's ability to distinguish between weed categories and healthy rice plants. The high testing accuracy of the models, especially ResNet-CBAM, suggests that these models can be applied at scale to large rice fields. The models' efficiency and accuracy make them suitable for both small and large agricultural operations.

3.1 Evaluation Metrics

Moreover, the performance of our proposed model was validated by employing the evaluation metrics mentioned in the classification report (Table 3). Our model's accuracy involves various metrics, including precision, accuracy, F1 score, and recall. Precision signifies the ratio of true positives to all positive predictions, while accuracy quantifies the accurately identified predictions out of the total predictions. Recall assesses the true positives relative to all potential positive predictions. The F1 Score, on the other hand, is computed as the harmonic mean of precision and recall. These measures are determined using the following formulas:

$$Precision = \frac{\text{True Positives}}{True\ Positives + False\ Positives} \quad (1)$$

$$Accuracy = \frac{\text{True Positives } + True\ Negatives}{Total\ Predictions} \quad (2)$$

$$Recall = \frac{True\ Positives}{True\ Positives + False\ Negatives} \quad (3)$$

$$F1 = 2 * \frac{Precision * Recall}{Precision + Recall} \quad (4)$$

Table 3. Performance evaluation metrics for the proposed model

Class	Precision	Recall	F1 Score
Healthy	0.97	0.96	0.97
Cyperus difformis	0.97	0.97	0.96
Echinochloa colona	0.96	0.97	0.96

Table 3 evaluates the proposed model's performance in classifying three distinct classes: Healthy, *Cyperus difformis*, and *Echinochloa colona*. For the "Healthy" class, the model demonstrates high precision (97%), accurately predicting this category 97% of the time. The recall stands at 96%, indicating the model's ability to identify 96% of "Healthy" images, resulting in an F1 score of 97%. The *Cyperus difformis* class exhibits similar precision and recall values of 97%, with a slightly lower F1 score of 96%. In the case of *Echinochloa colona*, the model maintains a precision of 96%, correctly identifying 97% of actual *Echinochloa colona* images, resulting in an F1 score of 96%. These results collectively signify the model's strong performance and accuracy in classifying the mentioned classes, essential for efficient weed detection in rice fields.

Figure 4 illustrates the training and validation loss and accuracy curves for proposed model. These curves demonstrates that the model is good at both making accurate predictions and not missing actual instances. ResNet50-CBAM model is reliable and consistent in accurately identifying different weed categories. Its high performance suggests that it's an effective and precise tool for automating the detection of two major weed species

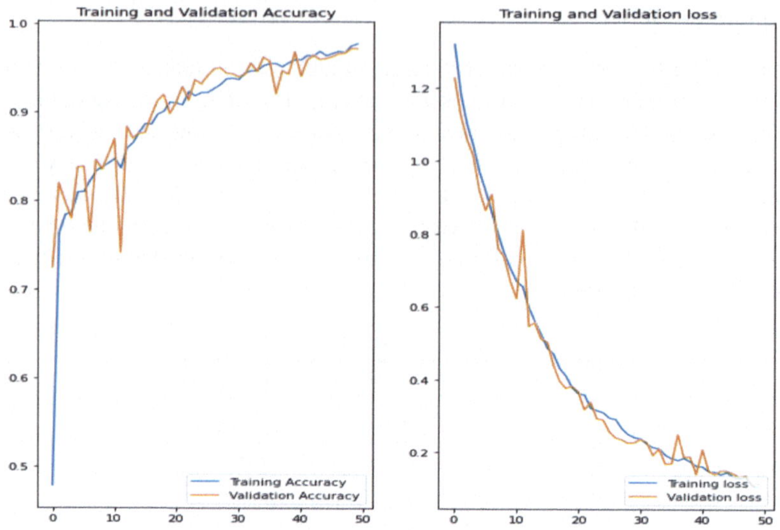

Fig. 4. Training and Validation loss and accuracy curves for the proposed model

in rice fields. Since among all the experimented CNN models here, ResNet50 has the second least number of parameters with 25 M. Other experimented models like VGG and InceptionV3 takes 140 M and 24 M parameters respectively. Although InceptionV3 takes 24 M parameters but has the network depth of 189 layers. So InceptionV3 is computationally complex in terms of number of layers and VGGNet is computationally expensive with 24 M parameters. This justifies the significance of ResNet50 over VGGNet and InceptionV3 with overall better performance, making it easily deployable in the mobile and portable devices.

4 Conclusion

This study addresses the issue of weed management in modern agriculture, particularly its impact on crop yield, quality, and the environment in the context of rice cultivation. Deep learning techniques, especially supervised learning, have demonstrated promising weed identification and classification outcomes. In this study, the authors have introduced a deep learning model for identifying two major weeds commonly found in India's rice fields. The model is developed using a pre-trained ResNet 50 architecture and a convolutional block attention mechanism at the end of the model to distinguish between weeds and rice crops. The model has achieved a testing accuracy of 93.81% with the attention mechanism and 97.40% without it. The experiments conducted highlight the potential of the attention mechanism in detecting variations in color, texture, and other features between weeds and healthy crops. Furthermore, the developed model can be seamlessly integrated into mobile applications, facilitating targeted weed management strategies for farmers. The implementation of our proposed model has the potential to improve agricultural practices, mitigate economic losses, reduce herbicide usage, and promote sustainable farming methods.

References

1. Bin Rahman, A.R., Zhang, J.: Trends in rice research: 2030 and beyond. Food Energy Secur. **12**(2), e390 (2023)
2. Nigam, S., Jain, R.: Plant disease identification using deep learning: a review. Indian J. Agric. Sci. **90**(2), 249–257 (2019)
3. Xu, K., Yuen, P., Xie, Q., et al.: WeedsNet: a dual attention network with RGB-D image for weed detection in natural wheat field. Precis. Agric **25**, 460–485 (2024). https://doi.org/10.1007/s11119-023-10080-2
4. Shahi, T.B., Dahal, S., Sitaula, C., Neupane, A., Guo, W.: Deep learning-based weed detection Using UAV images: a comparative study. Drones **7**(10), 624 (2023)
5. Nigam, S., et al.: Deep transfer learning model for disease identification in wheat crop. Eco. Inform. **75**, 102068 (2023)
6. Nigam, S., Jain, R., Marwaha, S., Arora, A.: Wheat rust disease identification using deep learning. De Gruyter (2021)
7. Haque, M.A., et al.: Deep learning-based approach for identification of diseases of maize crop. Sci. Rep. **12**(1), 6334 (2022)
8. Deng, Y., Yang, G., Xie, Z., Yu, J., Jiang, D., Huang, Z.: Effects of different weeding methods on the biomass of vegetation and soil evaporation in eucalyptus plantations. Sustainability **12**(9), 3669 (2020)
9. Ashraf, T., Khan, Y.N.: Weed density classification in rice crop using computer vision. Comput. Electron. Agric. **175**, 105590 (2020)
10. Hu, W., et al.: Review of deep learning-based weed identification in crop fields. Int. J. Agric. Biol. Eng. **16**(4), 1–10 (2023)
11. Zheng, Y., Zhu, Q., Huang, M., Guo, Y., Qin, J.: Maize and weed classification using color indices with support vector data description in outdoor fields. Comput. Electron. Agric. **141**, 215–222 (2017)
12. Espejo-Garcia, B., Mylonas, N., Athanasakos, L., Fountas, S., Vasilakoglou, I.: Towards weeds identification assistance through transfer learning. Comput. Electron. Agric. **171**, 105306 (2020)
13. Gerhards, R., Christensen, S.: Real-time weed detection, decision making and patch spraying in maize, sugarbeet, winter wheat and winter barley. Weed Res. **43**(6), 385–392 (2003)
14. Zhao, X., Wang, X., Li, C., Fu, H., Yang, S., Zhai, C.: Cabbage and weed identification based on machine learning and target spraying system design. Front. Plant Sci. **13**, 924973 (2022)
15. Rosle, R., et al.: Weed detection in rice fields using remote sensing technique: a review. Appl. Sci. **11**(22), 10701 (2021)
16. Peng, H., Li, Z., Zhou, Z., Shao, Y.: Weed detection in paddy field using an improved RetinaNet network. Comput. Electron. Agric. **199**, 107179 (2022)
17. Xu, K., et al.: Precision weed detection in wheat fields for agriculture 4.0: a survey of enabling technologies, methods, and research challenges. Comput. Electron. Agric. **212**, 108106 (2023)
18. Nigam, S., et al.: Wheat disease severity estimation: a deep learning approach. In: International Conference on Internet of things and Connected Technologies, pp. 185–193 (2021)
19. He, K., Zhang, X., Ren, S., Sun, J.: Deep residual learning for image recognition. In: Proceedings of the IEEE Conference on Computer Vision and Pattern Recognition, pp. 770–778 (2016)
20. Woo, S., Park, J., Lee, J.-Y., Kweon, I.S.: Cbam: Convolutional block attention module. In: Proceedings of the European Conference on Computer Vision (ECCV), pp. 3–19 (2018)

Intrusion Detection System Using Deep Learning Techniques for Internet of Medical Things (IoMT)

Naveen Saran[1(✉)], Nishtha Kesswani[3], and Ravi Saharan[2]

[1] Department of Computer Science, Central University of Rajasthan, NH-8, Bandarsindri 305817, Rajasthan, India
naveen.saran90@gmail.com

[2] Department of Computer Science and Engineering, Central University of Rajasthan, NH-8, Bandarsindri 305817, Rajasthan, India
ravisaharan@curaj.ac.in

[3] Department of Data Science & Analytics, Central University of Rajasthan, NH-8, Bandarsindri 305817, Rajasthan, India
nishtha@curaj.ac.in

Abstract. The Internet of Things (IoT) has tremendously impacted people's daily lives in recent years. Researchers and corporations also show interest in diverse IoT applications, such as the Internet of Medical Things (IoMT). In various aspects, security issues are the most important area of IoT, specifically for IoMT. The healthcare industry's primary concern is safeguarding patient data from numerous intrusion attacks. It is essential to develop an effective and secure Intrusion Detection System (IDS) for IoMT. In this research article, we have proposed an efficient Intrusion Detection System (IDS) to identify intrusion attacks on the Internet of Medical Things (IoMT). We have developed a robust Intrusion Detection System (IDS) by employing a variety of Deep Learning (DL) techniques such as Simple Recurrent Neural Networks (S-RNN), Long Short-Term Memory Networks (LSTM) and Gated Recurrent Units (GRU) on a publicly available healthcare dataset. The experiment's findings showed very high accuracy rates of intrusion detection using Simple Recurrent Neural Network (S-RNN), Long Short Term Memory Network (LSTM), Gated Recurrent Units (GRU) classification techniques compared to the state-of-art techniques. The overall accuracy of the experimented Intrusion Detection Systems was more than 99 % for all the implemented DL classification techniques.

Keywords: Internet of Medical Things (IoMT) · Intrusion Detection Systems (IDS) · Machine Learning (ML) · Deep Learning (DL) Techniques · Smart Health Care Sector · Healthcare Data Security

N. Kesswani and R. Saharan—Contributing authors.

D. Pastor-Escuredo et al. (Eds.): ICDLAIR 2023, LNNS 1001, pp. 752–763, 2024.
https://doi.org/10.1007/978-3-031-60935-0_65

1 Introduction

Millions of devices are linked together by the Internet of Things (IoT), that incorporates numerous services onto a single network without any human intervention [1]. The term IoT refers to a sensor network that consists of several powered, miniature devices that communicate with one another over wired or wireless internet or a computer network architecture to transmit the data for additional processing. Smart industries, smart cities, smart ecosystems, smart vehicular networks, and other functional areas are a few examples of IoT networks. International Data Corporation (IDC) forecasts that there would be around 41.6 billion connected IoT applications creating 79.4 Zetta-Bytes (ZB) of informative data compared to the expected human population of 8.1 billion, in 2025 [2].

In our daily lives, both at work and at home, the healthcare environment IoT applications like smart healthcare devices, remote health monitoring system, and personal health assistants gadgets are becoming essential. These Internet of Medical Things (IoMT) associated with the remote healthcare monitoring applications generate a substantial amount of continuous data from vital signs and other signals like, Electro-Encephalo-Gram (EEG), Electro-Cardio-Gram (ECG). Nowadays, people are more concerned about the security of their personal healthcare information, and several nations have already established legal protections for data security. However, incidents of intrusion attacks on healthcare data poses a significant threat to smart healthcare system. We want to ensure the security of health data from numerous intrusion attacks [3]. A major healthcare problem could result from modification in the data by the attacker, such as a mistaken diagnosis, delay in an emergency, or other health issues that could result in human life loss [4]. Consequently, the study focuses on the security of the medical data generated through IoMT in smart healthcare system.

IoMT has a major impact on how the healthcare sector develops. Furthermore, physical devices that now monitor vital signs using a range of sensors and transmit real-time healthcare data to the medical staff over the internet have become major source of medical information breach. Figure 1 depicts the patient data processing from healthcare monitoring sensors to smart healthcare data server and processing units via the network control systems. IoMT devices are easily hack-able, easy to attack, and capable of being managed remotely to create IoT-based botnets. Sensitive data leaks, violations, and infringements occur as a result of these attacks in the larger IoT-enabled system. Denial-of-Service (DOS), Ransomware, Distributed Denial-of-Service (DDoS), and Botnet attacks are some of the common attacks on IoT-based healthcare devices. As a result, an Intrusion Detection Systems (IDS) has become the best platform for defending the security, privacy, and integrity of the smart healthcare system. Since the first IDS was created, efforts have been made simultaneously to increase its effectiveness. It handles an enormous volume of data, which has an impact on system performance. The study of Machine Learning (ML) based healthcare applications is expanding quickly. Some of the ML models have already been utilised in Intrusion Detection Systems (IDS) for security issues, while others have been used in the surveillance of numerous diseases [5].

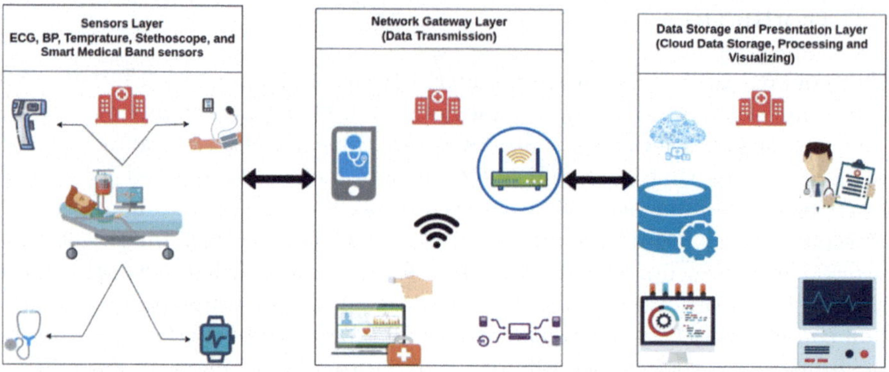

Fig. 1. Internet of Medical Things (IoMT) enabled smart healthcare system

In this research work, we have introduced an Intrusion Detection Systems (IDS) using three different Deep Learning (DL) techniques named as Simple Recurrent Neural Networks (S-RNN), Long Short-Term Memory Networks (LSTM) and Gated Recurrent Units (GRU) on the publicly available benchmark IoT Healthcare Dataset to detect the intrusion attacks on healthcare information collected from multiple IoMT healthcare devices in smart healthcare system. The above mentioned DL techniques are types of neural network architectures designed to handle the sequential time series data generated through numerous IoMT sensor devices [6]. In order to assess the effectiveness of all implemented Deep Learning (DL) classifier algorithms and choose the best DL classifier for Intrusion Detection Systems (IDS) in smart healthcare system, we also carried out a comparison of the resultant outcome derived from all DL classifier algorithms. The experiment produced data with a very high rate of accuracy. Following is a summary of this paper's significant contributions:

1. Implemented three different Deep Learning (DL) algorithms Simple Recurrent Neural Network (S-RNN), Long Short Term Memory Network (LSTM), Gated Recurrent Units (GRU) on the publically available benchmark IoT Healthcare Dataset to develop an effective and robust Intrusion Detection Systems (IDS) for Internet of Medical Things (IoMT) in smart healthcare system.
2. An analytical comparison was also carried out to assess the effectiveness of all implemented Deep Learning (DL) classifier algorithms.
3. Evaluate the best Deep Learning (DL) classifier algorithm out of all implemented DL techniques to develop a highly accurate Intrusion Detection Systems (IDS) for IoMT in smart healthcare system.

Rest of the paper is organised as follows: We briefly discussed about the literature that emphasize the already existing Intrusion Detection Systems (IDS) techniques for IoMT in Sect. 2. The dataset we utilised to carry out the different DL classification approaches is described in Sect. 3. A brief introduction of the implemented DL techniques in our research work is discussed in Sect. 4. Section 5 consists of the system configuration and experimental setup information. Preprocessing, Research model and performance metrics of the research work are addressed in Sects. 6. Results and analyses of this research work are mentioned in Sects. 7. Finally, Sect. 8 discusses the conclusion and future scope of the research work.

2 Related Work

Current section provides brief description about earlier research work done on Intrusion Detection Systems (IDS) for Internet of Medical Things (IoMT) in smart healthcare system, but most of the reviewed papers fail to solve the issue of intrusion attacks on IoT healthcare dataset.

To identify intrusion attacks in Internet of Things (IoT) network, Sah *et al.* [7] focused on COVID-19 decease early prediction using grid search cross validation approach and Monte Carlo simulation to justify the accuracy on publicly available dataset records of indian cases. Vishwakarma *et al.* [8] presented a Two-stage Intrusion Detection System (TIDS) from remote local access and user to end point access. The Naive Bayes (NB) classifier is the initial stage of intrusion detection in suggested model. The records that pretend to be normal are then passed to the k-means classifier, which is the secondary stage of the classifier. Sah *et al.* [9] tried to build a COVID-19 prediction analytical technique. For prediction of the COVID-19 confirm and active cases they used Prophet, ARIMA, and stacked LSTM-GRU models on indian healthcare dataset. Shyla *et al.* [10] proposed the Nesterov-Accelerated Adaptive Moment Estimation-Stochastic Gradient Descent (HNADAMSDG) algorithm to evaluate the performance of Intrusion Detection Systems (IDS). When compared to existing classification methods like Logistic Regression (LR), Ridge Classifier, and Ensemble algorithms, the suggested technique yields results with a high degree of accuracy. Sah *et al.* [11] used deep learning models like Recurrent Neural Network (RNN), Long Short-Term Memory (LSTM) and Gated Recurrent Units (GRU) to predict the forthcoming COVID-19 cases. Choudhary *et al.* [12] present a hybrid classification approach to detect multi-class attacks in the IoT environment. The suggested model makes use of Principle Component Analysis (PCA) and Linear Discriminant Analysis (LDA) to extract the important features and decrease the amount of features from the IoT Healthcare dataset. A combination of Neural Network (NN) and Support Vector Machine (SVM) classifiers was employed to help boost the Detection Rate and lower False Alarm Rates. Said *et al.* [13] proposed an Anomaly Detection System (ADS) for smart medical system by implementing two modules Intrusion Detection Component (IDC) and Event Detection Component (EDC) for network attacks and to identify online health scenarios. Choudhary *et al.* [14] created the Correlation and Regression based network Intrusion Detection System (CRIDS) to identify suspicious activities in the IoT network. For bi-variate, multi-variate, and distribution functions, this CRIDS method is useful. This method implies that each data collection may have two or more properties. The author found higher True Positive

Rate (TPR) in results as compared to the most recent approaches. Gadekallu *et al.* [15] proposed a block-chain based model to secure the healthcare data generated from smart healthcare applications. Amoon *et al.* [16] offered a Swarm Optimized based Artificial Immune System Approach (SOAIS) on IoT healthcare dataset in their research article. Choudhary *et al.* [17] proposed a hybrid intrusion detection method to work on the detection of sinkhole and selective forwarding attacks in the 6LoWPAN environment using Routing Protocol for Low power and Lossy Networks (RPL) Protocol. Alrashdi *et al.* [18] proposed a Fog-Based Attack Detection (FBAD) framework using an Ensemble of Online Sequential Extreme Learning Machine (EOS-ELM) to detect the malicious activity. The proposed distributed framework outperform the centralised framework by reducing the time to identify the attack in IoT network. Almulhim *et al.* [19] an efficient light-weight and robust group-based authentication scheme smart IoT-based healthcare system to securely channelize the healthcare sensor nodes and nursing (base) station. Kim *et al.* [20] followed Long Short-Term Memory Recurrent Neural Network (LSTM RNN) model with nadam optimizer approach for Intrusion Detection Systems (IDS) in network security. The proposed IDS model generates high intrusion Detection Rate (DR) and low False Alarm Rate (FAR). Khemissa *et al.* [21] proposed a light-weight authentication scheme to authenticate both, sensor enabled healthcare devices and base station to secure the patient health information.

3 Dataset Description

In this research work, we implemented S-RNN, LSTM, and GRU techniques on IoT Healthcare Dataset. The dataset includes both regular and IoT attack traffic records [22]. The dataset was implemented on a 2 bedded ICU simulation where 9 different types of patient monitoring smart healthcare devices and one control unit referred as Bedx-Control-Unit (BCU) were used to generate the patient health record under different traffic flows like attack or normal. The implemented dataset consists of 1,88,694 records with 52 different types of features under 3 different types of traffic classes a.) Attack class, b.) Environment monitoring class, and c.) Patient monitoring class. The Normal class (Environment monitoring class and Patient monitoring class) and Attack class of IoT healthcare dataset holds 1,08,568 and 80,126 records respectively. The dataset has 4 different types of attacks namely, MQTT Distributed Denial-of-Service (MDDoS), MQTT Publish Flood, Brute Force, and SlowITE [23].

4 Experimental Setup

The Internet of Medical Things (IoMT) environment is made up of many medical sensor applications that operate on a single network and offer a variety of healthcare services. We require a powerful Intrusion Detection Systems (IDS) to protect the IoMT devices against intrusion attacks in smart healthcare system. We used a computer system with the following specifications to develop a reliable IDS: Ubuntu 20.04.4 LTS operating system, 8 GB of RAM, an Intel Core i7-6700 CPU running at 3.40 GHz, and a 1 TB hard drive. Python code was written using Jupyter Notebook. We completed the research work by implementing three different DL approaches covered in the previous section on the benchmark IoT Healthcare Dataset.

5 Research Methodology

To identify the intrusion attacks in IoMT enabled smart healthcare system we processed the IoT Healthcare Dataset through the our proposed three different Deep Learning (DL) Classifiers techniques. The proposed model is schematized in 3 phases 1.) Data Pre-processing, 2.) DL model building and 3.) Result generation. Figure 2, demonstrates all the necessary steps taken to process the given dataset from initial network traffic data load to identify the dataset class whether it is Normal type data or attacked type to overcome the issues of intrusion attacks in IoMT enabled smart healthcare system.

5.1 Preprocessing

We have developed a robust Intrusion Detection Systems (IDS) using the IoT Healthcare Dataset for IoMT devices. And for this we need to pre-process the dataset to identify the intrusion attacks more effectively. To pre-process the given dataset, primarily we classify the dataset into Normal and Attack classes by denoting 0 as normal and 1 as attacked type data. Further, we drop the null values and reduce the dataset features from 52 to 42. And Extracted the best features and targeted features. We Label Encoded the selected best features out of all remaining features of the given dataset. Further, we split the pre-processed dataset into train data and test data using best features and targeted features of the given dataset and normalize the data using Min-Max Scaler. To outcast the over-fitting problem we used the Generative Adversarial Network (GAN) model on trained dataset.

5.2 Research Model

The pre-processing of the IoT Healthcare Dataset was meant to increase the quality of dataset by reducing the inappropriate features and records and retrieving the best & targeted features. To build the Deep Learning (DL) model we applied Simple Recurrent Neural Network (S-RNN), Long Short Term Memory Network (LSTM), Gated Recurrent Units (GRU) techniques on pre-processed data of given IoT Healthcare Dataset. Algorithm 1 shows all the major steps taken to develop the proposed IDS system. The best DL classifier technique for Intrusion Detection Systems (IDS) in IoMT was found after performing the overall result analysis comparison to assess the performance of all applied DL classifier techniques. We calculated precision, recall, F1-score, and overall Accuracy of the proposed system. Finally, a comparison table (Table 1) was created between proposed and state-of-art techniques proposed of earlier researchers comparing the output values produced by the Deep Learning (DL) techniques.

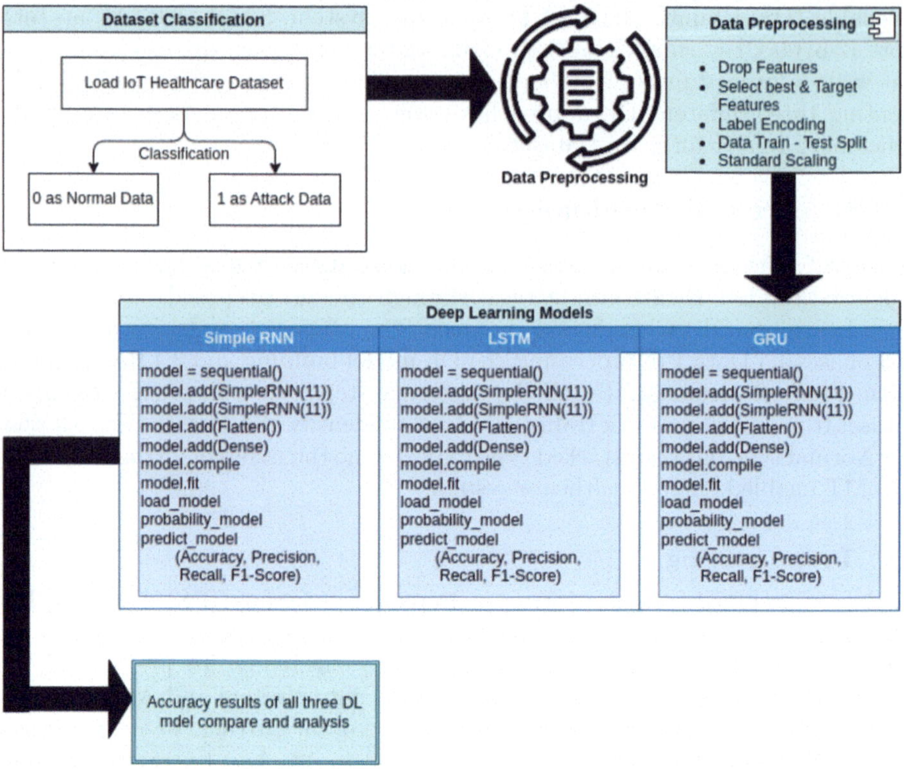

Fig. 2. Internet of Things Healthcare Data Processing Model

5.3 Performance Metrics

The False Positive Rate (FPR) and True Positive Rate (TPR) technique are used to gauge performance. FPR is chosen as a Failure or Warning Rate. It results from several instances of records being incorrectly identified. The Detection Rate (DR) is calculated by dividing the overall number of positive instances by the overall number of positively detected positive occurrences (DR). Recall or True Positive Rate (TPR) are other names for it. In Eqs. 1 to 4, the performance metrics are displayed.

$$Precision\% = \left(\frac{TP}{TP+FP}\right) * 100 \tag{1}$$

$$Recall\% = \left(\frac{TP}{TP+FN}\right) * 100 \tag{2}$$

$$F1 - Score\% = \left(\frac{2TP}{2TP+FP+FN}\right) * 100 \tag{3}$$

Algorithm 1. Algorithm of Proposed Methodology for IDS in IoMT

Input: Path to IEEE Healthcare Dataset (`path_to_dataset`), data pre-processing methods, GAN, and Deep Learning models.

Output: Accuracy (Acc), Precision (Prec), Recall (Rec), and F1-Score (F_1) of Deep Learning models.

1: **Load the dataset from the specified path:**
$D \leftarrow \text{load\_dataset}(\texttt{path\_to\_dataset})$

2: **Perform exploratory data analysis:**
For each feature f in Dataset D:
- **Compute basic statistics:** Mean (μ_{f_i}), Standard Deviation (σ_{f_i}), Median (median_{f_i}), Mode (mode_{f_i}).
- **Compute value counts:** (count_{f_i}) and display unique values (unique_{f_i}).
- **Visualize data distributions** using bar charts and pie charts.

3: **Data Preprocessing:**
- **Handle missing values:** by dropping rows with missing data. $D' \leftarrow D$ where rows with missing values are removed.
- **Feature Selection:** Select a subset $S \subseteq D'$ based on criteria (e.g., variance, correlation).
- **Encoding and Scaling:** For categorical features $C \subset S$ using label encoding $E(C)$.
- **Apply MinMax scaling:** $S_{\text{scaled}} = \frac{S - \min(S)}{\max(S) - \min(S)}$.
- **Data Splitting:** Split S_{scaled} into training and testing sets ($X_{\text{train}}, X_{\text{test}}, y_{\text{train}}, y_{\text{test}}$).

4: **Define and Train GAN Architecture:**
Define generator (G), discriminator (D), and Optimize G and D using adversarial loss $\mathcal{L}_{\text{adv}}$.

5: **Train Machine Learning Classifiers (GB,AB,RF,DT,NB):**
For each ML classifier model $M_k \in \{\mathbf{GB}, \mathbf{AB}, \mathbf{RF}, \mathbf{DT}, \mathbf{NB}\}$:
- Train M_k on ($X_{\text{train}}, y_{\text{train}}$).
- Evaluate M_k on ($X_{\text{test}}, y_{\text{test}}$) using Accuracy $\text{Acc}(M_k)$, F1 Score $F1(M_k)$, and Confusion Matrix $\text{CM}(M_k)$.

6: **Train Deep Learning Models (RNN, LSTM, GRU):**
Define RNN, LSTM, GRU model architectures.
For each DL model $M_{\text{DL}} \in \{\mathbf{RNN}, \mathbf{LSTM}, \mathbf{GRU}\}$:
- Train M_{DL} and evaluate using metrics: Accuracy $\text{Acc}(M_{\text{DL}})$, Precision $\text{Precision}(M_{\text{DL}})$, Recall $\text{Recall}(M_{\text{DL}})$, F1 Score $F1(M_{\text{DL}})$.

7: Plot training and validation loss and accuracy.

8: Save the best-performing models and their evaluations.

$$OverallAccuracy\% = \left(\frac{TP + TN}{TP + TN + FP + FN}\right) * 100 \tag{4}$$

6 Results and Analysis

In this research work we have implemented S-RNN, LSTM and GRU Deep Learning (DL) techniques to improve the accuracy of Intrusion Detection Systems (IDS) in smart healthcare system. We compared the proposed IDS system with

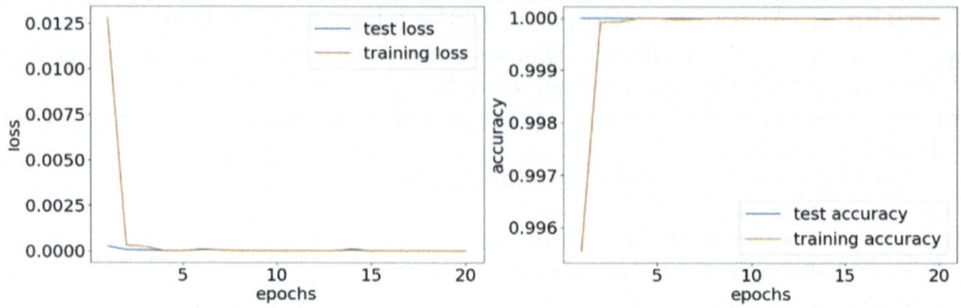

Fig. 3. Simple-RNN Loss and Accuracy Graph

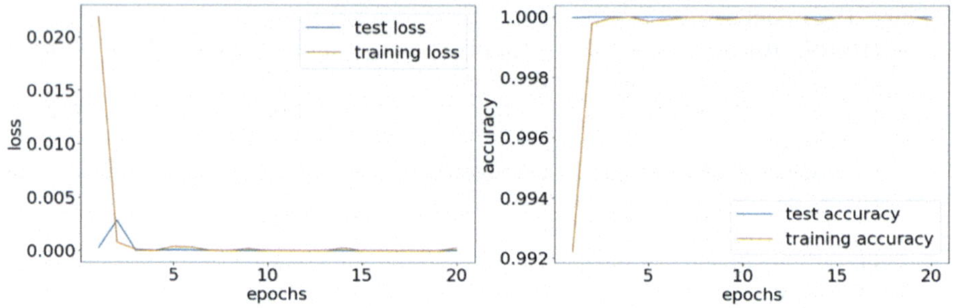

Fig. 4. LSTM Loss and Accuracy Graph

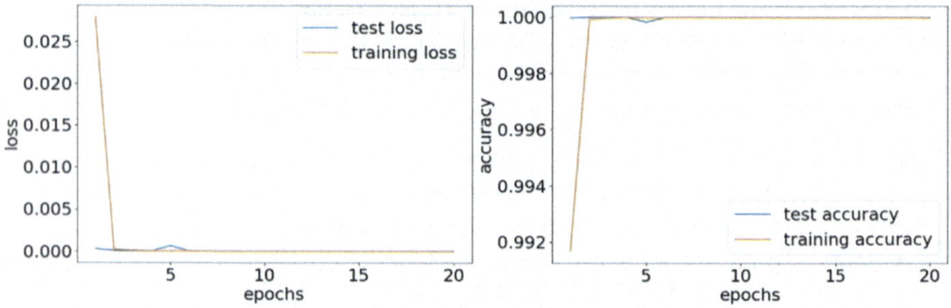

Fig. 5. GRU Loss and Accuracy Graph

state-of-art techniques for the given dataset and came up with better result outcomes. On the basis of accumulated results we found that the system's overall accuracy was higher using Long Short Term Memory Network (LSTM) and Gated Recurrent Units (GRU) performs accurately detect the intrusion attacks on given dataset. Figure 3, 4, and 5 show the Loss and Accuracy Graph of Simple-RNN, LSTM and GRU algorithms respectively.

The result comparison is also made in Table 1, between the proposed model with some other techniques offered by researchers in their earlier proposed

Table 1. Comparison of state-of-art IDS techniques with proposed IDS technique

References	Dataset	Precision (%)	Recall (%)	F1-Score (%)	Accuracy (%)
Tabassum *et al.* [4]	Simulated Data	99.70	98.90	99.10	99.87
Amoon *et al.* [16]	Healthcare Data	99.75	...	...	99.66
Alrashdi *et al.* [18]	NSL-KDD	...	97.09	...	98.19
Kim *et al.* [20]	KDD Cup99	97.69	98.95	...	97.54
Proposed	Healthcare Data	100.00	99.99	99.99	99.99

research work as discussed in literature survey section of this research article on IoMT enabled smart healthcare system using same and distinct datasets.

7 Conclusion and Future Scope

The Internet of Medical Things (IoMT) performs the healthcare operations easily and in an intelligent way. Medical sensor devices connected in an IoT environment are vulnerable to the intrusion attacks and other security issues. It is crucial to create an Intrusion Detection Systems (IDS) that is reliable and secure, capable of detecting several classes of intrusion attacks in an IoT environment, as well as feasible to figure out any new type of potential intrusion attack on a healthcare dataset. The IoT Healthcare Dataset is used to carry out the operations. We have proposed an intelligent Intrusion Detection Systems (IDS) that consists of three different Deep Learning (DL) techniques namely S-RNN, LSTM and GRU to address the issue of intrusion attack in IoMT healthcare sensor devices. The dataset consists of 52 features holding 4 classes of intrusion attacks. Almost all Deep Learning (DL) algorithm implemented in the experiments achieved the total accuracy more than 99%. The main objectives of this research are to determine the most efficient DL technique and to identify intrusion attacks in the IoMT environment. Whereas, we found that Long Short Term Memory Network (LSTM) and Gated Recurrent Units (GRU) obtained the overall accuracy for the proposed Intrusion Detection System (IDS) as 100% and accurately detect the intrusion attacks on the given IoT healthcare dataset. By applying these DL techniques on lightweight IoMT devices, the proposed IDS system may generate some overhead issues such as computational and memory overhead, energy consumption, data management and heat dissipation and all such overhead issues may carried out in the future research work of this field. Further, we can extend this research work by incorporating some new dataset holding newly generated intrusion attack classes by incorporating classification techniques such as Federated or Transfer Learning to further identify the newly developed intrusion attacks and already existing attacks in IoMT enabled smart healthcare system.

References

1. Islam, N., et al.: Towards machine learning based intrusion detection in IoIslam, networks. Comput. Mater. Contin **69**(2), 1801–1821 (2021)
2. Atlam, H.F., Wills, G.B.: IoT security, privacy, safety and ethics. In: Farsi, M., Daneshkhah, A., Hosseinian-Far, A., Jahankhani, H. (eds.) Digital Twin Technologies and Smart Cities. IT, pp. 123–149. Springer, Cham (2020). https://doi.org/10.1007/978-3-030-18732-3_8
3. Ambarkar, S.S., Shekokar, N.: Toward smart and secure IoT based healthcare system. In: Dey, N., Mahalle, P.N., Shafi, P.M., Kimabahune, V.V., Hassanien, A.E. (eds.) Internet of Things, Smart Computing and Technology: A Roadmap Ahead. SSDC, vol. 266, pp. 283–303. Springer, Cham (2020). https://doi.org/10.1007/978-3-030-39047-1_13
4. Tabassum, A., Erbad, A., Mohamed, A., Guizani, M.: Privacy-preserving distributed ids using incremental learning for IoT health systems. IEEE Access **9**, 14271–14283 (2021)
5. Ashraf, E., Areed, N.F., Salem, H., Abdelhay, E.H., Farouk, A.: Fidchain: federated intrusion detection system for blockchain-enabled IoT healthcare applications. In: Healthcare, vol. 10, p. 1110 (2022). Multidisciplinary Digital Publishing Institute
6. Yamak, P.T., Yujian, L., Gadosey, P.K.: A comparison between ARIMA, LSTM, and GRU for time series forecasting. In: ACAI 2019, pp. 49–55. Association for Computing Machinery (2020)
7. Sah, S., Surendiran, B., Dhanalakshmi, R., Yamin, M.: Covid-19 cases prediction using SARIMAX model by tuning hyperparameter through grid search cross-validation approach. Expert. Syst. **40**(5), 13086 (2023)
8. Vishwakarma, M., Kesswani, N.: A two-stage intrusion detection system (TIDS) for internet of things. In: Troiano, L., et al. (eds.) Advances in Deep Learning, Artificial Intelligence and Robotics. LNNS, vol. 249, pp. 89–97. Springer, Cham (2022). https://doi.org/10.1007/978-3-030-85365-5_9
9. Sah, S., Surendiran, B., Dhanalakshmi, R., Mohanty, S.N., Alenezi, F., Polat, K., et al.: Forecasting COVID-19 pandemic using prophet, ARIMA, and hybrid stacked LSTM-GRU models in India. Comput. Math. Methods Med. **2022**, 1556025 (2022)
10. Shyla, S., Bhatnagar, V., Bali, V., Bali, S.: Optimization of intrusion detection systems determined by ameliorated HNADAM-SGD algorithm. Electronics **11**(4), 507 (2022)
11. Sah, S., Kamerkar, A., Surendiran, B., Dhanalakshmi, R.: Predicting the trends of COVID-19 cases using LSTM, GRU and RNN in India. In: Satapathy, S.C., Peer, P., Tang, J., Bhateja, V., Ghosh, A. (eds.) Intelligent Data Engineering and Analytics. SIST, vol. 266, pp. 459–470. Springer, Singapore (2022). https://doi.org/10.1007/978-981-16-6624-7_46
12. Choudhary, S., Kesswani, N.: A hybrid classification approach for intrusion detection in IoT network. J. Sci. Ind. Res. **80**, 809–816 (2021)
13. Said, A.M., Yahyaoui, A., Abdellatif, T.: Efficient anomaly detection for smart hospital IoT systems. Sensors **21**(4), 1026 (2021)
14. Choudhary, S., Dey, A., Kesswani, N.: CRIDS: correlation and regression-based network intrusion detection system for IoT. SN Comput. Sci. **2**(3), 1–7 (2021)
15. Gadekallu, T.R., Manoj, M., Kumar, N., Hakak, S., Bhattacharya, S., et al.: Blockchain-based attack detection on machine learning algorithms for IoT-based e-health applications. IEEE Internet Things Mag. **4**(3), 30–33 (2021)

16. Amoon, M., Altameem, T., Altameem, A.: Internet of things sensor assisted security and quality analysis for health care data sets using artificial intelligent based heuristic health management system. Measurement **161**, 107861 (2020)
17. Choudhary, S., Kesswani, N.: Cluster-based intrusion detection method for internet of things. In: 2019 IEEE/ACS 16th International Conference on Computer Systems and Applications (AICCSA), pp. 1–8. IEEE (2019)
18. Alrashdi, I., Alqazzaz, A., Alharthi, R., Aloufi, E., Zohdy, M.A., Ming, H.: FBAD: fog-based attack detection for IoT healthcare in smart cities. In: 2019 IEEE 10th Annual Ubiquitous Computing, Electronics & Mobile Communication Conference (UEMCON), pp. 0515–0522. IEEE (2019)
19. Almulhim, M., Islam, N., Zaman, N.: A lightweight and secure authentication scheme for IoT based e-health applications. Int. J. Comput. Sci. Netw. Secur. **19**(1), 107–120 (2019)
20. Kim, J., Kim, H., et al.: An effective intrusion detection classifier using long short-term memory with gradient descent optimization. In: 2017 International Conference on Platform Technology and Service (PlatCon), pp. 1–6. IEEE (2017)
21. Khemissa, H., Tandjaoui, D.: A lightweight authentication scheme for e-health applications in the context of internet of things. In: 2015 9th International Conference on Next Generation Mobile Applications, Services and Technologies, pp. 90–95. IEEE (2015)
22. Hussain, F., et al.: IoT healthcare security dataset (2021)
23. Hussain, F., et al.: A framework for malicious traffic detection in IoT healthcare environment. Sensors **21**(9), 3025 (2021)

Detection and Grading of Diabetic Retinopathy from Fundus Images by Applying Transfer Learning

Md Aasif Raza[1(✉)] and Krishan Berwal[2]

[1] Department of Computer Engineering, National Institute of Technology, Kurukshetra, India
rmdaasif77@gmail.com

[2] Military College of Telecommunication Engineering, Mhow, India
kk@nitkkr.ac.in

Abstract. Diabetic Retinopathy (DR) is a common complication of diabetes that can cause damage to the veins in the retina, resulting in significant vision loss. Timely treatment can help prevent this vision loss, but due to delays in diagnosis and a shortage of ophthalmologists, patients often experience vision problems before receiving proper care. Therefore, it is crucial to detect DR at an early stage. Automating the analysis of diabetic retinopathy is essential for addressing these challenges. While deep learning has achieved high accuracy in binary classification, its performance in multi-stage classification, particularly in the early stages of the disease, is less impressive. To address the abovementioned problem, Densely Connected Convolutional Networks are proposed for detecting Diabetic Retinopathy in retinal images. The proposed method has two cases (case 1- Binary classification, i.e., DR and No DR, and case 2 Multilevel classification, i.e., classification with five stages of DR). Also, the proposed method is tested with a publicly available retinal image dataset, i.e., APTOS. The accuracy of binary and multiclass classifications is 99% and 95% for the APTOS. The accuracy of the DR screening system is improved due to preprocessing.

Keywords: Diabetic retinopathy · Transfer learning · Fundus Image

1 Introduction

Elevated blood glucose levels brought on by inadequate insulin secretion or synthesis by the pancreas cause the chronic disease known as diabetic retinopathy. From 1980 to 2014, there were approximately 422 million people with diabetes worldwide, a significant rise in the case of diabetes during the previous three decades. The liver, heart, kidneys, and joints are just a few of the organs that diabetes affects. The eyes are also one of them. It is one of the main reasons for blindness for people under 50 [7]. The condition known as diabetic retinopathy (DR) is an effect of diabetes caused by a blockage of blood arteries that supply the eye. It causes swelling and fluid leaking, which may cause significant eye damage. Early identification is essential to avoid the problems linked to chronic illnesses like diabetes. The development of aberrant blood vessels in the retina caused

D. Pastor-Escuredo et al. (Eds.): ICDLAIR 2023, LNNS 1001, pp. 764–773, 2024.
https://doi.org/10.1007/978-3-031-60935-0_66

by DR may result in scarring, bleeding, and eventually blindness. DR accounts for 2.6% of blindness cases worldwide.

Physicians often use retinal imaging to diagnose DR by identifying different lesions, such as microaneurysms (MA), hemorrhages (HM), soft and hard exudates (EX) [6], and venular beading.

- Microaneurysms (MA) are tiny, circular red spots on the retina that are early signs of DR, whose size is smaller than 125 μm and have well-defined edges.
- When huge irregular-shaped patches more significant than 125 μm are seen on the retina, hemorrhages (HM) are detected. Hemorrhages, unlike microaneurysms, may be divided into two categories: flame and blot.
- Hard exudates, caused by plasma leakage, appear as yellow patches on the retina. They have well-defined edges and are found in the retina's outer layers.
- Soft exudates are whitish lesions with an oval form that develops on the retina due to nerve fiber swelling.
- Venous beading is when the retina's venules' irrational constriction and dilation.
- Intra-retinal microvascular abnormalities (IRMAs) are unnatural alterations to the retina's tiny blood vessels.

Ophthalmologists classified DR into three categories: Normal, non-proliferative diabetic retinopathy - NPDR, and proliferative diabetic retinopathy - PDR, based on the degree of abnormality. Further NPDR ratings include mild, moderate, and severe [9] as illustrated in Fig. 1.

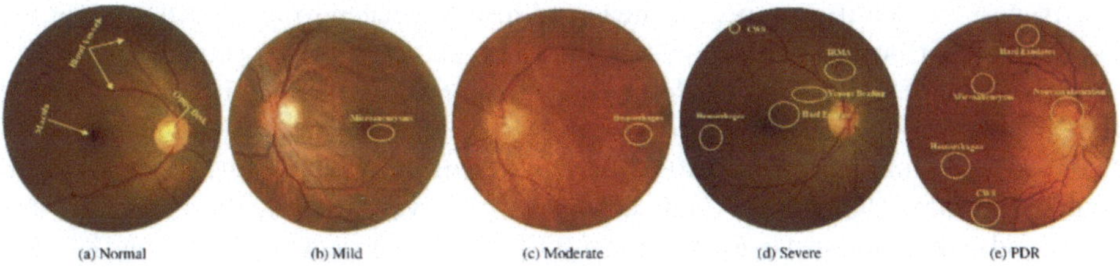

Fig. 1. Different stage of DR.

2 Motivation

The most significant reason for blindness and loss of vision in the world, especially for those with diabetes, is diabetic retinopathy (DR). Early detection and immediate treatment are essential to avoid severe eye damage and enhance patient outcomes. The manual identification of diabetic retinopathy is a very tedious and time-consuming method. Deep learning methods are now becoming practical tools for analyzing medical images, and they can entirely change how diabetic retinopathy is recognized. Convolutional neural networks (CNNs) are a type of deep learning algorithm. Deep learning techniques have shown outstanding performance in many image identification tests. We are using the deep learning model's ability to learn complicated patterns and features from big datasets; they can simplify and improve the accuracy of diabetic retinopathy diagnosis.

3 Literature Survey

Studies about categorizing diabetic retinopathy (DR) fall into three groups, i.e., binary, multiclass, and hybrid (classification with lesion localization). 73% of the papers under consideration employed binary classifiers that were either DR or non-DR, whereas 27% used several phases. Studies that categorize DR lesion location and DR detection and grading using fundoscopic images fall under the heading of "hybrid classification." The problem of DR was addressed creatively by Sraddha Das et al. [1]. The suggested approach comprises extracting the branching blood veins from fundus scans of the eye using preprocessing and segmentation methods. A convolutional neural network (CNN) was trained using specific layers for each class to categorize the retinal pictures. The findings showed that it performed better than traditional methods with fantastic accuracy and precision on the DIARETDB1 dataset of 98.7% and 97.2%, respectively. Padmanayana et al. [2] developed a method where retinal pictures are divided into no diabetic retinopathy and diabetic retinopathy categories using the proposed CNN architecture. The Gaussian-blur filter method is used to apply a Gaussian blur filter to retinal pictures to make them smooth. Ben Graham's preprocessing technique [15] is also used to improve the lighting in the photographs. The proposed model has been evaluated with 86% sensitivity, 96% specificity, and 88% accuracy rate on the Kaggle dataset.

Renato et al. [3] presents a comprehensive framework for diagnosing and prescribing diabetic retinopathy (DR) to prevent vision loss. For enhanced segmentation, a modified U-Net architecture named DR-UNet was utilized. The 84.50% dice similarity coefficient and 87.10% intersection over union segmentation accuracies showcased the framework's effectiveness in identifying specific areas and lesions related to DR. Furthermore; the classification performance reached 99.20%. Aziz et al. [5] proposed a novel hemorrhage network called HemNet. It was introduced for hemorrhage classification and compared with other deep models. The performance of HemNet was evaluated using two datasets, DIARETDB121 and DIARETDB022, based on sensitivity, specificity, precision, and accuracy metrics. HemNet's overall accuracy (AC) was 97.12%, the best of all the models and nearly identical to AlexNet's 97.08% AC. However, with 87.25% and 89.58% accuracy values, respectively, ResNet50 and AlexNet surpassed HemNet. Compared to LeNet-5 and VGG-16, HemNet's accuracy of 86.43% was greater. These findings show that HemNet performs exceptionally well identifying and categorizing hemorrhages on the DIARETDB0 dataset.

Muhammad, et al. [8] presented an ensemble model for diabetic retinopathy (DR) detection that utilizes a hybrid neural network approach. This unified approach used an entropy enhancement technique based on discrete wavelet transform. Kobat et al. [9] presented a method for classifying DR pictures that extract deep features from the images by utilizing transfer learning with the DenseNet201 architecture. They created a revolutionary technique that combined horizontal and vertical patch divisions after being inspired by popular image recognition techniques like the vision transformer and multilevel perceptron mixer, which employed patch division for multilevel deep feature extraction.

Jena et al. [10] presented a novel approach to diabetic retinopathy diagnosis. Their research developed a method that classified diabetic retinopathy lesions into four categories: standard, microaneurysms, hemorrhages, and exudates. Their approach began with segmentation using U-Net. The extracted features were then subjected to a two-step classification process. First, CNN was employed to analyze and categorize the lesions. Following this initial step, an SVM was utilized to refine the classification results further. The results were impressive, with the APTOS dataset achieving a remarkable 98.6% accuracy in diabetic retinopathy identification and the MESSIDOR dataset achieving an accuracy rate of 91.9%. Moreover, the method demonstrated excellent accuracy rates of 96.9% for exudate detection in the APTOS dataset and 98.3% in the MESSIDOR dataset. Islam et al. [11] presented a groundbreaking AI model for diabetic retinopathy (DR) detection. Their approach hinged on the application of supervised contrastive learning. The model yielded remarkable results, boasting an accuracy rate of 98.36% and an impressive AUC score of 98.50%. Furthermore, they employed a CNN architecture, using Transfer Learning and deep learning techniques to achieve the final classification. To assess the severity of DR using fundus images, Islam et al. employed supervised contrastive learning within the Xception CNN model, utilizing the APTOS and Messidor-2 datasets. Their results were highly encouraging, demonstrating an accuracy rate of 98.36% and an AUC score of 98.50% for binary classification. In multiclass classification, the model achieved an accuracy of 84.3% and an AUC score of 93.8%.

4 Proposed Work

A pre-trained CNN model is employed to apply transfer learning in our research. The DR detection process uses color fundus retinal pictures. The retinal pictures are improved, preprocessed, and using a model it is categorized in case1 or case2, i.e., proposed method has two cases (Case1- Binary classification, i.e., DR and No DR, and Case2-Multilevel classification, i.e., classification with five stages "NO DR," "level1", "level2", "level3", "level4" of DR) also, in proposed method publicly accessible retinal image datasets are used for testing, i.e., APTOS.

4.1 Dataset Description

In this method, a Retinal image dataset is used which is available publicly, i.e., APTOS. Experienced ophthalmologists annotate these images. In the APTOS dataset, 3662 images are used and split it into 80% training and 20% testing.

4.2 Preprocessing

The photographs within the dataset under consideration exhibit a need for more homogeneity due to their collection from rural portions of India under varying settings. Preprocessing was therefore required in order to improve the pictures. Here, in the preprocessing phase, the approach for retinal image analysis is to perform the following steps:

- Conversion to Grayscale: To simplify the analysis while retaining information.
- Center Cropping: To remove black borders in retinal images.
- Resizing Image: Retinal images may be of various sizes; resizing is essential.

- Enhance finer details: Gaussian blur filter was applied to improve the overall contrast of the image for noise reduction and tackle different illumination problems, i.e., background color differences and lighting conditions.

Ben [15] was applied by applying these preprocessing steps to enhance image quality further. These steps aim to enhance the visibility of critical retinal structures, reduce unnecessary variations, and ensure consistent and standardized inputs for subsequent stages of the algorithm, as shown in Fig. 2.

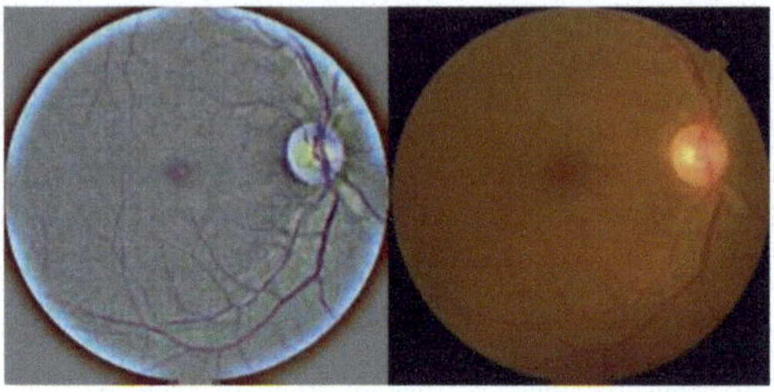

Fig. 2. Original vs preprocessed retinal image.

4.3 Data Augmentation

Data augmentation expands the dataset by adding additional data points through modifications or model-generated samples. Both model performance and data collection costs are improved. It is commonly used in deep learning applications to reduce over-fitting; parameters are shown in Table 1.

Table 1. Data Augmentation Hyper-parameters with values

Data Augmentation Hyper-parameters	Values
Rescale	1./255
Shear range	0.2
Zoom range	0.2
Horizontal flip	True
Vertical flip	True
Rotation range	20
Width shift range	0.1
Height shift range	0.1

4.4 Implementation

Implementing our work involves several key steps. First, we preprocess the images and split them into training and test data in an 80:20 ratio. Next, we introduced the Keras modules, including ImageDataGenerator, Model, Dense, and Flatten. The code was designed to work with pre-trained models, allowing for flexibility in architectural choices. The block diagram of the proposed model is shown in Fig. 3.

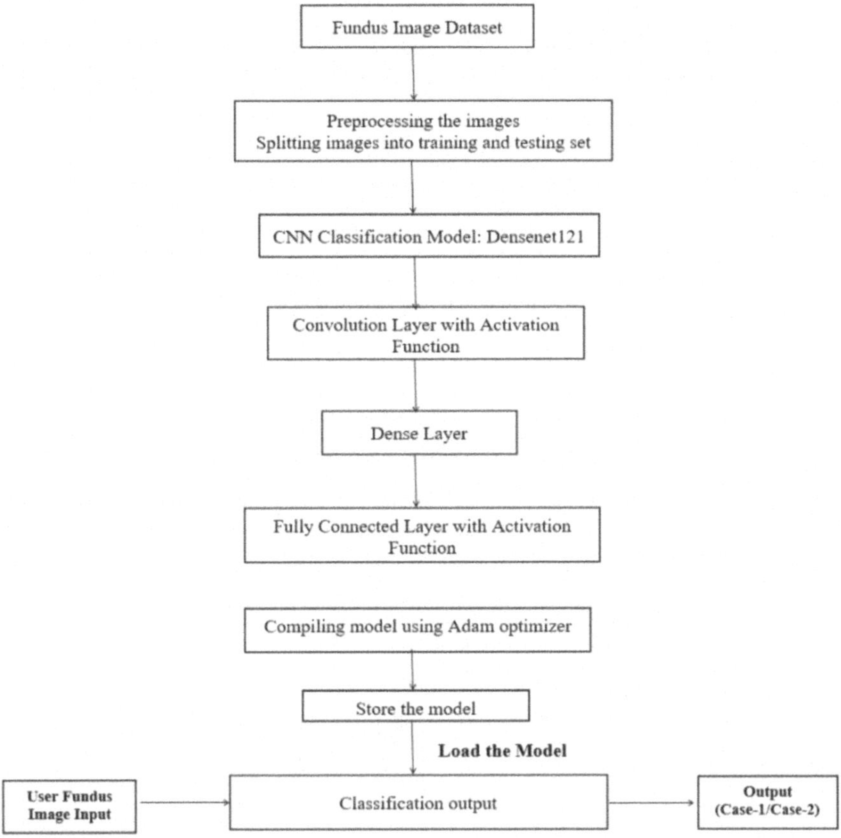

Fig. 3. Block Diagram of proposed model.

We specified the image size as 320 by 320 for the Densenet121 architecture, which was used to create grayscale images. The Densenet121 model was imported with weights from ImageNet, and the top layers were removed. A dropout of 0.5 was added at the last layer. We then determined the number of output labels based on the classes in our training data set. A Dense layer was created, and a final layer with a SoftMax activation function was added to match the number of classes in our output layer. For case-1, we added a layer with two nodes; for case-2, we added a layer with five. We compiled the ADAM optimizer and accuracy-based optimization metrics with learning rate = 0.0001, decay = 0.000001, and loss as categorical cross-entropy. Case-1 has total parameters = 7,039,554, trainable parameters = 6,955,906, non-trainable parameters = 83,648 and for case-2 total parameters = 7,042,629, trainable parameters = 6,958,981, non-trainable parameters = 83,648. To train our model, we took a batch size of 32. The class_mode was set to categorical and trained the model with the model.fit() method with 30 epochs for cause-1 and 50 epochs for case-2.

5 Results and Analysis

Metrics like accuracy, F1 score, recall, and precision for the model evaluation is used and calculated these metrics using the confusion matrix, which consists of four characteristics.

Accuracy: The model's performance in all classes is gauged by its accuracy. It is calculated as the ratio of correctly guessed samples to the total samples.

$$Accuracy = \frac{True\,Positive + True\,Negative}{True\,Positive + True\,Negative + False\,Positive + False\,Negative} \tag{1}$$

Recall: The proportion of positive observations was identified correctly as positive relative to all positive samples. When recall is higher, there are more positive samples discovered.

$$Recall = \frac{True\,Positive}{True\,Positive + False\,Negative} \tag{2}$$

Precision: Accuracy (incorrect or correct) is measured as the ratio of correctly categorized Positive instances to all Positive samples. Precision is helpful in determining how accurate the model is when it declares that an instance is true.

$$Precision = \frac{True\,Positive + True\,Negative}{True\,Positive + False\,Positive} \tag{3}$$

F1-Score: The F1 score is the harmonic mean of recall and precision.

$$F1Score = 2 \times \frac{Precision \times Recall}{Precision + Recall} \tag{4}$$

5.1 Results and Comparison

This section presents the findings from the suggested deep learning approach for DR. Using a CNN classifier, the DR grading is based on the retinal image's impacted level. The pictures challenge dataset, or the APTOS color fundus images, is used to test the suggested model. The models' recall, precision, accuracy, and f1-score values are shown, along with a description of their performances. Figures 4 and 5 depict the confusion matrix for cases 1 and 2, respectively, summarizing the effectiveness of a classification method.

Results of the APTOS dataset in case-1, where we got the accuracy value of 99% in less than 30 epochs, and in case-2, where the accuracy value is 95% in just less than 50 epochs.

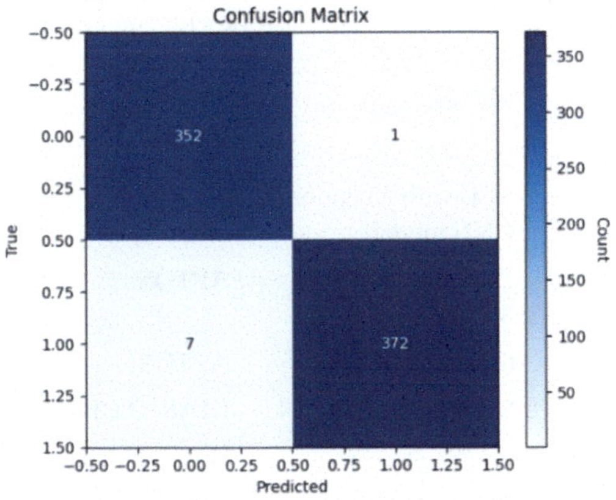

Fig. 4. Confusion matrix for Binary Classification.

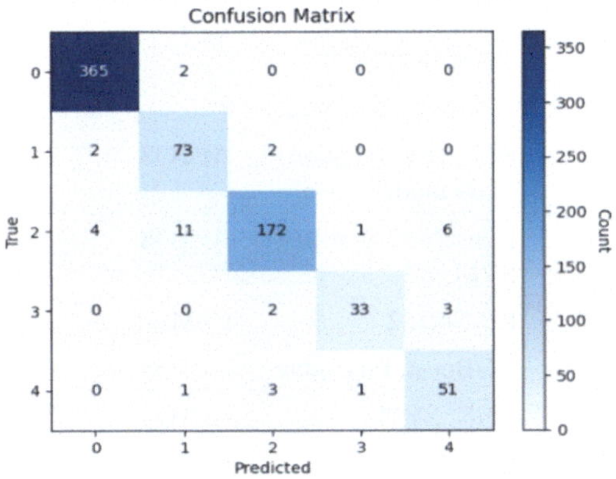

Fig. 5. Confusion matrix for Multiclass Classification.

5.2 Performance Comparison

The state-of-the-art in automatically classified DR is shown in Table 2. Regarding case-1, i.e., binary classification, It displays a performance comparison between the proposed and the current models using accuracy values. Notably, the APTOS dataset in instance 1 yielded findings with an accuracy of 99%. In contrast, the closest result models, i.e., Robiul et al. [11] and Padmanayana et al. [2], are 98.36% and 94.6%, respectively; for the same dataset, APTOS tells the proposed model outperforms those. The results of case 2, i.e., 95%, are also better than those of Robiul et al. [11] and Kobat et al. [9], i.e., 87.43% and 84.3%, respectively, for the same dataset APTOS that tells our proposed model outperforms those.

Table 2. Comparison of proposed work with contemporary work

Study	Method	Dataset	Result
Binary			
Robiul et al. [11]	SCL with Xception CNN model	APTOS	ACC = 98.36
Kobat et al. [9]	DenseNet201 with SVM	APTOS	ACC = 92
Padmanayana et al. [2]	CLAHE, CNN	APTOS	ACC = 94.6
Sraddha Das et al. [1]	Segmentation, CNN	DIARETDB1	ACC = 98.7
		Hospital dataset	ACC = 97.4
Uppamma et al. [4]	Taylor AVO, SqueezeNet	IDRiD	ACC = 94.2
Tamoor Aziz et al. [5]	HemNet	DIARETDB0	ACC = 97.12
Proposed	Ben Graham [15], DenseNet121	APTOS	ACC = 99
Multiclass			
Robiul et al. [11]	SCL with Xception CNN model	APTOS	ACC = 84.3
Kobat et al. [9]	DenseNet201 with SVM	APTOS	ACC = 87.43
T.M. Usman et al. [12]	ResNet152	CSME	ACC = 94.40
Sikder, N et al. [16]	XGBoost, Ensemble	APTOS	ACC = 94.20
Tariq, M et al. [17]	CNN, SVM	APTOS	ACC = 93
Xiaoling Luo et al. [13]	Long-Range units into the CNN	EyePac	ACC = 84
A.E. Minarno et al. [14]	Ben Graham [15], EfficientNet-B7	APTOS	ACC = 89
Proposed	Ben Graham [15], DenseNet121	APTOS	ACC = 95

6 Conclusion and Future Plan

Our project focuses on both the binary classification as well as multiclass classification of Diabetic Retinopathy (DR) detection using the APTOS dataset. We have employed a Convolutional Neural Network (CNN) architecture, DenseNet121, to train the model. The model allows us to input an image and obtain a binary and multiclass classification prediction. This model achieved a test accuracy of 99% for binary classification and 95% for multiclass classification.

In future, we propose several avenues for improving the model's performance and accuracy. These include training the model with real datasets from hospitals, using large datasets like Eyepacs, and explore segmentation techniques to improve the quality of the model especially, for multiclass classification.

References

1. Das, S., Kharbanda, K., Suchetha, M., Raman, R., Dhas, E.: Deep learning architecture based on segmented fundus image features for classification of diabetic retinopathy. Biomed. Signal Process. Control **68**, 102600 (2021)
2. Anoop, B.K.: Binary classification of DR-diabetic retinopathy using CNN with fundus colour images. Mater. Today: Proc. **58**, 212–216 (2022)
3. Maaliw, R.R., et al.: An enhanced segmentation and deep learning architecture for early diabetic retinopathy detection. In: 2023 IEEE 13th Annual Computing and Communication Workshop and Conference (CCWC), pp. 168–175. IEEE (2023)
4. Uppamma, P., Bhattacharya, S.: Diabetic retinopathy detection: a blockchain and African vulture optimization algorithm-based deep learning framework. Electronics **12**(3), 742 (2023)
5. Aziz, T., Chalie, C., Srijidtra, M.: Deep learning-based hemorrhage detection for diabetic retinopathy screening. Sci. Rep. **13**(1), 1479 (2023)
6. Atwany, M.Z., Abdulwahab, H.S., Mohammad, Y.: Deep learning techniques for diabetic retinopathy classification: a survey. IEEE Access **10**, 28642–28655 (2022)
7. Gani, V.K.U.A., Shanmugasundaram, N., Thiyagaraj, N.: Recent advancements in automated screening techniques for diabetic retinopathy. In: 2022 3rd International Conference on Communication, Computing and Industry 4.0 (C2I4), pp. 1–5. IEEE (2022)
8. Imran, M., Ullah, A., Arif, M., Noor, R., et al.: A unified technique for entropy enhancement based diabetic retinopathy detection using hybrid neural network. Comput. Biol. Med. **145**, 105424 (2022)
9. Kobat, S.G., Baygin, N., Yusufoglu, E., Baygin, M., Barua, P.D., et al.: Automated diabetic retinopathy detection using horizontal and vertical patch division-based pre-trained DenseNET with digital fundus images. Diagnostics **12**(8), 1975 (2022)
10. Jena, P.K., Khuntia, B., Palai, C., Nayak, M., Mishra, T.K., Mohanty, S.N.: A novel approach for diabetic retinopathy screening using asymmetric deep learning features. Big Data Cogn. Comput. **7**(1), 25 (2023)
11. Islam, M.R., et al.: Applying supervised contrastive learning for the detection of diabetic retinopathy and its severity levels from fundus images. Comput. Biol. Med. **146**, 105602 (2022)
12. Usman, T.M., Saheed, Y.K., Ignace, D., Nsang, A.: Diabetic retinopathy detection using principal component analysis multi-label feature extraction and classification. Int. J. Cogn. Comput. Eng. **4**, 78–88 (2023)
13. Luo, X., et al.: A deep convolutional neural network for diabetic retinopathy detection via mining local and long-range dependence. CAAI Trans. Intell. Technol. (2023)
14. Minarno, A.K., et al.: Classification of diabetic retinopathy disease using convolutional neural network. JOIV: Int. J. Inform. Visualizat. **6**(1), 12–18 (2022)
15. Graham, B.: Kaggle diabetic retinopathy detection competition report. UoW **22** (2015)
16. Sikder, N., Masud, M., Bairagi, A.K., Arif, A.S.M., Nahid, A., Alhumyani, H.A.: Severity classification of diabetic retinopathy using an ensemble learning algorithm through analyzing retinal images. Symmetry **13**(4), 670 (2021)
17. Tariq, M., Palade, V., Ma, Y.: transfer learning based classification of diabetic retinopathy on the Kaggle EyePACS dataset. In: The 3rd International Conference on Medical Imaging and Computer-Aided Diagnosis (2022)

Indian Sign Language Bilateral Translator

Shivansh Katoch[1](✉) and Krishan Kumar[2]

[1] National Institute of Technology, Hamirpur, India
shivanshkatoch123@gmail.com
[2] National Institute of Technology, Kurukshetra, India

Abstract. Deaf and hard of hearing individuals face significant challenges and hardships in their daily lives due to their hearing loss, which significantly impact their quality of life in a negative way. Even though a lot of research has been done in sign languages used in countries like Britain and America, little has been done in Indian Sign Language. Hence, we have created a robust bilateral Indian Sign Language Translation system in order to convert Indian Sign Language to text and vice versa. The first model takes input in form of English text, which is further parsed to structure grammar representation on which grammar rules of Indian Sign Language are applied. The sentence is pre-processed and a concatenated video file will be generated for the sentence. For the second model, we have used LSTM model to predict gestures provided as input. To lessen the struggles faced by those who are deaf or hard of hearing, society must work toward greater accessibility, inclusivity, and understanding. For this reason, this research aims to enhance sign language detection and makes use of a production system to promote communication between deaf and hearing individuals.

Keywords: Indian sign language (ISL) · parsing · stemming · lemmatization · MediaPipe · LSTM

1 Introduction

Communication is a critical component of human connection, enabling the exchange of thoughts, ideas, and emotions. It serves as the cornerstone for interpersonal relationships, mutual understanding, and teamwork. It makes it possible for people to convey their needs, impart knowledge, and engage meaningfully with others, but the manner in which we do so depends on our upbringing, education, society, and point of view.

For those who are not hard of hearing, spoken languages are the main form of communication. They rely on auditory cues, including spoken words and voice tones, to communicate and comprehend others. However, it becomes extremely challenging for those who are deaf or hard of hearing to communicate with the majority. With reference to Table 1, about 2.68 Cr people, or 2.21% of India's 121 Cr population, are classified as "disabled" according to the 2011 Census [1]. In India, 20% of people with disabilities are unable to move, 19% are blind or visually impaired, another 19% are hearing impaired and 8% of people have several impairments as shown in Fig. 1. Although we have made

D. Pastor-Escuredo et al. (Eds.): ICDLAIR 2023, LNNS 1001, pp. 774–788, 2024.
https://doi.org/10.1007/978-3-031-60935-0_67

great efforts and continue to make great efforts to modernize the nation's technology infrastructure, the needs of the DHH (deaf or half of hearing) population have been utterly ignored.

Table 1. Disabled Population in India.

Population, India 2011			Disabled persons, India 2011		
Persons	Males	Females	Persons	Males	Females
121.08 Cr	62.32 Cr	58.76Cr	2.68 Cr	1.5 Cr	1.18 Cr

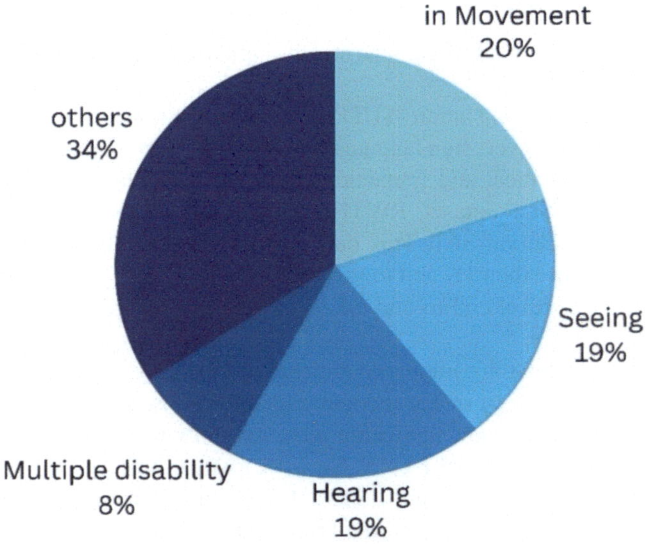

Fig. 1. Disability Statistics as per Census 2011 in India.

A lot of attention has been paid to other sign languages since they are simpler to study, such as those used in America. However, there hasn't been a lot of study done on Indian Sign Language (ISL) because there are variances in signs and regional dialects due to the vast linguistic environment of India. Moreover, it involves both static and dynamic gestures, single as well as double handed gestures, and the hands involved in gesturing can also have complex movements. Despite the fact that various methods and equipment have been developed over time, they lack affordability and accessibility.

With the advances in the field of machine learning, deep learning and computer vision, systems for recognizing sign language that utilize machine learning algorithms to translate sign language into spoken or written language or vice versa can help bridge the communication gap between the deaf and hearing communities. The authors of this research hope to address issues that deaf and hard of hearing persons face. The Indian Sign Language Bilateral Translator concentrates on translating English text into Indian

Sign Language (ISL) and vice versa more precisely than the current models. The various salient features of our work are:

- Unlike existing methods, our English to ISL conversion project makes use of a YouTube player for the visualization, which gives us the impression that we are speaking with a real person rather than an animation.
- The ISL to English text conversion system uses the MediaPipe Holistic and LSTM model to identify dynamic hand signs more accurately and easily.
- The suggested work can function independent of Signer and environment conditions.
- The proposed work outperforms the existing state-of-art models by giving almost 100% accuracy, precision and recall.

2 Related Work

Depending on the nature of sign language and the signs, different authors have employed different methodologies.

The Sign Language Translation (SLT) problem was initially presented by Camgoz et al. in [3]. It seeks to convert sign language films into spoken English while accounting for the intricate grammatical and syntactic structures of sign language. They collected the first "Continuous SLT dataset, RWTH-PHOENIX Weather 2014T" to assess the performance of their Neural SLT. This dataset includes more than 67K signs from a sign vocabulary of more than 1K and more than 99K words from a German vocabulary. Their frame and gloss-level end-to-end tokenization networks received scores of 9 and 18 respectively.

Deep Learning technology improvements have made it possible to automate image recognition using a variety of image recognizing algorithms. Convolutional neural networks, for instance, have made significant advancements in the field of deep learning recently. A CNN (Convolutional Neural Network) was utilized by G. Jayadeep et al. in [7] to extract picture features, and LSTM (Long Short Term Memory) was used to classify these movements and convert them into text.

By combining the Bag of Visual Words model (BOVW), SURF (Speeded Up Robust Features) features, and classification models like Support Vector Machines (SVM) and Convolutional Neural Networks (CNN), Katoch et al. suggested a method for deciphering Indian Sign Language alphabets and numbers in [5]. Using Pyttsx3, the Google voice API, and other tools, the system provides predicted labels in both text and audio.

In the Automated Speech to Sign Language Conversion using Google API and NLP approach by Bharati et al. [10] speech is first acquired as input. It is transformed to text using Google API. Infected portions of the free text retrieved afterwards are eliminated using NLP. A matching process is carried out for each word or character against the sign language video database. The matched videos are then retrieved from the database and joined together to create a new video.

Sruthi C.J. and Lijiya A. [11] created a sign-recognizing model named "Signet", using the CNN technique, supervised learning, and succeeded in learning all 24 ISL static alphabets. The suggested method seems to be effective at recognizing ISL alphabets making it a useful tool for speech-impaired people who need communication.

K. Saija [12], proposed a model to convert speech into English text that uses the Hidden Markov Model. Using Wordnet, an American linguistic database, and pre-processing, the resulting text creates an Indian Sign Language Gloss. Stems, tokenization, selective stop word removal, and punctuation removal are a few of the pre-processing techniques used.

Even though there has been a lot of research on Indian Sign Language recognition in recent years to bridge the gap between the Indian DHH community and the general public, there is still more work to be done to address issues with gesture diversity, complex hand movements, and lighting.

3 Proposed Work

3.1 English Text to Indian Sign Language Translation

When communicating with others, sign language makes use of physical cues such body language, hand gestures, and facial emotions. The authors of this article have created a model that make use of videos for specific words combined to translate the text language into sign language. The model consists of following steps:

An English parser to analyze texts in English
A module for rearrangement of sentences based on ISL grammar rules
A Stopword eradicator
Stemming to determine each word's roots and synonym replacement for words not found in dictionaries.
A module for video conversion.

Dataset Collection. We have used the http://www.indiansignlanguage.org/ and "Indian Sign Language RKMVU-CBE" YouTube channel [2] for accessing the videos of words and storing them with their respective URLs of the videos as key-value pair in the Python dictionary.

Algorithm Design. The authors have used "en_core_web_sm" model which is a pre-trained English language model from the spaCy library.

The "en_core_web_sm" model carry out operations like tokenization (dividing text into discrete tokens), part-of-speech tagging (assigning grammatical labels to tokens), dependency parsing (figuring out the syntactic relationships between words), named entity recognition (finding named entities like people's names, places, etc.), and more. After processing the input from the model, the text is converted into Subject-Object-Verb structure because ISL uses the Subject-Object-Verb structure in addition to a variety of negative and interrogative sentences, whereas English texts adhere to the Subject-Verb-Object structure. Unwanted words are then eliminated because ISL only uses words that have meanings and does not utilize any linking verbs, articles, or other helpful terms and output is forwarded to the lemmatization module, which breaks down each word into its root form. Synonyms are used in place of words that are not in the dictionary.

The flowchart of the algorithm design can be seen from the Fig. 2. These major steps are as follow:

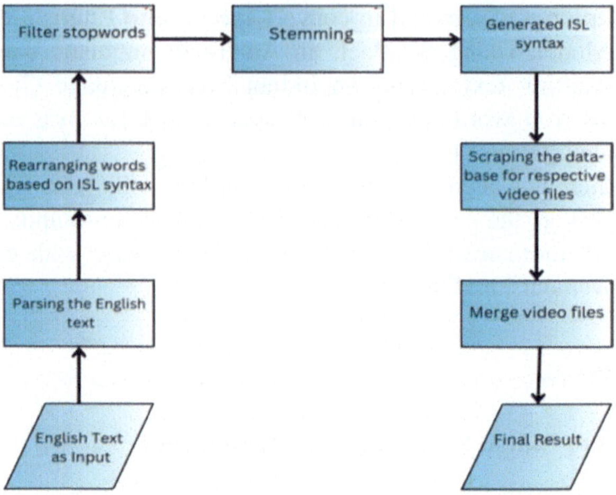

Fig. 2. Major Steps in proposed system-1

Parsing the English Text. The input English text is analyzed and parsed in this stage using pre-trained model. It entails dissecting the text into its component sentences, phrases, and words in order to examine the grammatical structure and relationships between them. Understanding the syntactic and semantic components of English sentences is made much easier by it. For example, as seen in Fig. 3, in the sentence "I like to read books," "I" is a personal pronoun (PRP), "like" is a verb present (VBP), "read" is a verb (VB), and "books" is a plural noun (NNP).

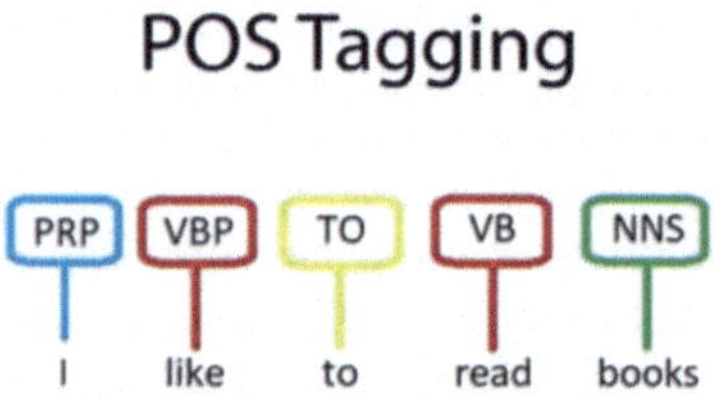

Fig. 3. Part of Speech Tagging.

Rearrangement of Sentences Based on ISL Grammatical Rules. ISL has its own unique syntax and sentence structures. Therefore, this module focuses on rearranging the parsed English sentences in accordance with ISL's grammatical structures and rules. This stage makes sure that the rearranged sentences adhere to the linguistic standards of ISL, which enables accurate and natural sign language generation. Table 2 shows how the grammatical rules of the English and ISL languages differ from one another.

Stopword Elimination. Stopwords are often used words that have minimal semantic value, such as "the," "is," "are," etc. Such stopwords can be eliminated from the sentences that have been processed using an eliminator module in ISL recognition. Eliminating

Table 2. Comparison of Grammar of English and Indian Sign Language.

Grammar Aspect	English Grammar	Indian Sign Language (ISL) Grammar
Word Order	Subject-Verb Object (SVO) order	Subject-Object-Verb (SOV) order
Verb Forms	Various verb forms and inflections used	No inflections used, root form of the verb
Dictionary Size	Large dictionary with extensive vocabulary	Limited dictionary with approximately 1800 words
Question Word Position	Question word at the start of interrogative sentences	Question word always at the end of the sentence
Auxiliary Verbs	Use of helping verbs in sentences	No use of linking verbs or helping verbs
Articles/Conjunctions	Use of articles and conjunctions in sentences	No use of articles or conjunctions
Adjective Position	Adjective placement before the noun	Not specified

stop words improves the accuracy of ISL translation by focusing on the more content-rich words and concepts while reducing noise. For instance, the Stopwords "is" and "a" in the sentence "this is a test" can be eliminated, as illustrated in Fig. 4.

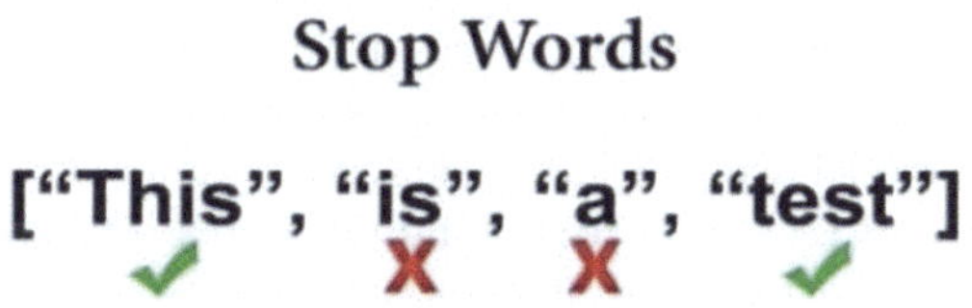

Fig. 4. Removal of Stopwords.

Stemming of Words. Words are stripped down to their fundamental or root form during stemming. Stemming can be used in ISL recognition to determine the roots of each word in the parsed English sentences. By standardizing the vocabulary, this stage makes the translation process easier. Additionally, a synonym replacement module can be used to locate relevant substitutes or comparable terms that have matching signs in ISL for words that are not included in the dictionary. As seen in Fig. 5, for example, the various inflections of word 'create' such as creative, creates, created, and creating can be reduced to their core word "create."

Video Generation Module. After completing the aforementioned steps, we have the ISL transformed text. The program then searches the datasets provided for matches for each word. This will be based on the fundamental string matching method between the labels

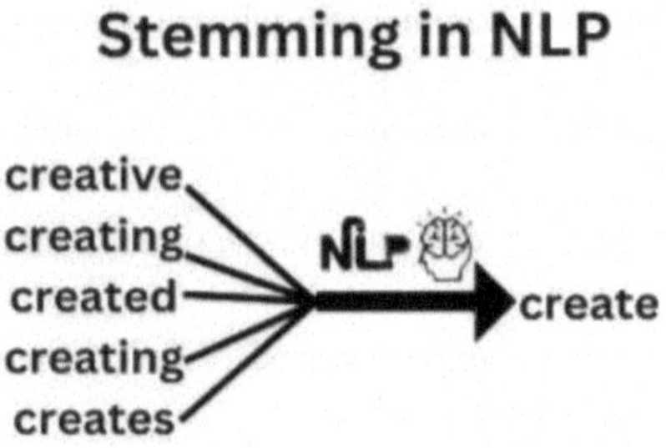

Fig. 5. Stemming of Words.

of the videos and the processed input text. Finally, a display of several videos is shown on the screen in a sequential order.

Output of the Model. The database will be having video for each and every separate words. The model's output will be a display of concatenated videos for each word in the ISL translated text.

3.2 Indian Sign Language to English Text

To develop a comprehensive landmarker for the human body, the authors combined parts of the pose, face, and hand landmarkers using the MediaPipe Holistic Pipeline. It produces 543 landmarks in total, including 21 hand landmarks per hand, 468 face landmarks, and 33 pose landmarks, in real-time.

In order to improve hand gesture identification, the authors manually collected the frames from the OpenCV with a webcam and utilized MediaPipe Holistic to monitor and display the salient features of the hand motions. We were then able to train our sign language recognition model using the array that contained the final key points for each class and video. We used LSTM layers to train the model. The output is fed into two fully connected layers with ReLU' activation, a dropout layer, and a SoftMax output layer for classification. Figure 6 shows the flowchart of the algorithm design for this system.

Dataset Creation. The data was gathered in real-time using a predefined set of sign language gestures. A total of 40 videos were recorded for each sign, each of which had 30 frames. We created a dataset with the words "hello," "good," "wish," and "thanks" (Fig. 7).

Features Extraction Using MediaPipe. MediaPipe Holistic accepts video from a camera or a previously recorded file. In order to use MediaPipe Holistic, the first step is to identify the person's stance within the video frame. It involves finding vital body parts such as head, shoulders, arms, hips, and legs. After evaluating the stance, the next step is to find the landmarks for the hands and face. After locating the landmarks on the hands, face, and body, MediaPipe Holistic integrates the key points to produce a collection of landmarks that correspond to the person's stance, hand, and face in the video frame.

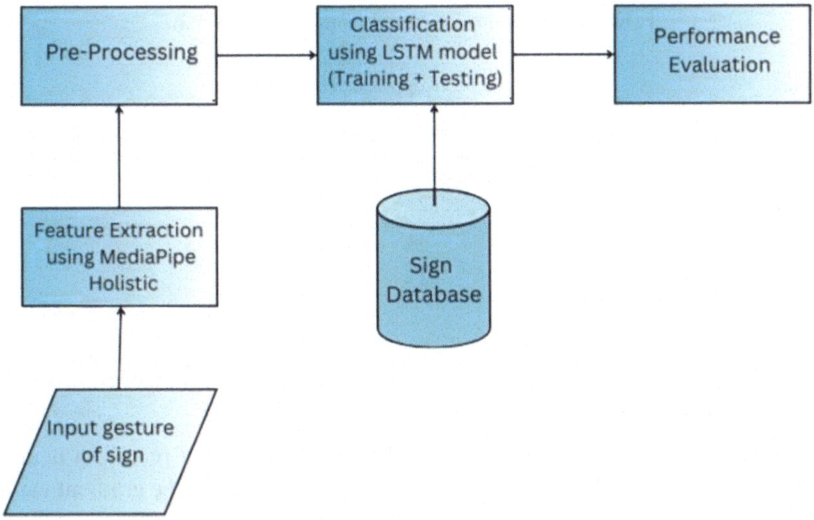

Fig. 6. Major Steps in proposed system-2.

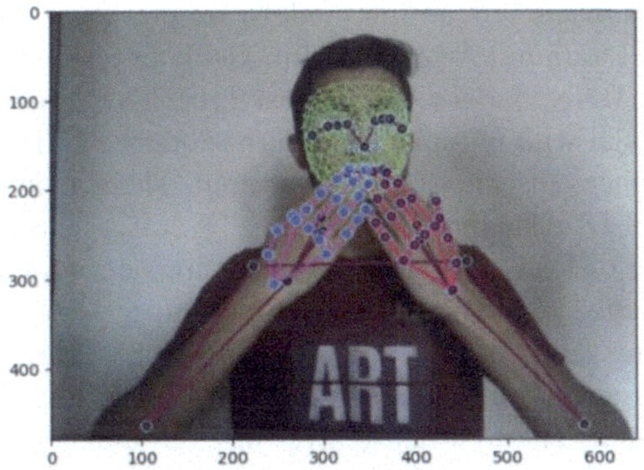

Fig. 7. Sample image of the dataset collection process.

A single coordinate space was created for each frame using a total of 1662 key points based on stance, hand, and face landmarks. To maintain the shape of the array, we passed zero values for any of these points that the model did not observe. When carried out on all frames of video, it enables precise tracking of person's movement. The key points extracted for different body parts by MediaPipe are given in Table 3. They are calculated in the three-dimension space: X, Y, and Z in addition to the visibility.

The number of extracted keypoints of hands is calculated as follows: keypoints in hand × Three dimensions × No. of hands = $(21 \times 3 \times 2) = 126$ keypoints. The number of extracted keypoints from the pose estimation is calculated as follows: keypoints in pose × (Three dimensions + Visibility) = $(33 \times (3 + 1)) = 132$ keypoints. The number of extracted keypoints from the face is calculated as follows: keypoints in face × Three dimensions = $(468 \times 3) = 1404$ keypoints. The total number of keypoints for each

Table 3. Keypoints extracted from each body part.

Body Part	Keypoints Extracted
Face	468
Left Hand	21
Right Hand	21
Body (Pose)	33
Total	543

frame is calculated as follows: keypoints in hands + in pose + in face = (126 + 132 + 1404) = 1662 keypoints.

LSTM Model. Long Short-Term Memory (LSTM) is a form of recurrent neural network (RNN) architecture LSTM networks are intended to address the gradient vanishing problem that might arise when trying to train conventional RNNs, as well as the shortcomings of standard RNNs in capturing and remembering long-term dependencies in sequential data.

Before training our machine learning models, we pre-processed the data to ensure its suitability for model training. A sequential model is initialized having the first three layers belonging to the RNN model while the last three layers are dense layers. The first LSTM layer to the model with 64 units has return sequence = True indicating that the layer will return sequences instead of a single output. The activation function used for the LSTM layer is "ReLU". The input shape of the layer is a sequence of length 30 with 1662 features. The use of LSTM model lessens the impact of vanishing gradients and speeds up the convergence of the network during training. A second LSTM layer with 128 units, also returning sequences is added. The third LSTM layer contains 128 units, but this time return sequences is set to False to indicate that it won't return sequences, but rather a single output. The first dense layer has 128 units, while the second dense layer has 32 units are added and a final Dense layer with the number of units equal to the number of actions utilizing the SoftMax activation function for probability distribution generation are set. Each dense layer is followed by dropout layer which randomly sets input units to 0 with a `rate` of 0.2, which helps prevent overfitting. The output of last layer represents the predicted probabilities for each action. The Adam optimizer is used for this model. Since this is a classification problem involving multiple classes, the activation function SoftMax and the categorical cross-entropy loss function are used.

4 Result

4.1 English Text to Indian Sign Language Translation

In the context of Indian Sign Language translation, the model generates ISL text from English text as input and searches the database for each word's matching video file. Then, all of these video files are combined to create a more orderly, educational, and understandable visual representation of Indian Sign Language.

Fig. 8. Output of the ISL Translation Model.

For instance, if we enter the text "We are playing volleyball." as input, the model generates ISL text like "We volleyball play" and then the video output for the ISL text is played by extracting the video of the words from the database using the Youtube IFrame API, as seen in the Fig. 8.

4.2 Indian Sign Language to English Text

Instead of using CNN and LSTM or other models, the authors used MediaPipe Holistic and LSTM because it requires less data to produce a hyper accurate model. Moreover, rather than having 30 to 40 million parameters in the neural network there are only about half a million parameters (with reference to Fig. 10) making it easier to train model. Since the neural network is much simpler, it is much faster when it comes for detecting in real time. Therefore, the authors employed LSTM model to identify gestures and signals. On the training set during the last epoch, we saw an overall accuracy of 93%, however the testing accuracy was higher than 99%. Figure 9 shows the output of the ISL recognition model.

Fig. 9. Output of the ISL Recognition Model

The Fig. 10 shows the model summary which includes type of model and the layers in the model used.

The Fig. 11 and Fig. 12 shows the model accuracy and loss plot signifying how well the model performs.

Performance Metrics. *Accuracy.* It is the simplest performance measure and is given by a ratio of correctly predicted observations to the total observations. In terms of true positives (TP), true negatives (TN), false positives (FP) and false negatives (FN), the formula of the accuracy can be written as:

$$\text{Accuracy} = \frac{\text{TP} + \text{TN}}{\text{TP} + \text{FP} + \text{FN} + \text{TN}}$$

```
Model: "sequential"
_________________________________________________________________
 Layer (type)                Output Shape              Param #
=================================================================
 lstm (LSTM)                 (None, 30, 64)            442112

 lstm_1 (LSTM)               (None, 30, 128)           98816

 lstm_2 (LSTM)               (None, 128)               131584

 dense (Dense)               (None, 128)               16512

 dropout (Dropout)           (None, 128)               0

 dense_1 (Dense)             (None, 64)                8256

 dropout_1 (Dropout)         (None, 64)                0

 dense_2 (Dense)             (None, 4)                 260

=================================================================
Total params: 697,540
Trainable params: 697,540
Non-trainable params: 0
_________________________________________________________________
```

Fig. 10. Model summary for LSTM for gesture dataset.

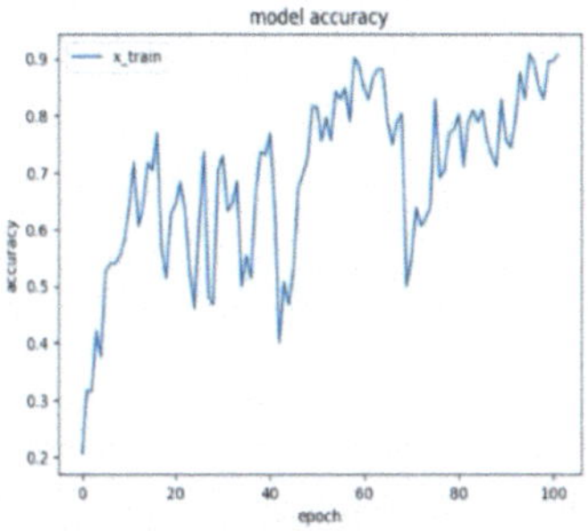

Fig. 11. Model's Accuracy Plot.

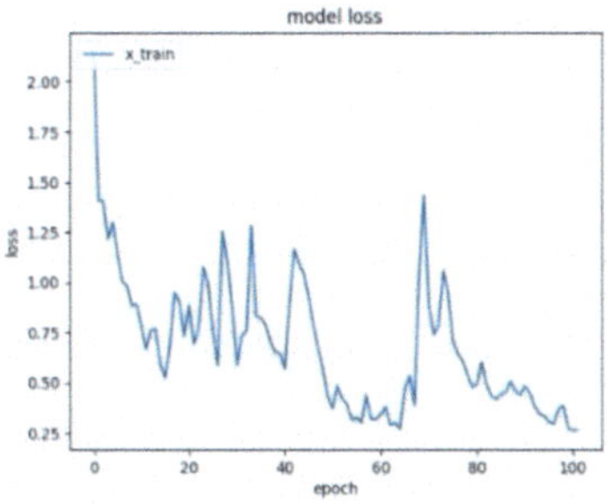

Fig. 12. Model's Loss Plot.

Precision. It is the ratio of correctly predicted positive observations to the total observations which are positive.

$$\text{Precision} = \frac{\text{TP}}{\text{TP} + \text{FP}}$$

Recall. It is the ratio of correctly predicted positive labels to the total no. of labels which are positive.

$$\text{Recall} = \frac{\text{TP}}{\text{TP} + \text{FN}}$$

F1 Score. It is the weighted average of precision and recall.

$$\text{F1 Score} = \frac{2*(\text{Recall}*\text{Precision})}{\text{Recall} + \text{Precision}}$$

Classification Report. It provides insights into how well the model performs for each class individually and gives an overall view of the model's performance. The classification report for our model can be seen by referring to Fig. 13 and Fig. 14. The accuracy, precision, recall and f1-score for all classes is 100% in the test dataset which demonstrates that our model performs better and produce great results.

Classification Report

	precision	recall	f1-score	support
0	0.94	0.79	0.86	39
1	1.00	1.00	1.00	36
2	0.82	0.92	0.87	39
3	0.97	1.00	0.99	38
accuracy			0.93	152
macro avg	0.93	0.93	0.93	152
weighted avg	0.93	0.93	0.93	152

Fig. 13. Classification Report of Train data.

Classification Report

	precision	recall	f1-score	support
0	1.00	1.00	1.00	3
1	1.00	1.00	1.00	1
2	1.00	1.00	1.00	1
3	1.00	1.00	1.00	3
accuracy			1.00	8
macro avg	1.00	1.00	1.00	8
weighted avg	1.00	1.00	1.00	8

Fig. 14. Classification Report of Test data.

Confusion Matrix. It is a tabular representation used to assess how well a classification model is performing. For a collection of data samples, it contrasts the model's predictions with the actual ground truth labels. In a classification task, the model's performance for each class is explained by the confusion matrix. From the Fig. 15, we can visualize the performance of model for the prediction of data samples.

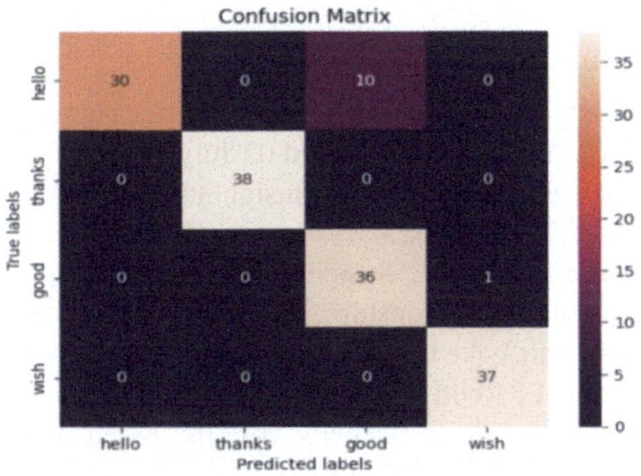

Fig. 15. Confusion Matrix of Train data.

5 Conclusion and Future Scope

We've made an effort to develop a model that will let persons with impairments express themselves clearly, allowing to build the communication gap between them and the rest of society. In this study, a bilateral model for recognising and translating Indian Sign Language was developed. The English text to ISLtranslator transforms English text into ISL text and creates a combined video file for every word's corresponding video file in the database. The visualization for our English to ISL translation project uses a YouTube player, giving users the impression that they are actually communicating with someone. By adding new words to its lexicon, the model's breadth can be expanded, leading to several advances. To improve communication, text-and-speech integration can be implemented on a project. Furthermore, Google Translate APIs can be used to facilitate translation from any regional Indian language to ISL. This would facilitate communication between someone who uses ISL and someone who is illiterate in English.

For ISL to English text conversion, we used the LSTM model for reconising signs in sign language. The LSTM performs effectively for time series events such as gestures sign recognisation. The suggested work function for both Signer and environment-independent Signers. The proposed work outperforms the existing state-of-art models by giving 100% precision, recall, and f-measure for test data. Future studies could look at how Indian Sign Language proficiency can be improved via the use of augmented reality (AR) and virtual reality (VR) technologies. Future studies might concentrate on enhancing the reliability and accuracy of systems that recognize sign language. By doing this, we can ensure that the DHH community is included in society and is not left behind in the digital era.

As a responsible society, it is our responsibility to guarantee that everyone has equitable access to technology. In order to achieve this, we must contribute to develop a technology specifically to meet the needs of the DHH community rather than concentrating only on methods to improve the lives of "normal" people.

References

1. Office of Chief Commissioner for Persons with Disabilities Homepage, http://www.ccdisabilities.nic.in/resources/disability-india. Accessed 02 July 2023
2. Indian Sign Language Portal. https://indiansignlanguage.org/. Indian Sign Language RKMVU-CBE. Accessed 29 June 2023
3. Liu, Y., Wang, R., Shan, S., Chen, X.: Structure inference net: object detection using scene-level context and instance-level relationships. In: IEEE/CVF Conference on Computer Vision and Pattern Recognition, Salt Lake City, UT, USA, pp. 6985–6994 (2018)
4. Velmula, K.R., Linginani, I., Reddy, K.B., Meghana, P., Aruna, A.: Indian sign language recognition using convolutional neural networks. In: Kiran Mai, C., Kiranmayee, B.V., Favorskaya, M.N., Satapathy, S.C., Srujan Raju, K. (eds.) Proceedings of International Conference on Advances in Computer Engineering and Communication Systems. LAIS, vol. 20, pp. 393–400. Springer, Singapore (2021). https://doi.org/10.1007/978-981-15-9293-5_35
5. Katoch, S., Singh, V., Tiwary, U.S.: Indian Sign Language recognition system using SURF with SVM and CNN. Array **14**, 10141 (2022)
6. Von Agris, U., Zieren, J., Canzler, U., et al.: Recent developments in visual sign language recognition. Univ. Access Inf. Soc. **6**, 323–362 (2008)
7. Jayadeep, G., Vishnupriya, N.V., Venugopal, V., Vishnu, S., Geetha, M.M.: Convolutional neural network based Indian sign language translator for banks. In: Proceedings of the 2020 4th International Conference on Intelligent Computing and Control Systems (ICICCS), Madurai, India, pp. 1228–1232 (2020)
8. Samaan, G.H., et al.: MediaPipe's landmarks with RNN for dynamic sign language recognition. Electronics **11**, 3228 (2022)
9. Monga, H., Bhutani, J., Ahuja, M., Maida, N., Pande, H.: Speech to Indian Sign Language Translator, School of Computer Engineering and Technology. MIT, Cambridge
10. Bharati, R., et al.: Automated speech to sign language conversion using google API and NLP. In: Proceedings of the International Conference on Advances in Electronics, Electrical & Computational Intelligence (ICAEEC). Elsevier (2019)
11. Sruthi, C.J., Lijiya, A.: Signet: a deep learning based indian sign language recognition system. In: Proceedings of the 2019 International Conference on Communication and Signal Processing (ICCSP), Tamil Nadu, India, pp. 596–600 (2019)
12. Saija, K., Sangeetha, S., Shah, V.: WordNet based sign language machine translation: from english voice to ISL gloss. In: IEEE 16th India Council International Conference (INDICON), pp.1–4 (2019). https://doi.org/10.1109/INDICON47234.2019.9029074
13. Pandey, S.: Automated gesture recognition and speech conversion tool for speech impaired. In: Reddy, A.B., Nagini, S., Balas, V.E., Raju, K.S. (eds.) Proceedings of Third International Conference on Advances in Computer Engineering and Communication Systems. Lecture Notes in Networks and Systems, vol. 612. Springer, Singapore (2023). https://doi.org/10.1007/978-981-19-9228-5_39
14. Indian Sign Language Research and Training Center (ISLRTC) Homepage. https://www.islrtc.nic.in/,last. Accessed 30 June 2023

A Review on Security Challenges: Cryptography and Blockchain for Internet of Things

Praveen Prakash[1(✉)], Priyanka[1], Lalit Kumar Awasthi[2], and Krishan Kumar[3]

[1] National Institute of Technology Hamirpur, Hamirpur, India
{praveen_phdcse,dr.priyanka}@nith.ac.in
[2] National Institute of Technology Uttarakhand, Srinagar, India
lalit@nituk.ac.in
[3] National Institute of Technology Kurukshetra, Kurukshetra, India
kk@nitkkr.ac.in

Abstract. In the current era of digital technology, we know the importance of security and we are worried about the security laps in the Internet of Things(IoT) fields. There is a lot of research going on in different fields of IoT, but when we think about IoT systems, security is more important and needs to be updated regularly. In this article, we discuss various security issues and emerging trends in IoT. This work serves as a valuable resource for gaining insights into potential directions for future research. Specifically, it focuses on the enhancement of security in IoT data and the utilization of blockchain to address privacy and security concerns within the IoT landscape.

Keywords: Internet of Things · Security · Cryptography · Blockchain

1 Introduction

In 1999, Kevin Ashton [1], a member of the radio frequency identification team, came up with the idea of the IoT, which has grown in importance over the years thanks to technological advancements and the capabilities provided to its users. Melody et al. in [2] the IoT has witnessed a major development and rise in communication and software technologies, which have impacted daily life and ushered in a new era of smart things, in which millions of things are interconnected through exchanging, communicating, and sensing data. The users can manage, plan, make decisions, and take appropriate action as necessary [3] due to these connected gadgets' constant analysis and information collection. Vermesan et al. [4] divided into three categories: technologies that govern user security and privacy, machines' ability to gather relevant information, and machines' ability to analyze consistent information.

D. Pastor-Escuredo et al. (Eds.): ICDLAIR 2023, LNNS 1001, pp. 789–795, 2024.
https://doi.org/10.1007/978-3-031-60935-0_68

2 Security Challenges in IoT

Security in today's distributed environment is an enormous challenge. There are several research published on security, but many new emerging fields have multiple challenges in security. One of the fields is IoT which has been booming in recent years. Here, we discuss various challenges related to IOT security. Device Heterogeneity and Resource Constraints [5]: IoT devices exhibit significant heterogeneity in hardware capabilities, operating systems, and communication protocols. Many of these devices are resource-constrained, lacking the computational power and memory required for robust security implementations. Data Privacy and Confidentiality [6]: The huge amount of data generated by IoT systems, often including sensitive information, presents substantial privacy and confidentiality challenges. Unauthorized access or data breaches can have severe consequences. Scalability and Network Security [7]: Securing large-scale IoT deployments presents scalability challenges. The dynamic and distributed nature of IoT networks requires sophisticated network security measures such as firewalls and intrusion detection systems. Identity and Access Management [8]: The secure authentication and authorization of devices and users is crucial with the increasing number of IoT devices. Weak IAM can result in unauthorized access and malicious activities.

3 Cryptography in IoT Security

The security of IoT devices and networks depends heavily on cryptography. IoT devices are frequently used in a variety of difficult conditions, leaving them open to numerous security risks. In IoT systems, cryptography contributes to data's secrecy, integrity, and validity. Here is a summary of how cryptography plays a role in ensuring security in IoT.

3.1 Data Encryption

Cryptography is pivotal in IoT security by ensuring data confidentiality and integrity. Data encryption techniques, such as Advanced Encryption Standard and Elliptic Curve Cryptography [9], are essential for securing data in transit and at rest. Some new developments in data encryption techniques. B. Umapathy, G. Kalpana [10] used UK symmetric block cipher to increase block complexity and reduce round substitutes using key expansion. However, work cannot reduce the encryption and decryption time without inconvenient the security and complexity. A combination of public and private clouds and is aimed at protecting sensitive data generated by IoT devices, the sensitive data is encrypted using Rivest cipher 6 and Fiestel encryption schemes and stored in a private cloud. At the same time, nonsensitive data is encrypted using the Advanced Encryption Standard algorithm and stored in a public cloud, and a multilevel authentication scheme with trusted authority is proposed in their paper [11] to protect cloud-stored data from unauthorized access. The recommended approach in this research [12] on protecting communications between devices and between devices and the cloud, guaranteeing the security and privacy of network entities, and implementing authentication and access control techniques. Zhang, Ke, et al. [13] introduce the concept of attribute-based encryption schemes as a solution to ensure data transmission security and fine-grained sharing of encrypted IoT data and use a lightweight and efficient data sharing and searching scheme that significantly reduces the computational cost for IoT devices and allows for multiple-keyword searching. The Probabilistic Super Learning model [14] is employed by this framework to manage secure data transmission efficiently. It analyzes the characteristics of the data received from IoT devices to ensure its safety. To encrypt and decrypt data, the framework uses the Elliptic Curve Cryptography model and the Random Hashing technique, ensuring the security of data storage and retrieval.

3.2 Authentication and Key Management

Authentication and key management [15] are critical components of IoT security. Cryptographic methods including digital certificates and Public Key Infrastructure, authenticate devices and establish secure communication channels. To enhance the security of cloud and IoT systems, Chandrasegar Thirumalai et al. [16] propose an efficient public key secure scheme. The RSA Public Key Cryptosystem is introduced as a secure scheme, and its workflow and security strength are explained. The document proposes a new Key Generation Scheme and presents experimental results and performance analysis comparing it to other RSA variants (Table 1).

4 Blockchain Technology for IoT Security

Blockchain technology[17] offers immutable data integrity by creating a decentralized, tamper-proof ledger. Transactions recorded on the blockchain are transparent, secure, and resistant to modification. Blockchain's decentralized nature[18] eliminates single points of failure, making it resilient to attacks. This feature enhances security in IoT applications where data and device trustworthiness are paramount.

A solution given by Wang, Haiyan, and Jiawei Zhang [19] to handle the challenges of managing massive amounts of data and communication is the Blockchain and Bilinear mapping-based Data Integrity Scheme for large-scale IoT data. This scheme effectively addresses trust issues that arise with third-party authorities (TPAs) by utilizing blockchain technology. The approach significantly enhances the efficiency of verifying data integrity for large-scale IoT data without the need for TPAs.

5 Integration of Cryptography and Blockchain

Hybrid solutions [20] that combine cryptography and blockchain offer robust security for IoT. Cryptographic methods can secure data and devices, while

Table 1. Existing Models of Cryptography and Blockchain for IoT Security

Author	Method	Objective
B.Umapathy, G. Kalpana [10]	New multi-layered cryptographic algo named as UK	Symmetric cryptographic method to design block complexity for cloud data security
Saleh Atiewi et al. [11]	RC6 and Fiestel AES algorithm trusted authority	IoT System Based on Multifactor Authentication and Lightweight Cryptography
Ke Zhang et al. [13]	lightweight and efficient data sharing and searching scheme	Attribute-based encryption (ABE) schemes as a solution to ensure data transmission security and fine-grained sharing of encrypted IoT data
Adil O. Khadidos et al. [14]	Random Hashing and AI-based Probabilistic Super Learning model	Secured data retrieval with attack detection and feature learning models
Chandrasegar Thirumalai et al. [16]	efficient and non-shareable Public Key Exponent Secure Scheme	Enhance the security of the cloud and IoT systems by proposing an efficient public key secure scheme
Haiyan Wang and Jiawei Zhang [19]	Blockchain and Bilinear mapping-based Data Integrity Scheme	Enhances the efficiency of verifying data integrity for large-scale IoT data
Anand Singh Rajawat et al. [22]	Blockchain, SHA256 hash algorithm, Consensus protocol	Blockchain-based model that utilizes the SHA256 hash algorithm to ensure the security and integrity of healthcare data
P Velmurugadass et al. [23]	Software Defined Network and Elliptic Curve Integrated Encryption Scheme	Cloud based Software Defined Network and Elliptic Curve Integrated Encryption Scheme is used to encrypt packets before they are transferred to the server

blockchain ensures data integrity and trustworthiness. Zhenqi Wang, Shaopeng Guan [21] proposed a data-sharing scheme to improve data security and reduce storage pressure on the blockchain. In Fig. 1, we can see how Cryptography Integration with Blockchain appears on page number 5.

A blockchain-based technique [22] using consensus protocols, the SHA256 hash algorithm, and priorities such as verifiable, appropriateness, extensiveness, uniqueness, and sturdiness was employed in the study focuses on guaranteeing the security and integrity of medical data in IoT gadgets, controlling data access and permissions based on access rights and policies entails integrating IoT devices into a blockchain system, suggested strategy seeks to overcome the difficulties associated with managing enormous amounts of data while maintaining data security in healthcare organizations.

P Velmurugadass et al. [23] design a cloud-based Software Defined Network by cloud server, authentication server, mobile nodes (IoT devices), open flow switch, blockchain-based controllers, cloud server, and investigator. The Elliptic Curve Integrated Encryption Scheme is used in the mobile nodes to encrypt packets before they are transferred to the cloud server.

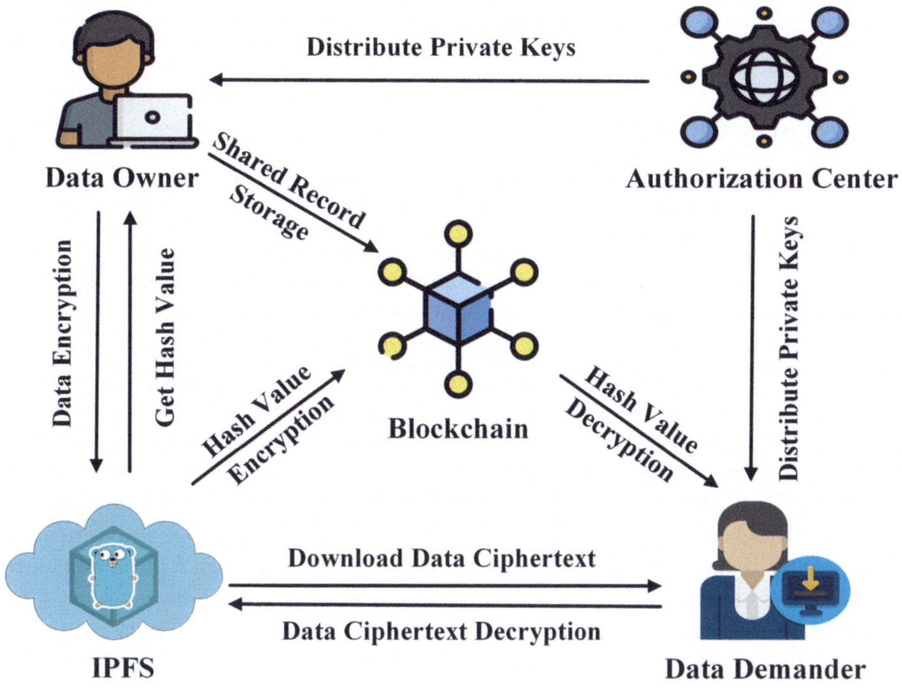

Fig. 1. Cryptography Integration with Blockchain

6 Future Directions and Challenges

The future of IoT security lies in advancements in cryptography and blockchain integration. Emerging trends include the development of lightweight cryptographic algorithms and the evolution of blockchain consensus mechanisms.

6.1 Emerging Trends

Quantum Resistant Cryptography: The development of quantum computing significantly threatens existing cryptography techniques. To protect IoT devices from potential quantum attacks, researchers are working hard to create quantum-resistant encryption algorithms. Homomorphic Encryption: As IoT devices generate and transmit sensitive data, preserving privacy becomes crucial. Homomorphic encryption allows computation on encrypted data without decryption, enabling secure data analysis and processing in IoT applications. Blockchain for IoT Security: Blockchain technology is being explored for securing IoT networks and transactions. It provides a tamper-resistant and decentralized ledger, enhancing data integrity and trust in IoT ecosystems.

AI-Driven Threat Detection: Utilizing artificial intelligence(AI) and Deep learning for anomaly detection and threat mitigation in IoT networks is becoming more prevalent. AI can help identify and respond to security incidents in real time.

6.2 Ongoing Challenges

Compliance challenges persist in the realm of IoT security. Addressing these challenges will be essential for widespread adoption. Limited Computing Resources: It is common for IoT devices to have restricted processing capabilities and storage space. Implementing robust cryptographic algorithms can be challenging due to resource constraints. Finding lightweight encryption methods that provide adequate security is a continual challenge. Key Management: Managing encryption keys for a vast number of IoT devices can be complex. Ensuring secure key generation, distribution, and storage prevents unauthorized access. Secure Bootstrapping: Establishing trust between IoT devices and networks, especially in uncontrolled environments, is challenging. Secure bootstrapping and onboarding procedures are needed to ensure the integrity and authenticity of devices. Blockchain Scalability: While blockchain offers tamper resistance and transparency, it faces scalability issues. Ensuring that IoT data can be efficiently recorded on a blockchain without network congestion is a challenge. Smart Contract Security: If IoT devices interact with blockchain through smart contracts, ensuring the security of these contracts is crucial. Vulnerabilities in smart contracts can lead to financial and security risks.

7 Conclusion

The security difficulties in the IoT ecosystem pose demand for innovative solutions. Cryptography and blockchain technology, when integrated effectively, provide a robust framework for enhancing IoT security. While challenges remain, continued research and development in this domain hold the key to a more secure and trustworthy IoT system. Our focus in the future will be on developing a decentralized system that is self-organizing and capable of responding to security threats.

References

1. Ashton, K.: That 'internet of things' thing. RFID J. **22**(7), 97–114 (2009)
2. Moh, M., Raju, R.: Machine learning techniques for security of Internet of Things (IoT) and fog computing systems. In: 2018 International Conference on High Performance Computing Simulation (HPCS). IEEE (2018)
3. Dash, S., Prusty, D.: Domain-specific IoT applications. In: Internet of Things: Enabling Technologies, Security and Social Implications, pp. 27–36 (2021)
4. Vermesan, O., Friess, P. (eds.): Internet of Things: Converging Technologies for Smart Environments and Integrated Ecosystems. River publishers (2013)

5. Anantharaman, P., Palani, K., Nicol, D., Smith, S.W.: I am joe's fridge: scalable identity in the Internet of Things. In: 2016 IEEE International Conference on Internet of Things (iThings) and IEEE Green Computing and Communications (GreenCom) and IEEE Cyber, Physical and Social Computing (CPSCom) and IEEE Smart Data (SmartData). IEEE (2016)
6. Miorandi, D., Sicari, S., De Pellegrini, F., Chlamtac, I.: Internet of things: vision, applications and research challenges. Ad Hoc Netw. **10**(7), 1497–1516 (2012)
7. Atzori, L., Iera, A., Morabito, G.: The internet of things: a survey. Comput. Netw. **54**(15), 2787–2805 (2010)
8. Alaba, F.A., Othman, M., Hashem, I.A.T., Alotaibi, F.: Internet of Things security: a survey. J. Netw. Comput. Appl. **88**, 10–28 (2017)
9. Xia, Q.I., Sifah, E.B., Asamoah, K.O., Gao, J., Du, X., Guizani, M.: MeDShare: trust-less medical data sharing among cloud service providers via blockchain. IEEE Access **5**, 14757–14767 (2017)
10. Umapathy, B., Kalpana, G.: A novel symmetric cryptographic method to design block complexity for data security. Comput. Electr. Eng. **104**, 108467 (2022)
11. Atiewi, S., et al.: Scalable and secure big data IoT system based on multifactor authentication and lightweight cryptography. IEEE Access **8**, 113498–113511 (2020)
12. Medileh, S., et al.: A flexible encryption technique for the internet of things environment. Ad Hoc Netw. **106**, 102240 (2020)
13. Zhang, K., Long, J., Wang, X., Dai, H.-N., Liang, K., Imran, M.: Lightweight searchable encryption protocol for industrial internet of things. IEEE Trans. Ind. Inf. **17**(6), 4248–4259 (2020)
14. Khadidos, A.O., Shitharth, S., Khadidos, A.O., Sangeetha, K., Alyoubi, K.H.: Healthcare data security using IoT sensors based on random hashing mechanism. J. Sens. **2022**(1), 8457116 (2022)
15. Roman, R., Zhou, J., Lopez, J.: On the features and challenges of security and privacy in distributed internet of things. Comput. Netw. **57**(10), 2266–2279 (2013)
16. Thirumalai, C., Mohan, S., Srivastava, G.: An efficient public key secure scheme for cloud and IoT security. Comput. Commun. **150**, 634–643 (2020)
17. Dorri, A., Kanhere, S.S., Jurdak, R.: Towards an optimized blockchain for IoT. In: Proceedings of the Second International Conference on Internet-of-Things Design and Implementation (2017)
18. Dorri, A., et al.: Blockchain for IoT security and privacy: the case study of a smart home. In: 2017 IEEE International Conference on Pervasive Computing and Communications Workshops (PerCom Workshops). IEEE (2017)
19. Wang, H., Zhang, J.: Blockchain based data integrity verification for large-scale IoT data. IEEE Access **7**, 164996–165006 (2019)
20. Zhang, A., Lin, X.: Towards secure and privacy-preserving data sharing in e-health systems via consortium blockchain. J. Med. Syst. **42**(8), 140 (2018)
21. Wang, Z., Guan, S.: A blockchain-based traceable and secure data-sharing scheme. PeerJ Comput. Sci. **9**, e1337 (2023)
22. Rajawat, A.S., Rawat, R., Barhanpurkar, K., Shaw, R.N., Ghosh, A.: Blockchain-based model for expanding IoT device data security. In: Bansal, J.C., Fung, L.C.C., Simic, M., Ghosh, A. (eds.) Advances in Applications of Data-Driven Computing, vol. 1319, pp. 61–71. Springer, Heidelberg (2021). https://doi.org/10.1007/978-981-33-6919-1_5
23. Velmurugadass, P., Dhanasekaran, S., Anand, S.S., Vasudevan, V.: Enhancing Blockchain security in cloud computing with IoT environment using ECIES and cryptography hash algorithm. Mater. Today: Proc. **37**, 2653–2659 (2021)

Current Approaches and Challenges in Medical Image Analysis and Visually Explainable Artificial Intelligence as Future Opportunities

Nandani Sharma and Sandeep Chaurasia(✉)

Department of Computer Science and Engineering, Manipal University Jaipur, Jaipur 303007, Rajasthan, India
nandani.219351015@muj.manipal.edu, sandeep.chaurasia@jaipur.manipal.edu

Abstract. Medical imaging plays a significant role in the disease diagnosis process. The automation techniques used today are helping medical professionals. The present methods' Black-box structure makes it difficult for them to be implemented in the medical field with reliability. As a solution of the black box nature, an option is to use the XAI method. In this review, a discussion is done on the available popular visual sources of medical diagnosis. Further, the discussion is extended to the current position of machine learning and deep learning in the medical image domain. The performance of deep learning and machine learning approaches in medical image analysis is covered in the paper that is being presented. There is also a thorough examination of the difficulties these methods present for processing medical images in the real-time implementation of artificial intelligence techniques. Present XAI approaches as potential remedies for problems with deep learning approaches. In the paper discussion is on the different techniques that are available for the explanation of the outcome of the artificial intelligence models. In the last the future of the XAI in the medical domain are discussed.

Keywords: medical image processing · machine learning · deep learning · Explainable Artificial intelligence · XAI

1 Introduction

Medical images (MI) are a piece of necessary information in the process of disease diagnosis by medical experts. Different images are used for diagnosis purposes such as histological images help in the diagnosis of cancer, fundus images are used for diabetes retinopathy, X-ray for fracture diagnosis, MRI scan image for tumors, CT scan for heart disease, and many more. With the evolution of deep learning, medical image automatic analysis is revolutionary. It is helping the

N. Sharma and S. Chaurasia—These authors contributed equally to this work.

D. Pastor-Escuredo et al. (Eds.): ICDLAIR 2023, LNNS 1001, pp. 796–811, 2024.
https://doi.org/10.1007/978-3-031-60935-0_69

healthcare sector break barriers in terms of accuracy and robustness to achieve efficient and robust computer-aided diagnostic systems [57].

Recently machine learning approaches have shown progress in the medical domain, due to the following variables brand new statical learning algorithms, an extensive volume of data, robust hardware availability, and last but not least cloud storage [3]. With the evolution of deep learning, medical image automatic analysis is revolutionary. The deep learning efficiency is measured in terms of accuracy, f1-score, precision, and recall. Deep learning is exhibiting extraordinary efficiency in terms of accuracy in machine medical image analysis. Deep learning models have demonstrated exceptional performance in Image Classification and disease Diagnosis using CNN for chest X-rays in [58], and classifying Cancer Cell Images using the vit-cnn ensemble model in [54]. Despite the favorable performance, the DL and ML approaches are facing challenges such as a lack of transparency of the output, unknown learning features, a low number of deployments, and the accuracy of a good performance evaluation technique.

The European government law named the European Union's General Data Protection Regulation (GDPR) motivates to explain the machine-generated decisions. The GDPR states that it "requires the right of patients to receive meaningful information about how a decision was rendered from the individual's data" [17,18]. This suggests that if the algorithm makes a decision, it should be included with an explanation. As the medical field is a high-risk field the decision made by automation is a black box. The explainable techniques motivate the legal and ethical face of Artificial intelligence models. Amerirad B. et al. [29] discovered that introducing a small bit of perturbation to the input of the deep learning model can classify the input into distinct class labels. The perturbation added is so negligible that it cannot be noticed with the human eyes.

The black-box nature of deep learning is a hurdle in the acceptance of automated results by medical persons. At the legal and ethical level and the trust building, the need for explanation with the profound learning results is also in request. Rajpurkar et al. in [43] performed a survey with chest X-rays one provided an explanation and the other not provided with an explanation. The results of the experiment show that 10 out of the 13 participating physicians (77%) had better diagnostic accuracy when assessing chest X-rays with an XAI explanation instead of no explanation. There is an urgent requirement for explainability decision-supported systems in domains such as medical, autonomous transportation, defense domain, and legal domains [13,14].

XAI full form is Explainable artificial intelligence. Explainable Artificial intelligence is a branch of artificial intelligence algorithm developed to explain the machine learning and deep learning model output for the end-users. Explainable Artificial Intelligence is in the growing stage there is no universally acceptable definition by scientists and researchers. The most justified definition found by Barredo Arrieta, A., et al.

"XAI proposes the development of machine learning techniques producing highly effective, explainable models which humans can understand, manage and trust" [22]. The current literature review is structured in the following sections: Sect. 2 discusses the general information about different medical images. Section 3 reviews deep learning and machine learning approaches for medical image analysis. In Sect. 4 what are the Challenges in DL and ML approaches faced in real-time implementations Sect. 5 reviews the visual XAI approaches in medical images. Section 6 concludes the paper.

The literature review aims to answer the following research questions: The research questions we aim to address are as follows:

RQ1: What are the common visual sources for the medical diagnosis available?
RQ2: How well the deep learning and machine learning algorithms are performing in terms of performance used in medical image analysis?
RQ3: What challenges have been reported for deep learning and machine learning algorithms in real-time implementation?
RQ4: Are visual XAI techniques a possible solution for the black-box nature of deep learning and machine learning?

2 Popular Visual Sources of Medical Diagnosis

In medical science variety of tools assist in the diagnosis of the disease, most common visual tools such as the X-ray, CT scan, and MRI are discussed in this section:

X-Ray:- X-rays are the most popular, and the oldest tools emerged in 1895, used for medical image diagnosis. X-rays are the special varieties of waves that exist in the electromagnetic spectrum. X-rays work on the idea that the amount of radiation being absorbed from different types of material is different. Bones absorb the X-ray in large amounts, whereas the fat and tissues absorb fewer rays and air does not absorb rays. So bones, tissue, and air are visualized as white, gray, and black in color respectively in the X-ray images [19,43]. The advantage of using X-rays is that it is cost-effective, easily available, no surgery required, and easily understandable. The disadvantage of using X-rays is the absorption of radiation and less information.

MRI:- The full form of MRI is Magnetic resonance imaging developed in 1972. MRIs contribute primarily to structural information inspection of the non-bony part of the body or the soft tissues of the body. MRIs are used to scan different anatomical positions. MRIs work on the concept that the orientation of the protons can be manipulated by applying a strong magnetic field, then the return of the protons to the stable state is measured. Some of the advantages of MRIs are that: provide higher-resolution images, and are safe compared to other scanning techniques used. The disadvantage is the cost of performance is higher, and unable to be utilized with the patients having pacemakers.

Ultrasound:- Ultrasound creates images of organs and structures within the body by employing sound waves of high frequency. Ultrasound waves were first used in the medical domain in the year 1956. Ultrasound provides high spatial resolution images, at a low cost and in a secure manner these are some of the advantages. The disadvantage of Ultrasound is the waves provide inferior penetration in bone or air and along with this is need expert radiologist for the interpretation.

CT Scan:- The use of CT scans in the medical domain started in 1971. The technology used for the CT scans is similar to X-ray scans except that CT scans penetrate the X-ray waves at a certain angle and from specific distances. The advantage of using the CT scan is that it provides a higher penetration depth view, and is less sensitive to the patient's movement. The disadvantage of the CT scan is that the patients are more exposed to higher radiation as compared to the X-rays (Table 1).

Table 1. Major diagnostic radiology image

S.No.	Image Name	Most Usages	Database	Link
1	ultrasound	Organs and tissues	HC18 dataset	[53]
2	fundus	eyes	MESSIDOR-2, IDRiD, and E-Ophtha.	[54]
3	X-ray	bones/chest	Covid-chestxray-dataset	[55]
4	CT scan	Any body part	MICCAI 2017 LiTS challenge dataset	[56]
5	MRI	Brain	PPMI	[57]
6	PET	Any body part	Alzheimer's Disease Neuroimaging Initiative database	[58]

Fundus Images:- The Fundus images anatomy location-specific tools used especially for the eyes. The fundus image camera was made available for the first time in 1926. The main use is for the diagnosis of Retinal diseases. The advantage of fundus images is that they are developed quickly and in a simpler manner. At any given time fundus images observe an extended retinal area. The fundus images are that are two-dimensional in nature so fail to acquire the depth of the image and find problems in finding abnormalities such as cotton wool sports.

PET:- PET stands for Positron Emission Tomography. The complete-body PET scanner was introduced in 1977. PET uses the technology employing isotopes where a couple of photons are produced in each annihilation [21].

3 Deep Learning and Machine Learning Approaches in the Medical Domain

The need for automation of the medical image analysis is been required for an easy and fast diagnosis process. Currently, artificial intelligence is contributing in the field at the medical domain with two broader approaches machine learning and deep learning. Here we review the machine learning and deep learning approaches for the evaluation perspective of the different types of images. Machine learning and deep learning approaches are mainly evaluated on accuracy, precision, recall/sensitivity, F1-score, AUC, and ROC. Luan et al. [50]

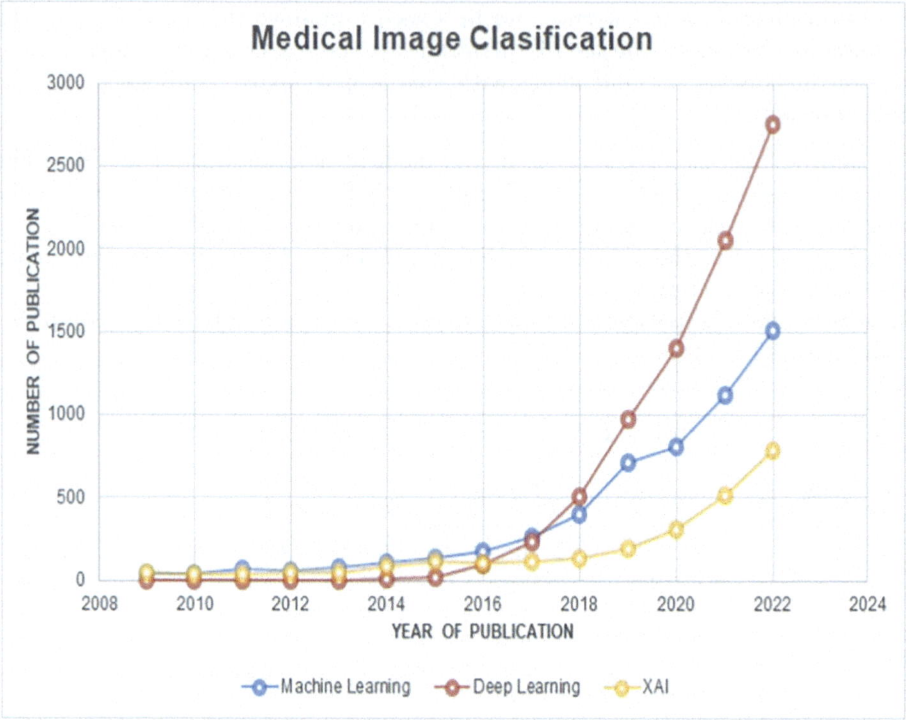

Fig. 1. Annual publication of MIC ML, DL, and XAI [Source:Scopus]

review of machine learning papers and it has been observed that for the evaluation of the machine learning model, accuracy is used in most of the papers as compared to the other evaluation techniques such as precision, recall/sensitivity, F1-score, AUC, or ROC.

Khan et al. [34] use five different models (SVM, KNN, Random Forest, Naive Bayes, and Decision Tree)) for the binary classification of normal and COVID-19 images. Achieved the accuracy of the 96%, 92%, 92%, 90%, and 82%, for the respective models. In the future plan to use different feature extraction techniques to improve performance accuracy. Hassanienet et al. [35] classify the X-ray images for binary labels of normal and COVID-19 affected using multi-level thresholding with the Support vector machine. The proposed model achieved an accuracy of 97.48%. Later we plan to modification of the hybrid ML model. Amin J et al. [36] use brain MRI to classify as cancerous and non-cancerous. Using the machine learning SVM architecture gets success in terms of the accuracy of 97.1%. Saleh et al. [37] analyzed the Lung CT-Scan with the machine learning hybrid CNN and SVM and achieved 97.91% accuracy. Javaid et al. [38] studied the skin Image utilizing SVM (Medium Gaussian), and Random Forest architecture and achieved the accuracy of 88.17% and 93.89% respectively.

The deep learning architectures achieve higher accuracy, as Jiang Z et al. [26] analyzed Lymph Cancer Cell Images using the ensemble model and achieved an accuracy of 99.0%. A diagnostic approach for acute lymphocytic leukemia, which could classify cancer cells and normal cells. Abbas A et al. [27] use X-ray classification of COVID and normal state the future work to add an explainability component to enhance the usability of the model. Apostolopoulos ID et al. [28] use different versions of the CNN architecture as VGG19, MobileNet, Inception, Xception, and Resnet-v2 and achieve respective accuracy of 97.96%, 97.40%, 92.85%, 99.10%, and 97.09%. The limitation is that the symptoms of the images are not visualized on X-rays. Rajpurkar et al. [30] use ChexNet of 121-layer for the detection of multiple diseases outperforms the previous achieved the 93.71%. Sawant et al. [31] use brain MRI with the LeNet-5 technique achieving 99%. In the future planning to use GPU and the interpretation for better understanding. Hosny et al. in [33] outperform several state-of-art in skin lesion classifications with an accuracy of 96.86%. for future direction need to provide more support to the clinical by explaining.

4 Challenges of Deep Learning and Machine Learning

In the present section challenges of the deep learning and machine learning approaches are showcased to keep in mind to help the new researcher to find the research gap. In this section, the challenges of implementing the solutions provided by machine learning and deep learning in the medical domain are discussed.

1. **Black Box Behaviour of Deep Learning:-** The Deep learning solutions are performing well in terms of accuracy [29] but their black-box behavior is a hurdle in the implementation in the medical area [51]. Xu et al. [12] raise the point that the result provided by Deep Neural Networks (DNNs) cannot be explained, neither by the neural network itself, nor by an external explanatory component, and not even by the system developer.
2. **Is accuracy an adequate criterion for evaluating model learning?** Many authors has raised the question of accuracy as a measure of evaluation of machine learning models, mainly in crucial areas such as facial recognition, automatic transport systems, financial transactions, and medical applications [13,22].
3. **Few Deployments of Machine Learning model:-** As shown in the Fig. 1. For a few years, the publication of deep learning and machine learning in the medical domain has been increasing rapidly, but the rate of deployment for the medical domain is very slow compared to the rate of publications. The possible reason pointed out is a lack of transparency and explainability [15,16].
4. **What features the Deep Learning models are learning?:-** As the deep learning models are black-box in nature it is difficult to identify which features of the input are influencing the final output of the models. It has been

observed by A.J. DeGrave et al. in [20] that the deep learning model does not rely on the COVID-19 medical pathology features but instead on learning text features imprinted on the x-ray. Lapuschkin et al. [13] observed that DNN looked at the contour of the actual horse, whereas the Fisher Vector model mostly relied on a certain copyright tag, that happens to be present on many horse images.

5. **Lack of Data:-** Jiao et al. [52] point out that one of the challenges of a deep learning model being implemented in real life is the scarcity of the available dataset. Even the available data are imbalanced [29]. In a case study, Caruana et al. [13] write that the AI model prediction for the "Pneumonia - Asthma" patient is completely different from the real situation. The model predicts that a patient with heart issues has less chance of death from pneumonia the a normal person. In a practical scenario, the patient has a higher risk of death. The model predicts this as there is less availability of such data (Table 2).

Table 2. Deep Learning and Machine Learning

Sno	Paper	Medical Speciality	Type of Image	Technique	Dataset	Accuracy	Description
Machine Learning							
1	Khan et al. [34]	chest	X-ray and CT scan	SVM KNN Random Forest Naive Bayes Decision Tree	different Kaggle dataset	**96%** 92% 92% 90% 82%	Use the five different models for the binary classification of normal and covid-19 images. To improve the performance in the future another feature extraction approach may be used
2	Hassanien et al., [35]	lung	X-ray	SVM	covid-chestxray-master	**97.48%**	user the SVM with the multi-level thresholding to classify the Covid-19 X-rays. **Later modification of hybrid ML model is used**
3	Amin J et al. [36]	Brain	MRI	SVM	Harvard, RIDER and Local	**97.1%**	Approach classify cancerous and non-cancerous Magnetic Resonance Imaging (MRI) use the different segmentation techniques to identify the candidate lesion. **But fail to highlight the segment contributes to the classification of the end user**
4	Saleh et al. [37]	Lung	CT-Scan	hybrid CNN and SVM	Chest CT-Scan Images	**97.91%**	The model perform well in term of accuracy and also aim to increase accuracy but doesn't talk about the implementation in real life
5	Javaid et al. [38]	Skin	Skin Image	SVM, and Random Forest	ISIC-ISBI	88.17% **93.89%**	The classification of the skin image is performed using the SVM and random forest
Deep Learning							
6	Jiang Z et al. [26]	Lymph	Cancer Cell Image	vit-cnn ensemble model	ISBI 2019 dataset	**99.0%**	With the ensemble approach classify the cancer cell image cancer cells and normal cells
7	Abbas A et al. [27]	Lungs	X-ray	ResNet 18	CXR dataset	**93.1%**	The model uses the ResNet for X-ray classification of covid and normal. state the **The scope of the XAI techniques to be implemented in feature work**
8	Apostolopoulos ID et al. [28]	chest	X-ray	VGG19; MobileNet; Inception; Xception; Resnet-v2	Cohen, RSNA, SIRM	96% 97.40% 92.85% **99.10%** 97.09%	Use the Different models for classification for covid-19. **The limitation is that the symptoms of the images are not visualized on X-rays.**

(continued)

Table 2. (*continued*)

Sno	Paper	Medical Speciality	Type of Image	Technique	Dataset	Accuracy	Description
9	Rajpurkar et al. [30]	Chest	X-Ray	CheXNet is a 121-layer CNN	ChestX-ray14	**93.71%**	The use ChexNet for the detection of multiple diseases **outperforms the previous state of the arts**
10	Sawant et al. [31]	Brain	MRI	LeNet-5	Diff. sources	**99%**	The use the LeNet-5 architecture for the brain cancer detection. And achieve higher accuracy. **In the future use GPU and the interpretation for better understanding**
11	Hosny et al. [33]	Skin	Skin image	Alex-net	MED-NODE Derm, ISIC	**96.86 %**	outperforms several state-of-art in skin lesion classification with accuracy 96.86 %. for future direction need to provide more **support to the clinical by explanation.**

5 XAI Model

For the challenges of the machine learning and deep learning approaches discussed in Sect. 4, several authors suggest the use of explanation artificial intelligence [47–49]. The domains with high risk such as self-driving vehicles, smart cities, security, finance, and medical, need to be trusted by the users. For the trust issues the Explainable artificial intelligence field of artificial intelligence are providing good support. In this section, the discussion on the XAI algorithms used in the field of medicine is shown in the table.

CAM. The Convolutional neural network is the most common neural network for computer vision. Zhou et al. [] use the CNN's forward pass for the classification and localization at a time. For localization, the purpose is to use the global activation pooling and the softmax in the case of the classification. CAM is used for the local visualization of symptoms learned by the different neural network architectures from the medical images. Rajpurkar et al. [5] generated CheXNeXt, a version of the CNN to detect the presence of 14 different pathologies concurrently. The abnormalities in the disease-affected X-Ray are highlighted using the CAM. Chakraborty et al. [9] use CAM to highlight the area of the brain MRI for the identification of Parkinson's disease diagnosis. Turk et al. [11] use the CAM to explain the results generated by the ensemble architecture using ResNet50, VGG19, InceptionV3, and MobileNet for tumor detection in MRI images.

Grad-CAM. Grad-CAM is the more generalized form of CAM. Grad-CAM is proposed for the local highlighting areas of importance in an image in the prediction process. Grad-CAM [] uses a gradient on the CNN layer for a specific class of prediction and provides the regions of significance in the images concerning the prediction. The explanation provided by the Grad-CAM is also utilized in the medical field analysis purpose. Brunese et al. [2] use the Grad-CAM to visually highlight symptoms of pneumonia, COVID-19, and healthy chest X-rays. Hagele et al. [4] implemented the Grad-CAM to emphasize the symptoms in the histopathological images after the classification was performed using the GoogLeNet architecture. Rajaraman et al. [6] visualize and explain deep learning predictions for pneumonia diagnosis on the pediatric chest X-ray using the Grad-CAM. Malhotra et al. [25] used the GradCAM to verify the correct learning of the COMiT-Net by focusing on the unhealthy region.

LIME. LIME is a visualization technique referring to locally interpretable model-agnostic explanations. The LIME works on the perturbation for the generated neighborhood of the instance to be a visualized superpixel of the learning of the model [19]. The LIME works on text, images, and tabular data. LIME is used in the medical domain by Rajaraman et al. [6] to highlight the pneumonia symptoms and Malhi et al. [7] visualize the bloody regions in the gastric endoscopy images.

LRP. LRP is the technique that operates from output to input in reverse order and utilizes the weights and activation generated in the forward pass. The output of LRP at the end is the visual contribution of the model. The contribution is specified in the relevance score. LRP is the Layer-Wise Relevance Propagation []. Zhang et al. [1] visually highlight the reason in the ultrasound image the growth of the fetus' head circumference during pregnancy with the help of LRP.

Prototypes. Prototypes use examples as explanations, which they described as 'this-looks-like-that'. The method reflected the case-based reasoning [32]. Uehara et al. [8] explain the Histology images decision of model of cancer and non-cancer classification by prototypical methods.

TCAV. Testing with Concept Activation Vectors (TCAV) idea is to incorporate directional derivatives to determine the intensity of a user-defined concept, such as the presence of whiskers, for picture classification in cat images. [B. Kim, M. Wattenberg, J. Gilmer, C. Cai, J. Wexler, F. Viegas, R. Sayres, Interpretability beyond feature attribution: quantitative testing with concept activation vectors (TCAV), in International Conference on Machine Learning, 2018]. Kim et al. [10] implement the TCAV for the medical domain global explanation of the internal state of the neural architecture for diagnosis of diabetic retinopathy with the help of fundus images.

Text Explanations. The textual description of the visual input is a multidomain field that includes the natural language processing and the computer vision combination. Some of the techniques used for text explanations are image caption generation, image question and answer, and image dialogs [46]. In the present paper, authors discuss the image caption generation only. The image caption is the technique in which a textual description of the image input to the model is provided. Biswas et al. [42] visual attention by the union of high-level features and the low-level features obtained from the object-specific salient regions. Text generation obtains the Bleu-2 score of 0.528. Xu et al. [43] a model that can fix its attention on salient objects while generating the corresponding words in the output sequence. For the medical domain, Yang S et al. [44] scored the BLEU 0.44 for the multi-view encoder and two-layer LSTM decoder architecture mode to generate the caption of the x-ray images. Gajbhiye G. et al. [45] implement the Multilevel Multi-Attention encoder-decoder approach by combining textual Attention to generate a medical report for different views of chest X-ray images.

In Table 3: It has been observed that the visual explanation techniques lack the evaluation measurements, whereas the explanation provided by the evaluation for the explanation techniques is done by the evaluation techniques such as BLUE score, ROUGE-L, METEOR, CIDEr, SPICE.

Table 3. XAI Models

Sno	Paper	Medical Speciality	Type of Image	XAI Technique and Evaluation	Dataset	Description
1	Zhang et al. [1]	Fetal	Ultra sound, MRI	LRP **NO**	HC18 dataset	fetus head circumference (HC) is a key biometric to monitor fetus growth during pregnancy. Using the CNN
2	Brune se et al. [2]	Chest	X-Ray	Grad-CAM **NO**	covid-chestxray-dataset, COVID-19, NIHCX	pneumonia, Covid-19, and healthy using CNN model and **accuracy equal to 0.97. Use the Grad-CAM to highlight the areas of interest in the X-ray that are symptomatic of the Disease**
3	Chen et al. [3]	Chest	X-ray	Grad-CAM **NO**	Chest X-ray 14 dataset	A novel lesion location attention-guided network named LLAGnet is used for the detection of lesions in X-ray. **Weakly supervised lesion localization with the Grad-CAM**
4	Hagele et al. [4]	Multiple	histopathological	Grad-CAM LRP **NO**	TCGA, Skin, Breast, Lung	use the GoogLeNet architecture for binary classification task of tumor tissue and **visual explanations tumor tissue at different granularity levels at patch level to pixel-wise**
5	Rajpurkar et al. [5]	Chest	X-ray	CAM **NO**	ChestX-ray8	Develop CheXNeXt, a CsNN to detect the presence of 14 different pathologies concurrently. **The abnormalities locations in the disease-affected X-ray are highlighted using the CAM**
6	Raja raman et al. [6]	Chest	X-ray	LIME, Grad-CAM **NO**	pediatric CXRs	Visualizing and explaining deep learning predictions for pneumonia detection in pediatric chest radiographs. The explanations generated by the **LIME show few false positive superpixels reported**
7	Malhi et al. [7]	Gastrointestinal	Endoscopy	LIME **use case-oriented evaluation**	Red Lesion Endoscopy data	Uses CNN for classification and LIME to explain which areas in gastral endoscopy images **contained bloody regions**
8	Uehara et al. [8]	Breast	Histo logy	Prototypes **NO**	Kylberg, PCAM	For the explanation purpose the prototypes technique is used, the model **finds similarity of the images to prototypical examples** learned from the training sets. Work on cancer and noncancer classification
9	Chakraborty et al. [9]	Brain	MRI	CAM **NO**	PPMI	For the classification 3D convolutional neural network (CNN) architecture is used for detection of Parkinson's Disease in the Magnetic Resonance Imaging (MRI) scans. **visual CAM is used.**

(*continued*)

Table 3. (*continued*)

Sno	Paper	Medical Speciality	Type of Image	XAI Technique and Evaluation	Dataset	Description
10	Kim et al., [10]	Eye	fundus	TCAV **NO**	-	It is a linear explanation of the internal state of neural networks, yielding a **global explanation** of the networks in terms of human-understandable concepts
11	Turk et al. [11]	Brain	MRI	CAM **NO**	-	The model's uses an ensemble model to binary classify. Lastly use **the CAM for the explanation of the image**
12	Matsu moto et al. [23]	Chest	X-RAY	Saliency map **NO**	chest X-ray	use chest X-ray for diagnosis of heart failure. High-level accuracy in medical images. **Ambiguous images misdiagnosed by model**
13	Winkler et al. [24]	Skin	dermo scopic images	grad descent backpropaga-tion **NO**	Skin Imaging Collaboration	**skin markings significantly interfered with the CNN's correct diagnosis of nevi by increasing the melanoma probability scores and consequently the false-positive rate**
14	Malhotra et al. [25]	Chest	Xray	Grad-CAM **NO**	ChestX-ray-14, CheXpert,	**The gradcam help to verify the correct learning of the COMiT-Net by focusing on the unhealthy region same as manual annotate**
15	Allaouzi et al. [39]	chest	X-ray	Retrieval-based models **BLEU, 0.56**	Own created	Hybride of VGGNet and ResNet is used for the feature extraction.**The report generation is performed using the semantically nearest pattern in the database**
16	Singh et al. [40]	Multiple	Multiple	Encoder decoder **NO**	VQA-Med	use **CNN for the encoder and LSTM for the decoder**
17	Park et al. [41]	-	X-ray	hierarchical (LSTM) **BLEU-1 0.3731**	MIMIC-CXR	Use resnet512 for image feature extraction and hierarchical LSTM for the text generation. **The text is not more specific to highlight the abnormal findings**
18	Biswas et al. [42]	Non-medical	Non-medical	Use both top-down and bottom feature **Bleu-2 0.528**	MSCOCO dataset	visual attention by the union of high-level features and the low-level features obtained from the object-specific salient regions. **Also use the re-ranking caption candidates using beam search decoders. Lack of deployment.**

6 Conclusion

The Explainable technique explained above can be used on a wider range of medical images such as Ultrasound [1], MRI [9], x-ray [2,5,7], histopathology [8], Endoscopy [7], and fundus [10]. The XAI techniques can be used in the medical domain irrespective of the anatomical locations such as skin [47], chest [39,48,49], brain [9,11], eyes [10], gastrointestinal [7], and Fetal [1]. To answer RQ1 (What are the common visual sources for the medical diagnosis available?) in the article the section (Popular Visual Sources of Medical Diagnosis discussion) is made on the technique of image available, dataset available locally, and advantages and disadvantages. The answer for the RQ2 (How well the deep learning and machine learning algorithms are performing in terms of accuracy used in medical image analysis?) is provided by the section (Deep Learning and Machine Learning Approaches in medical domain) in which the current machine learning models as well as the deep learning models. Not only the model discussion but also the accuracy achievements in the medical diagnosis are discussed. The section (Challenges of Deep Learning and Machine Learning) is capable of answering the RQ3 (What challenges have been reported for deep learning and machine learning algorithms in real-time implementation?) to support the answer the challenges discussed are black box natural of the deep learning, the higher accuracy does not assure the increase in the model deployment, lastly, the lack of medical data availability is also a challenge. Explainable artificial intelligence as a possible solution for the challenges of deep learning and machine learning is discussed in the section (XAI Model) as the answer to the RQ4 (Are visual XAI techniques a possible solution for the black-box nature of deep learning and machine learning?). In section (XAI Model) different methods are discussed such as CAM, Grad-CAM, LIME, LRP, Prototypes, TCAV, and Text Explanations. It has been observed that the publically available dataset for the medical image is limited. The evaluation of a model's goodness based on accuracy alone does not assure the deployment of the models. The XAI is one of the solutions but the visual method of XAI automatic evaluation is lacking to date. Future research options evolution matrix for the visual explaining techniques. The explanation techniques will be of great help to medical people in identifying the new symptoms of the disease that have not been observed previously. With the explanation provided by the techniques.

References

1. Zhang, J., Petitjean, C., Ainouz, S.: Segmentation-based vs. regression-based biomarker estimation: a case study of fetus head circumference assessment from ultrasound images. J. Imaging **8**(2), 23 (2022). https://doi.org/10.3390/jimaging8020023
2. Brunese, L., Mercaldo, F., Reginelli, A., Santone, A.: Explainable deep learning for pulmonary disease and coronavirus COVID-19 detection from X-rays. Comput. Methods Progr. Biomed. **196** (2020). https://doi.org/10.1016/j.cmpb.2020.105608
3. Chen, B., Li, J., Lu, G., Zhang, D.: Lesion location attention guided network for multi-label thoracic disease classification in chest X-rays. IEEE J. Biomed. Health Inform. **24**, 2016–2027 (2019)
4. Hägele, M., et al.: Resolving challenges in deep learning-based analyses of histopathological images using explanation methods. Sci. Rep. **10**(1), 6423 (2020). https://doi.org/10.1038/s41598-020-62724-2. PMID: 32286358. PMCID: PMC7156509
5. Rajpurkar, P., et al.: Deep learning for chest radiograph diagnosis: a retrospective comparison of the CheXNeXt algorithm to practicing radiologists. PLoS Med. **15**, e1002686 (2018)

6. Rajaraman, S., Candemir, S., Thoma, G., Antani, S.: Visualizing and explaining deep learning predictions for pneumonia detection in pediatric chest radiographs. Med. Imaging Comput. Aided Diagn. (2019)
7. Malhi, A., Kampik, T., Pannu, H., Madhikermi, M., Framling, K.: Explaining machine learning-based classifications of in-vivo gastral images. In: Proceedings of the International Conference on Digital Image Computing: Techniques and Applications, DICTA 2019. Institute of Electrical and Electronics Engineers Inc., Department of Computer Science, Aalto University Finland, Finland (2019). https://doi.org/10.1109/DICTA47822.2019.8945986
8. Uehara, K., Murakawa, M., Nosato, H., Sakanashi, H.: Prototype-based interpretation of pathological image analysis by convolutional neural networks. In: Palaiahnakote, S., Sanniti di Baja, G., Wang, L., Yan, W. (eds.) ACPR 2019. LNCS, vol. 12047, pp. 640–652. Springer, Cham (2020). https://doi.org/10.1007/978-3-030-41299-9_50
9. Chakraborty, S., Aich, S., Kim, H.C.: Detection of Parkinson's disease from 3T T1 weighted MRI scans using 3D convolutional neural network. Diagnostics **10**, 402 (2020)
10. Kim, B., et al.: Interpretability beyond feature attribution: quantitative testing with concept activation vectors (TCAV). In: Proceeding of the 35th International Conference on Machine Learning, ICML 2018. International Machine Learning Society (IMLS), pp. 4186–4195 (2018)
11. Turk, O., Ozhan, D., Acar, E., Akinci, T.C., Yilmaz, M.: Automatic detection of brain tumors with the aid of ensemble deep learning architectures and class activation map indicators by employing magnetic resonance images. Zeitschrift für Medizinische Physik (2022)
12. Xu, F., Uszkoreit, H., Du, Y., Fan, W., Zhao, D., Zhu, J.: Explainable AI: a brief survey on history, research areas, approaches and challenges. In: Tang, J., Kan, M.Y., Zhao, D., Li, S., Zan, H. (eds.) NLPCC 2019. LNCS, vol. 11839, pp. 563–574. Springer, Cham (2019). https://doi.org/10.1007/978-3-030-32236-6_51
13. Lapuschkin, S., Binder, A., Montavon, G., Muller, K.R., Samek, W.: Analyzing classifiers: fisher vectors and deep neural networks. In: Proceedings of the IEEE Conference on Computer Vision and Pattern Recognition, pp. 2912–2920 (2016)
14. Bento, V., Kohler, M., Diaz, P., Mendoza, L., Pacheco, M.A.: Improving deep learning performance by using Explainable Artificial Intelligence (XAI) approaches. Discov. Artif. Intell. **1**, 1–11 (2021)
15. Wiens, J., et al.: Do no harm: a roadmap for responsible machine learning for health care. Nat Med. **25**(9), 1337–1340 (2019). https://doi.org/10.1038/s41591-019-0548-6. Epub 19 Aug 2019. Erratum in: Nat Med. **25**(10), 1627 (2019). PMID: 31427808
16. Lipton, Z.C.: The mythos of model interpretability. Commun. ACM **61**, 36–43 (2018)
17. European Parliament. Regulation (EU) 2016/679 of the European Parliament and of the Council of 27 April 2016 on the protection of natural persons with regard to the processing of personal data and on the free movement of such data, and repealing Directive 95/46. OJEU 59, 29 (2016)
18. Amerirad, B., Cattaneo, M., Kennett, R.S., Luciano, E.: Adversarial artificial intelligence in insurance: from an example to some potential remedies. Risks **11**(1), 20 (2023). https://doi.org/10.3390/risks11010020
19. Ribeiro, M.T., Singh, S., Guestrin, C.: Why should I trust you?" explaining the predictions of any classifier. In: Proceedings of the ACM SIGKDD International

Conference on Knowledge Discovery and Data Mining, pp. 1135–1144. Association for Computing Machinery, New York (2016). https://doi.org/10.1145/2939672.2939778
20. DeGrave, A.J., Janizek, J.D., Lee, S.-I.: AI for radiographic COVID-19 detection selects shortcuts over signal. Nat. Mach. Intell. **3**(7), 610–619 (2021)
21. Yeom, S.-K., et al.: Pruning by explaining: a novel criterion for deep neural network pruning. Pattern Recogn. **115** (2021)
22. Jung, Y.J., Han, S.H., Choi, H.J.: Explaining CNN and RNN using selective layer-wise relevance propagation. IEEE Access **9**, 18670–18681 (2021)
23. Matsumoto, T., et al.: Diagnosing heart failure from chest X-ray images using deep learning. Int. Heart J. **61**(4), 781–786 (2020)
24. Winkler, J.K., et al.: Association between surgical skin markings in dermoscopic images and diagnostic performance of a deep learning convolutional neural network for melanoma recognition. JAMA Dermatol. **155**, 1135–1141 (2019)
25. Malhotra, A., et al.: Multi-task driven explainable diagnosis of COVID-19 using chest X-ray images. Pattern Recogn. **122**, 108243 (2022)
26. Jiang, Z., Dong, Z., Wang, L., et al.: Method for diagnosis of acute lymphoblastic leukemia based on ViT-CNN ensemble model. Comput. Intell. Neurosci. **2021** (2021)
27. Abbas, A., Abdelsamea, M.M., Gaber, M.M.: Classification of Covid-19 in chest X-ray images using DeTraC deep convolutional neural network. Appl. Intell. **51**(2), 854–864 (2021)
28. Apostolopoulos, I.D., Mpesiana, T.A.: Covid-19: automatic detection from X-ray images utilizing transfer learning with convolutional neural networks. Phys. Eng. Sci. Med. **43**(2), 635–640 (2020)
29. Dhar, T., Dey, N., Borra, S., Sherratt, R.S.: Challenges of deep learning in medical image analysis-improving explainability and trust. IEEE Trans. Technol. Soc. **4**(1), 68–75 (2023)
30. Rajpurkar, P., et al.: CheXNet: radiologist-level pneumonia detection on chest X-rays with deep learning. arXiv/abs/1711.05225 (2017)
31. Sawant, A., Bhandari, M., Yadav, R., Yele, R., Bendale, M.S.: Brain cancer detection from MRI: a machine learning approach (tensorflow). Brain **5**(04), 2089–2094 (2018)
32. Chen, C., Li, O., Tao, D., Barnett, A., Rudin, C., Su, J.K.: This looks like that: deep learning for interpretable image recognition. In: Wallach, H., et al. (eds.) Advances in Neural Information Processing Systems, vol. 32, pp. 8930–8941. Curran Associates, Inc. (2019)
33. Hosny, K.M., Kassem, M.A., Foaud, M.M.: Classification of skin lesions using transfer learning and augmentation with Alex-net. PLoS ONE **14**, e0217293 (2019)
34. Khan, N., Ullah, F., Hassan, M.A., Hussain, A.: COVID-19 classification based on Chest X-Ray images using machine learning techniques. J. Comput. Sci. Technol. Stud. **2**(2), 01–11 (2020)
35. Hassanien, A.E., Mahdy, L.N., Ezzat, K.A., Elmousalami, H.H., Ella, H.A.: Automatic X-ray Covid-19 lung image classification system based on multi-level thresholding and support vector machine. medRxiv (2020)
36. Amin, J., Sharif, M., Yasmin, M., Fernandes, S.L.: A distinctive approach in brain tumor detection and classification using MRI. Pattern Recogn. Lett. **139**, 118–127 (2020)
37. Saleh, A.Y., Chin, C.K., Penshie, V., Al-Absi, H.R.H.: Lung cancer medical images classification using hybrid CNN-SVM. Int. J. Adv. Intell. Inform. **7**(2), 151–162 (2021)

38. Javaid, A., Sadiq, M., Akram, F.: Skin cancer classification using image processing and machine learning. In: 2021 International Bhurban Conference on Applied Sciences and Technologies (IBCAST), pp. 439–444. IEEE (2021)
39. Allaouzi, I., Ben Ahmed, M., Benamrou, B., et al.: Automatic caption generation for medical images. In: Proceedings of the 3rd International Conference on Smart City Applications (SCA 2018) (2018)
40. Singh, A., et al.: Medical image captioning using optimized deep learning model. Comput. Intell. Neurosci. **2022**, 9638438 (2022). https://doi.org/10.1155/2022/9638438
41. Park, H., Kim, K., Park, S., Choi, J.: Medical image captioning model to convey more details: methodological comparison of feature difference generation. IEEE Access **9**, 150560–150568 (2021)
42. Biswas, R., Barz, M., Sonntag, D.: Towards explanatory interactive image captioning using top-down and bottom-up features, beam search and re-ranking. KI-Künstliche Intelligenz **34**, 571–584 (2020)
43. Xu, K., et al.: Show, attend and tell: neural image caption generation with visual attention. arXiv/abs/1502.03044 (2015)
44. Yang, S., Niu, J., Wu, J., et al.: Automatic medical image report generation with multi-view and multi-modal attention mechanism. In: 20th International Conference on Algorithms and Architectures for Parallel Processing, ICA3PP 2020, vol. 12454, pp. 687–699 (2020). https://www.scopus.com/inward/record.uri?eid=2-s2.0-85092700542&doi=10.1007%2f978-3-030-60248-2_48&partnerID=40&md5=847b95b721a38473839d4fd2881ed768
45. Gajbhiye, G., Nandedkar, A., Faye, I., et al.: Automatic report generation for chest X-ray images: a multilevel multi-attention approach. In: 4th International Conference on Computer Vision and Image Processing, CVIP 2019, vol. 1147, pp. 174–182 (2020). https://www.scopus.com/inward/record.uri?eid=2-s2.0-85083724536&doi=10.1007%2f978-981-15-4015-8_15&partnerID=40&md5=e28028f91eaaf6e681e9fa1574c112b0
46. Cambria, E., Malandri, L., Mercorio, F., Mezzanzanica, M., Nobani, N.: A survey on XAI and natural language explanations. Inf. Process. Manag. **60**(1), 103111 (2023)
47. Lysaght, T., Lim, H.Y., Xafis, V., Ngiam, K.Y.: AI-assisted decision-making in healthcare: the application of an ethics framework for big data in health and research. Asian Bioeth. Rev. **11**, 299–314 (2019). https://doi.org/10.1007/s41649-019-00096-0
48. Caspers, J.: Translation of predictive modeling and AI into clinics: a question of trust. Eur. Radiol. **31**, 4947–4948 (2021). https://doi.org/10.1007/s00330-021-07977-9
49. Kundu, S.: AI in medicine must be explainable. Nat. Med. **27**, 1328 (2021). https://doi.org/10.1038/s41591-021-01461-z
50. Luan, H., Tsai, C.C.: A review of using machine learning approaches for precision education. Educat. Technol. Soc. **24**(1), 250–266 (2021)
51. Mukhamediev, R.I., et al.: Review of artificial intelligence and machine learning technologies: classification, restrictions. Opportunities Challenges Math. **10**(15), 2552 (2022)
52. Jiao, Z., Hu, P., Xu, H., Wang, Q.: Machine learning and deep learning in chemical health and safety: a systematic review of techniques and applications. ACS Chem. Health Saf. **27**(6), 316–334 (2020)
53. https://www.kaggle.com/code/buffyhridoy/efficientnet-hc/input

54. https://www.kaggle.com/datasets/mariaherrerot/idrid-dataset
55. https://www.kaggle.com/datasets/prashant268/chest-xray-covid19-pneumonia
56. https://www.kaggle.com/datasets/andrewmvd/lits-png
57. https://adni.loni.usc.edu/data-samples/access-data/
58. https://www.kaggle.com/datasets/vikasukani/parkinsons-disease-data-set

Early Detection and Classification of Zero-Day Attacks in Network Traffic Using Convolutional Neural Network

Mahendra Pratap Singh[1(✉)], Virendra Pratap Singh[1], and Maanak Gupta[2]

[1] National Institute of Technology Karnataka, Surathkal, Mangaluru 575025, India
{mp_singh,virendra.223cs002}@nitk.edu.in
[2] Tennessee Technological University, Cookeville, TN, USA
mgupta@tntech.edu

Abstract. In a Zero-Day cyberattack, attackers exploit a software vulnerability for which the software vendor is unaware or has not released a patch. This can make it difficult for organizations to protect their systems until a patch or mitigation is developed. To stay ahead of these evolving cyber threats, it's critical to keep up to date with the latest threat information and to remain vigilant. Traditional methods for detecting and classifying zero-day attacks often require session-wide features, which can be challenging to implement. This paper presents a novel approach for detecting and classifying Zero-Day attacks in network traffic. Specifically, we present a framework composed of a 1D Convolutional Neural Network (1D-CNN), which involves minimal preprocessing and directly leverages raw network data as byte sequences to learn features, eliminating the need for complex feature extraction. To test the effectiveness of our proposed approach, publicly available network traffic datasets encompassing various malware families are used. Results show that the proposed approach is significantly effective in detecting and classifying Zero-Day attacks, empowering organizations to combat evolving cyber threats.

Keywords: Network traffic · Malware · Zero-Day Attack · Deep Learning · Early Detection · 1D-Convolutional Neural Network

1 Introduction

The internet has completely changed how we communicate, conduct business and work. It has enabled us to connect billions of devices worldwide, making connectivity seamless and easy. Nevertheless, with this exponential growth comes a surge in cyber threats and vulnerabilities, particularly malware infiltration into network traffic. Malware, any software program or code designed to exploit, harm, or gain unauthorized access to computer systems or networks, comes in many forms, such as Viruses, Trojans, Worms, Ransomware, Spyware, and Adware. The goal is to disrupt, damage, or control targeted systems or steal sensitive information. Malware in network traffic is malicious software that is

D. Pastor-Escuredo et al. (Eds.): ICDLAIR 2023, LNNS 1001, pp. 812–822, 2024.
https://doi.org/10.1007/978-3-031-60935-0_70

transmitted over computer networks. It can be spread through various network communication protocols, such as email, web browsing, file sharing, instant messaging, or direct network attacks. Malware can be disguised as legitimate software, documents or embedded within files. As internet usage continues to rise, so does the vulnerability of computer networks to malware attacks.

Signature-based techniques [15] have always been the most common method for fighting network traffic malware. This approach involves comparing network packets against a database of known malware signatures, which allows for detecting previously encountered threats. While this method effectively identifies known malware, it must be flexible enough to handle new or modified threats. Furthermore, it has a high false positive rate and limited scalability, so the signature database must be updated frequently.

As cybersecurity threats evolve and become more sophisticated, one promising approach is machine learning (ML) methods, which can automatically analyze network traffic and identify patterns that may indicate malware activities. Several ML algorithms have shown particular promise in this area, including Support Vector Machines (SVMs), XGBoost, Decision Trees (DT), and Random Forests (RF). By leveraging these algorithms, cybersecurity professionals can improve their detection rates while reducing the number of false positives, which can help minimize the impact of malware attacks and keep sensitive data safe from harm. Overall, using ML techniques for malware detection represents a significant step forward in the fight against cybercrime. As threats evolve, we must explore new approaches and leverage the latest technologies to stay one step ahead of attackers and keep our networks secure.

This paper explores a new method for early detection of unknown malware in network traffic. Our approach uses Deep Learning, specifically CNN, which surpasses the limitations of traditional methods and Machine Learning techniques. Using CNN without requiring manual feature engineering, we can automatically extract essential features from raw network data. CNNs are well-known for their impressive performance in image recognition tasks, and their ability to learn hierarchical representations makes them an ideal choice for detecting complex malware traffic patterns.

Our approach offers several benefits over other methods. By training a CNN on a large dataset of benign and malicious network traffic, we aim to improve accuracy, reduce false positives, and enhance the detection of evolving threats. Furthermore, CNN models can be optimized for real-time processing, enabling swift and accurate malware detection in high-speed network environments. The novel contributions of the work presented in this article are summarized below:

- We present a novel approach to detect and classify Zero-Day Attacks using CNN.
- We also present an experimental analysis of the proposed approach using publicly available datasets to show the effectiveness of our approach.

The remainder of this paper is organized as follows. Section 2 provides literature on malicious network traffic detection. The overall system architecture

Table 1. Comparison of Related Work

Paper	Proposed Approach	Dataset
Lim et al. [2]	ResNet-Based CNN	UPC Dataset [16]
Marin et al. [3]	1D-CNN, LSTM	USTC_TFC2016, ISCX2012
Zhang et al. [4]	1D-CNN	USTC_TFC2016
Hwang et al. [5]	1D-CNN	USTC_TFC2016, Mirai-RGU [18]
Lu and Wang [6]	CNN	CICMalDroid2020 [19]
Yeo et al. [7]	CNN	USTC_TFC2016
Fox and Boppana [8]	2D-CNN	USTC_TFC2016
Bar and Hajaj [9]	SimCSE	USTC_TFC2016, ISCX2012

is presented in Sect. 3. Section 4 presents the experimental evaluation. Finally, Sect. 5 concludes the article and gives directions for future research.

2 Related Work

This section provides a brief overview of the studies related to our proposed work by analyzing recent publications from the past few years.

Researchers have proposed several deep learning-based approaches to detect and classify network traffic, as shown in Table 1. For example, Lim et al. [2] proposed a ResNet-based CNN approach to classify network traffic by analyzing packet-level data. At the same time, Marin et al. [3] presented a DeepMAL method for detecting and classifying malware using two 1D-CNN models on raw packet traces. Additionally, Zhang et al. [4] analyzed the CPU usage of their 1D-CNN model by varying the input packet size, and Hwang et al. [5] investigated various input configurations for their 1D-CNN-based analysis. Lu and Wang [6] proposed a methodology called F2DC for classifying Android malware using raw traffic data and neural networks. At the same time, Yeo et al. [7] presented a novel method for malware detection using a CNN to analyze packet flow. In addition, Fox and Boppana [8] introduced a method to detect malicious network flows with minimal preprocessing, and Bar and Hajaj [9] presented a novel traffic detection method that uses simple contrastive learning of sentence embeddings (SimCSE) to learn from raw packet data and classify traffic as malicious or benign. Although these methods show promising results, further research is needed to evaluate their effectiveness on encrypted malicious traffic and their real-world reliability.

In [10], the authors apply decision trees and SVMs machine learning for network intrusion detection, aiming to overcome the limitations of traditional signature-based systems, especially against zero-day attacks. However, machine learning models often face high false positive rates, including these, potentially causing resource wastage and desensitizing operators to real threats. On the other hand, a novel method for classifying encrypted traffic is introduced in [11] that involves extracting the first 500 unencrypted bytes of payload as key data

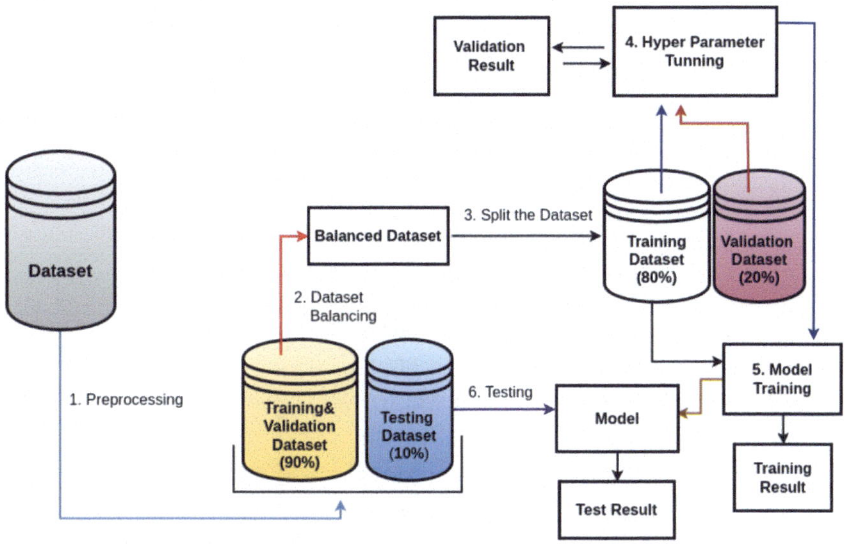

Fig. 1. Proposed Framework for Zero-Day Attacks Detection and Classification

and adding a length threshold identifier to the payload header. However, the method depends on the first 500 bytes being unencrypted, limiting its applicability to some encryption protocols like TLS 1.3.

Wang et al. [12] proposed Deep Hierarchical Network (DHN) for detecting malicious traffic at the packet level. While DHN excels in detection, it requires substantial training data, which can be resource-intensive to gather and label. In contrast, Chen et al. [13] presented a 2D-CNN-based Network Intrusion Detection System (NIDS) that shows 96.55% and 99.56% accuracy with features and raw data input, respectively. It can only detect known malicious traffic and cannot handle encrypted traffic. On the other hand, Xu et al.[14] proposed a model for detecting malicious network traffic by combining CNN and Long Short-Term Memory (LSTM). While achieving high detection rates, the model requires extensive training data and is challenging to deploy in resource-constrained real-world scenarios.

In the next section, we present a framework to address the limitations of various methods.

3 Proposed Approach

This section presents a framework and describes its components in detail. We also discuss the dataset, its split ratio, and its use as training, validation, and testing datasets.

With the framework shown in Fig. 1, we aim to detect and classify Zero-Day attacks in network traffic as soon as they occur to prevent the entire system from being compromised. The framework consists of several components, each with its unique function, and their description is as follows.

Table 2. Statistics of Different Traffic Categories in the USTC-TFC2016 Dataset

Traffic Category	Number of Instances
Cridex	16,386
Geodo	40,947
Htbot	6367
Miuref	13,481
Neris	33,791
Nsis_ay	6069
Shifu	9634
Tinba	8504
Virut	33,103
Zeus	10,970
BitTorrent	7517
FaceTime	6000
FTP	101037
Gmail	8629
Skype	6321

3.1 Pre-processing

In this work, we considered two datasets: USTC-TFC2016 [1] and CTU-13 [17]. The USTC-TFC2016 dataset comprises ten types of malware and normal traffic in pcap format captured in actual and simulated environments. We used the USTC-TFC2016 dataset for unknown malicious network traffic detection and classification, and the statistics of different malware categories in it are shown in Table 2. We also leveraged the CTU-13 dataset for testing our model, which contains 13 types of malware.

Preprocessing is critical for improving a model's efficiency and accuracy. This involves cleaning the dataset, which includes identifying flows and creating feature vectors. For flow identification, we split the pcap file into bidirectional flows. To represent real-world malicious network traffic, we truncate the Source IP address, Destination IP address, Source port, and Destination port information. This is because attackers often use a limited number of internet IP addresses in small testbed-created datasets. We then extract the initial 98 bytes of the IP packet, convert them into integers, and store them in the packet vector. We truncate the remaining bytes if the packet size is more than 98 bytes. If it is less than 98 bytes, we pad zeros to ensure the desired size. The final feature size would be (n, 98) if the number of packet vectors concatenated together is n, where n is the number of packets per bidirectional flow.

3.2 Dataset Splitting, Balancing, and Hyper-parameter Tuning

The dataset was split into training, validation, and testing. The model was trained using the training dataset and validated using the validation dataset.

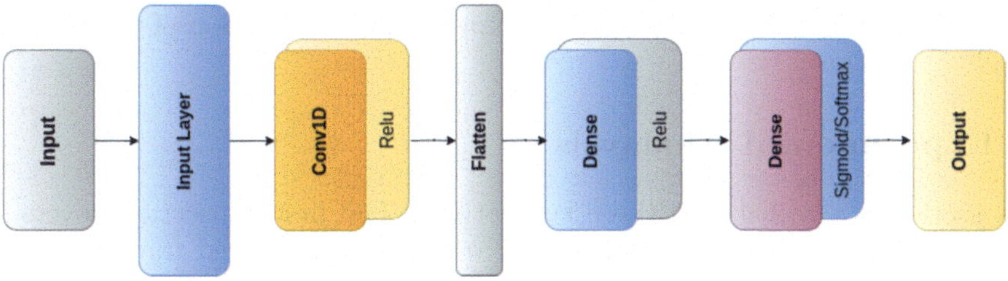

Fig. 2. Proposed 1D-CNN model

Finally, the model was tested on the testing dataset, which contained unseen instances. Initially, the dataset was split into a training and testing set in a ratio of 90:10. The training dataset was balanced using SMOTE (Synthetic Minority Oversampling Technique) and divided into 80:20 proportions to create training and validation datasets.

When training a model, it's essential to adjust certain variables called hyper-parameters to achieve the highest level of accuracy. This process, known as hyper-parameter tuning, involves testing various possible values for these variables before settling on the best-suited option for our data and algorithm. To streamline this process, we utilized the Optuna framework, which helped us identify the most effective hyper-parameters for our model. By fine-tuning these variables, we can ensure our model is as accurate and effective as possible.

3.3 Model Training, Validation, and Testing

The model shown in Fig. 2 comprises several layers that work together to process input data. The input layer defines the shape and dimensionality of the data that the network expects to receive. It does not perform any computations but instead passes the input data to the next layer. The Conv1D layer is used for processing sequential data and applies filters to the input data using a sliding window. This layer is beneficial for capturing temporal dependencies and patterns in sequential data. The Flatten layer converts multidimensional input data into a one-dimensional form and is typically used when transitioning from a convolutional layer to a fully connected layer. Finally, the Dense layer is a fully connected layer where each neuron is connected to every neuron in the previous layer. Each connection has a corresponding weight, and the output of each neuron is calculated by taking a weighted sum of the inputs, applying an activation function, and passing it to the next layer. These layers work together to create a powerful model for processing input data.

The model's performance is assessed during each training epoch by evaluating its accuracy using the validation dataset. This dataset is unbiased and comprises examples the model has not seen during training. In the final stage, the model is tested against a previously unseen dataset derived from the original to evaluate its behavior in real-world scenarios.

In the next section, we present the experimental evaluation of the proposed model.

4 Experimental Analaysis

This section is divided into three subsections. The first subsection covers the hardware and software requirements for testing the proposed framework. The second subsection outlines the evaluation metrics used to assess the performance of the proposed framework. Lastly, the third subsection evaluates and compares the proposed model to existing approaches.

4.1 Experimental Setup

All the experiments were performed on a Linux 64-bit operating system with Intel(R) Core(TM) i5-1035G1 CPU 1.00 GHz, 1190 Mhz, 4 Cores, 8 Logical Processors, and 8 GB of RAM. In the experiments, Scikit-Learn, TensorFlow, and Numpy libraries were used to implement the 1D-CNN model in Python programming language. Also, to select the best values of hyper-parameters to build our model, the Optuna framework is used, and values that give the best accuracy are chosen for each model.

4.2 Evaluation Metrics

For evaluating the proposed model, True Positive (TP), True Negative (TN), False Positive (FP), and False Negative (FN) measures are considered. Using the above measures, we calculate accuracy, precision, recall, and F1-score to evaluate the performance of our models.

- **Accuracy:** It gives the total number of true predictions over all the predictions made by the model.

$$Accuracy = \frac{(TP + TN)}{(TP + TN + FP + FN)} \tag{1}$$

- **Precision:** It gives the total number of true positive predictions on all positive predictions made by the model.

$$Precision = \frac{TP}{(TP + FP)} \tag{2}$$

- **Recall/Sensitivity:** It gives the total number of true positive predictions on all actual positive predictions.

$$Recall = \frac{TP}{(TP + FN)} \tag{3}$$

- **F1-Score:** It gives the harmonic mean of precision and recall.

$$F1 - Score = 2 * \frac{(Precion * Recall)}{(Precion + Recall)} \tag{4}$$

Table 3. Zero-Day Attacks Classification Using USTC-TFC2016 Dataset with 1 Packet Per Flow as Feature

Class	Accuracy (%)	Precision (%)	Recall (%)	F1 Score (%)
Credix	100.0	100.0	100.0	100.0
Geodo	100.0	100.0	100.0	100.0
Htbot	100.0	100.0	100.0	100.0
Miuref	99.99	100.0	99.99	99.99
Neris	100.0	100.0	100.0	100.0
Nsis_ay	93.67	92.18	100.0	95.92
Shifu	100.0	100.0	100.0	100.0
Tinba	100.0	100.0	100.0	100.0
Virut	100.0	100.0	100.0	100.0
Zeus	99.95	99.95	100.0	99.97

Table 4. Zero-Day Attacks Classification Using USTC-TFC2016 Dataset with 2 Packets Per Flow as Features

Class	Accuracy (%)	Precision (%)	Recall (%)	F1 Score (%)
Credix	100.00	100.00	100.00	100.00
Geodo	99.99	99.99	100.00	99.99
Htbot	99.92	99.92	100.00	99.96
Miuref	100.00	100.00	100.00	100.00
Neris	100.00	100.00	100.00	100.00
Nsis_ay	99.50	99.35	100.00	99.67
Shifu	99.99	99.99	100.00	99.99
Tinba	100.00	100.00	100.00	100.00
Virut	100.00	100.00	100.00	100.00
Zeus	99.91	99.91	100.00	99.95

Table 5. Zero-Day Attacks Detection Using CTU-13 Dataset

Feature Config.	Accuracy (%)	Precision (%)	Recall (%)	F1 Score (%)
1 packet	98.68	97.90	98.96	98.41
2 packet	99.72	99.79	99.53	99.66

4.3 Experimental Results

We utilized 1 packet and 2 packets per flow to evaluate our model as input. This evaluation was conducted on the USTC-TFC2016 dataset containing 10 different malware classes. Additionally, we performed binary classification using the same USTC-TFC2016 dataset, encompassing the malicious network traffic from 10 malware classes and 5 benign classes. Moreover, we incorporated the CTU-13 dataset to assess the model's performance in detecting Zero-Day attack scenarios. This comprehensive analysis encompassed the malicious network traffic of 10 malware classes from the USTC-TFC2016 and CTU-13 datasets, enabling a thorough evaluation of the model's capabilities. Deep Learning experiments split the dataset into training, validation, and testing. Then, the model was fit into the train data to predict the test data. We used USTC-TFC2016 and CTU-13 datasets to build and test the models.

One packet and two packets per flow were used as features to train the model for multiclass and binary classification. Tables 3 and 4 show the performance evaluation results for the models under the Zero-Day attack scenario on the USTC-TFC2016 dataset. We trained our model using the USTC-TFC2016 dataset and tested our model for the Zero-Day attack scenario on the CTU-13 dataset. Table 5 shows the performance evaluation results.

Based on the experiment's findings, it was discovered that the accuracy for all classes, except for "Nsis_ay", reached almost 100% when using one packet per flow as a feature. However, the accuracy for the "Nsis_ay" class significantly improved from 93.67% to 99.50% when utilizing two packets per flow as features. This means that two packets per flow are enough to detect malicious traffic in the Zero-Day scenario effectively. The experiment also analyzed the CTU-13 dataset containing encrypted and unencrypted traffic. The model was trained using the USTC-TFC2016 dataset, and the accuracy achieved with one packet per flow as a feature was 98.68%. However, increasing the number of packets per flow to two as features improved the accuracy to 99.72%. This shows that the model effectively detects malicious traffic, even when dealing with encrypted data.

5 Conclusion

This paper presented a 1D-CNN model that uses raw packet data sequences as input to detect and classify Zero-Day attacks. For two packets per flow as features, our approach achieved almost perfect accuracy for all the classes and precision for all classes except for the Nsis_ay class. However, even in the case of Nsis_ay, our model demonstrated significantly higher accuracy rates than other models, reaffirming its ability to detect Zero-Day attacks effectively. Moreover, our model was tested on the CTU-13 dataset, which includes both encrypted and unencrypted traffic, emphasizing its versatility and efficacy in detecting malicious traffic in various network environments.

In terms of future scope, there are two promising directions for further research in the field of malicious traffic detection. Firstly, testing the model on datasets with TLS1.3 encryption and, secondly, incorporating mixed datasets encompassing various forms of malicious traffic, including those encountered in 5G networks.

Acknowledgement. This work is partially supported by the US NSF grants 2025682 and 2230609.

References

1. Wang, W., Zhu, M., Zeng, X., Ye, X., Sheng, Y.: Malware traffic classification using convolutional neural network for representation learning. In: 2017 International Conference on Information Networking (ICOIN), pp. 712–717 (2017)
2. Lim, H., Kim, J., Heo, J., Kim, K., Hong, Y., Han, Y.: Packet-based network traffic classification using deep learning. In: 2019 International Conference on Artificial Intelligence in Information and Communication (ICAIIC), pp. 46–51 (2019)
3. Marín, G., Cañas, P., Capdehourat, G.: DeepMAL - deep learning models for malware traffic detection and classification. In Data Science - Analytics and Applications, pp. 105–112 (2021)
4. Zhang, W., Wang, J., Chen, S., Qi, H., Li, K.: A framework for resource-aware online traffic classification using CNN. In: Proceedings of the 14th International Conference on Future Internet Technologies (CFI 2019), pp. 1–6 (2019)
5. Hwang, R., Peng, M., Huang, C., Lin, P., Nguyen, V.L.: An unsupervised deep learning model for early network traffic anomaly detection. IEEE Access **8**, 30387–30399 (2020)
6. Lu, T., Wang, J.: F2DC: Android malware classification based on raw traffic and neural networks. Comput. Netw. **217**, 109320 (2022)
7. Yeo, M., et al.: Flow-based malware detection using convolutional neural network. In: 2018 International Conference on Information Networking (ICOIN), pp. 910–913 (2018)
8. Fox, G., Boppana, R.V.: Detection of malicious network flows with low preprocessing overhead. Network **2**, 628–642 (2022)
9. Bar, R., Hajaj, C.: SimCSE for encrypted traffic detection and zero-day attack detection. IEEE Access **10**, 56952–56960 (2022)
10. Thaseen, I.S., Poorva, B., Sai, U.P.: Network intrusion detection using machine learning techniques. In: 2020 International Conference on Emerging Trends in Information Technology and Engineering (ic-ETITE), pp. 1–7 (2020)
11. Ma, Q., Huang, W., Jin, Y., Mao, J.: Encrypted traffic classification based on traffic reconstruction. In: 2021 4th International Conference on Artificial Intelligence and Big Data (ICAIBD), pp. 572–576 (2021)
12. Wang, B., Su, Y., Zhang, M., Nie, J.: A deep hierarchical network for packet-level malicious traffic detection. IEEE Access **8**, 201728–201740 (2020)
13. Chen, L., Kuang, X., Xu, A., Suo, S., Yang, Y.: A novel network intrusion detection system based on CNN. In: 2020 Eighth International Conference on Advanced Cloud and Big Data (CBD), pp. 243–247 (2020)
14. Xu, G., Zhou, J., He, Y.: Network malicious traffic detection model based on combined neural network. In: 2022 6th Asian Conference on Artificial Intelligence Technology (ACAIT), pp. 1–6 (2022)
15. Petliak, N., Klots, Y., Titova, V., Cheshun, V., Boyarchuk, A.: Signature-based approach to detecting malicious outgoing traffic. In: 4th International Workshop on Intelligent Information Technologies and Systems of Information Security (IntellTSIS 2023), pp. 486–506 (2023)
16. Carela-Español, V., Bujlow, T., Barlet-Ros, P.: Is our ground-truth for traffic classification reliable? In: Passive and Active Network Measurement Conference, pp. 98–108 (2014)
17. García, S., Grill, M., Stiborek, J., Zunino, A.: An empirical comparison of botnet detection methods. Comput. Secur. **45**, 100–123 (2014)

18. McDermott, C.D., Majdani, F., Petrovski, A.V.: Botnet detection in the Internet of Things using deep learning approaches. In: 2018 International Joint Conference on Neural Networks (IJCNN), pp. 1–8 (2018)
19. Mahdavifar, S., Kadir, A.F., Fatemi, R., Alhadidi, D., Ghorbani, A.A.: Dynamic android malware category classification using semi-supervised deep learning. In: 2020 IEEE International Conference on Dependable, Autonomic and Secure Computing, pp. 515–522 (2020)

Correction to: Safeguarding Ecosystems and Efficiency in Peer-to-Peer File Sharing Systems: An IoT-Inspired Approach to Pollution Mitigation

Ritu

Correction to:
Chapter 21 in: D. Pastor-Escuredo et al. (Eds.): *The Future of Artificial Intelligence and Robotics*, LNNS 1001, https://doi.org/10.1007/978-3-031-60935-0_21

In the original version of the book, the affiliation of the author of chapter 21 "Ritu" has been changed to "Chandigarh College of Engineering, Chandigarh Group of Colleges, Jhanjeri-140307, India."

The updated version of this chapter can be found at
https://doi.org/10.1007/978-3-031-60935-0_21

D. Pastor-Escuredo et al. (Eds.): ICDLAIR 2023, LNNS 1001, p. C1, 2024.
https://doi.org/10.1007/978-3-031-60935-0_71

Author Index

D. Pastor-Escuredo et al. (Eds.): ICDLAIR 2023, LNNS 1001, pp. 823–825, 2024.
https://doi.org/10.1007/978-3-031-60935-0

The manufacturer's authorised representative in the EU is Springer Nature Customer Service Centre GmbH, Europaplatz 3, 69115 Heidelberg, Germany. If you have any concerns regarding our products, please contact ProductSafety@springernature.com

Printed and bound by CPI Group (UK) Ltd, Croydon, CR0 4YY

01/10/2026

02225893-0018